# Human Body

## A Wearable Product Designer's Guide

# Human Body

## A Wearable Product Designer's Guide

Karen L. LaBat, PhD.
Karen S. Ryan, M.D.
Illustrations by Le Wen

CRC Press is an imprint of the
Taylor & Francis Group, an **informa** business

CRC Press
Taylor & Francis Group
6000 Broken Sound Parkway NW, Suite 300
Boca Raton, FL 33487-2742



CRC Press is an imprint of Taylor & Francis Group, an Informa business


Printed on acid-free paper

International Standard Book Number-13: 978-1-4987-5571-9 (Hardback)



**Library of Congress Cataloging-in-Publication Data**

Names: Labat, Karen L. (Karen Louise) author. | Ryan, Karen (Medical doctor)
Title: Human body : a wearable product designer's guide / authored by Karen Louise LaBat and Karen S. Ryan.
Description: Boca : Taylor &Francis, 2019. | "A CRC title, part of the Taylor & Francis imprint, a member of the Taylor & Francis Group, the academic division of T&F Informa plc."
Identifiers: LCCN 2018047989| ISBN 9781498755719 (hardback : alk. paper) | ISBN 9780429055690 (e-book) Subjects: LCSH: Human anatomy. | Physiology. | Fashion design. | Clothing and dress.
Classification: LCC QM26 .L33 2019 | DDC 612--dc23
LC record available at https://lccn.loc.gov/2018047989

**Visit the Taylor & Francis Web site at**
**http://www.taylorandfrancis.com**

**and the CRC Press Web site at**
**http://www.crcpress.com**

# *Contents*

# *Preface*

Wearable products surround us every day of our lives, and many designers are engaged in developing the products. A wearable product is evaluated by the person wearing it—does it enhance the body, is it comfortable, does it function as anticipated? In other words, did the designer meet the product design goals? From the start, our goal as we wrote *Human Body: A Wearable Product Designer's Guide*, was to present wearable product designers with realistic, reliable knowledge of human anatomy and function from a design perspective. And, importantly, *Human Body: A Wearable Product Designer's Guide* seeks to provide a common language for designers and collaborators from many disciplines.

We hope to inspire a fresh approach to your design process and to provide something of value for all, from novice designer to seasoned professional, resulting in wearable products that enhance health, performance, safety, and pleasure. May this book expand your current knowledge and serve as a starting point for projects for years to come.

We approach the topic of designing for the human body from two different backgrounds and perspectives. Karen L. LaBat is professor emerita with experience in conducting product design research focused on applying human body dimensions to wearable products. Karen S. Ryan is a medical doctor with expertise in the body's musculoskeletal and neurological systems, with a master's degree in apparel design. We focus on the human body, but through different lenses—the lenses of our educational backgrounds and experiences. We found common interests and concerns while working together in the Human Dimensioning© Laboratory at the University of Minnesota.

We have structured this book in a manner we believe will be most useful, engaging, and intuitive for designers. To understand the human body as the sum of many structures and functions, start with Chapter 1 and work your way through this book beginning to end. Other features of this book may be particularly useful for designers. Anatomical and product terms are defined in an extensive glossary. The first time a term is mentioned in the text it is italicized. We describe the term in the text and give a succinct definition in the glossary. References are collected alphabetically in a separate section—a listing of old and new sources which we hope will serve to provide details beyond the scope of this book. Anatomical features are the foundation for developing the form and fit of a product. Understanding how to landmark and measure the body based on those features is an essential design tool. The text gives examples of landmarking and measuring the body with product applications, while the appendices describe landmarking and measuring methods and terms in detail.

Chapters begin with a brief introduction and a list of "Key Points" that will be covered in the chapter. We worked closely with our illustrator to develop original illustrations which clearly demonstrate body features and concepts related to wearable products. Body scans of real people are used to contrast with ideal images of the body presented in many design books. Basics of designing for the body in motion are incorporated throughout this book. Range of motion is described and illustrated with line drawings that apply directly to wearable products.

We have searched the medical and product design literature, and present many topics not available in other wearable product design books. Examples include: challenges of designing for the intricate structures of the hand and the foot, fitting and protecting the female and male genitals, developing products to manage incontinence, measuring and fitting the female breast, and accommodating changes related to pregnancy.

We have enjoyed planning and completing this book. We come away from the experience with a new appreciation of the form and function of the human body and of wearable product designers' creations and insights. We wish you well in your exploration of the human body as the foundation for wearable product design. We believe that the human body can provide unlimited inspiration for designing innovative wearable products.

# Acknowledgments

We wish to thank all of the people who supported this book project. We extend special thanks to our exceptionally talented illustrator, Le "Lettie" Wen. Lettie's technical knowledge, creativity, and problem solving allowed us to convey so much information visually. Thank you, Jane Ryan, the master of all things "APA," for your technical editing skills, and thank you, Cindy Renee Carelli, CRC Press Executive Editor, and Erin Harris, CRC Press Editorial Assistant, for your fast responses to all of our questions.

We thank Bryan Moloney, our project manager at Deanta Global, for assistance with production details.

We acknowledge financial support from the *USDA Agricultural Experiment Station* for funding several research assistantships through the years of research and writing, and to the *University of Minnesota Imagine Fund* for supporting the work of our illustrator.

Thank you to the numerous University of Minnesota Human Dimensioning© Laboratory graduate student research assistants who contributed research, ideas, and enthusiasm over many years. Our thanks go to the many people who read, edited, and commented on sections and/or whole chapters of this book, from the proposal stage to the final draft: Robert Anderson, Missy Bye, Robin Carufel, Crystal Compton, Lucy Dunne, Sheila Egan, Steve Erickson, Rachael Granberry, Linsey Griffin, Deepti Gupta, Heather Hettrick, Brad Holschuh, Dong-Eun Kim, Nokyeon Kim, Young-A Lee, Paula Ludewig, Saemee Lyu, Penny Madvig, Otto Mayer, Marlys McGuire, Angie Miller-Foley, Changhyu (Lyon) Nam, Juyeon Park, Mary Riley, Gregory Sheehan, Carol Salusso, Susan Sokolowski, and Rebecca Van Amber. Thank you, Nak Koh, for your hospitality during our work sessions with Lettie at your home. And thank you Kathy Guiney, Tara Faricy, and Julie Hillman for administrative help.

---

I thank my family, friends, and colleagues who supported the project. A special thank you goes to husband, Joe, who listened to early ideas for this book on a long road trip to Montana in 2007. Thank you, daughter Nicole, for hugs and baked goods, and son Sean and daughter-in-law Mary, for support from afar. Thank you, Craig, for texting every week—is it done yet? Thank you to many colleagues who lent encouragement along the way; with special thanks to Sherri Gahring, Missy Bye, and Su Sokolowski. Thank you, co-author Karen S. Ryan and illustrator Lettie Wen for your tireless and dedicated work on this book.

*Karen L. LaBat*

I thank the many teachers, classmates, colleagues, and students who asked me thought-provoking questions throughout my education and life, especially mentors Jacob Polta, Elaine Adams, Ross Shoger, Frederic Kottke, Glenn Gullickson, Jr., Miland Knapp, and Carol and Titus Bellville. To my family (including my two loving granddaughters) and friends, past and present, who have been there when I needed you, thank you, too. To Karen L. LaBat—thank you for inviting me to join you in this endeavor. And to Lettie Wen, thanks for your patience and your friendship through our many months of work on this project.

*Karen S. Ryan*

# *Authors' Biographies*

**Karen L. LaBat**, PhD, is Horace T. Morse Distinguished Professor Emerita, College of Design, University of Minnesota. She was instrumental in establishing the Wearable Product Design Center and the Human Dimensioning© Laboratory at the University of Minnesota. Her research focus is product development for human health and safety and study of the body to improve function and fit of wearable products. Research funding sources include: National Science Foundation, 3M, U of MN Clinical and Translational Science Institute, U of MN Institute for Advanced Study, and others. She has published in *Applied Ergonomics; Australasian Medical Journal; Ergonomics; Clothing and Textiles Research Journal; Journal of the Textile Institute; International Journal of Fashion Design, and Technology*; and other journals. She was co-editor for the *Clothing and Textiles Research Journal Focused Issue on Fashion and Health,* co-editor of the *Clothing and Textiles Research Journal Focused Issue on Research and Teaching in the 21st Century,* and an associate editor for *Fashion and Textiles—the Journal of the Korean Society of Clothing and Textiles,* and served on the editorial board of the *Journal for Health Design.* She was co-chair of the U of MN, College of Design, *Fashion and the Body 2016 Symposium.* Her awards include: International Textiles and Apparel Association Distinguished Scholar, U of MN College of Design Outstanding Research Award, Lectra Innovation Award for Faculty Research, and U of MN College of Design Innovation and Mission Advancement Award. Dr. LaBat received the 2013 U of MN *Horace T. Morse Award for Excellence in Undergraduate Teaching* and was inducted into the Academy of Distinguished Teachers.

**Karen S. Ryan**, MD, MS, has been a Research Associate with the Human Dimensioning© Laboratory, University of Minnesota, since she completed her Masters of Science in Apparel in 2006, with a project on designing apparel for women with posture changes from osteoporosis. Her research grew from her longstanding interest in apparel design and her clinical and teaching practice as a specialist in Physical Medicine and Rehabilitation (PM&R). Her clinical practice focused on musculoskeletal issues, electromyography (EMG), and neurological rehabilitation, with a strong emphasis on patient education. She has served on the faculty of the Department of Physical Medicine and Rehabilitation, University of Minnesota Medical School, including as the Director of the Electromyography Service and as Residency Training Director. She has been an Oral Examiner for the American Board of Physical Medicine and Rehabilitation and the American Board of Electrodiagnostic Medicine. She has made numerous PM&R presentations on a wide range of topics and in many venues, from patient family education sessions to University of Minnesota Television to national PM&R and EMG meetings. She has presented her design and health research in local, national, and international meetings and published in the medical and apparel design literature.

# *Illustrator Biography*

**Le "Lettie" Wen** graduated from Tsinghua University, China with a Masters of Literature. She began work on illustrations for this book as a graduate research assistant in the Human Dimensioning© Laboratory at the University of Minnesota. She received a master's degree in Graphic Design from the University of Minnesota in 2017. Her graduate thesis, titled "Human Anatomy for Wearable Product Designers," combined knowledge from the fields of anthropometry and medical illustration. She developed a unique style to portray the interactions of the human body and a product. She plans to continue work as an illustrator, with an emphasis on anatomical illustration.

# 1

# *The Human Body as the Foundation for Wearable Product Design*

The human figure serves as the foundation for wearable product development. An understanding of human structure (*anatomy*), human function (*physiology*), and the actions of natural mechanical forces and energy in and on the body (*biomechanics*) can generate products that are compatible with complex bodies. Wearable product designers identify questions of "Who, What, Why, When, Where, and How" and apply *product component* knowledge to find a solution—a wearable product—that serves an individual's needs.

Enrich and expand the *design process* by learning anatomical, physiological, and biomechanical terms. As terms (anatomical and product) are introduced in this book, they are italicized. Italicized terms, with definitions, are collected in the glossary. As you design a wearable product, use anatomical terms to reference (a) specific body structures, (b) sections of the body, (c) body processes, (d) relationships between body parts, and (e) body/product/environment interactions.

Anatomical knowledge can help you look at your *user group* or *target market* from a new perspective, see design problems in a new frame of reference, modify where and how you place your product on the body, and create innovative designs. Build anatomical, physiological, and biomechanical knowledge to set the stage for successful relationships between the human body and products and for advantageous interactions of the body and product with the environment.

**Key points:**

- The human body is the basis for wearable product design.
- The human body is wonderfully complex and variable.
- Anatomical terms can help visualize body structures and relationships.
- Anatomical terms provide a standard for team communication in the design process.
- Environments influence wearable product designs.

- The body can be characterized by its structure, function, and response to natural mechanical forces.
- Wearable products serve as buffers between the body and the environment.
- Wearable products must move with the body.
- Wearable product materials and structures influence body function.
- Variations in body structure complicate wearable product *fit* and sizing.

Words matter! Anatomists, professionals who study the human body, use an international language of terms defined by the structures of the body (Federative Committee on Anatomical Terminology, 1998). FCAT terms, some including links to illustrations, are available online (http://terminologia-anatomica.org/en/Terms).

For purposes of this book, a *wearable product* is defined as anything that surrounds, is suspended from, or is attached to the human body. In some cases, wearable products are inserted into the body. Many products fit this definition—from *fashion apparel* to *medical devices.* While external medical devices like wearable *blood pressure monitors* and *heart event monitors* may be placed by a professional, they are included because they are worn for significant periods—a day or longer. Products like *hearing aids* and *birth control diaphragms* are placed in the body by the wearer.

When designers incorporate knowledge of body *shape, form,* and *size* in their design thinking, they can improve everyday wearable products like apparel. Shape is the two-dimensional outline (silhouette) of the three-dimensional body (form), viewed from the front, back, or side. Each person's body has characteristic curves, planes, and angles. Product size reflects an individual person's body dimensions.

Designers often plan a wearable product by looking at the body in a *static* pose. However, understanding the body in motion, the *dynamic* body, is important too. Gemperle, Kasabach, Stivoric, Bauer, and Martin (1998) developed design guidelines for best placement of wearable forms on the static and dynamic three-dimensional body. As you work on a design, accommodate body movement.

The designer's job becomes more challenging as wearer demographics change and product categories expand. Population age shifts call for modification of product sizes, shapes, and forms to meet wearers' changing needs. Global markets demand careful study of cultural differences in body ideals and of real body forms that influence wearable product designs. Wearable products for *hazardous environments* and medical and healthcare settings require in-depth knowledge of the human body. Understanding anatomy, physiology, and biomechanics prepares designers to meet new design challenges.

## 1.1 Body/Wearable Product/Environment

Two givens, the body and the environment, influence wearable product designs. Where do you begin in order to deal with the complexity of the body in the environment?

- Gather knowledge of human body form, function, and movement.
- Evaluate the environment. What are the environmental conditions? Are any of them hazardous?
- Manipulate *wearable product components*—structure, materials, and fit—to meet the needs of the body and the demands of the environment.
- Relate the structure and placement of the wearable product to the body and environment.

See Figure 1.1 for a simple illustration of the human body within the environment; the wearable product is the mediating variable. Understanding these factors is essential for the wearable product designer.

Anatomy is defined as "the study of structure" (McKinley & O'Loughlin, 2006, p. 2). The root of the word anatomy comes from Greek, meaning to "cut up" or "cut open," harking back to the early pioneers in human body discovery as they *dissected* human remains to discover the mysteries of the body. Anatomists, medical students, and scientists use dissection today to study the parts of the body and the relationships of the parts. Understanding body structure is crucial when planning and implementing how a wearable product will be placed, attached, and work with the body.

Body structures suggest function; however, a separate scientific field, physiology, focuses on body functions and activities. Although this book does not provide in-depth explanations of physiological functions, basic body functions that affect design choices are covered as *body systems* in Chapter 2. Wearable products can hinder, enhance, or protect physiological functions. Laing and Sleivert (2002) delve into the effects of clothing and textiles on human performance (including physiology) in a comprehensive, thoughtful, and extensively referenced critical review. They specifically look at wearable product *ergonomic* requirements, design, fit, and material issues.

Over-heating in a hot and humid atmosphere illustrates how complicated body/product/environment interactions can be. Clothing might inhibit the body's *evaporative cooling* by limiting air flow that could evaporate perspiration from the body surface. Clothing may reflect or shield the body from the sun's heat. And *wicking fibers* may help cool the body by moving moisture away from the skin. The process of body *thermoregulation* involves the *circulatory system, respiratory system,* and the skin, part of the *integumentary*

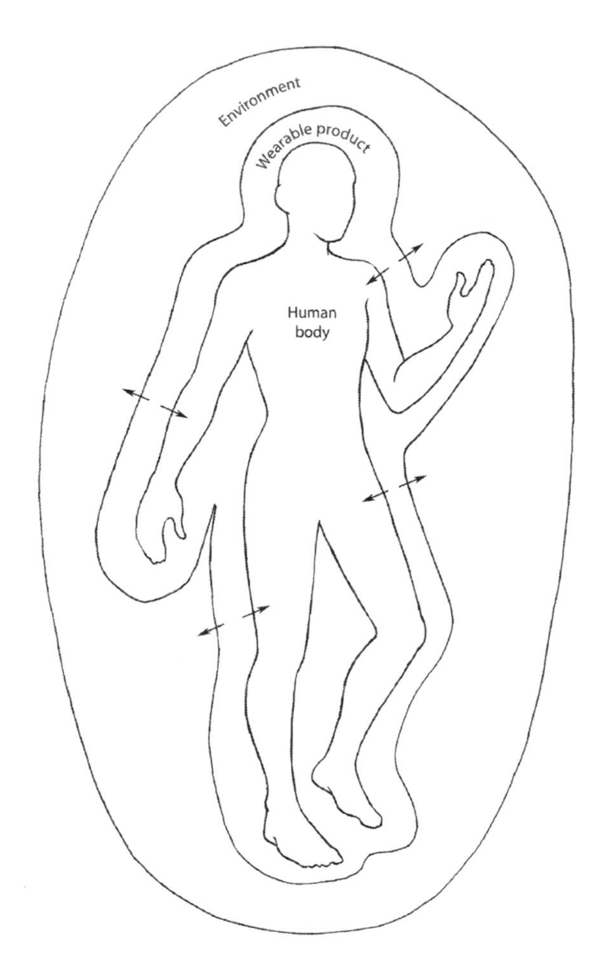


**FIGURE 1.1**
Human body/Wearable product/Environment interactions.

*system*—all discussed in Chapter 2. For a more detailed discussion of wearable products and thermoregulation see Laing and Sleivert (2002).

Body structures move in the environment and in relationship to each other according to the laws of mechanics. Biomechanics helps explain (a) the forces acting on the body, (b) the body's *center of gravity* and *base of support*, (c) *muscle work* and *power*, and in summary (d) movements of the body in the environment (Everett & Kell, 2010). The mechanical nature of the body is most evident in the *skeletal system* and the *muscular system*, the body systems providing a body's frame and motion. The human body is not well-suited to carrying loads. *Bipedal mobility* (upright walking), alone, produces significant stresses to the *skeleton*. Designing a backpack to suspend weight on a moving body requires applying biomechanical principles. Chapter 4 discusses some of the challenges of backpack design.

### 1.1.1 Environment

Designers consider where, when, and how their products will be used. A major focus of this book is the body's relationship to physical environments. However, the *social/psychological environment*—expectations for personal appearance in a social/cultural setting—affects product acceptance. Physical environments can be non-threatening everyday surroundings, the situations most of us encounter every day. Environments can also be challenging; temperature extremes require wearable products that protect body functions. Extremely hazardous environments, like sites that have been contaminated with hazardous chemicals or nuclear waste, or areas where infectious disease is out of control, call for full body protection. Whatever the environment, understanding the human body is essential when designing wearable products.

#### *Everyday Environments*

Clothing that most people wear every day is designed to meet mass market needs and often falls into the fashion apparel category. In these settings, physical environments are not extreme, but understanding the social environment is important. All types of apparel, as well as *accessories* like hats, gloves, and shoes, are used in everyday-wear environments. See Chapter 3 for specifics on designing for the head and neck, Chapter 7 for information on hands and wrists, and Chapter 8 for details on feet and *ankles.*

The fashion industry uses an ideal image of the human body based on selected body measurements, a standardized company *manikin,* or a live *fit model* believed to represent the target market. Wearable products based on the standards can lead to consumer dissatisfaction, especially for those who do not fit the ideal. Mass market demands are blamed for these problems. Companies want to sell as much product as possible with the least cost, so product shapes are simplified and numbers of sizes are reduced. With an emphasis on simplification for economy, manufacturers do not incorporate information about forms, functions, and sizes of real bodies.

#### *Sports and Athletic Environments*

Designing apparel and equipment for sports and athletic activities requires comprehensive knowledge of body structure, function, and movement in the sport-specific environment. The requirements of the sport or athletic activity, characteristics of the playing environment, and interactions with other athletes who are part of the environment need to be addressed. Consider the similarities and differences in wearable requirements of swimmers and downhill skiers. Each activity is typically performed in a specialized setting and the athletes, amateur and professional, perform specific and often difficult physical feats. Professional athletes may have more detailed and stringent requirements and a governing body or league may be involved in what can or can't be worn (LaBat & Sokolowski, 2012). What the athlete looks like—social environment requirements—also plays a role as some sports require a long lean body line while athletes in other sports, like hockey, want a bulky and imposing look.

### *Medical and Health Environments*

Wearable medical products like *blood pressure cuffs*, thermometers, and *heart monitors* may be worn in different settings; some in a hospital with medical personnel as part of the environment, some in a home or social setting where the wearer makes decisions about product use. Medical monitoring devices for home use are slightly different from hospital models, so the person can apply the device without assistance and easily read and understand the output. In addition to knowledge about body structure and function, medical product designers need information about medical and health conditions. Durfee and Iaizzo (2015) discuss the medical device design process, from understanding the opportunity, through *prototype* testing, clinical trials, regulatory approval, and developing a business plan.

Healthcare products cover a wide range—from monitoring devices to disposable *incontinence* products (diapers) for infants or adults. Some of these products require interaction with at least one other person; infant diapers are changed by an adult and devices used in a medical setting are applied and the output read and recorded by a medical professional. Some products like prescription eyeglasses are designed to improve body functions, while other products, *prosthetics* and *orthotics,* may replace body parts and/or assist with function. Cross-over products under the *wearable technology* category—wrist bands, arm bands, belt-mounted *pedometers*, and watches—combine health monitoring with sports activities. Designers have tended to concentrate on the functional aspects of healthcare and medical products, while viewing design *aesthetics* as secondary.

### *Theatrical Environments*

Theater *costume* must be designed with many factors in mind. A stage setting comes with challenges and restrictions: intense light, limited space, and interactions with other actors in the theatrical environment. The costume designer primarily is concerned with the visual presentation of the actor's character. But each actor is a human body with fit, function, and *comfort* requirements. Historic theatrical costume often requires precise fit. Think of designing an actress's period costume for the 1850s. The designer may want to present an authentic image on the stage, while giving the actress a comfortable garment that allows *full breath*—unlike the actual garments of the period. Theater designers with working knowledge of the body's form, functional motion, and range of sizes can design not only beautiful but working stage costumes that meet the aesthetic and functional needs of the actor and the environment.

#### *Hazardous Environments*

Designing *personal protective equipment (PPE)* for use in hazardous environments requires detailed knowledge of normal body functions as well as the environmental threats. Most often PPE refers to clothing and equipment (like respiratory masks) that protect a person from one or more natural or man-made environmental risks. When PPE—for example pesticide protective apparel, *asbestos abatement suits,* and *bio-hazard suits*—is used in industrial settings the suits must fit a wide range of body types and sizes. Specific hazards dictate distinct design features. Protective gear fit is important and may be crucial to a worker's health and survival. Accommodating task-related work movements helps maintain the integrity of the apparel/equipment ensemble in the hazardous environment.

Think about the requirements for a *space suit* worn outside the International Space Station (ISS). Space suit design requires a thorough understanding of the basic elements of wearable products, from fit to structure to materials, and of each body part, function, and movement—in addition to the multitude of threats in the space environment. See Chapter 9 for further discussion of PPE and an analysis of space suit design.

### 1.1.2 Human Body

The other given for the wearable product designer is the human body. The human body is often thought of as a uniform structure. In fact, human bodies come in a wide range and variety of forms and functional capabilities. The wearable product designer can improve designs and find inspiration for new designs not only by studying facts about the human body but by carefully observing real people and real bodies.

#### *Body Observation Exercise*

You may be taking the features and functions of your own body for granted. To test your awareness of your own body, try this simple design exercise. Look at your left hand. Really study it. First, start by looking at the *palm* side of your hand. Notice the variation in the color of your skin—it's not just flesh-toned, but a whole range of colors and shades. Note the texture of your skin—where it's smooth and rough—feel the differences with your right hand. Look at the wrinkles of your palm—some deep and prominent, some tiny and seemingly random. Look at the *blood vessels* running under your skin's surface. If you stretch your skin slightly, you might be able to see how blood vessels run from your wrist, through your palm, and into your fingers.

Notice the shape of your palm and the shapes and form of your fingers in relationship to each other. Which finger is the longest? the shortest? Are your fingers long and thin or short and chubby? Do you use common names for your fingers—*thumb, pointer, middle, ring,* and *pinky*? Now turn your hand over and use the same process to examine the top of your hand. How are the palm and top of the hand the same? Different? Why do your *knuckles* on the top of your hand have such wrinkly skin? Note the shape, color, and texture of your fingernails.

How does your hand move and function? Form your hand into a fist. How do your fingers work with your palm? What are the positions of each finger and the thumb? Did you place your thumb under your other fingers or on top of your fingers? Pick up a glass or other rounded object with your left hand. How does your hand work to do this? What do you feel with your hand—pressure? texture? temperature? What makes it possible for you to grasp something? Now try a pen or pencil and write or draw something. If you are left-handed this may be easy. If you are right-handed—not so easy. Observe how your fingers bend and work and move. Next, try picking up a smaller object like a pin or needle. Did your fingers function differently? What sensations did you feel with your fingertips? Was this task easy or difficult?

Now ask a colleague or classmate to do the same exercise with you. Tell the person you are in the exploration phase of design. Even if the person is your age, race, and *gender*—it's likely there will be differences in what you observe. You can extend this exercise further by studying the hands of people of different ages and genders. The next time you are in a public place, focus on hands and note what you see. Start a sketch book of hands with notations on your observations.

Now, consider this challenge. Design a *universal* hand covering that is warm in a cold climate, beautiful, comfortable, and provides excellent mobility; in other words, the wearer can pick up a glass, write or draw, and pick up a small object like a pin or needle. And while you are at it—make it one-size-fits-all.

### Human Body Diversity: Opportunity and Challenge

Humans come in many forms and sizes due to the variety of forms that skeleton, muscle, and *fat* allow. Typically, people are seen as short or tall or somewhere in between. People are described as "skinny" or "fat." Each individual's proportions and distribution of physical features contribute further to the diversity in human shapes. Apparel and other related industries have relied on the concept of an *ideal body* when designing products. The ideal body, male and female, varies culture to culture. Eicher, Evenson, and Lutz (2008) state, "Considerable consensus exists on what is the most beautiful in body build and everyone understands what the standards are even when they are not explicitly stated" (p. 349).

The ideal figure is often defined by the relationships of one segment of the body to another. The Greek nude female figure has been used as the classical standard of the ideal, with emphasis on perfection in proportion. For instance, the female figure should display the same unit distance between breasts, from breasts to *navel* and from navel to division of the legs. The Golden Ratio or Golden Mean (approximately 5:3) has been seen as visually satisfying for centuries and has been used to judge beauty in body proportion, apparel designs, and architecture (Finch, 2003).

Head length has been used as the unit of measure when defining body proportions. According to DeLong, the head can become dominant as the unit of visual measure, "with the head length of a particular body used for measuring the lengths of the remainder of the body" (1998, p. 108). As illustrated on the left side of Figure 1.2, the average female body is 7 ½ heads tall. In contrast, the fashion industry stretches vertical proportions to unrealistic lengths with fashion illustration figures using a body height that may measure anywhere from 9 to 12 head lengths (Figure 1.2, right side).

The ideals for athletic bodies are often based on requirements for excelling in a sport. Basketball players are tall, power weight lifters are muscular, and figure skaters present a long, elegant body line. While ideals may be useful, simplifying our view of the world and in some ways making design more efficient, relying on ideal forms may lead to missed opportunities: human needs that could be addressed with products designed for real human bodies. As population demographics change, "body demographics" also change. While increasing obesity rates are not a healthy trend, the need for wearable products for larger people is undeniable. As a melting pot of nationalities, the United States presents particular challenges for designers trying to design wearable products for people of many body types and sizes.

Bodies change through the aging process. Aging populations, worldwide, will open new wearable product opportunities. The aging body develops obvious changes in body shape and form, size, and *posture*. Sometimes changes are subtle, sometimes drastic. By understanding the body through a lifetime, designers can better address specific needs and concerns. Focused attention to the changes in aging bodies may produce better and more successful products. Watch for examples of changing anatomy through the lifetime throughout this book.

### Factors Affecting Body Diversity

Many factors affect the shape, form, and size of a human body. Skeletal structure can determine if a person is *long-waisted* or *short-waisted*. Long-waisted means that the upper *torso* is long in proportion to the lower body, and short-waisted means the opposite. Most apparel companies assume that upper torso and lower torso are ideally proportioned and produce apparel accordingly. Separates in apparel, tops and bottoms designed to be worn together, are one method of addressing variance from the ideal allowing more or less

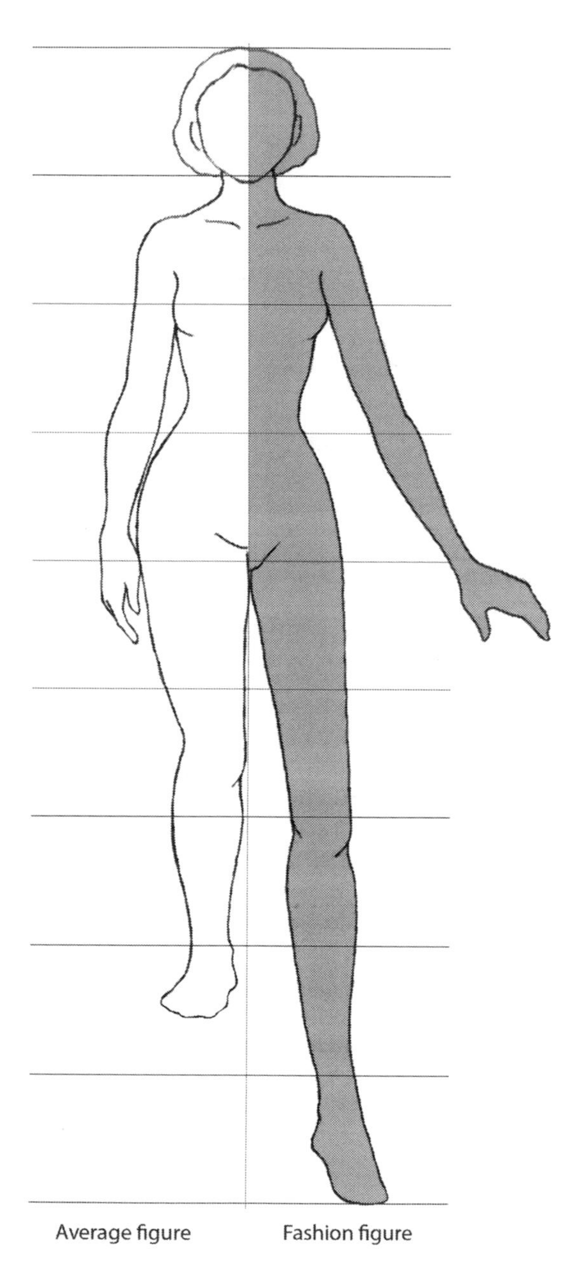


**FIGURE 1.2**
Head length as measure of body proportions: 7 ½-head average figure (left side), 9-head fashion figure (right side).

overlap at the mid-torso and providing adequate fit for a wide range of people. Fitting pants to a pendulous mid-torso is difficult. Personal preference may be to wear the *waistband* below the belly or to hike up the waistband above the belly—each giving a unique aesthetic look.

Differences in muscle definition and bulk also affect the size and fit of products. Consider how a shirt is designed and sized to fit a man. Two men may have similar torso measurements, but very different neck measurements and/or different arm lengths. Men's dress shirts are offered in sizes labeled with neck size and sleeve length, sometimes called a *bi-dimensional sizing system* (Chan, 2014). The neck is the primary measurement and arm length is the secondary measurement. Two men with similar arm lengths could have very different neck measurements and markedly different torso shapes, forms, and bulk if one of the men is a body builder with muscular, sloping shoulders, and a large neck and chest. Differences in posture will also affect fit and placement of a shirt collar on the body. Gender, health and fitness status, diet and nutrition, and ethnicity also lead to variations in body form and size.

### 1.1.3 Wearable Product as Mediator between Environment and Human Body

Designs for the human body must take the complex interactions of the body and environment into account. The wearable product mediates the interactions of the body and environment. The designer must consider the physical and social/psychological settings of product use. One may dominate, but typically both influence design choices. In a formal social setting like a wedding, appearance and aesthetics are major concerns. What are the expectations of participants? What should be worn in that setting? In a challenging physical environment like outer space, physical demands outweigh social concerns. How can the human survive and work in the most extreme conditions?

Conflicts between social/psychological demands and physical needs often present design challenges. Many women love to wear high-heeled shoes, placing social appearance needs above comfort. Years of wearing shoes that put the foot in an unnatural position, subjecting the ball of the foot to excessive pressure (often compressing the toes to a point) eventually, if not immediately, leads to foot pain. Men wear dress shirts and ties that constrict the neck, possibly affecting visual performance (Langan & Watkins, 1987). Designers are challenged to accommodate social expectations: what people "should" wear. Even when staying within the expected appearance and structure of items like high heels and men's shirts and ties, designers can improve products if they have basic working knowledge of the human body.

Do shoe designers typically know that there are 26 bones in the foot that function independently and in concert to provide stability and mobility for the body? Shoes can be designed with *anthropometric* shaping and material selection to stabilize and cushion the foot. Men's dress shirt designers should understand the basic anatomy of a man's neck. Which spinal structures control neck stability and flexibility? What measurements should be used to *draft* a shirt collar and shape the neckline seam? Menswear designers should understand the neck's skeletal structure, muscles, and the blood vessels passing through the neck (to and from the head) to design shirts that provide comfort and fit. A designer can begin to address many design challenges by matching the design and structure of the shirt to the form and function of men's bodies.

When designing a wearable product, the designer makes important decisions about the structure of the product, the materials that will be used to make the product, and some method of fitting the product to the body. These three product elements—structure, materials, fit—also interact affecting the function of the product. For most products, the permutations of these elements can seem endless. However, developing selection criteria early in the design process can smooth the transition from concept to product.

## 1.2 Wearable Product Design: Anatomy and Product Interaction

After developing a design concept for an environment, the first steps in designing a wearable product are developing its shape and form and relating the key parts of the product to key body parts and features.

### 1.2.1 Complex Body to Simple Forms to Product Structure

The human body is complex. To begin planning *product structure,* start by viewing the body as simple, two-dimensional, shapes. Think about the shape of the neck and shoulder area. It resembles a coat hanger you find in your closet. The stem of the hanger relates to the neck, while the extensions of the hanger form the sloping T-shape of the shoulders. Obviously, the real body has more padding or "flesh" than a hanger, so transform the hanger shape to a padded body form in your imagination. If you continue in this vein, you will find the human body can be viewed as linked simple three-dimensional forms like cylinders, spheres, and cones—a simplified body (Figure 1.3).

Although the body may be envisioned as simple forms, it is composed of complex structures. The skeletal system lends structure to the body. Muscle and fat layer on the skeleton to develop the human form. The *nervous system,* muscles, and bones of the skeleton collaborate to produce motion. The bony skeleton and the body's outermost layers: skin, hair, fingernails, and toenails protect other body structures. Visible or easily felt anatomical structures are most relevant for wearable product designers. Constricting, poorly fitted products may impede body functions. But deeper anatomical structures,

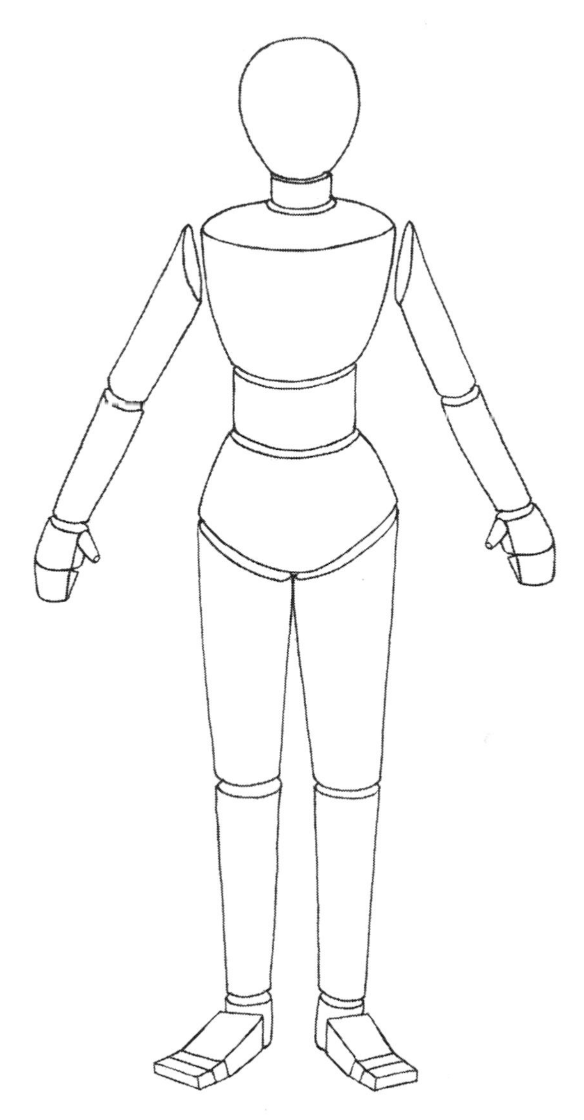

**FIGURE 1.3**
Human body as simple forms.

physiological processes, and biomechanical principles also have design implications. A product designed to provide *thermal protection* must work with the body's internal methods of heating and cooling. Building products to function with body structures is a design challenge.

Products may be structured as simple forms that mimic the forms of the simplified body. A *basic straight skirt* is a tube that wraps around the cylinder of the lower torso and encompasses the tapered cylinders of both legs. The top of the tube-shaped skirt is reduced in diameter using shaping methods, like *darts* or *pleats* or *gathers*, to fit to a waist measurement smaller than the hip. (A dart is a triangular-shaped fold with the base at the waist pinching in excess fabric and coming to a point at a body fullness to shape the tube of fabric to the curved body.)

Many wearable products use *two-dimensional planar materials*—textiles, films, foams—that are flexible and pliable. To form a three-dimensional wearable product the material is cut into shapes and joined (sewn or bonded) together at strategic locations. Product structures for close-fitting garments use designated *key points* to help relate the product to body *landmarks* and to indicate where to join product pieces to one another. Figure 1.4 illustrates a basic upper torso or *bodice* apparel *pattern* with key points that relate to key landmarks on the body.

Because designers often design on a manikin in a pose similar to the standard *anatomical position*, face forward and figure erect (Figure 1.5), the joinings are typically positioned to be less visible from the front view of the body. For example, a shirt or blouse is joined or seamed at the top of the shoulder and at *side seams* between front and back sections of the garment. These seams or join areas are best positioned by understanding the body features that relate to the product sections. Many apparel designs focus on designing for the front of the body while considering the back and sides of a garment as an after-thought. Some garments, like bridal gowns, place greater emphasis on design for the back.

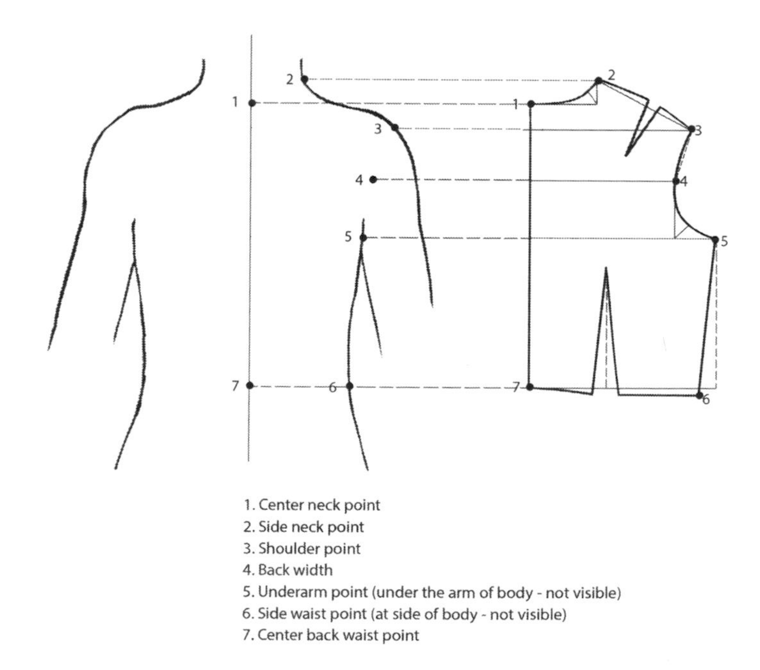


**FIGURE 1.4**
Fitted bodice back pattern; key points related to body landmarks.

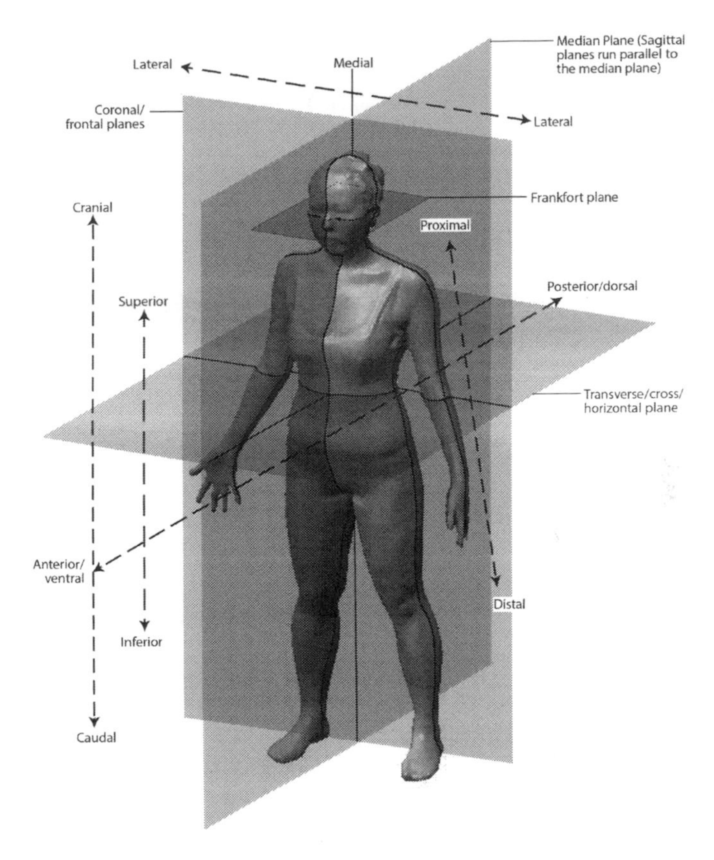


**FIGURE 1.5**
Scan of a person in anatomical pose with orienting planes and descriptions. (Courtesy of the Human Dimensioning© Laboratory, University of Minnesota.)

Structuring a product for some body areas is difficult, for example, the transition areas between the limbs and torso. Landmarks serve as key points in transition areas. Landmarks, in simple terms, refer to the body as the "land" with recognizable topical features that can be "marked" as reference points and used when taking measurements, shaping, and fitting a wearable product to a body. A garment for a transition area like the shoulder, where the cylinder of the arm fits into the inverted cone of the torso, needs to cover both geometric forms and be shaped to adapt to the motion and associated changeable *polyhedral* form of the shoulder. The cylinder of a *fitted sleeve* is attached to the product body by co-locating the intersection of the top of the sleeve with the torso *shoulder seam* and a landmark on the body called the *shoulder point*. In anatomical terms, the shoulder point is the *acromion*, the most lateral bony protuberance of the shoulder blade (*scapula*). The acromion can be obscured by flesh, muscles and fat, making it difficult to locate precisely. Read Chapter 4 on the upper torso and limbs for details on identifying the acromion and related product key points.

For some products knowing precisely how to relate a part or feature of the product to a body anatomical feature is essential. An eyewear frame and lens must be carefully located in relationship to the pupil of the eye to provide the best vision correction and, thus, proper eyewear fit (See Chapter 3 for information on fitting eyewear). Support structures like *braces* and *splints* should relate directly to the body features that need support and any included or adjacent *joints* that must move. Monitoring devices must be carefully positioned in relationship to internal anatomical structures to detect body signals.

### 1.2.2 Product Materials Selection

Selecting appropriate materials for a wearable product helps ensure a successful design. Wearable product designers can choose from a wide range of materials including textiles, rigid and pliable plastics, metals, and more. These basic materials are available in many variations leading to an almost endless assortment of choices. For example, *polymers* can be structured as film or foam—with totally different *performance characteristics.*

Build basic knowledge of materials used for your product category. To research material characteristics and stay up-to-date on new materials access internet sites such as *Materials Connexion* (https://www.material-connexion.online) and *Transmaterial*, an online catalog of new materials (http://www.transmaterial.net). *Smart materials* research and development is opening a new realm of possibilities for wearable product designers. Van Langenhove (2013) describes *smart textiles* as "textiles that can monitor man and his environment and react in an appropriate way" (p. 3). Because smart materials may change and adapt to the body and the environment, designers working with these materials need to have in-depth knowledge of the human body.

#### *Materials Selection Exercise*

To test your materials selection skills, consider the material characteristics needed for a form-fitting, comfortable rehearsal garment that covers a dancer, neck to ankle. Aesthetic and performance needs must be met. First, consider the physical movements of the dancer in the environment. The dancer will be active—twisting, bending, stretching—and the garment must move with the body. Comfort would also be a requirement—non-irritating, perspiration absorbing, and not constricting. Appearance is no doubt important to the dancer, and perhaps the instructor who observes the dancer. Ease of keeping the garment clean, probably meaning washable—is usually a concern. After developing a list of desired *product characteristics*, consider the broad category of materials that could be used. Textiles are used for most dancewear, but body paint or tattoos might be an option. Materials selection often involves compromises. The best product is designed by carefully considering the range of possible materials, narrowing the choices, and then *prototyping* to ensure that you made the best choice.

#### *Materials Selection Method*

Although many designers use years of experience or intuition to select materials for a product, a systematic approach to sorting and evaluating materials can be useful. Bye and Griffin (2015) present a Wearable Product Materials Research Model, which gives designers a structured method for making materials choices for a product. Such an activity may help designers eliminate inappropriate materials before going through a costly prototyping process. With the model, the designer answers a series of questions focused on how the materials might perform in a wearable product, with yes/no responses: (1) Does the material bend or allow movement? (2) Does the material hold shape and structure? (3) Can the material be joined to itself or other components (sewn, welded, etc.)? (4) Can the material support closures? (5) Will the material address user expectations and perceptual qualities needed for product success? "Yes" answers allow you to proceed to the next step, while "no" answers require that you loop back to an earlier step in the selection process. You can use the model or revise it to meet your design objectives.

As you develop your materials selection approach, consider how well materials mediate the body/environment interaction. For example, in a hot and humid environment, will the material help maintain a healthy *core temperature* by directing heat and moisture away from the body? Conversely, in a cold environment, will the material prevent heat loss from the *body core* where the essential *organs* of life are contained and protected? Models can help designers think through the combinations of materials needed for products used in extreme or dangerous environments. After determining the performance characteristics needed for a product, a method to sort through and evaluate materials is not just useful, but necessary.

### 1.2.3 Fitting the Product to the Body

Fit, how a wearable product or garment conforms to the body, affects product usefulness and comfort. A product may be described as having "poor fit," meaning a mismatch of product to body. Or a product may have "good fit" when structure and materials selection work together to produce a product that approximates the human form. Wearable products need to fit a human body or—for many products—a whole range of human forms and sizes. Therefore, determining how to fit the product structure and materials to the body is a necessary step in developing wearable products.

### *Static Body/Product Fit*

Wearable products are supported, in varying degrees, by the body. How do designers make a product fit? It is tempting to try to replicate the surface of the body, allowing only minimal space between the body and the product, much like applying upholstery to a piece of furniture or dipping your hand in warm wax and then letting it cool. Scan technology makes it possible to "skin" a three-dimensional form, removing the surface and flattening it into two-dimensional shapes and pieces. But this approach does not work for most wearable products. Why not? Our bodies are not static structures—the curves and planes of the body vary with voluntary movements and automatic activities, such as breathing and *digestion*.

A product may conform to the body or, at the other extreme, make contact at only a few body points (Figure 1.6). Consider a speed skating suit that conforms closely to the body. In the skater's characteristic crouched body posture, it contributes to a smooth and *aerodynamic* form. These garments apply shaping methods to two-dimensional materials or use pliable, stretchy materials. The suit is in such close contact with the body that it may compress the body, altering body form. A *sandwich board*, a prop for a human walking-talking advertisement, is suspended at two points, from the shoulders, and does not conform to the body at all. Most wearable products fall somewhere between exact replication or compression and minimal contact.

### *Fit "Ease" and Style "Ease"*

A wearable product must be tight enough to stay on the body and perform its intended function but can't be so tight that it impedes activities or body

**FIGURE 1.6**
Contrasts in fit: Speed skating suit and sandwich board.

functions or causes discomfort. Everyday apparel is designed to shape to the body with *fit ease* and possibly *style ease* added to the garment. In other words, most wearable products do not conform precisely to the size and form of the human body. Fit ease is defined as "length or width added to a body measurement to insure comfort, mobility, and smoothness in a garment" (Minott, 1978, p. 6). In some cases, for instance *compression garments* like *shapewear* and *girdles*, the fit ease may be a negative value. Style ease is length or width, in addition to the minimum fit ease, that produces a desired style or garment shape.

A basic straight skirt that conforms to a body with a 30-inch waist and 38-inch hip circumference would typically have a waistband with an additional ½ inch to 1 inch added to the band to allow the person to move and breathe, and an additional 2 inches at the hip to allow bending, sitting, and leg movement. A gathered skirt, sometimes called a *dirndl*, for the same person would have the same waistband measurement, but the skirt may include an additional 30 inches or more of fabric at the hip as style ease. Figure 1.7

**FIGURE 1.7**
Gathered (dirndl) skirt superimposed on a basic skirt with darts.

illustrates a full gathered skirt superimposed on a basic skirt with darts. The two garments demonstrate the differences between a form-fitting garment and a garment that does not conform to the body. Gersak (2014) has written an excellent explanation of types of ease and considerations for ease allowance choices for different designs.

### Dynamic Body/Product Fit

While designing for the body in the anatomical position is a good starting point, it is necessary to consider motion and the various physical postures and poses assumed while wearing a product in order to design a comfortable and functional product. Formal clothing and tailored garments are designed for a person in a static standing position. Look at the scans of a person wearing a fitted *tailored jacket*—in a standard pose and in a motion pose (Figure 1.8). While the jacket lies smoothly on the body in the anatomical pose, wrinkles and strain on the fabric are quite noticeable in the scan of the woman with one arm raised.

Designs for wearable products such as *athletic apparel* and protective clothing require features to allow as much movement as the wearer needs. Data on human body *range of motion (ROM)* has been collected and used in medical settings for a long time. *Human factors* researchers and *ergonomists* have conducted studies of the body in extended poses for use in design of automobiles, interiors, and furniture. Use of range of motion information in wearable product design, especially apparel, is fairly new. *3D scanners,* associated software programs, and *motion capture systems* are changing how wearable product designers use motion data in design development. Range of motion for body regions is described throughout this book and Chapter 9 presents basics on motion capture technology.

### Fit and Comfort

Fit comfort is subjective. One person may prefer products that conform to and perhaps constrict areas of the body, while another may want less-conforming or even very loose products. Ashdown and DeLong (1995) explored individual preferences for loose and tight clothing and made some suggestions on meeting the challenge of fit comfort. Some differences can be accommodated with a range of sizes in a well-designed *sizing system.* Designers need to know the likes and dislikes of the people they are designing for, although discerning fit comfort for a large target market is difficult. Meeting individual comfort needs is possible.

*Tactile* or *sensorial comfort* (how satisfactory a product feels to the wearer) includes the complex interaction between the surface characteristics of product materials and the wearer's skin. It also includes *perception* of the weight of the product materials and total product at a given point on the body or on the body as a whole. Materials can be perceived as comfortable or uncomfortable.

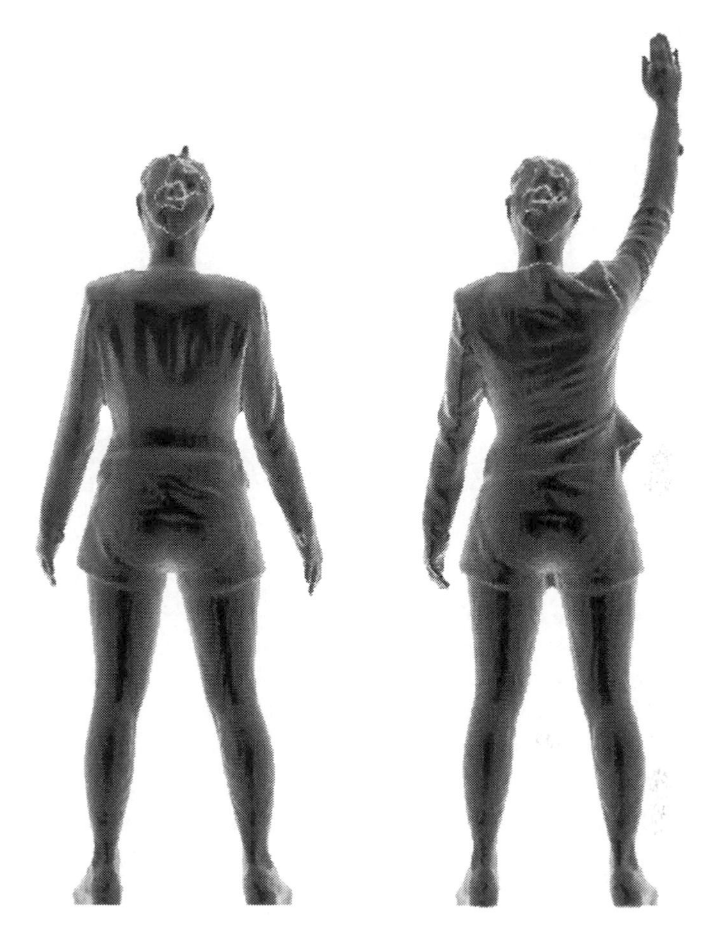

**FIGURE 1.8**
Woman wearing a fitted jacket in standard pose (left) and motion pose (right). (Courtesy of the Human Dimensioning© Laboratory, University of Minnesota.)

Wearers may describe products as soft, smooth, scratchy, cool, or clammy. An understanding of skin structure and function (see Chapter 2) can help designers increase product tactile comfort.

Wearer comfort does make a difference. In order to be accepted and used as directed, wearable medical products must meet the wearer's physical and social/psychological comfort needs. Personalized *prosthesis* designs for *amputees*, which make the prosthesis look high-tech or beautiful, can increase acceptance, furthering a well-fitted prosthesis' usefulness as a substitute limb. With an understanding of human anatomy, body function, and body movement—in the context of the wearer's environments—a designer can make educated decisions on shaping the product structure to the body, selecting appropriate materials, and fitting the product to the body to meet individual needs.

## 1.3 Wearable Product Design: Prototype to Multiple Sizes

Most wearable products start with a prototype fit to one person who represents all product users—the *sample size* or *model size*. Then, to prepare for mass production, dimensions for multiple sizes are calculated. This is the typical process for apparel. Some medical devices, like blood pressure cuffs, were first developed and sized for adults and then down-sized for children and infants. Developers seek sources of information on the range of sizes in the target market, smallest user or customer to the largest.

How do designers do this? There are three possible methods: (a) estimate, (b) use the company's database of measurements, (c) apply measurements acquired from a reliable anthropometric study. Estimation is based on readily available information, perhaps measurements of people in the company who might represent the target market or user group. However, estimation usually results in products that do not represent the range of people who will actually wear the product. Some companies have a database of measurements developed from numerous sources over the years. The advantage of this method is that the customer is familiar with the company's product sizing and fit. When designing a new product, a good approach is to use a reliable database of anthropometric measurements. A database from a large-scale study may be purchased or an entrepreneur designer with knowledge of anatomy and product design may develop a small-scale anthropometric study.

### 1.3.1 Assessing Body Shape and Size: Anthropometry

Roebuck (1995, p. 1) describes *anthropometry* as a method to establish the physical geometry of the human body. The name derives from *anthropos* meaning human and *metrikos* meaning measuring. Anthropometric studies are conducted to collect measurements from a representative sample of people. They can be large scale, sampling a wide range of people, or small scale and tailored to a company's needs. An example is the Civilian American and European Surface Anthropometry Resource Project (*CAESAR*®). Bougourd and Treleaven (2014) present an excellent overview of anthropometric studies conducted in many countries using 3D scanning methods. They also discuss planning national surveys, databases, and future trends for such studies. Data sets from these studies can be purchased and limited data is available free of charge for some audiences.

Croney (1971), in one of the first books on designers' use of anthropometry, stated, "Measuring from fixed points on the unclothed figure forms the bedrock of all anthropometric work" (p. 70). Wearable product designers can learn a lot about human body variation by conducting a small-scale anthropometric *pilot study* of people who will wear a product. The process of *landmarking* and measuring several people will give you a better idea of the measurements you might use from a large anthropometric study.

After concluding your study, use the measurements you collected to design or size a product. Your pilot study may help you decide how to modify measurements from a database. For example, data sets do not usually include a head circumference measurement that encompasses the ears. If you are designing a hat that covers the ears, how could you adapt a basic head circumference measurement? You could measure a sample of people taking the standard head circumference measurement (around the head just above the ears), and a second measurement encompassing the ears. Then calculate the percentage difference and use the percentage as an increase for the standard measurement from the database. The percentage increase is likely small, but making the adjustment may make the product more comfortable. Materials characteristics, like knit stretch percentage, also affect product dimensions. A rigid material requires more precision in shaping and sizing than a stretchy material.

### 1.3.2 Size and Sizing Systems

Chan (2014) presents an overview of *apparel sizing systems* developed from anthropometric data. He explains how *key* or *control measurements* are used to develop sizing systems. Apparel is perhaps the best example of a wearable product category that requires a sizing system. For instance, apparel designed for women in the U.S. ready-to-wear market is sometimes produced in 12 sizes, with Misses sizes 2 to 24 as *size designators.* Some products are offered in small, medium, large, and extra-large and may be expanded from extra-small (XS) to extra-extra-large (XXL). These products have forgiving structures that accommodate a range of body sizes. T-shirts are often sized using small/ medium/large size designators, because the style is a simple T-shape and made of stretchy knit fabrics. Faust (2014) discusses the historical development and current practices in size designations and labeling; modern systems are based on simplification and expediency. Consumers are often frustrated by the confusing array of size labels and sizing methods. At one time the U.S. government developed standards for apparel size categories (LaBat, 2007). The industry adopted the standards. While these standards are still available, the measurement data is from the 1930s and 1940s. So, some apparel companies develop their own systems based on target market demographics.

#### *Pattern Grading*

A prototype is finalized, and a perfected pattern is drafted from the prototype—the *sample size pattern. Grading* is the method of systematically increasing and decreasing measurements from the sample size pattern to fit the company's range of sizes. Figure 1.9 illustrates a graded apparel bodice front pattern. Key points that relate to body landmarks are projected out for larger sizes and pulled in for smaller sizes. For example, if the sample size *bust* or chest (body) measurement is 76 cm (30 in.) and the next largest size is 81 cm

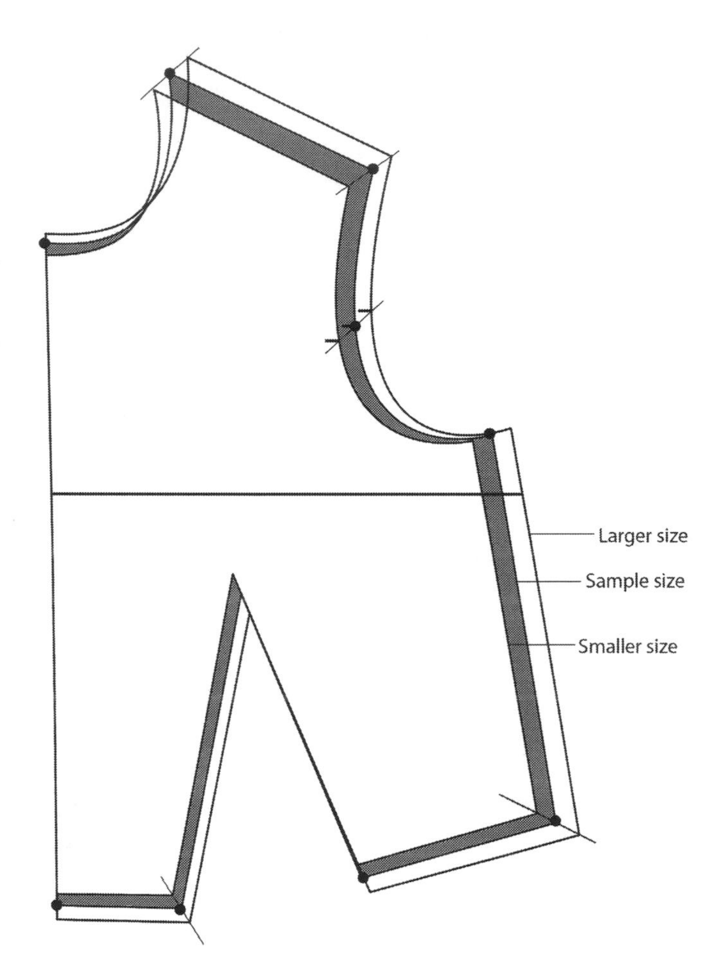


**FIGURE 1.9**
Graded bodice front with sample size (shaded), one size larger, and one size smaller; dots at key points.

(32 in.), then the related product *pattern dimension* must increase a total of 5 cm (2 in.). The 5 cm (2 in.) increase is distributed over the number of pattern pieces in the garment structure. Four pieces increase at each key location 1.27 cm (½ in.) each. The increase and decrease increments are determined in three ways: (a) using standard *grading tables*, (b) applying a company's proprietary *grade rules* or (c) conducting an anthropometric study of a target market and calculating grade increments based on data. Interpreting anthropometric data into useful grade rules can be difficult. The grading process is a further flattening of three-dimensional body data into a two-dimensional shape that must adequately fit a range of body sizes. Schofield and LaBat (2005) present an overview of the background, advantages, and pitfalls of the grading process.

### *Size Selection*

Size designators help customers select a product size. Companies may provide a chart to guide apparel size selection. The chart is based on the assumption that the customer knows his/her body measurements. Most charts include: bust/chest circumference, *waist circumference*, hip circumference, and perhaps back waist length. *Pantyhose* size charts typically use weight and height as the basis for product selection. And weight is the size designator for infant diapers. Footwear size designators differ country-to-country.

The appendices (A through H) present details on landmarking and measuring procedures used to develop product dimensions, determine a sizing system, and/or calculate grading increments. Chapters 3 through 8 describe sizing, landmarking, and measuring techniques for use for products specific to the body region.

## 1.4 Landmarking and Measuring the Body for Product Design Applications

Accurate body measurements are the basis for product dimensions and good product fit, and measurement accuracy relies on precise placement of body landmarks. A well-defined landmarking and measuring plan helps ensure accuracy and consistency. Body landmarks for anthropometric studies refer to somewhat stable, easy to locate points on the human body which are common to all people. The landmarks relate to key points on the product. Landmark locations are based on anatomical features, most often solid bony structures, or slightly flexible features composed of *cartilage*, like the tip of the nose. Fleshy features, like the tip of the nipple or the bottom of the ear lobe, also serve as landmarks.

Measurements (circumferences, lengths, and two-dimensional widths and depths) are distances relating one or more landmark to another. For example, "shoulder point to shoulder point across the back" measures the distance along the body surface from one bony feature to another. Shoulder width measures the *planar* distance between the same two landmarks. Measurements vary person-to-person and depend on each person's skeletal size and amount of fleshy tissue—fat and muscle. A petite woman and a large man have left and right shoulder points, but the span between the points is different, so the measurements are different. All anthropometric measuring methods require good working knowledge of the body, the vocabulary to describe common anatomical structures, and critical observations of each person being measured.

If you are conducting an anthropometric study to collect measurements for a product, have the person being measured wear undergarments appropriate for use with the product—often their own underwear. What a person wears will affect the collected data (Kim, LaBat, Bye, Sohn, & Ryan, 2014). Standard undergarments, like bodysuits for women and shorts for men, are usually issued to participants in large-scale anthropometric studies. Note that stretch bodysuits like the one worn by the model in the photograph in Figure 1.10 may compress the body thus affecting measurements. If you are using data from an anthropometric study, try to find out what the participants wore and determine how what they wore could affect the measurement data you are using.

The person's body position and breath cycle can affect measurements. Foot placement and the relationship of one foot to the other affect height measurements and can affect other measurements like hip circumference. Total

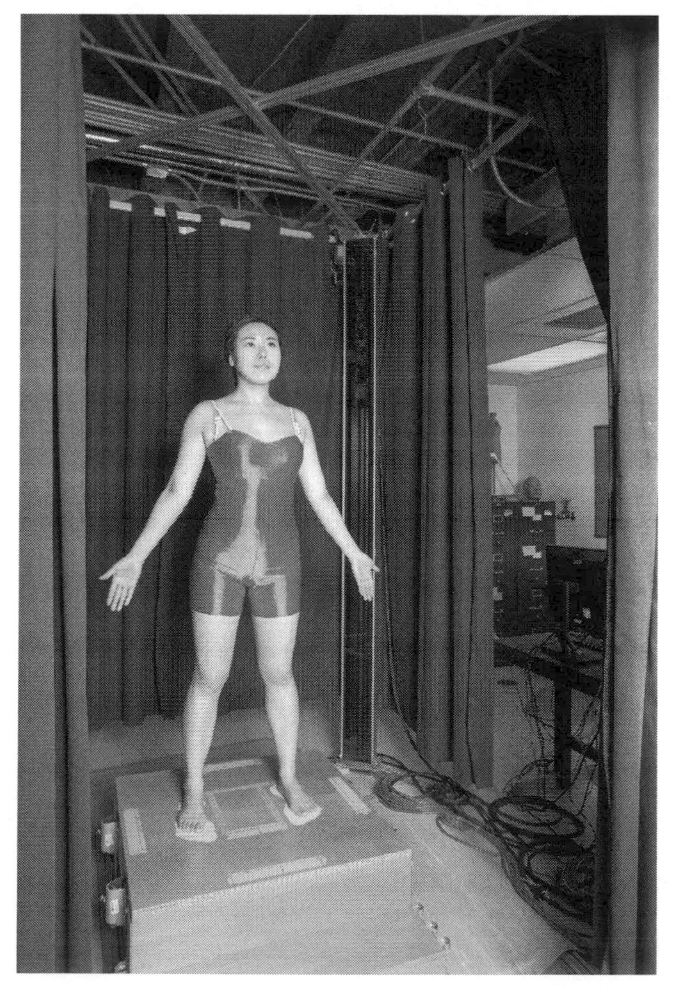

**FIGURE 1.10**
Woman in bodysuit, in 3D body scanner. (Courtesy of the Human Dimensioning© Laboratory, University of Minnesota.)

height measurement for a person with feet spread apart will be different from the height measurement taken with ankles touching. Decide the best body position for the measuring procedure and ensure that each person is positioned in that *stance* when being measured.

Observe a person as they inhale and exhale. Note how the body changes through this breath cycle, typically affecting circumference measurements. If breath cycle will affect a product's dimensions and comfort, determine the best point in the cycle to take a measurement. Body scan procedures often include a protocol to coach people through a *breathing cycle* to take the scan at full inhalation. This protocol assumes that full inhalation will capture maximum torso circumferences. Think through each step of the measuring process and develop a written step-by-step guide that is followed by every person taking measurements.

### 1.4.1 How to Locate Landmarks on the Body

Anatomical features provide the basis for locating landmarks, which, in turn, relate to wearable product key points. Most landmarks require demarcation with a marking pen or an adhesive marker. A marking pen is used to draw a point or a short line. Easily removed adhesive markers, small dots of paper, can be used but are less precise. The waist is usually defined (landmarked) with a cord or narrow belt. See Chapter 6 for guidelines on locating the waist and measuring waist circumferences. For some products, the person is asked to position the waist circumference cord or belt.

The *landmarking process* involves three steps:

1. Look—assess the body for visible key characteristics.
2. Touch—feel the body for key locations, often related to skeletal features.
3. Approximate—after establishing visible and *tactile points*, approximate locations of other landmarks.

#### *Look*

Combine the first step of observing a person (look) with explaining the landmarking and measuring procedure. Observation gives the designer the opportunity to put the person being measured at ease. The designer needs to keep the sensibilities of the person in mind, respecting personal space and modesty as much as possible. The procedure may also involve asking the person about product position preferences, for example, the most comfortable position for a pants waistband. Establishing rapport is useful in obtaining the most accurate and consistent measurements.

Visualizing the product on the person's body is necessary for some products, especially when key points of the product do not relate directly to an easily identified body feature. Some landmarks must be placed at areas of the body that are not well defined. *Bust point* describes the apex of the female breast or bust and is often landmarked. Bust, not an anatomical term, refers to the breast mound. Bust and bust point are clothing terms. Bust point relates to the anatomical feature of the *nipple* visible on the nude female and male bodies as a coloration difference sometimes demarked by a visible central tip. This anatomical feature is obscured when a woman wears a *bra*, so bust point (nipple tip) is approximated as the apex of the breast mound.

#### *Touch*

*Palpation* is a medical term used to describe the method of using the sense of touch to identify or examine internal structures (McKinley & O'Loughlin, 2006, Glossary, p. G-4). Key points on the body that you can feel, tactile points, may be stable and easily identified landmarks. Although this section (1.4.1) focuses on establishing landmarks, in practice landmarking and measuring are done sequentially. As you work, tell the person what you will do throughout the procedure and ask permission to touch them. When practical, ask the individual being measured to help locate landmarks, position, and hold the measuring device (tape measure) in or over areas such as the bust points and *groin* region. Ask the individual being measured to pass the tape measure around their body when making circumferential measurements, to help avoid crowding in close to the subject. Tactile points most often relate to skeletal structures and muscles and are common to all human bodies. Their stability makes them good reference points for determining similarities and differences in form and size, person-to-person.

The back neck point is often used as a landmark to co-locate the center back of a product (key point) to the body feature. The skeletal formation for back neck point is the *cervicale*, the "tip of the *spinous process* of the seventh *cervical vertebra* which is visible and easily palpated when the person bends their head forward" (Kouchi, 2014, p. 87). A landmark drawn or placed on the skin over a tactile point can move relative to underlying anatomical features, if the subject changes position. Ensure that the person is in the preferred stance before taking a measurement. After a tactile point is located and marked, other points can be located relative to it, using visual clues such as skin folds or other means of approximation.

#### *Approximate*

Some landmarks and measurements are an approximation of the best locations on an individual. One of the most difficult areas to measure is the waist—often used for designing apparel and other wearable products. The waist circumference measurement can be taken in various positions of the mid-torso (see illustration 6.5, in Chapter 6). The measurement location is related to each wearer's personal preference. Locating key landmarks for the waist measurement requires approximation and knowledge of the application of the measurement to a product. A discussion with the person can help in locating rather ambiguous and personal preference areas. The designer decides landmark location based on product function and aesthetic preferences. If you are using data from a large-scale anthropometric study, learn how landmarks were located and measurements taken for that study.

### 1.4.2 Measuring Based on Landmarks

Measurements are taken after landmarks are established. A measurement is the distance from one landmark to another. Sometimes a measurement combines distances among several landmarks, for example, *neck base circumference* is measured from the center back neck landmark, to the right side neck point landmark, to the center front base of neck landmark, to the left side neck point landmark, and returning the tape measure to the center back neck landmark—joining all four landmarks. Measurement procedures are described in detail in the appendices. Bye, LaBat, and DeLong (2006) reviewed methods of collecting anthropometric measurements. They classify the methods as: (a) linear, (b) multi-probe, and (c) body form.

#### *Linear Measurements*

*Linear measurements* are distances over or around the contour of the body. A flexible, non-stretchy, tape measure is used to measure the circumference or distance over the body between two or more landmarks. Linear body measurements are collected from a sample of people to form a database of measurements for size increment determination. *Apparel drafting methods* have been used for centuries. Formulas, based on traditional concepts of body form, are used to reduce three-dimensional data (the human body) to two-dimensional geometric shapes (the pattern). Pattern drafting formulas are based on theoretical proportional relationships between body segments. Linear measurements from one individual are used to draft the shape and dimensions of a prototype pattern. In theory, if the body that is measured matches the draft's proportional image, the resulting pattern will fit that person. However, direct relationships between two-dimensional data and three-dimensional product forms are not assured (McKinney, Gill, Dorie, & Roth, 2017). See Figure 1.11 for an example of a bodice draft based on body measurements.

#### *Multi-Probe Methods*

*Multi-probe methods* measure the distance between two points across a plane: body width in a *coronal* plane, body depth in a *transverse* plane, and

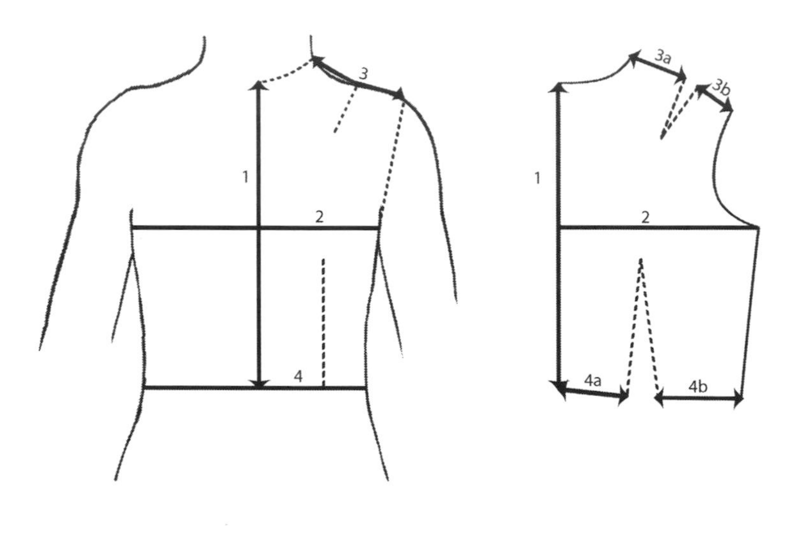


**FIGURE 1.11**
Draft of back pattern based on body measurements.

body height in the *median* plane. Multi-probe planar measurements can be used with linear measurements to determine relative proportions of the body, front to back and to develop three-dimensional product proportions. *Anthropometers* and *calipers* are used to measure widths, depths, and heights. See Appendix A, Figure A.1 for an illustration of a standing anthropometer, a hand-held anthropometer, and a caliper. An anthropometer has two probes, one placed at one landmark on the body and the opposite probe placed at a landmark a distance from the first. Total standing height is measured using a *standing anthropometer*. Widths and depths are measured with a hand-held *beam caliper*. For example, the length of the arm from shoulder point to elbow can be measured by matching a probe on the beam anthropometer to the shoulder point and then extending the device to match the opposite probe to the elbow point. Some areas of the body use specialized equipment. Small *spreading calipers* are used to measure dimensions of the head, face, and hands (see Appendix A, Figure A.1). A *Brannock™ device* is used to measure foot length and breadth and is often used in shoe stores to determine shoe size (see Appendix G, Figure G.2).

### Body Form Methods

*Body form methods*, like 3D body scanning, record information about the surface, shape, and volume of the body. See Figure 1.10 for a photograph of a person in a body scanner. 3D body scanning is augmenting, and sometimes replacing, manual measuring methods. Traditional linear and two-dimensional planar measurements can be extracted from a body scan. The science of body form measurement is advancing as 3D body scanning methods and technologies improve. The power of 3D scanning is the ability to assess body shape, proportion, and posture. New ways of assessing the body with scan data are changing the way designers collect and use information about the body. Scanning also gives designers a better understanding of human body similarities and differences. Figure 1.12 shows outlines of scans of a standard manikin used for designing apparel and three real people all with body measurements for bust, waist, and hip that closely match the measurements of the manikin. You can readily see differences in body shapes, proportions, and postures that may affect product form and fit.

3D body scanning is also changing how designers shape wearable products to the body using *virtual draping*. This technique lets the designer fit two-dimensional patterns to a 3D scan of an individual or a fit model. This technology saves time and material resources. However, the designer still needs to know body landmarks based on knowledge of anatomy, related key points of the product, and how 3D scanner software places landmarks on the body. Accurately describing body feature locations requires looking at the whole body, somewhat like a map or globe.

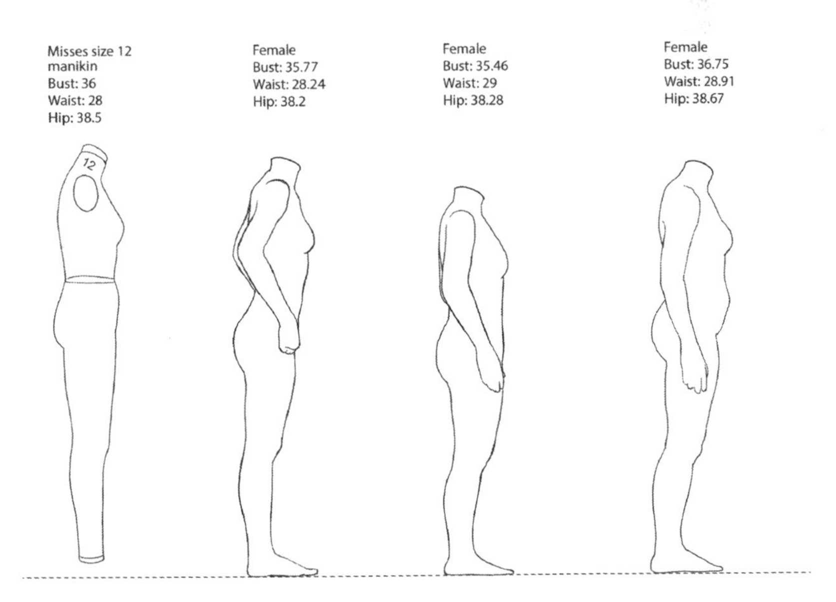


**FIGURE 1.12**
Misses size 12 manikin and scan outlines of three women with Misses size 12 body dimensions. (Courtesy of the Human Dimensioning© Laboratory, University of Minnesota.)

## 1.5 The Human Body "Map"

Discussion of anatomical structures and related product features requires clear communication of specific location and relative location of one feature to another. Think of the human body as a three-dimensional geographical map, like a globe, but with a distinct front and back. When trying to locate a city or feature on a map or globe you use locator aides to pinpoint the location. Similarly, medical professionals use terms to help reference body part locations. These terms are useful for designers. Directional language—based on anatomical terms—for a body "map" will be used throughout this book. Leonardo da Vinci, one of the most well-known artists and designers, described and illustrated the human body using geographical mapping terms (Lester, 2012).

Like a geographical map with latitude and longitude and directions of north-south and east-west, the body can be segmented using planes to distinguish left from right, front from back, upper from lower. Figure 1.5 illustrates the anatomical position; standing, with the palm facing forward and thumb pointed outwards (model's right side). It also shows the anthropometric position with the palm facing the body (model's left side). This slight difference in hand position could influence measurements used in designing a product.

The planes are defined as:

- The *median plane,* or *midsagittal plane,* is the midline plane through the long axis of the body dividing the body into equal left and right halves.
- *Sagittal planes* run through the body front to back and are parallel to the median plane.
- The *coronal* or *frontal plane* runs through the body from side to side, perpendicular to sagittal planes, dividing the body into unequal front and back parts. A coronal plane through the shoulders, or alternatively through the geometric center of the body, is commonly used as a reference plane.
- A *transverse, cross, or horizontal plane* is perpendicular to the long axis of the body and is used to create a body cross-section. The *horizontal* plane bisects the body at the mid-section into upper (cranial), and lower (caudal) parts.
- The *Frankfort* (sometimes spelled *Frankfurt) plane* is a plane passing through the lower edge of the left bony eye socket and the upper margin of each ear canal. Ideally, this plane lies parallel to the floor. It is close to parallel to the floor in the normal head position in the standing person. It is sometimes referred to as the auriculo-orbital (ear-eye) plane.

The following paired anatomical terms relate to the planes:

- *Cranial* refers to a structure being closer to the head.
- *Caudal* refers to a structure closer to the feet.
- *Superior* places a part higher than another structure.
- *Inferior* places a part lower than another structure.
- *Anterior* or *ventral* refers to a structure being more in front, toward the chest and belly, than another structure.
- *Posterior* or *dorsal* refers to a structure being more in back of, toward the body back, than another structure.
- *Medial* refers to a structure closer to the median plane than another structure.
- *Lateral* refers to a structure further away from the median plane than another structure. May also refer to a side view of the person.
- *Proximal* is often used in reference to the limbs (arms, legs) and refers to a structure being closer to the median plane or attachment of the limb to the torso than another. You might think of "proximity to."
- *Distal* is often used in reference to the limbs (arms, legs) and refers to a structure being further away from the median plane or attachment of the limb than another structure. You might think of "distant from."

Designers can use these planes and terms to clearly communicate their product ideas in relationship to the body. These planes and terms are used throughout this book when describing parts of the body and when describing one part of the body in relationship to another.

## 1.6 Design for the Human Body

The body is the foundation for wearable product design. Chapter 2, on human body systems, introduces how body functions affect product design choices. In Chapters 3 through 8 the body is segmented into body regions to look at relationships between wearable products and body structures, functions, and movements. Chapter 9, on design for the whole body, presents design examples that incorporate relationships between whole body wearable products and many body systems and structures.

# 2

# *Body Systems: The Basics*

Basic anatomical terms and functions of the components of each body system are essential knowledge for wearable product designers. Products may be designed to enhance, assist, restrict, and/or replace some elements of systems. Everyday clothing can serve many purposes—modesty, warmth, comfort—and most people prefer clothing that enhances their body. Eyeglasses may enhance a person's appearance while correcting vision. Compression garments can be used to assist the body's *lymphatic system*. A brace is used to restrict a body part, and a prosthetic replaces a body part. Learning how to observe, evaluate, and measure system components in static and dynamic forms will facilitate approaches to designing for human anatomy with all of its integrated systems.

**Key points:**

The systems and *tissues* described in this chapter include:

- Skeletal system: Body frame
- Muscular system: Movers
- Nervous system: Sensing, directing, and coordinating
- Fat: Form, fuel storage, and insulation
- Respiratory and circulatory systems: Exchange and transporting fuel
- Lymphatic system: Body cleanser and defender
- Digestive system: Energy conversion and waste disposal
- Urinary system: Fluid regulator
- Reproductive system (*genitalia*): Pleasure center and multiplication unit
- Integumentary system (skin, *nails*, hair): Body covering and protection
- Endocrine system: Regulating and balancing body functions

The human body, starting with a single cell from each parent, grows and develops into a complicated structure of many tissues. Some tissues organize into organs, for instance *kidneys* or *lungs*. Organs and tissues work together as systems for example: the nervous system, the *digestive* system. For maximum health and performance, human body systems must synchronize complex internal interactions at all levels: cells, tissues, and organs. Wearable product users benefit when designers understand body systems that directly interface with the environment, are accessible from the body exterior, and affect the outer surface and structures of the body.

All systems are sources of similarity and variation. The systems share characteristic developmental sequences. Starting before birth, throughout the body, tissues differentiate before distinct organs and systems develop. Tissues may be hard or soft—consider the differences between a bone and an eyeball. Some tissues, like skin with hair and fingernails, grow continually. Other cells, like those in the brain and the heart, have little potential for proliferation after the body matures. Those organs have limited capacity to heal if injured. Muscle cells change form and size as they act, and muscle contours change beneath the skin. Bones—although they can change in density—are rigid and more constant in location and form.

Body systems influence each other and external forces also act on all components of the ever-changing body. Despite obvious differences between tissues throughout the body, all body cells respond to two major influences: (1) nutrients, waste products, and chemical messengers continuously bathe the cells in a biochemical "soup" and (2) "... mechanical signals [forces] operate in tandem with biochemical cues to properly coordinate cell behavior and pattern tissues" (Chanet & Martin, 2014, p. 317). Atkins and Escudier (2013) define two of the forces that can affect the body: *compression force*—pushing; and *tensile force*—pulling. These forces also influence design strategies.

*Mechanotransduction* is the process by which mechanical signals affect the form and actions of all parts of the body. Products worn for prolonged periods have the capacity to alter anatomy—perhaps in an unintended manner. Consider a ring finger that has changed its form from years of wearing a wedding ring. Alternately, think of a cast for a broken wrist. The cast must protect the broken bones from bending forces. However, in the weeks to months it is worn to immobilize the wrist, the bones lose strength and mass (Ruggiu & Cancedda, 2015) and the muscles that move the bones shrink from disuse/*atrophy* (Wisdom, Delp, & Kuhl, 2015).

In the early 1990s, scientists began to explore the ways that cells convert mechanical stimuli (compressive, tensile, and bending forces) into biochemical signals to "sense their physical environment and adjust their structure and function accordingly" (Isermann & Lammerding, 2013, p. R1113). Throughout life, the body continually replaces itself, healing damaged areas with varying success, and maintains a generally consistent form. Because external mechanical forces impact body structure, wearable product designers have the opportunity and responsibility to determine if, and how, mechanical stimuli from products change cells and the tissues made up of those cells, and thus body function.

Wearable products like women's body shapers can produce a degree of wide-spread body compression. Some wearable products, like support hosiery, compress a body part. Tight-fitting body-shaping garments, snug wet suits for diving, stretch tights, and leotards for dance compress the body in different ways and to varying degrees. Belts, collars, or any encircling wearables need to grasp the body. What happens to the body as a result of sustained compression? Narrow bra shoulder straps, supporting the weight of the breast, can create grooves in soft tissues at the top of the shoulders. Will the belt on a pair of trousers have a similar effect on the waist?

Gravity serves as a compressive force on the body. It is an important positive mechanical force in human health. Mechanical loading from gravity helps maintain bone strength by balancing bone formation and breakdown in weight-bearing bones. Astronauts returning from long stays in space, where they are mostly gravity-free, often lose bone density and are prone to broken bones (Sibonga, Spector, Johnston, & Tarver, 2015). Can a wearable product exert gravity-like compression on an astronaut's body to slow deleterious effects of long periods of weightlessness?

Stretching, putting tissues under mechanical tension, elongates *connective tissues* (soft supporting and connecting tissues of the body). A joint (where two or more bones make contact) immobilized in a cast after a bone break is stiff when the cast is removed. Progressive gentle stretching with a splint (a thin piece of rigid material to maintain a fixed body position) and specific exercises aim to restore normal *ligament* and *tendon* length and normal joint motion, but achieving flexibility may take far longer than the period of immobilization (Akeson, Amiel, Abel, Garfin, & Woo, 1987). Keep mechanical forces, as well as anatomy and system functions, in mind as you design products.

To design successful wearable products, it is important to first examine how basic body systems interact with wearable products. The skeletal system (skeleton) forms the frame of the body; the muscular system and fat (a specialized connective tissue) add contours to the frame. Muscles burn *carbohydrates* for energy, and fat provides energy storage and insulation. The muscular and skeletal systems work together to produce body movement. The nervous system contributes direction and coordination for systems throughout the body and plays a key role in body motion. The skin, hair, and nails of the integumentary system are protective coverings. The circulatory and respiratory systems work together to circulate oxygen to the body and to remove waste products. The lymphatic system plays an essential role in draining surplus tissue fluid into the bloodstream and helps fight infection. The digestive and *urinary* systems manage nutrients and fluid balances, and provide waste elimination. The *reproductive* system, physically intertwined with the urinary system, produces babies. The *endocrine* system supplies hormonal controls for multiple body functions.

The skeleton, muscles, fat, and skin have the greatest significance for wearables. The circulatory, respiratory, and nervous systems are necessary in sustaining life. Portions of these three systems are close to the body surface, thus are vulnerable to the environment while giving easy access for body monitoring. The lymphatic system gets special attention relative to malfunctions and wearable products developed to alleviate the malfunctions. The essential tasks of the digestive, endocrine, urinary, and reproductive systems related to wearable product design are also introduced in this chapter.

## 2.1 Skeletal System: Body Frame

The skeletal system or skeleton, the framework for the body, imparts "sameness" to all humans with its arrangement of segments and symmetries (see left sides of Figures 2.1 and 2.2). The skeleton contributes to the variety and range of human body types and forms; an array of short to tall to average. There are similarities in types within a race and a gender, but also vast differences. The similarities create a starting place for designers who are shaping and sometimes developing multiple sizes for a product. The skeleton establishes the base for shaping, segmenting, and suspending products. However, the skeletal differences person-to-person set up challenges. Can you make a single product that accommodates and/or facilitates, maybe even enhances, the functioning of the shortest user to the tallest user? Can you design a one-size-fits-all product that serves the needs of all?

The basic components of the skeleton that relate to wearable product design are described in this chapter—including key information on ligaments and joints and on their relationship to stability and motion. Skeletal details for each body region are in the related body segment chapters. Learn how to identify bony structures that guide functional positioning of a product on the body and acquire information as to how the product can work with the skeleton in its static and dynamic forms. Use the language to communicate skeletal formations to consistently shape and develop a range of product sizes.

Common design terms are presented in relation to anatomical sites. For instance, apparel designers often refer to the vague *"shoulder point"* at the *armscye* (garment armhole). The shoulder point is the lateral tip of the acromion, a bony prominence surrounded by muscles and fat. Designers can learn to accurately locate this landmark and improve the fit of product components like sleeves and shoulder pads.

### 2.1.1 Overview of the Skeleton

The adult skeleton consists of 206 linked bones that give the body both stability and the ability to move. Parts of the skeleton are impact-resistant structures that encase and protect vital organs. For example, the *skull* (the bones of

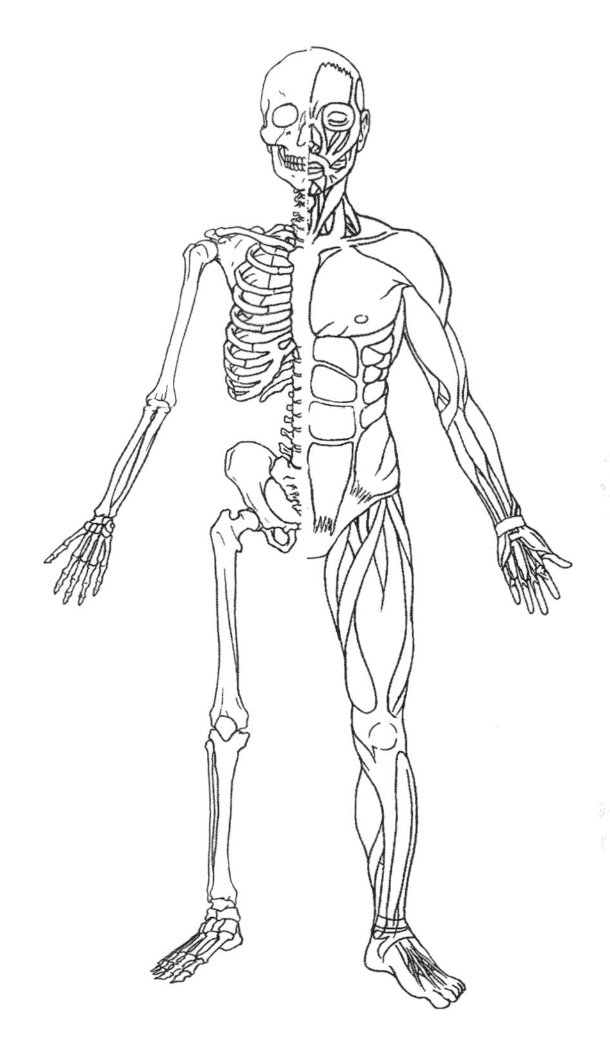

**FIGURE 2.1**
Skeleton (left side) and muscles (right side), anterior view.

the head) surrounds and safeguards the brain. The *rib cage* shields the heart and lungs. The *bony pelvis* contains the intestines and bladder. In women the bony pelvis forms a "cradle" for the unborn child. The female bony pelvis is broader than the male's to facilitate birth. While the skull (without the jaw), ribs, and pelvis have relatively little motion, the bones of the *spine* (backbone) and *limbs* (arms and legs) are essential to mobility. The spine or *vertebral column* protects the *spinal cord;* much like the skull protects the brain. With its segmented structure, the vertebral column allows us to survey the surrounding environment and reposition the head and the body. In adults, there are 33 individual *vertebrae* (bony segments of the vertebral column)

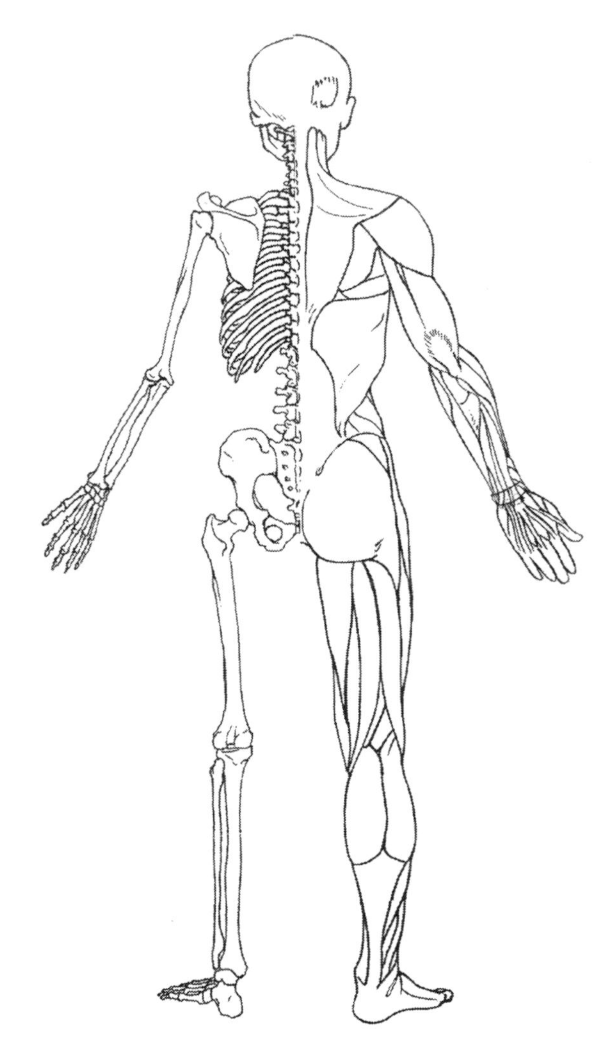

**FIGURE 2.2**
Skeleton (left side) and muscles (right side), posterior view.

named by the body region where they are found; the upper 25 are movable (Moore, Agur, & Dalley, 2011).

The limbs supply large movements over wide arcs, as well as small and precise actions. Limb structures become increasingly more elaborate the farther they are from the torso (trunk of the body). The single bone of the *arm* (between the shoulder joint and the elbow), with the two bones of the *forearm* (between the elbow and wrist), and the small bones of the wrist and hand, all act together to help perform fine manipulations. The leg, parallel to the arm in complexity, serves a far different purpose, delivering support and mobility in three dimensions.

### 2.1.2 Materials of the Skeleton: Bone Types, Cartilage, and Ligaments

The body has two basic bone types: *cortical* and *trabecular* bone. Cortical bone is structurally strong, dense, and found near bone surfaces. Trabecular bone is generally interior and has a spongy lattice form which holds some of the blood forming bone marrow. The long bones in the arms and legs are primarily cortical bone with areas of trabecular bone near the joint surfaces. At birth, infants' skeletons are composed of bone and cartilage (a resilient, firm connective tissue) that will be replaced with bone as the child develops. Areas of cartilage near the ends of the bones, known as growth plates, allow the bony skeleton to expand throughout childhood and into adolescence. Bone types, a long bone example, and bone maturation are illustrated in Figure 2.3. Some areas of the adult skeleton are made of cartilage. Feel

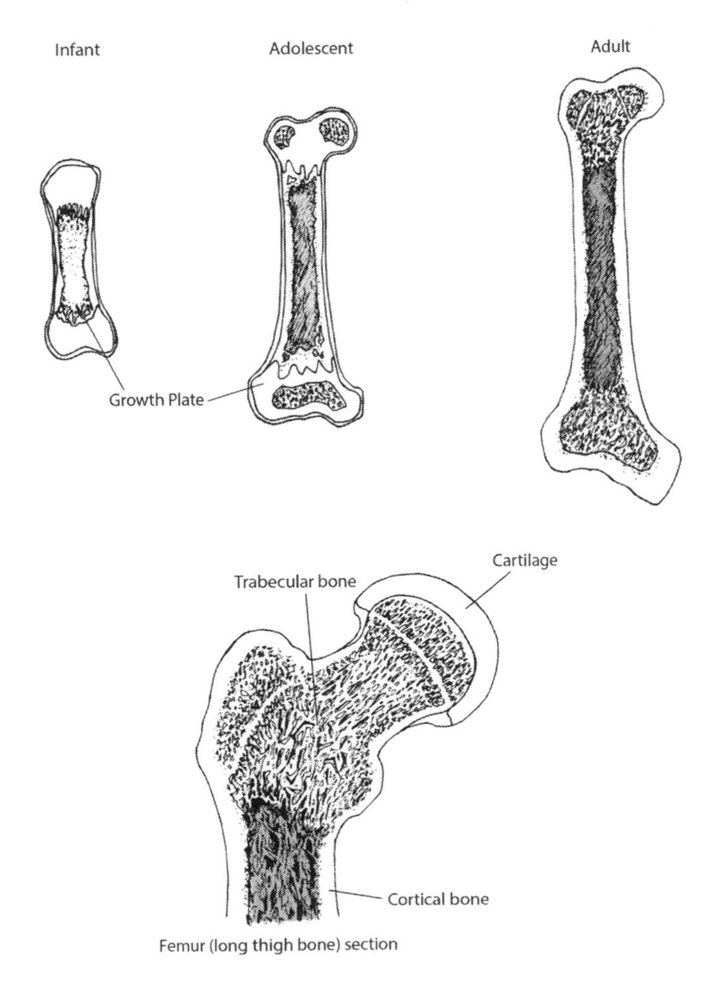


**FIGURE 2.3**
Bone growth plates, bone types, and cartilage.

the tip of your nose. You will notice that it is quite flexible compared to the bony *bridge* of your nose. Cartilage in your nose supplies flexibility. Cartilage can also provide a cushion where bone meets bone in a joint. Ligaments are strong bands of connective tissue that link bones across joints.

### 2.1.3 Linking Body Segments: Joint Types

The spinal column joins the head, the upper torso, and the lower torso. Understanding the structure and function of the spinal column is essential because it forms the "backbone" for designing many wearable products. Because most wearable products are constructed for either the upper body (Chapter 4) or the lower body (Chapter 5), we consider the skeleton divided at the "*waistline*" (Chapter 6) but also look at the spinal column in its entirety in Chapter 4. Limbs join indirectly to the spine. Arms link to the spine via the rib cage at the shoulder joint and legs connect to the spine via the pelvis at the hip joint.

These *articulated* (interconnected) parts must coordinate to facilitate human body motion. As skeletal transition points from one body segment to the next, joints raise challenges in designing products that accommodate motion and facilitate maximum function. Designers need to know joint locations, and how these joints move to ensure that wearable products move with the body. As with other body structures, joint names frequently combine names of the two bones that form the joint, with the proximal bone first in the combined name. For example, the *radiocarpal* joint of the wrist is the intersection of the *radius* in the arm and the *carpal bones* at the wrist. In general, there are cartilage coverings on the opposing surfaces in limb joints. Most joints are movable. At each joint, ligaments not only supply stability but also allow flexibility. Wearable products may be designed to help stabilize a joint, to prevent stretching of the ligaments. Too much joint restriction can interfere with natural function. Joints, as intersections of body segments, can be used as body landmarks. Body landmarks, in turn, correspond to key points of wearable products.

Joints can be classified by the materials of the joint or by their function or motion. There are four joint types based on key materials: *fibrous, cartilaginous, bony,* and *synovial.*

Adjacent bones bound by *collagen* (a specific connective tissue protein) fibers are fibrous joints. The bony segments of the *cranium* and face are joined by fibrous joints called *sutures.* Teeth are anchored in their sockets by the collagen fibers of the *periodontal ligaments* (Saladin, 2007). Teeth are slightly mobile, a fact that allows realignment with braces.

The *tibiofibular joint* and the *radioulnar joint* are *syndesmosis joints*, specialized fibrous joints. Densely organized collagen structures, *interosseous membranes,* span the distance between the *tibia* and *fibula* (bones of the lower leg) and between the radius and *ulna* (bones of the forearm). In the forearm the radioulnar joint allows you to move your hand palm up to palm down and back.

Cartilaginous joints, found where adjacent bones are bound together by cartilage, may be immovable or partly movable. *Synchondrosis joints* are held together by *hyaline cartilage* (a cartilage with few fibers). The joint of the first rib with the *sternum* (breast bone) is an example. A *symphysis joint*, such as the *pubic symphysis*, is a slightly movable joint held together by *fibrocartilage* (a cartilage with many fibers). You can locate this skeletal structure by tracing a line from your *umbilicus*, or belly button, straight down until you feel a bony ridge. In women this joint can flex during childbirth giving the baby's head and shoulders more space to exit the body.

Bony joints were fibrous or cartilaginous joints at an early stage of development. These joints fuse over time and do not move. These skeletal areas, not often considered when designing a product, may need protection during some activities. The median line of the *frontal* (forehead) bone and fused growth plates in adult long bones like the *femur* (in the *thigh*) are bony joints.

In synovial joints the ends of the adjacent bones are covered with *hyaline articular cartilage* and lubricated by *synovial fluid*. The whole joint, with the fluid, is enclosed in a fibrous bubble called a *joint capsule*. Synovial joints are the most common joint type in the body and are the most mobile. There are several types of synovial joints which differ in geometry and function: *ball-and-socket, condylar, plane, saddle, hinge,* and *pivot*. See Figures 2.4, 2.5, and 2.6 for illustrations of these joints. Learning the basic function and locations of each type is useful for designers because the joints must be considered when accommodating motion in a product.

The shoulder and hip are ball-and-socket joints. The superior end of the upper arm bone (*humerus*) is a smooth hemispherical round head (ball) which fits into the saucer-like socket of the shoulder blade (scapula). Likewise, the ball of the thigh bone (femur) fits in the cup-like socket of the pelvis. These joints move in all directions as well as rotate; the shoulder more than the hip.

Condylar (also called *ellipsoid*) joints have an oval convex surface of one bone that articulates with an elliptical depression of another. An example is the radiocarpal joint of the wrist. Condylar joint motion is similar to ball-and-socket joint motion, but the extent of motion is more limited.

A plane (*gliding*) joint is formed of complementary slightly concave and convex bone surfaces that slide across each other (Figure 2.5). The joints between the *articular processes* (interconnecting bony sections) of adjacent vertebrae of the spine represent this joint type. Bending the neck forward, these joint surfaces glide on one another. The joints between the small bones at both the wrist and the ankle are also gliding joints.

A saddle joint is formed with two bone surfaces that are saddle-shaped (concave on one axis and convex on the perpendicular axis). One such joint is the *first carpometacarpal joint* (CMC) between the wrist and the thumb that allows you to move your thumb toward your little finger.

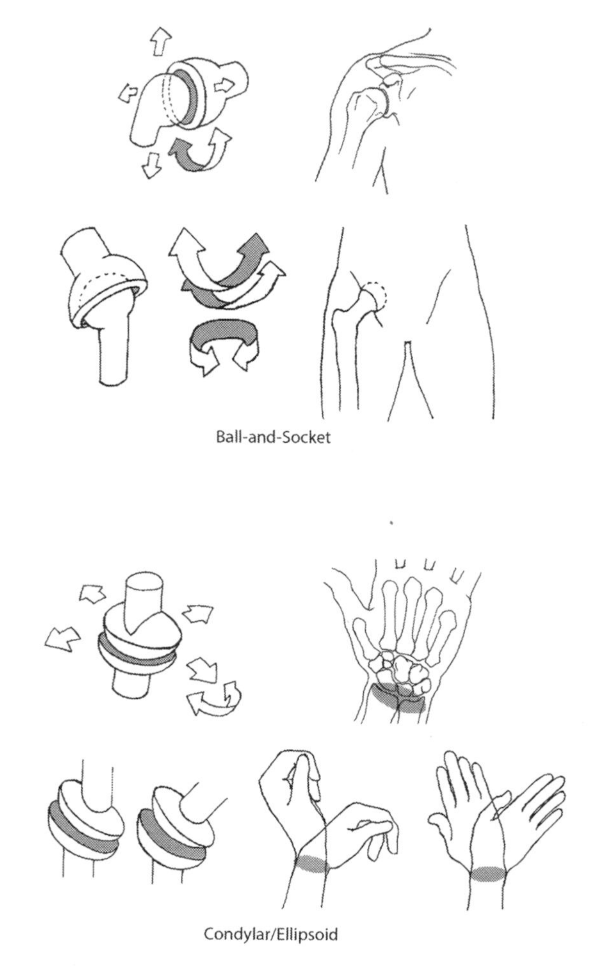


**FIGURE 2.4**
Joint types: Ball-and-socket and condylar (ellipsoid).

Hinge joints flex and extend in only one plane (Figure 2.6). The *interphalangeal* joints, the two joints closest to the tip of the fingers and toes are pure hinge joints. Both the elbow and knee primarily move with a hinge motion. When you bend and straighten your elbow, you will see there is a limit to the joint extension, much like a hinge. The *humeroulnar* joint, between the humerus, the large proximal bone of the arm and the ulna, one of the two bones of the forearm, controls this movement. The knee has a similar motion, but is sometimes described as a condylar joint, because the end of the thigh bone, the femur, has two condylar surfaces (knuckle-like prominences). In addition to flexing and extending, knee motions include "gliding, rolling and a rotational component" (Moore et al., 2011, p. 385).

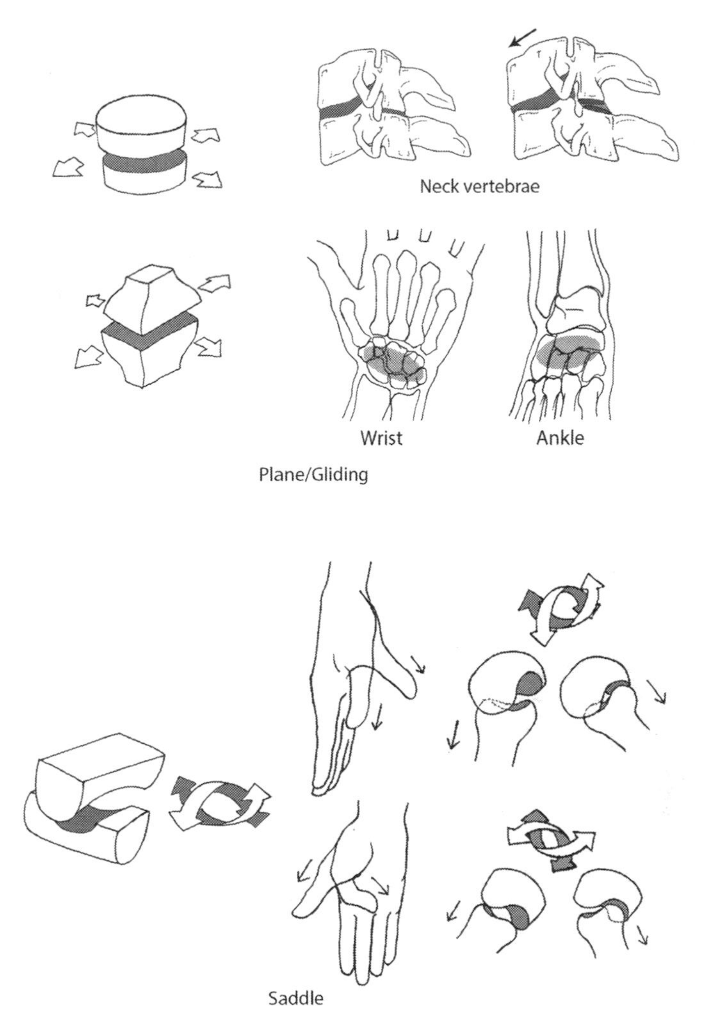


**FIGURE 2.5**
Joint types: Plane (gliding) and saddle.

The pivot joint allows rotation of one bone relative to another. A projection of one bone fits into a ring-like ligament of a second bone, allowing the first bone to rotate perpendicularly (see Figure 2.6). You can pivot your head from left to right because the first two vertebrae at the top of the spinal column allow a significant pivotal motion, although you cannot rotate your head 360 degrees. You can rotate your hand into a palm up/palm down position because of the pivot joint at the elbow between the two bones of the forearm, the radius and the ulna, stabilized by the interosseous membrane between them.

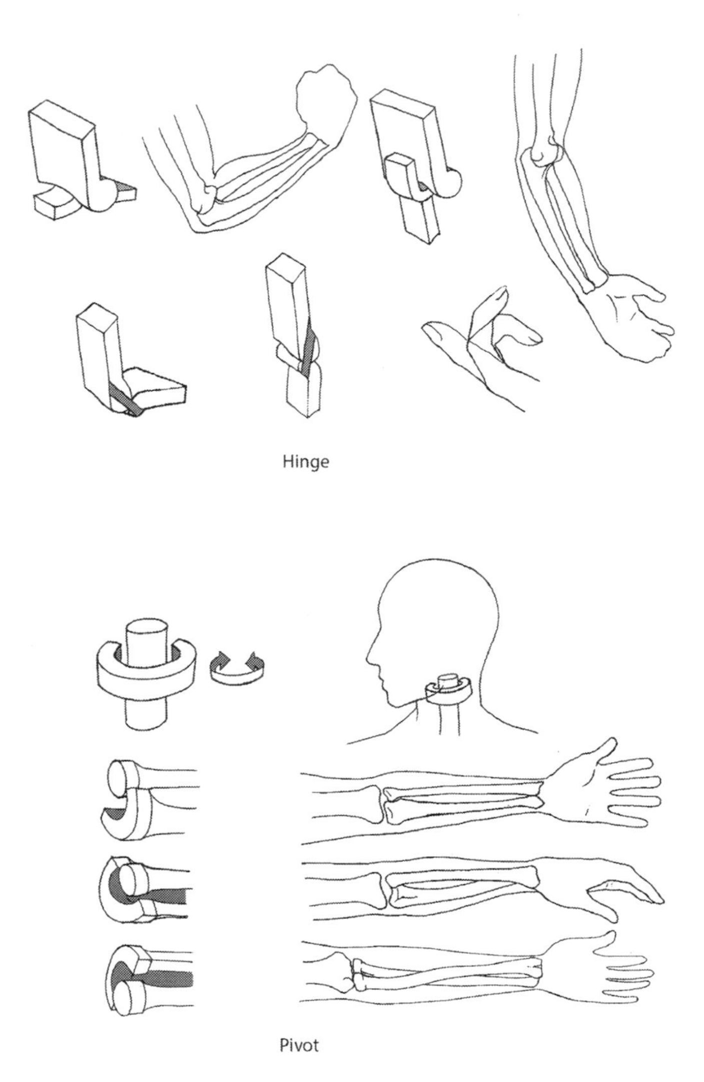


**FIGURE 2.6**
Joint types: Hinge and pivot.

## 2.2 Muscular System: Movers

Muscles form a dynamic body system with many functions. They contribute to body form, produce changes in body position, and maintain posture. Walking, a complex learned activity, efficiently integrates muscle activity in the trunk, legs, and arms—see details about *gait,* how we move, in Chapters 5 and 9. Postural muscles must contract continually to maintain a person's standing or sitting posture. *Exoskeletons,* frames that fit over the body and legs, are wearable products intended to supplant weak or paralyzed walking and postural muscles. Muscles support portals to the interior of the body that open and close and help keep internal organs in place. They aid communication through facial expressions and even modulate sound pressure in the middle ear. *Nerves* course into and out of muscles to furnish electrical connections from and to the brain. Messages from the brain direct and control contractions while electrical feedback to the brain carries information about the positions of joints and tension within the muscle. Blood vessels carry oxygen and nutrients to the muscles to supply the fuel needed for muscle contraction and carry away heat and chemical wastes generated by muscle activity. Muscles generate heat when they contract and respond to cold by raising "goose bumps" and producing shivers.

### 2.2.1 Overview of Muscles

Muscle names often have Latin roots. The names may give an indication of the muscle's form; *biceps*—having two parts. The name *abductor digiti minimi* describes the task of the muscle. It moves the little (*minimi*) finger (*digiti*) away from the central axis of the hand (*abducts*). The term *pectoral* designates a general location, on the chest, the location of the *pectoralis* muscles. Muscle names sometimes include Latin words to indicate the locations of the muscle's *origin* (fixed or mostly stationary muscle end) and *insertion* (muscle end which is more movable during muscle contraction). For example, the *brachioradialis* muscle's origin is a moderately long section of a surface ridge on the humerus, the large bone in the upper arm (in Latin, *brachium*). It inserts on the distal radius, the large bone of the forearm, which it moves. The brachioradialis helps to bend the elbow.

### 2.2.2 Muscle Types

There are three types of muscle: *cardiac muscle, smooth muscle,* and *skeletal muscle.* Cardiac muscle is involuntary, keeping the heart pumping without conscious effort. Smooth muscle, found in the stomach and intestines, urinary system, blood vessels and reproductive organs, is also involuntary. Smooth muscles move blood throughout the body, push food through the body in the digestion process, dispose of wastes through *defecation* and *urination,* and push a baby from the body in childbirth. Layers of skeletal muscles, which perform specific actions, cover the skeleton throughout the body. Some skeletal muscles are superficial and easily palpated, or felt, while others are deep within the body. Most are *striated* (literally "striped") muscles that attach to the skeleton, although some attach to other organs such as the eye or even to the skin. They are considered *voluntary* because a person can consciously use them, for instance, to pick up an object. Circular bands of muscle, *sphincters,* contract to open and close passages throughout the body, like the mouth opening. The striated muscles at the mouth controlling entry to the digestive tract and the muscles controlling waste elimination from the digestive and urinary tracts may be closed or opened voluntarily.

### 2.2.3 Muscle Configuration and Action Rely on Microscopic Changes

When superficial striated muscles change configuration, product appearance and function can be affected. For a clearer understanding of how muscle form changes, study microscopic changes—a type of "sliding" between filaments in the muscle. Look at the right side of Figure 2.7 to see a whole muscle at rest, while contracting to shorten, and contracting to lengthen. The volume of the muscle remains constant while the diameter changes relative to the muscle length. Now look at the illustration from left to right to observe the five muscle structure levels, from submicroscopic (not visible with microscope) to microscopic to whole muscle: (1) a representation or "concept" of *myofilaments* in a *myofibril,* (2) a myofibril, (3) a *striated muscle fiber,* (4) a *muscle fascicle,* and (5) the whole muscle.

On a submicroscopic (1) and microscopic level (2 and 3) striated muscles are a picture of structural and chemical complexity (Kimura, 2013, Chapter 12). Starting at the left of the figure (level 1), note the changes in the relationship of the *actin* (illustrated with solid black lines) and *myosin* (illustrated as changing cylinders). Actin and myosin are two different *contractile protein myofilaments* (threads of protein capable of contracting). These myofilaments,

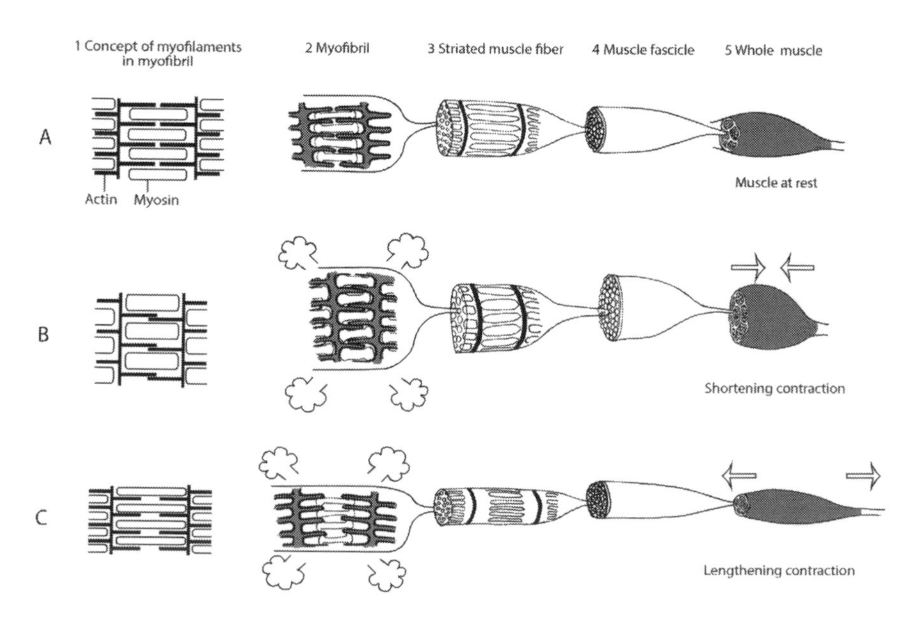


**FIGURE 2.7**
Microscopic level of striated muscles and muscle action.

which cannot be seen with an ordinary microscope, are the basis of the *sliding filament theory of muscle contraction* (Huxley & Hanson, 1954). Non-contractile protein myofilaments (not illustrated) provide a framework for the contractile protein myofilaments. They all join to form myofibrils (level 2). The non-contractile elements give a resting striated muscle fiber (collection of myofibrils) a stable configuration and consistency (level 3).

Other cell structures are part of the muscle fiber, like *nuclei* (plural of *nucleus*, the control center of a cell) and connections from the nervous system to individual striated muscle fibers. *Nerve signals* and chemical changes within the muscle fiber trigger movement of the actin and myosin (levels 1 and 2) and the muscle fiber contracts. Individual muscle fibers (level 3) are separated by thin connective tissue coverings. Groups of muscle fibers surrounded by another outer layer of fine connective tissue strands form muscle fascicles (level 4) which in turn are bound together by more connective tissue into a muscle (level 5). When muscle fibers (level 3) contract at the same time throughout a muscle, the whole muscle contracts and its contours change (level 5). This muscle activity offers an opportunity for development of products to monitor muscle function.

In simple terms, muscles convert sugar (carbohydrate) calories into mechanical energy and heat with the aid of oxygen to produce the microscopic changes of muscle contraction. Starting from the rest position (Figure 2.7-A), these chemical reactions drive the building and destruction of cross bridges between actin and myosin myofilaments, sliding them, to shorten (Figure 2.7-B) or lengthen (Figure 2.7-C) the muscle. These chemical reactions release heat and waste products. Consider the work of the biceps, the major muscle on the front of the upper arm. The muscle contracts and shortens (and bulges beneath the skin) as you bend your elbow and lift weight (Figure 2.7-B). It contracts and lengthens, if you use the muscle to control the movement of weight as you straighten your elbow and lower the weight (Figure 2.7-C).

### 2.2.4 Muscles and Energy Consumption

Muscles play a big role in keeping the body at a healthy temperature. If you exert a great deal of muscular effort working or exercising, you burn large numbers of calories and you feel warm. Excess heat is dissipated when you *perspire* (sweat) and the perspiration evaporates from the skin's surface. Wearable products may promote or prevent this natural process. The complexities of thermal balance and thermal comfort are further explored in more detail in many of the body region chapters.

### 2.2.5 Muscles and Body Form

The muscles of most importance for designers are those closest to the surface. Refer to Figure 2.1 (right side) for the anterior view of the major superficial muscles and Figure 2.2 (right side) for a view of the posterior superficial muscles. The sliding muscle filaments initiate the form changes when a muscle is activated, as in the biceps example. Muscle configurations and sizes vary throughout the body, according to their actions and their origins and insertions. Muscles may be relatively tubular, such as the biceps. The *trapezius* in the back-shoulder area and the *pectoralis major* in the chest are fanlike and move the arm/shoulder mechanism on the torso. The four-part *quadriceps* of the thigh, which straightens the knee, is a *compound* muscle.

Some muscle features are relatively consistent from one muscle to the next. Muscles generally have a red or pink, "meaty" *muscle belly*, which is relatively thicker and at the muscle center. Muscle fibers form the belly. A tough thick band—a tendon; or a flat sheet—an *aponeurosis*, anchors the muscle (most often to bones) on either end. These two anchoring structures are formed from opaque, non-contractile connective tissue. They are the accumulation of all the fine connective tissue strands and coverings described in the five levels of muscle structure. While tendons are sturdy connections that facilitate motion, they are vulnerable to damage. *Tendonitis* is an *inflammation* (a combination of pain, swelling, and warmth in the tissues) or irritation of a tendon caused by repetitive motions or damage from minor impacts. If an athlete ruptures his/her *Achilles tendon*, which attaches the calf muscle to the heel bone (*calcaneus*), the result is pain and immediate loss of strength. Wearable products have been developed as aids in *rehabilitation* (the restoration of satisfactory function) for both tendonitis and tendon ruptures.

Muscles help to fill out the human form, while contributing to the motion and control of body parts. Muscles assume different configurations at rest and at work, and muscle volume varies according to the degree of muscle use. The frequency and degree of muscle activity change the composition and structure of a muscle. Consider the differences in physique between a sedentary adult and a body builder. Each has the same number of muscles, in the same locations, but the visible difference in the form and bulk of the muscles is due to the degree of exercise the muscles have had recently. Admittedly there are genetic and environmental, particularly nutritional, influences on muscle mass and development, but all humans have the same basic muscular structure.

Exercise programs can be tailored to keep muscles in good working condition or to produce muscle bulk and firmness. Body building programs can create visible definition of separate muscles (Figure 2.8). Working the *rectus abdominis* muscles creates the abdominal "six-pack." A person who works to build upper body strength can so change the configuration and bulk of muscles of the neck, shoulder, and arms that standard shirt/blouse sizes do not fit well. Increased *deltoid* muscle volume broadens the shoulders; pectoralis major development enlarges the chest. Inactivity in the extreme, from paralysis or prolonged periods of time in bed due to illness, can result in muscle atrophy (muscle wasting or deterioration).

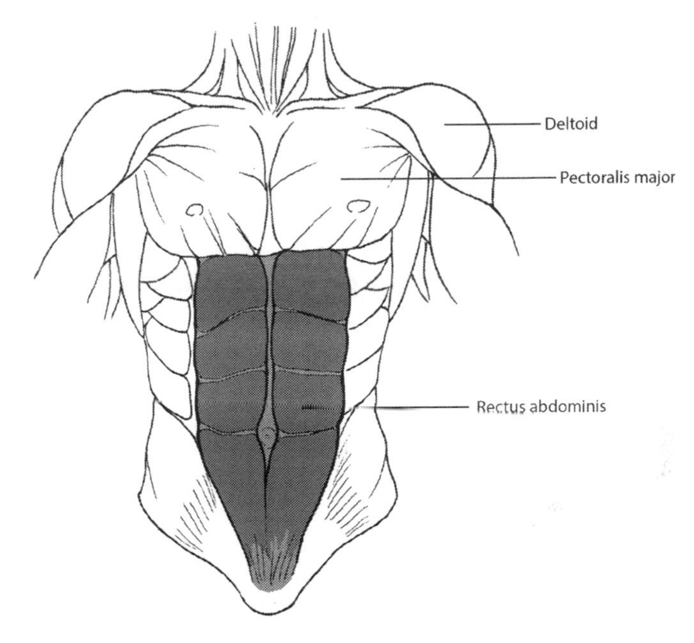


**FIGURE 2.8**
Upper torso muscles including defined rectus abdominis muscle ("6-pack"), deltoid, and pectoralis major.

Designers and fitness professionals frequently talk about "toned bodies," in relation to body form. But the definition of *muscle tone* (also called *muscle tonus*) relates to a normal responsiveness of muscles to nervous system input, not to size or form. Bernstein gives a concise definition of [muscle] tonus as "not a condition of elasticity, but a condition of readiness" (1967, pp. 111–112). He goes on to point out that tonus is a condition that reflects all the connections from the brain to the muscle, which is related to coordination and readiness "as a state is to an action" (1967, pp. 111–112). A toned body might be considered a trained body but should not be confused with a bulky or an exercised body. See Section 2.6 to consider how *adipose tissue*, commonly called fat, contributes to body form.

### 2.2.6 Supplements to the Skeletal and Muscular Systems

The term *fascia* describes a loose and then denser connective tissue, which acts much like "glue" to hold different tissues together. The superficial loose layer attaches the base layer of the skin to deeper structures. Farther into the body, the superficial fascia transforms into deep fascia, which is dense and more organized. The deep fascia connects the superficial fascia to the fibrous coverings of other internal structures, including muscles. It creates divisions between functional groups of muscles, the *intermuscular septa*. It also surrounds individual muscles.

*Bursae*, the plural of *bursa*, are small fluid-filled pouches located between adjacent muscles, between bone and skin, or where a tendon slides over a bone. Bursae allow smooth motion of the body parts as they work together. *Bursitis* is inflammation of a bursa and is often seen adjacent to joints or other structures that are moved repetitively. The nervous system, described in the next section, is an essential component of body stability as well as body motion, as it monitors and initiates the interaction of muscles with the skeleton.

## 2.3 Nervous System: Sensing, Directing, and Coordinating

The nervous system is crucial in all bodily functions. Wearable products may be designed to protect portions of the nervous system. A splint, to restrict wrist motion, may be used to treat *carpal tunnel syndrome*, which results from pressure on, or excessive motion of, a major nerve at the wrist, the *median nerve*. When damage to the nervous system has limited physical capacities, exoskeletons may facilitate walking for people with spinal cord injuries, strokes, or other neurological disease (Bortole et al., 2015; Pehlivan et al., 2014). A poorly designed wearable product may impair nervous system functions. A heavy backpack with shoulder straps that compress the nerves traveling from the neck to the arm can limit muscular function, cause *numbness* (loss of sensation), or create pain in the shoulder, arm, forearm, and/or hand. The nervous system is always in operation—receiving and sending messages to the body. The nervous system (Figure 2.9) can be categorized by anatomical structure or by function.

### 2.3.1 Anatomical Structure: Central Nervous System and Peripheral Nervous System

Anatomically, there is (1) the *central nervous system* (CNS), the brain and spinal cord inside the skull and vertebral column, and (2) the *peripheral nervous system* (PNS), nerve tissues connecting the brain and spinal cord to the rest of the body, outside of the skull and spinal column.

The CNS is composed of the brain, *brainstem*, and the spinal cord (see the central shaded areas of Figure 2.9). Components of the skeletal system provide crucial protection for the CNS. The skull encases and protects the brain/brainstem; and the upper two-thirds of the vertebral column houses and protects the spinal cord centrally, in the *vertebral canal*. If a portion of the CNS is damaged some or all functions of the body will cease. Find more information on protecting the parts of the CNS in Chapter 3 on the head and neck and in Chapter 4 on the upper torso and limbs.

The PNS includes *cranial nerves* extending from the brain to structures in the head and body, and *spinal nerves* extending from the spinal cord into

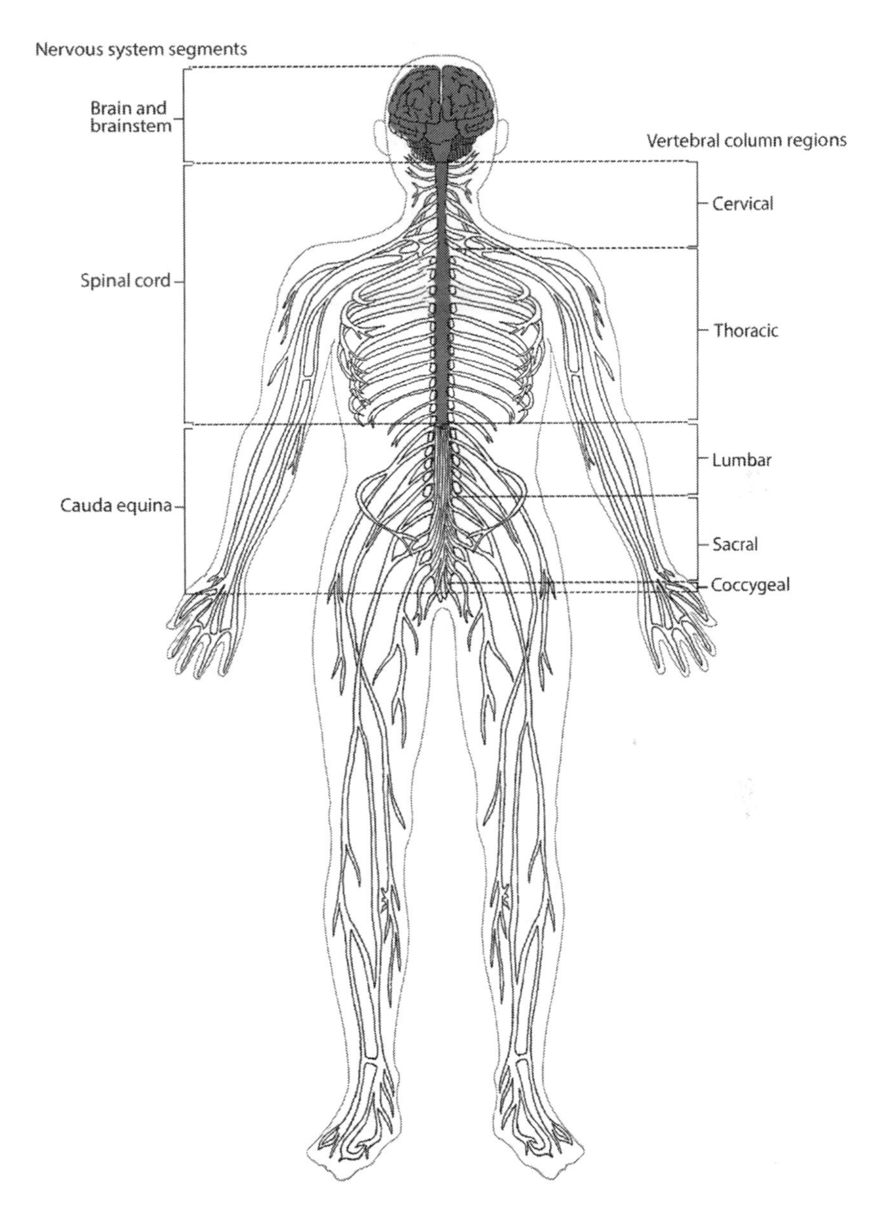


**FIGURE 2.9**
Central and peripheral nervous system, anterior view.

the neck, torso, and limbs. There are 31 spinal nerve pairs. They are named and numbered for the spinal region of origin (*cervical, thoracic, lumbar, sacral,* and *coccygeal*) and level (1, 2, etc.). After leaving the spinal canal, individual nerve fibers in the spinal nerves regroup into different combinations as they travel through *nerve plexuses* within the neck and torso. The interlaced nerve fibers in the plexuses look somewhat like a railroad switching yard with multiple junctions. *Peripheral nerves,* the bundles of regrouped nerve fibers from multiple spinal nerves, leave a plexus to travel to specific areas of the neck, torso, and upper and lower limbs. *Ganglia,* clusters of nerve cells located outside the CNS, may be part of the *sensory* or *autonomic nervous systems* discussed in the next sections. Read Chapter 4 for information relating to the spinal and peripheral nerves arising from the lower cervical, thoracic, and upper lumbar regions, and Chapter 5 for the remaining lumbar, sacral, and coccygeal spinal and peripheral nerves. The spinal nerves and ganglia of the PNS are generally protected within the body and not affected by wearables. Peripheral nerves are more vulnerable, particularly to direct or sustained pressure. Prototype trials of wearable products may include assessing restriction of the peripheral nerves through physical measurements and/or through questioning the wear testers about possible weakness, level of comfort, possible pain, or change of sensation from wearing the product.

### 2.3.2 Functional Divisions: Sensory and Motor

Functionally, the nervous system also has two components: a *sensory* division and a *motor* division, or the input and output channels of the nervous system. The sensory division is responsible for collecting sensory information via *receptors* (specialized structures in skin or other peripheral tissues) in the PNS and transmitting this information to the CNS. Parts of the brain and spinal cord interpret the signals. Sensory information from the special senses of vision, hearing, smell, and taste travel to the brain via the cranial nerves. The rest of the sensory division is divided into *somatic* (related to structures of the body wall) components and *visceral* (related to the internal organs) components. The somatic senses are senses a wearable product may directly activate: touch, pain, pressure, vibration, temperature, and *proprioception* (the sense combining position in space, balance, and/or movement in the muscular and skeletal systems). Visceral senses include pain and stretch of organ walls. Visceral senses may be affected by wearable products, but we are not as consciously aware of these senses.

The motor division is responsible for transmitting motor impulses from the CNS to muscles or *glands* (groups of cells or organs producing secretions). The motor division also includes both CNS and PNS components. The motor division is subdivided into a somatic motor component and an *autonomic* component. Somatic motor nerves are called voluntary like their associated muscles. The autonomic nervous system is called the involuntary system since we do not consciously decide on the actions of these motor nerves which control smooth muscle, cardiac muscle, and glands, including sweat glands. The autonomic nervous system also responds to stress, producing the "fight or flight" response.

Sensory and motor nerves are essential to musculoskeletal motion. The regions around joints are particularly well supplied with nerves which collect sensations of movement, muscle tension, and joint position, contributing to the sense of proprioception. They transmit electrical impulses to the brain, where a network of interconnections interprets the information from multiple sources in the limbs, torso, and head (including the eyes and ears), as well as various parts of the brain itself. Signals are then sent, via the motor nerves to the muscles, programming coordinated shortening and/or lengthening contractions, to move the body in either a deliberate (voluntary) or *reflex* (involuntary) manner.

## 2.4 Stability and Motion: Interactions in a Neuro-Musculo-Skeletal System

The configurations of the individual bones and joints, combined with the ligaments which span the joints, give the skeleton some resistance to collapse. Nerve-monitored and nerve-activated muscles, plus the body's fascia, work with the skeleton to hold the body steady or to make it move. Wearable products may be designed to assist with either operation, with care to avoid impeding natural function. Stability and motion required for an activity will affect design decisions. Some speed sports require wearable products that are aerodynamic; closely fitted to the body, but non-constricting. Full-body protective clothing, like bio-protective suits, should be designed to allow the worker to move as fully as possible—to stretch, reach, and bend. The suit can't be so loose that it impedes work, but it can't be so tight that the seams split or the fabric tears, allowing hazardous materials inside. Ankle braces may be designed to assist the body's stabilizing features for the short or long term. Understanding the static and dynamic anatomy of a body area and/or the whole body will assist designers as they design products to tie in stability or motion.

### 2.4.1 Body Stabilization

Skeletal muscles attached to the skeleton must work together to stabilize the entire body and body parts and to maintain body positions. The nervous system monitors and coordinates the muscular activity required to keep the body in a state of balance against the forces of gravity. The amount of muscle contraction needed, in a standing or seated position, can be quite small when the skeleton is in an idealized erect posture (Basmajian & De Luca, 1985). Muscles form a body wall, surrounding and supporting the internal organs located in the abdomen. Abdominal muscles also help to stabilize the multi-segmented spine and the bony *pelvic girdle*. While products can be designed to assist stability or good posture, products should not substitute for healthy use of postural muscles.

### 2.4.2 Body in Motion

Just as the nervous system helps regulate stabilizing activities, it is involved in body mobility. Skeletal muscles and bones generate motion through an intricate and sophisticated lever and pulley system. Muscles work together to produce desired body motion, from large movements such as running to fine movements such as threading a needle. One muscle frequently works in tandem with another muscle which opposes the primary motion or action of the first muscle. For example, when the *triceps* (the large muscle on the posterior upper arm) contracts and shortens to straighten the elbow, the biceps, its opposing muscle, contracts and lengthens. The movements of gait are addressed in more detail in Chapters 5, 8, and 9.

Designers often think of the human body in a static pose, much like the anatomical position, when designing a wearable product. However, humans are almost always in motion and, for many situations, incorporating design features to accommodate or stabilize motion is critical. You have read about anatomical features that make motion possible: interactions of bones, ligaments, muscles, and tendons, and the linkages formed by joints; along with the brain sending movement commands to and collecting sensation from all parts of the body. Basic methods of observing and recording motion, describing movements, and measuring the body in motion follow in this chapter. Chapters on body regions elaborate on how each segment moves and include examples of designing for stability and motion.

Most of the motions discussed in this book are voluntary motions that require a smooth interplay of brain activity, nerve signals, and muscles acting on the skeletal structures. As you observe the human body, think of the parts of the body as segments in a three-dimensional frame of reference. The body in the anatomical position (Chapter 1, Figure 1.15) is the starting point and motion is described related to that position. The sagittal, coronal, and transverse planes of the anatomical figure act as reference planes for body motion.

#### *Body Links*

Begin by picturing the human body as segments linked together at joints. Motion at a linkage point may be discussed in terms of movement around an *axis of rotation* (the center of a joint). A body segment can be represented as a straight line attached to one or two centers of joint rotation. Figure 2.10 shows a simplified linkage system to use for mapping the movement of one body part relative to another. By visualizing the joint locations within the body, you have a basis for measuring motion around that link. Expanded information on motion at specific joints is included in Chapters 3–9.

#### *Describing Body Motion*

Several terms are commonly used to describe the body in motion and many resources give definitions and applications (Everett & Kell, 2010; Kottke &

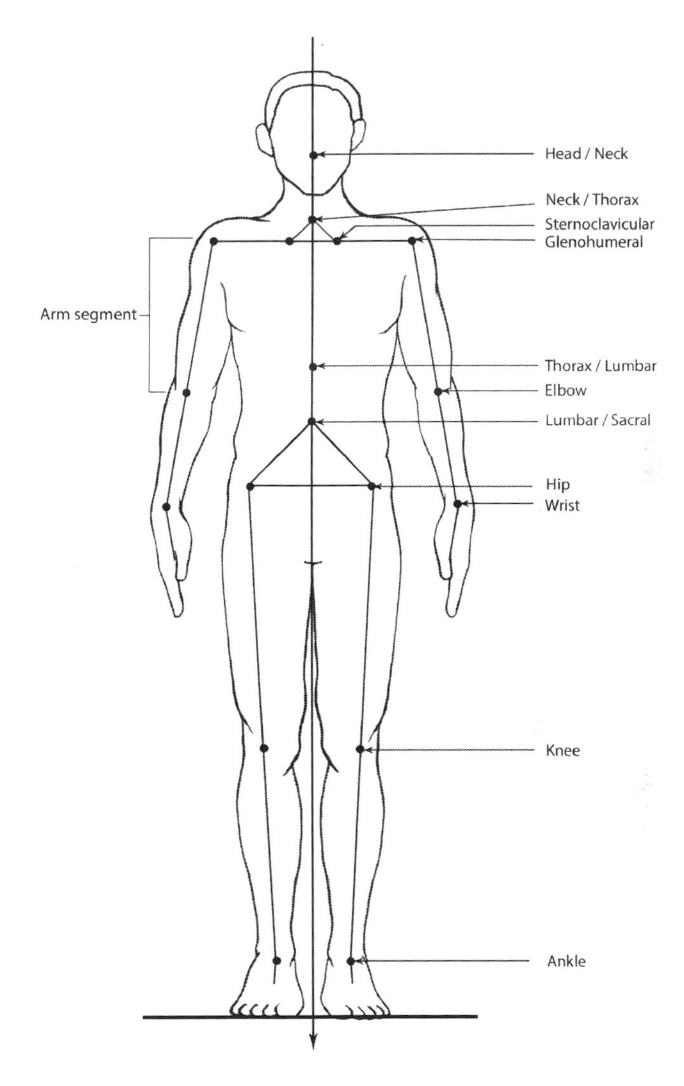


**FIGURE 2.10**
Body linkage system: Joint rotation centers and arm segment example.

Lehman, 1990; Langley, Telford, & Christensen 1980). Some definitions are quite lengthy and complex while others are short and simple. Common body motion descriptors that can be applied when designing a product or describing product functions related to body motion are:

- *Flexion*: Bending a limb at a joint, decreasing the angle between anterior or posterior parts of the body. Examples: Bending the arm at the elbow bringing the forearm toward the upper arm; bending the leg at the knee, bringing the calf toward the thigh.

- *Extension*: In simple terms, the opposite of flexion; straightening the joint. Example: Moving the forearm away from the upper arm so that the arm forms a straighter line.
- *Hyperextension*: Extending the joint beyond the natural extended position. Example: When the forearm is extended from the upper arm so that the angle at the elbow goes beyond the straight or 180-degree angle.
- *Abduction*: Movement of a body part away from the midline of the body. Example: Swinging the entire leg to the side away from the body with the axis of motion at the hip.
- *Adduction*: Movement of a body part toward the midline of the body. The opposite of abduction. Example: Moving the abducted leg back toward the neutral position or in front of or past the other leg.
- *Lateral rotation*: Turning away from the midline of the body. Example: Rotating the head to look to the left or right. This movement, for the arm, at the shoulder, and for the leg at the hip may alternatively be called *external rotation*.
- *Medial rotation*: Turning toward the midline of the body. Example: Returning the laterally rotated head to face anteriorly. This movement at the shoulder and at the hip may alternatively be called *internal rotation*.
- *Circumduction*: A movement combining flexion, abduction, extension and adduction; resulting in a continuous circular movement of a limb. Example: Rotating the entire arm in a circle with the shoulder joint as the axis. The motion at a ball-and-socket joint is complex and you must consider how the joint moves in multiple planes.

Additional terms specific to various body regions are explained in the related chapters. *Pronation* and *supination* describe specialized movements of the hands and/or feet. *Eversion* and *inversion* refer only to foot/ankle motions. *Protraction* and *retraction* also apply to specific joints, including the articulation of the jaw to the skull, the skull to the neck and shoulder movement on the torso.

### 2.4.3 Range of Motion

Joint range of motion (ROM) reflects the extent of an individual's flexibility around a joint center, the axis of rotation, in a specified plane of motion. Measurement of ROM is an art which requires consistent technique and practice in order to obtain reproducible results. Wearable product designers may not need to produce precise measurements of ROM, unless they are working with a totally customized product, but they do need an understanding of general principles of joint motion, the methods used to document ROM, and the standards used for ROM recording systems. A *goniometer,* a simple tool with two arms and a friction pivot point, with a 360-degree protractor, is commonly used by medical professionals, along with very specific protocols, to document ROM. Figure 2.11 shows measurement of elbow flexion with a goniometer, in a sagittal plane.

Briefly, to measure movement at a joint, such as shoulder abduction/adduction (Figure 2.12) follow this process: (1) look at the motion of the arm from the anatomical position; (2) place the goniometer pivot point over the shoulder joint center, in a coronal (frontal) plane in front of or behind the body; (3) align the 0-degree mark of the 360-degree circle up in the direction of the top

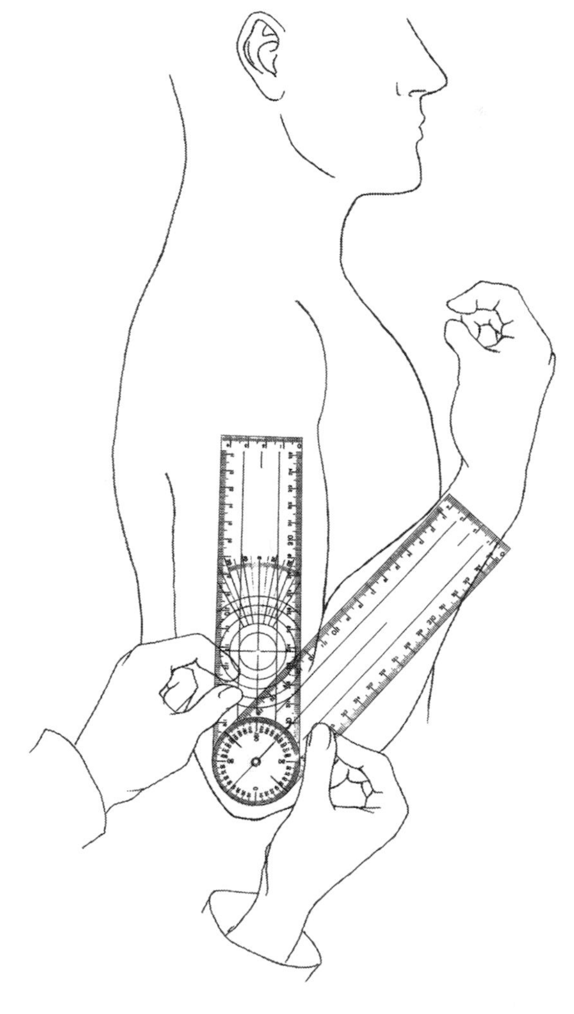

**FIGURE 2.11**
Measurement of elbow flexion with goniometer, in a sagittal plane.

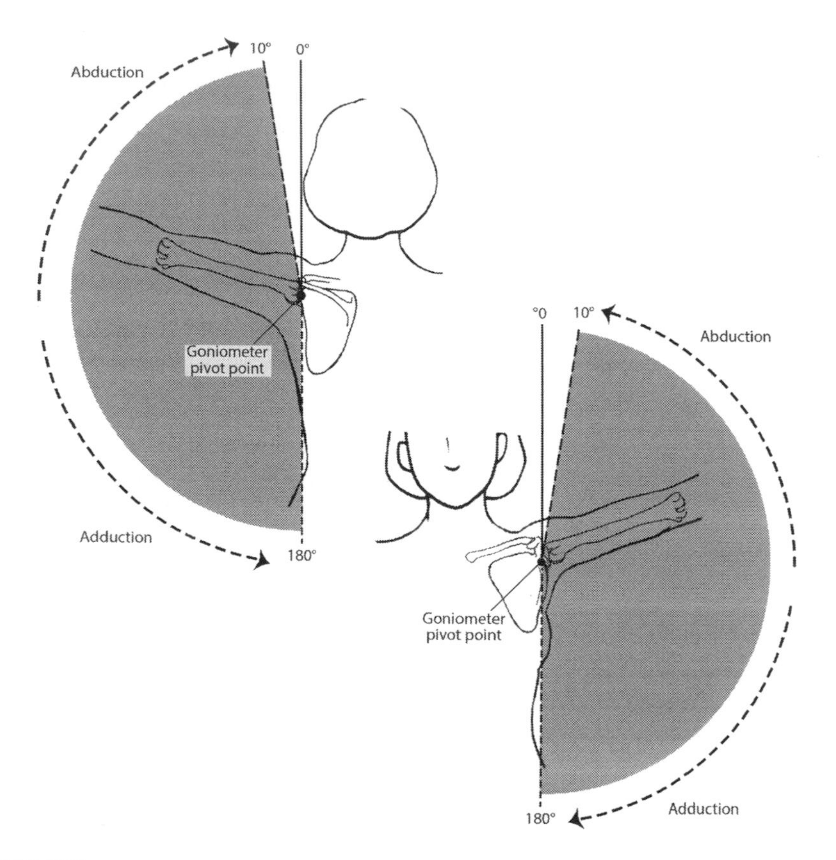


**FIGURE 2.12**
Shoulder abduction and adduction range of motion.

of the head; (4) move the goniometer arms to measure the arm's full motion; (5) record the joint's motion in degrees. Study Figure 2.12 and follow shoulder abduction-adduction ROM from 10 degrees to 180 degrees in a coronal plane. When the arm is abducted—raised up and outward from the side of the body above the head and aligned with the ear (thumb pointed toward the midline)—the smallest angle between the arm and the parasagittal plane is normally 10 degrees. When the arm is adducted—the arm lowered and at the side of the body—the angle is normally 180 degrees. The U.S. National Center on Birth Defects and Developmental Disabilities, Centers for Disease Control and Prevention (2011) produced a useful training resource for documentation of ROM—a series of videos demonstrating the techniques of ROM measurement for a number of major joints.

This approach works well for large joints and body segments. The disadvantage is that you can only measure one joint at a time. ROM is usually described for a pair of actions, specifying a particular joint, plane, and the extent of the movement for the actions. Motions that occur in a transverse (horizontal) plane, such as lateral/medial rotation of the head/neck on the torso, are usually measured as deviations from a perpendicular plane through the neutral joint position. Many disciplines collect motion data for varying purposes. Recording systems and reporting formats may vary. When looking at ROM numbers, find out which system was used to collect and report the data in order to interpret and use results effectively.

The mobile anatomical structures in each chapter are paired with examples of products that might accommodate or stabilize the motions. If you are designing a product requiring hip abduction motion, read Chapter 5 on lower torso and limbs and use the illustration of hip abduction ROM as a starting point for your design project. Although a pair of numbers is used to describe the limits of the normal joint ROM, there are sometimes wide variations for ROM. Differences may arise due to age, health status, and gender. For example, a person with generalized *osteoarthritis,* degeneration of bone and cartilage in many joints, may have significantly restricted ROM.

Keep in mind that ROM examples in this book were collected from several sources. Most range-of-motion studies were conducted by the military, so subjects were young, healthy, and almost always male. Each individual, even if healthy and the same age, has different capabilities and characteristic movements. When you observe, document, and apply motion to a design, try to find product testers who represent your intended product user or select a database that as closely as possible represents your intended users.

## 2.5 Design for Motion

Follow a step-wise approach to incorporating motion or stability into a wearable product.

- Observe characteristic movements.
- Describe the movements.
- Select movements essential to the motion of the activity.
- Measure critical movements performed during the activity or use data on the normal ROM for the target market.
- Incorporate essential movements into a design.
- Determine success of the product when worn in motion.

### 2.5.1 Observe Characteristic Movements

When movement or stability is essential for a product category, the first step is to observe potential wearers of the product performing key activities. For a designer's purposes motion is the sum of separate movements making up an activity. Successful sports activities require coordination of many movements. The designer can observe competing athletes, as a casual observer, and then go on to observing and measuring the movements of the sport with one or several athletes. The designer might also want to observe movements of people who interact with the wearer to determine if they change the wearer's movements and total motion.

### 2.5.2 Describe the Movements

In early design stages, very simple descriptions or stick figure drawings based on what the designer knows about body linkages may be sufficient. Determine which body areas are most involved in the motion. Focus on those areas and note the variety of movements in the motion. How does the torso move? How does the head move? How do limbs move in relationship to the torso? Are there movements that go beyond typical ranges of motion with risk of injury?

Consult texts and online sources to help develop a deeper understanding of the movements you are analyzing. Entire research fields are devoted to studying human body movement. Kinesiology is the scholarly study of human movement; and biomechanics, within the field of kinesiology, is dedicated to the study of motion and what initiates motion (Everett & Kell, 2010; Hamill & Knutzen, 2003; Knudson, 2007; Kottke & Lehman, 1990; Langley et al., 1980).

### 2.5.3 Select Essential Movements

A sport requires specific movements and good form in order for the athlete to be successful. Essential movements make up an activity's motion. Some of these can be considered critical. Once isolated and quantified, they can be accounted for in a design. Sport activity may require extending some parts of the body, sometimes to extremes, while flexing other parts. While the segments of the body are in motion, range of motion at body links can be observed. For some activities, a product should limit range of motion to maintain health and safety of the wearer. For example, a knee brace may be designed to prevent hyperextension or other angular movements that could stretch ligaments and tendons out of normal range.

### 2.5.4 Acquire the ROM and Linear Excursion of the Critical Movements

After describing movements that are critical to the activity or activities, either find the appropriate data set or measure the movements. To manually measure, start with the body in the anatomical position or a specified static pose. The person performing the motion is asked to stop at key points of the motion so the measurer can determine a length or an angle and record it. Lengths can be measured with a standard tape measure. Joint angle data can be measured with a goniometer (see Figure 2.11). For each critical movement, define the full ROM of the joint and plane of the action. With manual measurement methods, the angles are measured in two-dimensional space. Multiple measurements, in the different anatomical planes, are necessary to represent the three-dimensional motion of the body.

Digital methods are also used to quantify motion. Compared to a goniometer, motion capture, or a "mo-cap" system (Figure 2.13), requires as much or more care in set up and data acquisition, but it can collect massive quantities of real-time joint movement information, from many joints at once, without stopping motion.

This sophisticated technology can analyze the great quantity of data obtained and extract useful joint range of motion from that data for application to wearable product designs. Designers in many fields are using motion capture equipment to aid product design or to evaluate a product's effectiveness in motion.

### 2.5.5 Incorporate Movement into a Design

When the body is in motion surface areas can expand or contract. As a simple demonstration, look at the "pointer" (*index*) finger of one hand. With the finger extended, measure the length of your finger on the dorsal (top) side from the base of the finger to the tip of the finger. Then flex (bend) the finger as much as possible and measure the dorsal side again from base of the finger

**FIGURE 2.13**
Motion capture system. (Courtesy of the Human Dimensioning© Laboratory, University of Minnesota.)

to the tip. Note that the length of the underside of the finger is changing in a different direction. If you need to cover and accommodate movement of the finger with a product, develop structure and materials variables that will expand and contract with the finger movement. Range of motion data can help you to understand the degree of mobility needed in your product.

### 2.5.6 Determine Success of the Product

Prototypes designed to facilitate or limit motion should be wear-tested to determine success. Computer simulations are possible with virtually draped products displayed on avatars. However, *in vivo* tests (with a live test subject) may be preferable as the designer can assess fit and motion visually and ask the wear tester how the product feels.

First establish a baseline. Assess the motion of a person when not wearing the product. Then assess motion while the person is wearing and moving in the product. Observe for accommodation and/or restriction of motion by observing wrinkles or strain on the product materials. Question the wearer about ease of moving. Ask how comfortable the product is while standing, sitting, and moving. Revise the product if needed and repeat assessments.

Motion capture systems can be used to quantify motion of a person wearing a product. The type of system, *passive marker* or *active marker,* will affect the approach and quality of data. Passive marker systems use reflective markers placed at the joint intersection. One caution; passive markers on a product may move independently from the body. Markers reflect light to cameras placed around the body. Passive marker reflection may be blocked by the product or parts of the body. Active markers, on the other hand, transmit electronic signals from the marker to a receiver. The markers can transmit through a product worn on the body and do not move with the product.

Thoughtful incorporation of range of motion parameters into a wearable product design, done well, can add immeasurable value. Benefits may range from improved aesthetics, to greater safety, to improved function for the wearer. From medical devices to sports equipment, from fashion apparel to personal protective equipment, wearable products that move with the body regions they encompass are needed and will be welcomed.

## 2.6 Fat: Fuel and Form

Fat, a term interchangeable with the anatomical term adipose tissue, is a specialized connective tissue that is distributed throughout the body. It stores energy, provides thermal insulation, fills space between other tissues, and helps shape the body. Fat, as an additional layer of tissue, can make defining and locating landmarks and body linkages more difficult.

### 2.6.1 Fat Types and Functions

There are two types of fat in humans: brown fat and white fat. Brown fat is a heat generating tissue found chiefly in fetuses, infants, and children. It is concentrated in fat pads at the upper back and shoulder and around the kidneys. Infants and children have a much higher ratio of surface area to volume than adults, with a resulting increased possibility of loss of body heat, so the layer of brown fat delivers thermal protection. White fat is more abundant. It also supplies thermal insulation, while supporting and cushioning some organs, like eyeballs and kidneys. Characteristic female body contours come from white fat deposits in the breasts and at the hips. Fat at the hips adds to the already wider dimension of the female pelvis. Too little fat can reduce female fertility. Females' greater body fat content, compared to males, assists in providing nourishment to the unborn child and adds calories for the nursing mother.

Both the distribution and the amount of fat on a body affect a person's shape and size. Although we often speak of fat in simple terms—we say people are fat, we diet to lose fat, excess fat on the body is unhealthy—distribution of fat within the body is complex. There are two primary locations of fat tissue; *subcutaneous adipose tissue* (SAT) lies just under the skin and *deep adipose tissue* (DAT) occupies internal body spaces. Amounts and distribution of both SAT and DAT influence a person's appearance. DAT volume, particularly in the abdomen, increases with psychological and/or physiological stress. Lack of sleep is a physiological contributor to increased fat.

### 2.6.2 Fat Patterning

According to Bouchard and Johnston (1988), SAT distribution is conventionally referred to as *fat patterning* to distinguish it from the accumulation of DAT. Although fat is distributed throughout the body, "subcutaneous fat in the normal adult has a predilection for residence in particular body areas" (Croney, 1971, p. 29). Fat distribution patterns change with age, and men and women accumulate SAT in different locations (Baumgartner, 2005). Gender differences begin to appear at adolescence. SAT in boys tends to grow around the trunk, while girls' SAT is more evident at the buttocks and thighs (Mueller & Wohlleb, 1981). The typical fat patterning sites for adult males and females, derived from several body composition studies (Croney, 1971; Garn, 1954; Garn, 1955; Škerlj, Brožek, & Hunt, 1953) are shown in Figure 2.14. Fat patterning for women includes fat in the breasts, triceps, lateral mid-torso, abdomen down to the pubic arch, the hip, buttocks, and thighs. Fat collection areas for men include the pectoral region, anterior and posterior mid-torso, and, to some extent, the upper thigh.

The relationship between the skin and subcutaneous fat structures also differs in men and women. Cellulite, the puckering skin changes often seen on women's buttocks and thighs, arises from specific anatomical features

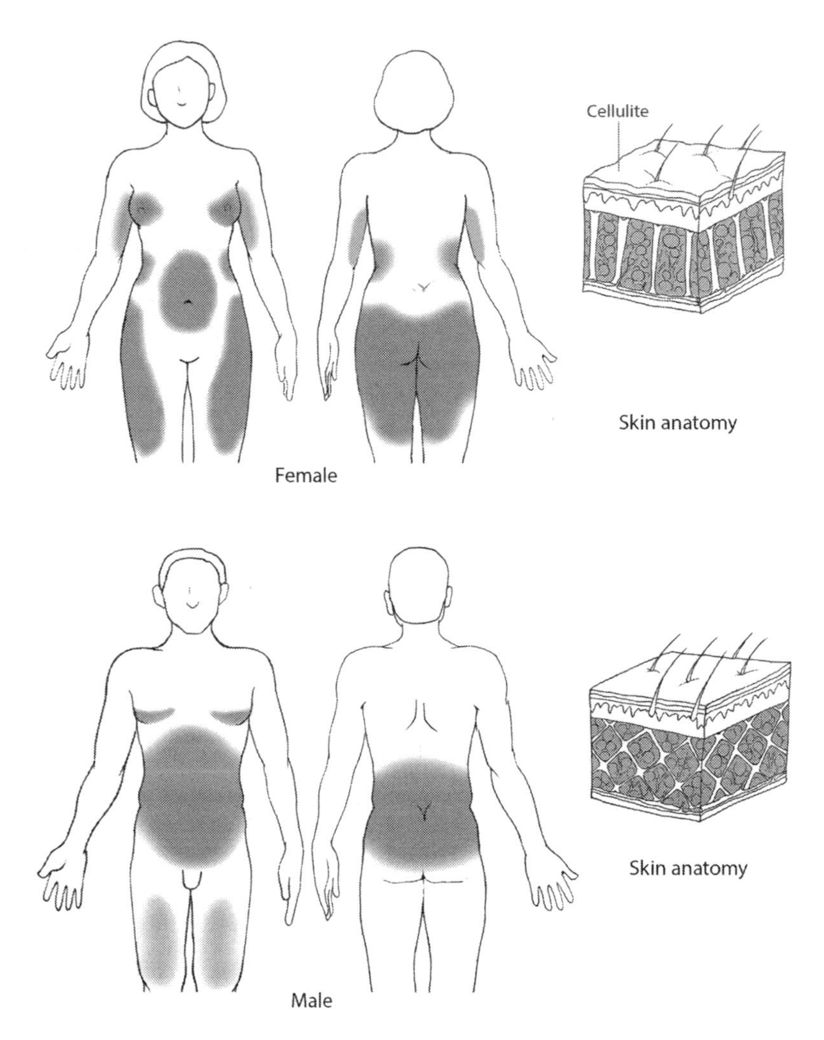


**FIGURE 2.14**
Fat patterning and cellulite, female and male differences.

(Khan, Victor, Rao, & Sadick, 2010). See Figure 2.14, right side, for an illustration of cellulite anatomy in adult men and women. Note that the female skin cross-section looks something like a tufted quilt as subcutaneous tissues pull the skin into a textured surface. In the male, skin cross-sections are networked with diagonal connective tissue giving a smoother surface.

Think of the body structures already discussed—skeleton and muscles—then add the padding of adipose tissue. Each person is born with a limited number of fat cells. Normal *adipocytes* (fat cells) range from 70 to 120 μm (micro-meters) in diameter but may be five times as large in obese people (Saladin, 2007). The volume of stored fat may vary, but the number of adipocytes is stable in an adult over time (Spalding et al., 2008). There is a "constant

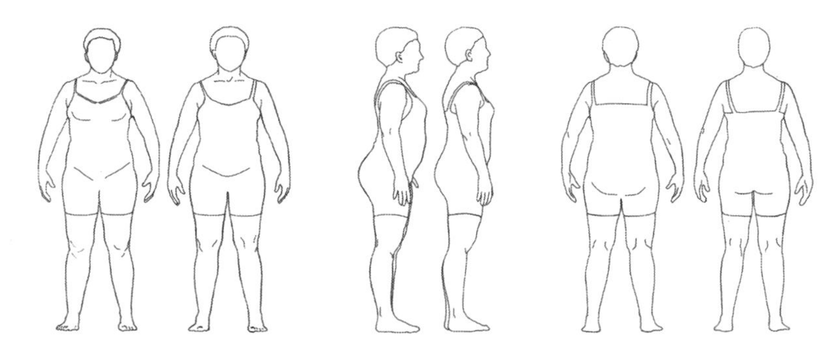

**FIGURE 2.15**
Female figure before (left) and after (right) 8% BMI reduction (figure outlines derived from body scans). (Courtesy of the Human Dimensioning© Laboratory, University of Minnesota.)

turnover of stored fat, with an equilibrium between synthesis and breakdown, energy storage and energy use" (Saladin, 2014, p. 66).

When energy use exceeds energy storage, whether from eating fewer calories and/or exercising more, adipocyte diameters shrink, and for this reason, total adipose tissue will decrease. With fat decrease, body weight goes down and body contours and size change. Figure 2.15 illustrates body changes due to weight loss with line drawings derived from body scans. The individual shown participated in a medically supervised 12-week weight loss study (Aldrich, Reicks, Sibley, Redmon, Thomas, & Raatz, 2011). Look at the before and after scans to note the 8% change in her body mass index (BMI). The typical fat accumulation sites, with more fat cells, show noticeable changes because the fat cells decreased in size. The areas of the body with fewer fat cells show less change. For example, the thigh has more adipose tissue than the calf. So, the after-weight-loss scan shows more change at the thigh and less change at the calf.

Exercise produces body form changes. If dietary intake increases to meet the increased energy needs of exercise, adipose tissue will remain stable, resulting in larger dimensions. However, if dietary intake remains stable or decreases, exercise may decrease body dimensions while increasing muscle bulk. Abdominal muscle exercise will increase support to the abdominal contents and the back. Waistline dimensions may decrease because those muscles become a natural "corset."

## 2.7 Internal Soft Tissue Systems: Respiratory, Circulatory, Lymphatic, Digestive, Urinary, Reproductive, and Endocrine

Designing a new and improved blood pressure monitor requires knowledge of the circulatory system. An appreciation of the structures and functions of the lower torso including the urinary, reproductive, and digestive systems is helpful for designers interested in managing incontinence, the inability to control the body's natural evacuations. The "internal soft tissue systems" influence design of wearable products because they are essential to whole body function and contribute much to how the body interacts with the environment. The importance of these systems, individually, to designers of wearable products varies, and scientific knowledge of these systems is also variable. While the circulatory system is perhaps best defined and understood, lymphatic structures, which are particularly important for medical wearable products, are definitely less well studied (Choi, Lee, & Hong, 2012).

### 2.7.1 Respiratory System

The respiratory system is made up of the lungs and air passages that bring in and expel air. It acts as an interface between the external environment and body tissues. The respiratory system and the circulatory system must work together to supply a basic element of life, oxygen, to the body. The lungs work in a bellows fashion as the diaphragm, a muscle below the lungs, helps in the expansion and contraction of the lungs in the action known as breathing. The respiratory system brings oxygen to the circulatory system where it can be transported throughout the body. The circulatory system returns *carbon dioxide,* a waste product of the body's oxygen and nutrient use, to the lungs where it can be expelled from the body. Respiratory system actions influence body form in the upper torso, the lower torso, and the waistline. A waistline measurement taken with a breath in, *inhalation,* will differ from a measurement taken after a breath out, *exhalation.*

### 2.7.2 Circulatory System

The circulatory system serves the entire body, including deep and superficial structures. It includes the *heart,* which pumps blood (a complex fluid) into *arteries* (Figure 2.16-A) and receives blood from *veins* (Figure 2.16-B). In each section of Figure 2.16 the deep structures are shown on the left and those which are more superficial are on the right. Blood consists of *plasma* (the liquid portion of blood) and *formed elements.* Plasma makes up about 55% of blood volume, and the formed elements about 45% (Kapit & Elson, 2014). All of the components play crucial roles. Plasma includes *antibodies,* infection fighting proteins. The formed elements are red blood cells, white blood cells, and platelets. The red blood cells are non-rigid, membrane-lined sacs of *hemoglobin,* which binds to oxygen, giving these cells their red color. White blood cells' major function is protection. There are several types of white blood cells with some responsible for fighting infections. Platelets help control bleeding.

The arteries, leading away from the heart, subdivide into successively smaller sections as they carry blood to the cells of the body. *Capillaries,* the

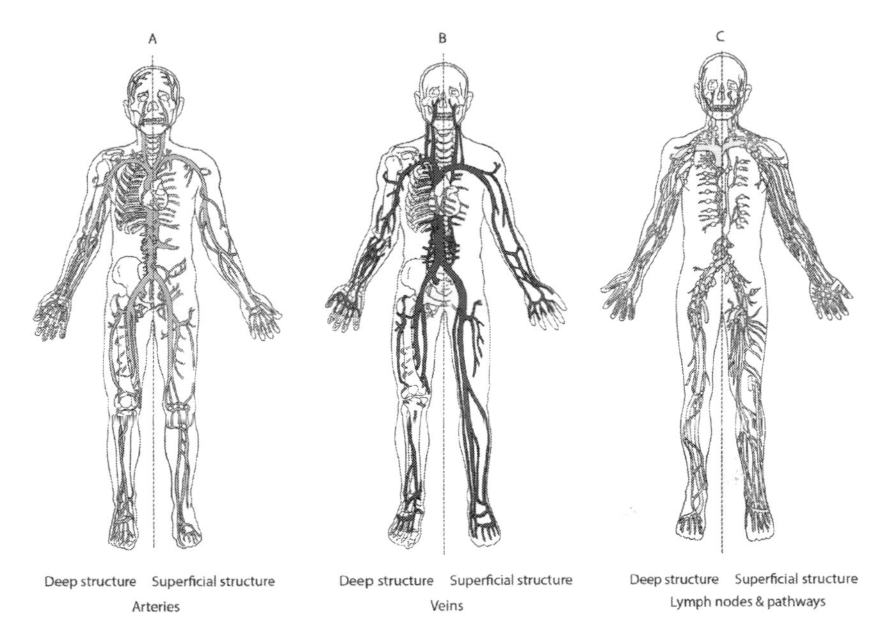


**FIGURE 2.16**
Circulatory system and lymphatic system: Arteries (A), veins (B), lymph nodes and pathways (C), with deep and superficial structures.

tiniest vessels of the circulatory system, lie in between cells in the tissues of the body. The microscopic capillaries bridge between the smallest arteries and tiny veins, which then join together to form increasingly larger veins to return blood to the heart. The thin walls of the capillaries allow gases and nutrients to move from inside the capillary to the body's cells beyond. Capillaries also collect cellular and gaseous waste to be carried out of the body.

There are two blood flow circuits made up of arteries, capillaries, and veins. The *pulmonary circulation* conveys oxygen-depleted blood from the right side of the heart to the lungs to take in oxygen and release carbon dioxide and then returns the oxygen-replenished blood back to the left side of the heart. The *systemic circulation* carries oxygen-rich blood from the left side of the heart to the body and returns oxygen-poor blood to the right side of the heart. We think of oxygenated blood as red and oxygen-poor blood as blue. Look at the underside of your wrist and you can easily see the superficial blue veins of the systemic circulation near the skin surface.

*Pulse* and *blood pressure* are markers of heart and circulatory system function. You can hear the heart beating in the chest cavity and feel its effects throughout the body as it pumps. *Pulse points* are locations on the body where the "pump" action or pulse of blood from the heart, through an *artery,* can be detected. Examples of pulse point areas are both sides of the inner wrist, inside of the elbow, the side of the neck, and the top of the foot

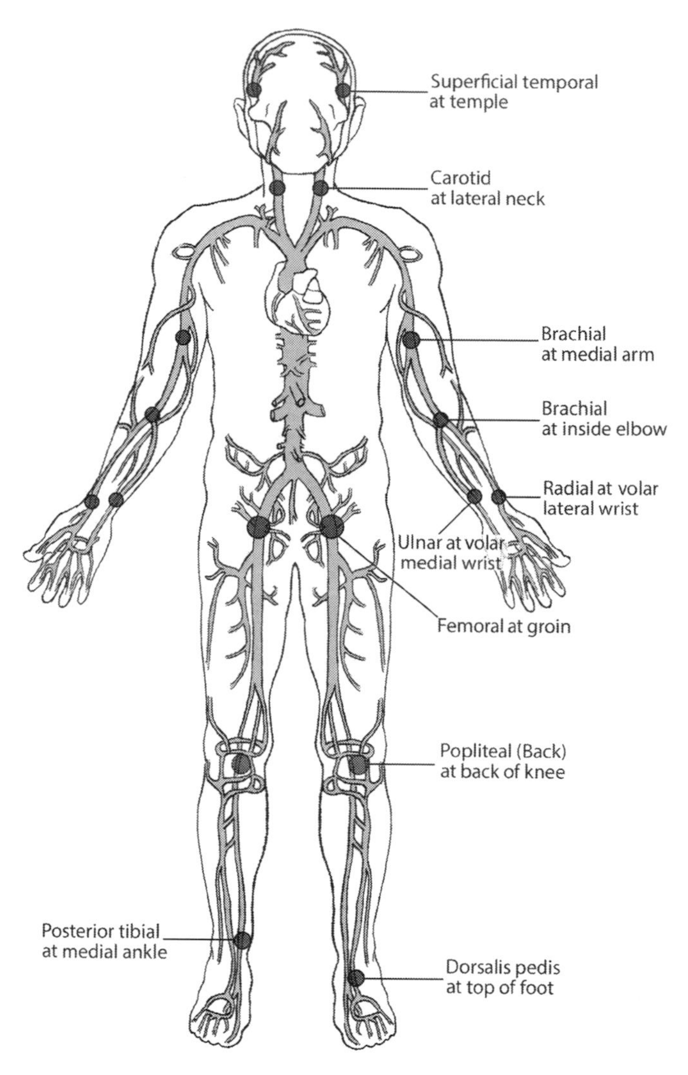


**FIGURE 2.17**
Arterial pulse points.

(Figure 2.17) When you feel the pulse (throbbing of the artery) count the number of beats, or pulses, in 6 seconds and multiply by 10 to calculate your pulse rate. Normal adult *resting heart rate* (pulse rate) is between 60 and 100 beats per minute. Many factors affect heart rate including illness, work load, environmental conditions, fitness, body position, and body size (Dewey, Rosenthal, Murphy, Froelicher, & Ashley, 2008). Medications can also affect pulse rate. *Pulse quality*—regularity, strength and variability—is an indicator of heart and circulatory system function (Felner, 1990). Wearable pulse monitors, as "fitness" trackers, are now widely available. Professional and amateur athletes sometimes track both their pulse and blood pressure to determine response to physical activity.

Blood pressure, sometimes called arterial blood pressure, is the pressure of circulating blood on the walls of the arteries. A peak pressure (*systolic*) is produced when the heart beats. When the heart muscle rests and the heart refills with blood (between beats) a lower, *diastolic,* pressure is evident. Arterial blood pressure, along with pulse, is a principal vital sign measured to assess both healthy people and people who are ill, under stress, or have experienced trauma. Wearable products have been designed to monitor blood pressure. You may have experienced having your blood pressure taken. The medical professional wraps a band called a cuff around your upper arm. The inflatable cuff expands putting pressure on the *brachial artery,* the major artery of the arm. As the cuff pressure is released the device (*sphygmomanometer*) detects the systolic pressure and then the diastolic pressure. The pressures are recorded in millimeters of mercury, for example 120/80 (spoken as "120 over 80"). The normal blood pressure for systolic is 90–119 and for diastolic 60–79. Pressures below those numbers indicate a state of hypotension and numbers above are described in four stages from prehypertension to hypertension (two stages) to hypertensive emergency (American Heart Association, 2016). Heart rhythm, heart muscle contraction strength, and heart rate can be measured simultaneously with *electrocardiography* (measurement of heart electrical activity), discussed in Chapter 4. *Pulse oximetry* is a non-invasive method for monitoring oxygen in the blood. Monitoring devices clip on the ear or the finger, or on the foot for infants. The device measures light absorption to determine the oxygen content of the pulsing arterial blood. The device must easily fasten to a part of the body that allows the light to penetrate through one side of the body and be picked up on the other side (Severinghaus & Honda, 1987).

### 2.7.3 Lymphatic System

At its simplest, the lymphatic system is composed of *lymph* fluid, vessels that carry the fluid, *lymph nodes* that act as filters, and lymphatic organs, for example, the *spleen.* Unlike the circular circulatory system, the lymphatic system is a "blunt-ended" linear system (Choi et al., 2012). It drains lymph—a collection of excess tissue fluids, leaked plasma proteins, and "debris from cellular decomposition and infection"—from the spaces between cells (Moore, Dalley, & Agur, 2014, p. 43). *Edema,* any collection of fluid in soft tissues of the skin and the tissue spaces beneath the skin, causes swelling. It increases pressure in the tissues. The tiny, thin-walled, *lymph capillaries* are the first segments of the lymphatic drainage vessels. They respond to changes in pressure, opening to drain the increased fluid (edema), along with wastes (Choi et al., 2012). Forbes (1938) defined the anatomy and drainage patterns of the lymphatic capillaries in the skin. See the left side of Figure 2.18 which illustrates the normal network of skin lymphatic capillaries.

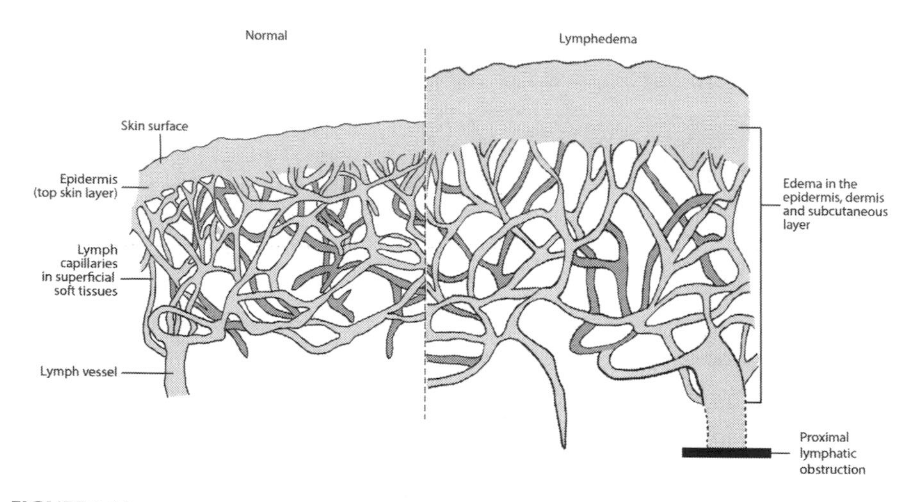


**FIGURE 2.18**
Cross-sections of normal skin and superficial soft tissue with lymph capillaries and skin and superficial soft tissues showing lymphedema.

In superficial tissues, the lymphatic capillaries are found near capillaries of the circulatory system. They eventually drain into larger, deep, lymphatic vessels which tend to follow the large vessels of the circulatory system. Refer to Figure 2.16 and compare the locations of the deep lymphatic vessels throughout the body (Figure 2.16-C) to the locations of the arteries and veins (Figure 2.16-A and 2.16-B). The lymphatic vessels empty the lymph back into the bloodstream at two large veins in the upper chest. Lymph must flow the full distance from the hands and feet to the central chest locations. Lymph is conveyed through the lymphatic structures by muscular actions of the lymph vessels, pressure differences created by breathing, movement of skeletal muscles, smooth muscle movements in the intestines, and arterial pulsations (Wittlinger & Wittlinger, 2004). The lymphatic vessels also have valves to help prevent lymph backflow (Forbes, 1938).

If tissue edema cannot drain naturally, *lymphedema* results. Edema can be caused by injury, poor health, inflammation, external pressure, or the force of gravity on the body. Lymphedema, appearing distally in the skin and subcutaneous tissues, is illustrated in Figure 2.18, right side. It occurs when there is lymphatic system obstruction from lymph vessel damage, lymph node removal, damage and/or inflammation, or infection in the lymphatic system (Zuther & Norton, 2013). Lymphedema may be permanent. It may be mildly uncomfortable to painful, and the amount of swelling can be minimal to extreme. Wearable products designed to try to minimize the extent of the swelling and discomfort have shown mixed results. Information on specific wearable products for lymphedema can be found in several chapters.

Look at the distribution of the lymph nodes illustrated in Figure 2.16-C. The lymph nodes are small, localized enlargements embedded in the vessels of the lymphatic system. Lymph nodes filter lymph fluid, removing solids; including bacteria, parasites, and foreign materials. They also produce blood cells to fight infections and cancers. Lymph nodes swell when they are challenged by infections or foreign materials. The spleen is a large lymphatic organ located in the upper left side of the abdomen. It produces blood cells. See Chapter 4 for further discussion on protection for this vital organ.

Understanding lymph flow patterns is important when designing to minimize or prevent lymphedema effects. Think of the body as a map (see Figure 2.19) separated into surface segments called *watersheds* (Forbes, 1938; Wittlinger & Wittlinger, 2004; Zuther & Norton, 2013). Note the landmarks (Figure 2.19) which determine the lines dividing the body's skin surface: the median plane of the body, the *clavicle* (collar bone), the *scapular spine* (the bony projection on the posterior side of the shoulder blade), the 2nd *lumbar* vertebra, and the navel (belly button).

The skin and superficial tissues of the head and neck, upper body and arms, and lower body and legs drain independently of one another. Similarly, the right and left halves of the body drain independently of each other. Edema fluid from the chest, arms, and head primarily drains through lymph nodes in the arm pits. Fluids from the pelvis, feet, and legs drain through lymph nodes in the groin (the fold between the abdomen and the thigh). Fluid from the right side of the head and upper body (the shaded area in Figure 2.19)

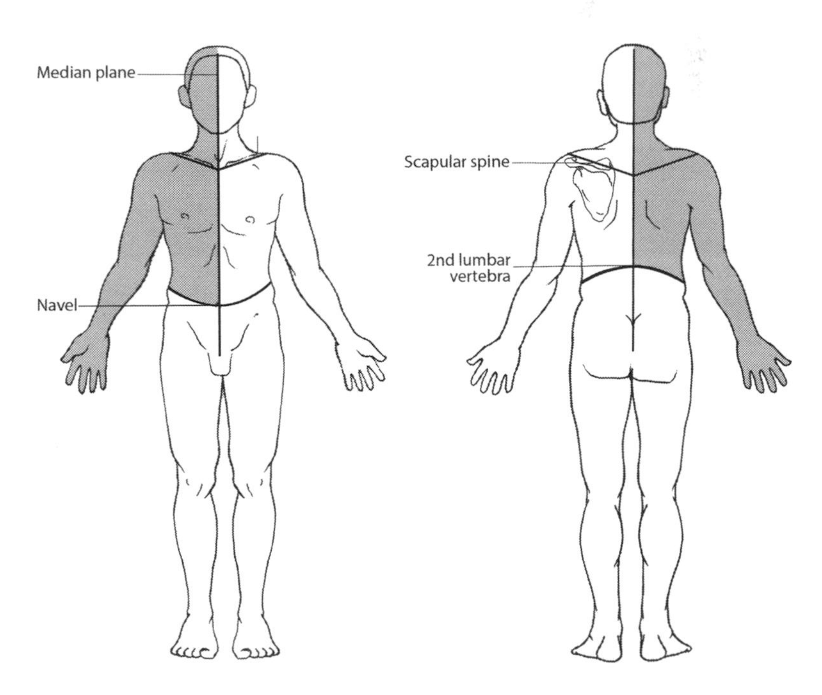


**FIGURE 2.19**
Lymphatic drainage patterns; landmarks and watersheds.

flows into a large vein in the upper right chest. Fluid from the rest of the body empties into a large vein in the upper left chest. These divisions are important when the lymphatic system develops obstructions resulting in lymphedema of a particular body part. As a disadvantage, the lymphatic system can spread cancer cells via the lymphatic watersheds. In general, lymph from the organs in the abdomen and pelvis and from deep tissues in the feet, legs, and thighs drains like the fluids from the skin of the legs and feet.

### 2.7.4 Digestive System

The digestive system is a continuous tubular pathway through the body (Figure 2.20). Food, water, and *toxins* entering the digestive system can be absorbed into the body through the selectively permeable lining of the "tube," a barrier similar to the skin. The body converts food to a useable fuel through a series of steps: (1) *mastication* (commonly called chewing), (2) chemical modification by digestive juices and by the bacteria of the *gut microbiome*—all of the microorganisms living inside the lower digestive tract (Sonnenburg & Sonnenburg, 2015), (3) absorption of the resulting simpler chemical compounds, followed by (4) elimination of non-digestible food and waste products (*feces*) through the *rectum* (final internal section of the digestive system) and *anus* (opening for passage of the feces at the lower end of the digestive system). Water needed by the body to maintain health also passes into the body through this system. While most things we eat support health; toxins, such as lead paint, can be absorbed and damage vital body structures and functions. This apparently simple tube and process is really quite complex. Several associated glands add *enzymes* (specialized proteins, produced by the body, which promote specific chemical changes) and fluids to the system. Smooth muscle contractions move the contents through the tube as it transforms from entry to exit. See Figure 2.20 for more detail on digestive inputs (left side), the steps of digestive action (center section), and an outline summarizing the digestive outputs from simple sugars to bacterial by-products and water (right side).

Wearable product designers develop products that must fit and function at the digestive system entry, the mouth and surrounding structures, or at the exit. Wearable products that are designed for the mouth area include mouth guards worn by athletes, scuba equipment, and protective masks. Diapers are, perhaps, the prime example of a wearable product for the waste products of the digestive system. Products that interfere with the flow of the digestive tract, like constricting waist "cinchers" and belts, may cause discomfort and can affect health.

### 2.7.5 Urinary System

The *urinary system* produces, transports, and eliminates liquid waste. It consists of two kidneys and *ureters*, the *bladder*, and the *urethra*. The kidneys filter waste

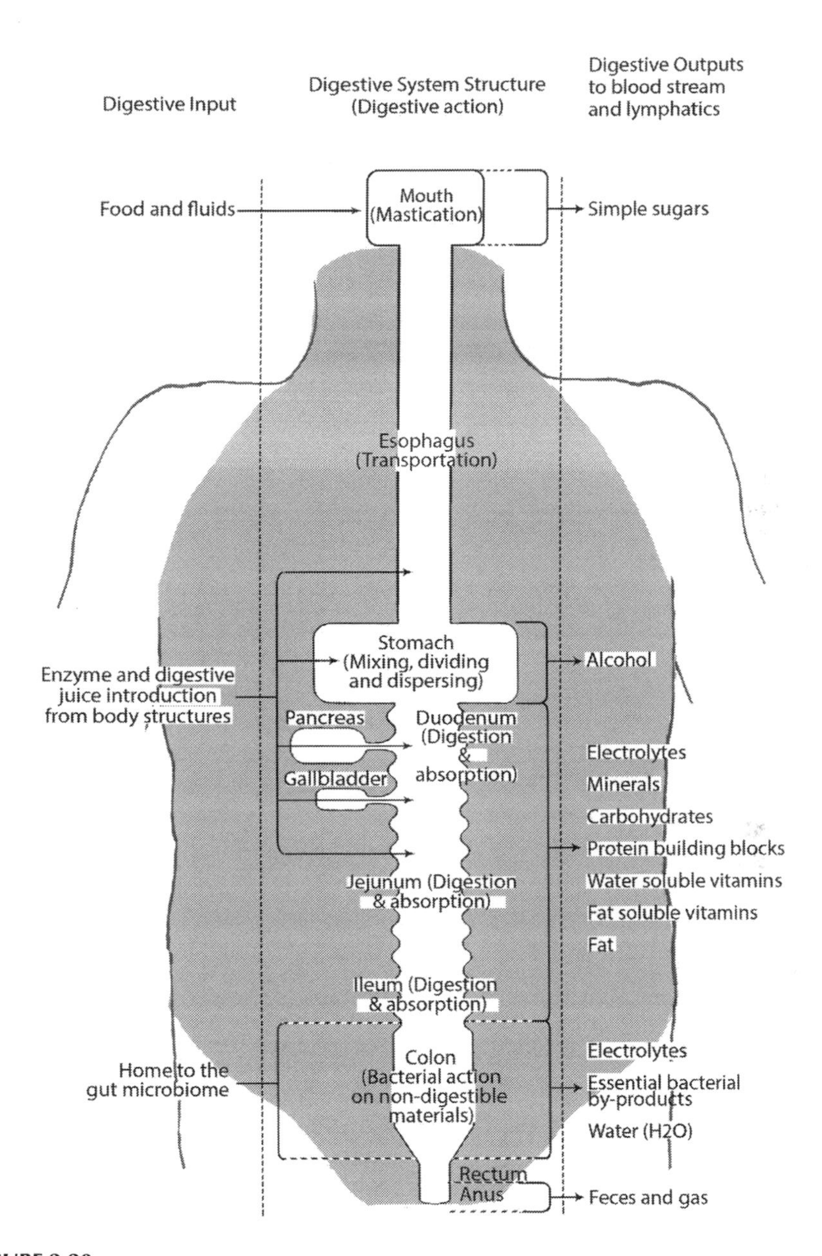


**FIGURE 2.20**

The digestive system "tube" and functions. (Modified from L. Mahan and S. Escott-Stump, Chapter 1: Digestion, absorption, transport, and excretion of nutrients. In *Krause's food and nutrition therapy* (12th. ed., p. 19) St. Louis, MO. Copyright 2008 by Elsevier Saunders.)

products, including excess water, from the blood to make *urine* (liquid body waste). Urine flows through the ureters, tubes connecting the kidneys to the bladder, the storage vessel for urine. After storage in the bladder, urine exits the body through the urethra, a tubular structure with a sphincter to control urine flow. In the female the urethra opens at the external border of the *vagina*, the internal female reproductive canal. When the bladder contracts, the sphincter opens and urine is expelled from the body. In males the urine flows through the urethra to the tip of the *penis*, the external male reproductive organ.

Products may be designed to protect urinary tract features and/or to meet needs of its natural functions. Athletic products are designed to protect kidneys from impact in contact sports. Everyday wearable products that are worn at the exit of the digestive and urinary systems include diapers for infants and underwear for men and women. Products designed for special needs include pads, garments, and drainage tubes for people who have limited or no control of urine elimination. Wearable product designers can consider the female urinary system as closely related to the digestive system and the reproductive system, discussed next, because they all exit the body in the same region. The unique features and combined functions of the male and female external organs of *reproduction*, the *external genitalia*, need special consideration when designing products for this region.

### 2.7.6 Reproductive System/Genitalia

Three components make up the reproductive system/genitalia for both males and females: *gonads*, ducts, and external genitalia. The *ovaries* are the female gonads. They reside deep in the pelvis and produce *ova* or eggs which carry the female's genetic material for reproduction. Male gonads, the *testes*, are part of the male external genitalia and are located in the *scrotum*, a muscular skin structure suspended from the body. Testes produce *sperm*, the structural vehicle for the male's genetic material in reproduction. Sperm development is temperature sensitive, a fact to be considered when designing products that cover the male genitalia. The egg and sperm are the key elements that merge to reproduce the species—make a baby. Males and females have ducts or tubes that transport eggs and sperm internally. The external genitalia, the penis and scrotum with testes of a male and *vulva* and *clitoris* of a female help connect partners in sexual intercourse to convey sperm to egg. If *conception* (merger of sperm and egg) occurs, the female reproductive system holds and nourishes the *fetus* (developing infant) and gives an exit path via the vagina for the baby's birth. Wearable products for both men and women are used as a means to prevent conception (Chapter 5).

### 2.7.7 Endocrine System

The endocrine system, a collection of glands located throughout the body, secretes *hormones* (chemical products that target specific organs or tissues in the body) directly into the bloodstream. *Insulin*, secreted by specialized cells of the *pancreas* (a digestive system component located near the posterior body wall of the abdomen), is an example of a hormone. It is necessary for the body to properly process and use sugars from the diet. A deficit of natural insulin or insulin function causes the disease *diabetes*. The endocrine system controls other essential body functions, like the female *menstrual cycle*, sperm and egg production, and pregnancy. The designer may not design specifically for the glandular structures of the endocrine system but may design products for the areas of the body that are affected by the system, such as feminine hygiene products to absorb menstrual blood.

## 2.8 Integumentary System: Coverage and Protection

Skin, hair, and nails—the upholstery and trims of the body—each play an important role in our interactions with the environment. The integumentary system is made of layers of specialized cells which cover the body surface. Integumentary tissues are replenished continually, but gradually, with new cells. The surface structures (skin, hair, and nails) are protein products of the cells, and not living tissues. Dead skin cells naturally and gradually flake off and some hair falls out on a daily basis.

Although we don't often think of skin as an organ, it serves crucial body functions as much as the heart and brain do. It protects, helps regulate temperature, and collects sensations among other functions. Many wearable products enhance or aid the integumentary system. Wearable products can also compromise this system.

### 2.8.1 Skin Structure

The skin is the largest organ of the body, covering approximately two square meters. The skin, with all its layers, is thick, heavy, and durable. Skin is composed of three layers, the *epidermis, dermis,* and *subcutaneous* layers. The epidermis is further subdivided, but only the top epidermal layer will be discussed here. Study the skin's layers and structures in Figure 2.21.

#### *Epidermis*

The epidermis, the outermost skin layer, looks like layers of flattened cells. The topmost layer of the epidermis, the *stratum corneum,* made of dead *keratinocytes*—skin cells which make the protein *keratin*—slowly flakes away as new cells replace the old from beneath. They contain a high concentration of the tough fibrous keratin. *Melanocytes,* another cell type in the epidermis, produce the dark skin pigment *melanin* which determines skin color. Skin color is genetically determined, but sunlight can stimulate the cells to

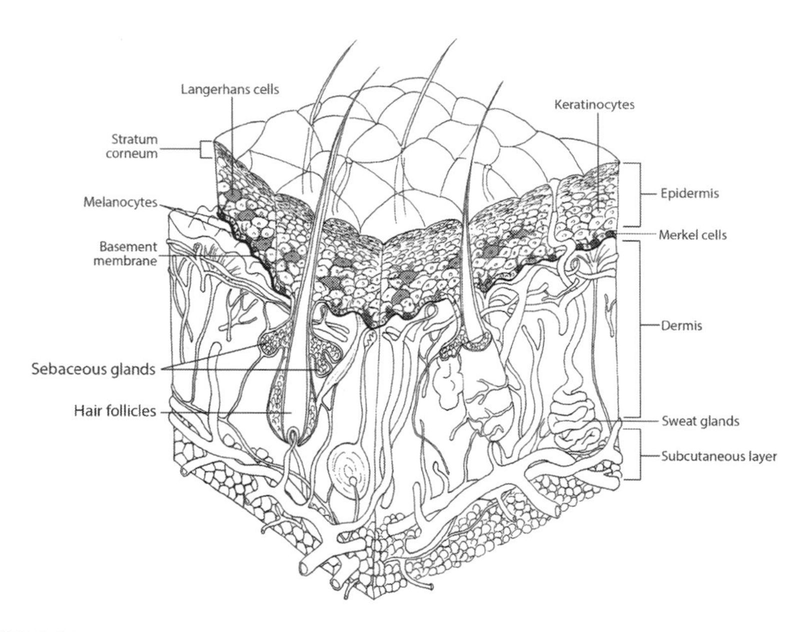


**FIGURE 2.21**
Skin anatomy, microscopic: Layers and structures.

produce more melanin, darkening or "tanning" the skin. The more melanin the darker the skin. However, melanocytes can also transform to become the dangerous skin cancer, *malignant melanoma.*

Components of the skin cells, especially in the stratum corneum, may be damaged in response to environmental stressors like *ultraviolet (UV)* light exposure (Anderson, 2012). Dangers of UV light exposure have been accepted and "In May 2000, the National Institutes of Health (NIH) added solar ultraviolet radiation and exposure to sunlamps and tanning beds to the list of identified carcinogens which are substances known to cause cancer" (Barrow & Barrow, 2005, p. 7).

*Langerhans cells,* a third cell type located in the epidermis, act to fight off microbes such as bacteria that threaten the body and *Merkel cells* are found at the junction of the epidermis and the dermis. These light touch receptors of the sensory nervous system are found in high touch body areas, like fingertips (Boulais & Misery, 2007). The epidermis is firmly attached to the dermis by the thin, ridged *basement membrane,* helping give the skin durability.

### *Dermis*

The middle layer of skin, the dermis, is mainly made up of connective tissue cells and their products, collagen fibers and *elastin* fibers, plus hair *follicles* (roots), *sebaceous* (oil) glands, and sweat glands. Collagen fibers give the skin a structural framework and strength. Elastin fibers, on the other hand, provide pliability and elasticity. The dermis contains networks of blood vessels, nerves, and lymphatics. Blood vessels in the dermis supply nourishment to the epidermis and the dermis. They play a vital role in temperature regulation, moving excess heat to the surface of the body where it can dissipate. Besides the touch-sensitive Merkel cells at the epidermal junction, there are nerves in the dermis which help in sensing temperature and pain. Lymphatic capillaries (discussed in Section 2.7.3 and seen in Figure 2.18) are located in the dermis.

### Subcutaneous Layer

The subcutaneous layer, the base layer of the skin is simpler in structure than those above it. It is the "subcutaneous" in the acronym SAT, the subcutaneous adipose tissue (Section 2.6.1). It is composed of adipocytes in a loose matrix of connective tissue, the superficial fascia (Section 2.2.6); with blood vessels, nerves, and lymphatic vessels serving all three skin layers.

### Nails and Hair: Associated Skin Structures

Toenails and fingernails are horny epidermal tissues made of keratin. These structures are generally less flexible than other soft tissues of the body. Nails are composed of plates of compacted keratinocytes, of the stratum corneum (Kapit & Elson, 2014). They are translucent, revealing capillaries in the dermis below. The proximal end of the nail, the *nail plate,* fits into a skin fold named the *nail fold*. The living nail plate grows and the nail extends distally from the nail fold.

Hair follicles, located in the dermis, are lined with epidermal cells. Living hair follicles grow hair, the keratinized filament that emerges at the surface of the skin. Hair is a strong, stable structure with layers of cells. Hair grows in cycles and, in humans, individual hair follicle growth cycles are independent of other hair follicles, so humans do not "shed" hair as most animals do. After a growing phase, human body hair passes through a two- to six-year resting phase before it is lost and replaced. For a more comprehensive discussion of human hair biology, see Ebling (1987).

Humans are distinct from other mammals because we are basically naked (Jablonski, 2006). While most mammals have a layer of protective hair on their bodies, the hair on the bodies of humans is very fine and often unnoticeable with the exception of some areas, like the head. Dermal layer sebaceous glands (only found associated with hair follicles) produce *sebum,* a combination of fatty compounds and cellular debris. The oily, waxy sebum flows along the hair shaft to the surface, to lubricate the hair and skin. *Sweat glands* also lie in the dermis and exit to the skin. *Apocrine* sweat glands are found in the armpits and groin. *Eccrine* sweat glands are found all over the body. Sweat glands produce a water-based solution of many different salts and body chemicals. Sweat from apocrine glands differs in composition from eccrine gland sweat. The composition of sweat and sebum produced by the skin has to be considered when designing products like hats and shirts with fitted collars that are in close contact with the skin and may easily pick up body oil and perspiration. Some materials are difficult to clean when they absorb body oils.

### 2.8.2 Integumentary System Function

The integumentary system is large and multifunctional. Skin keeps our bodies from shriveling to a dried-up mass. It holds a stable body temperature to let other systems work efficiently. It keeps us in touch with the world around us. Skin secretions include *pheromones* and breast milk. Skin manufactures the vitamin we need for strong bones and stores excess energy. It wraps everything up into an integrated, functioning body.

#### *Protection*

Skin, as a protective barrier, is water resistant, both shedding water from the body surface and keeping the body from drying out by preventing evaporation of crucial bodily fluids. Refer to the skin anatomy illustration. Starting from the surface, the epidermis and dermis, laminated together, help safeguard the inner body from abrasion and puncture. Skin shields other body tissues from ultraviolet radiation. The upper layer, the epidermis, which varies from one-half to four millimeters (mm) thick has "qualities that makes the epidermis sound more like a revolutionary new type of carpeting than a natural material" (Jablonski, 2006, p. 11). The skin can sometimes protect the body from microbes. However, several types of microbes live on the skin without harm and have beneficial effects (Grice & Segre, 2011). This useful microbe layer is called the *skin microbiome*.

The skin presents a large area of contact for harmful chemical or biological substances. When the skin is cool, dry, and intact it is relatively impermeable. Some protective clothing is designed to prevent harmful chemical or biological substances from coming in contact with the skin. Pesticide protective suits and asbestos abatement suits are examples of wearable products used in specific settings. The typical method of protection is to completely encase the body with a dense textile or film. Unlike sun protective products, these products don't have to meet aesthetic demands, so are often made of utilitarian non-woven textiles or coated materials. Watkins and Dunne (2015) provide an excellent overview of designing protective clothing for many hazardous environments.

Hair gives thermal protection and shields the scalp from sunlight. Wearable product designers may have to consider means of accommodating hair. Some external medical monitoring devices may require that a portion of the skin be shaved to provide close contact of a sensor to the skin. Fitting a protective head covering to a person with a shaved head will be much different from fitting a person with very long or dense hair.

The skin does not do a particularly good job of protecting the body from excessive heat and substances which cause burns. Wearable products are designed for burn prevention and to help minimize damage from burns. Oven mitts made of thick, non-conductive materials are an example of a product most people have in their kitchens to protect themselves when handling hot pots and pans. Blast furnace suits are made of textiles with a metallic coating to reflect high radiant heat away from the worker. See Section 2.8.4 for specifics on the damage done to skin by burns, including from ultraviolet radiation.

Humans use fingernails and toenails for aesthetic adornment. Other essential functions of these features can affect design choices. Fingernails and toenails protect the sensitive mobile fingers and toes and assist with grasp. They can be used for defense, for scratching an opponent. A relatively unique problem, *delaminated* or detached fingernails, reported by astronauts wearing protective space gloves is particularly challenging. The enclosed space of the glove with no mechanism to dissipate heat creates a moist atmosphere which, when combined with abrasion of the fingernails against the inside of the glove, causes this painful condition (Opperman, Waldie, Natapoff, Newman, & Jones, 2010).

### *Temperature Regulation*

Skin blood vessels, sweat glands, and hair help to maintain normal body temperature (37° C, 98.6° F). Blood vessels control the amount of blood reaching the body surface, dilating when the body is too warm, constricting when the body is chilled. When the body warms, or when emotions trigger sweating, sweat discharged onto the skin evaporates, cooling the skin. Fluid loss from sweat can be surprisingly high, particularly in hot temperatures.

Hair, or fur, on animals acts as a source of thermal insulation. Our relatively naked bodies lack that protection. We experience a primitive automatic response to cold temperatures when hair stands erect to offer some insulation by trapping dead air in the space between hair and skin. This "goose bump" response, often accompanied by shivering, helps to generate some heat. Humans can't rely on their scanty hair for insulation, so clothing is worn to improve thermal protection.

### *Skin as a Sensory Organ*

The skin is the largest sensory organ of the body, communicating signals to the body from the environment. Skin is a remarkably sensitive interface. Nerve structures attached to hair and skin sense the world around us; from the pressure of a too-tight waistband, to a very delicate, but irritating, touch of a tiny fiber such as wool. Pain receptors in the dermis serve a protective function for the whole body. *Dermatomes* and *cutaneous* nerve distributions are two different ways of representing patterns of sensory innervation of the skin. A dermatome is an area of the skin that is innervated by sensory fibers from a single spinal nerve (Moore et al., 2014, p. 50). *Dermatome maps* have been developed to illustrate which spinal nerve serves which skin area. For example, a spinal nerve from the middle of the neck innervates an area of the posterior neck and anterior neck and shoulder skin. Medical personnel are particularly interested in dermatomal patterns to isolate injuries within the central nervous system.

In the peripheral nervous system, up to three separate spinal nerves may contribute to a specific peripheral nerve in an arm or leg. Cutaneous sensory nerve distributions show the sections of skin served by individual peripheral sensory nerves. For example, the median nerve, a major nerve to the muscles of the hand also provides sensation to the skin on the thumb, index finger, middle finger, and the middle-finger side of the ring finger, primarily on the palmar aspect. The area boundaries for both dermatomal and cutaneous sensory nerve maps are inexact. Cutaneous nerve distributions can provide designers useful guidance for product placement, to prevent inadvertent injury to the more vulnerable peripheral nerves. Illustrations of these patterns can be difficult to locate, although Netter (1986, pp. 118–127), Gray (1966, pp. 971, 972, 996, 1000, 1003), and Jenkins (2002, pp. 120, 200, 291, 292, 347, 348) include cutaneous nerve maps. Cutaneous nerve distributions for the arms, hands, and legs are illustrated in Chapters 4, 5, and 7.

### Exocrine Function of Skin

Sweat glands deliver their secretions to the skin surface so are called *exocrine glands*. The watery fluids and electrolytes in sweat are vital to body functions. If the body sweats profusely, it is important to replace fluids and electrolytes. The apocrine sweat gland also produces and releases pheromones, body chemicals which communicate between persons and modify behavior (Kohl, Atzmueller, Fink, & Grammer, 2001). Pheromones produce the scents that attract us to our mates. The exocrine *mammary* (milk producing) glands of the breast are modified apocrine sweat glands.

### Manufacture and Storage

Skin cells manufacture vitamin D from chemicals brought to the skin by the bloodstream. *UVB*, one of the ultraviolet ray types found in sunlight, triggers this process. Vitamin D is important for the body, helping build strong bones and teeth. Sun protective measures block this function. Fortunately, the skin can produce vitamin D with short daily periods of sun exposure. In regions of the world with less regular daily sunshine people are often vitamin D deficient. A simple blood test can measure vitamin D levels in the body. People who require extensive sun protection, or who don't get enough sun exposure, may need to take a vitamin D supplement.

The subcutaneous layer of the skin provides a sometimes unwanted reservoir of energy—in the form of SAT. The dermis layer also holds, in association with its function in maintaining body temperature, a variable volume of blood. Blood can be redirected from the skin to other body systems, such as the muscular system or the digestive system. You may have noticed that you feel "cold" and reach for an extra layer of clothing after a meal. That chill comes when warm blood is diverted from your skin to the digestive system organs.

#### *Skin as a Wrapper*

Skin envelops the body. Skin prevents the loss of body fluids by enclosing all of the body structures. Skin also physically supports and helps to hold components of other systems in place. The structures we can see and feel through the skin—the superficial muscles, tendons, veins, and arteries—stay in place, in part, due to skin support.

### 2.8.3 Healthy Skin Characteristics

Skin helps to maintain the body's internal stability, with no apparent effort. It neither restrains nor encourages activity. Skin is just there. Skin, more than many other body tissues, has two important qualities. It can *regenerate* (restore itself) and it has marked *resilience* (recovery from shape and/or length change).

#### *Skin Regeneration*

Given the extent of skin coverage and the importance of skin as a protective barrier, humans are fortunate that skin heals efficiently; whether dry, cracked skin in winter or a surgical incision. While new skin cells regenerate the skin on an ongoing basis, wound healing is a complex, multi-step process. Put simply, skin heals itself first by creating a temporary closure for a wound when blood coagulates to form a scab. Secondly, skin cells build a new structure as the permanent closure. Medical science has developed products to assist wound closure, but healing remains a challenging problem (Boateng & Catanzano, 2015).

#### *Skin Resilience*

Healthy skin has a remarkable ability to stretch, as well as maintain its integrity and strength. Collagen fibers and keratin sustain a strong, healthy skin structure. Elastin fibers—think of elastic textile fibers like spandex—allow skin to stretch and rebound as we move or as we gain and lose weight. Some wearable products, like swim "skins" worn for competition, perform like skin, accommodating motion, while imparting modesty.

### 2.8.4 Damaged Skin: Burns

Skin protects the body as a tough, water resistant covering; however, the environment poses threats it cannot completely ward off. Fortunately, wearable products can be designed as an additional layer of protection.

#### *Ultraviolet Light: Sunburn*

Sunlight, despite its benefits, can be detrimental. Ultraviolet rays emitted by the sun are made up of *UVA, UVB,* and *UVC* rays. UVC is absorbed by the atmosphere and doesn't reach us. UVA and UVB have long-lasting effects on human skin causing irreparable damage and, in some individuals, lethal changes to the skin. The bright red appearance of sunburned skin comes from increased blood flow, which also produces skin edema.

UV skin damage can be avoided with several strategies: absorbing, reflecting, or blocking the rays. The challenge in designing UV protective clothing is to cover as much exposed skin as possible while providing sufficient ventilation to avoid discomfort or over-heating. Materials selection is important. Textiles with a compact, dense weave or knit structure block sun and are most effective. Some textile finishes are engineered to absorb UV rays. Sun hats with broad brims effectively protect the top of the head, face, nose, and ears.

#### *Other Burns*

Skin can be damaged from external/environmental burn sources like hot liquids, heated cooking utensils, stoves, an iron, or open flame. Chemical and electrical burns can be severe. Some protective functions remain in burned skin and help protect the deeper tissues from infection, but burns will typically leave a lasting visible mark on the skin, a *scar,* due to damage to the skin structure. Persons with severe burns may wear plastic face masks (Parry et al., 2013), compression garments (Engrav et al., 2010), or specialized products, like splints, that apply constant, even pressure to the damaged skin in attempts to prevent uneven scarring.

#### *Burn Classification*

Burns are classified by severity (MayoClinic.org, 2017):

- *First-degree:* Minor burns that affect the epidermal layer of the skin, with redness and pain for several days to a week, which usually heal without scars. Sunburn is a standard example; however, repeated sunburns can cause long-term damage.

- *Second-degree:* A deeper burn, involving both the epidermal and dermal skin layers. The skin may appear red or pale and/or wet, with *blisters*, pain, and swelling. Deep second-degree burns frequently cause significant scarring. Blistering sunburns are second-degree burns.
- *Third-degree:* Burns through the two top skin layers which reach into the subcutaneous, the deep, third skin layer. The tissues may be blackened. These burns may be painless due to nerve damage and permanent numbness may result. The skin may look stiff, waxy white, and leathery. Scars from third degree burns tend to be rigid and thick. Scars crossing joints can markedly limit joint motion.

### 2.8.5 Inflamed, Irritated Skin

#### *Contact Dermatitis*

Wearable products may cause skin irritation, most often from the material (textile, film, fur) used to make the product. Individuals vary in their sensitivity to materials; some people are highly sensitive while others have no reaction. When severe, the irritation is called contact *dermatitis.* According to the Mayo Clinic, "contact dermatitis is a red, itchy rash caused by a substance that comes into contact with the skin. The rash isn't contagious or life-threatening, but it can be very uncomfortable" (MayoClinic.org, 2017). Contact dermatitis from adhesives, plastic, and metals is likely to influence the viability of wearable technology. Contact dermatitis is divided into two categories: irritant contact dermatitis (nonspecific skin damage from friction and chemicals like detergents) and allergic contact dermatitis (poison ivy and fragrances). Allergic contact dermatitis tends to get worse over time. Irritant contact dermatitis goes away when the irritant is removed.

*Textiles and the Skin* edited by Elsner, Hatch, and Wigger-Alberti (2003) presents a wealth of information on interactions of textile products and human skin. Experts explain how textile additives, like detergents, dyes, and formaldehyde, can cause allergic reactions. The authors state that the incidence of textiles causing skin irritation is very low; however, for those who suffer with the conditions it is an important consideration in materials choice.

#### *Atopic Dermatitis*

*Atopic dermatitis,* also called *eczema,* is a chronic relapsing skin disease (Drucker et al., 2017). It involves an itch-scratch cycle and can be mild to severe. Wearable products may cause eczema: fabrics, clothing, rubber or latex gloves, and shoes. Designers of everyday apparel should have at least a basic knowledge of the raw materials and finishes used in their products to try to avoid undue skin changes. Wearable products can play roles in both protecting the skin and irritating the skin.

## 2.9 Body Systems: Conclusion

Wearable product designers have developed products that play an important part in human safety, health, and healing. Designers also develop products that affect everyday comfort and pleasure. Armed with this basic information concerning relevant body systems, you can go on to consider designing for regions of the body. Start at the top, the cranial region, in Chapter 3 on Designing for Head and Neck Anatomy and work your way through the body regions in Chapters 4 through 8 before returning to design for the whole body in Chapter 9.

# 3

# *Designing for Head and Neck Anatomy*

Headwear and neckwear products are varied in purpose and coverage. Headwear and neckwear fit on or around any part of the head and/or neck. Some products fit inside the ears or mouth. Designing products for diverse markets—fashion, protection, medical—requires different types and degrees of anatomical knowledge. If you are designing eyeglasses or a face mask; bony facial features, eye function, and *facial nerve* and blood vessel pathways are important. If you are designing helmets, both how the brain reacts to impact and how the skull resists impact are of paramount importance.

**Key points:**

- Basic anatomy of the head and neck:
  - Bones of the skull protect the brain and sensory organs.
  - Anatomical structures allow motion of the jaw and neck.
  - Soft tissue structures add individual character.
  - Circulatory, nervous, lymphatic, and respiratory systems are vulnerable.
  - The vital neck connects head and body.
  - The head and the neck locate the senses in space.
- Headwear and neckwear opportunities and challenges:
  - Protection and enhancement.
  - Aesthetics and function.
  - Coverage and movement.
- Fit and sizing:
  - Bones determine product shape and size.
  - Compressible neck soft tissues must be accommodated.
  - Numerous body landmarks coordinate with product features.

Consumer products worn on the head include hats and helmets of all types, hoods attached to coats and jackets, headphones, and fashion eyeglasses. Headwear may be designed to prevent damage from external sources: hazardous chemicals, gases, or particles; noise or excessive pressure (deep ocean dive activities), water in the airway, or UV light causing sunburn. Medical products to assist body functions include prescription eyeglasses, hearing aids, noise cancelling headphones, and masks or nasal prong tubes for supplemental oxygen. Helmets for football, hockey, baseball, and biking are designed to try to prevent damage due to impact—from other players, the playing field, or objects like baseballs and hockey pucks.

Neckwear is also diverse in design and purpose. Collars, ties, and scarves are fashion neckwear products. Orthopedic and sports braces, encompassing the neck to varying degrees, either restrict neck motion or support the head. Protective devices for the neck may be suspended from protective headwear. Baseball and softball catcher throat guards and hockey goalie mask attachments protect the neck from impact. Some of these products surround the entire neck, others cover one vulnerable segment. Anatomical structures, functions, and motions of the upper and lower neck are covered in Sections 3.8 and 3.9.

In this chapter, study the head and neck—from the inside out. With a basic understanding of the anatomy of the head and the workings of the brain a designer is better equipped to design headwear. Begin by learning the basic elements of the brain as the control center for the body. Then add knowledge of the head's skeletal structure—its form and function—with distinct facial features and the skull as a protective container for the brain. The spine, in the neck, similarly protects the spinal cord. Visualize the muscles, cartilage, fat, peripheral nerves, blood vessels, skin, and hair that complete the picture.

A medical principle is "first, do no harm." This should be a design principle for any product, but especially when designing products for the head and neck. A product may compromise nerve function or blood flow if vital points are compressed. Knowledge of head and neck anatomy and purpose opens new opportunities for product innovation. Designers may be able to enhance the senses, help decipher the mysteries of the brain, and improve control of body functions.

## 3.1 The Brain: Central Control

The central nervous system includes the brain and spinal cord. It controls and coordinates all systems throughout the body. Human *cognitive functions*—thoughts, emotions, personalities, memories, dreams, and plans for the future—reside in the brain. The brain controls movement and interprets the world via touch, sight, smell, taste, and hearing. Anatomical structures and functions in the head provide natural protection for the brain. Headwear may be designed to try to provide additional protection for the fragile central nervous system. Protecting the *cerebral hemispheres* of the brain and the spinal cord in the neck is a focus of design for sports.

### 3.1.1 Brain Segments and Functions

*Neurons,* the cell type found in the brain as well as in other nervous system structures, transmit chemical and electrical messages to control functions. There are an estimated 100 billion neurons in the brain alone. The complexity of the brain and the brain's interactions with systems throughout the body have long interested and challenged researchers. New tools explore brain activities, from cognitive processing to the apparent seamless integration of body functions. Wearable products are being used to further this research. For example, skull caps or head bands with sensors that detect brain activity can provide a non-invasive method of studying the brain. Skull caps, acting as a brain-computer interface, can harness brain activity and manipulate mechanical devices, like a robotic arm (Meng et al., 2016). Non-invasive stimulation of nervous system components also provides significant opportunity for wearable medical product designers (Lefaucheur et al., 2017).

Look at the major segments of the brain: cerebral hemispheres, *cerebellum,* and brainstem (Figure 3.1). The paired right and left cerebral hemispheres make up the major portion of the brain. The illustration shows the left hemisphere from the lateral side and the right hemisphere from the medial side. Some functions are more exclusively located in one hemisphere. The right hemisphere controls *motor function*—movement—on the left side of the body and the left hemisphere controls the right side. Registration of sensory inputs is also crossed, right body to left brain and left body to right brain. People are sometimes called "left brained" when they demonstrate predominant verbal skills as the left hemisphere usually has more influence on verbal functioning. "Right brained" individuals show strong visual, spatial, and/or musical

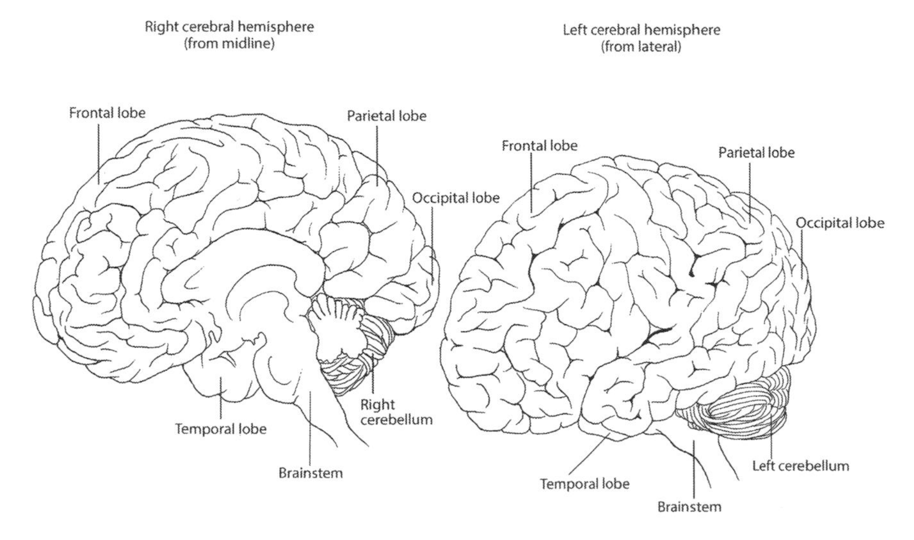


**FIGURE 3.1**
Major anatomical regions of the brain.

capacities suggesting the right hemisphere has more influence. There are individual exceptions to this right-left brain orientation.

Each cerebral hemisphere is divided into four anatomical regions: *frontal, temporal, parietal,* and *occipital lobes,* containing neurons with specialized functions. The neurons of the frontal lobe facilitate intellectual functions like reasoning, abstract thinking, and language formulation; emotional behavior and feelings; smell, memory, and voluntary movement. The parietal lobe is a primary center for body sensory awareness including taste, three-dimensional perception, and the image a person forms of his/her body. Hearing and language interpretation are centered in the temporal lobe. Vision is an occipital lobe function. Structures deep in the cerebral hemispheres function as emotional and major memory processing centers.

The cerebellum, the center for coordination, lies below the cerebral hemispheres toward the back of the head. The cerebellum makes up about 10% of the volume of the brain, but holds over 50% of the neurons (Saladin, 2007, p. 514). The brainstem lies below the cerebral hemispheres and in front of the cerebellum. It provides centers for head and face motor control, facial sensation; special senses of hearing, equilibrium, and taste; and regulation of automatic body functions. The vital brainstem is a vulnerable connector to the spinal cord, acting as a physical anchor for the "floating brain."

### 3.1.2 The "Floating Brain" and Concussions

The brain is a soft organ with a gel-like consistency. It is loosely anchored in the skull via the brainstem and spinal cord. The skull (cranium), when undamaged, protects the brain from crushing and penetrating injuries. The brain has a hydraulic mechanism that cushions the brain inside the skull. *Cerebrospinal fluid* (CSF), a colorless liquid, fills the space around the central nervous system (CNS) inside the skull and spine. It fills the spaces within the brain, the *ventricles* (see Figure 3.2). The CSF makes the brain buoyant in the skull. The human brain weighs about 1500 gm. (3.31 lb.), but when suspended in CSF its effective weight is only about 50 gm. (0.11 lb.) (Saladin, 2014, p. 402). This flotation mechanism keeps the brain from resting heavily on the bumpy and, in some places, sharp-edged floor of the cranium (brain cavity of the skull).

The brain, floating in the skull, is protected from jolting movements and minor impacts. If head motion is violent, due to direct blows to the head or to a "whiplash" motion of the head, the flotation mechanism can be insufficient. In these instances, the brain, tethered by the brainstem and spinal cord, can bounce off the interior of the skull. With enough force, it may strike first one interior side of the skull, then suffer a second blow at the opposite bony surface. *Concussion* is a disturbance of brain function caused by the brain shaking in the skull. A concussion can be mild to life-threatening. Several studies (Amen et al., 2011; Crisco et al., 2011) found that repetitive mild concussions can have serious and lasting consequences for mental function. There is more work to

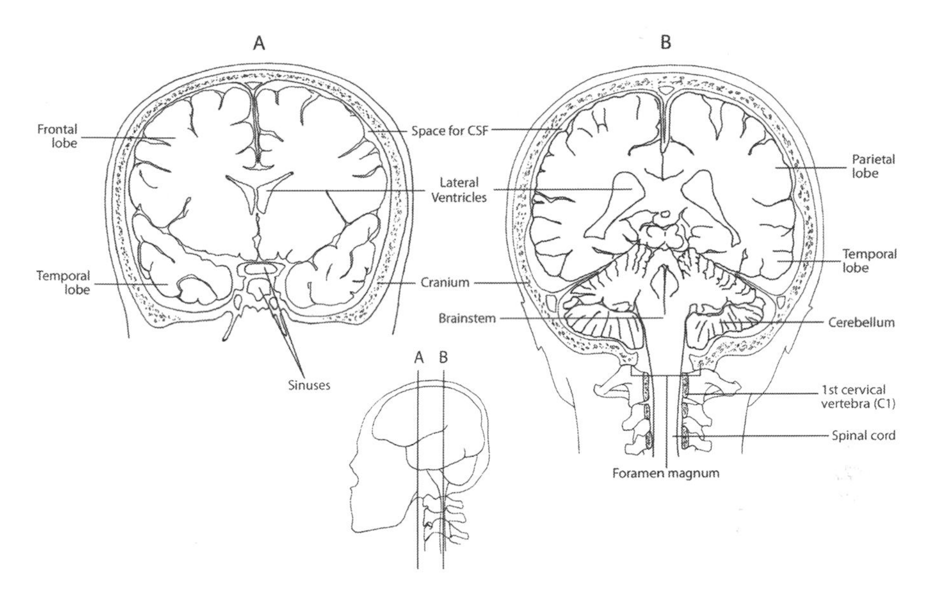


**FIGURE 3.2**
"Floating brain" in the cranium, CSF, and ventricles, two coronal views, from posterior perspective.

do on designing head protection, as a statement released after deliberations of the 4th International Consensus Conference on Concussion in Sport, stated that helmets may decrease mechanical stresses to the brain, but do not prevent concussion (McCrory et al., 2013). Gould, Piland, Krzeminski, and Rawlins (2015) give an overview of protective headgear for sports; descriptions of current technologies, test methods, and future design directions.

Awareness of long-term effects from concussion injuries in contact sports like football, hockey, soccer, and boxing is growing from grade school through pro sports (Marchi et al., 2013). Discussions and controversies about changing game rules to limit head injury are ongoing. At this time well-designed protective headgear is a must-wear part of the contact sports uniform. Protective headwear is equally important in leisure activities like bicycling, motorcycling, and snowmobiling. Helmets do help to prevent skull fractures and should be used for that purpose (McCrory et al., 2013).

## 3.2 Bones of the Head: Protection, Facial Form, and Stable/Mobile Jaw Bones

The head consists of *cranial bones, facial bones,* and the *mandible,* the lower jaw bone. Cranial bones form the cranium, the portion of the skull that encases and protects the brain and sense organs. The facial bones form the face providing each person with much of their characteristic facial appearance. The facial bones, with the lower jaw, also protect the entrances to the digestive and respiratory systems. Designers of protective headgear can apply knowledge of these anatomical features when designing and fitting products. Designers of other headwear can use the dimensions of these bony structures to guide sizing.

### 3.2.1 Cranium

The interior of the skull is more detailed and intricate then just one hollow space. The skull has several hollow areas, each with a specific function. Major superficial skull cavities are the *eye sockets* or *orbits, ear canals, oral* (mouth) *cavity,* and *nasal* (nose) *cavity.* The inside of the cranium is shaped around these cavities. The largest internal hollow area, the *cranial cavity,* acts as a container that holds the brain along with the CSF. The *sinuses* are hollow cavities near the respiratory passages and the cranial cavity. The spinal cord passes through a large opening at the base of the skull, the *foramen magnum.* These details are also seen in Figure 3.2. Additional small openings or passages throughout the skull provide spaces for nerves and blood vessels as they connect from inside the head to the superficial structures of the head.

Study Figure 3.3 (left side) and then look at your face and feel the features of the bones of your skull. Feel the top of your head, your cranium or skull

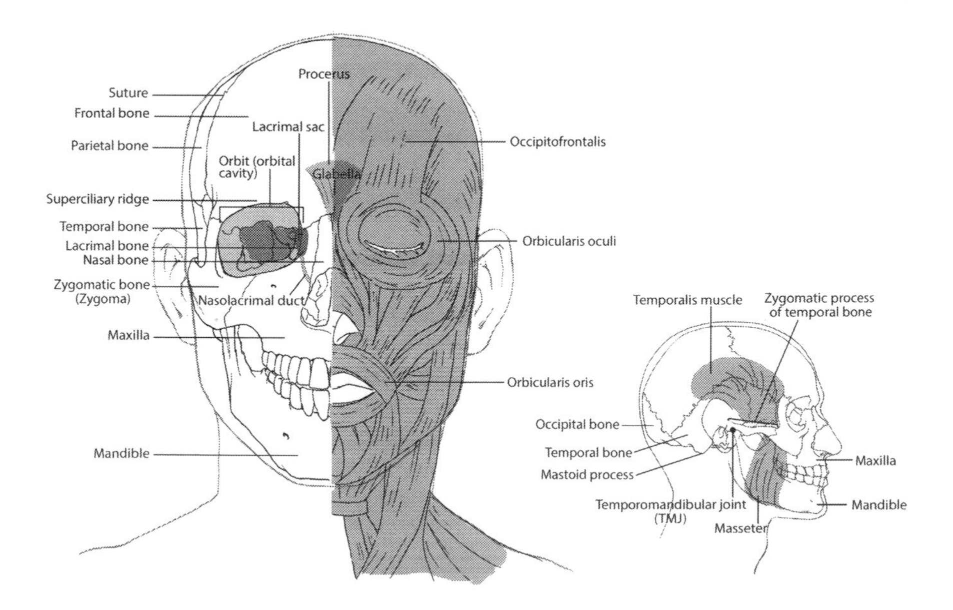


**FIGURE 3.3**
The head, anterior view: Bones of the skull (left side) and muscles (right side); structures relating to the temporomandibular joint, lateral view.

cap. Place both hands on your head and feel its shape and size. It is made up of several bones that mesh together with "seams" called sutures. As a product designer, you might think of the separate bones being stitched (sutured) together. The sutures allow the cranium to grow and expand with the brain through childhood and then incompletely fuse in adulthood. You can't feel the sutures, but you can imagine their placement. As you cup the top of your skull with both hands, the bones—one on the left and one on the right—at the top of your skull are the *parietal bones* which are joined to a single bone forming the back and much of the bottom of your skull called the *occipital bone*. Now run your fingers along the bone just above your eyebrows, this is the *frontal bone* that forms your forehead and the top front section of your skull. You can trace your brow arches or *superciliary ridges* of the frontal bone above your eyes. The *temporal bones* on either side of the head form your temples. The cranial bones lie over the corresponding lobes of the cerebral hemispheres.

While the bony structures of the skull provide protection in everyday situations, external blows can damage the bones of the skull. Helmets worn during sports add an extra hard shell of protection to the head. The U.S. Consumer Product Safety Commission (CPSC) has a helpful table listing a wide variety of sports, the appropriate helmet type, and the standards which relate to each helmet type (CPSC, 2014). Unyielding molded forms that shape to and/or cover the face and jaw, like a fencing mask, take the first blow then spread impact forces throughout the rigid material. They transfer some of the force to the cushioning that underlies the outer shell. The cranium acts in much the same way; impact is spread across the skull and then absorbed to some extent by the cerebrospinal fluid.

### 3.2.2 Facial Bones

The forehead of the face overlies the frontal bone. Other facial bones include the *lacrimal* (tear) bones, the smallest bones of the skull, which form the medial part of the eye socket. The slight indent between your eyebrows, just above the bridge of your nose, is a landmark spot called the *glabella*. The upper part of your nose, the bridge, is formed by the left and right *nasal* bones, facial bones that provide specific facial character. As you feel the length of your nose, you will notice that from the mid-nose to the tip the structure is quite flexible because it is made of cartilage.

In contact sports the nose often needs impact protection. Grids or grills on helmets provide good blunt force protection while allowing air circulation as the player breathes. But a grill or grid does not prevent small objects from striking the face. Helmets with full face protection using clear solid molded plastic may provide more protection from small objects, but can fog as players exhale warm, moist air. Small perforations can be strategically drilled into the face guard to provide ventilation; however, too many vent holes will weaken the plastic. Mesh fencing masks are designed to protect the eyes and face from the blades and tips of the foil, sabre, and épée, but restrict vision to a degree. Finding the balance between protection and wearability, including comfort, is always a challenge for designers.

The orbit, the skull cavity for the eye, is comprised of portions of several cranial and facial bones. This complex structure may help to reduce risk to the eye, as a break in any one of the bones can leave a majority of the cavity intact. The *zygomatic* or cheek bone (also called *zygoma*) is the lateral prominent facial bone next to the eye.

### 3.2.3 Jaw: Fixed Maxillae and Movable Mandible

The lower portion of the face, the jaw, provides protection as well as the motion essential for chewing and talking. The *maxillae* are the paired stationary upper jawbones which you can feel as a U-shaped structure holding your upper teeth. Each *maxilla* reaches all the way to the lower border of the eye and abuts the nasal bones medially and the zygomatic bone laterally. The mandible (lower jawbone), also U-shaped, is separate from the skull and holds the lower teeth. The *temporomandibular joint* (TMJ), the articulation between mandible and temporal bone, just in front of the ear canal, allows movement of the lower jaw so that you can talk and chew (inset, Figure 3.3). Keep the movement of the lower jaw (up, down, forward and back, and side-to-side) in mind when designing headgear that is strapped under the chin and/or worn while eating or talking.

The mandible, the strongest bone of the face, is susceptible to both impact and deterioration with age. Although most designers focus on external wearable products, wearable devices or appliances are also worn in the mouth, to replace teeth, such as dentures, or to protect and/or stabilize the maxillae and mandible in relationship to each other. TMJ splints or "night guards" worn when sleeping treat painful misalignment and degeneration of the TMJ joint. Mouth guards are often worn by athletes participating in contact sports to prevent damage to the teeth and mouth either from collisions or from upper and lower teeth clashing together during impact. Mouth guards may be custom crafted and fitted; somewhat adjustable guards can be purchased over the counter.

## 3.3 Head Muscles and Fat: Motion, Expression—Further Defining Individuality

Cranial and facial bones and the jaw are layered with muscles. Facial muscles help define who we are, adding further detail, along with fat, on the skull. As with all muscles of the body, some we control by thinking (voluntary muscles) and some work reflexively. We can choose to chew, moving the mandible and muscles of the mouth, tongue, and lips. We unconsciously blink, distributing *tears* over our eyeballs to keep them moist.

### 3.3.1 Muscles of the Head

Muscles of the mouth facilitate first steps in digestion as food enters the mouth through the lips. The tongue, muscles controlling the jaw (Figure 3.3, inset), and the bony structures of the mouth manipulate food helping the teeth to chew it into bits that can be swallowed. Consider muscle locations and functions when designing any headwear and avoid restricting muscle motion.

Muscles give us the ability to smile, frown, blink, and wink. Figure 3.3 (right side) shows only the outer layer of facial muscles. Facial muscles of expression originate on the head bones and insert into the superficial fascia of the skin. Contracting these muscles makes the skin move producing unique expressions. These muscles rely on signals from the *seventh cranial nerve* (cranial nerve VII), the facial nerve. When the *occipitofrontalis* muscle contracts it raises the eyebrows and horizontally wrinkles the skin of the forehead. The *procerus* muscle runs over the bridge of the nose, between the eyebrows, and produces horizontal wrinkles on the nose. It may be activated when you frown. The *orbicularis oris* muscle, circling the mouth, provides the ability to pucker the lips or form an O-shape. With this and other facial muscles a person can smile, whistle, kiss, and create equal pressure around a drinking straw. For specific applications you may need to research additional information on the deeper muscle layers of the face.

### 3.3.2 Fat Compartments of the Face

Fat literally rounds out the facial features over the bones and muscles. According to Rohrich and Pessa (2007), the face is "partitioned into multiple, independent anatomical compartments" (p. 2219). See Figure 3.4 for the approximate locations of the facial and neck fat compartments. The forehead has three separate sections of fat. The upper and lower eyelids hold envelopes of fat. The cheek area also includes three separate fat compartments. Pilsl and Anderhuber (2010) identified additional distinct fat compartments in the chin and upper neck. You know from the discussion of adipose tissue in Chapter 2 that numbers of fat cells do not increase or decrease with weight gain/loss, but facial and neck conformation can change as fat cells increase or decrease in size. Prolonged focused pressure can produce fat atrophy, causing visible indentations in fatty tissues, including in these compartments. Think of the creases sometimes seen in the side of the face and head when a person takes off close-fitting eyeglasses.

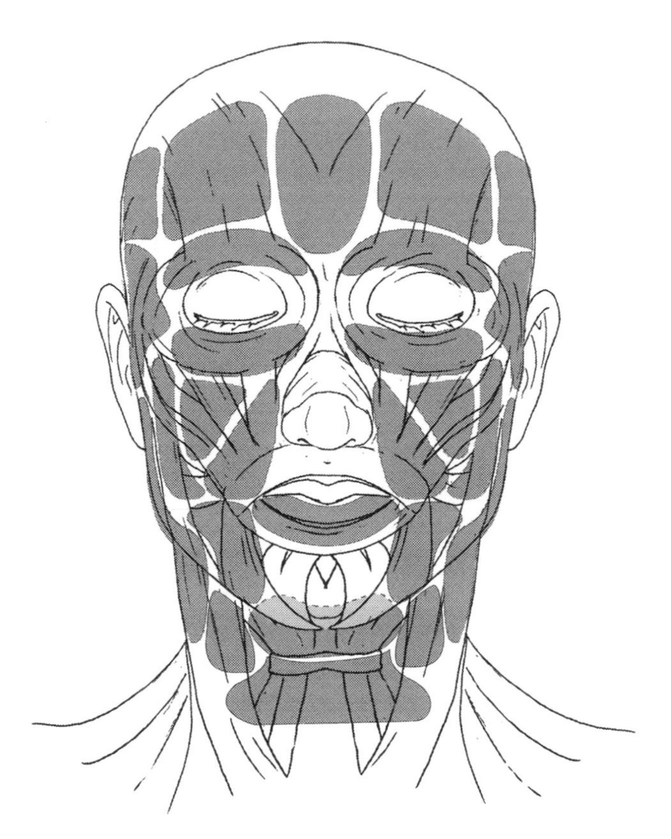

**FIGURE 3.4**
Facial and neck fat compartments.

## 3.4 Superficial Nerves and Blood Vessels in the Head: Functions and Design Cautions

An extensive network of nerves and blood vessels run throughout the head and neck. Review the basics of the systems in Chapter 2. Consider head and neck nerve and blood vessel locations and pathways when designing headwear and neckwear.

### 3.4.1 Nerves: Selected Vulnerable Nerves and Nerve Locations

Recall that sensory nerves convey sensation from receptors to the brain. Head and neck nerves pertinent to designers mostly carry sensation from the skin. Motor nerves carry messages from the brain to muscles. Nerve tissue, like the brain, is relatively soft. Pressure can damage nerves at the

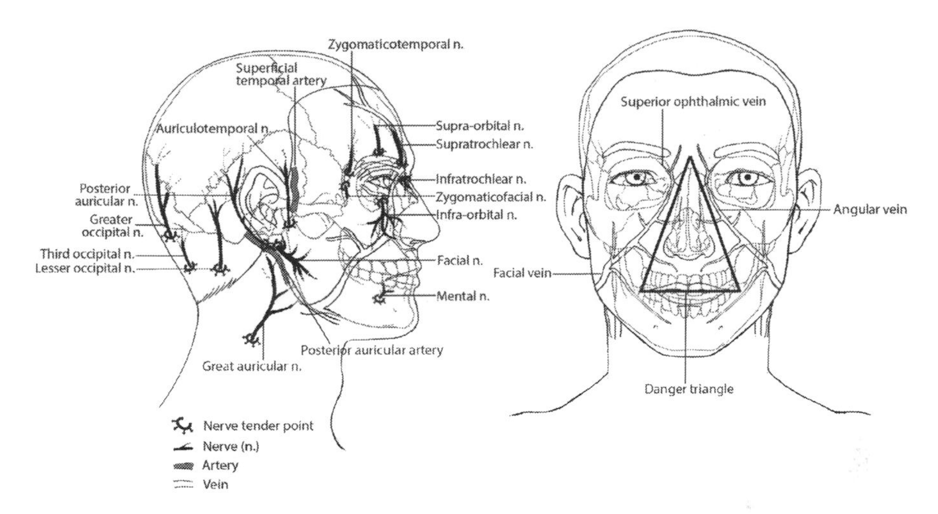


**FIGURE 3.5**
Facial/head nerves, veins, and arteries pertinent for headwear design; danger triangle of the face.

points where they transit through the skull or where they pass from the skin through muscles or connective tissue (Figure 3.5, left side). Protect these sites from undue pressure to avoid causing pain or numbness in the associated *innervated* areas of the face, *scalp*, and neck (Moore, Agur, & Dalley, 2011, pp. 521–522). Palpate your scalp, neck, and face to find the tender points in each of these regions to better understand the nerve locations. Some points are easier to find than others and sensitivity and degree of vulnerability varies among individuals. When designing headwear, use discomfort at tender points as a guide and avoid pressure at those areas.

The nerves, both sensory and motor, vulnerable to pressure are found in several different regions of the head and neck. Use the list below as a guide to locate these nerves.

- Back of the head
    - The *greater occipital nerve* (sensory) from the skin into the posterior neck muscles.
    - The *third occipital nerve* (sensory) toward the side of the head.
    - The *lesser occipital nerve* (sensory) in the upper neck between the ear and the back of the head.
- Near the ear
    - The *great auricular nerve* (sensory) below the ear and the *jaw line*.
    - The *auriculotemporal nerve* (sensory) behind the TMJ and traveling upward in front of the ear.

  - The facial nerve (motor) leaving the skull in front of the ear near the TMJ. It controls the muscles of facial expression (muscles not shown in illustration).
  - The *posterior auricular nerve* (motor) running to the muscles of the scalp behind the ear (muscles not shown in illustration).
- Lateral to the eye
  - The *zygomaticotemporal nerve* (sensory) going upward.
  - The *zygomaticofacial nerve* (sensory) traveling downward.
- Above, below, and medial to the eye
  - The *infratrochlear nerve* (sensory) on the side of the nose at the bridge.
  - The *supraorbital nerve* (sensory) just above the eye, between the nose and the center of the eye.
  - The *supratrochlear nerve* (sensory) running upward between the infratrochlear and supraorbital nerves.
  - The *infraorbital nerve* (sensory) below the eye between the nose and the center of the eye.
- Lateral to the mouth
  - The *mental nerve* (sensory) just below the corner of the mouth.

### 3.4.2 Arteries: Two Vulnerable Arteries and Artery Locations

The deep arteries of the head are well protected, and many surface arteries of the face and scalp are generally resistant to pressure, in part, because they are relatively protected by muscle and fat. Two arteries, the *superficial temporal* and the *posterior auricular,* can be affected with pressure (Figure 3.5, left side). The superficial temporal artery passes over the *zygomatic process* (Figure 3.3, right side), a part of the temporal bone which articulates with the zygoma in front of the ear. Use gentle pressure at this site, just in front of the ear to feel the artery pulse. Pressure on the artery may cause pain and restrict blood flow, both negative effects. However, pressure applied to this area with a wearable device has been used to treat migraine headaches (Cianchetti, Serci, Pisano, & Ledda, 2010). The posterior auricular artery, lying next to the posterior auricular nerve, about 1 cm (0.5 in.) behind the ear can be compressed with curving eyewear *bows* that reach the posterior auricular nerve tender point.

### 3.4.3 Veins: Vulnerable Veins and Vein Locations

Many head veins are near the skin surface and may be visible. As discussed in Chapter 2, venous drainage starts at capillaries, moving to ever larger vessels. Think of the venous drainage system like tributaries (the White Nile, the Blue Nile) flowing into a major river (the Nile River). Unlike river systems, veins frequently have branches that extend to more than one venous vessel. Therefore, compression of superficial veins to the point of obstruction is usually less concerning than arterial or nerve compression, because venous blood tends to have more than one drainage pathway (Moore et al., 2011, p. 26).

There is one important case where an extra venous drainage path contributes to a health risk. Two veins, the superficial *angular vein* and the deep *superior ophthalmic vein* can transmit infection from the central face to inside the cranium (Figure 3.5, right side). The angular artery runs near the angular vein between the corner of the eye and the side of the nose. Find the approximate position of the angular vein (and artery) by feeling for the angular artery pulse with gentle palpation along the upper lateral nose.

The infection risk has been known since the mid-1800s. Blood can flow in either direction in the angular vein; gravity moves the blood toward the neck, but there is no mechanical barrier to flow toward the brain (Zhang & Stringer, 2010). If the predominant venous drainage away from the central face toward the facial vein is blocked, blood will move toward the brain via the superior ophthalmic vein, near the top of the nose. Because of this risk, the nose and upper lip are descriptively named the *danger triangle of the face.* Although not a common problem, if compression obstructs the angular vein drainage from the danger triangle, and there is a major infection within the triangle, infected blood can "back up" into a major venous structure next to the brain. These serious infections are particularly concerning where antibiotic treatment is not readily available (Varshney, Malhotra, Gupta, Gairola, & Kaur, 2015). Use a precise design plan to avoid localized points of pressure on the angular vein.

### 3.4.4 Blood Vessels: Heat Conduits

Blood vessels not only carry blood, they also carry body heat. Numerous blood vessels carry blood to and from the brain. The scalp, the skin surface over the cranium, has a rich supply of blood vessels. This arrangement offers the opportunity to use the head as a site for cooling via thermal conduction. Researchers in the University of Minnesota (U of MN) Human Dimensioning© Laboratory and the U of MN Extreme Environments Lab experimented with a cap design (Figure 3.6) for astronauts to provide whole body temperature control (Kim & LaBat, 2010). Cool water circulating through tubing in the cap carries excess heat away from the head through a conduction heat transfer mechanism. A similar approach has been used in products designed to reduce chemotherapy-induced hair loss (van den Hurk et al., 2012). Conversely, wearing a hat in cold weather can help prevent overall heat loss by blocking radiation, convection, or conduction from the warm surface of the head to the cooler atmosphere.

**FIGURE 3.6**
Liquid cooling ventilation garment (LCVG). (Courtesy of the Human Dimensioning© Laboratory, University of Minnesota.)

## 3.5 Eyes, Ears, Nose, Mouth: The Senses

The head houses key sensory organs and wearable products are designed to enhance and/or protect these features. Eyes provide crucial information about the world. Many people wear eyeglasses or contact lenses to correct vision, and eye protection is required for many work and sports or leisure activities. Ears can be protected from the cold with earmuffs, shielded from loud noise with ear plugs, and used to hold earbuds that direct sound into the ear canal. The nose is a suspension point for eyeglasses, a means of collecting scents for information processing, and an entry point to the internal respiratory system. The mouth and nose can both be used as portals for devices to supply oxygen or air to the lungs. *Taste buds* on the tongue (the sensory receptors for the sense of taste) are sensitive to the flavor of any materials used in products worn within the mouth. Mouthpieces can be designed to be held in place by the mouth and/or jaw for respiratory assistance devices.

### 3.5.1 Eyelids, Eyelashes, Eyebrows

Eyelids function to protect, lubricate, and cleanse the eyeball. See Figure 3.7 for a cross-sectional view of structures of the eye. The muscle that opens and closes the upper eyelid allows the eye to blink, sweeping the moisturizing tear layer across the surface of the eye. The *orbicularis oculi* (Figure 3.3, left side), threaded through not only the eyelids, but also the other skin overlying the orbits, is the squinting muscle.

Using wind tunnel technology, Amador et al. (2015) documented that eyelashes help keep particles out of the eye and divert air away from the surface of the eye, decreasing moisture loss. Studying 22 different mammals, not including humans, they found the optimal length for eyelashes to be approximately one-third the width of the eye. Glasses and other eyewear need to have enough clearance from the eye and eyelashes to allow natural function of the eyelid. False eyelashes, generally intended as an aesthetic enhancement and applied with an adhesive just above the natural eyelash line, come with some apparent risks. They may exceed the optimal length for lashes, may reduce the clearance between the eye and eyewear, and the adhesive and materials used in the false eyelash may irritate the skin.

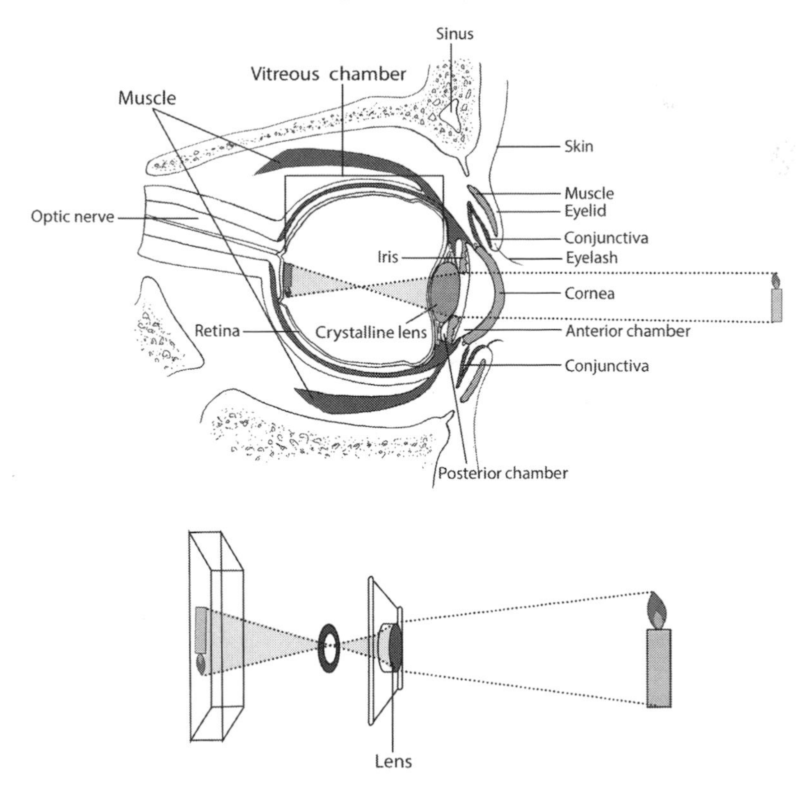


**FIGURE 3.7**
Anatomical structure of the eye, lateral view; comparison to simple camera.

Eyebrows are facial characteristics and to some extent protect the eye by helping keep perspiration out of the eyes. Prominent eyebrows functionally increase the depth of the orbit in the skull and may help shield the eyes from light, wind, and dust. Eyeglasses may be designed to emphasize, conceal, or mimic the shape of the eyebrow for aesthetic purposes.

### 3.5.2 Eyes and Sight: Visual Messages to the Brain

The eye is an approximate sphere with a 2.5 cm (1 in.) diameter (Forrester, Dick, McMenamin, Roberts, & Pearlman, 2016, p. 13). Refer to Figure 3.7 to study eye structures. When you look at a person's face you see just a portion of the white *sclera* of the eyeball surrounding the colored *iris* and the dark central *pupil*. The *cornea*, the clear front window of the eye, lies in front of the iris and the structurally deeper *crystalline lens*. The sclera, but not the cornea, is covered by the *conjunctiva*, a thin *mucous membrane* which is the outermost layer of the exposed eyeball and contiguous with the inner layer of the eyelid. It effectively prevents the migration of small *foreign bodies* or liquids into the eye socket and the cranium. The eyeball is not one fluid-filled balloon-like structure. It actually has three chambers. The *anterior chamber* lies between the cornea and the iris and bulges slightly under the upper and lower eyelids. The smaller *posterior chamber* lies between the iris and the crystalline lens. The fluid in these two chambers is called the *aqueous humor*. The third chamber, the *vitreous chamber*, between the crystalline lens and the *retina*, at the posterior of the eye, holds a more viscous fluid, the *vitreous humor*.

The eyes rest in the orbits and are cushioned with fat. Six small muscles originating on the cranium insert on each eye to move it. These muscles are innervated by three separate nerves. Like the bony complexity of the orbit, the multiple muscles and nerves serve as back-ups to each other. In case of injuries to an isolated muscle or nerve, the others preserve some function for the eye as a whole. The eye muscles, acting together, allow us to scan our environment—up, down, and side to side, as well as diagonally—using both eyes and without moving the head. The size and shape of the cranium determine the position of the eyes relative to other facial features, such as the distance from the center of one eyeball to the other and the position of the eyes relative to the bony bridge of the nose. Fitting and positioning eyewear for maximum visual capacity requires an appreciation of the functions of the eye.

The eye transforms entering light into stimuli which are interpreted into meaningful information by the brain. Good vision requires a functioning physiological optical system including the *refraction* (bending) of light by the crystalline lens to focus the image on the retina (Forrester et al., 2016, p. 269). Think of the eye as a "biologic camera" (Figure 3.7).

Seeing is a complex process. Light passes through the transparent cornea, the aqueous humor, and on through the center of the iris, the colored circular disk in the center of the eye. The amount of pigment in the iris determines a person's eye color. Blue eyes have the least pigment, brown eyes the most (Sturm & Frudakis, 2004). The iris acts like the diaphragm of a camera opening and narrowing to let in more or less light. You can see your iris in action by looking in a mirror, turning off the light, waiting a minute, turning the light back on, and watching the reaction of your iris. In the dark your iris opens to let in more light. When you turn the light back on your iris reacts by closing down and admitting less light.

The light passes through the opening at the center of the iris called the pupil to the crystalline lens. The crystalline lens focuses the admitted light on the retina at the back interior surface of the eye (Millodot, 2004, pp. 102–103). The retina contains millions of light-sensitive cells called *cones* and *rods* which collect the images for the brain to process. These cells contain chemicals that are activated by light. Cones are concentrated toward the center of the retina and are responsible for vision in a lighted environment, for example daylight, and for color perception. Vision in this center area called the *fovea* is the sharpest. Rods are located outside of the fovea, and do not detect color, but are very light sensitive so are responsible for low light or night vision. Unlike a camera, the eye rather commonly fails to produce sharp visual images. Fortunately, wearable products can be designed to improve sight.

### 3.5.3 Focus on Eyewear: Function and Fitting

Eyewear is a large design category. It encompasses devices to correct deficient eyesight, products to protect the eyes from all types of environmental threats, eyewear to augment vision and alter other cognitive functions, and fashion eyewear. Millodot (2004) defines eyeglasses/eyewear/spectacles as "An optical appliance consisting of a pair of . . . lenses mounted in a frame or rimless mount, resting on the nose and held in place by sides extending towards or over the ears" (p. 289).

#### *Eyewear: Function*

We commonly think of eyewear as glasses and contact lenses to correct vision. If the eye is not properly focusing light on the retina, corrective lenses can alter the path of light through the eye. Corrective lenses may be prescribed by an eye specialist or purchased off-the-shelf, such as common reading glasses. Optical corrections can be integrated into protective eyewear as well as into the lenses in eyeglasses and contact lenses. Placed directly on the surface of the eye, a contact lens has to float on the cornea without damaging it.

Eye protection may be as simple as tinted lenses or as involved as a full face and head mask. Good sunglasses protect the eye from bright sunlight and UVA and UVB rays. UVB is the most damaging to the eye. Industrial processes, such as welding, frequently cause eye injuries (Lombardi et al., 2005). Protective eyewear for welding must provide extensive coverage and fit well to prevent harmful light rays, heat, large and/or small objects, wind, fluids, or chemicals from making contact with the eyes. Contact lens wearers in industrial settings with specific hazardous airborne chemicals need specialized protection, such as indirectly vented goggles, to prevent contamination of the tear layer on the eye (Peate, 2007).

Safety is the priority when designing all protective eyewear. Vinger (2000), in his comprehensive discussion of eye protection in a wide range of sports, also provides references to more general eye safety standards from major standard-setting organizations such as the *American Society for Testing and Materials* (ASTM). In order to work, protective eyewear must fit well and counteract hazard. It should cover the eye area as efficiently as possible while not restricting sight. Lateral shielding diverts particles and wind away from the eye but can obscure peripheral vision. Cushioning on the edges of wearable products such as swim goggles is intended to seal the area around the eye to prevent substances from seeping under the eyewear and into the eye, but may cause discomfort over sensory nerves. To help insure stray bits and pieces don't reach the eye, the gap between safety eyewear and the face needs to be minimal. A small tool designed to measure the gap (U.S. Patent No. 9,507,175, 2016) can either help refine a design or facilitate a custom fit for safety eyewear. Understanding skeletal, muscular, vascular, and neurologic structures of the face and head helps designers meet individual needs for protection. Selecting materials with flexibility to conform to differences in facial structures or providing a selection of eyewear styles to accommodate anatomical variations helps to fit a variety of people.

Increasingly, eyewear is being developed to augment vision in low light, smoke-obscured, or high-speed environments, or to selectively block specific light colors from entering the eye. Head-mounted displays (HMDs) have small optics in front of one eye or both eyes, augmenting and modifying visual information for the wearer, effectively expanding the visual experience. HMDs are used by military personnel and in medical and video game settings. New applications are being developed and HMDs are trending to smaller optical head-mounted displays (OHMDs). Stroboscopic glasses have been used to enhance visual motor training for athletes (Appelbaum, Schroeder, Cain, & Mitroff, 2011). Specialized lenses, from contact lenses to various spectacles to goggles, selectively filter the colors reaching the eye in attempts to strengthen our natural visual capacities, even, in some cases, to decrease color blindness. Promising research in mood disorder treatment employed eyeglasses with lenses tinted to specifically block blue light (Henriksen et al., 2016).

Cosmetic and aesthetic aspects of eyewear have grown in importance as corrective eyewear has become an integral part of the fashion industry. Many factors: facial shape, scale of the eyewear relative to the size of the face, proximity of the frame to the bony features of the face, all contribute to eyewear aesthetic choices (DeLong & Daly, 2013, pp. 124–125). Frame and lens shapes enhance positive features or minimize perceived negative features. Facial features such as the width and boundary of the bony portion of the nose affect both the width of the bridge of the frame and the shape of the portion of the lenses and the frame near the nose (Abel, 1939). *Bespoke* (custom-made) eyeglasses emphasize aesthetics of the frame, while meeting the physical requirements of visual correction. Some bespoke manufacturers use 3D head scans and then make minute adjustments to the frame and lens for a custom fit. Given the use of 3D printing of eyeglass frames, bespoke frame materials may be more unusual and expensive than mass market frames. Aesthetic concerns are not limited to fashion eyewear. Even though eye protection is mandated in most risk situations, people want to look good.

### *Fitting Eyewear*

Fitting eyeglasses, or any eyewear, to the structure of the face and head while correcting vision is a complex procedure. The eyeglass frame has two important functions: holding the lens and correctly positioning it in relationship to the eye to provide the *refractive correction* needed to restore maximum visual acuity. Variations in bone structure change the position of the ears relative to the plane of the eyes, the distance between the eyes, and the shape of the bridge of the nose. Therefore, the nose and ears, as the support points for the frame, are key anatomical elements in positioning the front of the frame. Variations in the eyeglass frame, especially the length of the side pieces (the bows or *temples*) and the width of the frame, accommodate the bony variations and create aesthetic differences. Common frame measurements include: *lens width, distance between lenses, lens height, temple length,* and *total width* (Figure 3.8). The *bridge* of the frame, the connecting section over the bridge of the nose, must correspond to the size and shape of the nasal bones. The bridge of a metal frame is supported by nose pads. In a plastic frame the bridge may simply rest directly on the bridge of the nose.

Ideally, the frame positions the lens in front of the eye with the same *vertex distance* used when determining the lens prescription during the eye exam. The vertex distance is measured along the line of sight between the cornea and the posterior surface of the eyeglass lens (Millodot, 2004. p. 338). The *interpupillary distance,* between the centers of the pupils, is also important when positioning the lens (Millodot, 2004. p. 88). These measurements are important because the lenses held in the frame must be shaped and positioned to correctly refract light through the center of the pupil of the eye, along the line of sight. If the refracted light doesn't line up, eyesight will not be corrected.

Eye motion allows correction for far and near vision in a single pair of glasses. The line of sight, when looking straight ahead, passes through the upper/upper-middle part of the lens in the frame. When looking downward, without tilting the head, the line of sight passes through the lower segment of the lens. Bifocal lenses incorporate lenses of different refractive power in the same lens; the upper portion for distance and the lower portion for near vision. Bifocals may be constructed with two distinctly separate lens

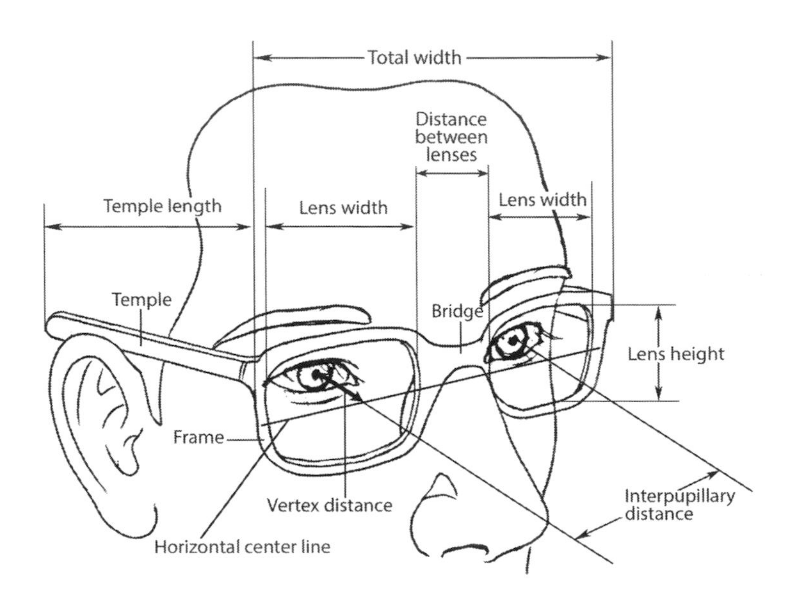


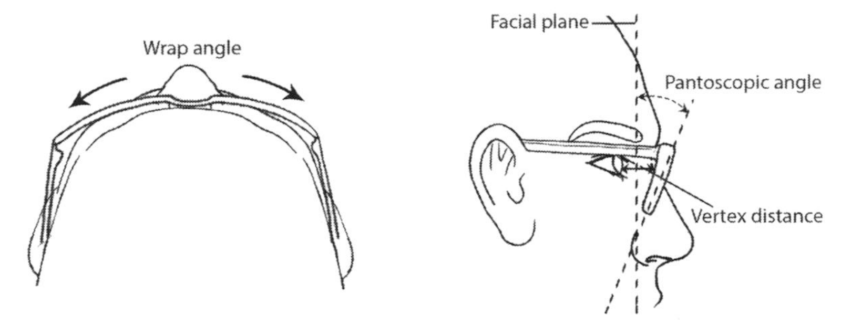


**FIGURE 3.8**
Eyeglass landmarks and measurements.

segments or as a "blended" lens, with a progressive change in the refraction and no clear delineation between the segments.

The temples of the frame attach to the outer margins of the frame and are supported by the ears. The angle of the temple to the plane of the lenses can be adjusted to modify the position of the frame. It is not unusual for a person's ears to be asymmetrically positioned on the head, in other words, on different horizontal planes. If so, each temple can be adjusted independently to align the lenses to the pupils of the eyes. Ears on a higher plane than the eyes will tend to tilt the lower section of the lenses toward the face, increasing the *pantoscopic angle* or *tilt* (Figure 3.8, lower right). The pantoscopic angle is the amount of tilt of the front of the eyewear frame—the spectacle plane—relative to the frontal plane of the face (*facial plane*) (Millodot, 2004. p. 20). Adjusting the pantoscopic angle helps keep the vertex distance, from the cornea to the lens, appropriate when the person looks downward to read or do other close work. This is most important with bifocal lenses. The *wrap angle* (Figure 3.8, lower left) is the curve of the lens section of the frame around the contour of the face. Wraparound prescription sunglasses or protective eyewear with corrective lenses may require an additional lens shape modification to maintain corrected vision throughout the field of vision. Curving a progressive bifocal corrective lens by wrapping the lens around the face expands the effective area of vision, improving visual experience (Han, Graham, & Lin, 2011).

The eyewear industry strives to meet standards for eyeglass frames and lenses established by organizations such as the *International Standards Organization* (ISO) and the ASTM. Eyeglass frame component parts (frames, temples, lenses) come in limited sizes. The lens width and distance between lenses varies between frame styles. Although there is no requirement that information be printed on the frame and no agreed upon order of listing the measurements, you can check the inside of the temples of prescription eyeglasses for sizes, typically given in millimeters.

Many types of eyewear are covered under ISO, ASTM, and other standards. Tang, Tang, and Stewart (1998a) present a detailed anthropometric approach to eyewear frame fitting, which can supplement the information in ISO standards. They identify multiple head and facial dimensions in three categories: measurements relative to the frame front, distances and angles in the bridge area where the nose bears the weight of the eyewear, and dimensions associated with the frame side (temple). Among many terms, they define the *horizontal reference line* (of the face), as an important line, tangential to the upper margin of the lower eyelid. This line is used to help establish a number of other anthropometric angles and reference points around the nose. The horizontal reference line is not the same as the *horizontal center line* as described in ISO 13666:2012—a line used as a reference for vertical measurements of the eyeglass frame itself. However, the horizontal reference line (of the face) may be used to align the facial structure to the frame/lens system, by relating it to the horizontal center line of the frame. In a companion paper Tang, Tang, and Stewart (1998b, pp. 296–298) present an approach for fitting eyewear frames considering ethnic facial differences. The discussion is specific to anthropometric data from a population of Hong Kong Chinese. The paper also cites several other international ophthalmic anthropometric studies and gives comparisons of measurement data from the sources. In a more recent publication, Eze, Uche, Shiweobi, and Mba (2013) provide data on selected eye-related anthropometric parameters in a Nigerian population.

Facial measurements and anatomical features, varying person to person, should mesh with the selected eyewear frame style. Then, the prescribed corrective lenses must be fitted into the frame and the frame adjusted to the individual's anatomy. 3D head and face scanning technologies have the potential to become an essential tool for fitting eyewear and in selecting an aesthetically pleasing frame for the individual. Shwom (2011) encourages eye professionals to take advantage of the progress in lens manufacturing by customizing corrective lens prescriptions, with careful attention to the details related to fitting.

### *Cautions in Designing Eyewear*

Abel (1939), in a never repeated work on eyeglass fitting, presents a discussion of head and facial anatomy that remains pertinent for designing eyewear today. He proposes developing a "correct" or "natural" eyeglass frame to conform to the anatomic structures of an individual's face and head. Like Tang et al. (1998a, 1998b), he acknowledges the challenges of fitting around the nose, but also points to issues related to the forehead, temple, ears, and specific nerves, arteries, and veins (seen in Figure 3.5). The following guidelines are largely derived from his work.

Eyewear nose pads need to be positioned so they do not rest on the infratrochlear nerve or the angular vein. Sustained pressure on the nerve will be uncomfortable. The risks of compressing the angular vein were discussed in Section 3.4.3. The section of the eyewear temple resting above and around the ear can cause pressure on the posterior auricular artery or the posterior auricular nerve, behind the ear (in Sections 3.4.1, 3.4.2). The eyewear temple may cause discomfort if resting on muscles in the area—the muscles used to wiggle our ears—especially if the muscles are well developed.

Some eyewear extends beyond the orbital areas—over the forehead, lateral face, or middle face. For these designs there are several other nerves that may be compressed. Five sensory nerves—the supraorbital, supratrochlear, infraorbital, zygomaticofacial, and zygomaticotemporal—should be considered. Straps running around the circumference of the head to secure eyewear may cause discomfort from or compromise the function of the greater, lesser, and third occipital nerves (nerve specifics are found in Section 3.4.1).

The *lacrimal apparatus*, the drainage system for tears produced to lubricate and clean the surface of the eye, is also vulnerable to compression by nose pads and eyewear. Tears collect at the medial corner of the eye, drain into the *lacrimal sac*, then the *nasolacrimal duct* (both seen on Figure 3.3, left side) and finally empty into the lower lateral portion of the nose. Compression of these anatomical features may be signaled by tears overflowing the eyes, without crying.

Products designed to be worn on the head should not interfere with eyesight nor compromise the function of other structures of the head. "If it (your product) hurts, don't do (produce) it" is a good guideline when you evaluate eyewear prototypes. Pain caused by pressure is likely to get worse, making a product unwearable.

### 3.5.4 External Structures of the Ears

Ear sizes and shapes vary: close to the skull, protruding from the skull, with long or short fleshy *ear lobes*. The ear, *auricle* in Latin, is also called the *pinna* meaning "wing" which makes sense when you consider the shape. The visible portions of the ears are made of skin and elastic cartilage and are attached to the head with collagen structures covered by skin (Figure 3.9). Feel your ear from the upper edge to the lower tip or ear lobe. The firm outer rim of the auricle is the *helix*. The flatter inner portion of the auricle next to the helix is the *scapha*. The small springy protrusion toward the center of the ear, near the opening of the ear canal, is called the *tragus*. This cartilage-based feature can help ear plugs stay in place and is sometimes used as a landmark for developing dimensions for headwear. You can bend it quite easily. Hold it in place for a few seconds, release it, and it will bounce back to its original shape. Because the auricle is thin and extends beyond the many heat-carrying blood vessels of the head, it is extremely susceptible to frostbite if exposed to sufficiently low temperatures. Ear "muffs" for warmth, sound protection, or sound enhancement should comfortably encase the pinna.

Ears serve a variety of functions. The *external ear* directs sound through the *external acoustic meatus* (ear canal) of the skull to the sound transforming (*acoustic*) structures of the ear and also protects access to the *middle ear*. The ears are a convenient place, in addition to the bridge of the nose, to perch eyeglasses and protective masks. Ears have been used for centuries as a place to adorn the body with clip-on jewelry or jewelry inserted through piercings. Due to the relatively poor blood supply, piercing the helix or scapha carries a risk of serious infections (Sosin, Weissler, Pulcrano, & Rodriguez, 2015). The fleshy ear lobes are vulnerable to tearing if jewelry in a piercing is so large that it catches on something or is very heavy (Niamtu, 2002).

### 3.5.5 Ears: Internal Structures and Functions

After passing through the external acoustic meatus, the sound energy reaches the *tympanic membrane*. This first acoustic structure is sometimes called the ear drum because it literally vibrates like the stretched head of a drum. When sound energy vibrates the tympanic membrane, it is converted to mechanical energy. As the mechanical energy enters the middle ear it encounters an amazingly complex arrangement of tiny bones—*malleus, incus, stapes*—which move to amplify and conduct the mechanical energy further into the ear. These bones are often descriptively called the hammer, anvil, and stirrup. Note these shapes in Figure 3.9. Ear infections most commonly occur in the middle ear and may cause both pain and problems with hearing. In the *inner ear* the mechanical energy is converted to electrical impulses by sensory cells and then travels to the brain for interpretation into meaningful sounds. Wearable products for the auditory system can be protective, sound enhancing, or pressure equalizing.

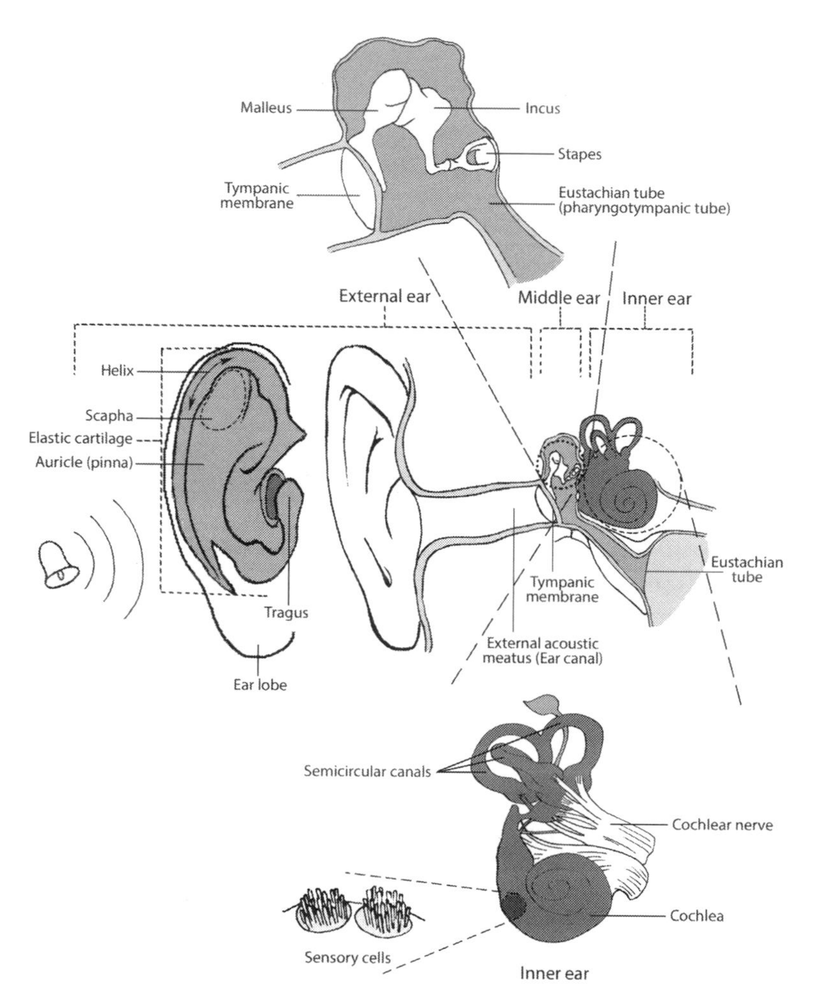


**FIGURE 3.9**
External, middle, and inner ear.

## *Protecting Hearing*

The anatomical features of the middle and internal ear are very delicate. When the mechanical energy transmitted through the ear drum and these tiny bones is excessive, as with long duration exposure to loud noises, or a single exposure to an extremely loud noise, the sensory cells of the inner ear can be permanently damaged. Therefore, protecting hearing in loud environments is important. A key approach is to prevent violent or excessive vibrations from reaching the tympanic membrane. In industrial settings, solid molded ear muffs that totally cover and seal around the external ear may be most effective. Less visibly obvious protection, ear plugs worn in the external acoustic meatus, may safeguard hearing from lower intensity noise.

Protective devices that are inserted into the ear canal may be off-the-shelf or custom fit. The effectiveness of both types of devices depends on a good fit as well as the materials selected to block the sound vibrations. Sound insulating materials that incorporate dead air space, like foams, are often the materials of choice.

### *Enhancing Hearing*

One purpose of sound enhancing equipment is to direct sound to the device wearer while limiting sound heard by other people. To do this the sound device is placed directly in the ear with earbuds or over the external ear with stereo muffs. These wearable devices attach to sound sources. One risk of directing sound in this way is that the sound waves may be so intense that the sensitive cells of the inner ear are damaged. Another danger is that the wearer's attention to sounds of the environment, for example traffic, may be blocked.

Another purpose of sound enhancing equipment is to amplify sound conducted to the tympanic membrane. Most of us are familiar with the acoustic stethoscope that doctors use to listen to lung, heart, and bowel sounds. Sound travels from a chest piece placed on the patient's body, through air-filled hollow tubes to ear pieces in the doctor's ears. The chest piece detects sound: one side is a plastic disc, a type of diaphragm, while the other side is a hollow cup called the bell. The diaphragm side amplifies high frequency sounds. When the bell side is used, rested lightly on the skin, low-frequency sounds are more easily detected (Felner, 1990, p. 64). Early mechanical or acoustic hearing aids operated on similar principles. Modern hearing aids electronically amplify different pitches, to try to improve speech recognition. Musicians rely on hearing as a crucial sense and have needs for both hearing protection and management of hearing loss (Chasin, 2009).

### *Maintenance of Balance*

The three *semicircular canals* of the inner ear, situated at right angles to one another, form an important part of the balance apparatus which helps to keep us upright and oriented to the force of gravity. Each of the canals responds to head movement in one of the major planes of the body: the sagittal, coronal, or transverse planes. Problems in the semicircular canals can produce a sensation of *vertigo*; that the environment around you is spinning with you at the center. The semicircular canals are not involved with hearing. Specialized goggles can be used in the diagnosis of balance problems.

### *Pressure Equalization*

If the pressure is unequal between the external and middle ears, the tympanic membrane cannot vibrate normally and hearing is impaired. The middle ear has a feature, the *eustachian tube*, which runs from the ear to the *nasopharynx*, the cavity in front of the upper spine behind the nasal cavity (see Figures 3.9 and 3.10). Contemporary literature often uses the term *pharyngotympanic tube*, which describes the pathway for this structure. The structure allows equalization of air pressure between the middle ear and the external acoustic meatus, on the other side of the tympanic membrane—relieving the sensation of "plugged up" ears. Graves and Edwards (1944) provide a comprehensive review of the structure, from anatomy to clinical significance.

Sudden equalization of pressure between the middle ear and the external acoustic meatus can result in a popping sensation, for example when airplane cabin pressure changes. Because infants have an underdeveloped eustachian tube a change in pressure may be quite painful. You may have suffered along with a screaming baby as an airplane descends and pressure changes. Bluestone (1996) suggests that crying may help an infant raise the pressure in the middle ear, to compensate for the change in exterior pressure. Products have been developed to try to address this problem (U.S. Patent No. 5,467,784, 1995).

### 3.5.6 Nose and Mouth: Airways, Smell, and Taste

Feel the tip of your nose. The rounded skin-covered cartilage structures at the sides of the nose are called *wings*. Run your finger along the central ridge on the underside of the nose, above your upper lip. This feature is the inferior aspect of the *septum*, a cartilage and bony structure which separates the left and right *nostrils* and nasal cavities. *Vibrissae*, hairs inside the nostrils, offer some protection from small airborne particles. *Olfactory cells* (the nerve cells responsible for the sense of smell) are found in the upper reaches of the nasal cavities and communicate information about odors to the brain via the *olfactory nerves* (Figure 3.10, upper section). The frontal lobe of the brain lies in the cranial cavity above the roof of the nasal cavities. The sinuses are cavities in the skull located above and below the eyes, as well as deep inside the skull (Figures 3.2, 3.7, and 3.10). The *pharynx*, commonly called the throat, is a muscular structure which funnels air, liquids, and solids to the appropriate locations in the neck. It includes the nose (nasopharynx) and mouth (*oropharynx*). In the neck, the *laryngopharynx*, the lowest region of the pharynx, is behind the voice box.

You can feel the internal features of your mouth (oral cavity) with your tongue. Run your tongue along the inner surfaces of your teeth; then feel the adjacent roof of your mouth (*hard palate*). If you move your tongue towards the posterior of your mouth, you can feel the transition from the hard palate to the *soft palate*. The nasal cavities lie directly above the hard palate. The oropharynx is the space inferior to the nasopharynx and posterior to the oral cavity. The tongue occupies most of the oral cavity. Figure 3.10 shows the tongue's size; it is a large muscular structure. Taste buds of different shapes are clustered on the tip, posterior sides, and very back of the tongue; all five

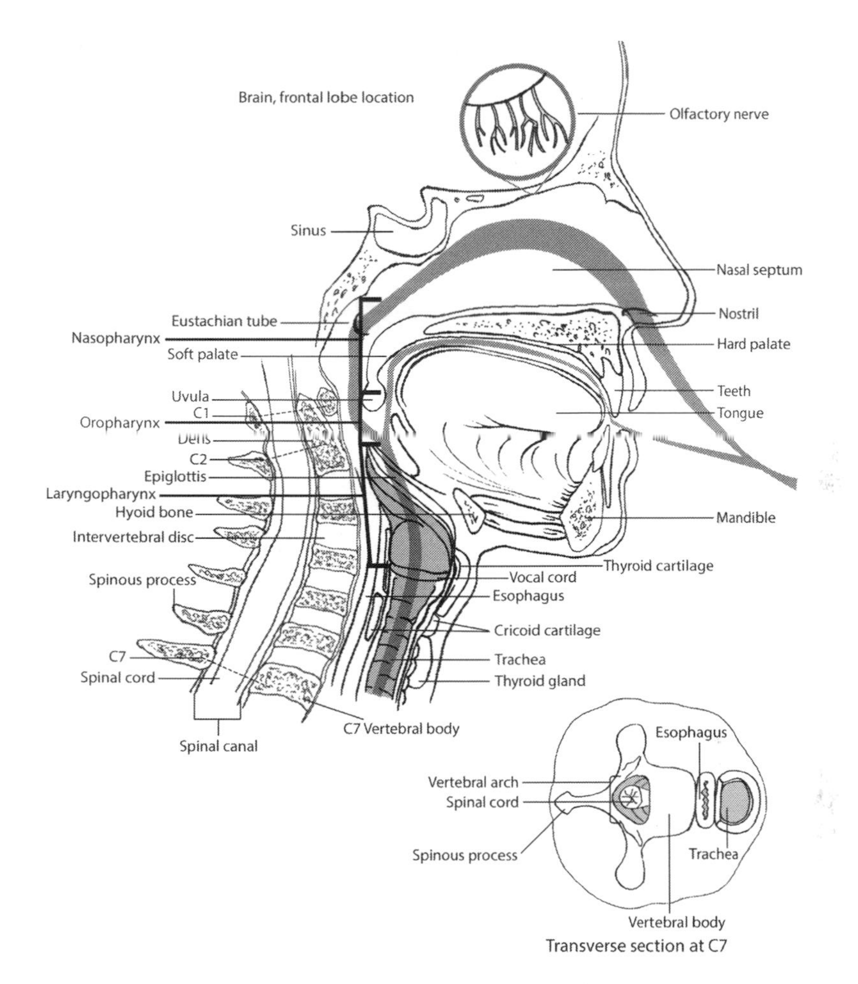


**FIGURE 3.10**
Nose, mouth, throat, and spine; sagittal section, transverse section at C-7.

taste sensations (sweet, sour, salty, bitter, and umami) are detected in each area. Avoid using materials with intense or disagreeable tastes, or which may leach chemicals into the mouth, for products like mouthpieces and nipples for infant's bottles. The *hyoid* bone sits at the base of the tongue muscle (Figure 3.10). It is unusual in that it does not articulate with any other bone. *Tonsils* are lymphatic structures in the mouth and pharynx. Tonsillar tissue helps defend the body against bacteria, viruses, and other foreign materials.

The functions of the mouth and nose are closely related. Air moving through the nostrils and nasal cavities flows past the olfactory receptors, allowing you to monitor odors in the environment. The olfactory receptors also enhance the sense of taste; detecting the aromas of food in the mouth. From the nose, the air moves into the nasopharynx and then the oropharynx. If the nasal airway is blocked with nose plugs, breathing through the mouth becomes necessary. When breathing through the mouth, air can dry out the tissues of the mouth as it flows back to the oropharynx. Think about your breathing when your nose is congested due to a cold or other viral infection. The nose, mouth, and sinuses help humidify and warm air as it enters the body. The sinuses help lessen the weight of the head and add resonance to the voice. Products pressing on the facial sinuses may cause discomfort.

The mouth, as the entry point to the digestive tract, initiates digestion through chewing and moistening food with *saliva*. The tongue assists with the formation of speech sounds and facilitates chewing and swallowing—two important digestive system activities. The hyoid bone props the entrance to the airway open (Moore et al., 2011, p. 594). The respiratory system and the digestive system meet in the oropharynx where the intricate interplay of muscles, cartilage, and the hyoid bone directs food into the digestive tract and air into the respiratory system. Collapse of the airway in this region during sleep results in *obstructive sleep apnea*. A continuous positive airway pressure (CPAP) machine, a treatment for sleep apnea, uses a facemask interface. Facemasks or devices worn in the nose are also used to assist respiration by directing oxygen-rich air to the lower respiratory system.

The *gag reflex* is an involuntary contraction of muscles in the pharynx. This reflex occurs when the posterior part of the tongue or another structure in the back of the mouth or throat is touched. It could be triggered by objects in the mouth such as dental X-ray film holders or trays for dental impressions. The gag reflex helps to prevent choking.

You may have had the unpleasant sensation of "swallowing the wrong way." If the oropharynx is not working in a coordinated fashion, *aspiration* can occur. Aspiration is the movement of food, liquid, or a foreign body from the upper airway into the lungs. It can cause pneumonia or death. Solids or liquids in the respiratory system lead to coughing or choking. Coughing is the body's self-protective response to try to force material out of the respiratory system and away from the lungs. Wearable products can help prevent substances, like water, from entering the lungs. Snorkel mouth pieces fit into the mouth, so a swimmer can breathe without water entering the airway.

Breath is essential to life and the air-flow pathways through the nose and mouth to the lungs need to be protected. Because small children tend to put things into their mouths and occasionally push objects into their nostrils, they are especially susceptible to choking. Choking can occur if an object or piece of food lodges at the back of the oral or nasal cavity, anywhere in the pharynx, or erroneously reaches the trachea. If a small garment element (a piece of trim or button) detaches with use and is small enough to enter the throat of a child, it becomes a choking hazard.

The U.S. Consumer Product Safety Commission issued a regulation to try to prevent death and injury from choking on, inhaling, or swallowing small objects a child may put in his/her mouth or nose (CPSC, 2017). The CPSC regulation is linked to ASTM F963 (2016a), which describes the test procedure for measuring "small parts." F963 was updated in 2016 and the new test

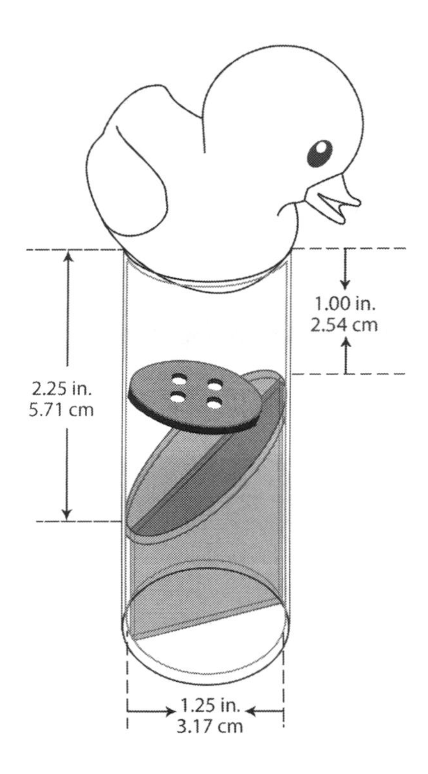


**FIGURE 3.11**
Small parts test device.

requirements became effective April 30, 2017. The test relies on knowledge of the approximate diameter of the pharynx of a young child. According to the test, a small part is defined as "any object that fits completely into a specially designed cylinder 5.71 cm (2.25 in.) long by 3.17 cm (1.25 in.) wide that approximates the size of the fully expanded throat of a child under three years old" (Figure 3.11). Baby and toddler clothes with potentially hazardous small parts, like attached buttons or bows, must be tested and meet the standard.

All of the senses of the head are essential for sustaining life and enjoying the world around us. Well-designed wearable products can protect, enhance, and/or amplify these senses.

## 3.6 Head and Neck Lymphatics: Fluid Conduits and Debris Collectors

The lymphatic system structures in the head and neck function in the same way as they do in the rest of the body. As in other parts of the body, head and neck lymph nodes are generally not evident from the surface. It is fortunate

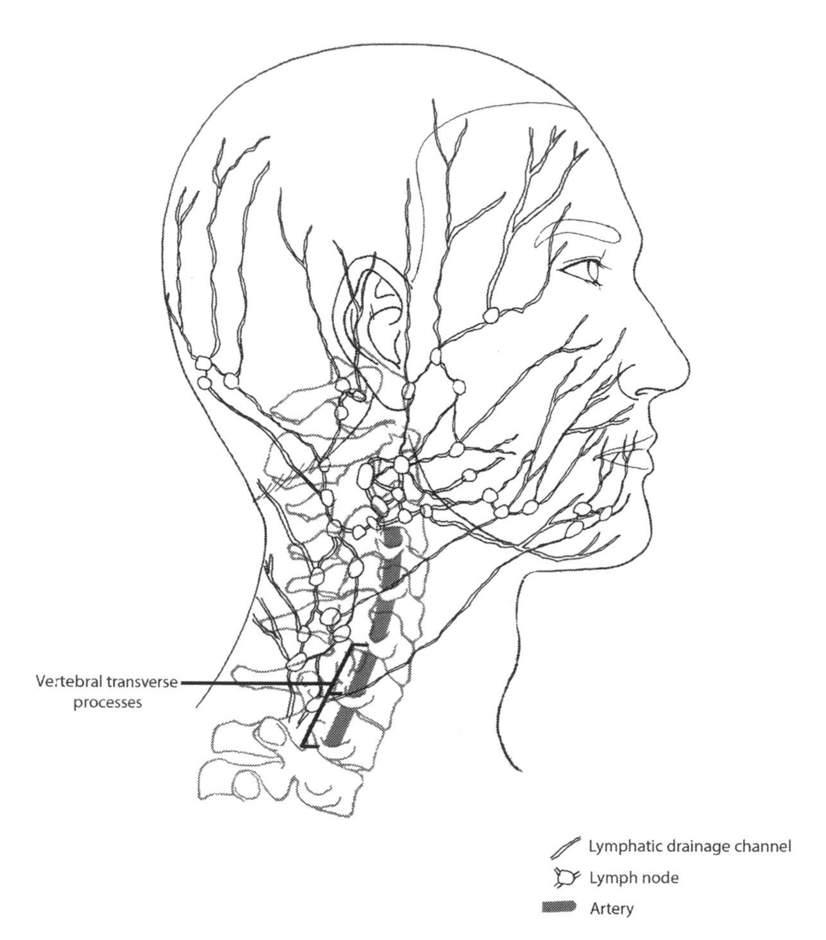


**FIGURE 3.12**
Head and neck lymphatic system and vertebral artery.

that there are numerous lymph nodes in the head and neck region (Zuther & Norton, 2013, pp. 10–11), because the mouth, nose, ears, and eyes are all possible entry points for *infectious agents*. When lymph nodes become enlarged in response to cancer cells, *autoimmune disease*, or bacterial, viral, or parasitic illness; lymph nodes can be palpated (Karpf, 1990, p. 711). See Figure 3.12 for locations of head and neck lymph nodes.

As in other parts of the body, fluid can accumulate in the spaces between the cells in the skin and subcutaneous tissues of the head and neck, causing swelling. Refer to Chapter 2, Figure 2.19, to see the lymphatic watershed drainage of the head and neck. Lymphedema can occur in the head and neck although it is less common than in other body areas. Diffuse pressure from a wearable product can influence the function of the lymphatic capillaries. Designing wearable products for treatment of lymphedema in this region is both controversial and a challenge because of the risk to important compressible blood vessels in the neck and the chance of impairing, instead of facilitating, the lymphatic drainage pathways in the neck (Zuther & Norton, 2013, pp. 199–208).

## 3.7 Skin and Hair: Encasing It All

Skin and hair cover the structures of the head and have aesthetic and functional roles.

### 3.7.1 Facial Skin and Mucous Membranes

The surface skin of the head transitions from the appearance layer—making up the complexion of a person—to mucous membranes that line the inside of the mouth, nose, and eyelids and cover the sclera of the eyes. Facial skin contains a wide variety of nerve endings which collect somatic sensory information to transmit to the brain. The lips are sensitive to touch, making kissing pleasurable and helping to facilitate nursing for infants. A slight touch to an infant's lips can result in a *sucking reflex*. Nerve endings in the mucous membranes serve the same function as nerve endings in the skin. In the eye they alert us to blink, if the eye is drying or if something lands on the surface of the eye.

### 3.7.2 Hair and Scalp

Hair, sometimes referred to as a person's crowning glory, also provides a natural layer of thermal insulation and a small amount of impact protection. Fine fibers like fiberfill and human hair trap air; this dead air space has insulating properties. The quantity, density, and—to some extent—texture and degree of curl of the hair determines the effectiveness of the thermal and impact protection. The scalp, underlying the hair, carries an extensive network of blood vessels and nerves. Small involuntary muscles, the *arrector pili*, attach to the hairs in the hair follicles. Stress, like cold temperature, activates the arrector pili muscles causing the hair to stand up, thus increasing protective insulation. Hair coverage and hair density affect how thoroughly heat is conducted or radiated away from the head. Hair volume and density are often ignored when designing headwear. Factoring in hair volume and/or density of hair would be nearly impossible for most hats sold at retail. Custom-made products might incorporate this consideration when measuring the person to fit a helmet or other protective headgear.

The skull and the tissues covering it provide many distinct landmark features that can be used when designing products. Similar structures are apparent for all humans, but wide person-to-person variations provide challenges when designing products that will fit and function for a target market.

## 3.8 The Neck: Connecting Head and Torso

The neck is the crucial connector between the head and the torso encasing sections of vital body systems. The flexible neck moves the head's eyes, ears, and nose; helping us follow a visual target, turn toward noise, detect the origin of a scent, or locate the source of a touch. The tissues of the neck protect the spinal cord, as well as other nervous system pathways between the brain and the body. The blood vessels of the neck carry oxygen and nutrition to the brain and waste products from the brain. Food enters through the mouth and travels through a tube to the stomach to begin distribution to the body; air enters the body through the mouth and nose moving through the neck to the lungs. Figure 3.10 shows the relationships of the deep structures of the neck. Figure 3.13 details structures closer to the surface.

### 3.8.1 Bones, Ligaments, and Muscles of the Neck

Bones, ligaments, and muscles provide structure, stability, and mobility for the neck.

#### *Vertebrae and Hyoid Bone*

The skull balances on the top section of the spine, a stacked column of seven cervical (neck) vertebrae. It is a little like stabilizing a bowling ball on a pile of blocks. The *condyles* of the skull's occipital bone articulate with the first cervical vertebra, C1, called the *atlas*. The atlas is named after the mythological Greek figure who carried the world on his shoulders. C1 has a unique shape; it is more ring-like than the lower vertebrae. C2 through C7 are structurally similar to one another, with a *body, arch,* and spinous process (see Figure 3.10, transverse section). C2 has an additional bony feature, the *dens*—also called the *odontoid process*—a tooth-like structure which projects upward from the body of C2 to articulate with the anterior arch of the atlas (Teton Data Systems & Primal Pictures Ltd., 2001). A ligament inside the anterior arch of the atlas braces the dens against the atlas. The dens acts as an axis of rotation for the skull on the spine.

Most bony neck injuries occur either near the skull or in the lower neck when normal range of motion has been exceeded, usually after forceful impact. The C1-C2 region is vulnerable to devastating or fatal injuries if the dens is fractured or wrenched away from the anterior arch of the atlas and then forced into the tissue of the spinal cord or lower brainstem. Lower neck

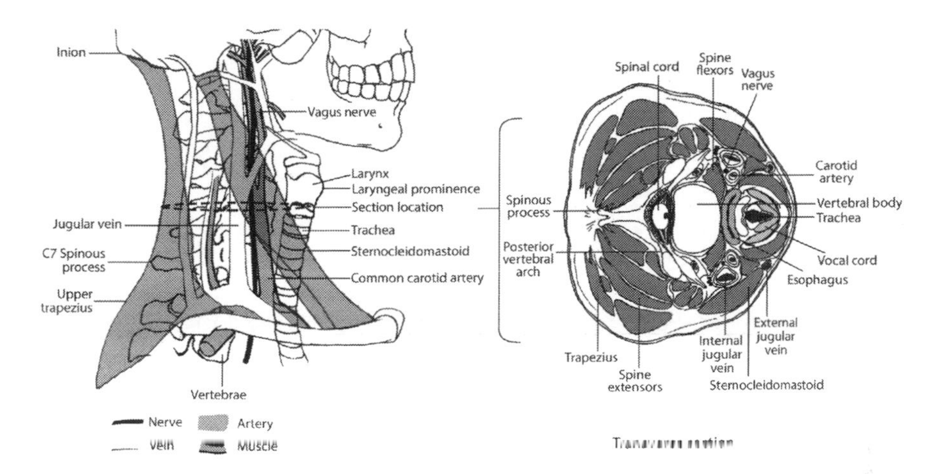


**FIGURE 3.13**
Neck nerves, veins, and arteries pertinent for neckwear design, transverse section at C6.

injuries can also result in spinal cord injuries. Cusick and Yoganandan (2002) discuss neck injury mechanisms, useful information when designing products to try to prevent or stabilize traumatic neck injuries.

The hyoid bone, at the base of the tongue muscle and the level of the C3 vertebra, functions as an attachment for anterior neck muscles. The posterior bony prominences of the vertebrae, the spinous processes (sometimes simply called the "spines" of the vertebrae), may suggest that the spine itself resides near the back of the neck. But the bulk of the spine is much more centrally located. See the transverse section of the neck in Figures 3.10 and 3.13 to explore the relationships of a vertebra to other internal structures. The spinous process of the seventh cervical vertebra is often used to mark the location of the base of the posterior neck. The *external occipital protuberance,* sometimes called the *inion* (a palpable landmark on the back of the occipital bone), lies at the junction of the skull and the neck, just above the intersection of the Frankfort plane and the median sagittal plane of the skull. The *mastoid process,* another bony prominence of the skull, is located just behind the ear lobe on each side of the head (Figure 3.3, lateral view).

### *Ligaments and Intervertebral Discs*

The neck would be a flimsy structure if the stacked vertebrae were its only major elements. The body of each bony vertebra in the neck is connected to the vertebrae above and below by a wide anterior ligament. A narrower, but still substantial, posterior ligament attaches the body of each vertebra to the adjacent vertebrae. Additional ligaments run vertically between the vertebral arches and between the spinous processes. There are intervening soft fibrocartilage structures, the *intervertebral discs,* between the vertebral bodies of all the vertebrae from C2 to the pelvis. See the bony vertebral components in Figure 3.10, transverse section, and intervertebral discs in Figure 3.10, sagittal section.

### *Muscles*

Several layers of neck muscles contribute to the intricate control of head movement. Many small muscles attach to the vertebral arches and bodies to help stabilize the column of stacked vertebrae. Refer to the cross-section of the neck in Figure 3.13 for an appreciation of the many muscle layers and shapes. Large muscles visible from the body surface—which are perhaps of most interest to designers as they are likely to interface directly with wearable products—include the upper trapezius muscle, the *sternocleidomastoid* muscle, and the *platysma*. See Figure 3.13 for locations of the sternocleidomastoid and trapezius muscles.

You can feel the trapezius contract on the back of your neck when you tilt your head back to look upward. If you turn your chin to look over your shoulder you can see and feel a strap-like muscle on the other side of your neck, running from behind your ear to your collarbone, near the *sternal notch*. That is the sternocleidomastoid. The platysma, a broad but thin sheet of muscle, originates at the jaw line (an arbitrary division between the face and the neck at the jaw) and runs down to and over the collar bone. Look in a mirror and clench your teeth to see several cords of contracted tissue, the platysma, on the front of your neck. Keep the shape and volume changes of these three muscles in mind when designing neckwear. The deeper neck muscles, which attach the cervical vertebrae to one another, to the head, and to the torso, tend to control small head position adjustments. Research details of these muscles if designing a product requiring fine head control.

This elegant structure of bones, ligaments, intervertebral discs, and muscles supports the head, yet is flexible. The strength and stability of these musculoskeletal structures vary person-to-person and change with age. They are particularly important when the head and neck need to support weight. Van den Oord, Steinman, Sluiter, and Frings-Dresen (2012) describe the challenges associated with developing properly balanced multifunctional military helmets. The vertebrae, ligaments, and muscles of the neck also help protect vital internal structures.

### 3.8.2 Nervous System in the Neck

The vertebrae provide a protective channel for the spinal cord, much as the cranium protects the brain. Study the transverse section of the neck (Figure 3.10). Locate the spinal cord as it passes through the *spinal canal,* the ovoid cylindrical space between the bony vertebral arches and the vertebral bodies. The paired, R and L, *vagus nerves* lie next to the *carotid arteries,* near the sternocleidomastoid muscle (Figure 3.13, lateral view). Other nerves originating in the brainstem as well as nerve branches from the cervical spinal cord innervate neck muscles and skin. They are relatively well protected as they travel through the tissues, often near small blood vessels.

The vagus nerve moderates many autonomic nervous system functions such as heart rate and digestive actions. Medical conditions, from depression (Daban, Martinez-Aran, Cruz, & Vieta, 2008) to epilepsy to asthma, have been treated with surgically implanted vagus nerve stimulators. More recently, electrical stimulation of the vagus nerve through the skin on the side of the neck (Ben-Menachem, Revesz, Simon, & Silberstein, 2015) has been tried to treat a variety of medical conditions. The nerve can be stimulated from the surface at this location, providing a new way to influence the nervous system. Knowledge of this combination of anatomy and technology presents opportunities to develop new wearable products to substitute for implantable vagus nerve stimulation.

### 3.8.3 Circulatory System in the Neck

Four major blood vessels, the carotid arteries and *vertebral arteries,* run through the neck carrying blood to the head and brain. The R and L carotid arteries are relatively superficial, need protection, and should not be constricted. You can feel the carotid artery pulse and monitor heart rate at the side of the neck at the level of the *thyroid cartilage* of the *larynx* (Figures 3.10 and 3.13). Palpate the movable cartilage in the front of your neck as you swallow. Then, move your fingertips to one side and, with light pressure, feel the rhythmic beat of your heart.

In contrast, the paired vertebral arteries (Figure 3.12), supplying blood to the back of the brain are relatively isolated from outside pressure or impact. Because they course through openings in the *transverse processes* on the R and L sides of the bodies of the cervical vertebrae, they can be injured by extreme or violent neck twisting or side bending motions. These four vessels, carotid and vertebral arteries, are interconnected at the base of the brain and provide redundancy for the circulation of crucial oxygen and nutrients to the upper central nervous system. If one is obstructed or damaged, the other three may compensate enough to preserve brain function.

The venous blood from the head flows back toward the heart through the four major veins of the neck, the R and L *internal jugular veins* and R and L *external jugular veins* (Figure 3.13). The veins lie adjacent to the carotid artery at the side and near the front of the neck.

Arteries and veins in the neck are vulnerable to minor and major injuries. Langan and Watkins (1987) identified neckties and shirt collars as products that can compromise carotid artery blood flow thus decreasing visual performance. Blunt or sharp trauma to these vessels—from motor vehicle accidents, industrial mishaps, or sports activities—can result in major injuries or death.

### 3.8.4 Air and Food Passages through the Neck

In everyday life, we are aware of air movement between the head and torso as we breathe. Similarly, we feel our stomachs fill as we eat.

#### *Larynx and Trachea*

Air travels to and from the nose and mouth via the larynx and *trachea* to the lungs and back (Figures 3.10 and 3.13). Cartilage of the two airway components helps shape, protect, and support the front of the neck. The larynx, or voice box, is the upper part of the semi-rigid cylindrical airway running the length of the anterior neck. The large thyroid cartilage forms a wall around the larynx. A portion of this cartilage, the *laryngeal prominence*, projects anteriorly and is commonly called the *Adam's apple*. This anthropometric landmark varies in size person-to-person and is more noticeable in men because its growth is stimulated with exposure to *testosterone*, a predominately male hormone. Ligaments and muscles hold the *laryngeal* cartilage framework, including the thyroid cartilage, in place.

The upper airway is protected from fluids and solids by movements of the *epiglottis*, a laryngeal cartilage just above the larynx (Figure 3.10), the structure that allows us to produce sound. Structures of the mouth (tongue, teeth, and lips) modify the sound produced by the vocal cords and larynx into intelligible speech. The vocal cords lie at the base of the larynx (Figure 3.10 sagittal section, Figure 3.13 transverse section), just above the *cricoid cartilage*, a ring of cartilage inferior to the thyroid cartilage which marks the beginning of the trachea. The vocal cords provide further protection from liquids and solids for the trachea, or windpipe, the lower section of the airway in the neck. The area between the thyroid cartilage and cricoid cartilage, the *cricothyroid membrane*, is also sometimes used as an anthropometric landmark.

The trachea connects the larynx in the neck and the lungs in the chest. You can feel the C-shaped cartilage rings of the trachea by running your fingers from under your voice box to the base of your neck. The trachea is approximately 12 cm (4.5 in.) long and 2.5 cm (1 in.) in diameter. The *thyroid gland* lies in the neck caudal to and somewhat lateral to the thyroid cartilage and the trachea.

The trachea is compressible and sensitive to pressure. If you push on your trachea in the front of your neck, you cough. With enough pressure on the trachea, the flow of air can be completely stopped, leading to loss of consciousness and, in the extreme, death. Small children, with a narrower trachea than an adult, are particularly vulnerable to airway obstruction. When designing products for children, avoid designs with parts that could accidentally wrap around the child's neck.

In 1997 ASTM issued a voluntary standard, *F1816-97 Standard Safety Specification for Drawstrings on Children's Upper Outerwear* to reduce the numbers of child deaths due to drawstrings becoming entangled with other objects. The standard applies to outerwear sizes 2T to 12, with an estimated age range of 18 months to 10 years. Rodgers and Topping (2012) reported the voluntary standard reduced the drawstring-related child mortality rate, 1985–2009, by an estimated 90.9%. Manufacturers identified an easy and inexpensive fix to address the drawstring safety hazard by substituting "alternative closures such as snaps, buttons, Velcro™, or elastic" (Rodgers & Topping, 2012, p. 654). For years, the U.S. Consumer Product Safety Commission (CPSC) considered that children's garments with neck or hood drawstrings presented a substantial risk of injury to children. They initiated recalls of such products. In 2011 the CPSC issued a final rule, *CPSC16 CFR Part 1120*, which states that products that do not comply with this ASTM voluntary standard are included in the CPSC substantial product hazard list. The CPSC can now order manufacturers, distributors, or retailers to repair or replace such products or refund the purchase price to a consumer. Items arriving from foreign manufacturers can be refused admission to the U.S. CPSC has a useful document, The Regulated Products Handbook, (CPSC, 2013) outlining regulated product requirements.

#### *Pharynx and Esophagus*

The pharynx, the space lying behind the nasal and oral cavities, and behind the larynx in the neck, extends between the trachea and the vertebral bodies to the level of the cricoid cartilage. The pharynx works with the epiglottis to move food into the digestive tract or air into the larynx. Food travels from the mouth through the oropharynx to the laryngopharynx, the section of the pharynx behind the voice box, then into the *esophagus* to the stomach. The flexible esophagus, a flat *fibromuscular* tube, links the mouth and the stomach. It expands to accommodate a mouthful of food after you swallow. The smooth muscles in the wall of the esophagus contract sequentially to squeeze the food through the esophagus to the stomach.

## 3.9 Head and Neck Range of Motion (ROM)

Understanding the relationship of ROM to the anatomy of the head and neck is important in designing headwear and neckwear. Some products stabilize or limit head/neck ROM. Others allow as much natural motion as possible. Wearable products can also unintentionally hinder motion. If a product modifies motion in some way, it also modifies the energy required to move or stabilize the head and neck.

### 3.9.1 Head, Jaw, and Neck Structures Facilitating Motion

Head, jaw, and neck structures move to varying degrees. The sutures between the skull bones and teeth in their sockets move miniscule amounts with the possibility of moving measurably over time. Products are designed to take advantage of this motion capability. Tooth motion allows gradual repositioning of misaligned teeth with braces. In the rare case when an infant's head is severely misshapen, wearing a rigid cap can gradually move cranial bones to normal positions.

The only major movable structure of the head is the mandible of the jaw, which needs to be unconstrained for talking, eating, and yawning. Two powerful muscles near the body surface that account for much of the mandible's motion are the *temporalis* and the *masseter* (see Figure 3.3, inset). Palpate the muscles as you clench your teeth—the masseter lies over the angle of the jaw, in front of the ear. The temporalis is located on the side of the head, in front of and extending up from the ear. These two muscles activate the temporomandibular joint (TMJ), assisted by smaller muscles inside the mouth to move the jaw in multiple directions. Jaw motions become painful or limited when there are problems at, or within, the TMJ.

The cervical vertebrae, with intervertebral discs between, move in relationship to each other and to the skull positioned at the top of the spine. Bogduk and Mercer (2000) detail cervical spine mobility, section by section. They divide the complex movement of the neck into four functional units, starting at the top. (1) C1, the atlas, cradles the occipital condyles of the skull. (2) The head and atlas rotate as a unit on the dens, the bony structure that projects upward from C2. (3) The C2 vertebra anchors onto the lower cervical vertebral section by a unique orientation of the articulating joints between C2 above and C3 below. (4) Each segment, C2-C3, C3-C4, etc. of the remaining five segments (C3-C7) move forward-backward and side-to-side. The vertebral bodies in this middle and lower region are saddle-shaped to facilitate the motions.

Many muscles help move the neck (Figure 3.13, transverse section). Small overlapping muscles immediately adjacent to the vertebral bodies supply both fine control and assist with bigger neck motions. Two major paired muscles move the head and neck; the R and L sternocleidomastoid and the R and L trapezius. These muscles change shape as they contract, possibly affecting product fit, position, and wearer comfort. Because of the elegantly interconnected anatomy of the head and neck, a person can either generate motion of the head or maintain a static pose by continually adjusting the skull's center of mass over the stacked vertebrae.

### 3.9.2 Jaw and Neck Range of Motion (ROM)

#### *Jaw ROM*

Products worn in the mouth or external to the jaw accommodate or limit jaw motion depending on purpose and activities. As with other products, gender and age affect jaw motion (Yao, Lin, & Hung, 2009). Some products are held in the mouth and rely on reciprocal actions of the upper jaw and the mandible; for example, opening the mouth to insert a mouthpiece and then applying pressure to hold it in place by closing the mouth around it.

Jaw motion is complex, involving multi-dimensional movements. The primary jaw motions for opening and closing the mouth are *depression* and *elevation* of the mandible. Depression is lowering the mandible. The upward motion is elevation—like an elevator going up. Palastanga, Field, and Soames (2002, pp. 566–568) discuss depression, elevation, as well as the more limited motions of the mandible including retraction (gliding the mandible backward) and protraction (gliding the mandible forward) and *lateral deviations* (side-to-side movement) of the mandible, needed for chewing. Speech also requires jaw elevation/depression and protraction/retraction. *Overbite* is the amount of overlap of the upper teeth in front of the lower teeth when the mouth is in an unstrained closed position.

Walker, Bohannon, and Cameron (2000) provide detailed instructions and illustrations for documenting jaw movements. The movements are relatively small, in the range of 45 mm (1.8 in.) for jaw opening, 20 mm (.79 in.) for total (R plus L) lateral deviation, and 7 mm (.28 in.) for the forward/backward motions. If you are designing a product with an element that is inserted into the mouth and/or held in place with the teeth, consult or collaborate with specialists in the dental field. When designing a product like a helmet with an integrated jaw protector, use the methods found in Walker et al. (2000) to measure mandible motion to help determine the clearance needed within the product.

### *Neck ROM*

Neck pain may be treated with wearable products that modify neck motion. Neck pain, a common problem, varies in degree and cause. Fatigue, muscle tension and strains, herniated intervertebral discs, and spine fractures can all cause neck pain. Products for neck pain are usually worn on or around the neck and cover a broad spectrum. A U-shaped pillow gently supports a comfortable neck posture. Soft collars which encircle the neck restrict some motion. A totally rigid plastic vest with projecting metal rods and a "halo" ring attached to the skull with screws is intended to prevent all movement of the neck (Guan & Bisson, 2017).

A product worn on the head or on the torso, which also covers a portion of the neck, either accommodates or limits neck motion. Headwear which extends to the neck will tend to influence rotational movements of the head and upper neck. Think of a cap that fastens beneath the chin or a hood on a jacket which makes it difficult to look right and left. A football player's neck collar or neck roll, resting on the shoulders, is worn to prevent a hyperextension or hyperflexion neck injury during a game (Figure 3.14). The design should preserve the player's ability to move, breathe, and see.

Professionals in a variety of fields are interested in documenting spine ROM, often with a focus on neck ROM. Measurement methods range from using simple mechanical tools to measure changes in the relationships of body segments (see goniometer in Chapter 2) to X-rays taken from multiple angles which can measure minute changes between adjacent bones. White and Panjabi (1978) gathered information and summarized spinal movements at every spinal segment. Bogduk and Mercer (2000), in a report specific to the neck, reviewed additional literature and pointed out controversies concerning ROM measurement techniques. Some sources, such as Cole and Tobis (1990), decline to list specific ranges for cervical motions, given the wide motion variations possible in a group of normal people. To best protect the neck with a wearable product, work to understand normal neck range of motion and methods to document it.

Because active muscles change in shape and volume, knowing which muscles produce a motion is also important when designing wearable products. Conley, Meyer, Feeback, and Dudley (1995) state, "Twenty pairs of muscles, acting about 37 joints, are responsible for movement of the head and neck in the sagittal, frontal [coronal], or axial [transverse] planes." However, three relatively large neck muscles, near the surface, are likely to be of greatest interest to designers. They are the sternocleidomastoid and trapezius muscles plus the *splenius capitis*, discussed in detail below.

Determining which muscles generate specific neck motions is at least as difficult as measuring neck ROM. Basmajian and DeLuca summarized studies which used *electromyography* to determine muscle activity associated with specific neck motions (1985, pp. 466–469). Electromyography uses surface *electrodes*, or a needle or fine wire inserted into a muscle, to study muscle activity. The number of muscles, overlapping nature of the neck muscles, and neck muscles' varied sizes, as well as the proximity of important nerves and blood vessels, limits electromyographic studies in the neck.

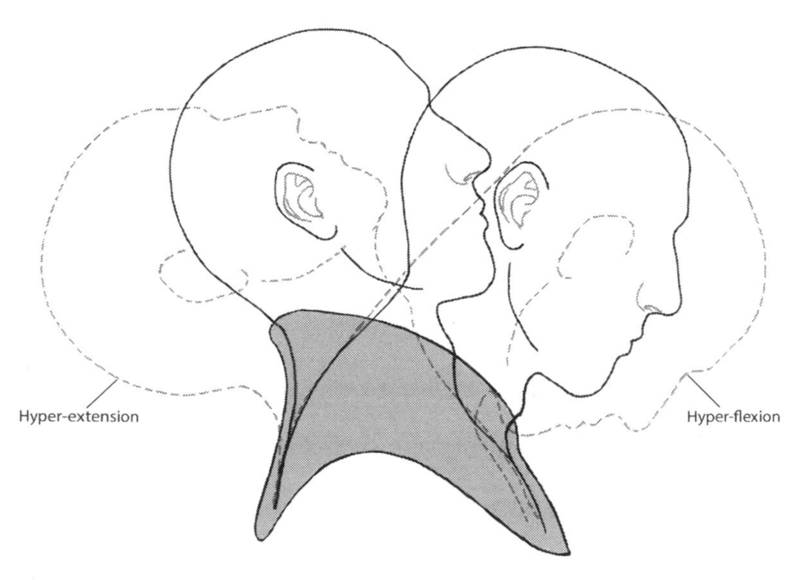


**FIGURE 3.14**
Football neck roll to prevent hyperextension.

The Conley research group used *magnetic resonance imaging* (*MRI*) to compare use patterns and intensity of use in neck muscles during neck motions (Conley et al., 1995). Since 1995 both techniques have been used to study neck muscle activity and neck range of motion relationships, but there still is variation among researchers and authors as to which muscles generate which neck motions.

The sternocleidomastoid, trapezius, and splenius capitis generate major head/neck motions in varying combinations and action intensities. Neck ROM has three major components: flexion-extension (including "nodding" of the head), lateral flexion (bending to the sides), and rotation (Bogduk & Mercer, 2000). Palastanga and Soames (2012) break ROM for these three motions into two categories: sub-occipital and lower neck. Additionally, head protraction-retraction ("gliding" the head forward and backward) contributes significantly to the complex three-dimensional movements of the head in space (Hanten, Lucio, Russell, & Brunt, 1991). Protraction-retraction can be either a gradual or postural response, or a rapid, "darting," motion (like a bird, as it pecks at a worm).

Flexion-extension is the forward and backward bending motion of the head and neck in the midsagittal plane. Look at the flexion-extension ROM (Figure 3.15, center left), and the muscles which control the motion (Figure 3.15 upper, center, and lower). The front to back "yes" nodding motion of the head (Figure 3.15, center right) accounts for about 20 degrees of movement. It occurs in the sub-occipital segment of the spine (between the skull and the atlas, C1, plus between C1 and C2). The R and L sternocleidomastoid and several small deep muscles originating on the upper cervical vertebrae and inserting on the skull create the nodding motion of the head on the neck. At each level below that segment (from C2-C3 to C7-T1) there are another approximate 10–15 degrees of motion, for 110 additional degrees of mobility. The total flexion-extension ROM from maximum forward motion to maximum backward motion is approximately 130 degrees.

Flexion-extension of the neck is a reciprocal motion which requires paired muscle actions. Section 2.4.2 (Body in Motion) pointed out that muscles frequently work in tandem with other muscles to carry out reciprocal motions. This is the case with neck flexion and extension. Neck flexion motion can be controlled, modified, or balanced by neck extensor muscles, and *vice versa*. In anatomical terms, mutually resisting muscles are called *agonists* and *antagonists*. Agonists are acting with shortening contractions while antagonists are opposing those contractions. The muscles moving the head and neck into flexion (or extension) act with shortening contractions while the muscles which extend (or flex) the head and neck simultaneously relax or work with lengthening contractions. Shortening and lengthening contractions were discussed in Chapter 2, Section 2.2.3 (Muscle Configuration and Action Rely on Microscopic Changes) and illustrated in Figure 2.7.

Neck flexion motion is naturally more limited than the extension motion. In other words, you can look up more fully than you can look down

(Palastanga et al., 2002, p. 520). In flexion, the sternocleidomastoid muscles on both sides of the neck act together with shortening contractions to bring the chin toward the chest (Basmajian & De Luca, 1985, p. 469). Recall the naming conventions for muscles. The sternocleidomastoid muscle originates from both the sternum and clavicle (*cleid* in Greek) and inserts on the mastoid process of the skull. With these origins and a head/neck flexion action,

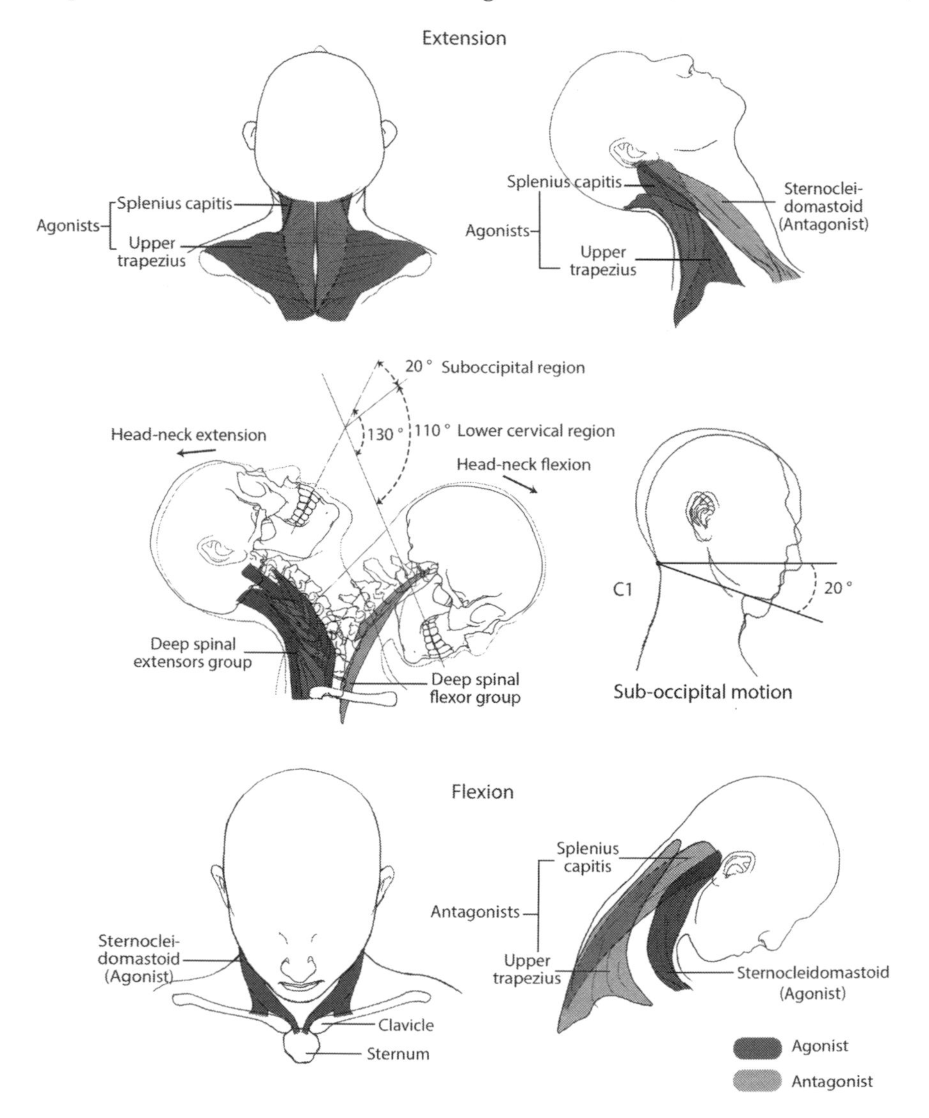


**FIGURE 3.15**
Head and neck flexion-extension range of motion, including nodding, with major associated contributing muscles. (Modified from Figure 4.61 (A) Total range of movement in the cervical region, flexion and extension, by N. Palastanga and R. Soames 2012, *Anatomy and Human Movement, Structure and Function*, 6th ed., p. 475. Copyright 2012 by Elsevier Ltd.)

the sternocleidomastoid muscles are the muscles most likely to come into contact with a collar or other wearable product encircling the neck. Many smaller deep muscles also assist with neck flexion (Figure 3.15, center left). With neck extension, the R and L trapezius (Moore et al., 2011, p. 309), at the back of the neck, and the slightly deeper R and L splenius capitis (Figure 3.15, upper) act along with several smaller and deeper muscles to tip the head backward (Basmajian & DeLuca, 1985, pp. 468–469). The trapezius and splenius capitis may also press against a product worn on the neck. Muscles producing these flexion or extension motions (agonists) act bilaterally, with shortening contractions. For example, when the muscles in the back (posterior) of the head and neck are shortening in extension (agonists), the flexor muscles in the anterior neck (antagonists) contract with lengthening contractions to control the motion, or else relax and stretch.

Lateral flexion, or side-bending of the head and neck toward the shoulder, occurs in a coronal (frontal) plane (Figure 3.16, lower). The C2-3 to C7-T1

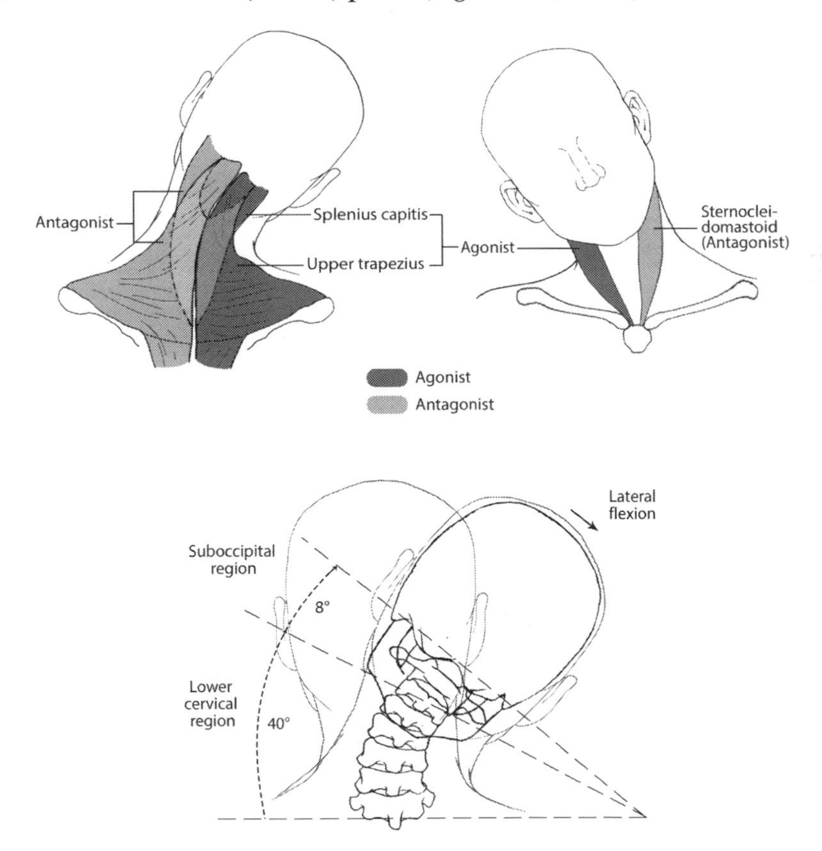


**FIGURE 3.16**
Head and neck lateral flexion range of motion with major associated contributing muscles. (Modified from Figure 4.61 (B) lateral flexion, by N. Palastanga and R. Soames, 2012, *Anatomy and Human Movement, Structure and Function*, 6th ed., p. 475. Copyright 2012 by Elsevier Ltd.)

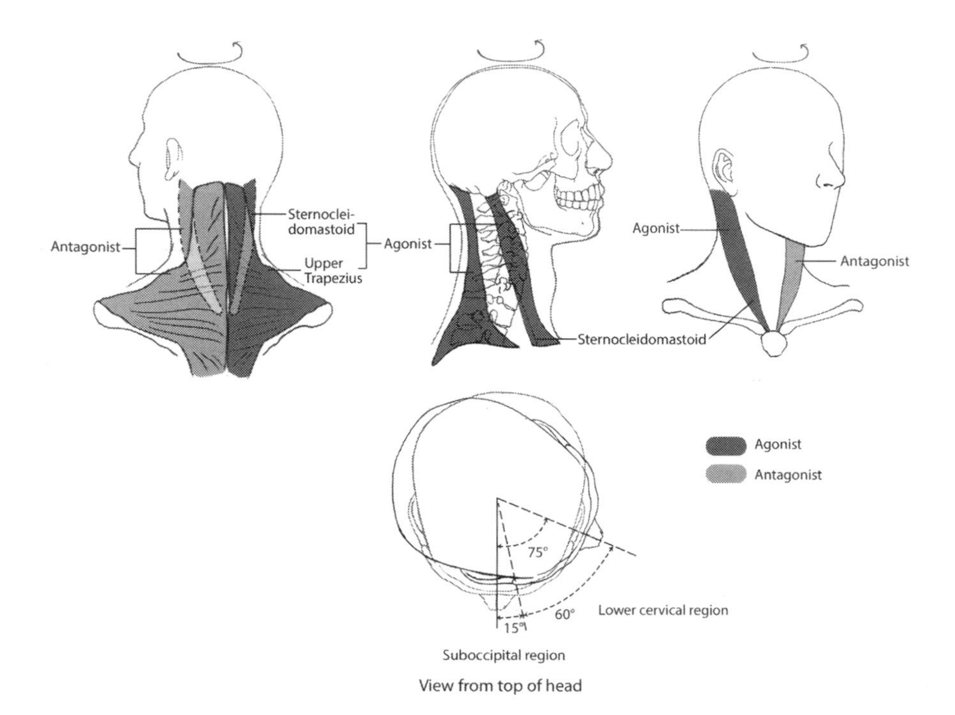


**FIGURE 3.17**
Head and neck rotation range of motion with associated muscles.

segments again provide much of the motion, about 40 degrees. Motion at the sub-occipital (skull-C1, C1-C2) segment adds about 8 degrees. The specific contours of the lower cervical vertebrae always add a degree of rotation to lateral flexion of the neck (Palastanga et al., 2002, p. 521). Although different authors credit different combinations of muscles with this movement, the sternocleidomastoid, splenius capitis, and trapezius muscles all likely supply major forces for this motion (Figure 3.16, upper), as well as for the return to an upright head and neck position (Kendall, McCreary, Provance, Rodgers, & Romani, 2005, pp. 148–149, Palastanga et al., 2002, p. 489). The R sternocleidomastoid and upper trapezius contract together, as agonists, in a shortening contraction, along with the R splenius capitis, to tip the R ear toward the R shoulder. The same muscles, on the L, control the motion with lengthening contractions and act as antagonists. The splenius capitis lends an element of neck extension when it laterally flexes the head on the neck (Teton Data Systems, & Primal Pictures Ltd., 2014, Head & Neck). To return the head to upright, the muscle actions reverse.

Head and neck rotation—the action of looking side to side or pivoting the head and neck with the head and neck held erect—occurs in a transverse plane (Figure 3.17). In contrast to flexion-extension and lateral flexion, the sub-occipital segment contributes most of this movement. Rotation of the C1 vertebra on the dens of C2 accounts for about 15 degrees of the total 75 degrees of motion. Products which extend downward from the head into the upper neck region can restrict this rotational movement. Products can be designed to intentionally limit this motion or poor design can inadvertently restrict the motion. The remaining 60 degrees of motion are divided among C3–C7. The upper trapezius and the sternocleidomastoid assist in rotating the head, but the rotation of the head is to the opposite side (the R sternocleidomastoid muscles turn the head to look left and *vice versa*).

Head protraction-retraction—the head moving in a forward-backward motion, without neck bending—occurs in a midsagittal plane (Figure 3.18). The motion includes both horizontal and vertical components. Head protraction-retraction, in daily movements, arises from actions in both the neck and the thoracic spine (Persson, Hirschfeld, & Nilsson-Wikmar, 2007). Head protraction can be gradual, a shift of resting head posture, or quick, as a "darting" motion. This motion can affect design choices. Wearable products that restrict head protraction are uncomfortable, like a T-shirt with a neckline which rises too high and tight on the front of the neck. If a product crossing the trachea is rigid, like a body brace, and the wearer moves quickly, the motion may cause more discomfort or, in the extreme, deformation of the trachea or larynx.

The degree of protraction-retraction motion varies according to the starting posture of the head and neck on the shoulders. If the upper back is rounded

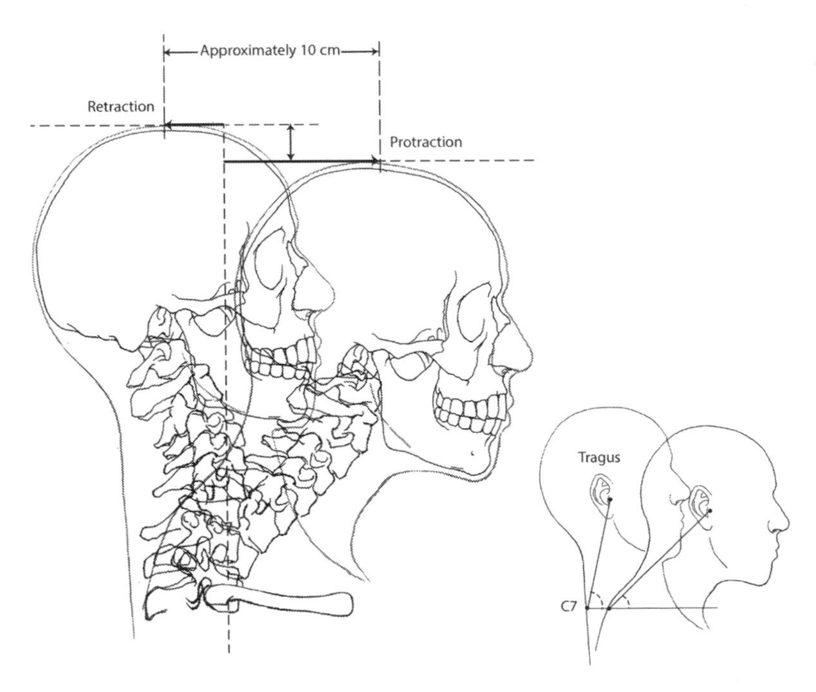


**FIGURE 3.18**
Head protraction and retraction range of motion.

and the neck has a degree of extra extension so the eyes can look straight ahead, protraction-retraction motion will be reduced. If the starting position is more erect, the protraction-retraction movement can be greater. Hanten et al. (1991) describe a simple method of documenting head protraction-retraction motion and resting head posture. They collected measurements of the horizontal movement of the head and found approximately 10 cm (4 in.) of protraction-retraction horizontal movement with variations by gender. They did not report the extent of the vertical component. Another method can be used to measure this motion (Braun and Amundson, 1989). Find the C7 spinous process. Establish a transverse (horizontal) plane at that level. Mark the tragus of the ear. Measure the angle between a line from the tragus to C7 and the transverse plane in full head protraction and full head retraction (Figure 3.18, inset). The angle will be greater in retraction than in protraction.

## 3.10 Head and Neck Landmarking and Measuring

Coordinate selected features of a head/neckwear design to key anatomical structures of the head and/or neck. Senses in the head—sight, hearing, taste, smell—require special attention. Review landmarking and measuring foundations in Appendix A, and details of landmarking and measuring the head and neck in Appendix B.

### 3.10.1 Landmarks and Measurements for Headwear and Neckwear Designs

The locations and numbers of landmarks and measurements for head/neckwear depend on: (a) intended purpose or use, (b) product material characteristics, and (c) extent of the product coverage. ASTM developed a general standard for a range of headgear (ASTM, 2015a). The standard incorporates testing for impact resistance, puncture resistance, and vision restriction among other concerns. ASTM also administers standards for specific headgear for sports like lacrosse, hockey, horseback riding, and more.

#### *Product Purpose*

Head/neckwear are designed for a variety of purposes, in the broadest sense ranging from aesthetics to function. Fashion hats and scarves characteristically fill an aesthetic need. Winter hats, warm scarves, and thermal face masks may combine function and aesthetics. Functional head/neckwear can be designed to protect, to detect and record functions of the brain, to enhance senses, or to compensate for sensory organ deficiencies. Fashion products usually require a simple landmarking and measuring scheme.

Products with critical protection functions and those used to enhance or monitor senses require precise landmarking and measuring procedures. Space suit helmet technology exemplifies the ultimate in protection, incorporating sophisticated mechanisms to support crucial functions like respiration. The space helmet and integrated face protection must fit the individual precisely to give best protection, while allowing full use of the senses.

Some head/neckwear are designed to monitor functions of the head. CPAP machines track attempts at breathing. Sense-enhancing wearables—like hearing aids or specialized eyesight enhancing products like those worn by dentists and surgeons—require an in-depth knowledge of hearing and eyesight and rely on a precisely-placed product to correlate with anatomical features. Virtual reality head-mounted devices alter a person's perceptions of his/her surroundings, requiring an understanding of eyesight and perception.

#### *Product Materials*

A stocking cap covers the head along the hairline, over the ears, and around the base of the skull. A turn-up cuff gives length adjustability. The design of the cap and the necessary landmarks and measurements relate to the stretch of the cap material. A simple stocking cap will require very few landmarks. A *bi-dimensional measuring scheme* is sufficient: (a) circumference of the head at the hairline below the inion and above the eyebrow level, and (b) a measurement from the tragus of one ear, across the top of the head, to the tragus of the other ear (bitragion coronal arc, see Figures B.3 and B.4 in Appendix B).

Traditionally hat blocks in various sizes are used to shape fashion hats (Figure 3.19). Some fashion hats use conformable materials, like buckram or felt, molded to hat blocks of different sizes (Henderson, 2002). Styles made of these materials provide some flexibility to fit a range of head shapes. In contrast, a helmet of rigid materials requires more landmarks to align protective features with the head and to develop a sizing system. Imagine a football helmet with face grid superimposed on head and neck illustrations (Figures B.1 and B.2 in Appendix B). The helmet and grid should accommodate the many head and neck structures and facial features. The helmet shell is rigid, which limits how well the structure will fit a range of sizes. Helmets often incorporate size-adaptable materials—foam, air pockets, or a webbing sling—within the shell. A landmarking and measuring plan must consider dimensions for both material layers.

#### *Product Coverage*

Headwear may encompass the total head and neck or cover limited areas. Even small products worn on defined areas of the head or neck may require very detailed and precise measurements. Examples include goggles for the eye region, respirators for the nose and mouth, and hearing protection for the

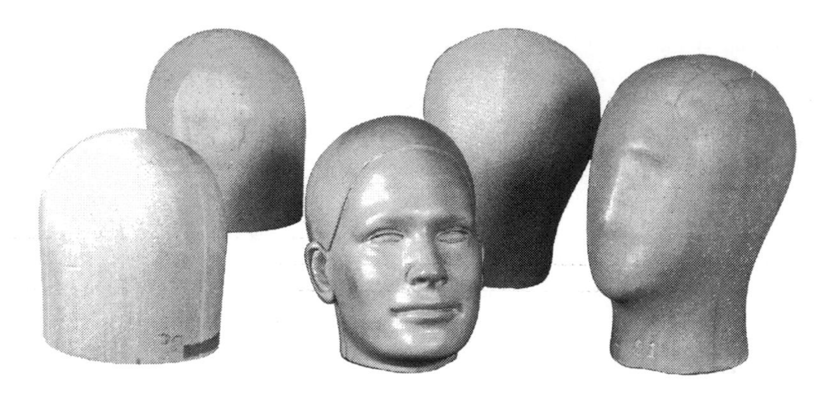

**FIGURE 3.19**
Hat blocks and headform for headwear shaping and sizing. (Photograph by David Bowers, University of Minnesota College of Design Imaging Lab.)

ears. Erickson et al. (2016) conducted a wear study to evaluate fit and comfort of sensors worn in the external acoustic meatus. They conclude "lack of knowledge about the anthropometric variability of the ear canal restricts the successful design of an earpiece that fits correctly and comfortably into the ears of various users" (p. 020966-2). Lee et al. (2016) studied ear shapes and dimensions of 100 Korean subjects (50 males and 50 females) and combined 3D scans and a casting method to determine the three-dimensional shape of the acoustic meatus. They identified detailed dimensions that can be used in designing various products inserted into the ear.

Products to protect eyesight must conform to the bony structures, muscles, and other tissues surrounding the eyes. Nose and mouth protection can range from a very simple cloth face mask filter for environmental particles to a complex self-contained breathing apparatus. The simple face mask comes in one size, is made of a flexible textile, and formed with accordion-like pleats so that the mask can shape around the lower face from the bridge of the nose to under the chin. Relatively long ties give size adaptability. Respirators require precise fit to the face to be effective. Shaped rigid materials paired with conformable materials can give a degree of good fit as well as comfort for the structures of the nose, mouth, and jaw. Although these products cover a limited area of the head, the need for good fit requires carefully selected and detailed landmarking and measuring procedures.

Other products cover a larger area of the head but require coordination with specific anatomical features. Protective sports helmets are designed to provide impact protection. However, a sports helmet also needs to allow full function of a player's senses, especially eyesight and hearing. Small openings incorporated into a hard-shell helmet to facilitate hearing must align with the acoustic canal, so that sounds can be directed into the ear.

### 3.10.2 Head/Neckwear Fit

Methods to analyze head/neckwear fit vary depending on product requirements. Review the discussion of fit ease and style ease in Chapter 1. Head/neckwear made of textiles incorporate fit ease and style ease. A method often used for fashion items is analysis of appearance—does the product look too big or too small? A man's fedora hat can have a comic appearance if the crown and circumference are too large for the man's head so the hat hangs below the eyebrow ridges. Another method is to ask the wearer how they think a product fits. Does the product feel too tight or too loose? Rigid headgear, like helmets, use quantitative methods to measure fit. *Standoff distance* (*SOD*), the measurement of the distance between the head and the inside surface of the helmet, is used to assess helmet fit (Meunier, Tack, Ricci, Bossi, & Angel, 2000). Finding the SOD for an individual that keeps the helmet in position, providing best protection while not being so tight as to be uncomfortable, is an important design skill. Ellena, Subic, Mustafa, and Pang (2016) developed a *helmet fit index* that incorporates standoff distance, *gap uniformity* (variations in SOD throughout the helmet), and *head protection proportion* (e.g. "2/3" or "3/4" of the head is protected by the helmet) to determine good fit. 3D body scanning allows precise quantitative measures of a person's head, the helmet, and the space between to assess fit.

The U.S. *Occupational Safety and Health Administration* (*OSHA*) issues standards for eye and face protection against chemical, environmental, radiological, or mechanical irritants. OSHA requirements state that eye/face protectors—in addition to providing adequate protection against specific hazards and being of safe construction—must be reasonably comfortable, fit snugly, and not unduly interfere with movements. OSHA's online "Eye and Face Protection eTool" gives guidance on selecting many products (OSHA, 2017). Tests were also developed to assess how well a product to block dangerous chemicals seals around the face. The OSHA Technical Manual (2017) describes a qualitative test procedure using a human tester who wears the mask while a harmless chemical with a noticeable scent is sprayed near the mask. If the person detects the scent, the product fails the test. OSHA and the *National Institute for Occupational Safety and Health* (*NIOSH*) provide standards for many types of protective headwear products.

Motion can drastically affect fit. Consider how motion of the head and/or neck could affect your design. If the distance from one landmark to another changes when wearing a product for typical activities, take multiple measures using stable landmarks. For example, when wearing a product fastened with a chinstrap, the wearer may lower the mandible (open the mouth) to talk, chew, or yawn. Two measurements may be needed to determine the length of the chinstrap; one measurement from the L TMJ, around the chin, to the R TMJ with the mandible fully elevated and another measurement with the mandible fully depressed. Incorporate length adjustability in the chinstrap to coordinate with the measurements.

Demographic factors affect head shape and size: gender, ethnicity, and age. Select measurements representative of your target market. The U.S. population, a melting pot of ethnic groups, presents challenges for head/neckwear sizing and fit as described in a study of variations in head and face shape of U.S. workers (Zhuang, Shu, Xi, Bergman, & Joseph, 2013). Although more women are joining military services, military headgear has traditionally been designed and sized for males. Female head shapes and sizes differ from males. When designing headwear for women, use head dimensions for females. As a helpful guide, White and Folkens (2005, pp. 386–387) review basics of gender-related skull structure differences.

As international marketing of wearable products continues to grow, incorporating ethnic differences in body shapes and sizes becomes essential to meet the needs of diverse populations. Ball et al. (2010) used 3D scanning to compare the head shapes of 600 Chinese males and 600 Caucasian males. They concluded that in general Chinese heads can be characterized as rounder than Western counterparts, flatter on the back of the skull and across the forehead. When designing for the Chinese market, use of anthropometric data from a database of a Western population will result in products that don't fit the intended user.

Similarly, when compared to an adult, the head of an infant or small child is proportionately larger than the body. Children's wear designers who consider the larger proportion of head-to-body when designing a garment, like an undershirt, that is donned and doffed over the head will produce more useable garments. Other differences throughout the life cycle may affect design decisions. Using age-related anthropometric data is essential for good fit.

### 3.10.3 Head/Neckwear Sizes and Sizing Systems

Sizes and sizing systems based on anthropometric data ideally represent the intended wearers of the product, ranging from the smallest to largest wearer. Such sizing systems are not widely available for headwear. And, there are no standard formulas to determine sizes for headwear products. Manufacturers' and retailers' sizing systems vary, so trying on a hat is often the best way to attain good fit. If a size chart is available, a head measurement in inches or centimeters directs a person to a size that should fit. Sizes are designated with a whole number and a fraction (6 ¼, 6 ½, 6 ¾, 7, 7 ¼, etc.), progressing from smallest to largest product size. Limited-size designators are also used, from "one-size-fits all" to variations of small, medium, large, and extra-large. A landmark and measurement plan using head circumference and skull arc measurements often works for design of these limited-size products. Protective headwear is sized and shaped using a standard manikin head called a *headform* (Figure 3.19, form with facial features) and 3D scanning can be used to produce digital headforms.

Neckwear sizing systems are limited. Men's long-sleeved dress shirt size indicators include a neck measurement and a sleeve length. U.S. shirt neck sizes are given in inches, while European and Japanese systems use centimeters. There is no apparent consistency in relationship of neck circumference and collar size. Some online retailers instruct purchasers to simply measure neck circumference and consult a size chart to select the best size. This method presumably includes an ease amount in the product. Other retailers specify measuring neck circumference over the Adam's apple while allowing enough room to insert two fingers to allow ease at the collar. This method incorporates the body measurement and a personal measure of ease that is incorporated in the size selection chart. Neckties can be found in two sizes, long and extra-long, based solely on the length of the tie with no indication of how the tie length relates to neck or torso dimensions.

### 3.10.4 Head/Neck Measurement Databases

Head and neck measurement databases are available from several sources. Be aware that not all databases include measurements for the head and neck. Also, be certain that the database you select represents the demographics of your target market. *ANSUR I* and *ANSUR II* (U.S. Army Anthropometry Surveys) are available on the internet but are based on a military population trending to a young, physically fit demographic. SizeChina is a large commercial database focused on the Chinese population.

Anthropometric data can be collected for specific headwear products. Some of the product-specific databases are large scale, while others collect data from limited samples. Zhuang and Bradtmiller (2005) collected head and face anthropometric data from 3997 subjects representing 91% of the U.S. population. Their goal was to develop a database of respirator users and to use the database to establish fit test panels "to be incorporated into the National Institute for Occupational Safety and Health's respirator certification and international standards" (p. 567). Abd.Latiff and Yusof (2016) propose a landmarking and measuring scheme to design and size ready-to-wear hijab for Malaysian women, ages 18 to 30. Their ultimate goal is to establish ready-to-wear hijab sizing standards.

When using a database, be sure that the measurements in the database are the measurements you need for your product. If you are designing a stocking cap, you most likely want a head circumference measurement that is not parallel to the floor, but spans the arc above the eyes, extends around the base of the skull, and covers the ears. You may not find an exact match in the databases you consult. For instance, ANSUR head circumference measurements are taken parallel to the floor with the tape placed flat to the skull just above the ear top (Clauser, Tebbetts, Bradtmiller, McConville, & Gordon, 1988). With a specific product design in mind and an understanding of the basics of anatomy, you can devise a measuring method and collect specific measurements for the product you are designing.

Sizing system deficiencies—no specific data available for a target market or the need to fit a wide range of people at a manageable cost—can be addressed by using size-adaptable features in a product. Baseball caps use adjustable features at the back of the cap, like ratcheted fasteners, slide fasteners, and elastic bands. A rigid headwear structure will not expand or contract to fit many people, so incorporating an inner liner with flexible materials allows fitting a wider range of people.

## 3.11 Designing for Head and Neck Anatomy: Conclusion

When designing products for the head and/or neck, consider how the product you design might enhance or protect the anatomy. Know the functions and vulnerabilities of the brain, brain stem, and the senses when designing any type of head and/or neckwear. A person's individual facial features convey their identity, so consider how anatomical structures can be highlighted with well-designed headwear. Eyewear goes beyond eyesight correction to produce aesthetic effects, protect, or even enhance normal visual function. The skull with surrounding muscles and fat will aid in determining shapes, detail placement, and conformation of a product. The neck, as the connector between the head and torso, serves many functions. Numerous landmarks useful to designers are found in the head, neck, and the immediately adjacent areas of the upper torso.

# 4

# *Designing for Upper Torso and Arm Anatomy*

Designers create a wide range of products for the upper torso and arms. Shirts and blouses, backpacks, bras, and sports protective wearables are examples. Upper torso products cover all or sections of the body between the neck and pelvis with or without coverage for all or a portion of the arm. Although the anatomical term arm refers to the upper limb between the shoulder and the elbow, unless otherwise noted, we adopt the common meaning for "arm" as the entire upper limb, incorporating the anatomical arm and the forearm (between elbow and wrist).

As the head is the control center for the body, the torso contains organs to pump and aerate blood and to process nutrients and water. Upper torso and lower torso structures and functions overlap. For purposes of this chapter, the upper torso includes the skeletal structures from the neck to the back of the pelvis at the rim and internal organs most pertinent for designers: heart, lungs, stomach, liver, and major portions of the small and large intestines, plus the kidneys and spleen. This chapter also discusses the entire spine from the head to the tailbone, and upper torso nervous system structures.

The upper torso is divided into two compartments. The *respiratory diaphragm*—a dome-shaped muscular structure which originates from the rib edges, sternum, and spine—separates the compartments. The chest, or *thorax,* lies below the neck and above the diaphragm. The belly or *abdomen* is below the diaphragm. For apparel products, the "waist" defines the lower boundary of the upper torso. However, the waist is a moving and changing target. It is an ill-defined and arbitrary demarcation, but an important landmark for some products. See Chapter 6 for guidelines for locating and measuring the waist related to product design.

**Key points:**

- Basic anatomy of the upper torso and arms:
    - Terminology: Thorax, abdomen, *pectoral girdle*, arm
    - Vital organs and systems: Heart, lungs, digestive (stomach, liver, parts of the small and large intestine), lymphatic (lymph collectors, lymph nodes and spleen), nervous system

    - Systemic components passing through, to, and from the lower torso and legs: spinal cord and nerves, blood vessels, and lymphatics
    - Bony structures for protection and mobility: Rib cage, thoracic and lumbar spine, and total spine relationships—cervical/thoracic/lumbar/sacral/coccygeal
    - Muscles, ligaments, and joints involved in spine, rib cage, shoulder, and arm motion
    - Breasts: Distinguishing female anatomical feature
- Wearable product opportunities and challenges for the torso and arms:
    - Coverage allowing movement
    - Protecting, monitoring, and controlling upper torso structures and functions
    - Aesthetics with function
- Fit and sizing:
    - Motion as a fit consideration
    - Gender requirements

Upper torso and arm products are designed for fashion and/or function. Fashion items, like men's and women's suit jackets, are often difficult to fit, especially at the complex shoulder and chest/bust anatomy. Fashion items may fit loosely or be form fitting. Some products—like compression sleeves, bras, body shapers, or corsets—are form-changing. Sports bras are designed to meet athletes' functional needs; many are fashionable. The upper torso can support and carry objects like backpacks and baby carriers. Upper body apparel may insulate or help cool the body. Protective athletic gear for the upper body is often designed to protect vital organs from impact. Products worn only on the arms, for example blood pressure cuffs, protective sleeves, and compression sleeves, meet specialized needs. Some wearables for the arms are paired with gloves or integrate a glove into the garment. Wearables for hands are addressed in Chapter 7.

This chapter begins with internal anatomy and goes on to describe external features pertinent to wearable product design. The upper torso contains key components of many body systems. The vital circulatory and respiratory organs are centrally located. The torso houses skeletal elements, including the bony connections of the arms to the rib cage, and muscles that provide postural support and motion. Male and female torsos demonstrate secondary sexual characteristics with variations in skeletal structure, muscle bulk, fat patterning, and breast formation. Important superficial structures, skin and body hair, encase and protect the body. The natural motion of the upper body, total spine, and arms complicates the design process for this body segment. Methods of landmarking, measuring, and fitting the upper torso and arms need to take motion into consideration.

## 4.1 Thorax: Overview

The heart and lungs, above the diaphragm, are fully contained within the bony and cartilaginous rib cage and the muscular body wall. Posteriorly, the vertebrae in the second section of the spine, the thoracic spine, articulate with the ribs. Look at a schematic top-view and anterior view of the thorax in Figure 4.1. In the top-view of the thorax (at the level of the fourth thoracic vertebra) see the horizontal relationships of the heart and associated blood vessels, the spine surrounding the spinal cord at the posterior, the lungs on either side of the heart, and the body wall with ribs and muscles wrapping around the sides of the body. The anterior view illustrates the vertical relationships of the major organs of the chest.

## 4.2 Circulatory System in the Thorax and Arms

The heart, approximately the size of a fist, is located centrally in the thorax. The heart directs blood to the body and lungs in the systemic and pulmonary circulations. As part of the systemic circulation, blood flow in the arms progresses from arteries to capillaries to veins.

### 4.2.1 Circulatory System Structure: Heart, Arteries, Veins

About two-thirds of the heart lies to the left of the median (midsagittal) plane (Figure 4.1). The tip of the inferior portion of the heart, the *apex*—what you may think of as the point in the romanticized valentine—is a muscular blunted point that lies above the diaphragm. Skeletal structures: the sternum (breast bone), the upper thoracic spine, and the ribs protect the heart.

Study the structures of the heart in Figure 4.2. The heart is a muscular organ with four chambers. The upper chambers *(atria)* collect blood from the body on the R, and from the lungs on the L, emptying into the more muscular lower chambers *(ventricles).* The ventricles are connected to the pulmonary and the systemic circulation by major arteries. The *pulmonary trunk,* which branches into the R and L *pulmonary arteries,* leaves the R ventricle to carry blood to the lungs. The *aorta* (the primary artery of the systemic circulation) arises from the L ventricle. The *coronary arteries* (Figure 4.1), the first arteries to branch from the aorta, carry blood to the heart muscle. The aorta then ascends in the thorax, curls over the pulmonary trunk, and sends arteries to the arms and head. The curved section is called the *aortic arch.* When the aorta turns caudally (downward in the anatomical position) it is called the *descending aorta.* The aorta passes to the lower body through an opening in the diaphragm near the spine. The major veins entering the atria of the heart include the *superior vena cava, inferior vena cava,* and four *pulmonary veins.*

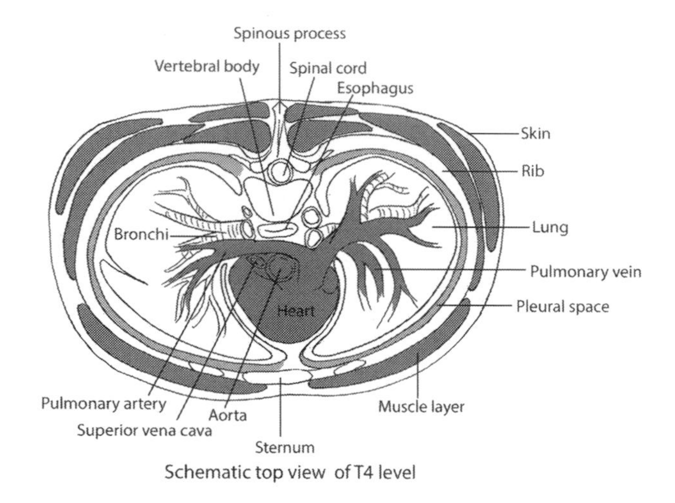


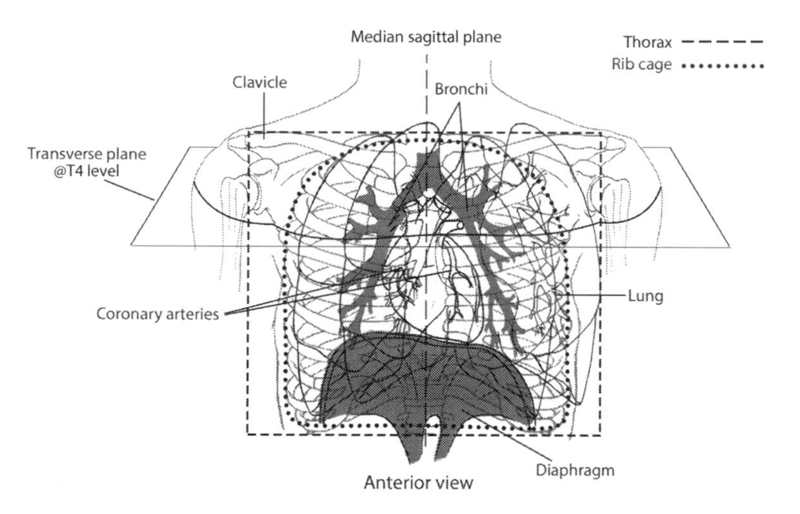


**FIGURE 4.1**
Relationships of heart, lungs, skeletal structures, muscles, and skin in the thorax; schematic superior view from the upper chest and anterior view.

### 4.2.2 Circulatory System Function: Energy Distribution, Waste Management, and Temperature Regulation

Look at the blood flow patterns indicated by the arrows in Figure 4.2. When the heart contracts and relaxes, it does so as a coordinated unit, sending blood from the ventricles to the lungs and throughout the body and filling the atria with blood from the body and lungs. Blood in the R heart is oxygen-poor but rich in nutrients absorbed from the digestive tract, having returned from the body. Blood from the R heart is pumped to the lungs, where it off-loads carbon dioxide and picks up oxygen, an essential body fuel. The oxygen- and

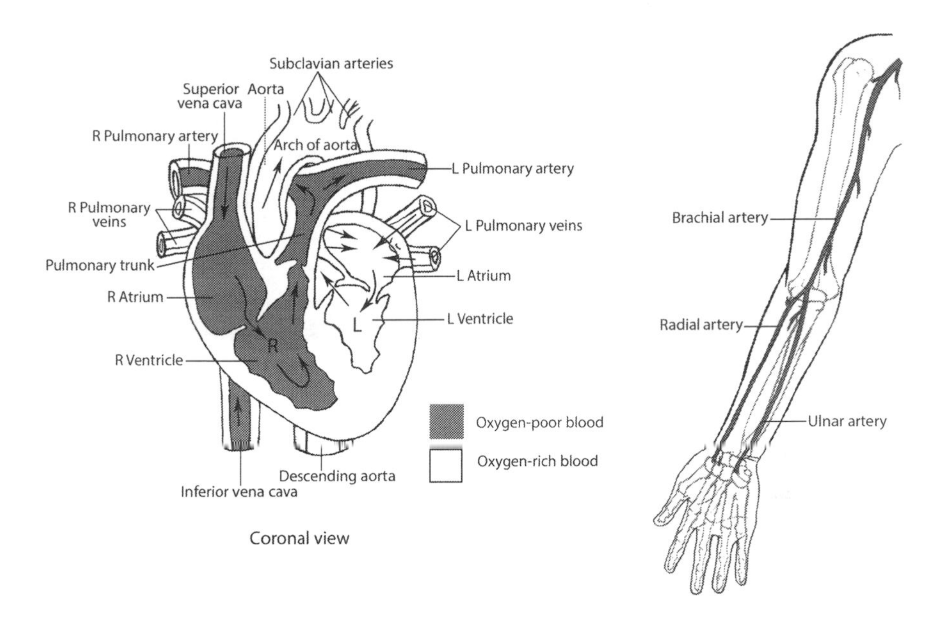


**FIGURE 4.2**
Heart chambers, circulation patterns, and major veins and arteries entering and leaving the heart, coronal view; arteries in the arm.

nutrient-rich blood returns to the L atrium from the lungs, flows into the L ventricle, then into the aorta, and to the body, supplying oxygen and nutrients to all the tissues. Blood flows through the dividing and subdividing arteries, propelled by the beating heart, to eventually reach the capillaries throughout the body. Nutrients, water, and oxygen—all essential for energy—leave the capillaries and enter the cells. Nutritional products of digestion and water absorbed from the digestive organs return with the blood through the veins to the R side of the heart. Waste chemicals and carbon dioxide from all the tissues also travel back toward the R side of the heart, for eventual removal from the body by the lungs (carbon dioxide), liver (chemical waste products), and kidneys (excess water and chemical waste products).

Serving the entire body, the circulatory system plays a role in controlling body temperature. Recall that muscles generate heat and the skin has a role in heat dissipation. Upper body garments may help regulate body temperature and provide thermal comfort. A coat with layers of fiberfill or down increases insulation, while a microfiber t-shirt can wick away perspiration giving a cooling effect. In many situations thermal balance is a whole body function.

### 4.2.3 Blood Vessels of the Arm

The arm has both arteries and veins. Arterial pulse points in the arm likely are of most interest to designers. The arterial blood supply for the arm arises from a large artery in the upper thorax. After this artery travels through the *axilla* (arm pit), it is called the brachial artery. Commonly blood pressure is measured over the brachial artery, as it passes along the medial aspect of the humerus above the elbow. The brachial artery provides branches to the tissues in the arm and forearm. Below the elbow joint, the brachial artery branches into the radial and ulnar arteries—two major arteries supplying the wrist and hand. The arteries run along the radial and ulnar sides of the forearm, and their pulsations can be felt at the *volar* wrist pulse points (refer to Figure 2.17 and Figure 4.2).

Blood returning to the heart from the arm flows in multiple veins. Arm veins are relatively easy to compress. Pressure on a single vein is seldom a problem because blood will find an alternate route through the arm. Product pressure encircling the arm or forearm can be a problem. A tight garment cuff or sleeve can cause the limb below to turn a bluish/purple color, become uncomfortable, and eventually painful as venous blood collects. If compression is continuous for a long period or is severe, it could cause permanent damage.

### 4.2.4 Documenting Circulatory System Function with Wearable Products

The heart, as a muscular pump, fills and empties with a regular rhythm. *Cardiac* (heart) muscle contractions, like all muscle contractions, involve chemical changes within the muscle fiber. The chemical changes create differences in the electrical energy inside and outside the cardiac muscle cells. Electrodes adhered to the chest can detect the heart muscle's electrical energy changes. When the electrode patches are connected by wires to a recording device, the signals can be recorded as an electrocardiogram (abbreviated as *ECG* or *EKG*). ECGs measure pulse rate, heart rhythm, as well as heart muscle contraction patterns and efficiency, to evaluate how well the heart is working.

Wearable technology for heart monitoring has a long history (Del Mar, 2005). After approximately 20 years of work on prototypes, Holter and Glassrock patented an ECG monitoring system to be worn during activity (Holter & Glassrock, 1965). Practical wearable heart monitoring devices, commonly called *Holter monitors*, were subsequently developed and marketed to provide *ambulatory electrocardiography*. Like many wearable medical products, ambulatory monitors were designed to accurately and efficiently capture medical information, with comfort and wearability as an afterthought.

Ambulatory ECG guidelines (Crawford et al., 1999) outline indications and technical requirements for ambulatory ECG monitoring, plus many challenges and limitations of the monitoring process. The size, weight, and method of suspension of the Holter monitor signal recording device have posed challenges for use. Newer technologies using miniaturization, increased flexibility of adhesive recording electrodes, and signal transmission techniques promise improvements (Raj et al., 2016). A lightweight ambulatory ECG monitor in an adhesive patch can now continuously record data from a single electrode on the chest wall for up to 14 days (Barrett et al., 2014). However, this long wearing time has the potential to cause problems. Adhesives can irritate the skin, and activities such as bathing may expose the monitor or conducting elements to moisture. Knit garments with integrated electrodes have been introduced. However, as Griffin and Dunne (2016) emphasize, proper sensor positioning is essential for obtaining accurate and useful data. Their analysis of fitting a knit sensing garment to a sample population, using an apparel grading system, illustrates problems to be solved. In November 2017, the United States Food and Drug Administration (FDA) approved a wristband system to record ambulatory EKGs (U.S. Food and Drug Administration, 2017).

As described in Chapter 2, blood pressure is the pressure of blood on the walls of the arteries generated by the force of the pumping heart. Typically, blood pressure is measured at a single point in time. An inflatable cuff attached to a recording device is wrapped around an arm or leg, inflated and then deflated, producing a blood pressure reading (*BP*). The cuff is then removed. *Ambulatory blood pressure monitoring* is used to record pressures over hours or days but is less common than ambulatory ECG monitoring. Stergiou et al. (2016) review the current limitations and future research and development needs for ambulatory BP monitoring.

## 4.3 Respiratory System in the Thorax

The lungs, the primary respiratory structure, occupy most of the volume of the thorax (Figure 4.1). They extend from the collarbones (clavicles) down to the diaphragm. The diaphragm bulges slightly upward to lie just below the lungs and heart. Abdominal organs nest into the space under the diaphragm. The chest wall and muscles of respiration work with the lungs to exchange waste gases from the body for oxygen from the air. Unlike the circular flow pattern of blood in the arteries and veins, the respiratory system is a linear structure—air first flows into and then waste gases leave the body through the same structures.

### 4.3.1 Respiratory System: Structure and Function

Several important features of the respiratory system are located in the head and neck: the respiratory control centers of the brainstem, and the respiratory tract structures of the nose, sinuses, pharynx, larynx, and trachea. Air is warmed and humidified in the head and neck, before it flows into the lungs. In the thorax, the tubular respiratory structures branch repeatedly.

The branching respiratory structure is called the *bronchial tree*. The trachea branches into the R and L main *bronchus* behind the sternum. The *bronchi* (plural of bronchus), like the trachea, are held open with cartilage rings. They direct air to all areas of the lungs through the bronchial tree. The bronchi reach into the *lobes* in the lungs; then into subdivisions of the lobes, the *segments;* and then into the *alveoli,* the tiny air sacs next to the pulmonary circulation capillaries. With the two main bronchi as trunks, the final branches of the bronchial tree are the *bronchioles*—the respiratory system equivalent of the capillaries of the circulatory system.

The bronchioles and clusters of alveoli are considered the respiratory region of the lung (Netter, 1980, p. 24), where the body completes the waste gas-oxygen exchange. A breathing or *respiratory cycle* includes a breath in, followed by a breath out. Quiet breathing, while a person is at rest, moves about .5 liter (30.5 in.$^3$) of air with each breath. Deeper breathing moves more air, farther into the structures of the lungs. As air moves back out through the bronchial tree, it carries carbon dioxide, other gases associated with health and disease, and water vapor from the body (Horváth et al., 2005). Wearable technology designers have the opportunity to improve methods of capturing exhaled compounds that are of medical interest. Water vapor and warm air expelled from the warm moist environment of the respiratory tree can cool the body. Think of "seeing your breath" on a cold day.

The lungs exchange air (Figure 4.3) with help from the "respiratory pump" (Trulock, 1990, p. 256). The bony and cartilaginous rib cage surrounding the lungs is surprisingly mobile. Joints within the ribs, between the ribs and the sternum, and between the ribs and spine allow chest wall movement in several directions. Three layers of muscles, the *intercostal* muscles, span the spaces between adjacent ribs. The pumping action of the respiratory cycle—moving air in and out—is generated by coordinated contraction and relaxation of the diaphragm, intercostal muscles, and abdominal muscles.

Lung/chest cavity volume is proportional to the radius and height of the relatively conical lungs. In an adult, the difference in lung volume between maximum inhalation and maximum exhalation is approximately 4 to 4.5 liters (244 to 275 in.$^3$) (Miller et al., 2005). Unrestricted rib cage motion and full relaxation and strong contraction of the diaphragm and abdominal wall are needed to produce that degree of volume change. The diaphragm flattens as it contracts, enlarging the chest cavity, causing negative pressure within the chest. This action leads to air intake or inhalation. As the intercostal muscles contract, the ribs move out and up—helping to stabilize the rib cage, preventing collapse, while allowing and assisting the chest cavity to expand. See Sections 4.4.1 and 4.14.4 for additional information on rib cage structure and rib cage motion. When the diaphragm relaxes, it moves up toward the lungs expelling air in exhalation. Abdominal wall muscle contraction against the abdominal contents forces more air from the lungs.

Lung volume changes affect the return of blood to the heart from the lungs. When the pressure in the chest is negative (during inhalation) blood

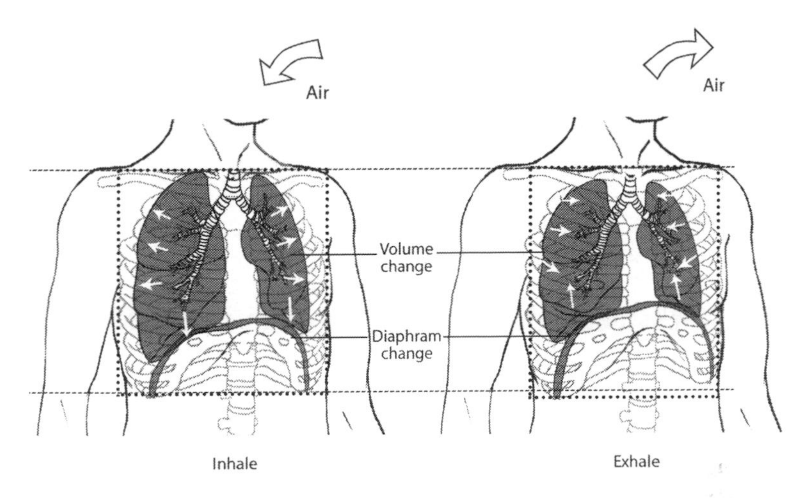


**FIGURE 4.3**
Lungs showing inhalation and exhalation with the associated diaphragm and volume changes of the lung and chest cavity.

pools very briefly in the lungs and lowers blood pressure (Brzezinski, 1990). Blood flow back to the heart from the lower body and legs, through the inferior vena cava, varies depending on how a person is breathing (Frisbie, 2016). Frisbie details the three common modes of normal breathing: (a) rib cage breathing with most of the motion in the chest, (b) diaphragm breathing with movement primarily where the chest and abdomen interface, and (c) belly breathing with fluctuations of the abdominal wall. Most of the time breathing is automatic and continues without conscious thought; however, a person can also breathe purposefully. Some wearable products limit one or more of the natural breathing processes. A tight bra band can restrict rib motion; a tight belt, waistband, or corset can hamper diaphragm movement indirectly and abdominal wall motion directly.

### 4.3.2 Focus on Products for Use with the Respiratory System

Wearable products have been designed to assist breathing in special cases. As noted above, abdominal muscle relaxation coordinated with diaphragmatic and intercostal muscle contraction facilitates inhalation and abdominal contraction can assist with exhalation. Most individuals paralyzed from the neck down (related to a cervical spinal cord injury) maintain some diaphragm function. But, because their intercostal and abdominal muscles are paralyzed, breathing can be severely impaired. Paralyzed individuals may use abdominal binders to mimic the pressure of the abdominal wall on the internal organs. These devices are designed to be used while seated and upright (West, Campbell, Shave, & Romer, 2012). However, chest, diaphragm, and abdominal relationships change when a person lies down. The chest cavity volume decreases due to gravity influencing and shifting organ positions (Moore, Agur, & Dalley, 2011, p. 84). Instead of being helpful, a binder may limit breathing when the person is in a horizontal position (Alvarez, Peterson, & Lunsford, 1981).

Other wearable products have been developed to actively assist breathing for people with neurological diseases, muscle diseases, respiratory disease, and injuries. Intermittent abdominal pressure ventilators are corset-like devices with an inflatable bladder which intermittently presses on the abdomen, pushing the diaphragm upwards and forcing air out of the lungs (Adamson, Lewis, & Stein, 1959). These devices assist exhalation in a seated position. When the device deflates, the abdominal contents descend and passively assist inhalation. Mehta and Hill (2001) review other wearable devices to assist breathing as well as masks used with ventilators which assist both inhalation and exhalation.

Medical personnel take special interest in monitoring lung function and volume after surgery, when breathing may be suboptimal due to pain or medications. Recordings of natural rib cage and abdominal wall motions have been used to monitor function. Non-invasive respiratory effort recording devices —specialized belts placed over the lower rib cage and/or over the abdominal wall just above the navel can provide useful information about pain control and/or lung function (Seppänen, Alho, Vakkala, Alahuhta, & Seppänen, 2017).

When designing a product that will be worn during physical exertion, extreme stress, or other activity requiring deep breathing, allow for significant chest and abdominal expansion. For example, full chest expansion is necessary for an actor or singer to be able to project the voice. Stage costumes should be designed to incorporate the full extent of the actor's breathing. If wearable products limit respiratory system function because they restrict rib cage motion or increase intra-abdominal pressure, they most likely will be uncomfortable. Discomfort at the chest or waistline may be used as a signal to further assess the effect of a design on respiratory function.

A cough is a protective respiratory system action. McCool (2006) describes the coordinated step-wise mechanism of an effective cough (Figure 4.4). First you inhale. Next the vocal cords close and the abdominal muscles and intercostal muscles contract, compressing the air in the lungs and increasing the pressure in the lungs and abdomen. Then, the vocal cords open and air rushes through the trachea to expel excess *mucus* or irritating particles like dust or small foreign bodies. Coughing is markedly impaired for individuals with cervical spinal cord injuries. Manual pressure on the abdomen or electrical stimulation of the abdominal wall muscles with electrodes applied to the skin can assist spinal cord injured persons when they want or need to clear their airways with a cough.

*Cystic fibrosis* (*CF*), a genetic disorder, affects the respiratory and digestive tracts. Everyone has mucus in the airways, but people with CF have very thick, sticky mucus that builds up. Ordinary coughing doesn't effectively remove the mucus. Figure 4.5 shows a vest, which wraps around the rib

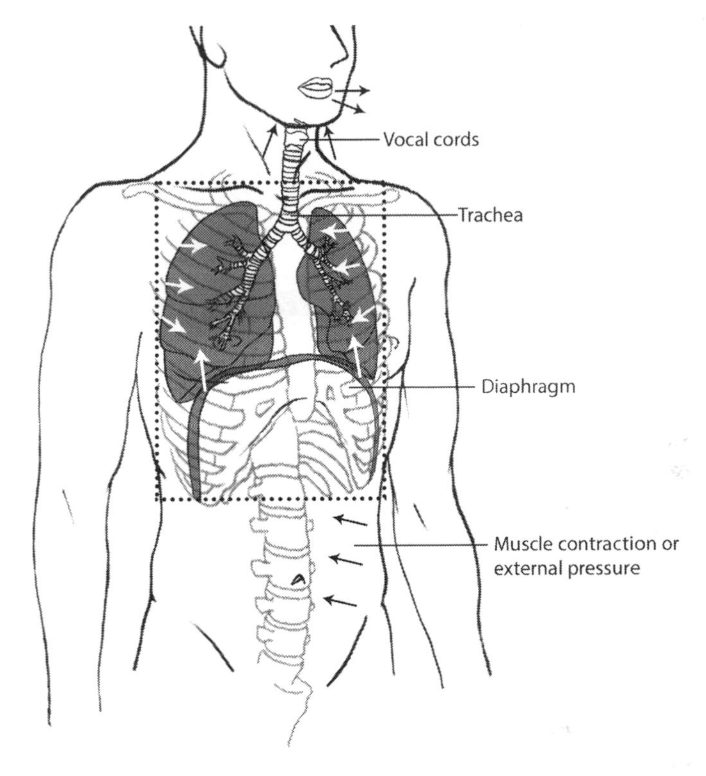


**FIGURE 4.4**
Key respiratory structures involved in a cough; tightening the abdominal muscles or providing external abdominal support to assist cough.

cage and vibrates, to help dislodge mucus. After the vest is worn for a short period, the mucus can be coughed up and expelled from the body (Lee, Lee, & Warwick, 2008; Warwick & Hanson, 1991).

## 4.4 Skeletal System in the Thorax and Arms: Protection, Mobility, and Reach

The framework of the body, the skeleton, can be divided into two sections: the *axial skeleton* and *appendicular skeleton* (Figure 4.6). The axial skeleton primarily offers protective coverings for vital body components; the appendicular skeleton in the thorax and arms allows reach. Both the axial and appendicular skeleton contribute to motion in the torso and arms and help to maintain balance in an upright posture over the body's center of gravity.

**FIGURE 4.5**
inCourage® Airway Clearance Therapy is a vest therapy system that uses air pressure and pulses to create compressions to the chest that help loosen, thin, and move mucus from lungs. (Courtesy of RespirTech®, St. Paul, MN.)

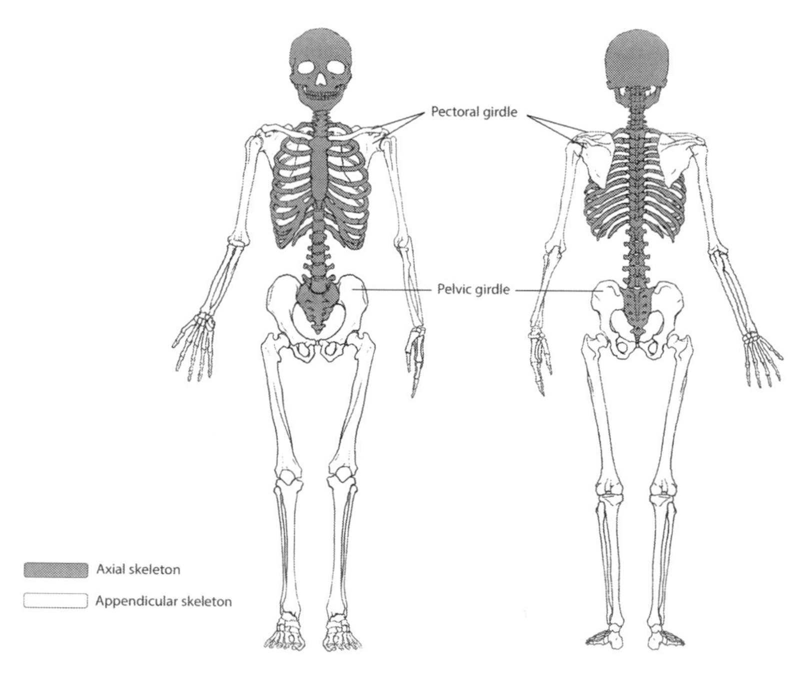


**FIGURE 4.6**
Axial and appendicular skeleton; pectoral and pelvic girdles.

### 4.4.1 Upper Torso Axial Skeleton: Structure and Function

The spine and rib cage form the axial skeleton in the upper torso. The rib cage, introduced earlier for its protective functions and as part of the respiratory pump, helps stabilize the thoracic spine. The pectoral (shoulder) girdle—the bones supporting the upper limb—are attached, or "appended" to the rib cage at the upper section of the sternum—the *manubrium*. The pectoral girdle and the arm/wrist/hand bones are part of the appendicular skeleton.

#### *Spine: Structure and Function*

The spine protects nervous system structures within the vertebral canal and, along with the rib cage, helps protect the heart and lungs. The lumbar spine in the mid and lower torso serves as a structural support for the muscular wall surrounding the abdomen. Recall from Chapter 3, the spine is made up of bones, the vertebrae, and discs that separate and cushion the vertebrae. The vertebrae in each torso region of the spine—thoracic, lumbar, sacral, and coccygeal—differ in shape and size. However, they have several features in common (Figure 4.7). A generic torso vertebra has an anterior cylindrical vertebral body. It has arches that protrude posteriorly and provide points of contact (the *facet joints*) with the arches of the vertebrae positioned above and below it. The arches meet and form spinous processes, the bony protrusions that you can feel up and down the spine in the center of your back. The vertebral arches cover the vertebral canal.

Vertebral shapes affect spinal movement and stability. The cervical, thoracic, and lumbar spine vertebrae are movable and become larger the farther they are from the head. The thoracic vertebrae connect with ribs; first ribs articulate with the first thoracic vertebra (T1), and so on. The bodies of the five lumbar vertebrae in the lower back, L1–L5, are sturdier than the vertebrae above, because they bear the accumulated weight of the head, neck, and torso (White & Folkens, 2005, p. 159).

The mechanical stresses on the sacral and coccygeal spinal elements are different from those in the thoracic and lumbar spine. The differences in stress lead to marked form variations of these lower vertebrae, compared to the relatively consistent forms of the thoracic and lumbar vertebrae. Below L5 there is little to no vertebral motion. The five sacral vertebrae are fused into a shield-shaped structure, the *sacrum* (Figure 4.8). Remnants of the S1 vertebral body and sacral vertebral spinous processes can be seen within the sacrum (Figure 4.7). The *sacroiliac joint*, between the lateral aspects of the sacrum and the *iliac* bones of the pelvis, joins the axial and appendicular skeletons in the pelvis. The tailbone or *coccyx* varies from one person to another with three to five coccygeal vertebrae. These very small vertebrae may or may not be fused together. The coccyx, fused or not, curls toward the front of the body from the tip of the sacrum. Injuries to the coccyx can be painful. Protective gear, such as football tailbone pads, have been designed to protect this vulnerable area of the spine (Figure 4.8).

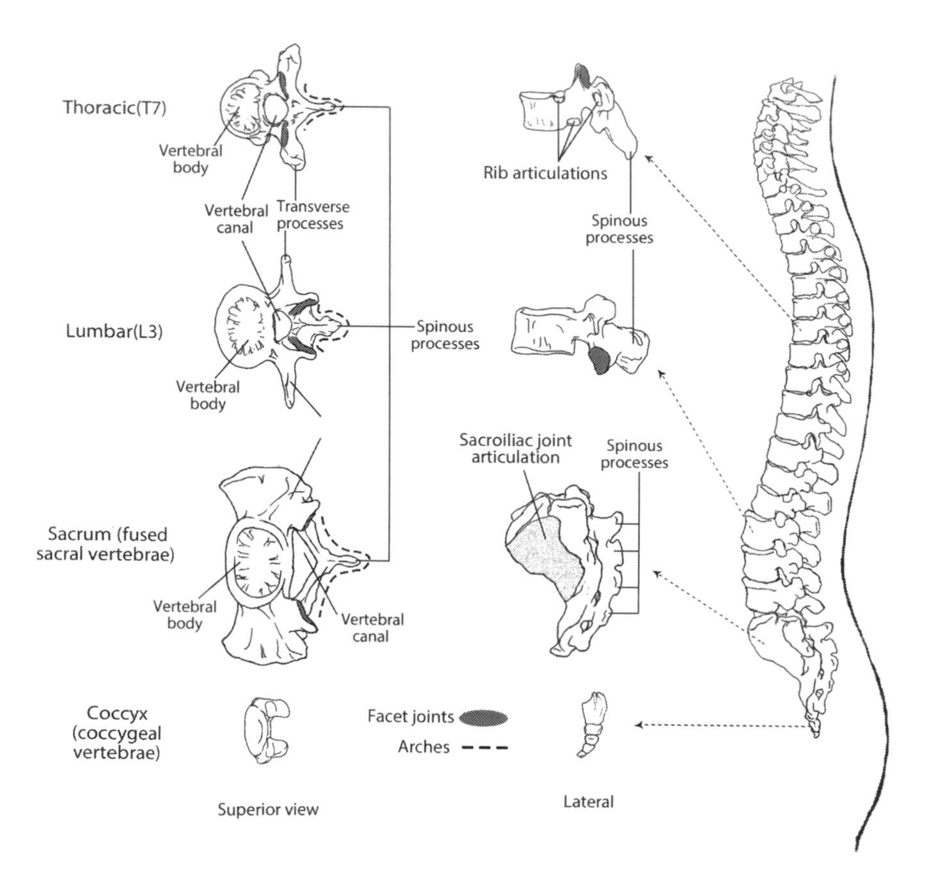


**FIGURE 4.7**
Thoracic, lumbar, sacral, and coccygeal vertebral structures.

Intervertebral discs lie between vertebrae in the cervical, thoracic, and lumbar regions, plus at the junction between L5 and the sacrum. Each disc and the two vertebral bodies on either side form a moving fibrocartilaginous symphysis joint (Figure 4.9), a kind of disc "sandwich." Intervertebral discs provide a physical spacer between the vertebrae in addition to transmitting loads and facilitating spinal flexibility (Urban & Roberts, 2003). When vertebrae shift with spinal column movements (even with normal motion), disc wear and tear occurs. New developments in light weight wearable products may offer solutions to slow degenerative spinal column changes. Thoracic spinal discs are not as easily damaged as lumbar discs, due to relatively restricted movements between thoracic vertebrae. Intervertebral discs account for about one-third of the height of the spinal column. In the lumbar spine, they are approximately 7–10 mm (.3–.4 in.) thick and 4 cm (1.6 in.) in diameter (Urban & Roberts, 2003, p. 121).

The intervertebral joints lose height due to age-related changes of the vertebral body, disc dehydration, the compressive force of gravity, and mechanical loading. Through the disc changes of normal aging, a person can lose 2.5–5 cm (1–2 in.) of body height. Aging and genetics contribute to degenerative change in the lumbar spine, but sustained mechanical loading of the spine also produces negative consequences (Urban & Roberts, 2003; Hadjipavlou, Tzermiadianos, Bogduk, & Zindrick, 2008). Lumbar degenerative disc changes are observed as early as 11–16 years of age, and severe degenerative changes are evident in almost everyone's lumbar discs by the age of 70. Males exhibit disc degeneration earlier than females (Urban & Roberts, 2003). Once the disc in the intervertebral symphysis joint has deteriorated, secondary degenerative changes develop at the vertebral edges and other spinal joints. Back pain, restricted motion, and compromise of the neural elements within and outside the vertebral canal can follow.

Although the movable vertebrae of the spine become increasingly sturdy from top to bottom, wearable products should not over-load the structure. Military research indicates that some carrying methods for loads of 35% of body weight are more comfortable than others (Legg & Mahanty, 1985). They found a load carried close to the body's center of gravity—with weight distributed in a body jacket or divided between a frameless front pack and a backpack with a frame—is more comfortable. However, study participants found it difficult to don and doff loads in the front pack/backpack-with-frame combination. Carrying the same load in a backpack alone—with or without a frame—or in a combination of a backpack with frame and weight suspended from a waist belt was less comfortable. Wearing loads in packs or jackets in any configuration restricted chest/respiratory function, some more than others, an important limitation to keep in mind.

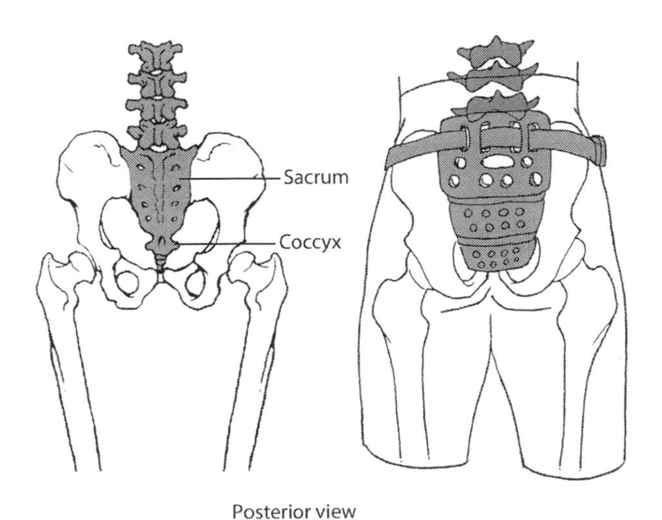


**FIGURE 4.8**
Sacrum, posterior view; coccyx pad for athletics.

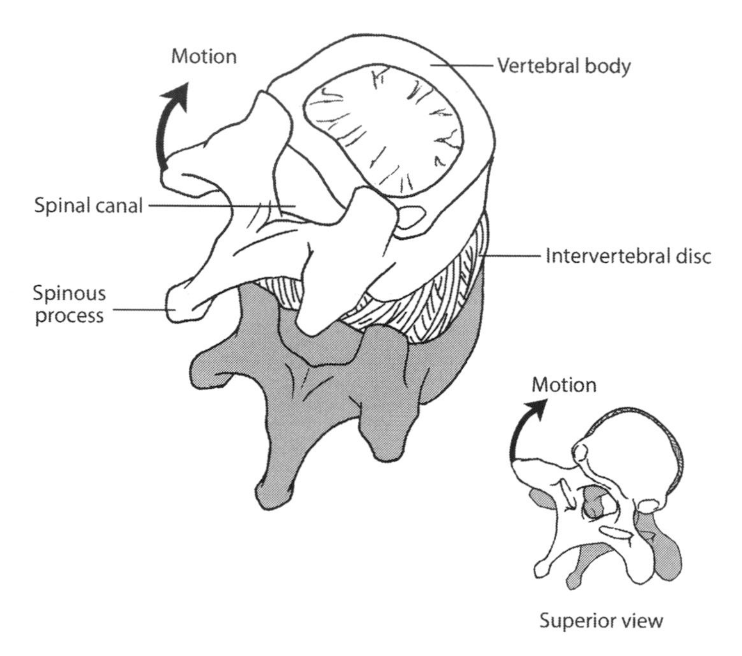


**FIGURE 4.9**
Fibrocartilaginous symphysis joint between two thoracic vertebrae and intervertebral disc/vertebral segment movement.

### *Rib Cage: Structure and Function*

The rib cage includes the sternum and the ribs (Figure 4.10). The sternum, a vertically oriented flat bone, is located front and center on the rib cage. Feel its solid structure running from your neck to your stomach. It has three parts: the manubrium, the body of the sternum, and the *xiphoid process*. The manubrium is somewhat trapezoidal in shape. You can feel the superior edge of the manubrium at the base of the neck. It forms a U-shape with the ends of the clavicles (collar bones) at the *sternoclavicular joints* and the origins of the sternocleidomastoid muscles. The edge of the manubrium is easily palpated and landmarked. The bottom of the U can be marked, or the two upper points of the U can be palpated and marked. Any of these locations can be used as landmarks for a product that fits the neck area. The body of the sternum, below the manubrium, is the major portion of the sternum. It is thinner and more vulnerable than the manubrium. Notches along the edge of the sternum are the locations of the rib *sternocostal* joints. The xiphoid process, at the inferior end of the sternum, is more delicate than the body of the sternum. Its shape can vary, person to person. You can feel some flexibility at the joint between the xiphoid and sternum. In children, the xiphoid is cartilaginous and in adults under 40, it may still be cartilage. With age, the cartilage transforms to bone, and in the elderly, it may be fused to the body of the sternum. The xiphoid marks the level of the inferior border of the heart and the central tendon of the diaphragm (Moore et al., 2011, p. 51). The joints between the manubrium, body of the sternum, and xiphoid process are slightly flexible fibrocartilaginous joints.

Look at the 12 pairs of ribs. Unlike the movable vertebrae, these bones are more delicate the farther they are from the head (White & Folkens, 2005, p. 185). As with the vertebrae, the form (shape and size) of the ribs (Figure 4.11) helps determine the motion of the rib cage. Structures vary between the 1st rib, ribs 2–10, and ribs 11 and 12. Each rib attaches to the thoracic spine at one or more movable *costovertebral joints* and then curves down and around toward the front of the body. The typical rib, like ribs 2–9, has a *head* which articulates with two adjacent vertebral bodies, a *tubercle* which articulates with the vertebral transverse process, an angled bony *body,* or *shaft,* with a *facet* (on the sternal end) which articulates anteriorly with a section of cartilage (shaded in Figures 4.10 and 4.11) (Palastanga, Field, & Soames, 2002, pp. 460–462; Talbot et al., 2017). Ribs 1–7 attach to the sternum with a simple bridge of cartilage. Ribs 8–12 are *false ribs,* because they do not attach directly to the sternum. Ribs 11 and 12 do not have tubercles or facets and are much shorter than the other ribs. They are *floating ribs* and do not attach to the sternum at all. Rib 12 is approximately one-fourth the length of rib 9.

Bellemare, Jeanneret, and Couture (2003) found significant structural differences in dimensions and configuration between the rib cages of men and women. These differences may relate to the structural accommodations for pregnancy in the female abdominal/pelvic region and should be considered when designing unisex upper torso wearables.

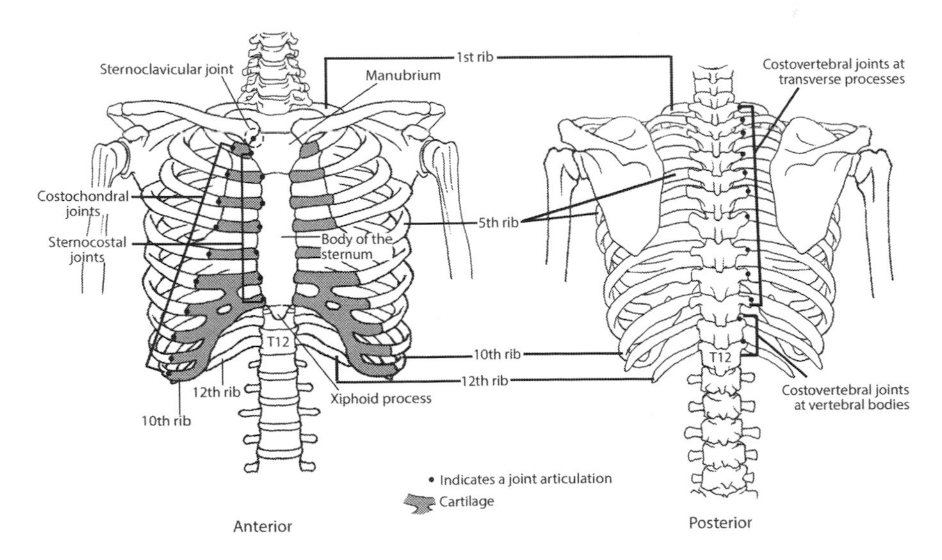


**FIGURE 4.10**
Skeletal elements of the thorax–axial skeleton (ribs and sternum) with associated joints; appendicular skeleton (clavicle and scapula) with associated joints.

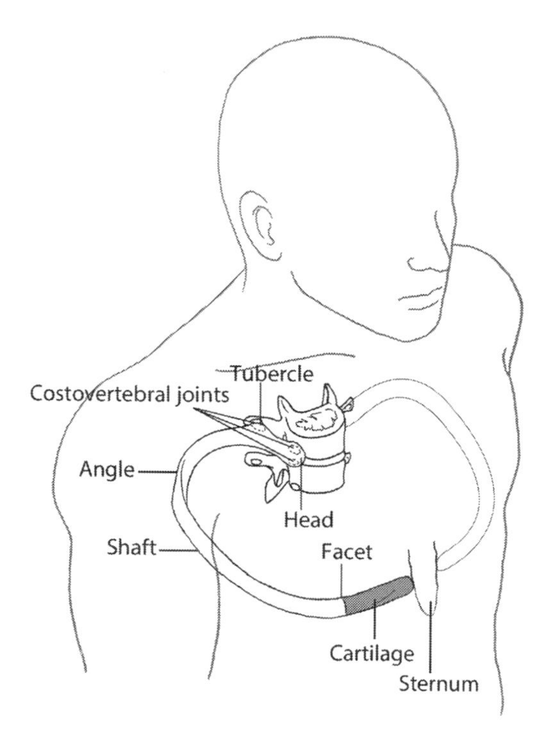


**FIGURE 4.11**
Typical rib structure and the costovertebral joints—the attachment of a rib to a thoracic vertebrae.

Ribs are vulnerable to direct chest wall trauma; the middle ribs are most commonly broken (Moore et al., 2011, p. 54). This may be a result of both shape and decreasing mass, compared to the upper ribs, and the fact they are anchored at both ends—to the spine in the back and to the sternum in the front—unlike the lowest ribs. Ribs often fracture at the angle of the rib (Carrilero, 2016, p. 54). Broken ribs are painful, and every breath or cough causes additional pain because of the associated rib cage motions. In addition, the sharp ends of broken ribs can cause significant injury to internal organs of the chest and/or abdomen.

Immobilizing the rib cage is undesirable (Carrilero, 2016, p. 57) and difficult. Recall the importance of rib motions for breathing and the importance of breathing and coughing for health. Innovative wearable approaches to preventing rib fractures may prove useful for: people with *osteoporosis* who are at risk for falls, soldiers, motorcyclists, all-terrain-vehicle riders, and athletes participating in contact sports. The floating ribs can be displaced by external pressure. As part of a medical study of the history of corsets and their medical and health impacts, Ludovic O'Followell (1908) documented with X-rays how a corset can distort the 11th and 12th rib positions. His study is accessible online from the Bibliothèque nationale de France.

The joints of the rib cage are essential for rib cage motions for breathing and coughing. Costovertebral joints are located posteriorly, and there are two kinds of rib joints in the anterior rib cage (Figure 4.10). The junction between the cartilage of the rib and the sternum is called a sternocostal joint or *chondrosternal* joint. The first sternocostal joint, between the first rib and the manubrium is not mobile (a synchondrosis joint). The other anterior rib cage joints are movable. The sternocostal joints for ribs 2–7 are synovial plane joints. The junction between the sternal end of each bony rib section and the cartilage connecting the rib to the sternum, directly or indirectly, is called a *costochondral joint.*

### 4.4.2 Focus on Products Protecting Thoracic Structures

Wearable products designed to protect the chest from impact and/or puncture can help prevent upper torso injuries ranging from chest wall bruises and punctures, to broken bones in the axial and/or appendicular skeletons, to rib joint and cartilage damage. Motocross chest protectors deal with the risk of impact from falls and from impact with the bike. Padded vests for fencers protect the chest from puncture. Participants in sports like hockey, lacrosse, and baseball may wear chest protectors to shield the thoracic wall structures and the vital organs of the thorax from impact. Chest protectors for baseball catchers protect not only the heart and lungs but also the upper abdomen.

Every year, 10 to 20 people in the U. S. suffer *commotio cordis,* cardiac arrest caused by a blow to the chest (Link, 2012). A projectile traveling at about 40 miles per hour must land directly over the heart, in the milliseconds between heartbeats, to stop the heart. Deaths have occurred despite wearing a protective vest. The National Operating Committee on Standards for Athletic Equipment (2017) issued a standard for chest protectors worn by lacrosse and baseball players, effective June 2018. Kumar et al. (2017) achieved effective chest wall protection with a novel pad of modest thickness in an animal model of *commotio cordis.* They propose further development of chest protector designs with the same materials for use on the playing field. Designers of bullet and blast resistant vests for police officers and military personnel face the challenge of designing a product that can stop bullets or bomb fragments and be comfortable for daily wear. Bullet-resistant vests can be made of rigid materials—ceramic shields placed into vests—or high-strength fiber textiles that are more comfortable and yet provide adequate protection. Although quite successful in protecting against penetrating injuries, body armor "flak" jackets can transmit force to the body, causing "behind armour blunt trauma" (Carr, 2016, p. 261). Couldrick (2004) states, "Good armour design is a matter of making the best of a bad situation" (p. 213). He notes that materials technology, threat, wearer, task, and environment are the pertinent considerations for body armor design. A vest that physically protects against bullet penetration but is too heavy, hot, or cumbersome will not be worn—offering no protection. Additionally, the designer has to balance financial cost with protection and ergonomic effectiveness. For a review of the features and design of bullet-resistant equipment, see Watkins and Dunne's (2015, Chapter 6) discussion of impact protective materials and products. Another functional military garment uses the rib cage as a support structure. A vest designed for combat radio operators incorporates a radio antenna in the vest materials, eliminating a visible antenna that projects above the soldier's head (Lebaric, Adler, & Limbert, 2001).

### 4.4.3 Spinal Curves

The spine is a complex three-dimensional structure. It helps determine the shape of the back and affects relationships with other skeletal regions. Think of the vertebrae as blocks that can be stacked in several ways, with variations possible in the sagittal, coronal, and transverse planes of the body. Study Figure 4.12 as you read about spinal curves. Recognize the position and magnitude of these curves when designing for the upper torso. In a fitted jacket or dress, fitting devices—darts or curved seams—shape two-dimensional fabric to the curves of the back. The healthy adult backbone has several opposing anterior-posterior curves between the base of the skull and the tip of the coccyx in the median sagittal plane (lateral view). Individual vertebrae incline forward or back. A thoracic convexity forms posteriorly (toward the back of the body). Cervical and lumbar convex curves form anteriorly (toward the front of the body). Viewed in a coronal plane (from anterior) the spine is straight, with no lateral curves. The vertebral spinous processes of the stacked vertebrae align, when viewed from above (superior view), with no twisting or rotation of one vertebra relative to another.

When the orientation of an individual vertebra or group of vertebrae changes within the spine, it can change the spinal curve in several ways. The vertebrae can develop: (a) altered inclinations in the sagittal or coronal plane and/or (b) rotations in a transverse plane. When the inclinations or rotations result from voluntary muscle activity, they contribute to the normal range of motion (ROM) of the neck, torso, and pelvis. Unusual spinal curves develop when vertebral orientations change from causes such as (a) fractures of the vertebral bodies due to osteoporosis, trauma, or other medical conditions, (b) variations in the shape of the vertebrae due to abnormal muscle function or posture during vertebral growth, (c) shifts in the body's center of gravity due to weight change or weight distribution changes as in pregnancy, and/or (d) muscle imbalances due to neurological disease.

Structural changes altering vertebral alignment produce exaggerated curves: *kyphosis,* an unusual or increased posterior convexity of the spine; *lordosis,* an unusual or increased anterior convexity; and *scoliosis,* a lateral

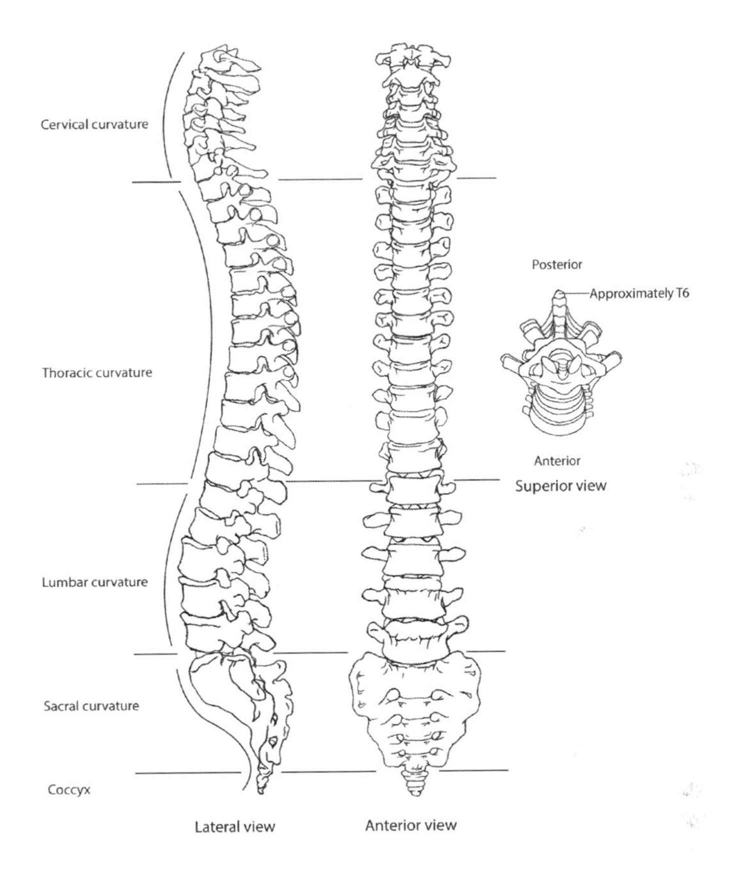


**FIGURE 4.12**
Normal spine; left lateral, anterior, and superior views.

deviation of the spine to either side (Figure 4.13). Scoliosis includes both vertebral inclinations and rotations in most cases. When the thoracic spinal curve changes, the alignment and shape of the rib cage are affected, because the ribs articulate with both the vertebral bodies and transverse processes of the vertebrae. Gorton, Young, and Masso (2012) documented postural and body surface changes associated with scoliotic spines with 3D body scanning. An altered curve in one section of the spine frequently, if not always, causes a compensating change in the curves of the other sections as the body tries to stay upright and maintain balance around the center of gravity. A person with lumbar lordosis and/or thoracic kyphosis may also have cervical lordosis. Note vertebral twists and inclinations as well as the associated postural shifts illustrated in Figure 4.13, and think about how altered spinal curves will affect product shape and fit.

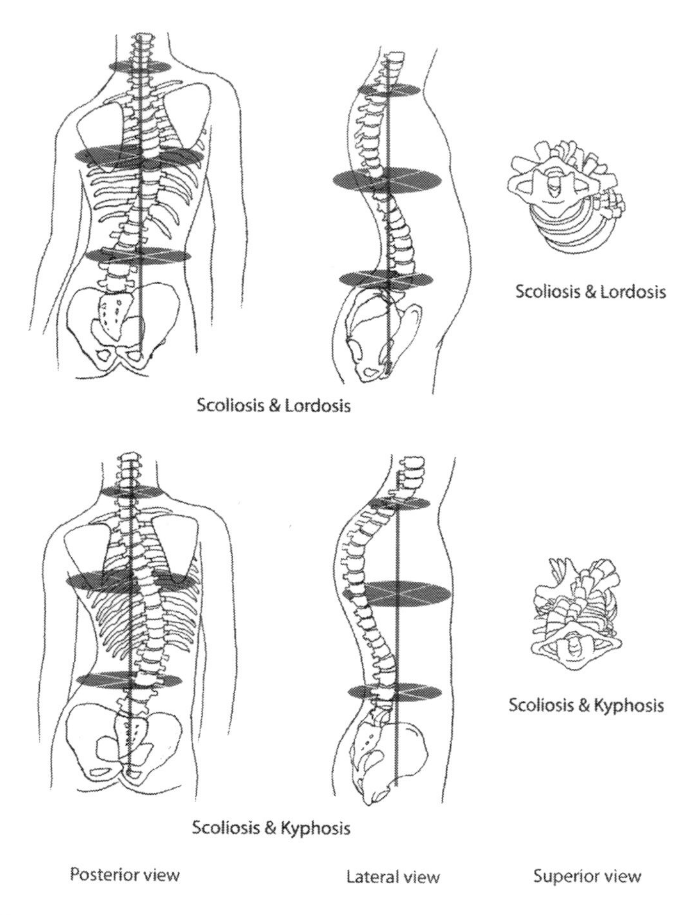


**FIGURE 4.13**
Spinal alignment variations; posterior, lateral, and superior views.

### *Kyphosis*

Kyphosis, most often seen in the thoracic spine, may be habitual or due to medical problems. Kyphosis is often related to *compression fractures* from osteoporosis. Osteoporosis, loss of calcium from the bones, causes structural weakness of the bones. It is frequently associated with hormonal changes of aging and is more commonly diagnosed in post-menopausal women than in men. Weakened vertebral bodies, subjected to bending forces or lifting activity, can fracture, developing a wedge-shaped vertebral body in the coronal and/or sagittal planes. Numerous wearable products, from weighted vests (Hakestad, Torstveit, Nordsletten, Axelsson, & Risberg, 2015) to wearable sensors (Dunne, Walsh, Hermann, Smyth, & Caulfield, 2008), have been developed to help correct kyphotic spinal curves.

### *Lordosis*

Lordosis is most frequently seen in the lumbar spine. It is sometimes called "swayback" and may be a compensation for thoracic kyphosis. Structural lordosis may be the result of osteoporotic lumbar vertebral fractures or other medical conditions. Pregnancy and obesity can shift the center of gravity resulting in lordosis. Muscle activity in the neck and lumbar spine produces changes in spinal lordosis and increased back extension increases lumbar lordosis. Wearing heavy gear like backpacks or body armor, or carrying heavy loads, may produce temporary lumbar lordosis and/or pain. Soldiers (n = 863) who wore body armor for four or more hours per day during their combat zone deployments reported an increased incidence of back, neck, and upper limb musculoskeletal pain (Konitzer, Fargo, Brininger, & Reed, 2008).

Al-Khabbaz, Shimada, and Hasegawa (2008) focused on muscle activity and posture in young men carrying backpack loads of 10%, 15%, and 20% of body weight. They reported increased back extension of approximately 3° with all load levels. Rodrigues, Domingues, Ferreira, Faria, and Seixas (2017) studied neck and low back muscles in six- to ten-year-old children who carried backpacks. They recommended limiting backpack weight to 10% of body weight. That translates to 2.7 kg (6 lb) including the pack for an elementary school child weighing 27 kg (59.5 lb). Brackley and Stevenson reviewed the literature in 2004 and reported approximately 90% of school children in the developed world use backpacks and the majority were carrying packs heavier than 10%–15% of their body weight. In one study 34.8% of students were routinely carrying over 30% of their body weight.

Carrying loads of any kind can produce vertebral orientation changes. Carrying a load asymmetrically affects both shoulder and spine posture and can compensate for, or aggravate, a natural posture imbalance (Lyu & LaBat, 2016). A pilot study in the University of Minnesota Human Dimensioning© Laboratory demonstrated posture change for an adult female carrying an infant weighing 3.1 kg (7 lb) in a front-positioned baby carrier (Figure 4.14). Infant slings that hug the baby close to the wearer's body, can cause both posture changes for the wearer and pose a risk to the infant. The U.S. Consumer Product Safety Commission (CPSC) advises parents and caregivers to be cautious when using infant slings for babies younger than four months, because slings can pose suffocation hazards (2017). The federal safety standard for infant sling carriers incorporated the voluntary standard developed by ASTM International, ASTM F2907-1 (2015b).

### *Scoliosis*

Scoliosis, technically an abnormal lateral curve of the spine in a coronal plane, occurs most often in the thoracic region. Because scoliosis invariably includes elements of kyphosis, lordosis, and rotations of the vertebrae

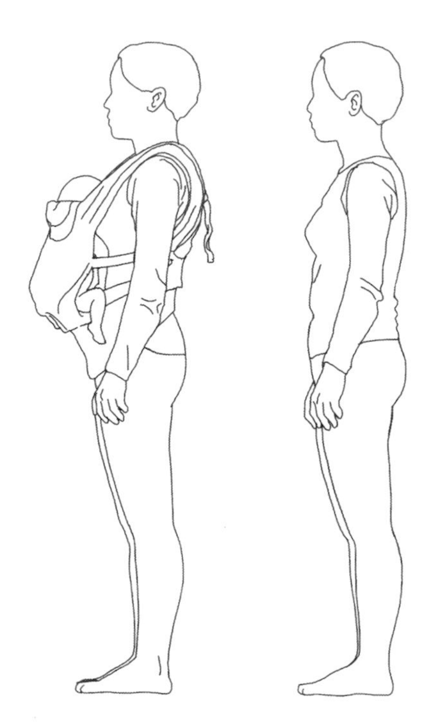

**FIGURE 4.14**
Spinal posture change when wearing a front-positioned baby carrier (outlines derived from body scans). (Courtesy of the Human Dimensioning© Laboratory, University of Minnesota.)

and ribs in transverse planes, it is often used, in non-medical language, to describe any distortion of the spinal curves. Scoliosis which develops without a known cause is termed *idiopathic* scoliosis. Occurring during growth, it is found more often in adolescent girls than in boys.

Corrective braces can be worn before skeletal growth is complete to attempt to straighten a scoliotic spine. The Milwaukee brace (Figure 4.15, upper), a well-known *thoraco-lumbo-sacral-orthosis* (*TLSO*) style, was developed by Blount and Schmidt in 1946 and remains a mainstay in the treatment of scoliosis. In general, bracing for scoliosis depends on application of transverse loads on the apex and both ends of the curve (Karimi, Ebrahimi, Mohammadi, & McGarry, 2017). A person with scoliosis may need to wear a brace for a year or more, 20–23 hours/day, so the brace should allow regular activities as much as possible. Spinal bracing presents three challenges: (a) achieving correction of abnormal spinal curves, (b) providing wearer comfort, and (c) either looking good or the brace being "invisible" to others.

A product related to braces for scoliosis is the TLSO worn after spinal fractures or spine stabilization surgery to try to fully immobilize the highly segmented and complex spine (Figure 4.15, lower). These full "body jacket" braces of rigid materials encircle the spine and rib cage and are worn relatively continuously, but for a limited period after surgery. Mobility and

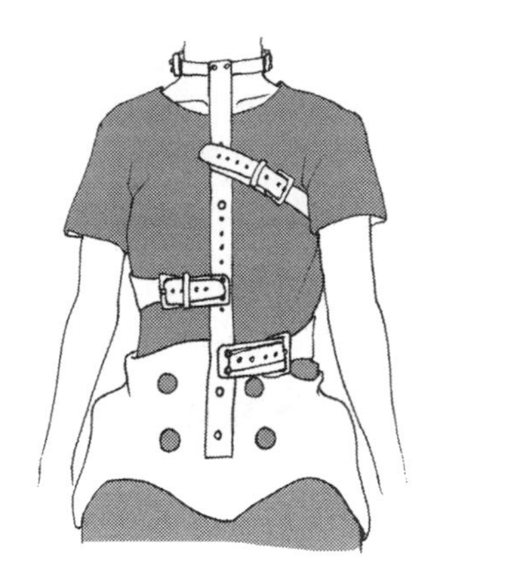
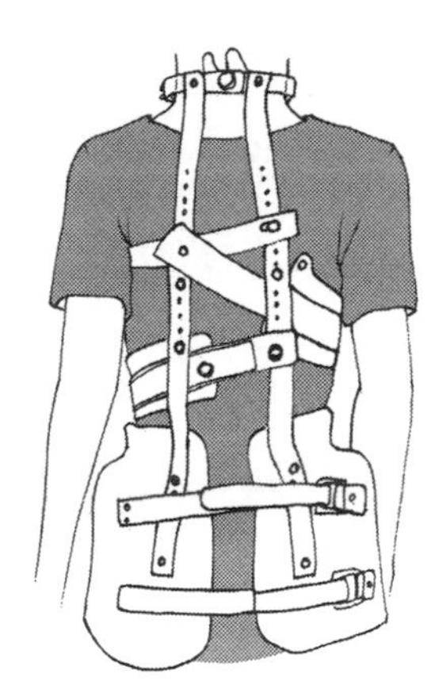

Milwaukee brace for thoracic scoliosis

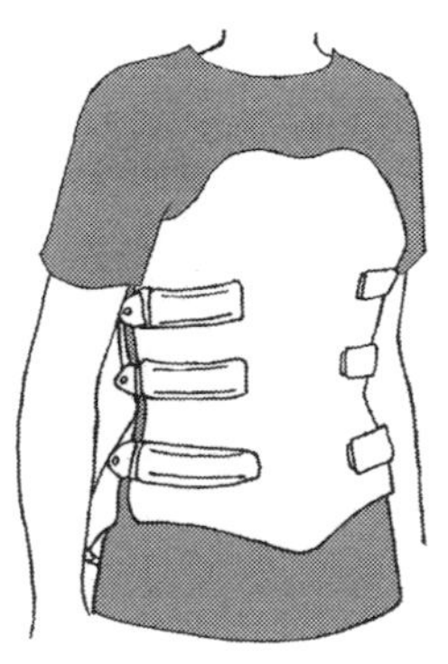
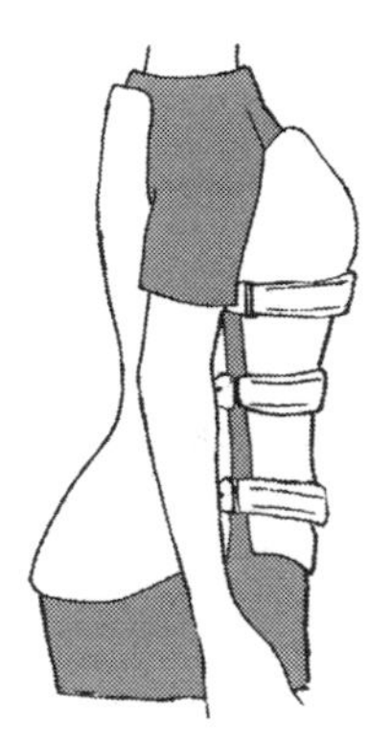

Body jacket

**FIGURE 4.15**
Milwaukee brace (typical) to treat scoliosis, and body jacket for post-operative/post-trauma spine immobilization. (Modified from "The influence of elastic orthotic belt on sagittal profile in adolescent idiopathic thoracic scoliosis: a comparative radiographic study with Milwaukee brace," Jun Jiang, Yong Qiu, Saihu Mao, Qinghua Zhao, Bangping Qian, and Feng Zhu. 2010. *BMC Musculoskeletal Disorders*, 11, p. 2. doi:10.1186/1471-2474-11-219. Copyright Jiang et al; licensee BioMed Central Ltd. 2010.)

activity are generally restricted during post-operative bracing, in contrast to routine daily activities that must be considered in scoliosis bracing.

### *Spinal Curve Variations: Wearable Product Design Considerations*

Spinal alignment and curvature affect body symmetry and how a product fits the body. Refer to Figure 4.13 to see rib cage and appendicular skeletal position changes related to spinal kyphosis, lordosis, and scoliosis. Spinal curvature effects may be subtle or severe. You may have noticed that you have one hip or shoulder higher than the other. This imbalance or asymmetry is

**FIGURE 4.16**
Shoulder pads in tailored jacket to give appearance of balanced shoulder line.

most likely due to small variations in the curve of your spine which throws other skeletal segments off-kilter. Most wearable products are designed for symmetrical bodies without concern for comfort or aesthetics for bodies that don't have an ideal posture. Shoulder pads can be used to achieve shoulder symmetry. A perfectly tailored, custom-made jacket will take asymmetries of the individual into account and incorporate shaping and padding to give a balanced appearance. Figure 4.16 illustrates a man's tailored jacket with different shoulder pad thicknesses to give the appearance of a level shoulder line. Spinal curves that vary from the ideal will also affect the fit and balance of lower torso garments. The hemline of a skirt will not hang parallel to the floor if even a mild spine imbalance causes one hip to be higher than the other.

### 4.4.4 Upper Torso and Arm Appendicular Skeleton: Structure and Function

A primary function of the upper limbs, the arms, is to maneuver the hands into the best positions to serve a person's needs. Our upright posture allows us to hold and manipulate objects with our upper extremities. We use our arms and hands to eat, gesture for emphasis or clarity of expression, and even to defend ourselves. Many upper limb motions are complex. For example, throwing a ball involves many "shoulder," elbow, wrist, and hand/finger joint motions.

Two major shoulder region functions are anchoring the mobile arm to the body and, at the same time, supporting the arm and any load the torso or arm carries. Because the "shoulder" is a common place for supporting wearable products, from everyday apparel to backpacks and baby carriers, wearable designs for the shoulder region need to address stability, mobility, and support.

The appendicular skeleton in the upper torso extends from the sternoclavicular joint between the collarbone and the manubrium of the sternum to the tips of the fingers. For designers, it is logical to divide it into four segments: (a) the pectoral (shoulder) girdle, (b) the *glenohumeral joint*, (c) the arm—the anatomical arm (upper arm) and forearm, and (d) the wrist and hand. The pectoral girdle, glenohumeral joint, and arm will be discussed in this section. Chapter 7 addresses the wrist and hand.

Bones, joints, ligaments, and muscles throughout this body segment connect the arm to the torso and serve to both support and move the upper limb. The skeletal structure of the pectoral girdle is vulnerable to excessive motion or load, as it relies only on connective tissues and muscles to hold the parts together while providing motion. Some shoulder region problems relate to skeletal issues, some to pressure on nervous system structures, and others to a combination of the two. The pectoral girdle and axial skeleton are intimately connected in structure and function. Upper torso and arm motion are similarly interconnected. See the motion discussions in Sections 4.14 and 4.15.

### *Pectoral (Shoulder) Girdle: Structure and Function*

The pectoral girdle is "a bony ring, incomplete posteriorly, formed by the scapulae and clavicles and completed anteriorly by the manubrium of the sternum" (Moore et al., 2011, p. 406).

It is shown in relationship to the axial skeleton and the arm in Figure 4.17. The pectoral girdle is stabilized by ligaments and supported and moved by upper torso and arm muscles. The spine and pectoral girdle align to allow arm and hand use for functional activities.

### *Clavicle*

The clavicles (collarbones) form part of the structure of the horizontal "clothes hangar" of the shoulder discussed in Chapter 1. "Collar" bone presumably refers to an apparel collar around the neck which extends to and covers a portion of the bone. The sternoclavicular joint—the articulation point for the clavicle and the manubrium of the sternum—is the only bony link between the axial and upper appendicular skeletal structures. It is a relatively complex synovial joint, with both ball-and-socket features and some hinge elements. The clavicle's relationships to the rib cage and arm are shown in Figure 4.17. It is supported by a ligament bridging the space between the R and L clavicles as well as by ligaments attaching the clavicle to the sternum and the first rib (Hollinshead & Jenkins, 1981, p. 79). The sternoclavicular joint allows moderate motion of the pectoral girdle in the transverse and coronal planes, but less movement in a sagittal plane. Because it can move in all three planes the clavicle can circumduct in the sternoclavicular joint, making "shoulder rolls" possible. This joint, somewhat surprisingly, tends to function well without many problems—testimony to the strength and versatility of the muscles and ligaments in the shoulder region.

The clavicle articulates distally with the acromion of the scapula at the *acromioclavicular joint*, a synovial gliding joint. The acromion is the bony

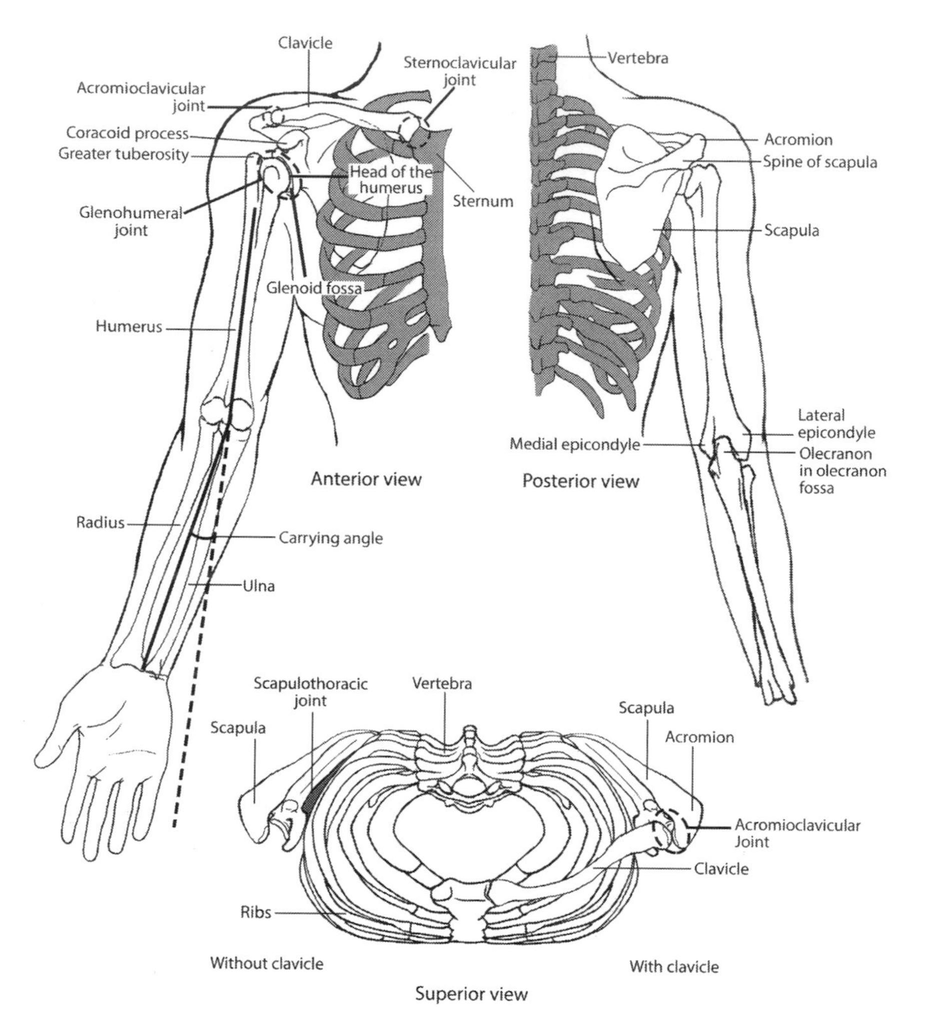


**FIGURE 4.17**
Upper torso and upper limb skeletal structures—including pectoral girdle structures: anterior, posterior, and superior views; carrying angle of the forearm.

upper shelf of the scapula that sits above the humeral head of the upper arm. You can palpate it as a ridge about 2 cm (.8 in.) in length—the size varies person to person. The acromial point, the lateral tip of the acromion, is an often-used landmark spot to position wearable product features. For example, the acromial point corresponds to the top point of the cap of a set-in sleeve (Figure 4.18).

The acromioclavicular ligament supports the acromioclavicular joint at the articulation, and three additional ligaments attach the clavicle to the scapula. The acromioclavicular joint is the weaker of the two clavicular joints. Impact can cause a "shoulder separation" when the distal end of

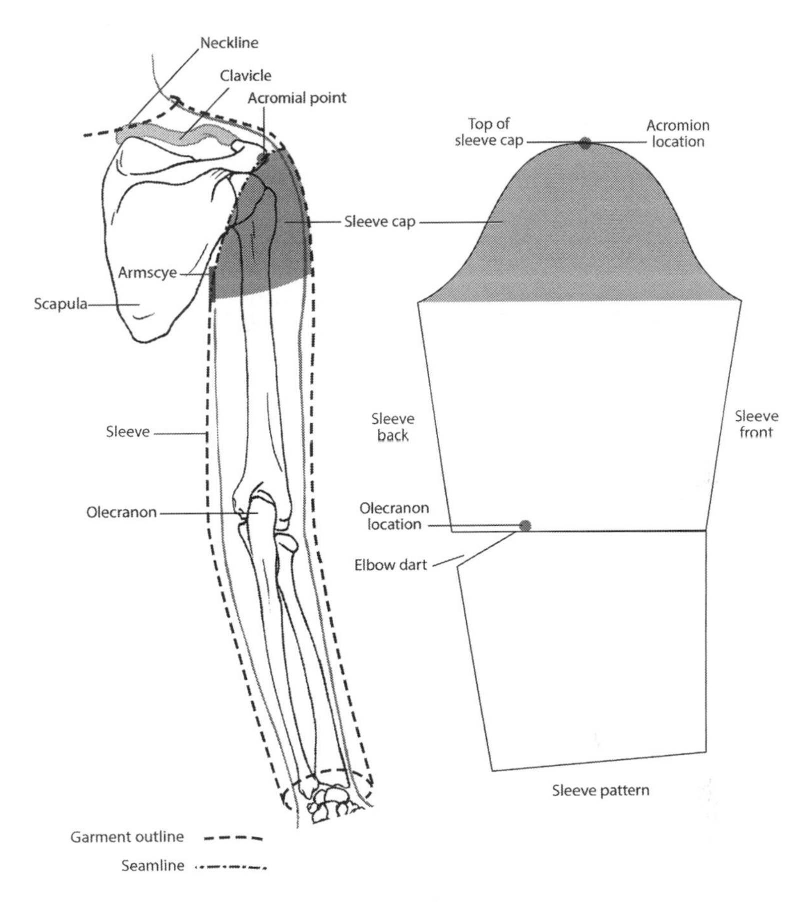


**FIGURE 4.18**
Shoulder and arm skeleton related to garment sleeve and armscye.

the clavicle rises above the acromion—causing a step-like elevation of the collarbone relative to the arm. A blow to the glenohumeral joint (connecting the arm to the scapula), or directly to the acromion can fracture the relatively long and slender clavicle. The most common causes of clavicular fractures are motor vehicle accidents and accidental falls (Postacchini, Gumina, De Santis, & Albo, 2002).

Breaking the clavicle, the one bony link controlling the humerus' and scapula's positions relative to the rib cage, severely limits arm use. Numerous designs have been patented for use after a clavicle injury, from simple *slings* to bandages in a figure-of-eight around both shoulders to complex, multipart arm supports. Simple slings are readily available, easily donned, and the best-tolerated common conservative treatment for clavicular fractures (Andersen, Jensen, & Lauritzen, 1987). A sling made from a triangle bandage must be carefully positioned to avoid additional injury from the sling. If the fastening knot is in the wrong location, pressure can injure the nerve to

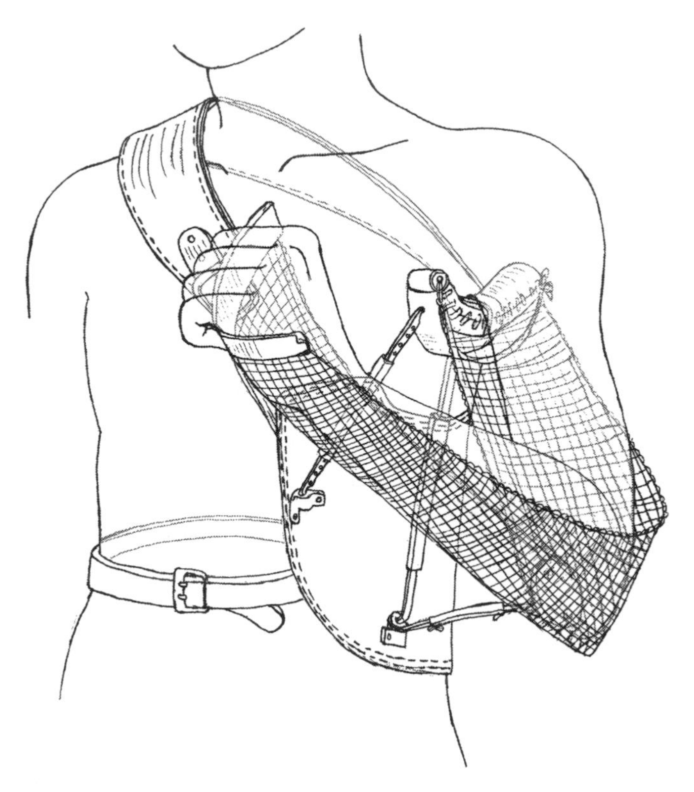

**FIGURE 4.19**
Historical orthopedic shoulder support; Cheatham's Clavicular Apparatus: U.S. Patent No. 890,842, 1908.

the trapezius muscle (Woodruff, 1950). An apparatus patented by Cheatham (Figure 4.19) features many parts including: a trough in front of the body to support the arm, an underarm crutch, a torso plate, strapping across the opposite shoulder and around the waist, as well as an adjustable bar to position the arm trough (U.S. Patent No. 890,842, 1908). Elements of this device have served as inspiration for many orthopedic supports for pectoral girdle and arm injuries. Unfortunately, all of these sling types fall short of effectively replicating healthy anatomy. A sling/shoulder support also complicates donning, doffing, and wearing garments.

### *Scapula*

The other bone of the pectoral girdle, the scapula, is relatively triangular in shape with a prominent ridge, the *spine of the scapula*, palpable from the body surface (refer to Figure 4.17). Several other features are located at the upper lateral angle of the scapula: the acromion, the *coracoid process*, and between these two structures, the *glenoid fossa*. The glenoid fossa is a key structure in the glenohumeral joint, the articulation between the arm and the scapula. The scapula curves to approximate the shape of the rib cage.

The curved scapula contributes to an important feature of the shoulder region—movement of the scapula on the rib cage. Multiple muscles hold the scapula in place on the back and several anatomical structures contribute to scapular motion on the rib cage (thorax). The concave interior surface of the scapula is covered with the *subscapularis* muscle. Another muscle, the *serratus anterior,* attaches to the scapula, and also covers the bones of the rib cage. The *subscapular bursa* is a fluid-filled sac that lies between these two muscle layers. The bursa facilitates smooth movement between the surfaces. This network of bones, muscles, and the bursa is called the *scapulothoracic joint*. Locate the scapulothoracic joint on the superior view of Figure 4.17. Understand that it is not a true joint because it has no bony articulation.

The muscles and adipose tissue layered on shoulder and arm bones of the skeleton can obscure these bony structures. The designer needs to visualize not only the entire shoulder/arm area in three dimensions, but also the intended function of the product—how the arm and shoulder will move with product use. Superimpose the design of the product on these structures to correctly design for, landmark, measure, and fit this area. Consider the task of the hockey equipment designer, the entire structure of shoulder to upper arm to lower arm requires well-designed protective gear that allows movement of the shoulder and arm. Fitting a suit jacket—as complex as that is, looks easy in comparison. Protective pad designs look like lobster shells with overlapping segments of external rigid materials lined with foam padding. The articulating pads provide a hard impact-resistant shell that can move to some degree with the body. Shoulder and arm protectors must also coordinate with gear worn to protect the chest and back of the torso.

### Glenohumeral Joint: Structure and Function

The humerus, the upper bone of the arm, articulates with the scapula at the *scapulohumeral* or glenohumeral joint (refer to Figure 4.17). This articulation joins the saucer-shaped glenoid fossa of the scapula and the hemispheric *head* of the humerus in a synovial ball-and-socket joint. The *glenoid labrum,* a lip of cartilage on the glenoid fossa, deepens the "saucer" a bit. A loose joint capsule surrounds the glenohumeral joint to contain the synovial fluid. Unlike the sternoclavicular and acromioclavicular joints, no ligaments span the glenohumeral joint to stabilize the humerus in the glenoid.

While the ball-and-socket structure of the glenohumeral joint is relatively simple, its function is very complex. Think of how you move your arms in multiple planes. Shoulders flex and extend in a parasagittal plane, so arms can reach in a wide range in front of, overhead, and directly behind the body. Starting with the arms hanging alongside the body, the joint can abduct and adduct the arm in a coronal plane allowing motion out to the side, into an overhead stretch and back down to the side. In addition, the joint can rotate medially and laterally in a transverse plane, allowing lateral reach as well as moving the arm into an across-the-body position. All these motions combined produce circumduction, a circular motion of the humeral head in the glenoid fossa of the scapula.

The glenohumeral joint "has more freedom of movement than any other joint in the body" (Stolzenberg, Siu, & Cruz, 2012). When upper torso muscle weakness alters glenohumeral joint function, the slings described earlier for use after injury do not work well. Designing a sling to compensate for muscle weakness, such as seen after a *stroke* paralyzes an entire side of the body, is even more challenging than supporting the arm after trauma. The degree of joint mobility tends to make the joint unstable when the shoulder muscles are weak.

When supporting/stabilizing muscles are weak the scapula changes position on the rib cage. The scapula rotates tipping the glenoid fossa downward. Then gravity, without the opposition of the muscles, tends to pull the humeral head downward, *subluxing* the humerus out of the glenoid fossa. This incomplete dislocation causes traction on the shoulder joint capsule, limiting not only the normal motion of the joint but frequently causing pain.

Smith and Okamoto provide a useful checklist for features of slings for individuals with such muscle weakness and shoulder joint *subluxation* (1981). The priority is restoring the humeral head position in the glenoid fossa. Other considerations include allowing for normal motions of the shoulder, arm, hand, and fingers; and helping to restore the scapula to a normal position. New slings better address the points on Smith and Okamoto's checklist (U.S. Patent No. 6,945,945 B2, 2005), but the problem of shoulder subluxation and pain after stroke, particularly when the problem is chronic, is not yet resolved.

Functional electrical stimulation (FES) of the pectoral girdle muscles has also been used to try to reduce shoulder joint subluxation and pain in stroke victims (Price & Pandyan, 2001). This type of wearable technology involves applying surface electrodes over muscles which bridge the glenohumeral joint and stimulating the muscles to contract intermittently for several hours a day, for several weeks. Although the joint subluxation tends to improve, at least temporarily, with this treatment, many people find the stimulation too uncomfortable, due to the sensitivity of nerve receptors in the skin, plus individuals and caregivers may have difficulty properly positioning the electrodes.

Applying stretchable "sports tape" to the skin of the arm and shoulder region in specific patterns is another approach to try to reduce the subluxation of the humeral head and shoulder pain after a stroke. This technique has also been used in patients with many kinds of shoulder pain. The mechanism(s) which may make taping work are unclear. Results have been mixed, varying from successfully decreasing pain, to modifying the degree and direction of humeral head subluxation, to simply decreasing pain with active motion immediately after application (Huang et al., 2017; Stolzenberg et al., 2012; Thelen, Dauber, & Stoneman, 2008).

### *The Mobile Shoulder Region: Wearable Product Design Considerations*

Shoulder region motions are important design considerations. The motions of the shoulder, with the attached upper arm, create many changing curves and planes. The curves of the spine, whether static or dynamic, influence both scapular position and scapular motion on the rib cage (Gorton, Young, & Masso, 2012). Scapular motions on the thorax (Figure 4.20) like elevation and depression, as seen with shrugging your shoulders, involve the pectoral girdle. Medial and lateral scapular rotations and/or scapular protraction and retraction on the rib cage accompany most arm motions.

For arm motions—reaching in front or back of the body, reaching overhead to the front or to the side of the body, reaching across the chest or behind the body in a horizontal plane—there is a summation and coordination of joint motions from the sternoclavicular joint, the acromioclavicular joint, the scapulothoracic joint, and the glenohumeral joint. The arm moves relative to the scapula and the scapula moves relative to the rib cage with a relatively consistent relationship, the scapulohumeral rhythm (Ludewig, Cook, & Nawoczenski, 1996).

Beyond the muscles of the scapulothoracic joint, other large, more superficial, muscles attach to the scapula and to the arm to control and direct

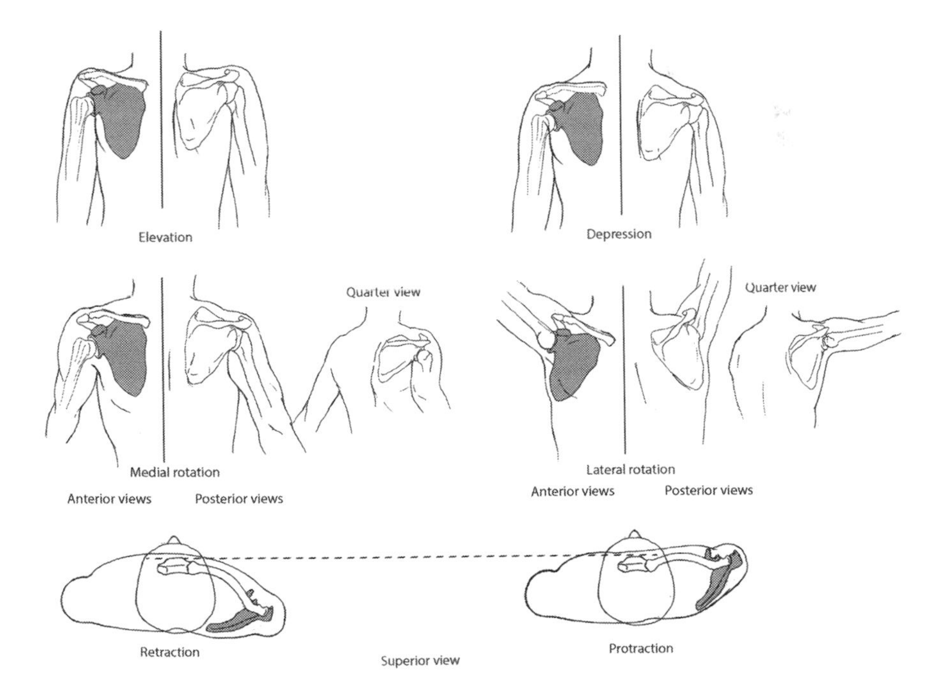


**FIGURE 4.20**
Scapulothoracic motions; anterior and posterior views of elevation/depression, medial/lateral rotation, protraction/retraction, and quarter views showing the scapulohumeral rhythm with medial/lateral rotation.

movement. See Section 4.5 for information on the muscles of this region and Sections 4.14 and 4.15 for discussion of motion.

The designer's challenge is to find a method to quantify the shoulder's changing shape and measurements. As the structures elevate, lower, flex, extend, rotate, and protract/retract the whole area becomes difficult to fit with a wearable product. Sohn and Bye (2014) conducted a pilot study of motion capture as a tool to determine body measurement changes that affect apparel pattern dimensions, focusing on body measurements used to draft a bodice back. Motion capture recorded measurement changes during a series of combined motions that they called "arm rotation." Measurements showed a decrease in shoulder width of 38.45% and increases of back width (16.08%) and back arc at the armpit (27.69%). The bony anatomy of the shoulder region and arm directly affect the fit of upper torso garments with or without sleeves. In addition to the relationship between the *sleeve cap* seam and the acromion (refer to Figure 4.18), the bony skeleton of the shoulder and arm affect fit of the neckline, shoulder seam, armscye, and sleeve of a garment.

### *Arm and Forearm: Structure and Function*

Continuing from the glenohumeral joint toward the hand, the bones of the arm include the humerus, the radius, and the ulna (refer to Figure 4.17). The humerus, the bone of the upper arm, is the longest and largest arm bone, with each end characterized by knobby protuberances. The *greater tuberosity*, a "knob" adjacent to the humeral head, along with the deltoid muscle and fat, forms the rounded contour of the upper arm. The fullness incorporated into a fitted sleeve cap is intended to smoothly shape over this contour transitioning from shoulder to upper arm. Variations in amount of muscle and adipose tissue affect the shape and size of the sleeve cap (refer to Figure 4.18). Two protuberances at the distal (elbow) end of the humerus, the *medial* and *lateral epicondyles*, sit on either side of an indentation, the *olecranon fossa* (refer to Figure 4.17, posterior view).

At its distal end, the humerus articulates with both bones of the forearm—the radius and ulna—at the elbow. The ulna is the longer of the two forearm bones. In contrast to the shoulder joint, the elbow joint is complex in structure (it has three synovial joints which share a single joint capsule), but simpler in function. The three joints are the humeroulnar (between the humerus and ulna), the *humeroradial* (between the humerus and radius), and the *proximal radioulnar* (between the radius and ulna) joints (refer to Figure 4.21). Because of the shape of the humeroulnar joint surfaces, when the arm is in the anatomic position (elbow extended and palm facing forward) the forearm is angled away from the body; in other words, it does not lie in a straight line with the upper arm. Seen in Figure 4.17, this lateral deviation is called the *carrying angle* (Jenkins, 2002, p. 117). Carrying angle is a medical term referring to how the forearm is positioned relative to the arm, not how a product is carried. The

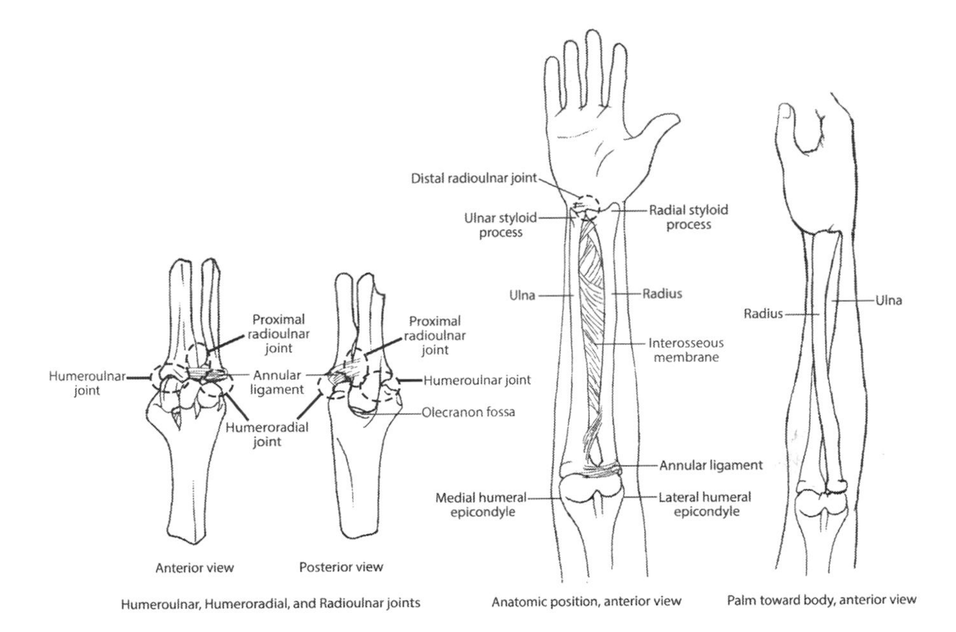


**FIGURE 4.21**
The three-joint elbow joint complex showing the annular ligament; proximal and distal radioulnar joints, forearm bones with interosseous membrane and key landmarks, and bony alignment for natural gesture of the hand.

carrying angle is usually greater than 15 degrees in women, but only 10 to 15 degrees in men (Moore et al., 2011, p. 483). The angle between the arm and forearm is smaller when the arm is in a more natural gesture with the palm facing the body (Jenkins, 2002).

When you straighten your elbow, the *olecranon process* of the ulna fits into the olecranon fossa of the humerus. With your elbow flexed (bent), you can feel the olecranon process, or olecranon, as the tip of your elbow. The olecranon is the landmark used to divide the upper arm and lower arm when drafting a fitted sleeve pattern. A fitted sleeve cannot be a simple cylinder. A dart in the back underarm seam is usually used to shape the sleeve contour, with the point of the dart directed to the olecranon. An elbow dart or darts "fit the natural bend in the arm and allow arm motion" (MacDonald, 2010, p. 39). Elbow pads for athletes or people at risk for falls can protect the olecranon but must allow elbow motion. Elbow pads may also be used to protect an inflamed bursa—a soft tissue swelling ranging in size from a golf ball to a softball—surrounding the olecranon. These pads should be comfortable and accommodate the swollen area.

Elbow joint motion includes flexion/extension and supination/pronation (Figure 4.22). If the elbow joint extends beyond 180 degrees, it is said to hyperextend. The humeroulnar and humeroradial joints act together as a hinge joint to provide flexion/extension motion. Product fit from shoulder to wrist will be affected by the degree of flexion/extension needed at the

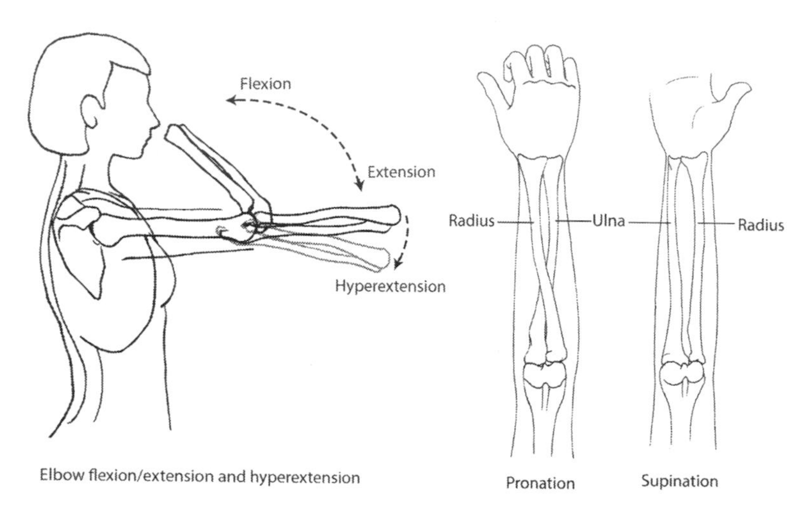


**FIGURE 4.22**
Elbow joint motions: flexion, extension, and hyperextension; forearm pronation, supination.

elbow, and how the arm is positioned relative to the torso. Determine how the wearer will hold the arm(s) at a point in time or the variation through time while wearing the product. Then shape the product accordingly.

A typical arm/forearm position of use for a wearable product, like a tailored jacket, is with the hand held in front of the body near hip level and palm facing posteriorly. Stand with your arm and hand in this position and flex your elbow to note how the angle at your elbow changes. Most tailored suit sleeves are shaped to fit over the flexed elbow. Directions in apparel pattern drafting books for locating a shaping device (dart or curved seam) to fit the elbow vary, however all directions specify flexing the arm to some degree (refer to Figure 4.18). The arm is then measured from the acromial point to the olecranon point to determine the location of the device.

Some products require consideration of the twisting motion of the radius and ulna. Can the product fit the arm closely and rotate with the arm or can the arm rotate within the product, for example move within a loose jacket sleeve? The radius and ulna articulate with each other at the elbow and the wrist in the proximal and *distal radioulnar joints* (refer to Figure 4.21). These two radioulnar joints are pivot joints which help rotate or "radiate" the radius around the ulna in pronation/supination, as in turning a screwdriver. In the anatomic position, the bones are parallel. As the hand turns over from palm up to palm down, the radius and ulna cross. The *interosseous membrane*, a sheet of tough connective tissue between the bones, helps to hold the forearm bones stable as they move. The natural gesture of the hand, with the palm turned toward the body, is a neutral pronation/supination position (Figure 4.21, right side).

Numerous ligaments cross the elbow joints. The *annular ligament* is a key part of the proximal radioulnar joint. Children, ages one to five, are vulnerable to dislocation of the proximal radioulnar joint, at the elbow (Choung & Heinrich, 1995). The head of the radius can be pulled out of the annular ligament if the elbow is extended and traction is applied to the joint, a condition called *nursemaid's elbow* or pulled elbow. Think of the adult who is pulling along a lagging child. A better solution to keep track of and/or manage a young child might be a harness with a lead.

The prominent bony protrusion on the little finger side of the distal forearm, the *ulnar styloid process* (Figure 4.21), is a landmark point. It is used to define the hem length or cuff location of a long-sleeved garment. To draft a long-sleeve pattern for women have the person stand with elbow slightly bent, place the 0-point of the tape measure at the point of the acromion, run the tape over the elbow point (olecranon), and measure to the ulnar styloid. If the sleeve includes a fitting dart at the elbow, the measurement is taken in two steps; from the acromion to the olecranon and from the olecranon to the styloid process of the ulna. For a man's shirt sleeve length, the measurement typically begins at the cervicale (the tip of the spine of the C7 vertebra) and tracks over the bent elbow, to the ulnar styloid.

The *radial styloid process* may also be used as a landmark spot. When designing products that encircle the wrist determine how and where the product will be worn and what activities need to be accommodated. Then consistently measure that area—below or incorporating the styloid processes of the ulna and radius—while considering any necessary movements of the radius and ulna during product use.

In addition to articulating with the ulna in the distal radioulnar joint, the combination of the radius/distal radioulnar joint articulates with the three proximal (of the eight) carpal bones to form the radiocarpal joint, "the" wrist joint (Jenkins, 2002, p. 181). Numerous ligaments bridge from the radius and ulna to the proximal carpal bones. The radiocarpal joint between the forearm and the proximal carpals is complex in both structure and function. As at the elbow, "the" wrist (radiocarpal) joint moves in flexion/extension and pronation/supination. The hand also deviates medially and laterally on the forearm at "the" wrist. See Chapter 7 for specifics on the more intricate distal upper limb: the wrist, hand, and fingers.

## 4.5 Upper Torso and Arm Muscles: Structure and Function

Recall from Chapter 2 that muscles attach to bones and are activated by nerves to produce movement. Muscles layer onto the upper torso contributing to an individual's contours, form, and size. Like the skeletal structures, the torso's muscular system is symmetrical, with paired R and L muscles. Upper torso muscles facilitate stability and motion throughout the many joints of the spine, rib cage, pectoral girdle, and arm. Upper torso muscles

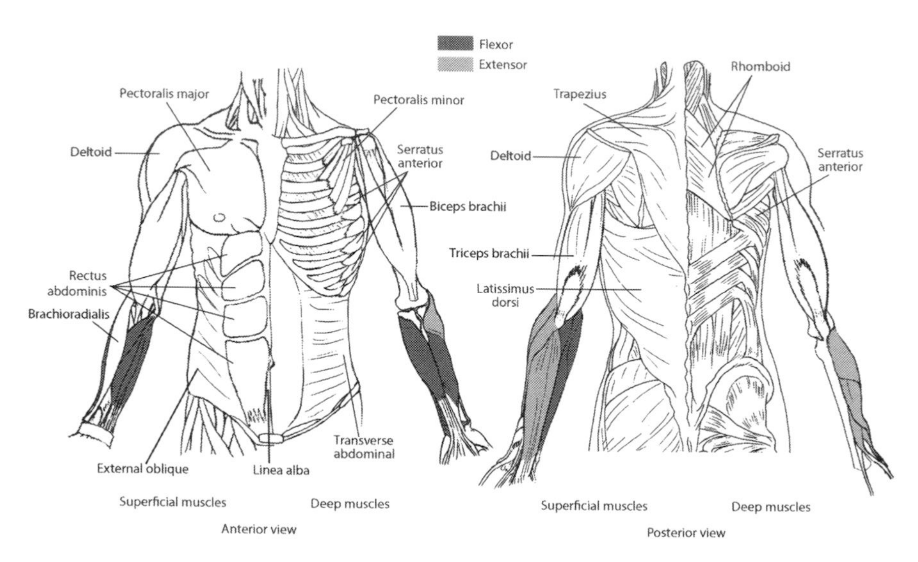


**FIGURE 4.23**
Torso and arm muscles—superficial and deep.

include both superficial muscles and many layers deep within the torso (Figure 4.23). Muscles acting on the arm and shoulder are probably of most interest to designers. They create body shape changes with activity, affecting wearable product fit and function. Basic descriptions of the more design-relevant upper torso and arm muscles follow.

To meet shoulder region fit challenges, try to relate the wearables to the underlying muscle groupings. For example, football shoulder pads cover muscles of the shoulder area, front and back; and must shape to, fit, and move with those muscles. Most upper torso wearable products suspend from the shoulders, and good fit over the muscles is necessary. Next, think of muscles surrounding the cylinder of the torso, a kind of natural girdle extending from the lower thorax through the abdomen and lower back. Although upper arm muscles work with shoulder muscles to act on the glenohumeral joint, they are not major contributors to shoulder motion or shoulder bulk. They primarily control the actions of the elbow and will be discussed with other arm muscles in Section 4.5.3. As with the muscles at the shoulder, designers have to consider the bulk of the arm muscles as they fit and size sleeves below the sleeve cap.

### 4.5.1 Shoulder Area Muscles

Grasp your shoulder near the glenohumeral joint with your hand curving over the top of the joint, and look at the muscles illustrated in Figure 4.23. Move your arm and shoulder to feel the muscles work on the front and back of your body. The deltoid muscle bridges between the shoulder area and the upper arm. It forms a smooth bowl-shape that originates from the edge of the acromion, the spine of the scapula, and the lateral clavicle. When designing an upper torso product, consider the size, shape, and contour of the deltoid along with the skeletal structure underneath. Large deltoid muscles often require a broader and higher sleeve cap to accommodate the muscle. A larger deltoid may also require a greater than standard upper sleeve circumference for a comfortable fit.

The major superficial upper back muscle, the trapezius, arises from the base of the skull and spinous processes in the neck and thorax. It inserts on the spine of the scapula, the clavicle, and the acromion. This large muscle shrugs the shoulders—up and down—and rotates the scapula on the rib cage. The other large superficial back muscle, the *latissimus dorsi,* originates from spinous processes in the thoracic spine, lumbar spine, and the sacrum plus the *iliac crest* of the pelvis. It inserts on the humerus, forming the posterior wall of the axilla. It extends, adducts, and rotates the humerus in the glenohumeral joint and raises the body toward the arms, for example during rock climbing activities (Moore et al., 2011, p. 425). Because of their many actions, these muscles change shape frequently.

Some deep muscles in the upper back interact to rotate the arm in the glenohumeral joint. The tendons of these muscles form the *rotator cuff,* which is vulnerable to wear and tear injuries because it lies between the acromion and the humeral head. Figure 4.19 illustrates a historical, but typical, orthopedic device worn to stabilize the shoulder after rotator cuff surgery. Shaping this type of product to stabilize or allow motion, while keeping the product in place where the shoulder and arm intersect, is difficult.

### 4.5.2 Upper Torso Muscles below the Shoulders, Anterior and Posterior

Muscles on the front of the body help to stabilize the axial skeleton, contribute to an erect posture, and deliver power and control for motions of the arms. The large pectoralis major muscle originates at the outer edge of the sternum and covers the upper thorax just below the clavicle. It forms the anterior wall of the axilla as it inserts on the humerus. It protracts the shoulder and adducts and internally rotates the arm in the glenohumeral joint. The pectoralis needs special design consideration. Body builders work to define a matched set of "pecs," which are often thought of as a male characteristic. Women, of course, have "pecs," but the pectoralis major muscle underlies the breast and is less evident in females, even when it is well developed. Fitting a tailored jacket to a body builder's form requires contour seaming with placement to shape from the convex curve of the chest muscles through the armscye. A constricting armscye or backpack straps that cross the pectoralis major at the axilla may restrict motion and be uncomfortable.

The major muscles below the pecs are the segmented rectus abdominis muscles, the abdominal "six-pack" (refer to Figure 4.23). These muscles span from the upper torso to the lower torso. They generate abdominal compression. The upper torso muscles at the back of the body play an important role in supporting and moving the upper torso. The *paraspinal* muscles run along the spine, stabilizing the axial skeleton and supporting upright posture. Layers of these muscles bridge between the vertebrae. They connect the spinous processes and transverse processes of the vertebrae, and also connect to the posterior ribs, the back of the head, and to the sacrum. The deepest cover only a few vertebral segments; the most superficial cover many. You can feel these muscles in bundles along both sides of the spinous processes as they work together. These muscles assist the skeleton in bending and twisting. The rectus abdominis muscles also support the lumbar spine along with the back muscles to help maintain upright posture.

The abdominal wall muscles cover the sides of the body. They extend into the front of the abdominal wall, beneath the rectus abdominis in the lower torso. Three layers of muscles make up the abdominal wall: (1) the deep *transverse abdominal,* (2) the intermediate *internal oblique,* and (3) the superficial *external oblique.* The transverse abdominal originates from multiple spine, pelvic, and rib structures and inserts into the anterior pelvic bones and the *linea alba* at the midline of the abdomen. The transverse abdominal muscle fibers run horizontally, so when the muscle contracts/shortens, it supports and squeezes the abdominal contents. The internal oblique muscles originate on the sides and back of the pelvic structures and insert on the front lower ribs. The external oblique muscles originate on the back lower ribs and insert on the linea alba and the iliac crest of the pelvis. Together, the abdominal oblique muscles flex and rotate the trunk, and support the abdominal contents. A designer might use the size and placement of these muscles for aesthetic inspiration when designing leotards for gymnasts.

These abdominal wall muscles provide support and stability to the core of the body. With the respiratory diaphragm and the pelvic floor (Chapter 5), they complete a cylinder of support. With products worn around the torso—braces or casts after injury, corsets or waist cinchers for aesthetics—the muscles lose strength quickly because these sheets of muscle have little natural bulk. They present an excellent example of "use it or lose it."

### 4.5.3 Muscles in the Arm

Muscles of the arm work with upper torso muscles to move the arm at the glenohumeral joint. Muscles in the arm also generate elbow movement, when they act on the bones of the forearm. Most of the muscles in the forearm act on the joints of the wrist, hand, and fingers. Like other regions, the upper arm and forearm have deep, internal, muscles and more visible, easy to palpate, superficial muscles. Deeper muscles are generally smaller, and they change less in volume with contraction and relaxation than the superficial muscles.

Muscles in each segment of the limb attach to and act on the joints at each end of the segment. The *biceps brachii* and *triceps brachii* are major muscles of the upper arm which flex and extend the shoulder and elbow with opposing lengthening and shortening muscle contractions (refer to Figure 2.7 and Section 2.2.3). The brachioradialis, in the proximal forearm, helps to

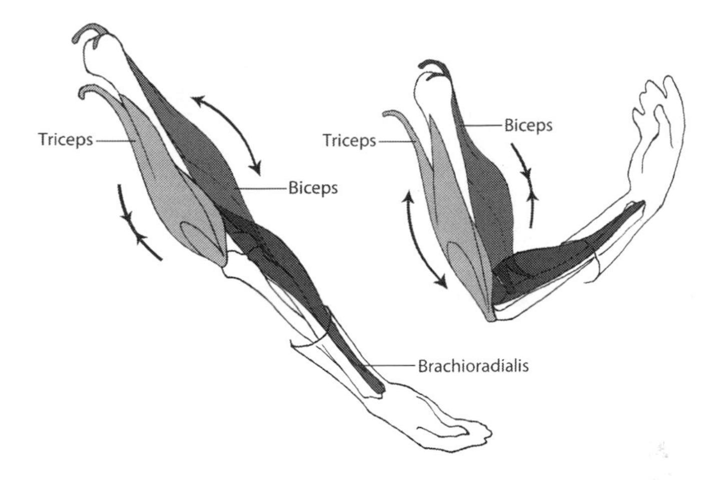


**FIGURE 4.24**
Elbow flexion and extension showing lengthening and shortening contractions of the biceps and triceps muscles and the location and action of brachioradialis.

flex the elbow. See these three muscles acting at the elbow in Figure 4.24. You are showing off these muscles when you flex your elbow in an exaggerated "strong man" pose.

Some relatively small muscles bridging the elbow help pronate and supinate the forearm. Other muscles which flex and extend the wrist, hand, and fingers are located together in the forearm. The muscles are grouped in layers below the elbow, as seen in the anatomic position of the forearm shown in the views of Figure 4.23. Groups of muscles in the medial volar forearm, the "flexor group," provide flexion actions for the radiocarpal joint, the bones of the hand, and the fingers. The muscle groups which act to extend the wrist, hand, and fingers are positioned on the dorsal lateral forearm, the "extensor group." Long tendons extend from the muscle bellies in the upper and middle forearm to the bones adjacent to the small joints of the fingers from both the flexor and extensor muscle groups.

To demonstrate how muscles of the upper arm and forearm change shape and bulk with contraction, try these activities. Rest your L hand on the middle of your R upper arm as you contract and relax the muscles. Feel the change in shape and bulk of the upper arm. Do the same with your forearm. Place your L hand on the muscles just below the elbow. Try these motions with your hand under your forearm and then on top of the forearm. Spread your R fingers out and then close them into a fist. Move your R hand up/down and around relative to your wrist. Pay attention to the muscle contractions and volume changes in the muscles on the top and bottom of your forearm. Consider how the changes in muscle bulk might affect a product worn over the upper arm or the forearm.

Decreased muscle strength in the arms creates significant disability and dependence in basic daily activities. Individuals with severe arm and/or hand weakness may not be able to feed themselves, complete basic grooming activities, draw, or keyboard. For children with *muscular dystrophy* (which causes significant muscle weakness), a wearable exoskeletal device, the WREX, offers life-changing help for them and their caregivers. The device consists of a two link, four degrees of freedom arm brace, controlled with linear elastic bands. It mounts to the child's wheelchair and provides enough assistance so that the child can move his/her arm against the force of gravity and perform basic tasks independently (Rahman, Sample, & Seliktar, 2004).

## 4.6 Nervous System in the Upper Torso and Arms

The skeletal elements and muscles of the torso and arms give form to the upper body and influence product shape. The nervous system helps make upper torso structures work, with and without a person's conscious effort. The upper torso contains elements of both the central and peripheral nervous systems, while the arms have peripheral nerves only. These central (CNS) and peripheral nervous system (PNS) structures serve both somatic (sensory and motor) and autonomic (*sympathetic* and *parasympathetic*) nervous system functions (review Chapter 2, Section 2.3).

The nervous system structures branch and subdivide in the body, but in ways that differ from the circular pattern of the circulatory system and the linear, in-out, pattern of the respiratory system. A nerve plexus is one way the nervous system directs different kinds of nerve fibers to the wide range of tissues they need to reach. Recall from Chapter 2 that a plexus is a complex of interlaced nerve fibers that looks somewhat like a railroad switching yard with multiple junctions. After leaving a nerve plexus, nerve fibers from several different spinal nerves usually contribute to the motor nerves for individual muscles and the cutaneous nerves for specific skin regions. This redundancy, an important organizing feature of the nervous system, helps assure function even if a section of the spinal cord or one or two spinal nerves are injured.

The parasympathetic nervous system has a few elements in the upper torso, but most of its structures are found in the lower torso, as well as in the head and neck. The upper torso is home to most of the sympathetic nervous system. The sympathetic nervous system affects many organ systems. It speeds up the heart. It makes hairs stand on end and increases sweat output. It tightens the sphincter muscles of the digestive and urinary systems, allowing them to fill with waste. Through these actions, the sympathetic nervous system prepares the body for "Fright, Flight, Fight, and Fill" (Palastanga, Field, & Soames, 2002, p. 654).

More than in the head and neck, wearable products can interfere with upper torso and arm nervous system function. Peripheral nerves connecting to the region's musculoskeletal structures, skin, and organs in the torso, are vulnerable to compression and injury. The spinal cord, the CNS structure connecting the brain to the rest of the body, and the sympathetic nervous system elements in the torso are less vulnerable than the peripheral nerves in the torso.

### 4.6.1 Nervous System Torso Components: Structure and Function

The spinal cord is a cylinder of nervous tissue extending from the brain stem through the entire cervical and thoracic spinal canal approximately to the level of the first lumbar vertebra. It is surrounded by cerebrospinal fluid (CSF). Like a fiberoptic cable with relay centers, it incorporates nerve cells and bundles of nerve fibers. Because the thoracic spine is less mobile than the neck, thoracic spinal cord injury (SCI) is less likely than cervical SCI.

When nerve fibers leave the skull or the spine, they become part of the PNS. Peripheral nerves in the upper torso and arms arise from nerve cells in either the brainstem or in the cervical and thoracic spinal cord. Two peripheral nerves come from the brainstem and influence structures in the thorax. The *spinal accessory nerve* controls the trapezius muscle. The vagus nerve (carrying parasympathetic nerve fibers) exits the skull and passes through the neck to reach and control the organs of the thorax and abdomen. Vagal nerve impulses slow the heart, stimulate smooth muscle contraction in the digestive and urinary system, and help relax sphincter muscles to allow the digestive and urinary systems to empty.

The spinal nerves from the upper neck to the tip of the spine contribute a variety of nerve fibers to peripheral nerves. Figure 4.25 illustrates a typical spinal nerve with labels for nerve structures. Spinal nerves have two *spinal nerve roots.* The spinal nerve roots are bundles of nerve fibers from the motor and sensory divisions of the central nervous system. A *spinal nerve ganglion,* an enlargement along the more posterior nerve root, contains sensory nerve cells, and lies in the opening between the vertebrae. The spinal nerve roots merge and then exit the spinal column between the vertebrae as a spinal nerve. The sensory nerve cells in the spinal ganglia communicate with sensory nerve fibers in the spinal cord and the sensory receptors throughout the body. Like the site for vagus nerve stimulation in the neck, the spinal ganglia offer possible locations for stimulating sensory nerves or blocking pain signals. *Transcutaneous electrical nerve stimulator* (TENS) units are devices used to treat painful conditions. TENS pads are sometimes placed over spinal ganglion sites, especially in the low back. Many persons report satisfaction with use of TENS for pain treatment; however controlled research trials have been unable to confirm that the product provides pain relief (Jones & Johnson, 2009).

Eight pairs of spinal nerves leave the spine in the neck; one pair above the first cervical vertebra and a pair below each of the seven cervical vertebrae, including the first. Spinal nerve numbering and vertebral numbering are therefore different in the neck. The numbers of nerves and vertebrae

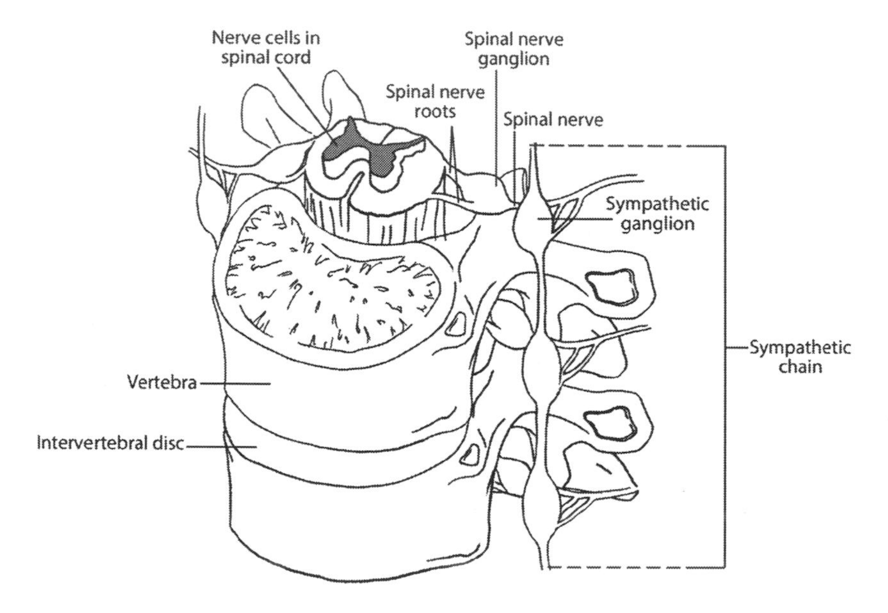


**FIGURE 4.25**
Schematic of central-peripheral nervous system junction: vertebra, spinal cord, spinal nerve roots, spinal nerve ganglion, spinal nerve, sympathetic ganglion, and sympathetic chain.

are equal in the rest of the spine, so this is not a problem below the neck. For most design use, it is sufficient to discuss the spinal nerves by their position relative to the numbered vertebrae, e.g., "the spinal nerve below (caudal to) the fifth cervical vertebra," or "below the third thoracic vertebra." If you are working with collaborators who discuss the cervical spinal nerves by number, take time to learn the nerve naming conventions.

Fibers from the first four cervical spinal nerves come together in the *cervical plexus*. They branch and subdivide into peripheral nerves with different names as they leave the plexus. One such nerve is the *phrenic* nerve, which is essential to breathing. Fibers from the cervical plexus, plus fibers from the spinal nerve caudal to the fourth cervical vertebra, form paired phrenic nerves. This helps assure at least a degree of continued respiratory function if one of the structures forming the phrenic nerves—the cervical spinal cord, spinal nerve root, spinal nerve, or even one phrenic nerve—is injured. The phrenic nerves pass through the lower neck and the thorax to connect to and innervate the diaphragm. Early research demonstrated that the phrenic nerves can be stimulated with a wearable external radiofrequency transmitter and an implanted receiver in the neck, to assist breathing in some cases of SCI (Glenn, Holcomb, Gee, & Rath, 1970).

Three kinds of specialized sympathetic nervous system structures are located at variable distances from the spine. These structures, which carry out specific transmission activities, include: (a) *sympathetic ganglia*—clusters of sympathetic nervous system cells, (b) connected sympathetic ganglia, the *sympathetic chain,* also called the *sympathetic trunk* (refer to Figure 4.25), and (c) sympathetic nerve plexuses which connect the internal organs to the CNS. Just as they carry somatic (sensory and motor) nerve fibers, the thoracic spinal nerve roots and the pairs of thoracic spinal nerves they form (plus the first two lumbar spinal nerve roots/nerves) carry sympathetic nervous system fibers between the spinal cord and the sympathetic ganglia. From the sympathetic ganglia, lying close to the vertebral bodies, the nerve fibers connect to components farther out in the body (other sympathetic nervous system cells, blood vessels, sweat glands, and internal organs). The sympathetic chain, like the thoracic spinal cord, is protected from all but extreme trauma. Many implantable devices have been developed to stimulate or block nerve signal transmission at sympathetic nervous system structures (Stanton-Hicks & Salamon, 1997). Non invasive wearable options may also be possible.

Understanding the location and structure of the *brachial plexus* is important when designing upper torso wearables (Figure 4.26). Spinal nerve fibers—motor, sensory, and autonomic nervous system fibers—from cervical spinal nerves 5 through 8, plus the first thoracic spinal nerve form the brachial plexus. They regroup into different fiber combinations as they move through the brachial plexus. As with the phrenic nerves, this fiber regrouping likely helps diminish localized injury effects. Leaving the plexus, the fibers branch into 16 peripheral nerves which serve the shoulder region, the arm (brachium), and hand. The brachial plexus is vulnerable to compression, and injury, between the clavicle and the first rib. Carrying loads in packs with straps across the trapezius muscle, above the clavicle, puts it at risk. Distributing weight in a vest or providing a waist belt to shift weight from the shoulder can help to decrease focal pressure on the brachial plexus.

Twelve pairs of spinal nerves leave the spinal cord in the thoracic spine. The first of the thoracic spinal nerves joins the brachial plexus as noted above. The remainder connect to muscles of the back, rib cage, and abdominal wall. They also collect sensation from the skin of the thorax, posterior torso above the waist, and abdomen.

Consider the bigger picture and how upper torso and arm wearable products can assist or inhibit nervous system functions. Central nervous system/spinal cord nerve fiber bundles and nerve cells work together to transmit motor impulses from the brain to the body and sensory impulses from the body to the brain. The peripheral motor nerves carry signals to muscles in the torso—the involuntary muscle of the heart, digestive organs, and other automatically functioning organs—as well as voluntary muscles helping to move the torso and limbs.

Cutaneous and deep tissue sensory receptors carry signals from the body back to the brain. Receptors in the torso can signal the brain that a wearable is too tight or is irritating the skin surface. While the skin on the torso is generally not as sensitive to stimuli as the skin on the palms and fingers, all torso areas have sensation. Some areas of the upper torso are more sensitive

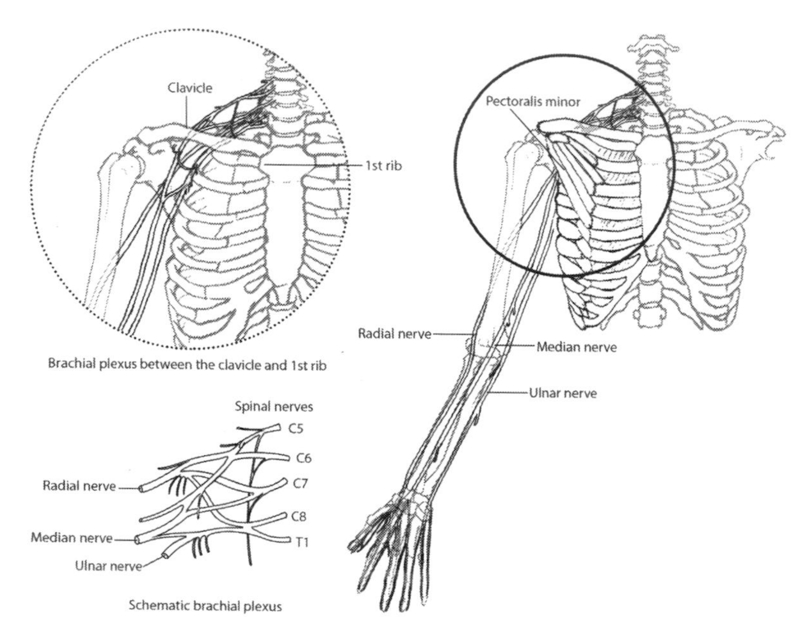


**FIGURE 4.26**
Brachial plexus schematic and anterior views at shoulder, showing interconnections of nerve fibers arising from the cervical spinal cord and ending as named peripheral nerves leading to the shoulder region and arm, and the relationship to clavicle; major peripheral nerves in the arm.

than others. The skin of the chest and breast, along with the belly, which has less bony protection, is more sensitive than the upper back. In the peripheral nervous system, an area of skin innervated by a single spinal nerve is called a dermatome. Thoracic spinal nerves (2–12) supply the upper torso dermatomes. If a spinal nerve or sensory nerves in an area of skin are damaged, nerves from adjacent dermatomes and/or cutaneous sensory nerves branch and grow into the damaged area to help restore sensation.

The central nervous system also has a way of dealing with loss of nerve function in the peripheral nervous system. It has the capacity to adapt and use alternate connections to restore lost functions, a property called *neural plasticity*. For example, a blind person can learn to read using Braille text. If the retina of the eye is damaged, the sense of touch substitutes for some eye functions. Machine-human interfaces have been developed to use central nervous system plasticity and sensory substitution for a variety of sensory deficits (Bach-y-Rita & Kercel, 2003). Collins and Bach-y-Rita (1973, pp. 296–298) developed wearable technology using mechanical or electrical stimulation on the skin and torso wall to help visually impaired individuals "see." Duvall, Dunne, Schleif, and Holschuh (2016) addressed problems related to autism spectrum disorders and attention deficit hyperactivity disorder. They developed an active hugging vest, a wearable product producing mechanical stimulation of the torso with a remote-control feature, to provide calming deep touch pressure.

A threat or dangerous situation triggers sympathetic nervous system activity. As the heart speeds up in response, blood diverts to the muscles and *sudomotor* nerves (sympathetic nerves to the sweat glands) cause a person to break out in a sweat. They stimulate both the eccrine sweat glands in the skin of the back and chest (and the arms) and the apocrine sweat glands in the arm pit (axilla), helping to regulate body temperature. Smith and Havenith (2012) mapped sweat patterns of female and male athletes finding some differences and suggesting that sweat mapping has important applications for sex-specific clothing design.

Severe spinal cord damage anywhere in the thoracic region produces *paraplegia*, paralysis of the torso and legs (loss of muscle control), plus loss of sensation. Thoracic level spinal cord injury also impairs the sympathetic and parasympathetic nervous system components of the autonomic nervous system including bladder, bowel, and sexual/reproductive functions, further discussed in Chapter 5, Lower Torso and Legs.

### 4.6.2 Peripheral Nerves in the Arm: Structure and Function

Peripheral nerves of the arm (Figure 4.26) emerge from the brachial plexus mixing motor, sensory, and sympathetic nervous system fibers. The motor fibers connect to all the upper limb muscles facilitating muscle contraction and motion. The radial, ulnar, and median nerves, the major peripheral nerves of the arm, and the locations of the muscles they innervate are shown in a schematic view in Figure 4.27. Peripheral nerves reach sensory receptors in deep tissues including muscles, joints, blood vessels, and other deep structures. They provide sensory feedback about muscle activity and information about body position and movement, allowing us to complete tasks requiring muscle power or fine control.

Sensory nerves also collect sensory information from receptors in distinct areas of the skin. Individual cutaneous nerves serve the illustrated shaded skin regions. The cutaneous nerves of the hand are most important. Impairment of sensation in the superficial radial, ulnar, or median sensory nerves changes our abilities to use and manipulate objects. These sensory areas are markedly different from the dermatomal distributions on the upper limb. Knowledge of the sensory nerve receptor areas has more utility in direct application to wearable products than placement of a product related to dermatomes.

Of the many peripheral nerves in the arm and shoulder region, only three are vulnerable to significant compression by wearable products: (a) the *radial nerve* as it spirals around the humerus, (b) the *ulnar nerve* near the olecranon (at the "crazy bone"), and (c) the median nerve on the anterior forearm, just below the elbow crease. These three nerves also run the full distance of the arm; two of them extend to innervate the hand muscles. If

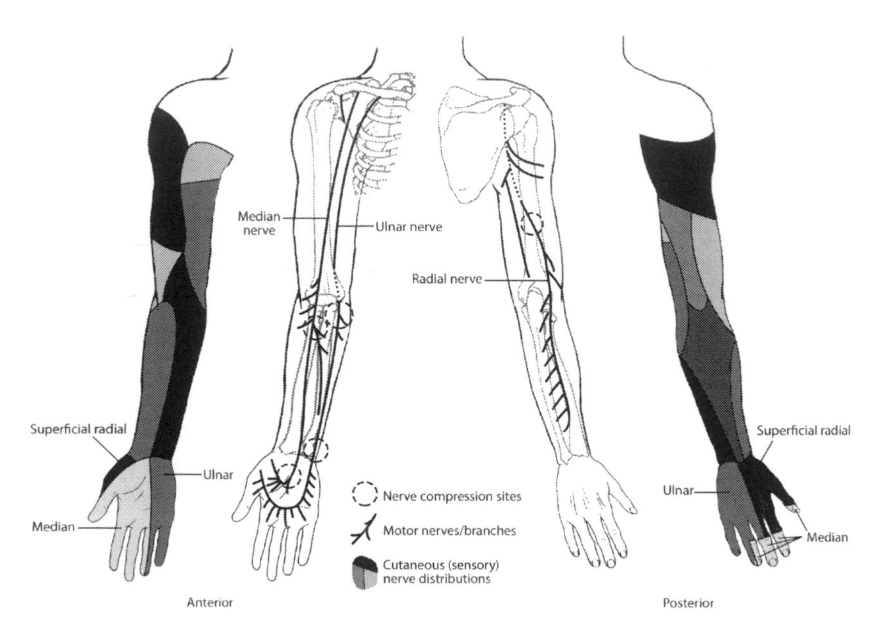


**FIGURE 4.27**
Major peripheral nerves in the arm, showing (1) schematic of motor nerve branches to individual limb muscles, (2) cutaneous (sensory) nerve distributions in the arm and hand, and (3) nerve sites vulnerable to compression.

wearable products cause muscle weakness, pain, numbness, or tingling in the arm or hand, that suggests nerve compromise. Wear test products that may apply pressure. If a product constricts the arm at any of these vulnerable sites, rework the design. Familiarize yourself with the cutaneous nerve distributions of the superficial radial, ulnar, and median nerves and understand that sensory change in the hand can signal an arm or forearm nerve compression problem.

## 4.7 Connecting Structures in the Thorax

The nervous system structures discussed above, while crucial, take up very little space. Circulatory, respiratory, and musculoskeletal structures occupy almost the entire volume of the thorax. Parts of two other body systems important to designers are also located in this region above the diaphragm. The esophagus, the hollow digestive system structure linking the mouth and the stomach (in the abdomen), and the *thoracic duct* and the *right lymphatic duct*, the final common pathways of the lymphatic system, pass through the thorax.

### 4.7.1 Esophagus

The esophagus is a tubular structure with smooth muscle in its wall. Its primary function is transporting food and liquid to the stomach. *Peristalsis,* intermittent smooth muscle contractions and relaxations, propel small portions of material through this hollow structure, somewhat like squeezing toothpaste from a tube. It is well protected in its central position (refer to Figure 4.1). The *esophageal hiatus* is a relatively small opening in the diaphragm, allowing the esophagus to travel from the upper body into the stomach. Because the esophagus is hollow and connected to the stomach, with no physical structure to prevent backflow, stomach contents can also be pushed upward into the distal esophagus. *Gastroesophageal reflux* is the medical term for this phenomenon—heartburn is the common term.

### 4.7.2 Thoracic Duct and Right Lymphatic Duct

The thoracic and right lymphatic ducts are also tubular structures passing through the thorax. The thoracic duct is the largest lymphatic vessel. It collects products of fat digestion in the abdomen as well as lymph from three-quarters of the body (Phang, Bowman, Phillips, & Windsor, 2014). It has numerous valves, to prevent backflow, and lies in front of the lower thoracic vertebral bodies. Like the esophagus, the thoracic duct passes through a hiatus in the diaphragm. The right lymphatic duct is the common drainage vessel for the R upper quarter of the body. It is located in the R upper thorax. The thoracic duct and R lymphatic duct empty into the "crotch" between two large veins in the upper L and upper R thorax—like putting a third pipe into a Y-junction between two other pipes. These lymphatic structures are well-protected within the body. To be effective, compression garments need to direct lymphedema fluid back to one of these two ducts.

## 4.8 Upper Abdomen: Digestive and Urinary Systems

Unlike the heart and lungs, which are fully contained within the bony rib cage above the diaphragm, the organs within the abdominal cavity are surrounded mostly by the muscular body wall and deep adipose tissue (DAT). Muscles and fat offer less protection than the rib cage but do accommodate the important normal mobility of the abdominal organs. The diaphragm is a muscular and movable roof to the cavity and the upper lateral abdomen is adjacent to the lower ribs. The upper abdominal organs include: the hollow stomach and gall bladder; the solid kidneys, liver, pancreas, and spleen (Figure 4.28). The consistency of the solid organs varies. Some, like the liver and pancreas, are relatively soft; the kidneys are firmer. Some organs are readily visible in illustration Figure 4.28 (anterior view) while some, hidden

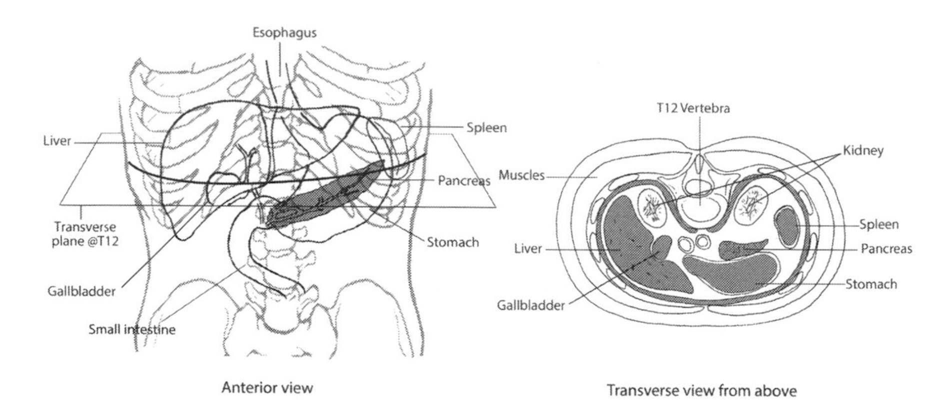


**FIGURE 4.28**
Solid internal organs in upper abdomen and stomach, transverse section at T12 and anterior view.

deeper in the body, are evident on the illustration's transverse section. The general horizontal relationships of the spine, internal organs, and the body wall with ribs and muscles wrapping around the sides of the body are seen in the transverse view. The gall bladder is embedded in the liver.

A relatively horizontal portion of the large intestine, segments of the small intestine, and parts of the supporting structure for the intestine, the *mesentery,* are also found near the lower ribs. The mesentery is a continuous connective tissue structure (Coffey & O'Leary, 2016). Blood vessels, nerves, and lymphatic vessels travel through the mesentery to the hollow small and large intestines. Large segments of small and large intestine and mesentery are also found in the mid torso and lower torso. The rectum, the most distal portion of the large intestine, fits almost entirely within the bony pelvis. There are 8.2 m (27 ft) of small intestine and 1 m (3 ft) of large intestine packed into the abdomen. Figure 4.29 shows the relationships of the intestines to the other abdominal organs and the exterior of the body.

If the ribs near the upper abdomen break, they can be a hazard to the nearby solid abdominal organs, the liver and spleen. Punctures and tears in these organs cause severe internal bleeding. The lumbar spine, posteriorly, provides a vertical support for the body above the pelvis, but not a great deal of protection for the abdominal contents. These vulnerabilities need to be considered when designing protective equipment, whether for military use or contact sports. With only the muscular body wall surrounding most of the abdomen, the digestive organs are vulnerable to penetrating injuries. If the large intestine is punctured, bacteria are released into the confined space of the abdomen, where they can cause life-threatening infections. Adequate torso coverage is essential for body armor designs to protect the vulnerable abdomen.

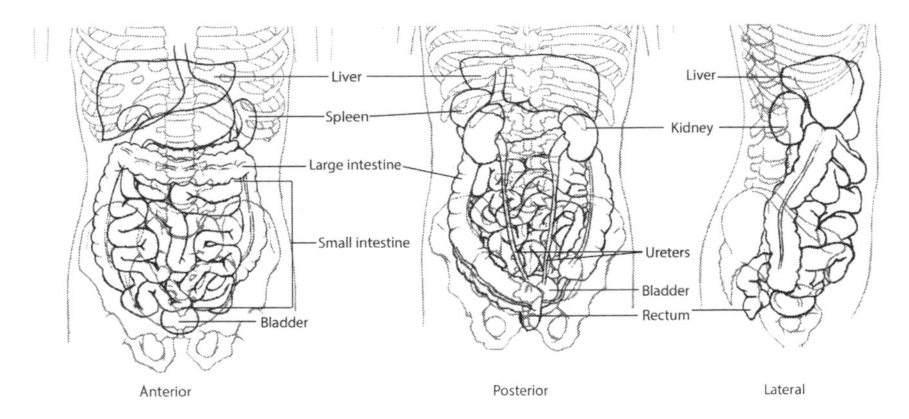


**FIGURE 4.29**
Abdominal organs: anterior, posterior, and lateral views.

### 4.8.1 Movable and Shapeable Abdominal Organs

Standard anatomy texts and anatomical models divide the abdomen into four quadrants with dividing lines through the umbilicus: R/L and upper/lower. This may give the impression that abdominal organs are solid and stationery, but that isn't the case. The abdominal organs below the diaphragm are variably pliable and can shift in surprising ways. "The whole abdomen behaves like a rubber bag, the contents of which can be displaced into any part by the squeezing of the muscles or in other more expressive words 'like a jumble of slithery things in a hole' (H. B. Fell)" (Barclay, 1932, p. 267). Barclay studied the stomach's position changes within the abdomen under several circumstances. He found that maximum exhalation increases intra-abdominal pressure elevating the stomach, laughing displaces the stomach, and stomach position changes when a person lying *supine* (horizontal, face-up) voluntarily pulls the belly in or pushes it out. Barclay also reports the large intestine seems as "acrobatic" as the stomach (p. 269).

Gravity also acts on internal organs, shifting them in various directions when a person changes positions, affecting external body shape and form. You can feel this shift in your abdomen. While standing, place one hand on your abdomen, then drop to a crawling position and feel the change in your abdomen. Anatomy text information is primarily based on studies of people or bodies in supine positions. Wearable product designers typically design for people in standing (vertical) postures—static or dynamic. Therefore, designers have to accommodate both the movable nature of the internal organs and the effects of gravity on the body. Keep the fluid nature of body form in mind and consider how these changes might affect product shape, fit, and position.

### 4.8.2 Digestive System in the Abdomen: Structure and Function

Digestion occurs primarily in the confined cylindrical space of the upper and lower torso. The principal digestive tract structures in the abdomen are the stomach, small intestine, and large intestine. Glandular abdominal organs supply chemicals to aid digestion. The primary function of the digestive system is to supply nutrition and fluid to the body. Review Figure 2.20 in Chapter 2, the schematic diagram of the digestive process.

When you think about something to eat, the stomach walls begin to secrete gastric juice to get ready to break food down (Netter, 1959, pp. 82–83). Food reaches the stomach after passing through the esophagus. The stomach collects food and liquid and uses muscular contractions to mix parts of a meal with highly acidic gastric juices. The stomach stores and partially digests the food before moving it into the small intestine.

The chemical process of digestion begun in the stomach continues in the small intestine. The intestine secretes high volumes of digestive enzymes and fluids into the intestine's interior. The intestine absorbs nutrients and fluids into the blood stream from the food material and digestive juices moving through the intestine. Normally, the volume of secretion is balanced by the volume of absorption. If you are *dehydrated* (your body needs more fluids), the balance will shift to extracting more fluid. If you are ill with an intestinal infection, the lining of the intestine may secrete normally but be unable to absorb fluids well, and you may have vomiting or *diarrhea*. Food that is not yet broken down continues to the large intestine, the colon, where the bacteria of the gut microbiome act on non-digestible materials. One of the byproducts of bacterial activity in the colon is gas, which can cause abdominal pain and distension if it is not expelled from the body. Essential bacterial byproducts and excess water are absorbed in the colon. Chapter 5 discusses the final digestive system actions, the storage of feces in the rectum and elimination of feces from the body.

Digestive tract volume varies according to the volume and movement of ingested portions of food and liquid plus the digestive juices added by the body and the gaseous products of the gut microbiome. Peristalsis in the hollow digestive organs facilitates both food breakdown and movement of material through the system. The stomach expands to hold the amount of food and liquid consumed. *Transit time* is the amount of time it takes food to move through the entire digestive tract. Transit time varies from individual to individual and from time to time, within one individual. When you are sick with the "stomach flu" with diarrhea, the transit time for food and liquid can be much faster than usual.

The liver, a large, relatively soft, but solid organ lies just below the diaphragm at the center and R side of the upper abdomen. It is the largest glandular organ of the body. The liver stores carbohydrates and removes chemical waste products from all over the body from the blood. It produces *bile*, a mix of chemicals to aid digestion of fatty foods. Like the stomach, the liver moves with diaphragm motion and the pull of gravity (Moody, Chamberlain & Van Nuys, 1926; Barclay, 1936, pp. 100–101).

Organ mobility and the volume changes of the digestive process alter abdominal form. Increased pressure within the abdomen—whether from digestive activity, anatomical structures, or from wearable products—affects the processes and comfort of digestion. Derakhshan et al., (2012) report a tight waistband, muscular abdominal compression, or obesity increase intra-abdominal pressure. Dodds, Hogan, Stewart, Stef, and Arndorfer (1974) state that increased intra-abdominal pressure slows the peristaltic waves in the esophagus, an undesirable effect. As Bowman (2014, p. 143) says, "A continuously flat stomach, while perhaps valuable for cultural purposes, is not necessary for physiological ones." As designers, evaluate the trade-offs between fashion and body function when designing products interfacing with the easily compressed tissues of the anterior abdomen.

### 4.8.3 Urinary System in the Abdomen: Structure and Function

The urinary system makes and excretes urine. The kidneys remove waste products and excess fluid from the blood, thus controlling the volume and composition of body tissue fluids. The kidneys, in the posterior upper abdomen (Figures 4.28 and 4.29), continuously filter the blood stream to make urine. From each kidney, urine flows into a *ureter*, a tubular structure connecting the kidney in the upper torso to the bladder, deep in the pelvis (Figure 4.29). Bladder structure, function, and the elimination of urine from the body will be addressed in Chapter 5.

Kidneys are about 11 cm (4.3 in.) in length, 2.5 cm (1 in.) thick and 5 cm (2 in.) wide. Because they are solid organs, their shape does not change. But, like several other solid and hollow abdominal organs, they are mobile. They can shift several centimeters caudally with deep inhalation or when the body moves from supine to an upright position (Netter, 1979, p. 2). Kidneys, located along the spine and near the posterior body wall, are vulnerable to impact. Body armor and athletic kidney pads used in contact sports can provide a degree of protection for the kidneys. These wearables should be positioned correctly, cover the body wall over the movable kidney's possible locations, and not move out of place during activity and contact. Keys to placement and stability include a product shaped and sized for the body, and a fastening system that facilitates product positioning.

## 4.9 Upper Torso Organs: Summary

The upper torso is the container for vital systems. The actions of the supporting bony and muscular structures in the thorax and abdomen are often complementary. In the thorax, the rib cage provides a flexible but relatively rigid shell to protect the heart and lungs. Wearable products for the chest region often are designed to augment the protection of the rib cage, but can limit chest expansion, causing either or both restricted function or discomfort.

Consider the contrasting effects of the abdominal wall structures. Abdominal wall muscle contractions push the abdominal contents up against the diaphragm raising the pressure in the lungs and in the stomach and lower esophagus. The same muscle contractions push abdominal contents down against the intestinal, urinary, and reproductive structures in the lower abdomen and pelvis. A person tightens the abdominal wall muscles to cough, sneeze, vomit, or to push while defecating, urinating, or giving birth.

In contrast, the soft structures of the abdominal wall accommodate shifts in organ position and volume changes in the urinary and digestive systems. These shifts add to the plasticity of body shape and form. The diaphragm, separating the two compartments, also accommodates reciprocal volume changes between the thorax and the abdomen. Wearable products for the abdominal region may be protective, but can increase intra-abdominal pressure, a possible detrimental effect.

## 4.10 Upper Torso and Arm Fat: Body Form, Energy Storage, and Insulation

Recall from Chapter 2 that the body's sum total of fat cells is stable, but the cells may vary in volume over time. Body fat (adipose tissue) volume, in general, and torso fat volume, in particular, increases in middle age and tends to persist in the torso into old age (Baumgartner, Heymsfield, & Roche, 1995). DAT in body cavities and subcutaneous adipose tissue (SAT), the fat you can pinch along with your skin, can account for 50% of body weight in very obese individuals (Trujillo & Scherer, 2006). Although not all that adipose is in the torso, a great deal is, and much is internal. The most obvious upper torso fat belongs to the female breast (Section 4.12). In the arm, SAT helps insulate the other tissues. Women tend to accumulate additional fat in the subcutaneous layer of the skin over the triceps muscles, more so than men.

Consistent with gender- and age-related fat patterning, Whitbourne (2002, p. 84) reports women accumulate fat around the torso, gaining 25%–35% in abdominal *girth* over their adult years, compared to 6–16% for men over the same time. While gaining torso fat, elderly persons tend to lose mass in the muscles of the shoulder region and arms. With these muscle and fat changes, torso dimensions may remain stable or even increase.

Fat in the torso and extremities stores energy, provides insulation, and contributes to body form. It also produces natural chemicals which influence organs and body functions (Trujillo & Scherer, 2006). In obese individuals, fat (DAT) occupies spaces between structures in the chest and in the abdomen and crowds other structures, including the hollow digestive organs, liver, and heart. These related changes have health consequences and can be seen in medical scans (Thanassoulis et al., 2010).

Body size and form are influenced by the contributions of both DAT and SAT. SAT added through weight gain increases bra cup size and torso circumferences, and creates fat folds on the abdomen. The increased torso fat and larger bust size, with no change of the underlying bony structures of the shoulder region, contributes to bodice and sleeve misfit of plus size women's apparel. Starting from a typical production dress pattern, Bye, LaBat, McKinney, and Kim (2008) developed custom-fit bodices for study participants, Misses sizes 6 to 20. Bye et al. (2008) then compared the perfected-fit pattern shapes with the standard patterns. Figure 4.30 illustrates the patterns. Note that the armscye of the custom-fit patterns does not increase in width and length as sizes progress, while the standard industry patterns uniformly drop and widen the armscye. This mismatch of armscye pattern shapes and dimensions to the anatomy of the upper torso, arm, and shoulder region is prevalent in industry practice. New approaches to shaping women's bodice patterns could reduce dissatisfaction with the fit of women's plus size garments.

## 4.11 Lymphatic System in the Upper Torso and Arms

As in the rest of the body, lymphatic capillaries drain spaces between cells in the torso and arm and return the lymph fluid to the vascular system (refer to Figure 2.19). The lymphatic capillaries associated with the breast (in the

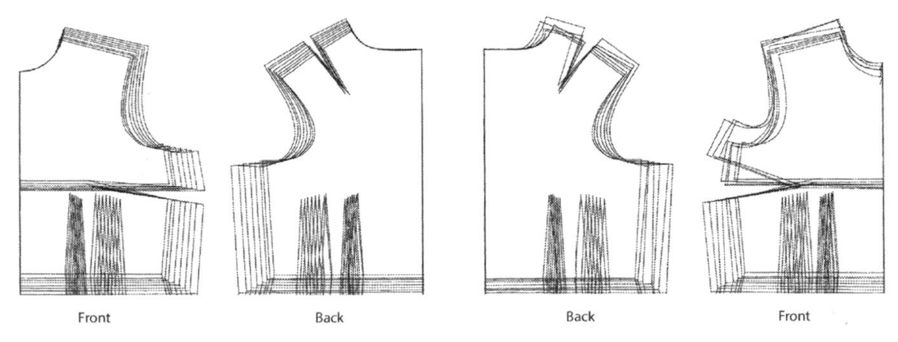


**FIGURE 4.30**
Traditionally graded patterns and custom-fitted patterns, stacked to illustrate armscye and bust dart differences. (Courtesy of the Human Dimensioning© Laboratory, University of Minnesota.)

skin covering the breast and in the breast tissue) drain to lymph nodes in the axilla and interior to the axilla. Similar structures drain lymph from the arm's tissues. These regional lymphatic structures are important to designers because of a breast cancer complication—lymphedema of the chest wall and arm. Compression garments can help treat breast cancer related lymphedema (Moseley, Carati, & Piller, 2006). Read more in Section 4.12.4.

The *spleen*, a solid, blood-cell forming organ in the upper abdomen is the largest lymphatic organ. Like the lymph nodes, it forms *lymphocytes*, specialized white blood cells to fight off foreign invaders. It also makes red blood cells. The spleen lies beneath the diaphragm, in the L upper quadrant of the abdomen, and is relatively close to the body surface, resting near the 9th to 11th ribs. The spleen normally moves along with the diaphragm. Splenic injuries are often associated with broken lower ribs. Wearable products that protect the ribs from fracture also help to protect the spleen.

The lymphatic capillaries of the skin and deep tissues of the arm drain into progressively larger lymph vessels as the lymph moves up the arm toward the torso. In the arm, as throughout the body, lymphatic vessels tend to track with the arteries and veins. The lymph flows through lymph nodes located in the *antecubital fossa*, the depression in front of the elbow joint, and in the axilla. The functions of the lymphatic system in the arm are no different than that of the lymphatic system elsewhere in the body. Lymph nodes in the arm, as in other body regions, are generally not palpable unless they are inflamed or activated.

## 4.12 Breasts: Structure, Function, Medical Issues, and Products

Breast tissue (Figure 4.31) adds to body shape and individuality. Both males and females are born with *mammary gland* tissue. The hormonal changes of puberty stimulate female breast development, preparing the mammary glands for milk production. Depending on genetics, hormones, and the monthly menstrual cycle, women's breasts vary in size. The following sections focus on women's breast structure, function, and products including *brassieres* (bras), a common upper torso wearable for women.

### 4.12.1 Breast Structure: Fat, Connective Tissue, Glands, and Surface Characteristics

Breasts—composed primarily of fat and connective tissue—vary in form, shape, and size person to person, and often side to side. The anatomical boundaries of the breast are the second or third rib superiorly and the *inframammary fold*, the crease below the breast; the sternum medially and the mid-axillary line laterally (Pandya & Moore, 2011). Two regions of breast tissue overlay thoracic wall muscles: (1) the cone-shaped body of the breast

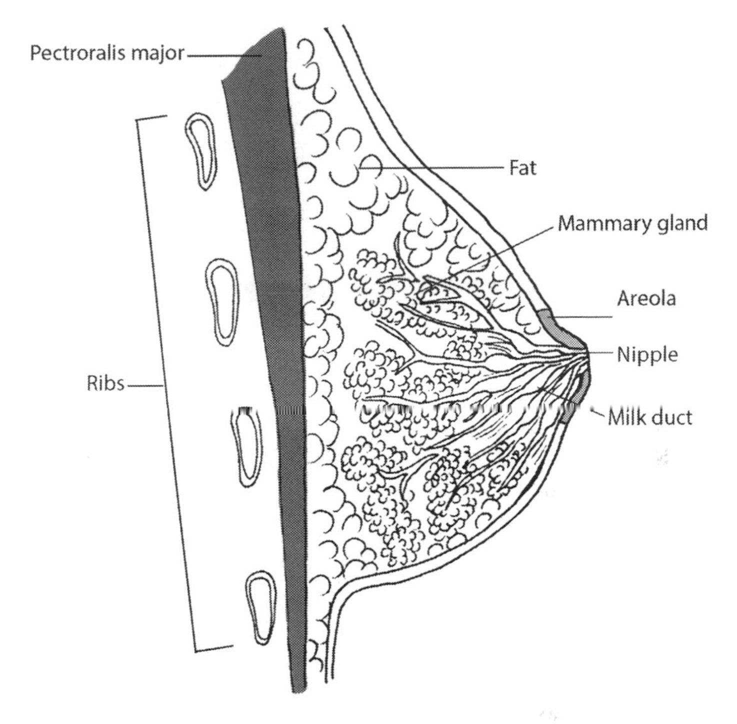


**FIGURE 4.31**
Breast cross-section.

with the nipple at its apex lies above the pectoralis major muscle and (2) an extension, the *axillary tail*, extends from the breast body into the armpit.

The nipple is surrounded by lobes of mammary gland tissue. Adipose tissue fills the spaces between the lobes. *Lactiferous ducts* (milk ducts) from the mammary tissue expand into cavities beneath the nipple (Pandya & Moore, 2011). On the surface of the breast, the nipple is surrounded by the circular, more densely pigmented zone, the *areola*. Sensory nerves in the skin of the areola are sensitive to cold, touch, and sexual arousal. The skin wrinkles and the nipple becomes erect in response to these stimuli. *Cooper's suspensory ligaments*, specialized connective tissue structures, pass through the breast tissue to loosely attach the breast to the dermis of the skin over the breast tissue and to underlying muscle. Support bras and sports bras are designed to restrict breast movement on the chest, to limit stress to these ligaments.

### 4.12.2 Breast Function: Specialized Exocrine Glands

At puberty, the female mammary gland enlarges, fat deposits in the breast increase in volume, the nipple elevates, and the pigment of the areola darkens. During pregnancy, the breast increases in volume and density and the areola gets darker. The areola's color changes may help the infant more clearly see and find the breast. These breast changes are associated with the developing capacity of the breast, as an exocrine gland, to produce and secrete milk.

*Lactation* is the formation and secretion of milk. Ramsay, Kent, Hartmann, and Hartmann (2005) studied the lactating breast using ultrasound. They found approximately twice as much glandular tissue as adipose tissue in the lactating breast, but also great variation woman to woman. Milk is said to be "ejected" from the lactiferous ducts and the cavities beneath the nipple, where it has collected, when it is secreted from the breast. Prime, Geddes, and Hartman (2007) report that milk ejection occurs not only during stimulation of the nipple but also reflexively. Conditioning of the milk ejection reflex, sometimes called *milk let-down*, is common and may be initiated by the mother thinking of her infant, in response to her infant's cry, or with the infant's *suckling* activities.

### 4.12.3 Anatomical Breast Changes Related to Breast Cancer

Breast cancer is the most frequently diagnosed cancer in American women (Siegel, Miller, & Jemal, 2016). Treatment often includes surgery to remove part or all of the affected breast. Breast cancer surgery ranges from *lumpectomy*—removing a mass of tissue including the tumor and a margin around the tumor, to *mastectomy*—removing the breast and a variable amount of adjacent tissue. Some women at high risk of developing breast cancer due to genetic status elect to have both breasts removed as a preventive measure. Lymph nodes may also be removed from the axilla. After recovery from surgery, breast cancer treatment may include radiation therapy and/or chemotherapy. Some women elect breast reconstruction surgery after mastectomy.

### 4.12.4 Anatomical Lymphatic System Changes Related to Breast Cancer

Review Figure 2.16-C, which illustrates the lymphatic system throughout the body. Notice the network of lymph nodes and connecting pathways throughout the upper torso. There are numerous lymph nodes near the sternum and clavicles, throughout the axillae, and adjacent to the breasts. Lymph from the breast drains into the lymph nodes of the axilla. Cancer cells from the breast can travel to the axillary lymph nodes and then to other parts of the body. When a malignant tumor is removed from the breast, it is common to remove and examine the first lymph node to receive cells from the breast to assess for possible spread of cancer. If the first lymph node is positive for cancer, more lymph nodes are removed.

Axillary lymph node removal interrupts the normal lymphatic drainage from the arm and anterior upper torso (refer to Figure 2.19) on the affected side. Lymphedema (refer to Figure 2.18, R side) with swelling (edema) of the arm, hand, and fingers may result. Chest wall lymphedema can also occur. The arm increases in weight and size due to the fluid build-up. The fluid limits joint mobility and creates mild to severe discomfort, depending on the volume change. Because lymphedema fluid increases the risk for infection in the arm, individuals with lymphedema need to protect the affected hand and arm. Cuts, scrapes, and insect bites—or any other trauma which creates a break in the skin—might result in infection.

Compression sleeves are often worn to try to control arm lymphedema and to relieve associated discomfort. Compression sleeves are slightly tapered knitted cylinders; wider in circumference at the axilla/biceps and tapering to a smaller circumference at the wrist. Some people also wear a gauntlet, a modified glove. Compression, in engineering terms, is defined as pounds/square inch. Health care professionals use a different convention, mm of mercury (Hg), to prescribe how much product compression is required. Garments often have "graduated compression"—reduced compression over bony prominences, nerves, or joints, such as the elbow. Compression varies between 15 and 40 mmHg (light to extra firm compression), but occasionally may go up to 50 mmHg. Compression garments can be difficult to put on due to their "skin tight" fit.

The anatomy of the shoulder and upper arm affects design approaches. Visualize the arm as a cone. The challenge is to keep the sleeve in place while providing the desired compression, avoiding blood flow restriction, and allowing full motion of the wrist, elbow, and shoulder. The larger circumference of the sleeve's upper cone may easily slide down to the narrower circumference of the lower arm. Also, a tubular sleeve does not extend to cover edema in the axilla and/or adjacent chest. Lymphedema pads and bra-like chest wall compression garments are sometimes used for chest wall lymphedema. New products are being developed to cover the torso and arm, but tend to be bulky. Custom-fit sleeves are available but must be constructed from accurate measurements and the sleeve must be correctly positioned each time it is worn to be effective.

### 4.12.5 Female Breast: Focus on Products

Form-fitting wearable products must be shaped to fit women with varying breast dimensions and shapes. Bras for many needs: everyday wear, maternity wear, to allow nursing an infant, sports activities, and to wear after breast surgery are primary examples. Products like uniforms and protective equipment for women working in previously male-dominated fields (construction trades, military service, and law enforcement) must also be shaped to accommodate breast variations. And chest/breast protection for female athletes should be shaped differently from products designed for men.

The bra is, perhaps, the most engineered and patented ready-to-wear garment, but bra designs still leave much to be desired. Yu, Wang, and Shin (2006) present three basic bra pattern development approaches: (a) adapting basic bodice patterns, (b) drafting from body measurements or an existing product, and (c) three-dimensional modeling on a digital form. They conclude that a more scientific basis for bra design is needed. Bra designers also need in-depth knowledge of breast anatomy. Good bra design is based on a pattern shape that fits the torso and breast, and selection of materials for comfort and support.

#### *Everyday Bras*

Many women begin to wear a bra as a pre-teen and wear one throughout their lives. Everyday bras are worn for breast support, modesty, and as foundations to enhance the aesthetic appearance of outer garments. As women age, beginning in their mid-forties, the skin over the breast progressively loses elasticity. As anatomical support of breast tissue decreases, the shape of the breast may change and "Women might therefore benefit from increased external breast support (i.e. a more supportive bra) with increasing age" (Coltman, Steele, and McGhee, 2017, p. 303).

#### *Maternity/Nursing Bras and Other Wearable Lactation Products*

Maternity bras are designed for wear during pregnancy to accommodate slightly enlarged, tender breasts. Wearable products for lactating mothers include: nursing bras with cups that can be opened, nipple shields, silicone covers to fit over the nipple and areola during nursing, and nursing pads to place in bra cups to absorb leaking milk. Keep the nipple's vulnerability in mind when designing these products. The nipple and skin of the areola can become irritated and cracked from the almost constant moisture exposure from milk and the baby's mouth plus the friction of the baby's suckling. Nursing bras should be non-abrasive, absorbent, and easily cleaned. Fetherston (1998) identifies nipple shield use as a risk factor for developing blocked milk ducts, and a tight bra as a risk factor for both *lactation mastitis*, a painful breast condition related to milk retention, and blocked milk ducts. Gordon (2015) conducted in-depth interviews with eight working mothers who were nursing infants aged one year and under. The mothers expressed functional and social needs when selecting apparel including: garments that allow breast access, provide support without being restrictive, camouflage for leaks and spills, coverage when breast-feeding in public, and comfort for painful nipples and breasts (p. 63).

#### *Sports Bras*

With minimal internal stabilizing structures, the breast moves freely as the body moves vigorously. Read more about breast motion in Section 4.14.4. Sports bras are worn to provide support and to prevent excessive and/or painful motion of the breasts. Haycock, Shierman, and Gillette (1978) and Lawson and Lorentzen (1990) filmed athletes' breast motion without and with support bras. They noted substantial vertical and lateral breast motion in the nude state and concluded support bras restrained motion. Haycock et al. (1978) surmised that forces during breast motion on Cooper's suspensory ligaments caused breast soreness and lengthening of the ligaments. Mason, Page, and Fallon (1999) found that a sports bra could reduce absolute vertical movement and the maximum downward deceleration force on breasts. Yip and Yu (2006) state that sports bras should function to control breast motion, reduce breast pain, and regulate body heat and moisture. Motion capture has replaced film to study breast movement; however, findings are much the same. Limiting breast displacement remains the major goal for sports bra design.

Compression and encapsulation are the two design approaches used to provide breast support and prevent motion (Yu & Zhou, 2016). The compression method relies on textiles with moderate to high elasticity to flatten breast tissue to the chest, applying equal pressure around the torso. The encapsulation method relies on non-stretchable, structured cups fitted closely to each breast. The cups are attached to a stabilizing chest band and shoulder straps. A sports bra chest band should expand easily with inhalation to allow enough air/oxygen intake to sustain high levels of activity.

Morris, Park, and Sarkar (2017) developed a sports bra to meet the needs of physically active breast-feeding mothers. Their focus group discussions and wear-test trials with eight participants identified needs for: good breast support while limiting compression, especially to sensitive nipples; easy breast access for nursing and milk pumping; and specific design features like soft wicking fabrics, wide straps for breast support, and designs that "look pretty" (p. 298).

Zhou, Yu, and Ng (2012) studied specific sports bra features to control breast motion. The study, limited to four participants wearing A and B cup sizes, may not apply to a wide range of body types. However, the research process considered many factors and used a state-of-the art motion analysis system. After studying seven styles, they concluded the best stabilizing features are: compression, short vest style with high neckline, inelastic cup seams, side slings, racer-back, a slightly elastic bound neckline to stabilize the upper breast boundary, and wide non-adjustable straps. They found that a center gore, lower breast "cradle," underwires, and padded cups were not effective. Improved sports bra designs can come from an understanding of breast motion mechanics, use of new materials, and extensive testing of combinations of bra features.

### *Mastectomy Bras and Other Products for Post-Cancer Treatment*

Wearable products for breast cancer patients and survivors are designed to accommodate both short- and long-term body changes related to breast surgery and/or radiation. Drainage tubes inserted under the skin during surgery drain excess fluid produced during healing. The liquid drains into a small flexible bulb that must be emptied several times a day for a week or two. Cho,

Paek, Davis, and Fedric (2008) developed and tested a hospital gown with pockets to securely hold the drains and features to access the drains.

After an initial post-operative period of healing, women can be fitted for a variety of mastectomy bras and products. Women who do not elect reconstruction surgery may choose to wear a breast prosthesis to replace the natural breast. A lightweight prosthesis that does not put pressure on healing tissues is sometimes worn for several weeks after surgery. After the surgical site is healed, a prosthesis of silicone with weight and size that replicates the natural breast can be worn. A mastectomy bra with a bra cup pocket holds the prosthesis (Figure 4.32). Gallagher, Buckmaster, O'Carroll, Kiernan, and Geraghty (2009) report experiences of women with post-breast cancer surgery products and services. Women using post-surgery products often reported dissatisfaction with the products and services related to the products, while service providers reported few problems (p. 560). The response differences suggest designers can develop more satisfactory products by working closely with survivors.

LaBat, Ryan, and Sanden-Will (2016) interviewed 52 breast cancer survivors. Common complaints about mastectomy bras and prostheses included: (a) warmth and perspiration between the silicone prosthesis and the chest wall, (b) shoulder/upper trapezius pain and fatigue (because the upper trapezius and clavicle become the main support for the weight of the prosthesis), and (c) unwanted motion of the prosthesis on the chest wall. Companies have developed prostheses with wicking channels incorporated into the prosthesis backing to attempt to address the thermal problems.

The issue of prosthesis weight can be difficult to address, as a prosthesis lighter in weight than the opposite breast (to reduce strain to the clavicle and trapezius) can impact balance and posture. Prostheses are available that adhere directly to the chest with an adhesive. These adhered prostheses may

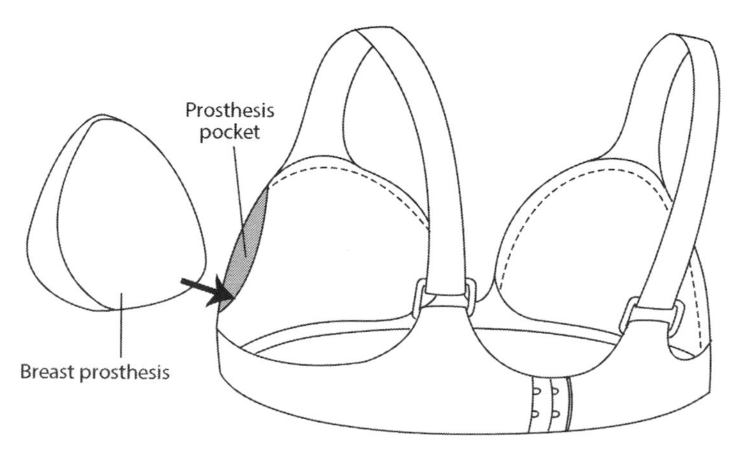


**FIGURE 4.32**
Mastectomy bra and breast prosthesis.

move more naturally with chest skin and muscles and help to decrease pull on the clavicle and trapezius, but dermatitis from the adhesive can become a problem. Is it possible to replicate the function of the Cooper's ligaments with a different approach to prosthesis design?

While studying shoulder pain in women, Ryan (2000) described the mechanics of breast suspension from a bra shoulder strap. He developed techniques to assess breast weight and found breasts could vary from 1 to 5.5 kg (2.2–12 lbs.). The weight of an object, such as a breast, is determined by volume and density (Kakalios, 2005, p. 26). The mechanics and solutions discussed by Ryan suggest new approaches to addressing the problem of prosthesis weight.

Breast reconstruction surgery is an option after mastectomy (Roses, 2005). If a woman has had reconstruction on one side of the body, breast asymmetry may be a problem as the reconstructed breast may not match the natural breast. Products, like partial bra cup fillers, have been designed to balance asymmetries. After reconstruction, some women report localized sensitivity and discomfort from clothing (LaBat, Ryan, and Sanden-Will, 2016).

Radiation may also be part of breast cancer treatment. Radiation may cause skin damage with mild to severe burns during treatment; and dry, itchy skin during or after treatment. Stevens and Cooper (1999) estimated that 30% of women treated with radiation develop some type of skin discomfort. They recommend several topical treatments and not wearing a bra until the condition subsides. Understanding textile/skin interactions after radiation could lead to improved bra designs for women who have had radiation therapy.

### Female Breast/Chest Protection

Protecting the female chest and breasts with a wearable product can be challenging. The problems are varied. Some relate to breast structure and others to breast/chest form, the underlying anatomy, or to the external threat to the body. Solutions to the problems are equally varied. Simple solutions for women who experience nipple soreness when running bra-less include placing tape over the nipples to prevent chafing and irritation; or wearing upper torso running gear made of absorbent, non-abrasive textiles. Contact sports require special attention. Protective products shaped and fit to the female torso and breasts are worn for fencing, softball/baseball, lacrosse, martial arts, and other sports. As more women participate in traditional male sports, wearable product research/development for women is needed.

Historically, body armor has been worn by men and made to fit men's torsos. As women entered military and public safety fields in greater numbers, the need for effective chest/breast and torso protection increased. Products, especially body armor to protect from bullets and bomb fragments, must fit and be positioned optimally on the body. The shape of the female torso must be considered. Boussu and Bruniax (2012) describe design of a lightweight bullet-proof vest for females. They suggest using 3D body scanning and specific pattern drafting techniques to custom shape vests for each woman. Such precise methods assure the vests fit individuals with equal pressure throughout the torso, increasing protection and comfort. They also discuss the benefits of using para-aramid or high-performance polyethylene yarns in two-dimensional textile structures and integrating them into the garment making process.

## 4.13 Axilla: Structure and Function

The female breast axillary tail extends to the axilla, otherwise the male and female axillary anatomy and function are the same. Sweat production in the axilla is a concern for wearable product designers.

### 4.13.1 Axilla Structure

The axilla or arm pit is the hollow between the arm and the body. The axilla, with the arm at rest at the side of the body, is a pyramidal space. When the arm is raised overhead, the fascia and skin of the axilla flatten, and the axillary space disappears. This kind of mobility can be difficult to accommodate in garments with set-in sleeves for athletes or orchestra conductors who require full overhead arm motions. A *gusset* (a diamond shape, cut on the bias) set into the armscye/underarm seam is one way to increase freedom of motion for the axillary region.

The glenohumeral joint forms the roof of the axilla and the pectoralis major and latissimus dorsi muscles form the bulk of the front and back walls. Serratus anterior, the deep torso muscle which helps to move the scapula on the rib cage, forms the medial wall and the humerus forms the lateral wall. Nerves, arteries and veins, and lymphatic vessels lying between the glenohumeral joint and the more superficial tissues (fascia and skin), pass through the axilla to and from the arm. The skin in the axilla has hair and numerous sweat glands.

### 4.13.2 Axillary Sweat Glands

There are three types of sweat glands in the axilla: apocrine, eccrine, and apoeccrine glands (Wilke, Martin, Terstegen, & Biel, 2007). After puberty, apocrine sweat glands become active. They discharge sweat into the hair follicles in response to emotional stimuli/stress and apocrine sweat is believed to exhibit a pheromone-like effect. Eccrine glands produce watery sweat to help lower elevated body temperatures. Apoeccrine sweat glands were first described in 1987. Reports indicate they sweat continuously, producing watery sweat at a higher rate than the other types, but what stimulates them and their functions is not fully understood (Wilke et al., 2007) Some authors debate the apoeccrine sweat gland's existence (Hu, Converse, Lyons, & Hsu, 2018; Piérard-Franchimont, Piérard, & Hermanns-Lê, 2017).

### 4.13.3 Perspiration/Sweat

Sweat is a chemical mix of interest to designers because of its association with body odor and perspiration stains affecting textiles worn in contact with the armpit. Apocrine sweat contains proteins, lipids, and steroids. It is thought to be mixed with sebum, a waxy substance secreted by sebaceous glands into hair follicles to lubricate the hair shaft. Eccrine and apoeccrine sweat is more watery than apocrine sweat. All sweat types likely contribute to stains on fabrics. Many apparel manufacturers test textiles for reactions to perspiration before using them in a product. A test developed by the American Association of Chemists and Colorists (AATCC) is an example. The test (AATCC TM15-2013) involves saturating a textile with a solution that replicates human sweat. After a period of time under specific conditions the textile is evaluated for color change and its likelihood of bleeding into fabrics that are worn next to the textile. A manufacturer may choose to test a red-dyed textile intended for a blouse design that may be worn over a white bra. If color change or dye transfer is not acceptable, a different textile may be selected.

Sweat contains scents related to body functions and diet (Wedekind, 2002). Axillary odor developing from the action of bacteria on sweat may be a problem. Body odors on apparel relate to the fiber type used in the garment (Callewaert et al., 2014). Polyester harbors different kinds of bacteria than cotton and appears to generate different and more offensive odors. Removable, washable or disposable, garment shields to absorb perspiration can prevent perspiration from saturating garments. Another design strategy is to incorporate armpit vents lined with vapor-permeable mesh to increase air circulation, allowing sweat evaporation.

## 4.14 Upper Torso: Spine and Rib Cage Motion

Incorporating motion into an upper body wearable product can increase function and improve comfort.

### 4.14.1 Overview of Spine Motion

Spine motion is generated by muscular actions. Many upper torso muscles originate from or insert on the spine. Recall the three primary planes that characterize body position—sagittal, coronal, and transverse planes. The

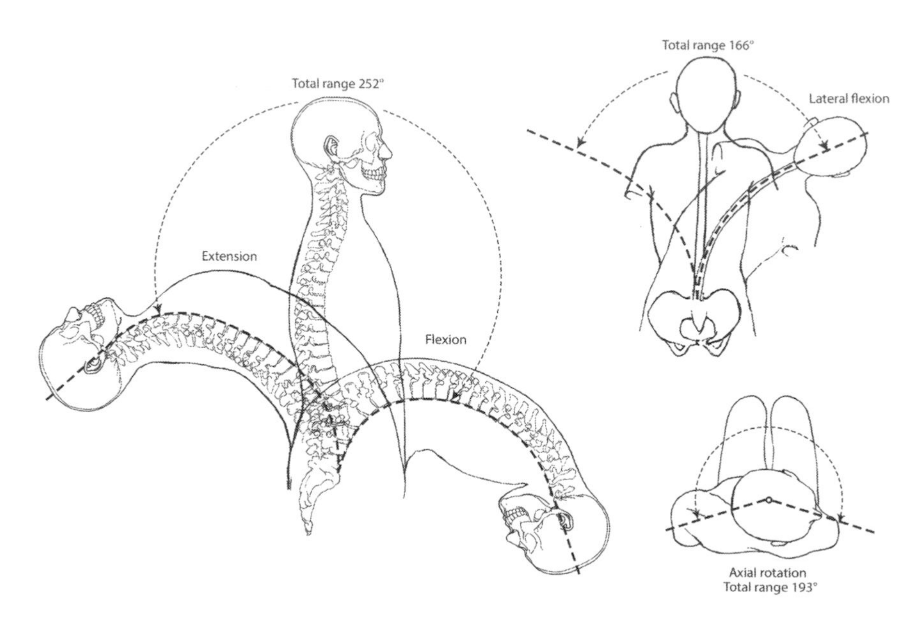


**FIGURE 4.33**
Spine range of motion: flexion/extension, lateral flexion, and axial rotation.

normal, upright, static spine position (refer to Figure 4.12, lateral/sagittal view) is characterized by balanced opposing curves. The spine is straight in the coronal plane (anterior view) and the transverse planes (superior view). Three basic movements characterize head/neck and torso motions: flexion/extension, lateral flexion (bending to the R and L) and back to upright, and axial rotation to the R and L around vertical axes of the vertebrae (Figure 4.33). Flexion and extension in the neck (cervical spine) and torso (thoracic and lumbar spine) involve changes in the inclination or tilt of the vertebrae in the midsagittal plane. Lateral flexion, bending to the side without twisting, occurs in all spinal regions, but the motion is slightly freer in the neck than the torso.

White and Panjabi (1978, p. 16) describe a ROM for each movable vertebral joint from the occiput of the head to S1 for the spine motions. From their data, approximate motions for the spine (cervical, thoracic, and lumbar segments combined) are: flexion/extension—252 degrees, lateral flexion—166 degrees, axial rotation—193 degrees. Motions of the cervical spine were discussed in Chapter 3. The rib cage moves with the thoracic spine, but also limits thoracic spine motion. The lumbar spine displays freer motion than the thoracic spine for some motions. Experience spinal axial rotation (twisting) by sitting on a bench and turning your torso side-to-side without sliding your buttocks on the bench. The intervertebral joint between the fifth lumbar vertebra and the sacrum is the last but a very mobile joint of the spine, with 20 degrees of flexion/extension motion (White & Panjabi, 1978). See Chapter 5 for the relationship of the sacrum to walking motions. If muscles of the neck, abdomen, and back (and related spinal joints) aren't used on a regular basis, motions may become restricted. When designing for people with limited motion, for example an elderly population, keep this in mind.

### 4.14.2 Thoracic Spinal Motion

The thoracic spine provides protection for the spinal cord while allowing a person to bend and twist. The costovertebral joints—(a) between the ribs and one or two vertebral bodies and (b) between the ribs and transverse process of the vertebral body—limit motion of the thoracic vertebrae especially for vertebrae T1 to T10 (refer to Figure 4.11). The ribs block motion, limiting side-to-side and forward and backward motion more than rotation. T11 and T12, with their attached floating ribs, move more like the lumbar vertebrae below them.

### 4.14.3 Lumbar Spinal Motion

Flexion/extension motion in the lumbar spine is freer than in the thoracic spine. To experience lumbar spine flexion, sitting or standing, tighten your abdominal muscles and try to "tuck" your tailbone in and under your torso. This action flattens the normal lumbar curve and tips the sacrum and coccyx, as well as the pelvis, toward the front of your body. You can also flex your lumbar spine while lying on the floor by pressing your low back into the floor. Lumbar spine flexion produces a stretch to the muscles positioned along the vertebral spinous processes. If you relax your abdominal muscles and push your abdomen/belly forward and arch your low back—sitting, standing, or lying down—you are extending your lumbar spine. Consider the variation in spine flexibility and extension when designing products that either need to move with the body, or that need to stabilize the body.

### 4.14.4 Rib Cage and Breast Movement

Rib cage motions facilitate lung function, vocal production, and coughing. Coordinated rib-vertebrae and rib-sternal motions expand the chest cavity with a breath in and reduce the volume of the chest cavity with a breath out (refer to Figure 4.3). Movements between individual ribs and the vertebrae and sternum are small (Figure 4.34). To help remember how individual ribs move up and out with breathing, think of the ribs as a series of bucket handles—attached to the sternum at the front of the body and to the spine at the back of the body (12 handles on each side). With each breath in, the 24 bucket handles lift upward a short distance. With each breath out, the 24 bucket handles return to their rest position. At the same time, the lower end of the sternum moves forward and up, like a pump handle, returning to its original position as you exhale.

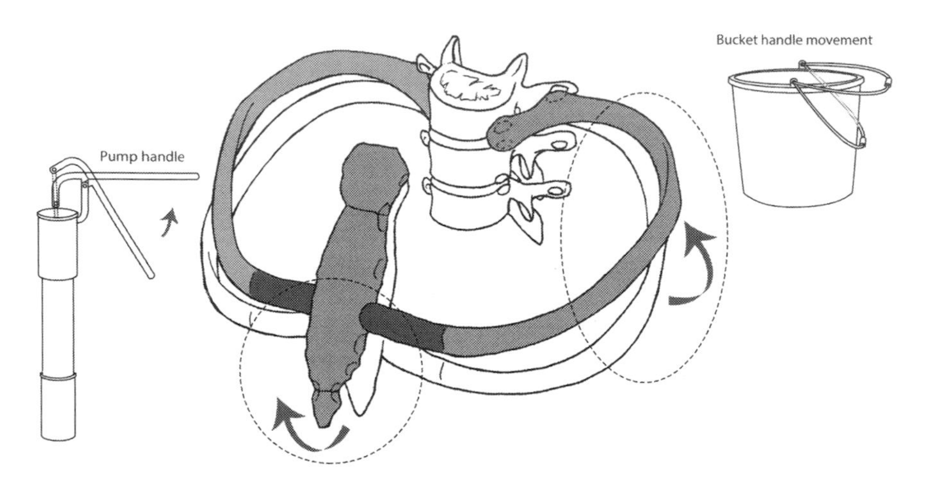


**FIGURE 4.34**
Rib motion.

Athletes may need to inhale and exhale deeply for peak performance. Layton et al. (2011) compared lung function in men and women and found chest wall motion differences during high-intensity exercise. They did not comment on what garments the subjects wore for the study, but illustrate a woman in a sports bra. Could the differences relate to the presence of the bra around the rib cage? A bra worn during quiet breathing, moving 0.5 liter of air, may have little impact on rib cage motion, but with effortful activity requiring breaths of 4–6 times that much, or more, a band around the chest may change respiratory function. The differences may also relate to the fact that female lung volumes are 10–12% smaller than males, in individuals of similar height and age (Bellemare, Jeanneret, & Couture, 2003).

Breast motion, particularly during vigorous activities, is an important consideration in sports bra design. All objects, breasts included, move according to how large they are, how dense they are, and how much force is applied to them. Like any soft tissue without skeletal support, the breast mounds move somewhat independently of the torso. Breasts rise and fall during running push-off/landing actions (Figure 4.35, center and right) and, to a lesser degree during less vigorous running and walking. They also move side to side as the torso twists; think about torso motions during running activities.

Quantitative breast motion data is not readily available. The breast motion studies reviewed in the section on sports bras used a limited number of participants. Breast motion analytical methods continue to evolve. Many variables affect breast motion: breast size and weight; motion of the torso, arms, and whole body; speed of motions; and impact forces—transferred or direct to the breast. Breast motion can be evident in sagittal and transverse planes. Sports bra effectiveness is often assessed by comparing breast motion: (a) nude or wearing a minimal support bra and (b) wearing a sports bra. Quantitative measures including change in position of the nipple landmark are an assessment option.

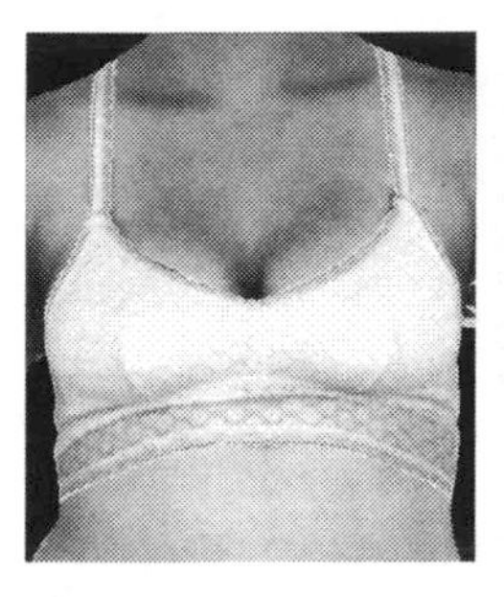
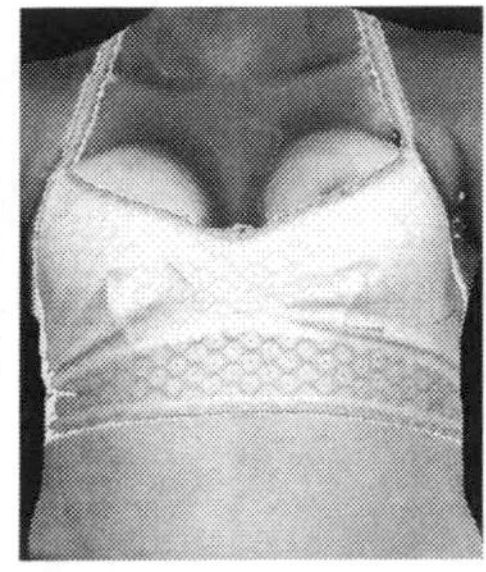
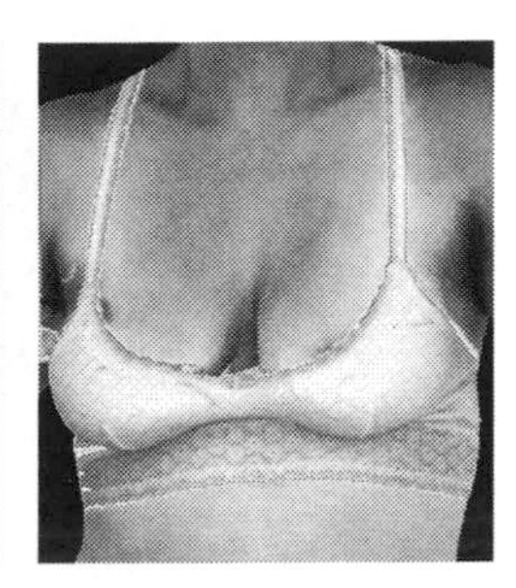

**FIGURE 4.35**
Temporal-3D Breast motion while jogging in a minimally supportive bra. Note the breast moves independently of the torso. (Extracted from "A Temporal-3D Torso Exercise with the 3dMDbody18.t System," 3dMD. 2017. A PowerPoint document prepared by 3dMD specifically for Professor Karen L. LaBat at the University of Minnesota, slide 2, for use in *Human Body: A Wearable Product Designer's Guide*, CRC Press, and may not be used in any other manner, distributed to or discussed with any third-party organizations without the prior written approval of 3dMD. Copyright 2007 3dMD. All rights reserved.)

## 4.15 Motions of the Shoulder and Arm

The upper torso and arms are made up of several complex three-dimensional parts that interconnect and move together. Look again at the body shapes illustrated in Chapter 1 (Figure 1.3) to recall how the upper body can be visualized in simple cone and tube shapes.

Products covering part or all of the upper torso should adjust to the motion of the shoulder blade (scapula). Scapular motion on the rib cage is only one element of the complex motion of the shoulder region. Rib cage motion also affects function and comfort of upper torso designs.

### 4.15.1 Shoulder Region Motion: Scapulothoracic and Glenohumeral

#### *Scapulothoracic Motions*

Scapulothoracic motions, particularly scapular protraction/retraction, produce significant changes in back dimensions. Scapular motions on the rib cage were described and illustrated earlier in this chapter (refer to Figure 4.20). The serratus anterior and *pectoralis minor* and major muscles protract the scapula; the rhomboid and middle section of the trapezius muscle retract it (refer to Figure 4.23 and see Figure 4.36). Recall that scapular protraction/retraction motions originate with the forward-backward motion of the clavicle in a relatively horizontal plane at the sternoclavicular joint. Numerical values are generally not available for sternoclavicular motion.

Back dimension change resulting from these motions can be seen in transverse 3D body scan sections, taken at the level of the acromial points. Measurements over the body, acromial point to acromial point, vary depending on torso and arm position. Figure 4.36 illustrates measurements (in centimeters and inches) taken from three scans of one adult subject. Compare the measurements taken in a natural relaxed position, in full scapular protraction, and in full scapular retraction. For this person, the back measurement increases 11.5% with full scapular protraction. With maximum scapular retraction, the measurement decreases 10.6% from the natural position.

Measurements across the chest at the level of the clavicle show different relationships. With scapular protraction, the chest becomes more concave and width decreases approximately 4.6%. This body change is not as crucial, as the garment will likely simply move somewhat farther away from the body. With scapular retraction, the chest becomes more convex and width increases approximately 1.5%. The differences between back and chest expansion/contraction with these two motions relate to the distance of the medial border of the scapula from the spine. The retraction cross-section shows that the skin over the back wrinkles as the scapulae move closer to the spine.

Accommodating back expansion in a garment can be especially important for athletes and to insure comfort in everyday clothing. As with all joint ROM, variation person-to-person is common. These observed measurement changes in one person suggest a need for study of upper body dimension changes related to scapular protraction/retraction, in a diverse population of subjects.

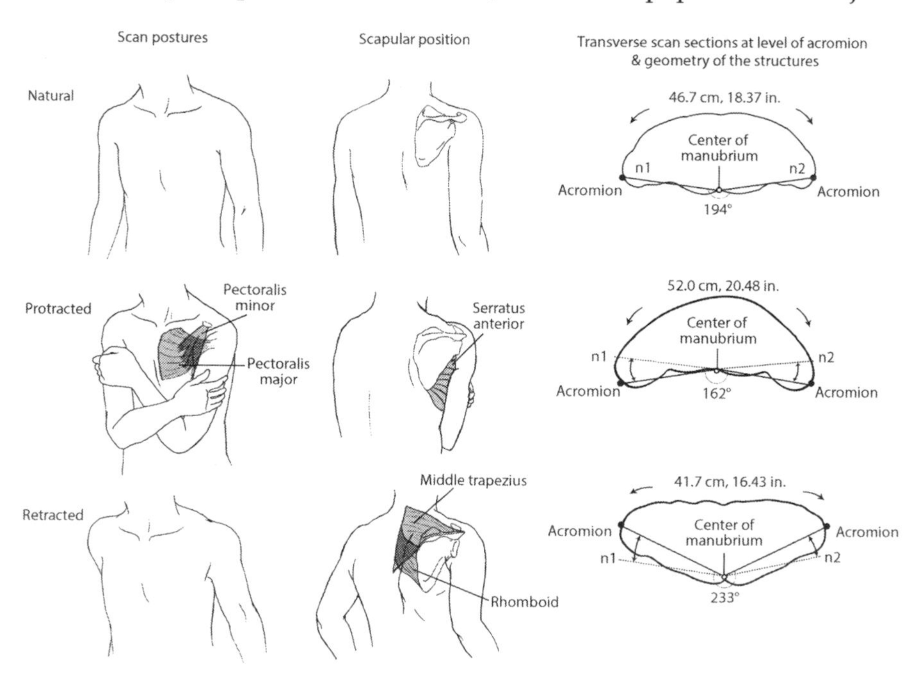


**FIGURE 4.36**
Sternoclavicular joint protraction/retraction motion (scapulothoracic motion), with muscles and transverse sections (body outlines derived from body scans). (Courtesy of the Human Dimensioning© Laboratory, University of Minnesota.)

### *Glenohumeral Motions*

Arm motions originating at the glenohumeral joint include: flexion/extension movements in a sagittal plane; abduction/adduction motion in a coronal plane; and—in a transverse plane—medial/lateral rotation of the humerus. Many muscles contribute to arm flexion (Figure 4.37). Pectoral girdle muscles: the deltoid, part of the pectoralis major, and the latissimus dorsi are large superficial muscles active in flexing and extending the humerus on the scapula. The flexion motion, starting with the arm hanging next to the body, is approximately 170 degrees. Arm extension from the same position is approximately 60 degrees. (Note the motions are based on a 360-degree system; see Section 2.4.3.) The biceps brachii contributes to the flexion motion. The triceps brachii contributes to glenohumeral joint extension.

Glenohumeral joint abduction moves the humerus up and away from the body in a coronal plane. The central section of the deltoid muscle is the major mover for arm abduction. The large muscles involved in pulling the humerus down and toward the body, adducting the arm, are the pectoralis major and latissimus dorsi (Figure 4.38). The arm moves 170 degrees from a fully adducted resting position, at the side of the body, to full abduction—with the arm and hand as high overhead as possible.

The standard terminology "medial/lateral rotation" for another glenohumeral joint motion assumes that the arm is adducted and the elbow flexed to 90 degrees; from there the forearm moves toward and then away from the body in a transverse plane. However, in many racket sports and throwing activities the arm is abducted to 90 degrees, and the forearm moves in a sagittal plane. The typical arm position for medial/lateral rotation for racket sports and the muscles which drive the motions are shown in Figure 4.39.

The muscles which execute the motion are the same, whether the arm is adducted or abducted. An intricately related set of muscles on the back and front of the chest holds the humerus in the glenoid fossa. Four small muscles, which make up the rotator cuff, attach to the front (rib cage side) and back (trapezius side) of the scapula. They help to rotate the humerus medially and laterally in the glenoid fossa of the scapula. Two trapezius-side muscles (infraspinatus and teres minor) rotate the humerus laterally. The posterior part of the deltoid also rotates the humerus laterally in the glenoid. It is the most superficial of this group of rotators.

The rotator cuff muscle on the rib cage side of the scapula (subscapularis), plus the fourth muscle, the teres major from the trapezius side, rotate the humerus medially. The large latissimus dorsi and pectoralis major also help to rotate the humerus medially. The anterior portion of the deltoid also acts as a medial rotator at this joint. The humerus laterally rotates approximately 90 degrees, when raising the hand on the bent arm up near the head. Medial rotation is usually limited to about 80 degrees. Throwing athletes, such as baseball pitchers, often have far greater lateral rotation than the average person, seen as they cock the pitching arm back before releasing the ball.

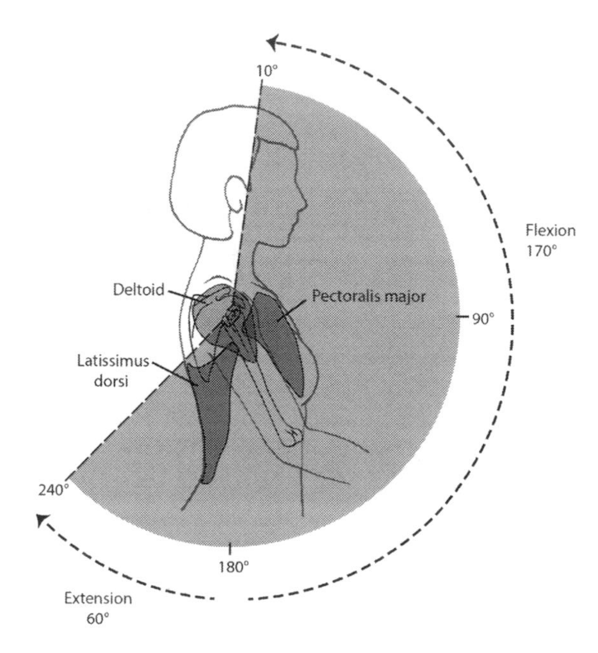


**FIGURE 4.37**
Shoulder range of motion, flexion/extension, with muscles.

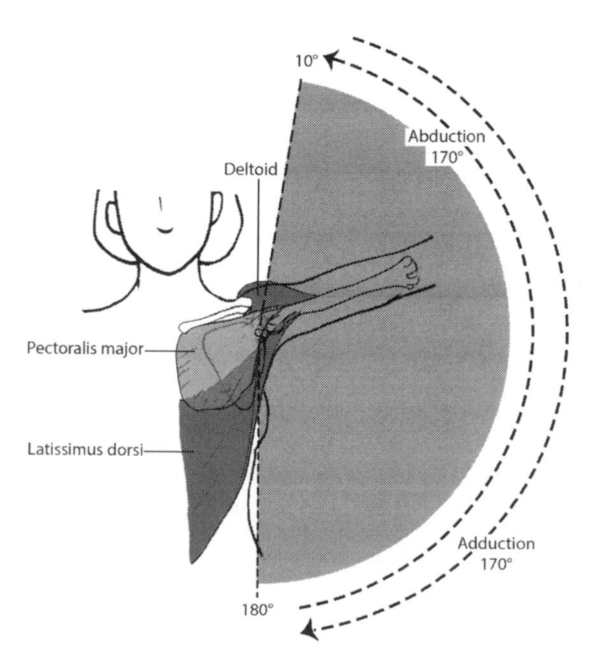


**FIGURE 4.38**
Shoulder range of motion, abduction/adduction, with muscles.

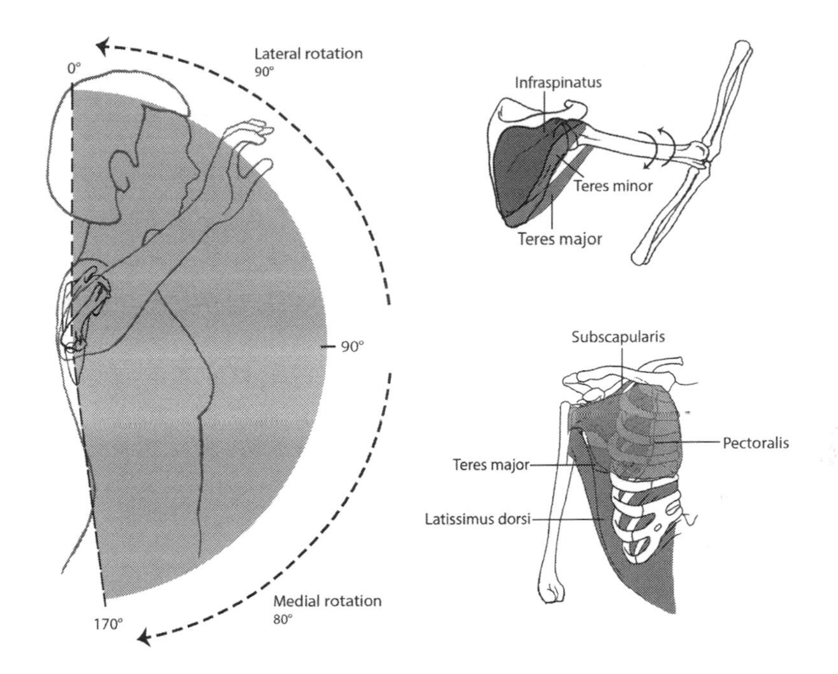


**FIGURE 4.39**
Shoulder range of motion, medial/lateral rotation, with muscles.

## 4.15.2 Elbow Motion

The arm has two paired motions at the elbow. The first, flexion/extension, is a hinge motion, bending and straightening the arm at the elbow. The second pair of motions is forearm pronation/supination, turning the forearm over and back.

Elbow flexion/extension as shown in Figure 4.40 begins with the arm at the side in the anatomic position (thumb pointed laterally and palm facing forward). The elbow, held at the body's side, bends and the forearm moves in a sagittal plane, approximately 150 degrees until the forearm presses against the biceps—the primary muscle acting to flex the joint. If the palm is facing toward the body when the elbow bends, the brachioradialis—in the forearm—helps to complete the motion. The triceps acts to extend or straighten the elbow and the forearm returns to the side of the body.

To understand pronation/supination at the elbow (Figure 4.41), sit at a table and place your forearm on the table. Pronation positions the hand palm-down, facilitating grasping. Supination puts the hand in a palm up position which can facilitate carrying. Use the first syllable of supination, "sup," to help remember that this is the "palm-up" position, which would allow you to cup your hand, as if to carry soup. The forearm positions for pronation/supination are fairly easy to describe. However, the bone and joint motions affecting the positions are complex. As the forearm and hand pronate, turning from palm

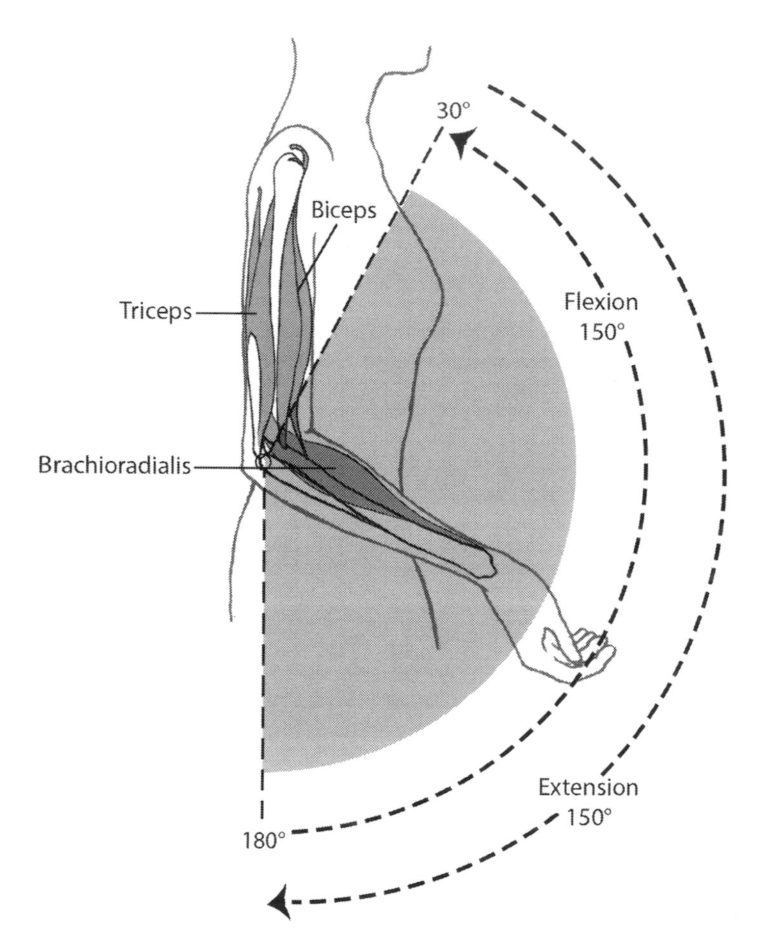


**FIGURE 4.40**
Elbow range of motion, flexion/extension, with muscles.

up to palm down, the distal end of the radius crosses the ulna and the bones of the forearm form an X. When the hand returns to palm up, or supinates, the bones of the forearm are parallel. These changes in bone position can affect product fit. Several small muscles around the elbow are active in pronation/supination, along with one muscle at the volar wrist. These muscles change minimally in volume as they act. The forearm and hand rotate approximately 90 degrees from the neutral position with each motion.

### 4.15.3 Radiocarpal Joint Motion

The radiocarpal joint is sometimes referred to as "the wrist joint." Muscle action across the radiocarpal joint (joint between the forearm bones and the hand) produces motion of the pronated or supinated hand relative to the forearm—flexion and extension and deviations toward the radius or ulna

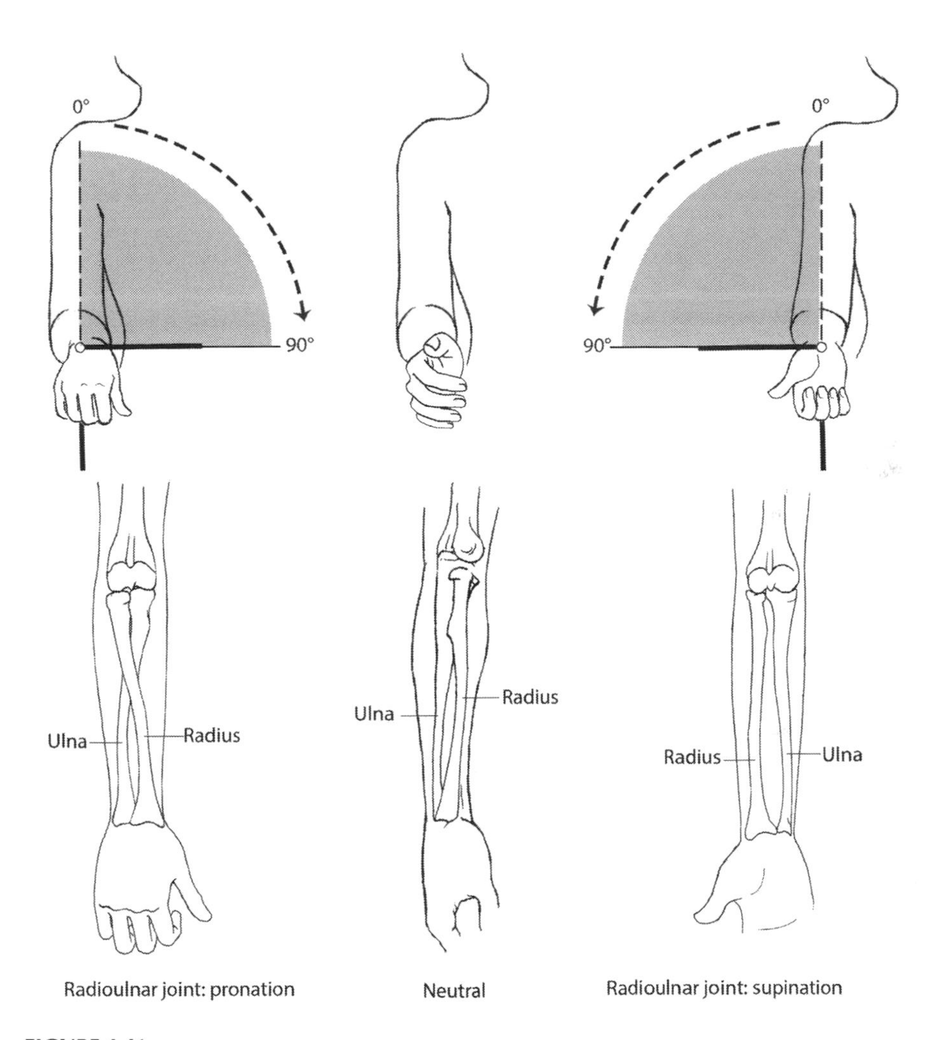


**FIGURE 4.41**
Elbow range of motion, forearm supination/pronation, with bone positions.

of the forearm. These motions are controlled by muscles in the volar and dorsal forearm (refer to Figure 4.23). See Chapter 7 for specifics on the radiocarpal joint.

### 4.15.4 Upper Torso Range of Motion: Conclusion

Products covering all or selected areas of the arms must move at the shoulder, elbow, and forearm. Understanding coordinated motion throughout the arm, shoulder, and rib cage and then applying that knowledge can improve designs. In a 2005 report, the International Society of Biomechanics proposed standardized terminology for joint motions of this region to facilitate communication among researchers (Wu et al., 2005). The details in that report may be useful for designers working on mechanical devices or assistive products mimicking specific joint motions. In an attempt to establish norms for a specific set of upper extremity motions, van Andel, Wolterbeek, Doorenbosch, Veeger, and Harlaar (2008) go a step further, describing the process, landmarks used, and 3D technology to assess upper extremity tasks involving the arm and pectoral girdle. In addition to the ROM activities described above, their participants completed four complex arm/pectoral girdle sequences used in daily activities.

## 4.16 Upper Torso and Arm Landmarking and Measuring

Upper torso and arm landmarking and measuring methods depend on many factors. With the wide array of products designed for the upper torso and arms, a comprehensive discussion of landmarking and measuring all products is not possible in this chapter. Information on unique aspects of upper torso and arm fit, and bra fit and sizing as a specialized upper torso product, are presented. While reading this section, refer to landmarking and measuring foundations in Appendix A, and details of landmarking and measuring the upper torso and arm in Appendix C.

### 4.16.1 Landmarks and Measurements for Upper Torso and Arm Designs

As in other body regions, the choice of landmarks and measurements for products for this region depends on product purpose, coverage, and materials. *Sizing in Clothing* (Ashdown, 2007) provides guidance for landmarking and measuring many types of apparel worn on the upper body. More resources are listed in Appendix A. When landmarking and measuring for specialized products—health monitoring devices, compression garments, braces and splints, impact protective vests, and more—search for guides specific to the product or alter general guidelines, based on your knowledge of the underlying anatomy.

Upper torso products are designed for function and/or fashion. Coverage varies for each product type. Some products, like tailored jackets, cover the torso and arms; and others, like a music playback device worn on the upper arm, cover a limited area. Materials are chosen to implement product purpose and to work with the body area(s) that are covered.

Tailored jackets require a detailed landmark and measuring process to fit the many curves and contours that must be accommodated with a close fit. The multi-contoured area of the glenohumeral joint and the axilla is particularly difficult to landmark and measure, because motion changes the relationships of the body parts. Individual morphological variations, especially around this area, make fitting a range of sizes difficult. Appendix C provides basic guidelines for landmarking and measuring the body to draft the garment armscye—the seam that encircles the glenohumeral joint/axilla in a *parasagittal* plane. The garment armscye may be visualized in two parts. With a person standing, and palms facing toward the sides of the body: (a) the upper armscye seam curves from the front of the body (at the juncture of the arm and the torso), over the acromion, to the back of the body (at the juncture of the arm and the torso); (b) the lower armscye is U-shaped, cupping below the axilla and joining with the upper armscye, front and back. The acromion may be thought of as a starting point for fitting the shoulder area of a tailored jacket. The acromion landmark is an anatomical feature common to all humans. However, the position varies depending on posture; and muscle and fat may obscure the bony feature.

Dimensions and placement of the upper armscye may be based, more realistically, on visualizing the seam on the body in relationship to the acromion. For example, if a jacket includes a shoulder pad the armscye seam is positioned a distance above and lateral to the acromion. Patternmaking directions for the lower armscye often state that the lowest point of the U-shape is approximately 2.54 cm (1 in.) below the axilla; however, as described earlier in this chapter, the axilla changes with motion, affecting how the lower axilla can be landmarked. Current body scanning methods cannot define the axilla area with the arm at rest, as it is obscured by the arm, and raising the arm to expose the axilla drastically changes its shape. New methods of defining the shape and dimensions of the axilla could lead to improved tailored garment fit.

Products other than tailored garments also use the acromion landmark as a reference point. Products requiring straps over the shoulder to support a product, like a backpack, may use acromion location relative to the neckpoint to determine optimum placement. Protective hard-shell products, like hockey and football shoulder pads, include a stand-off distance from the body (see stand-off applied to helmet designs in Chapter 3). The product relates to the acromion location and the glenohumeral joint to allow motion.

Apparel is usually gender specific, with unisex apparel in the ready-to-wear market and prevalent for military apparel. This chapter presented some of the problems when designing unisex apparel due to upper body skeletal and soft tissue gender differences. Upper body unisex apparel often uses simple, oversized T-shapes that do not require precision fit at the chest/bust and armscye. Key dimensions for simply-shaped men's upper body wearables are the neck circumference (discussed in Chapter 3), chest circumference, and combined back width and arm length—from the cervicale, over the acromion, to the olecranon or to the ulnar stylion. Upper arm and wrist circumferences may also be measured. Instructions for measuring upper arm circumference vary from measuring the largest circumference (flexed or relaxed) which depends on development of the biceps and amount of fat; to locating the measurement half-way between the shoulder point and the elbow point. Both methods can result in measurement variations. For most products measuring the largest circumference makes sense to ensure that the product will fit the largest person. Products worn only on the upper arm, such as music playback and physical activity monitoring devices, rely on close fit and compression to stay in place rather than product shaping or suspension.

Various materials are used in products to fit the upper body. Protective wearable materials are often stiff or bulky and don't conform to the body because they have limited shaping capability. Chest protectors for baseball catchers made of thick materials with sectioning allow only partial contouring to the body curve of the crouching catcher. Basic length and breadth measurements suffice, if the smallest and largest person can be covered and protected adequately from the neck to below the crotch.

Designers are accustomed to working with materials that have stable and predictable physical characteristics. New materials are being developed that have changeable characteristics. Branson and Nam (2007) describe innovative materials with the potential of custom-adjusting to an individual's body, requiring a new approach to fit and sizing. *Auxetic materials* are available for use in impact protective wearable products. These materials are somewhat flexible but react to high-speed impact by changing material properties from soft and flexible to stiff and protective (Branson & Nam, 2007). The designer has to predict how size and fit change as the material changes. Carneiro, Miereles, and Puga (2013) state that auxetic textiles are being developed to not only increase energy absorption, but also to improve drape to shape to the body.

### 4.16.2 Focus on Bra Fit and Sizing

Bra manufacturers try to fit many breast sizes and shapes without custom fitting each woman. Manufacturers use body measurements in two ways when developing a bra design: (a) to draft bra pattern dimensions and (b) to convey size designations to the consumer. Three basic body dimensions are used for these purposes, each relating to a specific garment component: (a) chest/rib cage circumference below the breasts relates to the bra band measurement, (b) breast volume relates to bra cup size, and (c) length over the shoulder from the top boundary of the breast to the bra band level at the back relates to bra strap length. Assumptions about other body and product dimensions are made from these three measurements. Data on over-the-shoulder length measurements related to bra strap length are not readily available, probably because many bra designs feature adjustable length straps. However; Chen, LaBat, and Bye (2010, 2011), identified bra straps slipping off the shoulder as a major complaint for women with sloping shoulders. Luk and Yu (2016) present a detailed guide to assessing bra fit on the body with guidelines for improving fit. Some of the suggestions may be useful when designing bras.

### *Chest/Rib Cage and Bra Band*

Determining the bra band dimension in the design process is based on a direct measure of the body, so measuring the rib cage at the appropriate band level is important. Most bra draft directions specify measurement of the total rib cage circumference directly below the breasts; often called the underbust measurement (at the inframammary fold). However, Burgess (2000) notes that women's rib cage proportions vary with some women being "broad-backed" in relationship to bra band sizes that do not distinguish proportional differences. In other words, they do not account for the differences between the width and positions of the breasts (anterior chest) and back (posterior chest).

Two methods are used to determine *bra band size designators*—the bra size a woman selects when purchasing a bra. The *imperial system* uses a formula applied to the underbust measurement calculated in inches. The bra band designator is calculated by adding 5 inches to an odd number measurement, or 4 inches to an even number measurement. Cup size is then determined by subtracting underbust from bust circumference, for example a one-inch difference is an A cup. Britain uses this system. The *metric* system—used by America and most European and Asian countries—is direct with the underbust measurement as the bra band size designator (Zheng, Yu, & Fan, 2006).

McGhee and Steele (2006) evaluated current bra measurement methods and point out an additional need is to correlate bra band size with the respiratory cycle. They suggest substituting a chest circumference measurement taken at the end of a voluntary exhalation, without any mathematical manipulation, to establish band size. However, given the importance of full chest expansion to provide maximum performance in athletic activities, this measurement method may not be appropriate for sports bra design. The percentage stretch of the bra band material will also affect fit and comfort. More research is needed in developing simple bra band measuring methods and size designator systems.

### *Breast Volume and Bra Cup Size*

An equally important, but difficult to quantify, dimension for determining bra design and fit is breast volume. Bra cups are sized and shaped to accommodate breast volume which varies woman to woman, depending on whole body size and amount of mammary gland and adipose tissue in the breast. Manufacturer's methods of determining bra cup shapes and dimensions are proprietary to each manufacturer. There are no standards for determining cup size related to breast volume (Zheng et al., 2006). In 1933, S. H. Camp and Company introduced letter cup sizing—A, B, C, D cups—as a reflection of breast size and pendulousness (Farrell-Beck & Gau, 2002, p. 72). Letter cup size designators have become the accepted method of conveying breast volume for size selection, and the number of cup sizes has increased to include sizes AA to G (Zheng et al., 2006, p. 48). The usual directions for selecting cup size are: (1) while wearing a bra, measure the fullest circumference of the breast to provide an estimate of breast volume, (2) measure the circumference of the chest directly below the breasts, (3) subtract the two measurements and (4) consult a manufacturer's chart to determine cup size. For example, a one-inch difference usually is an A cup, a three-inch difference is a C cup. This method of labeling bra cup size is used by many manufacturers, but each company determines how cup size and shape relates to the designators. Many manufacturer's size charts state that sizing is approximate and that larger women may have to try on several sizes to get good fit.

Research on breast volume with applications to bra design may lead to new methods of sizing and shaping bra cups. Research is conducted for cosmetic and reconstructive surgery and beginning to appear for design applications. Many of the medical studies use equipment and analysis methods that are not used by designers; and applying the findings to bra design is not a goal of the research. Adapting research results for bra design may be possible with collaboration between the two fields.

Shin (2014) reviewed several breast volume methods used by surgeons. Many methods focused on defining and then surgically replicating the "perfect" breast, having questionable application to bra design for most women. Pechter (1998), a plastic surgeon, suggested a measurement technique using a mammary hemi-circumference measurement, measuring the nude breast in a supine position from the lateral boundary, over the nipple, to sternum midline. He correlated the measurement to bra cup sizes, for example a seven-inch measurement corresponded to an A cup. However, he did not provide guidelines on determining band size. Coltman, McGhee, and Steele (2017) developed a method to document breast volume using 3D breast scanning with the woman in a *prone* (horizontal, facing downward) position, breasts freely suspended. This technique improved the accuracy of breast volume measurements for women with large, *ptotic* (drooping) breasts; however, guidelines for applications to bra sizing were not described.

A letter in the *Journal of American Society of Plastic Surgeons* (Tadisina, Frojo, Plikaitis, & Bernstein, 2016) suggests that plastic surgery residents familiarize themselves with the basics of bra sizing to better communicate with patients. However, the medical field also finds bra sizes as used in the apparel field confusing and inconsistent.

McGhee and Steele (2006) used an adapted water displacement device to measure breast volume of 104 women, and compared volume to the size of professionally fitted bras for each participant. They found a range of breast volumes for one bra size, and that volume of any cup size does not correspond to one band size. The study suggests that better directions for bra size selection should be developed to help women understand bra band and cup size determinations.

New methods for bra design applications use computer technologies. Chen and Wang (2015) developed a method to measure breast volume using mesh projections from 3D point cloud data from body scans. They verified results by comparing the mesh projection results to results from a standard water volume displacement method. They also proposed a method of determining breast boundaries that has applications in body scanning (see descriptions in Appendix C).

One major difficulty is that current scan-generated body models are rigid, inelastic, and incompressible—much different from actual body tissues. Cai, Yu, and Chen (2016) conducted preliminary research to develop a method to simulate bra fit. They state that to realistically simulate bra fit, new strategies need to be developed. They advise "using experimentally derived data from high-resolution body scans, biomechanical analysis of the viscoelastic properties of breasts, and tests to measure the mechanical properties of multilayer bras" (p. 151). Keep in mind that the *viscoelastic properties* of breasts vary from individual to individual and in any one individual over time. Variables include: proportion of mammary gland tissue to fat, cyclic changes in hormones, and mammary gland tissue activity related to pregnancy and breast-feeding.

The interacting variables of body, materials, and product shaping for bra design demonstrate the necessity of in-depth knowledge of all the components involved in designing an aesthetically pleasing, comfortable, and functional product.

## 4.17 Designing for the Upper Torso and Arms: Conclusion

When designing wearable products for the upper torso, arms, and spine, incorporate skeletal variations, vulnerabilities, and capabilities in your designs. Consider the functions, vulnerabilities, and variabilities of thoracic and abdominal organs. Keep the normal mobility of the organs in mind. Avoid products that either constrict, continuously displace, or otherwise limit natural organ functions. Incorporate motion into a design to the extent possible for axial and appendicular skeletal functions. Build the design on the body, for the body; choosing the details of the underlying anatomy that are appropriate for the design goals.

# 5

# *Designing for Lower Torso and Leg Anatomy*

Lower torso wearables include products varying from skirts and pants of all styles to highly personal products. Wearable products for the lower torso often incorporate features for both privacy and protection related to the anatomy, with its structures for elimination and reproduction. Day-to-day lower torso garments serve many functions, and generally shield the external genitals from view. Other wearable products serve to protect the genitals from direct trauma or infection, manage *menstrual* bleeding or incontinence, or prevent pregnancy.

Lower torso fashion wearables range from "second skin" garments like women's tights to a dirndl skirt (see Chapter 1, Figure 1.8). Wearable medical products for the lower torso vary from *ostomy* bags to collect feces after removal of sections of the intestines to medication patches. There is an equally wide range of products and devices for the legs, from pantyhose, socks, and leg warmers to trauma protection, knee braces, and prosthetic limbs. Some wearables for the legs, like shin guards, may extend to cover a portion of the ankle. Shoes and other wearables for the feet are described in Chapter 8.

Products that cover the lower torso and the legs must transition from a cylindrical shape to two cones, becoming a *bifurcated garment*. Fitting a bifurcated garment requires locating the waistline (discussed in Chapter 6). Pants cover the lower torso and extend as two tubes of varying lengths, pants legs, to cover the body's legs. Products that cover only the leg, or are attached to the leg, are difficult to shape and fit due to the leg's tapering conical form. We use the common term "leg" to mean the entire lower limb, both the thigh (proximal limb section) and the anatomical *leg* (distal limb section).

**Key points:**

- Basic anatomy of the lower torso and legs:
  - Terminology: leg, *lower torso region,* bony pelvis, pelvic girdle, pelvis, *pelvic outlet, pelvic diaphragm, perineal region*
  - Systems and vital organs: reproductive (female: uterus, ovaries, and vagina; male: testes, prostate gland, and penis), digestive (parts of the small and large intestine, appendix, rectum, anus), lymphatic (lymph collectors and lymph nodes)
  - Systemic components passing through, to, and from the lower torso and legs: blood vessels, spinal nerves and peripheral nerves, and lymphatics

  - Bony appendicular skeletal structures for protection and mobility: pelvis and legs
  - Muscles, ligaments, and joints involved in spine, pelvic girdle, and leg motion
  - Key soft tissues: male and female external genitals
- Wearable product opportunities and challenges for the torso and legs:
  - Coverage allowing privacy, protection, and movement
  - Compensating for impaired lower torso functions
  - Aesthetics with function
- Fit and sizing:
  - Motion as a fit consideration
  - Gender requirements

If you place your hands "on your hips" you can identify the top of the lower torso region. A horizontal plane through this edge lies approximately at the level of the belly button or navel. Structures inferior to this level are sometimes discussed in terms of being "in the pelvis." Garment waistlines, as discussed in Chapter 6, often generate a different defining plane between the upper and lower torso. Many lower torso anatomical structures and functions are gender specific and affect wearable product design structures and materials.

## 5.1 Lower Torso: Overview

Bones, muscles, and connective tissues of the lower torso form a container for the internal organs (Figure 5.1). The external genitals are, perhaps, the most obvious feature of the lower torso for designers; however, gender differences in the bony pelvis—shape, width, and hip joint location—also affect product designs. The bony and muscular structures of the lower torso must also coordinate with the actions of the legs. Sometimes, the lower torso is referred to as "the hip" —a term which is both nonspecific and insufficient to describe the many structural features and functions of the lower torso. For purposes of this chapter, the anatomy of the lower torso which begins at the superior edge of the bony pelvis in the posterior and lateral torso will be described in anatomical terms.

Carefully defining the anatomical terms used to describe lower torso structures facilitates communication, including communication between medical and design professionals:

*Lower torso region*: The portion of the torso between the top of the bony pelvis to and including the external genitals, but excluding the legs

*Bony pelvis*: The ring formed by the sacrum and coccyx of the spine with the R and L "hip" bones, at the slightly mobile sacroiliac joints (see details of "hip" bones in Section 5.5.2)

*Pelvic girdle*: The R and L hip bones joined at the *pubic symphysis* in the front of the body, the lower torso equivalent of the pectoral girdle (clavicle and scapula)

*Pelvis*: The bowl-shaped cavity of bone, muscle, and soft tissues to hold and cradle internal organs of the urinary, digestive, and reproductive systems, bounded by the anterior abdominal wall

*Pelvic outlet*: The opening in the bony pelvis allowing the urinary, digestive, and reproductive systems to pass to the body's exterior for expulsion of waste products or birth of a child

*Pelvic diaphragm*: The sling of muscles spanning between the pelvic bones, above the perineal region, and supporting the internal organs (see Section 5.6.1)

*Perineal region*: The skin and soft tissue structures between the legs including external genitals and anus; lying in a more or less transverse plane

*Anatomical crotch*: The angle between the lower torso and the medial proximal thighs, extending from the pubic bone anteriorly to the buttocks crease posteriorly

*Product crotch*: The product section covering the anatomical crotch—in a variety of product shapes and sizes

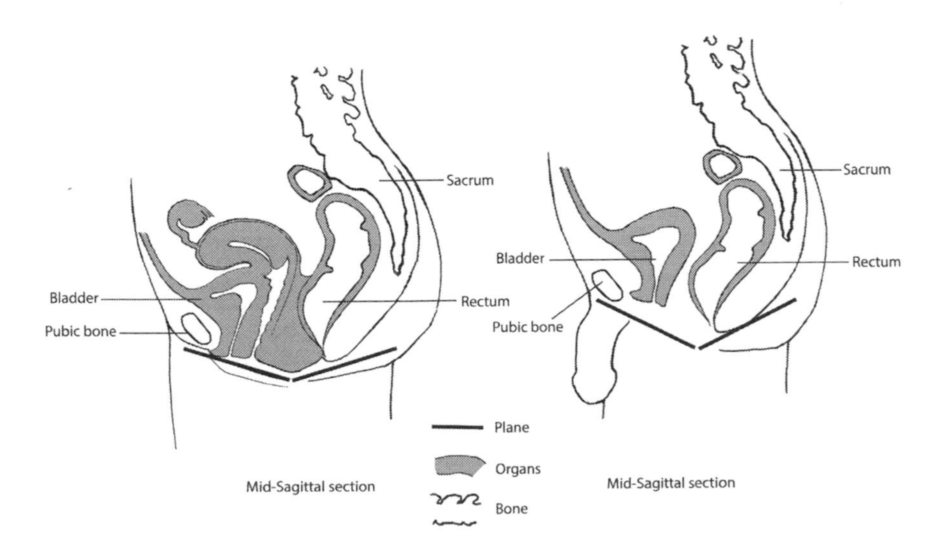


**FIGURE 5.1**
Overview of lower torso, female and male.

The external genitals of the perineal region, and the muscle and fat at the top of the thigh influence both the shape of the *crotch* of men's and women's undergarments and pants and the volume of space available beneath the perineal region for absorbent products. See Chapter 6 for discussion of pants' crotch seams and shapes related to waistline placement.

Because of the overlap of structures and functions between the upper and lower torso, much of the internal anatomy of the lower torso was introduced in Chapter 4. The digestive system and the urinary system span the upper and lower torsos. Many products described in this chapter relate to managing urine and feces, including incontinence. The reproductive system, located in the lower torso only, has both internal and external components and shares some structures with the urinary system. The lower torso also contains elements of the skeletal, circulatory, nervous, and lymphatic systems in addition to fat and muscles. Sections 5.2 through 5.4 describe the lower torso region's digestive, urinary, and reproductive organs.

## 5.2 Digestive System in the Lower Torso

The small intestine, large intestine, rectum, and anus—the hollow digestive organs in the lower torso region—continue the digestive process begun in the upper torso (Figure 5.2). For details on digestion, the body's process of extracting nutrients and water from food and liquid, review Chapter 4, Section 4.8 and Chapter 2, Figure 2.20. The small intestine loops back and forth throughout the abdomen and pelvis, with no structural division between upper and lower torso. It attaches to the large intestine in the R lower quadrant of the abdomen near the R lateral pelvis.

### 5.2.1 Large Intestine: Structure and Function

The ascending, transverse, and descending colon of the large intestine are shaped like an inverted "U." In the anterior view of Figure 5.2, the ascending colon is on the L and the descending colon on the R. The transverse colon spans between these segments in the upper torso abdominal compartment. The *vermiform appendix,* commonly called the *appendix,* is a skinny sac-like structure pouching out to the side at the beginning of the ascending colon on the R side of the body. The appendix is perhaps best known for getting infected and requiring surgery to remove it.

At the distal end, the descending colon makes an S-shaped curve to connect with the rectum in the posterior central pelvis. The rectum ends in the anal canal and anus, where the digestive system exits the body, controlled by the muscular *anal sphincter.* The circular sphincter serves to open and close the anus, releasing or retaining the feces. Infants wear diapers until

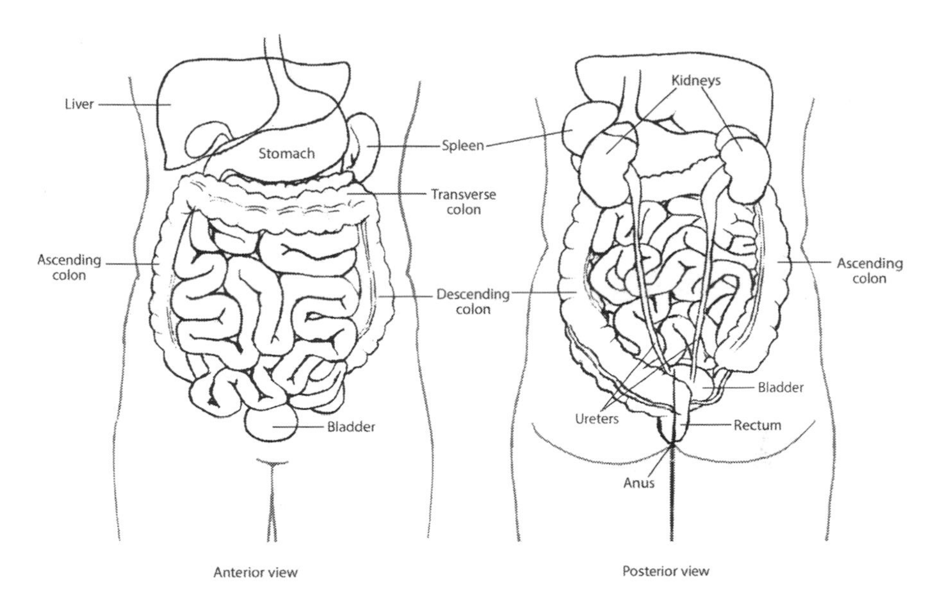


**FIGURE 5.2**
Digestive and urinary systems in the torso.

their nervous systems allow them to control the sphincters. Poor control of the anal sphincter can require an adult to wear an *incontinence management product* (IMP) a padded underwear-like garment to collect and absorb feces and/or urine. The anus lies in the perineal region (Figure 5.3) in front of the coccyx (tailbone). The anus' location is identical in men and women. In the anatomical position, the plane of the anus is tilted somewhat posteriorly as seen in Figure 5.1 (Drake, Vogl, & Mitchell, 2015, p. 502). It is useful for IMP designers to know that the anus lies in a different plane than the other openings in the perineal region (see Sections 5.3 and 5.4).

### 5.2.2 Digestive System: Evacuation of Feces

Once the small and large intestines have finished their digestive actions, the non-digestible waste, a variable amount of liquid, and large numbers of bacteria from the gut microbiome, in the form of feces, collect in the distal colon, rectum, and anal canal (Figure 5.4). The bacteria of the colon play essential and beneficial roles in digestion but can cause serious illness if they invade the urinary tract. Careful perineal region hygiene after *bowel movements* and use of clean and absorbent underwear helps to prevent urinary tract infections (UTIs). Hand washing after changing diapers or IMPs and after using the toilet also helps keep gut bacteria from spreading beyond the intestine.

The evacuation of digestive waste from the body as a bowel movement is the final step in the digestive process. *Stool* (feces, *fecal material*) is eliminated from the body with defecation as a voluntary action, at a socially appropriate

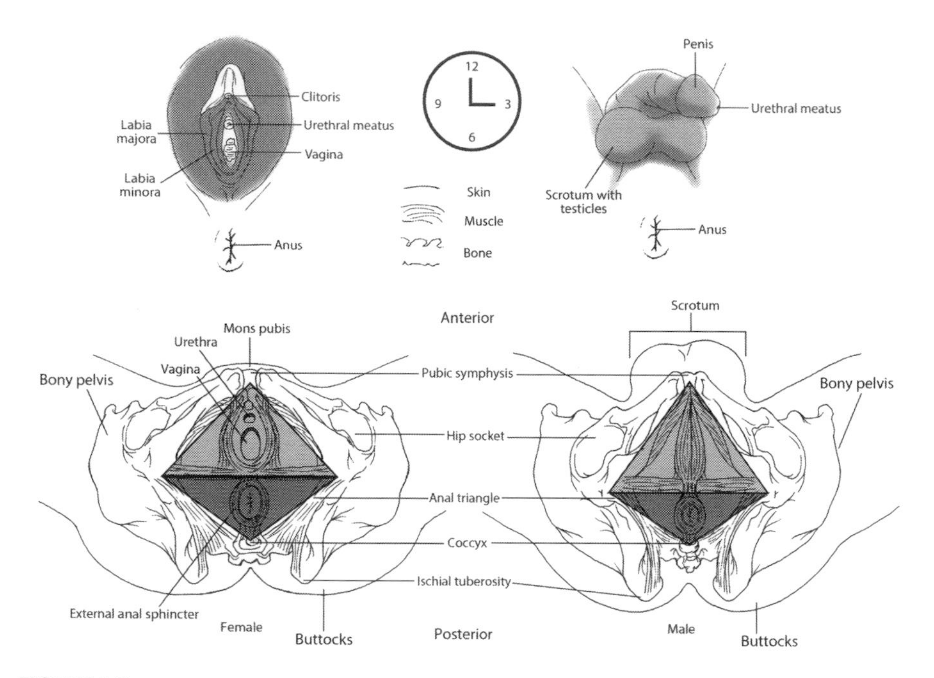


**FIGURE 5.3**
Female and male external genitals, inferior views: Relationships of the digestive, urinary, and reproductive system outlets to the female and male bony pelvis.

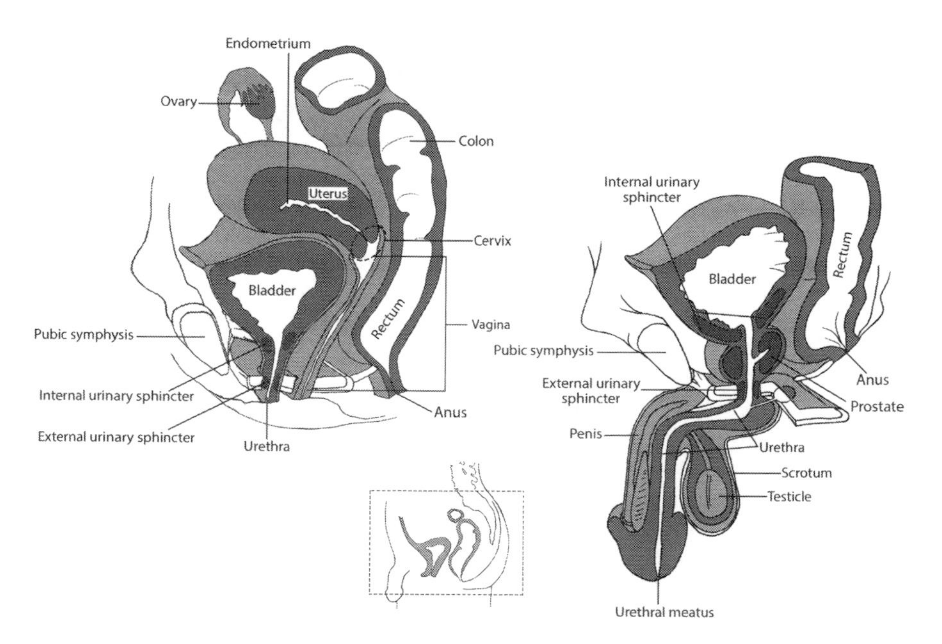


**FIGURE 5.4**
Quarter view of pelvis, female and male, with digestive, urinary and reproductive structures.

time and place. Children learn to control their bowel evacuations (and bladder emptying) in a process called toilet or "potty" training. As they gain control of both their bowel and bladder, toddlers transition from traditional diapers requiring an adult's assistance to self-managed underwear—disposable or washable *training pants*.

The consistency of the feces partially reflects how well the body is hydrated. Soft feces indicate an excess of liquid in the body. Diarrhea, frequent watery bowel movements, indicates liquid reabsorption in the colon is poor, due to disease or illness. Design of diapers, IMPs, and training pants requires an understanding of the process of elimination, of the possible consistencies of the eliminated feces, and of both stool volumes and speeds of rectal emptying. Moisture and fecal soil collected in a diaper or IMP can cause skin irritation, rashes, and infections, so these garments also need to allow easy removal and replacement as well as facilitate cleaning of the perineal region.

Wearable product size, strength, absorbency, moisture barriers, infection prevention, and odor control are factors to consider when designing these products (Wiesemann & Adam, 2011). Successful containment of a rapidly evacuated large liquid stool requires different design features than collection of a small, dry, and hard stool expelled over a longer period of time. Unfortunately, stool characteristics can vary without warning so effective designs need to address both extremes of stool quality. Experienced caregivers for children in diapers or adults using IMPs recognize the importance of erring on the side of extra absorbency and effective barriers against leakage.

The defecation process is of considerable medical interest, as good health depends on regular bowel emptying. Studies of the physics of defecation in mammals varying in size from cats to elephants (Yang, LaMarca, Kaminski, Chu, & Hu, 2017) may help shed light on human defecation. The studied mammals produce cylindrical feces as do humans and there are physical similarities between human intestines and these mammalian digestive tracts. Among their findings: (a) defecation in mammals depends on mucus to aid the physical movement of feces through the rectum and anal canal, (b) "The diameter of feces is comparable to that of the rectum, but the length is double that of the rectum, indicating that not only the rectum but also the colon is a storage facility for feces" (Yang et al., 2017), and (c) the defecation time is relatively constant at 12 seconds, +/− 7 seconds. These findings may help designers of diapers, *maximum absorbency garments* (*MAGs*) used in space travel, and IMPs to better understand the time frame for the process of defecation, and to predict fecal bulk.

The ease of passage of feces from the rectum provides information about the colon's mucus layer. If the mucus layer is absent or thinned (Yang et al., 2017) or the body has aggressively reabsorbed water to try to balance low fluid intake, *constipation* results. With this condition, feces may be dry, hard, and/or difficult to eliminate. Constipation can cause *hemorrhoids*, dilation of veins, a common medical problem. Hemorrhoids occur within the anus and/

or in the canal above it and occasionally the mucous membrane of the anus cracks, causing a painful *anal fissure*. *Suppositories*, waxy bullet-shaped cylinders containing bowel stimulants and lubricants, have been developed to treat constipation. Suppositories are medications, not wearable products, but a specially formed suppository may be considered a wearable product. The product features a side vent to allow passage of bowel gas. It is designed to be inserted and worn in the anus for four to six hours at a time to treat hemorrhoids or anal fissures (Vail, 2007).

Some medical conditions and treatments change how waste is eliminated, requiring specialized products. Surgical removal of a large part of the intestine due to cancer or other bowel diseases means the body needs an alternative pathway to eliminate waste. The current approach is to surgically connect the remaining intestine to an *ostomy*, an artificial opening in the body wall. Ostomies are commonly located somewhere in the rectus abdominis muscle. The ostomy has no sphincter-like control, so the intestinal contents flow unimpeded with each intestinal wall contraction and are collected in a pouch attached to the abdomen with an adhesive seal. The challenge for ostomy bag/collection device designers is to (a) create a collecting device that will adhere securely to the body without causing irritation, (b) develop a pouch of enough volume to collect semisolid stool for several hours, (c) provide a convenient way for the wearer to empty the ostomy bag both at home and away from home, (d) control the odor associated with the fecal discharge, and (e) provide a bag that will be relatively inconspicuous under clothing, whether the ostomy is positioned above or below the person's waist (Wasserman & McGee, 2017). Until new medical solutions are found, improving ostomy bag/collection devices is a major challenge.

## 5.3 Urinary System in the Lower Torso: Structure and Urine Production, Collection, and Release

Understanding the relationships and positions of the digestive, urinary, and reproductive openings in the perineal region is important when designing many products—underwear, swimwear, diapers, and more. The liquid urinary waste product, urine, may also need to be managed with a wearable product. Some products are designed to contain both feces and urine, while others are designed to absorb or collect urine only. Urine is produced by the kidneys (Chapter 4, Section 4.8.3), collected in the bladder, and voluntarily or involuntarily expelled from the body. Involuntary loss of urine (urinary incontinence) increases in adults with age and with obesity. A major concern when designing products to manage urine output is preventing leakage, so understanding urinary tract structure, urine production, and *bladder capacity* is crucial for designers of these products.

### 5.3.1 Urinary System: Structure

Urine drains from the kidneys in the upper abdomen, into the ureters passing through the upper, middle, and lower abdomen to reach the urinary bladder (refer to Figure 5.2). The bladder is a hollow, muscularly-walled but stretchy organ. As illustrated in a relatively empty state in Figure 5.4, it is found low in the lower torso region behind the pubic symphysis, the bony structure at the bottom front of the abdomen. Urine production and flow to the bladder is continuous. It is anatomically difficult to completely empty the bladder; at least 5–10 ml (0.17–0.34 oz) will remain after each *void* (single bladder emptying). The bladder is supported within the pelvis by a fascial covering and held in its anterior location by *suspensory ligaments.*

#### *Anatomical Bladder Features*

As the bladder fills with urine, it can increase to 100 times or more its emptied capacity. To accommodate this expansion, the upper surface of the bladder rises, along the anterior abdominal wall, extending as high as the level of the umbilicus (Moore, Agur, & Dalley, 2011, p. 226). Wearables—body shapers, bustiers, or tight belts and pants—that compress any segment of the abdominal wall will limit the bladder's natural ability to enlarge, resulting in either a more frequent need to empty the bladder or displacement of the bladder into another part of the lower torso region. Wearable product fit over the lower abdominal wall may be affected by the urine volume in the bladder. In extreme cases of bladder overfilling, displacement, or structural abnormalities, urine can be forced back into the ureters, which can damage kidney function or cause kidney infection—if the urine has bacteria in it.

#### *Male and Female Urethra and Urethral Meatus: Differences*

The *urethral meatus* is the external opening of the urethra, the tube connecting the bladder to the outside of the body (Figure 5.4). Differences in urethral length and the location of the urethral meatus between men and women affect the design of incontinence products and urine collection devices. The female urethra is short, approximately 2.5 cm (1 in.). The male urethra extends approximately 22 cm (8.7 in.) on its course through the *prostate gland* (an internal male reproductive structure) and the penis (Moore et al., 2011, p. 232). In young adulthood, the male's walnut-sized prostate weighs about 20 g (0.7 oz). As men age, *hypertrophy* (enlargement) of the prostate or the development of prostate cancer can distort the urethra and impede the flow of urine. Urinary sphincters, functioning much like the anal sphincters, constrict the urethra to help prevent urine leakage and allow urine flow when voluntarily relaxed. The male and female urinary sphincters are located at different points along the urethra. Urine is normally free of bacteria. However, bacteria can enter the bladder through the urethra, at the urethral meatus. Bacteria from the perineal skin, underwear, or from feces contained in IMPs can invade the bladder and cause infections more easily in women than in men. In women, the urethral meatus is located near the front of the perineal region inside the *labia*, just in front of the *vaginal opening*. In men it is located at the tip of the penis. The location of the urethral meatus is a primary reason women squat or sit to void and presents a continuing challenge for design of urine collection devices for female pilots and astronauts. Attaching an external urine collection device directly to the region of the female urethral meatus remains an unsolved problem, so devices are designed to encompass a larger area around the urethral meatus and the external genitals. *Female urination devices* have been developed to allow women to stand to void—a useful action while hiking, camping, or using dirty bathroom facilities. For example, Parks and Block (U.S. Patent 7,682,347 B2, 2010) developed a silicone funnel-like device that fits over the labia and directs a stream of urine out in front of the body when a woman voids from a standing position.

#### *Urinary Retention and Urinary Incontinence: Catheters for Bladder Drainage*

Prostate volume changes are the largest single cause of *urinary retention*, the inability to fully empty urine from the bladder. Abnormal bladder sensation, bladder wall muscle weakness, as well as other obstructions also cause urinary retention (Selius & Subedi, 2008). Urinary retention is seldom a complete inability to void; a person may pass a moderate volume of urine but retain up to several hundred ml of urine in the bladder. Consistently retaining more than 100 ml (3.4 oz) of urine after voiding or being unable to void large amounts of urine—i.e. over 1500 ml (50.7 oz)—are both considered significant retention needing treatment.

A *urinary catheter*, a tube to drain urine, can be inserted through the urethra to relieve retention. A catheter left in place for a period of time is an *indwelling catheter*. The catheter is attached to tubing which attaches to a urine collection bag suspended from the body in some manner. The catheter, tubing, and collection bag as a unit are a wearable product. An indwelling catheter left in place for a period of time increases risk of urinary tract infection (Warren, 1997). The wearer or a care giver must drain the urine periodically from a valve at the bottom of the bag. Although catheter designs have changed little in over 100 years (Warren, 2001), it is possible that improved designs or materials used in catheters, collection tubes and bags may decrease infection risk.

Indwelling catheters are also placed to manage urinary incontinence, most often for patients in nursing homes or other care facilities. In general, men experience incontinence less often than women (Milsom et al., 2014). The risk of male incontinence increases (a) after surgery for prostate enlargement, (b) after prostate cancer surgery, and (c) with age.

Fortunately, for men, a *condom catheter* is an alternative to help manage urinary incontinence (Figure 5.5). Instead of placing a catheter into the bladder, a pliable *condom*-like sheath which fits over the penis is connected to tubing which runs to a wearable urine collection bag ranging in volume from

250–1000 ml (8.5–34 oz). An anti-reflux valve prevents back flow into the tubing. The bag is typically attached to the medial leg between knee and ankle with two straps that adjust to fit around the top of and bottom of the calf. The weight of the bag increases as the bag fills with urine, making bag attachment and suspension increasingly difficult with long periods of wear. The entire device can be concealed under loose-fitting trousers; however, easy access to the emptying valve is necessary.

Compared to an indwelling catheter, this device appears to decrease the risk of UTI (Warren, 2001). Keeping the sheath in place can be challenging. Temperature differences and normal variations in blood circulation to the penis trigger shape and size changes. Condom catheters sometimes slip off the penis, demonstrating the common design problem of anchoring wearables on cylindrical or conical forms without causing constriction. Kinks in the connecting tubing prevent urine passage and cause back-pressure and leaks.

Saint et al. (2006) compared use of hospital patient condom catheters to indwelling catheters. They found fewer medical complications with the condom catheter and fewer complaints of discomfort, restriction of movement,

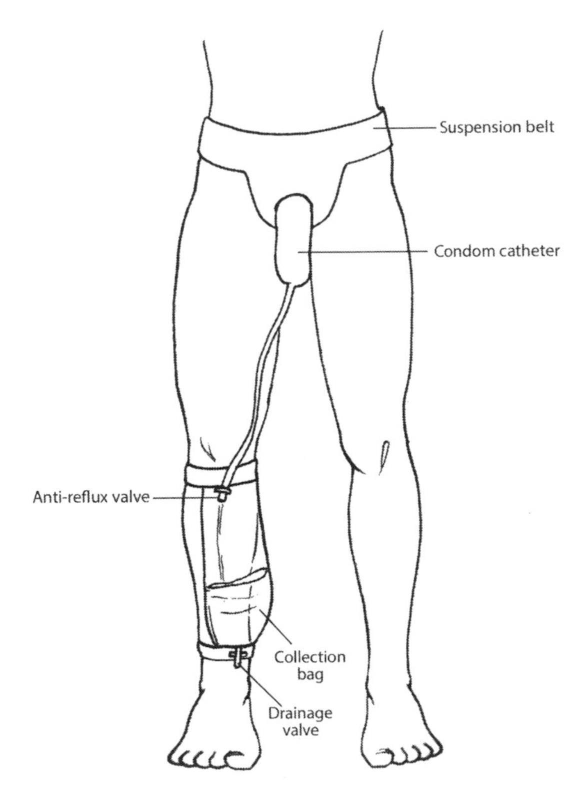


**FIGURE 5.5**
Condom catheter for male urine collection.

pain, embarrassment, or inconvenience. Military pilots, male astronauts, and men working extended hours without access to a bathroom can also use wearable urine collection devices similar to medical condom catheters. Non-hospitalized mobile patients may find any catheter with a leg bag restricting and uncomfortable.

### 5.3.2 Urine Production

Many factors influence the amount of urine produced by the kidneys: (a) fluid volume consumed; (b) kind of fluid, food, and/or medication consumed; (c) environmental temperature and humidity; (d) intensity of work or exercise; (e) sweat production; (f) pulse and blood pressure; (g) position of the feet (higher or lower) relative to the heart; and (h) sleep quality and duration. The same factors influence the design of incontinence products, for example products worn during a sport activity will be different from products worn during sleep.

Amundsen et al. (2007) recorded total daily urine production and volume of urine per void of 161 women—ages 19.6 to 81.8 years —with no history of urinary system problems. Urine production and bladder function varied greatly. Urine output over 24 hours ranged from 437 ml to 3861 ml (14.8 oz to 130.6 oz), with average urine output of 1730 ml (58.5 oz). Single void volumes ranged from 10 ml (0.3 oz) to 1000 ml (33.8 oz). The average void was 245 ml (8.3 oz). The women emptied their bladders, on average, seven times per day, with a range of two to 13 voids. Parsons et al. (2007) studied 92 men and 161 women ranging in age from 19.6 to 84.6 years, to assess differences in urine production and frequency of bladder emptying between day and nighttime hours. Like Amundsen et al. (2007), they found large variations in urine production and void measurements in these participants with no history of urinary problems. Significant for product design, they found that the urinary system adjusted output and bladder emptying during sleep. In general, each void volume was larger at night compared to day in both men and women, but more so for men. With increasing age, total urine production tended to increase at night for both men and women. The variations in urine production, void volume, and voiding patterns demonstrate the need for a wide selection of gender-specific bladder management product sizes, styles, and absorbency rates.

To better understand differences in urine output and bladder emptying patterns consider the following comparisons. A young man, enjoying a fall afternoon drinking beer and playing soccer with his friends will produce a different volume of urine than an elderly woman who is sitting and knitting for the same amount of time in her air-conditioned home, drinking water, with her feet resting on the floor. Even if they both drink the same volume of liquid, the young man may produce a large amount of urine due to the *diuretic* effect of alcohol—despite being physically active. The elderly woman, relaxed and comfortable, may not make much urine at all, particularly if she tends to develop swollen feet when her feet are lower than her heart.

The same two individuals may also have very different urine outputs overnight, despite eating the same evening meal (without additional diuretic foods, beverages, or medications). The young man may sleep well throughout the night and empty his bladder of moderately-concentrated urine in the morning. The elderly woman may have trouble getting comfortable, tossing and turning, and go to the bathroom two or more times through the night. Her swollen feet and poor sleep contribute to her urinary system's production of a larger volume of more dilute urine during the night. She may need to wear an IMP at night, if she has difficulty controlling her bladder or walking to the bathroom quickly. What specific features would you include in an IMP for the woman?

### 5.3.3 Urine Collection: Bladder Capacity

The bladder's sensory receptors help monitor bladder fullness. The terms bladder capacity or *maximum bladder volume* describe the total amount of urine a person can tolerate holding in the bladder. After infancy and early childhood, the bladder is controlled voluntarily and emptied by urination. If a person urinates (empties the bladder) before becoming uncomfortable, the void volume will usually be less than waiting until maximum capacity is reached. Bladder capacity varies over the lifespan. Adult bladder capacity can vary widely for an individual over time and, as reported by Wyndaele (1998), person to person and between men and women. Bladder capacity tends to diminish after age 50 (Van Haarst, Heldeweg, Newling, & Schlatmann, 2004).

Adult bladder capacity can be measured. Bladder capacity is a different measure than the void volumes noted above in the Amundsen et al. (2007) and Parsons et al. (2007) studies. Void volumes in their studies were based on the person's desire and choice to empty the bladder, through the day and night. To determine bladder capacity, an individual is asked to hold his/her urine as long as absolutely possible. Then he/she empties the bladder and the urine volume is measured. Like urine volume measured when a person chooses to empty, the bladder capacities measured with this method are variable. In a study of 60 normal adults, the average maximum bladder volume was 590 ml (20 oz), with a broad range from 17 to 760 ml (0.6 oz to 25.7 oz) (Brouwer, Eindhoven, Epema, & Henning, 1999). The range would have been wider, but two volunteers with capacities of 1400 ml (47 oz) and 1600 (54 oz) were excluded from the study due to technical issues. This information further supports the importance of designing MAGs and adult IMPs in a range of sizes and styles.

Bladder volume is also relevant for design and production of diapers. Obviously, infants and children have smaller bladder capacities than adults. Kaefer et al. (1997, p. 2261) developed equations to estimate bladder capacity in children, for two age ranges. For children less than two years old: age (years) + 2 = capacity (ounces). For children two and older: age (years)/2 + 6 = capacity (ounces). Thus, a normal ten-year-old child is estimated to have a bladder capable of holding 325 ml (11 oz). In their research, Kaefer et al. (1997) also found that girls tend to have slightly larger bladders than boys of the same age. Designers of infant and toddler lower torso garments need to consider both the volume of the diaper (and diaper contents) and how caregivers will change soiled diapers.

### 5.3.4 Urine Release

Normal urination occurs when the pelvic diaphragm muscles and the internal urethral sphincter relax, followed by voluntary contraction of the bladder wall muscle. Bladder emptying is prompt with a strong continuous flow (Lukacz et al., 2011, p. 1029). Urine flows from the bladder through the urethra and past the external urethral sphincter to exit the body. A full bladder normally triggers the urge to void. Bladder irritants can also create an urge to empty—including: (a) concentrated urine, (b) some foods and spices, (c) medications, or (d) a bladder infection. Fear or anxiety can also trigger an urge to urinate. When the bladder wall muscle contracts in response to any stimulation, the sphincters normally relax to allow bladder emptying. However, the urinary sphincters, as well as the pelvic diaphragm, can be voluntarily activated to help constrict the urethra to try to stop urine flow.

Bladder control and emptying problems are common. In addition to urinary retention, discussed in Section 5.3.1, voiding difficulties, which tend to increase with age for both men and women, include *overactive bladder (OAB)* and three classes of urinary incontinence: (a) *urgency urinary incontinence (urgency UI),* (b) *stress urinary incontinence (stress UI),* and (c) *mixed urinary incontinence (mixed UI).* The much longer male urethra may make men less prone to incontinence than women—from all incontinence causes (Irwin, Kopp, Agatep, Milsom, & Abrams, 2011; Tennstedt, Link, Steers, & McKinlay, 2008). The volume and duration of the urine loss with each of these problems can vary, from person to person as well as from one incontinent episode to another. A best-case scenario for IMPs for urinary incontinence would be gender-specific products to effectively collect variable urine amounts over variable periods of time. Extrapolating from 2008 data; Irwin et al. (2011) estimate that worldwide, as of 2018, 423 million people will suffer some form of incontinence (p. 1132).

OAB is defined by Lukacz et al. (2011) as an urgent need to urinate (p. 1027). The condition may be associated with urgency UI—leakage of small or large amounts of urine. Urgency UI affects women somewhat more than men. Stress UI, a problem 12 times more frequent for women than for men, occurs when increased intra-abdominal pressure from coughing, sneezing, or laughing negates the urethral sphincters' and pelvic diaphragm's actions and urine is forced from the bladder (Irwin et al., 2011, p. 1134). Obese women, with increased abdominal girth, experience stress incontinence more frequently than women with lower BMI's (Dwyer, Lee, & Hay, 1988). A large waist circumference—one measure of obesity—may correspond to increased intra-abdominal pressure, which in turn produces UI (Tennstedt et al., 2008). Mixed UI, combined urgency and stress incontinence, impacts women four times more often than men (Irwin et al., 2011, p. 1134). New and improved IMP designs focused on specific types of UI will improve quality of life for many people, especially women.

A rapid or slow bladder emptying time suggests a need for an IMP absorbent material with a matched absorbency rate. Although little information is available for humans; Yang, Pham, Choo, and Hu (2014) report uniformity in normal bladder emptying times for very small to extremely large mammals. They report, "all mammals above 3 kg [6.6 lb.] in weight empty their bladders over a nearly constant duration of 21 ± 13 s [seconds]" (p. 11932). How is it possible for an elephant and a cat to empty their bladders at similar rates? Yang et al. (2014) applied principles of fluid mechanics to analyze bladder emptying. They found "differences in bladder capacity are offset by differences in flow rate, resulting in a bladder emptying time that does not change with system size" (p. 11936). They calculated bladder volume as 4.6 ml/kg body mass, daily urine production as 26 ml/kg body mass and observed an average mammal's urination frequency as 5.6 times/day.

Applying their calculated values to humans, a 70 kg (154 lb) person would be expected to produce 1820 ml (61.5 oz) urine/day (slightly less than a 2-liter soft drink) and void 322 ml (10.9 oz) (small water bottle) six times/day. These averages are in the range of those documented in humans by Amundsen et al. (2007) and Parsons et al. (2007). There is no apparent anatomical reason the 21 seconds +/− 13 seconds rate of bladder emptying would fail to apply to normal people. This kind of data can help designers gauge how quickly a diaper, MAG, or IMP needs to soak up urine, how much volume each product unit must contain and control, and the number and frequency of product changes needed throughout a day.

### 5.3.5 Incontinence Management Products

Bladder health and function impact people's daily lives from birth to extreme old age. An inability to control elimination of urine (and/or stool) creates significant disability—chiefly related to social embarrassment. Diverse disposable products, from diapers for newborns to different types of IMPs to meet the elderly's bladder and bowel regulation problems, have been used since the 1970s. Product designs combine materials and barrier, transport, and absorbent layers into structures to contain liquid and solid waste, manage stool bulk, minimize odor, and to keep wetness and feces away from the skin and the urinary and reproductive system structures of perineal region.

IMP design features, material choices, product shape, and positioning on the body differ depending on specific uses (DeMarinis, Kaschak, & Newman, 2018). A range of fibers absorb body waste. Absorbent materials include natural and synthetic fibers in various lengths and forms: cellulose, including mechanically and chemically modified cellulose, polymers, and superabsorbent polymers (Wiesemann & Adam, 2011, p. 317). Granular superabsorbent polymers trap fluid into a gel as it is absorbed (Das, 2014, p. 79). Both men and women wear pads or absorbent full coverage lower torso garments to absorb urine. Male/female anatomical differences require different shapes and positioning of the product components. Product size, product materials, layering methods, amount of padding, and method of conforming to body contours need to vary relative to overall body size, urine output, and wear occasion. Will the individual need help to doff/don the product? The fastener style must meet the requirements of the wearer and/or caregiver. New materials or methods of wearable construction may further improve health, hygiene, comfort, and sustainability. The designer's challenge is to find the best combination of design elements to meet the needs of the target market.

Disposable diapers and IMPs are multi-layered (Figure 5.6). Layer structures range from non-wovens to fluffy loose layers, from bonded materials to films. Layers (from nearest the body outward) include: (1) a top sheet, or coverstock—to keep the skin surface dry, (2) an acquisition and transport layer to move fluids quickly to (3) the absorbent core layer, and (4) a back sheet to prevent fluid leakage to outer garments while allowing vapor migration to reduce skin wetness (Das, 2014; Wiesemann & Adam, 2011). The top sheet, acquisition and transport, and back sheet layers are usually very thin. Das (2014) explains the diaper layer material selection rationale (pp. 77–79). Overall, the weight and thickness of absorbent hygiene products have decreased significantly since the late 1980s (Wiesemann & Adam, 2011, p. 331). Microporous structures help reduce skin irritation and infection by reducing

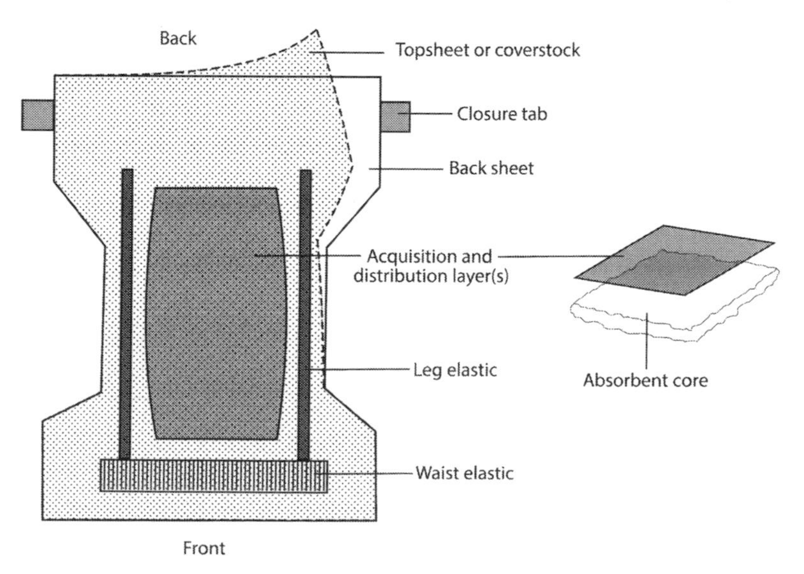


**FIGURE 5.6**
Typical baby diaper: Shape, features, and layers.

humidity near the skin. Runeman (2008) reviews skin conditions related to use of incontinence products—including the risk factors of frail skin of newborns and the aged. Skin health depends on managing interactions between (a) individual wearer characteristics, (b) skin care provided to the IMP user, and (c) the IMP (Runeman, 2008, p. 46).

Product shape and style vary according to the wearer's needs. Infant and baby diapers include: (a) tiny sizes for premature babies and newborns, (b) baby and toddler products in a range of sizes, and (c) toddler "pull-ups" intended for toilet training. Adult incontinent wearable products include: (a) all-in-one taped products for severe urinary and/or fecal incontinence; (b) pads for light, medium, or heavy incontinence worn as inserts in a special pants or in everyday underwear; and (c) male pouches for mild urinary incontinence, worn with a holder or with body-conforming underwear (Wiesemann & Adam, 2011, p. 318). Materials and structures used in incontinence collection products are similar to some feminine (menstrual) hygiene products discussed in Section 5.4.4.

In addition to the shape, size, and absorbency of an IMP, designers need to consider what a person, infant or adult, will wear over the product—its appearance, fit, and function. A bulky, padded look is acceptable for infants and toddlers, while adults want comfortable and fashionable garments with as little volume as possible in the seat and hip area—all while maintaining absorbency and avoiding leakage. Ambulatory adults can usually don and doff clothing to change an IMP. Adults with limited mobility or flexibility may need assistance in changing IMPs. Infant and toddler soiled diapers are changed by a caregiver, so easy access to the diaper is important. Current diaper access methods include crotch seams that open with fasteners such as snaps and hook-and-loop tape. Exploration of other access and fastener methods is warranted.

The European Disposables and Nonwovens Association (EDANA) and the International Nonwovens and Disposables Association (INDA) are trade associations providing information on baby diaper, incontinent, and feminine hygiene products industries. Some information is available free of charge from these associations via the internet. Members have access to more extensive information including market reports.

### 5.3.6 Protective Products for the Perineal Region

The male external genitals, serving both urinary and reproductive functions, are vulnerable to injury due to their anterior location and position between the legs. Consequences of penile, scrotal, and testicular injury range from problems of pain, swelling, and bruising to partial or total genital amputation and/or complete loss of function. The most extreme injuries occur in the context of war. Flak jackets, which adequately cover the upper abdomen, and leg armor tend to leave the penis and testicles somewhat exposed, even with an optional groin protector, because of the need for leg movement (Matic et al., 2006, p. 75;

Thompson, Flaherty, & Morey, 1998, p. 141). Consequently, war-time genital injuries still occur. Military veterans with serious genital trauma in addition to amputation of one or both legs rate genital injuries as more impactful than limb loss (Lucas, Page, Phillip, & Bennett, 2014). Designing military clothing to better protect the perineal region is challenging, but important.

Male genital protection strategies differ in less extreme conditions. Runners, bicyclists, baseball catchers, and competitors in contact sports have different perineal protection needs. Runners may choose to wear no support other than a *boxer brief. Athletic supporters*, commonly called *jock straps*, are worn for more genital support. The basic jockstrap features a knitted textile pouch to support the genitals. The pouch is held in place with a combination of a waistband and straps fastened to the band and to the pouch (Figure 5.7). For additional protection, especially in contact sports, the pouch may be constructed to hold an anatomically shaped rigid protective shell, an *athletic cup*, to cover both the penis and scrotum. The cup's edges may be padded to try to reduce impact-related bruising and trauma to skin and nearby soft tissues. Ventilation holes in the cup help to decrease heat build-up and allow some evaporation of sweat.

Female external genital anatomy, less vulnerable than the male's, may benefit from padding over the perineal region for specific activities. Male and female bicyclists may wear bike shorts with padding positioned to cushion the perineal region from bicycle seat pressure and abrasion. Textiles worn next to the perineal region should be soft to prevent abrasion and irritation, non-allergenic to prevent skin reactions, and either absorbent or wicking to prevent moisture build-up. Product shapes, materials, and joining methods can be chosen to minimize friction between the perineal region and undergarments during athletic activities.

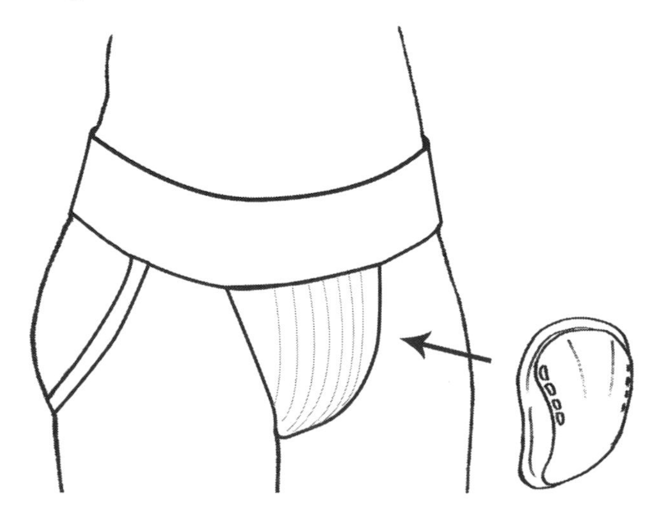

**FIGURE 5.7**
Male athletic supporter and athletic cup.

## 5.4 Reproductive System: Structures, Function, and Wearable Products

Human reproductive functions and sexual pleasure are integrated in the lower torso in the male and female reproductive systems. This association has influenced the design of many wearable products used with the reproductive structures. Most reproductive system wearables fall into five broad categories: (a) infection prevention (b) pregnancy prevention, (c) *fertility* promotion, (d) menstrual hygiene management, and (e) *maternity* products.

The reproductive systems become functional at *puberty*, sometime in adolescence. Physical changes related to puberty affect lower torso wearable product designs. Both genders develop pubic hair. Changes in oil production and increased perspiration can affect product materials choices. Products designed specifically for the male reproductive system may be prophylactic, meant to prevent pregnancy in a female partner or to prevent disease in the wearer or a partner. Everyday products for the male lower torso, especially underwear, may produce negative effects on fertility. Products designed specifically for females can also be prophylactic, to prevent pregnancy or disease. Other products related specifically to female anatomy and function include products designed to manage menstrual flow, maternity clothing to accommodate the pregnant figure, products to compensate for pregnancy-related skeletal changes, and internal supports for *pelvic organ prolapse*.

### 5.4.1 Male Reproductive System: Structures, Function, and Wearable Product Considerations

Designers of reproductive system wearable products need an understanding of reproductive system components: the anatomical structures and reproductive system functions. Male reproductive system structures can be separated into internal and external components. The major internal male reproductive component, the prostate gland and its relationship to the urinary system, was described in Section 5.3.1. It is not related to reproductive system wearable products.

#### *Penis and Testes: Male External Reproductive Structures*

The external genitals: (a) the testes or *testicles* within the scrotum and (b) the penis are the most evident male reproductive system features (seen in Figure 5.4). The penis and testicles require accommodation in men's wearable products. The penis is a key element for both reproduction of the species and sexual pleasure, which bonds partners.

The penis changes shape and size from *flaccid* to *erect* in response to sexual stimulation (or sometimes reflexively with a full bladder). As a male anticipates or initiates sexual activity, sympathetic and parasympathetic nervous system signals control blood flow to and within the penis. It enlarges and becomes firm. An *erection* prevents urine release from the bladder during *ejaculation*, the rhythmic forcible expulsion of *semen* (sperm plus other reproductive system fluids) from the penis. When partners connect in *sexual intercourse* and the male ejaculates, the semen travels from the urethral meatus of the penis into the vagina, the internal female reproductive canal, or the anal canal—depending on penis placement (refer to Figure 5.4). Other terms for intercourse include *copulation* or *coitus*. The male's semen, as well as the secretions of the vagina and anal canal, can transmit viruses and bacteria. To decrease the risk of a *sexually transmitted disease* (*STD*) some type of impervious barrier must be put between the penis and the vagina or anus and anal canal prior to every connection in intercourse.

Schneider, Sperling, Lümmen, Syllwasschy, and Rübben (2001) measured penis lengths and circumferences in flaccid and erect states in two groups of men (ages 18–19 and 40–65). Penile length in the flaccid state varied from 5 to 14 cm (2.0 to 5.5 in.) and width varied from 2.2 to 4.2 cm (0.9 to 1.6 in.). Erect penis length ranged from 10 to 19 cm (3.9 to 7.5 in.) and width at the base of the penis measured from 3 to 5 cm (1.2 to 2 in.). Herbenick, Reece, Schick, and Sanders (2014), had 1661 men, 18–60+ years measure themselves. They reported that erect penis length varied from 4 to 26 cm (1.6 to 10.2 in.) and erect mid-penis circumference varied from 3 to 19 cm (1.2 to 7.5 in.). The penis also changes size in response to environmental temperature changes. In cold weather, when the body core temperature drops, blood is diverted from the penis to the core with an associated decrease in penis size. Conversely, when the core temperature rises, blood can be shunted to the penis, resulting in some size increase.

The testicles produce sperm, the cells carrying male DNA, daily. The penis delivers sperm during sexual activity. Although the penis and the testicles are near each other in the perineal region, sperm travels a circuitous path through several internal structures, including the prostate gland, which adds fluids to the sperm, before it is released into the urethra of the penis. Like penile measurements, testicular size varies man to man: 3.6 to 5.0 cm (1.4 to 2 in.) length, 1.5 to 2.4 cm (0.8 to 0.9 in.) depth, and 2.5 to 3.3 cm (1.0 to 1.3 in.) width (Sakamoto et al., 2008). The scrotum, the skin-covered muscular pouch that holds the firm, smooth testicles, changes size in response to environmental temperature changes. The scrotal muscles contract as the body cools, pulling the testicles closer to the torso, to keep them from cooling too much. And, when the core temperature rises above normal, the scrotal muscles relax fully, allowing the testicles to move as far as possible away from the torso. These testicle and scrotum size and position variations complicate fit issues for men's reproductive structure wearables and for lower torso garments.

### Condoms: Male Wearable Products for Prevention of Pregnancy and Sexually Transmitted Disease

A condom is a sheath worn over the erect penis to collect ejaculated semen and act as a barrier between sexual partners. The condom is worn to prevent pregnancy and/or STDs. Condoms were available, reliable, and inexpensive by 1869 due to advances to rubber technology (Bullough, 1981, p. 104). The colloquial name "rubber" derives from the new rubber materials used for condoms at that time. Today most condoms are made of latex or latex substitutes. Condoms of lambskin or other animal membranes do not prevent STD transmission.

Given the risks of HIV/AIDS and other STDs, proper condom use and fit are important. Male condoms are commonly available in three sizes: (a) snug fit, (b) large, and (c) extra-large. For both infection prevention and pregnancy prevention, condom fit and integrity are important, but fit is often left to an individual's trial-and-error. If the condom tears or is painfully tight when placed on the erect penis, it is too small. If it slips around or falls off in the midst of sexual activity, it may have been used incorrectly or is too large. Online resources are available to determine proper condom size (CondomSizes.org, 2018). Researchers who have studied condom use and failure recommend producing condoms in a greater range of sizes to improve effectiveness and acceptability (Reece et al., 2008; Smith, Jolley, Hocking, Benton, & Gerofi, 1998). Read more about condom use and acceptance in Section 5.4.3.

### Male Wearable Products and Fertility

Lower torso garment materials and designs affect testicular temperatures. The body tries to maintain testicular temperatures cooler than the body core, presumably to protect sperm production and, thus, male fertility—the ability to father children (Durairajanayagam, Sharma, Du Plessis, & Agarwal, 2014, p. 106). In addition to scrotal muscle contraction and relaxation, scrotal skin and testicular vascular structures use universal *heat transfer mechanisms*: (a) convection, (b) radiation, (c) evaporation of sweat from the skin, and (d) counter-current heat exchange from the testicular arteries to the testicular veins (Durairajanayagam et al., 2014, pp. 105–106). Although research suggests scrotal cooling helps sustain male fertility, it has not been definitively proven (Nikolopoulos, Osman, Haoula, Jayaprakasan, & Atiomo, 2013). However, if a wearable product enhances *heat dissipation*, it will supplement the body's natural cooling abilities.

Multiple environmental and lifestyle factors contribute to testicular heat stress: ambient temperature, underwear style, the insulating effects of outer clothing, postural position, occupation, sauna use, and sport cycling activities. Durairajanayagam et al. (2014) provide an extensive review of these factors (pp. 112–118) and suggest elevated testicular temperature may become a useful component in a program of male *contraception*. In contrast,

Laven, Haverkorn, and Bots (1988) studied 56 male infertility patients and suggested that men sleeping "warm" in underwear or under down quilts or electric blankets, or who wear tight clothes are at risk for producing lower-quality sperm (p. 141).

Some cancer chemotherapy drugs affect sperm production. Cis-platinum, one chemotherapy drug used to treat testicular cancer, damages the testes' sperm production capabilities. Aminsharifi et al. (2016) reported experiments in male mice using scrotal cooling to protect fertility during cis-platinum administration. Scrotal cooling may function like the cooling used to try to prevent chemotherapy-related hair loss for breast cancer patients.

Many researchers have studied ways to provide testicular cooling. Jung, Schill, and Schuppe (2005) studied 20 men with low sperm counts. They found signs of improved sperm production after cooling the testicles with circulated air overnight for 12 weeks. Relying on the work of Jung et al. (2005), Laven et al. (1988), and others, Shoemake developed specialized underwear fitted with pockets to hold chilled gel packs to cool the testes during waking and sleeping activities (U.S. Patent Application 0,137,316, 2014). Using such a wearable to provide cooling in the perineal region may also prove helpful after vasectomies, minor urinary tract surgeries, or to decrease swelling after perineal region soft-tissue trauma. More research in this area is warranted.

### 5.4.2 Female Reproductive System: Structures, Function, and Wearable Product Considerations

Female reproductive system structures, like male structures, can be separated into external and internal parts. Although the internal female reproductive structures lie deep inside the pelvis, they are important to lower torso wearable product design, especially during pregnancy. External female reproductive structures in the perineal region, smaller and less obvious than their male counterparts, are important to designers of *menstrual management products* (*MMPs*), IMPs, underwear, swimwear, and tights or pantyhose.

#### *Vaginal Opening, Labia, and Clitoris: Female External Reproductive Structures*

The female reproductive structures in the perineal region are not as visibly evident as the male's, but they affect product design, shape, and, fit. (Review the external genital features in Figures 5.3 and 5.4). The *mons pubis*, near the external genitals, is the mound of fatty tissue overlying the pubic symphysis. The female external genitals are the vaginal opening, the *labia (minora and majora)* along each side of the vaginal opening, and the clitoris, the sensitive structure which lies anterior to the vaginal opening and inferior to the pubic symphysis. As a female anticipates and initiates sexual activity, sympathetic and parasympathetic nervous system signals direct blood to these structures and they enlarge somewhat.

Female external genital dimensions and spacing between the urethra, vagina, and anus in the perineal region should affect pattern shapes, positioning, and sizes of women's underwear, IMPs, and MMPs, but the data from the few studies that are available has not been collected with product design in mind. Krissi et al. (2016) documented external genital measurements in 32 premenopausal women whose BMIs ranged from 16.6 to 33.9 kg/m$^2$. Lloyd, Crouch, Minto, Liao, and Creighton (2005) measured female external genitals and their relative positions in 50 premenopausal women but neither they nor Krissi et al. (2016) provide information regarding any relationship of external genitals size to pelvis or hip size.

Measurements from both studies demonstrate wide variations between individuals, much like male external genital measurements, with relatively broad ranges for both structural dimensions and the distances between structures. Some distance measurements in the list below are given with reference to a standard non-digital clock with a 360° face—with the anterior edge of the vagina at 12 o'clock (0°) and the posterior of the vaginal opening at 6 o'clock (180°). Data from both studies (Krissi et al., 2016; Lloyd et al., 2005) were combined to give feature measurement ranges and spacing of the features. The data regarding feature spacing may be of more use to designers than the size of the structures. Refer to the upper left of Figures 5.3.

- Clitoral length: 0.5–3.5 cm (0.2–1.4 in.)
- Clitoral width: 0.2–1.0 cm (0.1–0.4 in.)
- Clitoris to urethral meatus: 1.0–6.0 cm (0.6–2.4 in.)
- Clitoris to anterior of vaginal opening (12 o'clock): 3.0–8.0 cm (1.2–3.1 in.)
- Labia majora length: 7.0–12.0 cm (2.8–4.7 in.)
- Labia minora length: 1.0–10.0 cm (0.4–3.9 in.)
- Labia minora width: 0.7–5.0 cm (0.3–2.0 in.)
- Urethral meatus to anterior of vaginal opening (12 o'clock): 1.0–5.0 cm (0.4–2.0 in.)
- Vagina, anterior wall to posterior wall, in perineal region: 1.0–4.0 cm (0.4–1.6 in.)
- Vagina, width, in perineal region 0.0–2.0 cm (0.0–0.8 in.)
- Posterior vaginal opening (6 o'clock) to center of anus: 1.5–7.0 cm (0.6–2.8 in.)
- Clitoris to anus: 8.5–17.5 cm (3.3–6.9 in.)

Additional research to relate these dimensions to total body size may be useful, but results may only underscore the large degree of normal variation found in genital structures. The variation suggests two different product design approaches: (a) enlarge product components (such as panty crotch panels covering the perineal region in all panty sizes) or (b) develop products (such as panties) with multiple size options to cover the perineal region—e.g. a size 6 panty with a S, M, or L crotch panel.

### *Ovaries, Uterus, and Vagina: Female Internal Reproductive System Organ Structures*

Internal female reproductive organs occupy a relatively small volume of the lower torso region (refer to Figure 5.4) but are important to designers of internal MMPs, products for prevention of pregnancy and STDs, and *pessaries,* internal supportive devices used to treat a displaced bladder, *uterus* or rectum.

The small paired R and L ovaries are the source of the mother's eggs (*ova,* plural of *ovum*) needed for reproduction. The ovaries are lateral to the centrally located muscular uterus. Unlike men, who release sperm during sexual activity, the ovaries release ova on an approximately 28-day cycle. Smith et al. (2011, pp. 160–161) note, in a non-pregnant pre-menopausal adult, the pear-sized uterus measures 8 to 10 cm (3.1 to 3.9 in.) in length. The *cervix,* at the stem-end of the pear-shaped uterus is 2 to 3 cm in diameter (0.8 to 1.2 in.). It opens into the vagina. The vagina helps direct the sperm to the cervix. If the female (1) is at the appropriate time in her menstrual cycle, (2) has released an ovum, (3) and one sperm successfully reaches and merges with the ovum, this is called conception (*fertilization*). Pregnancy results if the fertilized ovum embeds in the *endometrium*—the mucous membrane richly supplied with blood, which lines the uterus. The endometrium develops into the *placenta,* which nurtures the fetus and connects it to the mother. As the fetus grows, the uterus expands, affecting maternal torso shape and size, as well as wearable product shape and size. After approximately 38 weeks of pregnancy, the vagina serves as the infant's exit path from the uterus.

If conception does not occur the endometrium is sloughed through the cervix, creating the monthly *menstrual flow* of blood and endometrial tissue. See Section 5.4.4 for details on MMPs. If endometrial tissue tracks toward the ovaries, during *menstruation* (instead of the cervix), it can attach to other structures within the pelvis, causing a painful condition: *endometriosis* (Leyland et al., 2010).

The vagina, lined with epithelial tissue similar to skin and the lining of the intestines, has considerable capacity to expand during intercourse, for passage of an infant during birth, or for insertion of a wearable product. Most of the time it is collapsed on itself. Barnhart et al. (2006) collected vaginal dimensions and structural orientation information from MRI scans of 28 sexually active healthy young women. They report the vaginal axis changes direction at the level of the muscular pelvic diaphragm (see Section 5.6.1). Barnhart et al. (2006) describe the collapsed vagina's cross-sections at the cervix level as either a "W" or an "H" shape. Pregnancy prevention products, tampons, and pessaries fit into this space. They provide internal vaginal dimensions: (a) length from the vaginal opening to the cervix varies from 4.1 to 9.5 cm (1.6 to 3.7 in.); (b) vaginal diameter at the level of the cervix averages 4.2 cm (1.7 in.); and (c) vaginal diameter, measured 1 cm (0.5 in.) internal to the vaginal opening, averages 2.6 cm (1 in.). Vaginal dimensions increase with sexual stimulation—by 2.5 to 3.5 cm (1 to 1.4 in.) in length and 3.75 to 4.25 cm (1.5 to 1.7 in.) in diameter near the cervix (Masters & Johnson, 1966, p. 74). Pendergrass, Reeves, and Belovicz (1991) describe a three-dimensional casting technique to document vaginal shape. Vaginal configurational information and dimensions can be used by designers of internal wearable products ranging from *contraceptive* devices to MMPs to organ supports.

### *Condoms, Diaphragms, Patches, and Vaginal Rings: Female Wearable Products for Prevention of Pregnancy and STDs*

A *female condom,* a female-controlled barrier product, is currently the only internal wearable product useable for both pregnancy and STD prevention. Alternatively called an *internal condom,* it is designed to be placed into the vagina or anus prior to intercourse. It features a tube of pliable impervious materials, a ring at the closed end to help hold it in place inside the body, and a ring at the open end to keep it from slipping into the body and to facilitate removal. A female condom's loose fit (compared to the close fit of a male condom) allows it to approximate vaginal expansion related to female sexual excitement.

Correct use of any wearable product used for pregnancy or disease prevention is essential. Part of the design process for these products should include product semantic consideration—does the product communicate how to use it and a plan for educating first time users. Artz et al. (2000), in a study of 1159 female STD patients, found that an intensive intervention program to promote female condom use decreased unprotected sex. The program included: educational materials, one-to-one education, hands-on female condom use training, and suggestions for sexual partner education about this method.

A diaphragm is another barrier-type wearable product designed for pregnancy prevention. It does not prevent STD transmission. The dome-shaped device fits over the cervix to physically block sperm from entering the uterus (see Figure 5.8-A). The pliable impermeable dome is attached to a flexible ring, so the woman can compress the ring to place the diaphragm over the cervix, deep in the vagina. The ring encircles the cervix and the rubber dome cups the cervical mound. A spermicide is applied to the diaphragm before insertion. The woman removes the diaphragm several hours after intercourse. A woman is fitted for a diaphragm by a medical professional.

Vaginal rings infused with medication similar to birth control pills (BCP) are an intravaginal non-barrier pregnancy prevention product, but do not protect against STDs. They are designed to be inserted by the wearer, worn for three weeks, and then removed by the wearer for one week, before inserting a new ring. They are designed to be left in place during intercourse.

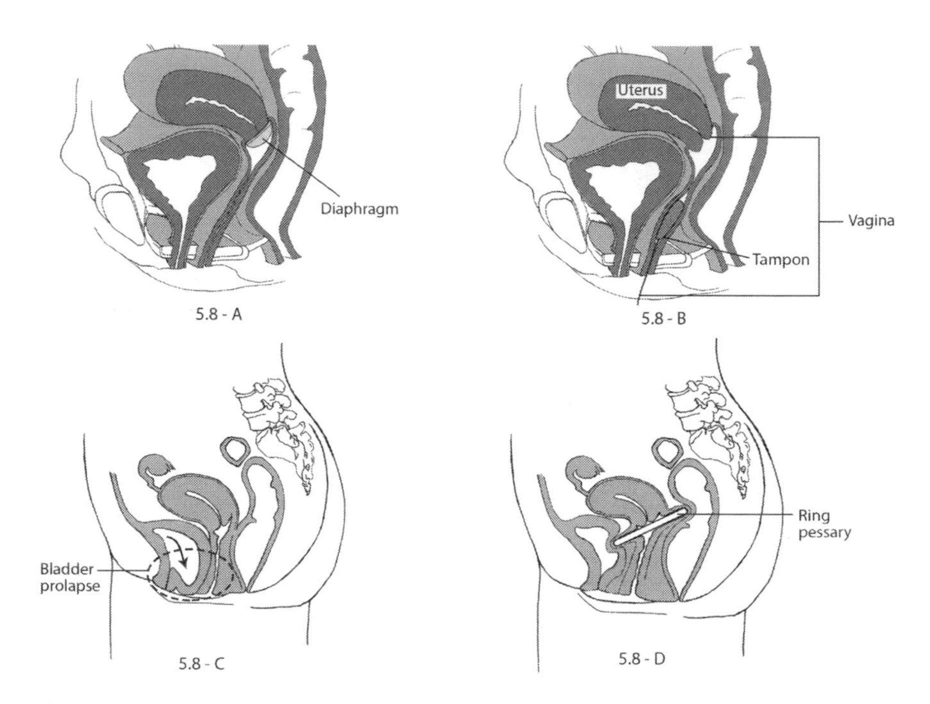


**FIGURE 5.8**
Internal female reproductive structures and products.

A longer-acting product, designed to be worn for one year, is being studied (Stifani, Plagianos, Vieira, & Merkatz, 2018). The authors identified factors leading to suboptimal use of vaginal rings and suggest strategies to improve compliance with product use instructions. Contraceptive patches worn on the skin contain the same kinds of medications as BCPs and vaginal rings. Worn for three weeks, then removed for one week, they are as effective as BCP and the vaginal ring for contraception, but occasionally cause skin rashes (World Health Organization, 2007).

Watnick, Keller, Stein, and Bauman (2018) studied use of a self-inserted 5.5 cm (2.2 in.) polyurethane intravaginal ring, a donut-shaped device, which time-releases medication for prevention of some major STDs, but not pregnancy. The ring, placed deep in the vagina, can be left in place during intercourse and menstruation (the monthly shedding of the uterus' lining in non-pregnant women) and provides a non-barrier female-controlled means of HIV/AIDS and STD prevention.

### 5.4.3 Wearable Contraceptive and STD Product Effectiveness and Acceptance

"For a contraceptive to be effective it has to be inexpensive, reliable, and available." (Bullough, 1981, p. 104). No currently available wearable product is 100% effective for pregnancy or STD transmission prevention. Trussell (2011) reports the pregnancy rate for contraceptive methods (including barrier and non-barrier wearable products) by several criteria: typical use, perfect use, and imperfect use. Among his conclusions: (a) Medication-based wearable contraceptives are most effective, ranging from 91–99.7%, depending on whether the product is worn with typical compliance or exactly according to directions. (b) Perfectly used male condoms can be 98% effective as a contraceptive, but as typically used, they are only 82% effective. Blunt-Vinti, Thompson, and Griner (2018) consider all barrier contraceptive methods to be ineffective because male condoms, female condoms, and/or diaphragms must be used for every instance of intercourse to be effective for pregnancy prevention.

Of the barrier methods controlled by females, female condoms have limited acceptability. Eldridge, St. Lawrence, Little, Shelby, and Brasfield (1995) questioned 178 women concerning preferences for male or female condoms. They found that women preferred condom use by the male partner. Research participants' dissatisfaction with the female condom included discomfort, the need for partner agreement for use, and interference with the sexual experience. Advantages of female condoms included that use was female-controlled and they were perceived to be safe and effective. Eldridge et al. (1995) suggest development of other safe barrier methods that are comfortable and convenient, female-controlled, and can be used without partner awareness.

When male or female condoms are used consistently and correctly, they significantly reduce, but do not eliminate, the risk of STD transmission (CDC, n.d.). In an attempt to find a more effective, but less obvious wearable to prevent HIV infection, Watnick et al. (2018) tested an HIV/AIDS preventive intravaginal ring in a group of 18 women for two weeks. The participants had limited interest in continued use of this female-controlled, HIV-preventive product, due to both individual and relationship concerns.

### 5.4.4 Menstruation and Menstrual Management Products

Menstruation, evidence of female reproductive maturation, is a very personal experience for women. Britton (1996) presents an anthropological perspective of menstruation: (a) its role in the emergence of female identity, (b) how social and cultural forces influence women's beliefs about menstruation, and (c) how women view their bodies. Despite the universal occurrence of this biological phenomenon, women's management of their monthly menstrual flow varies according to ethnic, cultural, religious, and social values (Buckley & Gottlieb, 1988; Dahlqvist, 2018). An understanding of cultural differences, menstrual flow patterns including the range of menstrual days within a cycle, quantity and consistency of menstrual flow, as well as vaginal shape and measurement data can guide designers as they consider shapes and sizes of products to be worn during menstruation. Women's needs for MMPs persist whatever the circumstances; design thinking may be able to improve the current responses to menstrual management needs during floods, earthquakes, armed conflicts, etc. (VanLeeuwen & Torondel, 2018).

### *The Biology of Menstruation*

Menstruation occurs approximately monthly from puberty to *menopause* in non-pregnant females. Due to the periodic nature of the menstrual cycle, menstruation is often referred to as a "period." Individual variations occur in (a) the interval between periods, (b) the duration of the menstrual period, and (c) the volume of menstrual flow. Use of oral, patch, or vaginal ring contraceptives can decrease the duration and volume of menstrual flow. Documenting the duration of menstrual flow is straightforward—ask women how long they bleed. Measuring menstrual flow volume is more difficult. Callard, Litofsky, and DeMerre (1966) collected menstrual blood in intravaginal cups which study participants removed in the research lab. They reported the heaviest flow on days 1 and 2, with a range of average flow (per woman) from 0.60 to 2.19 ml/hr (0.02 to 0.07 oz/hr). Toxqui, Pérez-Granados, Blanco-Rojo, Wright, and Vaquero (2014) administered a questionnaire to assess numbers of MMPs women used per menstrual period. They used a formula to combine product numbers with manufacturers' reported product absorbency data to calculate total menstrual blood loss. They noted correlations between the calculated menstrual blood loss and blood tests which reflect blood loss in general. Munro (2012) reported normal values including: (a) menstrual periods lasting from 4.5 to 8 days and (b) total menstrual blood loss per woman per menstrual period ranging from 5 to 80 ml (0.17 to 2.7 oz) (p. 231).

Heavy menstrual bleeding, defined as more than 80 ml/period, presents special challenges for MMP users and designers. Women with very heavy bleeding may also unpredictably pass large blood *clots*. Heavy menstrual bleeding is sometimes considered a symptom and sometimes a diagnosis, but such heavy bleeding affects women's social and work activities and health due to the degree of their blood loss (National Collaborating Centre for Women's and Children's Health (UK), 2018). If the start of a woman's menstrual period is unpredictable and bleeding is very heavy on the first day, she must constantly carry or have immediate access to a day's supply of MMPs. If a woman is unsure if her MMP will keep her from staining her clothes or the chair she sits on, she may limit her activities on heavy flow days. New methods of managing menstrual hygiene are especially needed for these women.

### *Menstrual Management Products (MMPs)*

MMPs are commercially available as feminine hygiene products. Products designed to absorb or collect menstrual flow include pads worn below the perineal region—*sanitary pads* (also called *napkins* or *towels*) and *panty liners* (or *shields*); and products inserted into the vagina—*tampons* and *menstrual cups*. Related products include menstrual panties to hold pads and liners securely in place, and products intended to relieve cramping and pain (*dysmenorrhea*) that may accompany menstruation. MMPs must be absorbent and/or retain fluid, stay in place, prevent odor, and be comfortable (Wiesemann & Adam 2011). Bullough (1985) summarizes MMP design strategies from 1854–1914, noting how bandaging technology (and surplus bandaging materials) from WWI led to production and acceptance of disposable sanitary pads in 1920. Much research published on MMPs has been conducted by industry researchers.

Sanitary pads are constructed in layers, much like IMPs (refer to Section 5.3.5), and shaped to fit inside the crotch of women's underwear—often with an adhesive backing to hold them in place. Panty liners are a smaller, thinner form of sanitary pad worn for light menstrual flow days, or to manage vaginal discharge, small urine leaks, and maintain underwear cleanliness. (see Section 5.11.2 for a discussion of non-menstrual vaginal secretions.) Because panty liners are thinner than sanitary pads, the inner core is made of a thin layer of rayon pulp fluff. New ultra-thin liner core materials made of super-absorbent fibers or granules wrapped in cellulose tissue are being developed by industry (Farage, Bramante, Otaka, & Sobel, 2007).

Bullough (1985) presents excerpts from a 1927 industry-commissioned report on the use of sanitary napkins in the U.S. Lillian Gilbreth, a pioneer in human factors and industrial management research authored the report. Her firm reviewed over 1000 questionnaires, conducted conferences on MMP design, and interviewed academics and young women. The report covered multiple topics: (a) women's opinions about menstruation, (b) information on dealing with menstrual flow, (c) a survey of available MMPs, (d) market penetration, (e) analysis of 57 different branded products, and (f) suggestions for improving sanitary napkins and marketing as well as a literature review and bibliography. The report provided the sponsors valuable information on this topic. While MMP technologies and women's opinions have no doubt changed since the 1920s, the need for good MMP design continues.

Modern tampons (refer to Figure 5.8-B), absorbent cotton/rayon plugs inserted into the vagina with a finger or removable applicator to collect menstrual flow, were patented by E. C. Haas in 1933 (U. S. Patent No. 1,926,900, 1933). Medical literature related to tampons began appearing by the late 1930s (Brander, 1942; Emslie & Wilkie, 1938). Tampons are removed from the vagina by pulling a string attached to the plug. Tampons eliminate the bulk of externally worn pads—a welcome feature for many women. Although Thornton (1943), in an industry-supported study, noted that women in the U.S. readily accepted this new MMP, tampon use is currently variable around the world. Ren, Simon, and Wu (2018) report that while most women in Europe and the United States use tampons, only 1.8% of Chinese women do. They explore multiple reasons for Chinese women's preferences for sanitary pads.

A menstrual cup is an MMP inserted into the vagina to collect, not absorb, menstrual flow. It is a reusable conical cup of silicone, rubber, or thermoplastic elastomer which can be folded for vaginal insertion, much like a diaphragm, to collect menstrual flow. The cup has an attached stem to facilitate removal. The inventor promoted the cup as an inconspicuous MMP (U. S. Patent No. 2,089,113, 1937). Menstrual cups must be cleaned after removal and before reinserting. Oster and Thornton (2011) compared traditional washable menstrual cloths to menstrual cups in a study with Nepalese school girls. They state the menstrual cups were popular with the girls and decreased time doing laundry. Madziyire, Magure, and Madziwa (2018) found, in a pilot study, that 54 sexually active young women (19–45 years) in Zimbabwe accepted and correctly used menstrual cups. A reusable MMP, compared to pads and tampons, is an attractive option because it can reduce costs and material use.

MMPs encompassing the lower torso region (from the waist to and including the external genitals) include menstrual panties and an acupressure garment intended to reduce menstrual period discomfort. A patent search reveals many patent applications for "menstrual" panties. Some of these panties are disposable and incorporate an absorbent pad. Other menstrual panties are reusable, typically featuring easy care materials that are leak-resistant. The body-conforming panties secure a pad under the perineal region.

Taylor, Miaskowski, and Kohn (2002) tested the safety and effectiveness of an acupressure garment designed to decrease primary dysmenorrhea pain, menstrual pain often described as cramps—in the lower abdomen, back, and/or thighs. The cotton/spandex brief exerts truncal compression of 1 to 2 psi. Three 3.8 cm (1.5 in.) dome-shaped latex rubber pads and one 7.6 cm by 12.7 cm (3 in. by 5 in.) pyramid-shaped pad are placed into pockets in the brief, with the curved sides toward the body. Pad position is based on traditional Chinese acupressure sites. Sixty-one women were randomly assigned to wearing the brief or their own underwear, over the course of two menstrual cycles. The acupressure brief wearers reported decreased use of pain medication and decrease in menstrual pain symptom intensity. This design demonstrates opportunities for adapting old therapeutic, non-pharmaceutical methods into new product applications.

Menstrual products may be used individually or in combination, depending on issues ranging from access to a private rest room with washing facilities (to allow product changes and hygiene) to rate of menstrual flow to cultural/social norms. Lack of information about both menstruation and MMPs, embarrassment about menstruation, and availability and cost of MMPs limit women's options. Each product has advantages and disadvantages related to convenience, product size and absorbency, sustainability and risks. Product features are important but not the only determining factor for MMP choice. A woman's age, daily activities, economic and political factors—including product availability, cost, and taxes on products

(Hartman, 2017)—affect choices (Kuhlmann, Henry, & Wall (2017). MMP designers need to factor in all these variables.

### 5.4.5 Vaginal Health and Intravaginal Wearable Product Use

Like on the skin and in the large intestine, there is a vaginal microbial community (VMC). Vaneechoutte (2017) describes the unique qualities of the human VMC and some of its microbial species. The natural bacteria of the VMC help shield vaginal tissues from bacterial infections (Taherali, Varum, & Basit, 2018). The VMC also maintains a healthy vaginal acidity (Godha, Tucker, Biehl, Archer, & Mirkin, 2018).

Reproductive system wearable products used in the vagina influence the VMC in various ways (Noyes et al., 2018). Wearable products, such as tampons, condoms, and diaphragms, can impact the vaginal microbiome and the vaginal lining. (See Section 5.11.2 for more information on the vaginal lining.) In an industry-supported study, Soper, Brockwell, and Dalton (1991) evaluated use of female condoms and diaphragms as contraceptive devices, looking for evidence of vaginal trauma and change of the VMC. Neither the female condoms nor the diaphragms caused trauma in the lower reproductive tract. However, diaphragm use changed the microbial community of the vagina over time.

VMC changes make the risk of *toxic shock syndrome* (*TSS*) a design and product use consideration. First described in women using tampons in 1980 (Shands et al., 1980), TSS is a rare but serious disease related to a toxin given off by some *Staphylococcus aureus* (*S. aureus*) bacteria which can grow in the vagina. Subsequent research prompted changes in the type and amount of absorbent materials used in tampons, with a significant reduction, but not elimination, of the risk of TSS (Schlievert, Tripp, & Peterson, 2004). Nonfoux et al. (2018) tested several tampon styles as well as menstrual cups, to determine effects on vaginal *S. aureus* and the toxin causing TSS. They concluded there is a risk of TSS while using tampons or menstrual cups. The primary precaution for users is to calibrate the size of tampon or cup to menstrual flow, as higher absorbency/higher capacity products—left in place for longer time periods, increase risk. They also recommended menstrual cup users alternate between two or more menstrual cups, sterilizing each cup by boiling after use to remove any *S. aureus* bacteria.

Raudrant et al. (1995) studied the effects of three different menstrual tampons on the vaginal lining, finding tampon use can cause vaginal tissue drying. This result, plus findings from an earlier study (Raudrant, Frappart, De Haas, Thoulon, & Charvet, 1989) led the authors to recommend (a) tampons be made with materials and a design to minimize vaginal drying and (b) tampons not be used other than during menstrual periods. Two more recent industry-supported research studies looked for tampon-related vaginal tissue and VMC effects. Hochwalt, Jones, and Meyer (2010) compared super-absorbent and ultra-absorbent tampons (both made with 100% rayon) and found no differences between the tampon types in either (a) the vaginal microbiome (suggesting no increased risk of TSS) or (b) the integrity of the mucous membrane. The second study compared film-covered tampons to fleece-covered tampons and reported similar findings (Chase, Schenkel, Fahr, Eigner, & Tampon Study Group, 2010).

Vaginal anatomy is important for both reproductive and sexual functions, and wearable products used in the vagina relate to those functions. Designers and the women who use intravaginal wearable products need to balance the complex set of (a) vaginal anatomical and physiological factors, (b) social and cultural factors related to reproduction and sexual expression, and (c) sustainability issues intertwined in these products.

### 5.4.6 Female Pelvic Organ Prolapse and Pessary Use

Pelvic organ prolapse (*POP*), a disorder seen only in women, may cause discomfort and disability in varying degrees. If the supporting connective tissues of the bladder, rectum and/or uterus fail, any or all of these organs can collapse into, and in extreme cases protrude from the vagina and beyond the labia majora. Figure 5.8-C illustrates a bladder prolapse. Problems eliminating urine and stool often accompany POP. Although specific causes for POP are unclear, it is a common problem which can be treated with exercise in mild cases and with pessary use in moderate cases. Severe cases often require surgical repairs. The incidence of POP increases with age, possibly related to loss of connective tissue elasticity throughout the body. Pregnancy with vaginal delivery of the infant is a risk factor.

Early pessaries, defined as any wearables or medications inserted in the vagina, date to prehistory (Oliver, Thakar, & Sultan, 2011). Wearable hard rubber ring-shaped medical devices inserted into the vagina to help stabilize and support the bladder, uterus, and/or rectum were used in Germany prior to 1843 (Bullough, 1981). The diaphragm evolved from these supportive devices in the mid-19th century. Modern pessary designs vary from relatively simple rings, with or without a central membrane, to cubes, to more complex three-dimensional forms. See a ring pessary in place in Figure 5.8-D. Most are made of silicone. Pessaries usually occupy more space in the vagina than the contraceptive diaphragm devices. Each style comes in many sizes. Pessaries decrease POP discomfort and help improve bladder and bowel emptying. They are intended for long-term wear. Even though women are fitted for a pessary by a medical professional, the wearer can remove and insert the devices to allow for sexual activity and cleaning.

### 5.4.7 Pregnancy

Pregnancy results when a sperm unites with an egg and it embeds in the endometrium of the uterus. The uterus expands in size to accommodate the developing fetus. As documented in the sequential 3D body scans of a pregnant woman (Figure 5.9), the lower torso/belly shape changes as the uterus expands (Sohn & Bye, 2015). The torso widens at the level of the umbilicus early in pregnancy, then develops an anterior bulge, which expands upward and outward—anteriorly and to the sides—relative to enlarging breasts. At this stage, the uterus may press on the stomach and cause heartburn. Later in pregnancy, the top of the enlarged uterus begins to protrude more anteriorly. As weight increases at the body anterior, the lumbar spinal curve can shift to allow the upper body to counterbalance the belly. As delivery approaches, the fetus "drops" and the uterine bulge tips more forward, seen in the scans as the lowering of the umbilicus relative to the breasts. At this stage, the pregnant uterus fills the lower abdominal cavity and can create discomfort as it presses against the bladder, bowel, and even the nerves of the lumbosacral and sacral plexuses (see Section 5.10.2 for details on these peripheral nerve structures).

Spinal posture changes as the weight of the fetus moves the body's center of gravity forward. Pregnant women frequently complain of low back pain and alter their activities because of that pain. LaBan and Rapp (1996, p. 474) state that low back pain in pregnancy, like most low back pain, has no clear single cause. They suggest posture change, pelvic girdle joint instability, plus hormonal, mechanical, and vascular factors contribute to low back pain in pregnancy.

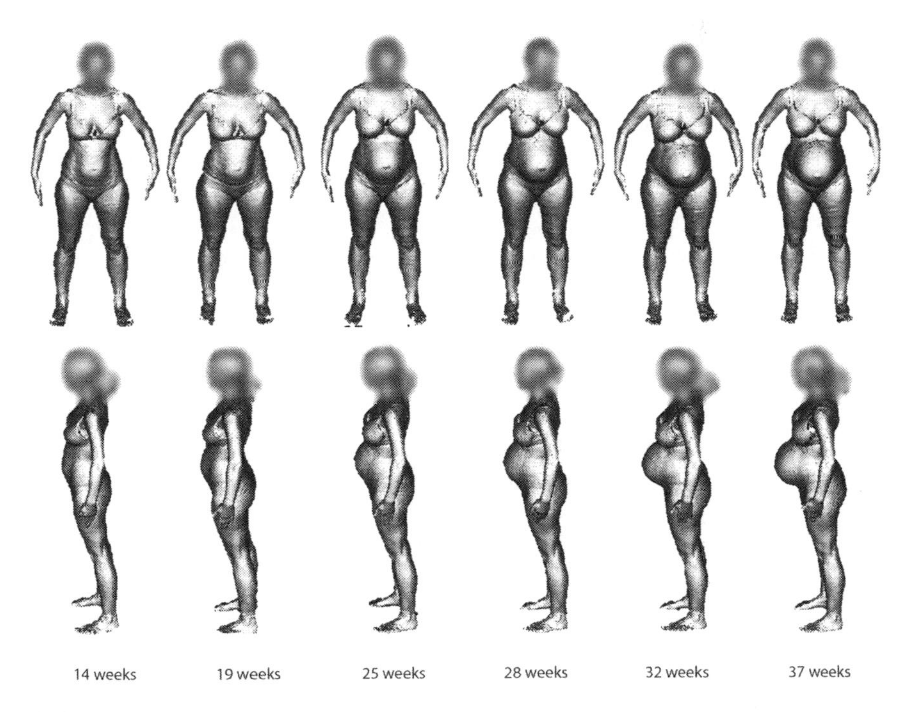


**FIGURE 5.9**
Progression of a pregnancy: 3D body scans in the same individual from weeks 14 to 37. (Courtesy of the Human Dimensioning© Laboratory, University of Minnesota.)

Due to multiple factors, low back pain in pregnancy is difficult to treat with a wearable product. Östgaard, Zetherström, Roos-Hansson, and Svanberg (1994) suggest pre-pregnancy fitness programs as a back-pain prevention plan.

Ho, Yu, Lao, Chow, Chung, and Li (2009b) interviewed 10 pregnant women using maternity supports. Sixty percent of the women discontinued maternity support use—one reason was perceived ineffectiveness. Other reasons included aesthetics, support style, inconvenience with donning, doffing, and support adjustment; safety, and itchiness. Ho, Yu, Lao, Chow, Chung, and Li (2009a), in a review of literature of maternity support belt effectiveness for low back and pelvic girdle pain raised issues relative to the designs. They concluded maternity support belts may have some benefits, with many caveats including needs for (a) careful clinical evaluation, (b) pairing a support with individualized exercises, and (c) ergonomics education (p. 1529).

In addition to considering biomechanical forces and torso shape and size factors, designers must deal with other changes related to pregnancy. Itching is a common complaint. Normal pregnancy-related itching, occurring in 20% of pregnancies (Geraghty & Pomeranz, 2011), and itching related to specific skin conditions in pregnancy (Panicker, Riyaz, & Balachandran, 2017) may be aggravated by tight clothing, heat and humidity, and fabric characteristics. Ho, Yu, Lao, Chow, Chung, and Li (2008) assessed thermophysiological, tactile, and movement comfort of 8 maternity support garments with 14 pregnant women. They related the results to fiber content—cotton and rayon were most comfortable—and fabric mechanical properties: soft, smooth, and breathable fabrics were preferred. They also considered ease of donning/doffing, and comfort with movement. Support products with a belt or brief design were more acceptable. Although they studied maternity support garments, their findings are applicable to other maternity wear such as panties or other daily wear in contact with affected skin.

Sixty-three percent of pregnant women develop *stretch marks* along skin tension lines on the breasts, abdomen, hips, buttocks, and/or thighs (Motosko, Bieber, Pomeranz, Stein, & Martires, 2017). Stretch marks are often associated with itching but the marks arise from changes in skin connective tissue fiber thickness and orientation. Many factors predispose women to stretch mark development (Geraghty & Pomeranz, 2011). Wearable product interventions likely cannot prevent stretch marks; however, the association of stretch marks with itching necessitates attention to materials used for maternity wearable products.

Other changes may develop during pregnancy. Significant changes occur in the bony pelvis to facilitate birth—see Section 5.5.2. Cellulite may increase (Rossi & Vergnanini, 2000) due to changes in connective tissue structure, fat cells, and the microcirculation of the skin—again independent of what one wears. Edema (swelling) in the legs during pregnancy reflects changes in the circulatory system. Compression stockings (see Section 5.10.1) may help (Muallem & Rubeiz, 2006), but may be too warm, and donning and doffing them late in pregnancy may be difficult.

### 5.4.8 Maternity Product Fit and Sizing

Maternity wear, the category of apparel designed for pregnant women, spans the lower torso and the upper torso. Increased breast and torso dimensions and postural changes influence maternity wear shapes, proportions, and materials. O'Brien (2005) traced what women wore during pregnancy from the medieval period to modernity, stating that throughout history and around the world "women's attire has of necessity been designed to adapt to the needs of pregnancy and breast-feeding" (p. 394). Until recently, women wore garments with expandable features like generous fabric, gathers, pleats, and ties at side seams designed to expand with and camouflage the pregnant body (Furer, 1967). Maternity wear as a separate market category was introduced in about 1910 (O'Brien, 2005, p. 395). Changing attitudes in the 1990s, especially in Western cultures, influenced maternity wear styles. Maternity wear now emphasizes and reveals the "baby bump," a popular marketing term for the expanding belly.

The degree and rate of belly expansion generally requires garment modifications by the fourth or fifth month of pregnancy (Kılıç, Tama & Öndoğan, 2014; Park & Lee, 2007; Sohn & Bye, 2012). Study the series of pregnancy scans in Figure 5.9 and visualize shaping a product to the changing figure. One of the first wardrobe changes may be selection of larger garments to accommodate increased bust girth. As pregnancy advances and the belly expands, a pregnant woman's apparel should accommodate increased torso girth. Suspending garments from the shoulders or using stretch materials to shape to the belly are functional design alternatives.

Designing for the lower torso is challenging as waistbands and mid-torso waistlines no longer function as methods to suspend lower torso garments. Stretch panels over the belly and increased front crotch lengths are typical maternity pants features to try to accommodate the larger belly. Yet, Kılıç et al. (2014) found that maternity pants fit is still a problem. Pregnancy's postural shifts also affect garment fit and balance. As posture changes, the hemline of a garment suspended from the shoulders progressively shifts from parallel to the floor to a slight backward tilt.

Women select everyday clothes for assurance, fashion, camouflage, individuality, and comfort (Tiggemann & Lacey, 2009). Given the relatively short period maternity clothing is worn, pregnant women often add affordability to that list. Park and Lee (2007) surveyed 50 pregnant women finding that, in addition to the cost of maternity wear, women were interested in designs that change with the body throughout pregnancy and garments that can also be worn after pregnancy. Sohn and Bye (2015), in a survey of 137 women, found they sought comfort and assurance, not camouflage, in maternity wear.

The limited research regarding maternity wear preferences looks bountiful compared to the very few studies collecting anthropometric data over the course of a pregnancy, especially studies comparing body changes to garment fit. Sohn and Bye (2012) used a case study approach to assess body changes for three pregnancies of two women. They tracked body change effects on the fit of a basic sheath dress. They used 3D body scanning to collect linear measurements and visually assessed body shape changes with midsagittal scan sections and transverse slices at the breast, midriff, waist, maximum belly, and hip. They virtually draped a dress to each body scan and evaluated changes in the garment pattern contours and dimensions. Their results indicate that body shape changes are as important as linear dimensions in fitting maternity wear. Although each woman's pregnant belly had a distinctive shape, the shape of the maximum belly became more elliptical, and the midriff became more rounded. Sohn and Bye (2012) concluded, after analyzing changes in the draped patterns, that different abdominal shapes lead to different pattern shapes. The study provides insights to fitting maternity wear; however, a much larger study is needed for application to mass-produced maternity apparel. As scan and virtual draping technologies become available, new methods of fitting and sizing maternity apparel are likely to develop.

Personal protective equipment (PPE) and uniforms that accommodate the body changes of pregnancy are needed for women who work in physically demanding and dangerous jobs and environments. Very little research is available on design strategies for these maternity categories. Pregnant workers wearing PPE and/or uniforms may need (a) accommodations for the shape and body size changes of pregnancy and (b) different thermal control options to prevent over-heating because, as pregnancy progresses, heat stress tolerance is reduced (Navy Environmental Health Center, 2007, p. 10). Manley (1997) suggested that if pregnant women resorted to wearing clothing that was either too small or too large, unsafe work conditions would result. After conducting an anthropometric measurement study of 90 pregnant women (46 completed the study), she urged more research and anthropometric data accumulation in order to design better fitting maternity protective clothing. The need for up-to-date anthropometric data of pregnant women persists, as well as new design approaches for maternity wear in fashion and protective apparel categories.

## 5.5 Lower Torso and Leg Skeletal System: Protection, Support, and Mobility

Recall the division of the skeleton into axial and appendicular segments. The sacrum and coccyx, the most distal sections of the spine, are part of the lower torso axial skeleton. The pelvic girdle and the leg (lower limb) bones make up the lower appendicular skeleton. Mechanical forces from loads of and on the head and upper torso and arms are transmitted down the spine to the sacrum and then through the pelvic bones to the legs and feet. Low back pain, a very common and often disabling condition, originates from both axial and appendicular skeletal structures in the lower torso. An improved understanding of musculoskeletal anatomy in the lower torso and legs may help designers improve wearable products to both treat and prevent low back pain.

### 5.5.1 Lower Torso Axial Skeleton: Structure and Function

The lower torso continuation of the axial skeleton differs from the rest of the spine in form and function. The vertebrae in this section of the spine are partially or completely fused and move very little relative to one another, acting instead as a foundation for the mobile spine—and a specific, key part of the bony pelvis.

The sacrum (refer to Section 4.4.1 description, Figure 4.8 illustration) with five vertebral elements fused into a shield-shaped bone, functions as a supportive base for the vertebrae above it. The lateral edges of the sacrum also serve as pivot points to allow subtle movements between the sacrum and the *iliac* portion (*ilium*) of the bony pelvis. These movements are necessary for walking and running. The coccyx or tailbone—three to five very small, variably fused vertebrae caudal to the sacrum—taper to a point above the rectum and anus.

Multiple ligaments span between the sacrum and the spinal structures above and below as well as between the sacrum and the pelvic bones. The ligaments bind the bones together, helping create a sturdy bowl to contain the digestive, urinary, and reproductive organs of the lower torso. Many muscles originating from the inner surfaces of the pelvic bones attach to the coccyx to offer a floor of support to these organs and facilitate motions occurring at the hip. The anterior and lateral abdominal wall muscles (refer to Section 4. 5.2) also connect to the pelvic bones to complete the muscular support system for the pelvis and the lower torso region organs.

### 5.5.2 Lower Appendicular Skeleton

The pelvic girdle and the leg/ankle/foot bones with their joints make up the lower appendicular skeleton (Figure 5.10). This chapter includes information on the pelvic girdle and the leg (thigh and anatomical leg). Chapter 8 discusses the foot.

#### *Pelvic Girdle: Structure and Function*

The pelvic (hip) girdle—the bones supporting the torso on the legs—is attached, or "appended" to the axial skeleton at the sacroiliac (SI) joint. Unlike the incomplete ring of the pectoral (shoulder) girdle, the components of the bony pelvis do create a ring-like structure.

#### *Ilium, Ischium, and Pubis: The Innominate Bone*

The ilium, *ischium*, and *pubic bones* form separately in each half of pelvic girdle of the developing fetus. A portion of each of these fetal bones extends into the hip sockets of the bony pelvis and the three bones mature and fuse into a single structure, the *innominate bone*, by adulthood. Many bony pelvic landmarks retain names indicating their origin from the fetal ilium, ischium, and

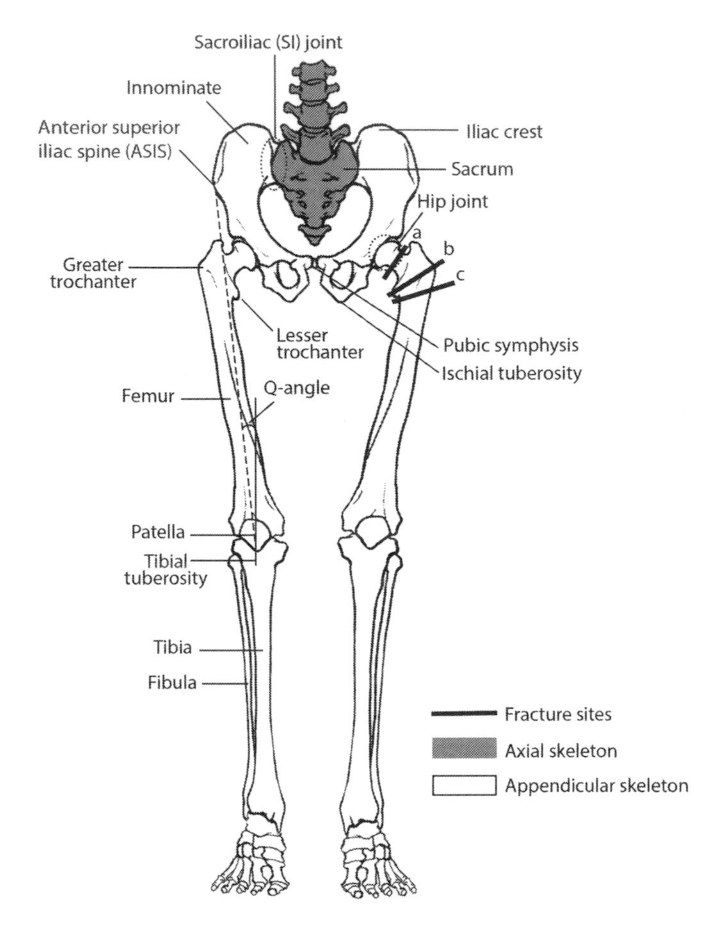


**FIGURE 5.10**
Lower axial and appendicular skeleton, anterior view.

pubic bones. The R and L innominates (also called hipbones), are joined to the sacrum at the SI joints and to each other at the pubic symphysis or pubic joint. The *ischial tuberosities* (refer to Figure 5.3) are the "sit bones," the bony contact points between the lower torso and a horizontal surface.

### *Sacroiliac Joint and Pubic Symphysis: Structure, Function, and Wearable Products*

The SI joint is a synovial joint with cartilage on both sides of the joint, but thinner cartilage on the articular surface of the ilium. SI joint surface area is approximately 17.5 $cm^2$ (2.7 $in.^2$) (Sizer & James, 2008). The articular surface of each ilium fits into/against the lateral aspect of the sacrum and the ilium transmits force from the spine to the legs. SI joint configuration influences how, and if, an individual can carry loads without back pain. Summarizing decades of research, Lee (2011, pp. 10–14) notes the mobility (and stability) of the SI joints depend on the specific shape of the sacrum: (a) the shape of the lateral aspects of the sacrum can be asymmetrical (R to L), (b) the body of the sacrum can be more or less triangular in a coronal plane—changing the angles of the SI joints to more or less vertical, and (c) the lateral aspects of the sacrum can be more or less rotated in a transverse plane, to position the SI joint surface in a more or less sagittal plane. Franke (2003) illustrates in a useful drawing (p. 23) secondary interconnected postural patterns associated with unstable SI joints—demonstrating the importance of healthy SI joint function within the musculoskeletal system. These structural and functional variations create uncertainty for designers of devices to help bear loads through the pelvis to the legs, and of products to maintain stability or to reinforce unstable SI joints.

While multiple muscles in the abdominal wall and lower torso region work together to support the SI joints and pubic symphysis, no individual muscles stabilize them directly. SI joint stability is key for load-bearing activities, with the sacrum "wedged" into the ring of the pelvis. However, when designing a product to enhance stability at the SI joint for load-bearing, remember walking and running require SI joint mobility. The innominate bones move small distances with normal hip flexion and extension, especially in gait activities. Designers who understand SI joint mechanics may be able to create custom devices in collaboration with medical clinicians to help individuals with SI joint stability challenges.

Gao, Sun, Goonetilleke, and Chow (2016), tested an on-hip load-carrying belt (HLCB). Fifteen healthy, young male participants reported less effort and demonstrated lower energy consumption while using the device; suggesting it created a mechanical advantage. The design appears to transfer weight directly to the innominate bones, while decreasing axial loading through the arms, shoulders, and spine. Such force changes, while decreasing stresses at the SI joint, may increase the normally large mechanical forces at the hip joint. Further study of devices which use innominate load-bearing—comparing forces on the SI joints and the hips—are warranted, especially for individuals with SI joint instability and/or hip pain or hip arthritis. New designs to shift forces away from the SI joint, without increasing hip joint load, may prove helpful for people who do heavy or repetitive lifting or have low back pain.

SI joint instability is a particular problem for some women in pregnancy. Although the pelvis is a relatively rigid support for lower torso organs, the pelvic joints and ligaments develop laxity in response to hormonal changes late in pregnancy, to facilitate the passage of the infant through the birth canal. The SI joints and the pubic symphysis become more mobile and the coccyx moves posteriorly (Moore, Agur, & Dalley, 2011, p. 211). LaBan and Rapp (1996) show significant pubic symphysis separation in an X-ray of a standing pregnant woman. They report separations up to 0.5 cm (0.2 in.). That degree of laxity can lead to pain, in and around the SI joints, pubic symphysis, and perhaps elsewhere in the low back and legs. Heckman and Sassard (1994) reviewed literature on low back pain and other musculoskeletal conditions in pregnancy. Snijders, Vleeming, and Stoeckart (1993) proposed a biomechanical SI joint self-locking model. They and Östgaard et al. (1994) suggest use of a *trochanteric belt*, combined with exercise, as treatment for pregnant women with SI joint pain. This product differs from most of the maternity support garments described by Ho et al. (2008, 2009a,b). It is relatively narrow and encircles the pelvis between the iliac crests/*anterior superior iliac spines* (*ASIS*) and the femoral *trochanters* (Figure 5.10) and is tightened to push the innominate bones and sacrum together. Lee describes SI joint dysfunction biomechanics and illustrates biomechanical effects of a pelvic girdle compression belt in U.S. Patent No. 7,037,284 (2006). Pelvic girdle compression belts can provide significant pain relief and decrease activity limitations. However, when using these belts pregnant women must be under the direction of a knowledgeable health care provider, to avoid compression to the fetus.

### Gender-Related Differences in Pelvic Bone Structure and Function

Characteristic female/male pelvic differences can be seen in the fetus as early as four months after conception (Gray, 1966, p. 247). Hormonal changes at puberty cause additional changes in the bone structure of the pelvis. The lower half of the female pelvic bone becomes broader/wider (Smith, Turek, & Netter, 2011, p. 8) and shallower (in a superior-inferior dimension) than the male's (Figure 5.11). The lateral upper rims of the pelvis, the hip sockets, and the ischial tuberosities are more widely spaced in females. These changes provide more space in the upper and lower portions of the pelvis for fetal growth and childbirth (Moore et al. 2011, p. 209). The angle below the pubic symphysis is larger in females. The *acetabula* (hip sockets), which face down (more so in women) and somewhat laterally, are larger in men (Gray, 1966, p. 245; White & Folkens, 2005, p. 394). These female/male pelvic differences affect garment fit from torso through the ankle, especially noticeable in bifurcated garments like slacks and jeans.

Sizer and James (2008) report three female/male SI joint differences: (a) instability occurs five times more commonly in women, (b) articular surface shapes differ, and (c) women's cartilage on the sacral articular surface is thicker. It is likely that b and c, along with hormonally induced laxity, lead to the higher incidence of SI joint instability in women. The position of the acetabulum contributes to female/male differences in the inclination of the *femoral shaft* relative to vertical—read more below.

### Pelvic Tilt and Wearable Product Design Considerations

Lumbar lordosis, the degree of concavity of the lumbar spine, varies from person to person and influences sacral position. The sacral inclination, relative to vertical, determines how much the innominate bones tilt forward or backward. The tilt of the pelvis changes the positions of the orifices of the perineal region—the urethral meatus, the vagina, and the anus—relative to the waist (Figure 5.12). For more discussion of *pelvic tilt* and *hip tilt*, see

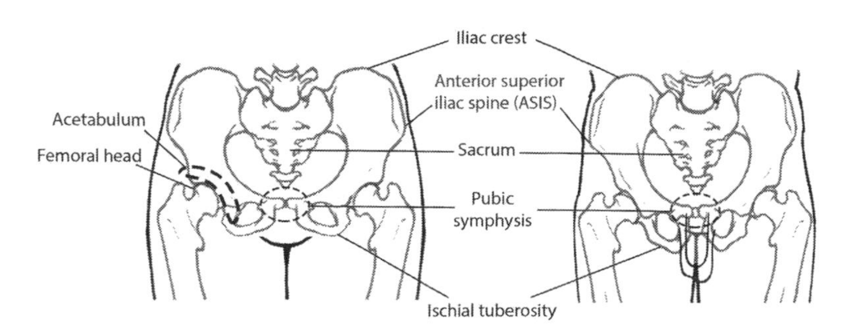


**FIGURE 5.11**
Comparison: Female and male pelvis.

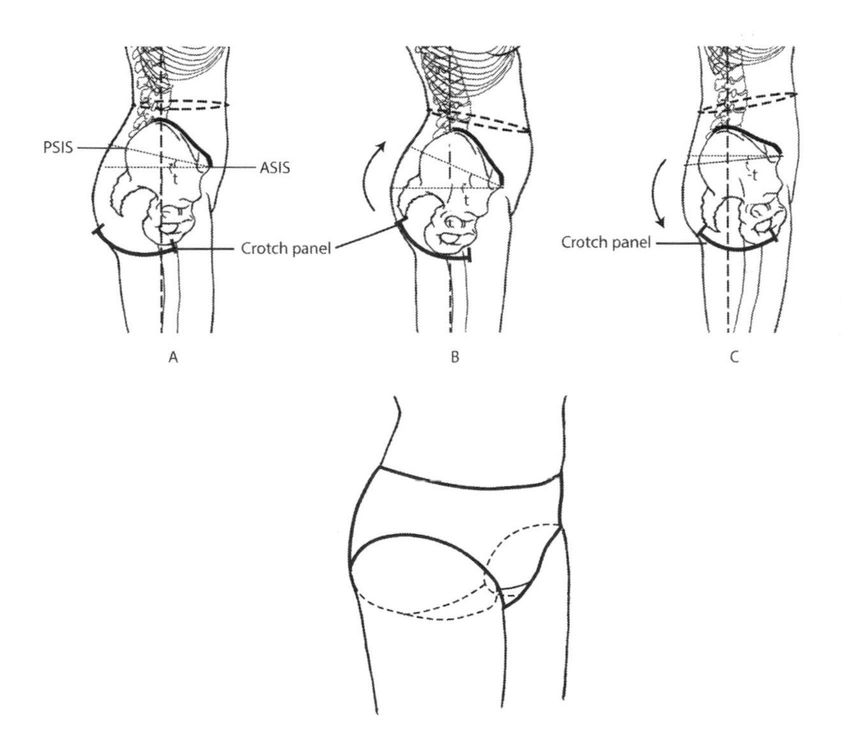


**FIGURE 5.12**
Woman's panty crotch panel location relative to pelvic tilt variations.

Section 6.9.1. Any product worn beneath the perineal region—underwear crotch panel, IMP, or MMP—should be positioned to collect discharges from the orifice(s). An alternative to custom design for pelvic tilt is to make the product section under the perineal region long enough to accommodate all pelvic tilt variations.

### *Hip Joint and Femur: Structure, Function, and Wearable Products*

As a designer, keep the weight-bearing and mobility characteristics of the hip joints in mind. The hip joint (see Figure 5.10) is a ball-and-socket joint, with a relatively deep socket and a prominent ball on the femur of the thigh—the *femoral head*. Ligaments between the pelvis and femur surround and stabilize the joint. A ligament within the joint joins the acetabulum to the femoral head. These features limit motion at the hip but make the hip a stable major weight-bearing joint.

With advancing age, (a) osteoporosis decreases bone strength and (b) balance problems increase vulnerability to falls. The combination often leads to hip fractures. Brauer, Coca-Perraillon, Cutler, and Rosen (2009) studied incidence and mortality related to hip fracture in the U.S., 1986–2005. Hip fracture incidence and chance of death within one year of a hip fracture, for men and women, declined somewhat between 1995 and 2005. However, several findings cause concern: (a) women—who live longer than men in the U.S—suffer hip fractures more than twice as often as men, (b) surgical treatment for hip fractures is expensive, (c) risk of death within the first year after hip fracture is significant—22% for women, 32% for men (Brauer et al., 2009), and (d) quality of life tends to deteriorate in the early months after a hip fracture (Randell et al., 2000). Prevention of hip fractures from falls can reduce costs, forestall premature deaths, and preserve quality of life.

The femoral neck spans between the femoral head and the *greater trochanter* —the bony prominence on the side of the thigh (see Figures 5.10 and 5.11). The femoral neck—fracture site **a** in Figure 5.10—is the most likely "hip" feature to break during a fall. Other fracture sites include the femur section between the lateral trochanter (greater trochanter) and the medial section of bone at the top of the femoral shaft (*lesser trochanter*)—fracture site **b** in Figure 5.10, and much less often, the area just below the trochanters near the top of the femoral shaft—fracture site **c** in Figure 5.10 (Zuckerman, 1996). The numbers of people affected, and the morbidity associated with a hip fracture present an important design opportunity: creation of better wearable hip fracture prevention devices.

Twenty years of U.S. government-funded research into fall prevention found that wearable hip protectors prevent injury (Sleet, Moffatt, & Stevens, 2008). But, for reasons varying from (a) wearer cognitive impairment, to (b) incontinence, to (c) concerns about appearance, discomfort, and inconvenience; hip protectors are poorly accepted and not often used, especially at home (Cianferotti, Fossi, & Brandi, 2015). Departing from designs using padding (Cianferotti et al., 2015) or auxetic materials (Yang, Vora, & Chang, 2018) over the trochanter, Raanan engineered a sensor-equipped inflatable air bag, worn at the waist, for prevention of femoral neck and intertrochanteric fractures (U.S. Patent No. 9,629,399, 2017). Citing a lack of consistent evaluation methods, the International Hip Protector Research Group developed recommendations for both biomechanical (Robinovitch et al., 2009) and clinical (Cameron et al., 2010) hip protector testing. Additional research employing these standards (for both efficacy of and compliance with hip protector use) may reveal the best design and education/marketing approaches, resulting in products acceptable to the target market and effective in preventing fall-related hip fractures.

Athletes in contact sports also experience trauma to the femur (both the trochanteric region and distally) and pelvic bony prominences. In addition to coccyx protectors, discussed in Chapter 4, incorporating auxetic materials or padding over the ischial tuberosities and greater trochanter helps decrease soft tissue and bony injuries for hockey players (Figure 5.13-A). Football pads protect the lateral iliac crest and coccyx as well as the trochanters. Pads inserted into football pants over the anterior thigh as well as the knee offer cushioning for the mid- and distal-femur.

### *Knee Joint, Patella, Tibia, and Fibula: Structure, Function, and Wearable Products*

The knee joint, at the distal end of the femur, creates a variety of problems for a wide range of people from adolescents to senior citizens, sedentary individuals to workers with physically demanding jobs, to athletes (Fairbank, Pynsent, van Poortvliet, & Phillips, 1984; Miranda, Viikari-Juntura, Martikainen, & Riihimäki, 2002; Peat, McCarney, & Croft, 2001). Although common, knee problems related to mechanical factors—including wear-and-tear—and injuries are complex. Obesity, which increases the weight borne by the knee, is a common contributing factor for many knee problems. And, as research on knee pain, injuries, osteoarthritis (OA), and treatment options demonstrates; knee issues raise difficult and often unsolved diagnostic and therapeutic questions. Many wearable products designed to address knee problems are available: protective knee pads, braces, taping systems applied at the knee, and orthotics inserted in/attached to shoes.

Weight-bearing and motion, the primary lower extremity functions, factor into most knee problems. Knee swelling and pain related to systemic illnesses such as rheumatoid arthritis occur when the joint's lining (the *synovium*) becomes inflamed. They are most often treated with medications and joint rest. The anatomy of the knee joint is not overly complicated, but knee function, within the *kinetic chain* of the lower torso and legs, is intricate (Palmitier, An, Scott, & Chao, 1991). Alterations in the stability, alignment, or range of motion (ROM) of the hip, knee, ankle, or foot joints or changes in the muscles controlling any of those joints can contribute to knee problems.

In simplest terms, the knee joint is a synovial hinge joint (refer to Figure 5.10), consisting of the two distal, rounded, cartilage-covered femoral condyles, which articulate with relatively flat condyles of the proximal tibia. *Medial* (*tibial*) and *lateral* (*fibular*) *collateral ligaments* (on the joint sides) help to stabilize the joint. A pair of ligaments (the *anterior* and *posterior cruciate ligaments*) bridge between the femur and the tibia, crossing in an X-shape within

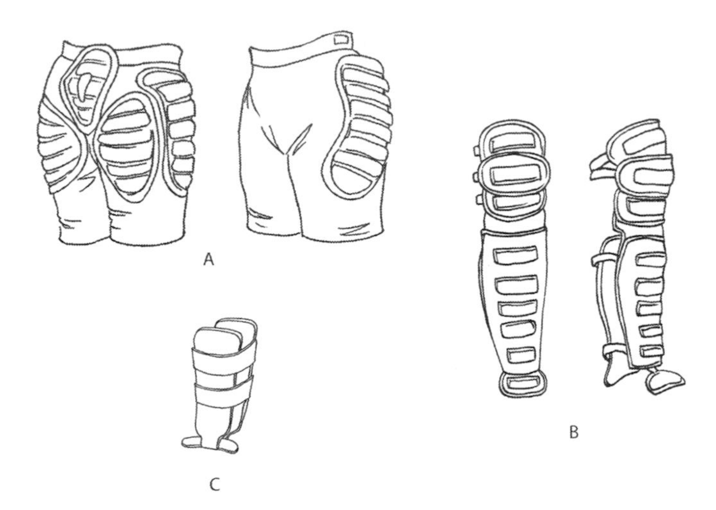


**FIGURE 5.13**
Product examples for skeletal protection in sport and daily activities: A: Pelvic bone hockey protectors, B: baseball catcher's knee/shin/ankle guards, C: stirrup style ankle brace.

the joint, and attaching to a non-articular area between the tibial condyles. Each femur has a *medial* and *lateral epicondyle* protruding from the condyles that can be palpated from the surface and which serve as landmarks for wearables. The *medial* and *lateral menisci* (plural of *meniscus*), partially movable crescent-shaped cartilage spacers between the femur and tibia, deepen the tibial articular surface (Jenkins, 2002, p. 277). Lip-like, they attach to the edges of the tibia along the medial and lateral aspects of the joint. The thin innermost portions of the menisci, particularly the medial menisci, are vulnerable to tears with knee trauma, and contribute to knee locking and/or buckling when damaged.

An anterior cartilage-covered section between the femoral condyles provides an articular surface for the underside of the *patella* (knee cap). The patella, suspended in the tendon connecting the quadriceps (the large muscle on the anterior thigh) to the tibia, tracks over the anterior distal femur in the *patellofemoral joint* of the knee joint (Wiberg, 1941). A joint capsule, reinforced by various connective tissue structures attaching muscles, ligaments, deep fascia, and tendons in the region surrounds both the tibial-femoral articulation and the patellofemoral joint. Tapping on the section of tendon distal to the patella, the *patellar ligament*, elicits the *patellar tendon reflex* (knee jerk) tested in routine medical exams. The patellar ligament attaches to the *tibial tuberosity*, the anterior prominence of the proximal tibia. The smaller of the two bones in the anatomical leg, the fibula, articulates with the proximal lateral tibia, but is not considered a part of the knee joint.

Patellofemoral joint pain (PFP) in adolescence and early adulthood and knee osteoarthritis (OA), in older adults, are wide-spread, multi-factorial, and not fully understood. PFP's diversity and ambiguity introduces new approach opportunities, including wearable products, but great difficulty in demonstrating intervention benefits. The design process requires a clear problem definition (LaBat & Sokolowski, 1999). The complex anatomy and function of the knee makes it difficult to do quality research on product function. A few randomized controlled trials (RCTs) have found some benefits from bracing for PFP. In 2004 the Van Tiggelen et al. study of a patellofemoral brace (a neoprene tube with modifications to place traction on the patella) found fewer test brace wearers (compared to a control group) developed anterior knee pain during intensive military training. They concluded that the brace prevented anterior knee pain from high physical demands but suggested more study to explain why. In another study, bracing of the patella, alone and in combination with home exercise, was found to reduce PFP (Lun, Wiley, Meeuwisse, & Yanagawa, 2005). Selfe et al. (2016) used a series of clinical tests to delineate three subgroups of patients with PFP—suggesting PFP is at least three separate problems and that targeted interventions may improve patient outcomes. Designers may be able to use this research to more clearly define PFP brace design puzzles. Collaboration between designers and health care providers is desirable and necessary.

Unlike PFP, painful OA has many physical manifestations. Altman et al. (1986) developed a comprehensive series of algorithms to classify knee OA to "assure consistency and improve communication" in research reports (p. 1048). They caution that the criteria may not measure therapy response (p. 1049), therefore application for determining product effectiveness is also uncertain. Bone change seen on X-ray has been used to diagnose OA and monitor disease progression but is unlikely to be useful for product evaluation.

As with PFP, wearable products for treatment or prevention of knee OA should be considered from clinical, functional, and economic points of view. Knee OA, when severe, can be treated with joint replacement, but it is expensive and the post-operative recovery and rehabilitation process is long and difficult. Efforts are being directed toward prevention of osteoarthritic change and/or treatment of milder knee joint changes; bracing, specifically designed to alter biomechanics, is often part of that process (Roos & Arden, 2016). Factors limiting patient compliance with knee bracing need to be considered in brace design (Moyer et al., 2015, p. 185). Until OA issues and OA interventions are more narrowly defined, product evaluations will be difficult.

On the other hand, Callaghan et al. (2015) studied the effects of wearing a patellofemoral knee brace for OA over a six-week period. Patients used the brace a mean of 7.4 hours/day, reported relief of knee pain, and demonstrated improvement of MRI-detected OA changes in the patella. Moyer et al. (2015), reviewing literature on biomechanical effects of bracing for medial compartment tibial-femoral joint OA, found biomechanical change with brace use. However, they also found uncertain clinical benefit and poor patient acceptance of long-term bracing. Cudejko et al. (2018) reviewed literature on soft knee sleeves without realignment properties and found benefit for users through pain reduction more than from improved physical function.

Sport-related knee injuries are common, widely studied, and can end an athlete's career. Given the knee's weight-bearing function and the motions possible at the knee, serious knee injuries occur in sports despite the extra ligaments at the knee. Anterior cruciate ligament (ACL) tears occur in a variety of sports. ACL injury mechanisms include: (a) taking a blow to the lateral knee or (b) pivoting the torso laterally relative to a landing foot, with a slightly flexed knee (Boden, Dean, Feagin, & Garrett, 2000; Krosshaug et al., 2007; Johnston et al., 2018). Women experience ACL tears more often than men, for unclear reasons. A blow to the lateral side of the knee can tear the medial collateral ligament; twisting and bending stresses at the knee often cause other damage, in the form of meniscal and ligamentous tears.

Athletes sometimes suffer a compound injury with tears of the anterior cruciate ligament, the medial collateral ligament, and the medial or lateral meniscus (Shelbourne & Nitz, 1991), resulting in a very unstable knee. Treatment for such an injury includes surgical reconstruction, an often long and arduous rehabilitation program, and a progressive return to activity and sometimes to sports. Once a mainstay of treatment, bracing following ACL repair has become less and less common. Wright and Fetzer (2007) maintain there is no post-operative or rehabilitative benefit from bracing after ACL repair. However, Chahla, O'Brien, Godin, and LaPrade (2018) mention post-operative knee bracing as well as bracing during return to sport in the context of multi-ligament repairs. This suggests bracing decisions depend on injury specifics, patient characteristics and progress, and medical and therapy professional evaluation. These divergent bracing practices underscore the evolving understanding of the intricacies of knee function and the importance of individualized knee injury treatment.

The tibia and fibula, bones of the anatomical leg, connect the knee and the ankle. The tibia bears body weight through the femur. Locate your tibia by palpating the inside of your leg just below your knee and the bony ridge down the front of the leg. The tibia articulates with a large bone of the rear foot, the *talus* (refer to Section 8.1.1 to learn more about the ankle joint). Feel the tibial *medial malleolus* at the inside of the ankle joint. Feel the other bone of the lower leg, the slender fibula, on the outside of your leg. Find the *fibular head,* the bony prominence on the outside of the leg, toward the back of your knee. This feature articulates with the tibia's lateral condyle in a minimally movable joint. At the bottom of the fibula, find the *lateral malleolus* at the lateral side of the ankle. The lateral and medial *malleoli* (plural of malleolus) help stabilize the ankle joint. Baseball catchers' articulated protective leg pads, which cover from the distal femur to the top of the foot, are a good example of a functional and protective wearable product for an entire section of the lower limb (Rosciam, 2010). See Figure 5.13-B.

The tibia and fibula share a minimally mobile syndesmosis joint, with upper and lower components, between the knee and ankle. A tear in the distal segment of the *syndesmotic ligament* is called a high ankle *sprain*, a different injury than the more common sprain at the lateral and/or medial malleoli (discussed in Chapter 8). Roemer et al. (2014) used MRI to study sport-related ankle sprains in young men and found about 20% included a syndesmotic injury. Williams, Jones, and Amendola (2007, p. 1199) note high ankle sprains are more common in "collision" sports (i.e. football, rugby, lacrosse) and sports whose athletes wear boots—hockey, skiing. Historically, this injury has been underdiagnosed and optimal treatment is not clear; however, it generally includes immobilization (Williams et al., 2007). Wearable devices for high ankle sprains include casts (fixed-in-place) or removable walking boots to stabilize a severe or recently injured syndesmotic joint. Stirrup style braces with good lateral support or ankle taping are used for minor injuries, sub-acute recovery periods, or to try to prevent recurrent injury (Figure 5.13-C).

It is difficult to determine the effect wearable products have in the treatment and prevention of high ankle sprains. Like knee problems, ankle sprains are complex, and available literature lumps all types of ankle sprains together. Lamb, Marsh, Hutton, Nakash, and Cooke (2009) found 10 days of ankle immobilization in a fixed-in-place below knee cast proved most beneficial for severe ankle sprain healing compared to a walking boot, ankle air splint, or tubular elasticized compression sleeve. All treatments were combined with use of crutches. Janssen, van Mechelen, and Verhagen (2014) reported bracing was superior to an exercise program for the prevention of self-reported recurrent ankle sprains, but participant compliance with the research protocol was an issue. Feger, Donovan, Hart, and Hertel (2014) documented decreased muscle activation with ankle brace use in a group of individuals who had chronic ankle instability after ankle sprains. Ankle brace use may prove to be a trade-off of benefits and detriments.

Many sports and activities use lower extremity protective gear to protect from bumps and bruises: rugby, football, hockey, baseball, soccer, skateboarding, basketball, volleyball, and lacrosse use sport-specific combinations of pelvic, hip, thigh, knee, shin, and ankle guards.

## 5.6 Lower Torso and Leg Muscles: Structure and Function

The muscular system below the waist has two separate but interrelated and equally important components: (a) the pelvic diaphragm, within the bony pelvis and (b) muscle groups active in posture control and motion associated with the lower appendicular skeleton. Connective tissues throughout the pelvis and lower extremities interconnect the two, and also connect to the upper torso and arms (Lee, 2011, pp. 40–44; Lee & Vleeming, 2007, p. 626).

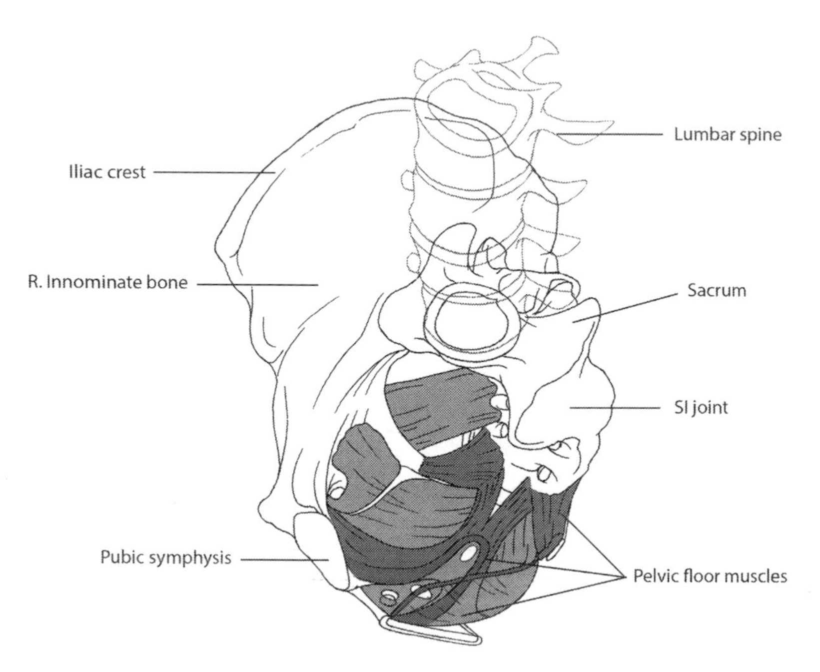


**FIGURE 5.14**
Muscles of the female pelvic diaphragm and sphincters.

### 5.6.1 Muscular Components of the Lower Torso

The muscular components in the lower torso can be subdivided by their deep or superficial locations, but they function together. Neumann (2017) presents a detailed biomechanical analysis of superficial and deep lower appendicular muscle actions on the hip joint (in sagittal, transverse, and coronal planes). His work is especially relevant to designers of specialized products such as hip braces and full-length prosthetic legs and leg braces.

#### *Deep Muscles: Pelvic Diaphragm and Sphincters*

Pelvic diaphragm muscles attach to the internal surfaces of the pelvis and form a concave "floor" of support for the lower torso organs. As part of that muscular sling, the sphincter muscles encircling the urethra and rectal canal actively control the elimination functions of the urinary and digestive systems (Figure 5.14). Other muscles originating deep within the pelvis are activated during reproductive system activities. The *cremaster muscle,* which surrounds the testes and contracts to elevate them within the scrotum is an example. The essential functions of these muscles were mentioned earlier in this chapter, but further details about these muscles are of limited importance for most wearable products. Products requiring in-depth anatomical knowledge of the pelvic diaphragm include:

(a) pessaries worn for support for pelvic organ prolapse, (b) contraceptive devices—the diaphragm and cervical cap—worn over the cervix, and (c) tampons (MMPs). These products stay in place better when the wearer has a strong pelvic diaphragm.

#### *Deep Muscles of the Lower Appendicular Skeleton*

Large paired muscles, the *psoas major* and *iliacus,* are found deep inside the lower torso (see Figure 5.15-B). The psoas originates deep inside the torso, from the anterolateral lumbar vertebrae. The iliacus originates from the interior aspect of the ilium. They merge into one muscle, commonly called the *iliopsoas,* as the muscles leave the lower torso. The iliopsoas attaches to the lesser trochanter of the femur (refer to Figure 5.10) and flexes the hip joint. You can feel the iliopsoas in action where it passes through the groin—the junction between the leg and your lower abdomen. While seated, palpate the area about half-way between your pubic symphysis and the side of your hip. Raise your leg and feel for movement of a narrow band of muscle, the iliopsoas. Tight products that encircle this intersection—such as underwear and swimwear elastic bands, as well as very tight jeans—may be uncomfortable, chafe, and could cause skin breakdown over the muscle.

*Psoas minor,* another paired but smaller muscle, also originates from the anterolateral aspect of a few lumbar vertebrae. It runs parallel to the psoas major and attaches to the top of the pubic bone, toward the hip joint. It helps rotate the anterior pelvis in an upward direction, tucking the tailbone under the body and flattening the lumbar spine. Taken to an extreme this action tilts the pelvis posteriorly (see Figure 5.12-C). In contrast, if you use paraspinal muscles (discussed in Section 4.5.2) to arch your low back—increasing your lumbar lordosis—you tilt the pelvis anteriorly (review Figure 5.12-B), stretching the psoas minor.

Several other small deep muscles of the appendicular skeleton extend from the interior bony pelvis to the femur and act to rotate the hip. You can observe their actions: (1) sit on a chair with your thighs and feet parallel, (2) lift one thigh about 1 inch off the chair and keep it in that position, (3) with your foot pointed straight ahead, twist the thigh in the hip joint to move your foot toward your opposite leg (medial hip rotation), and (4) twist the thigh the other way to move your foot laterally (to your side) to show lateral hip rotation. Some authors substitute "internal" for "medial" and "external" for "lateral" when discussing hip rotation. While essential for these motions, the hip rotators do not contribute significant surface bulk to affect product dimensions.

#### *Superficial Pelvic Girdle/Femoral Musculature: Anterior, Lateral, and Posterior*

Unlike the deep muscles, the superficial lower torso muscles can have major effects on product shape, size, and function. The skeletal components of

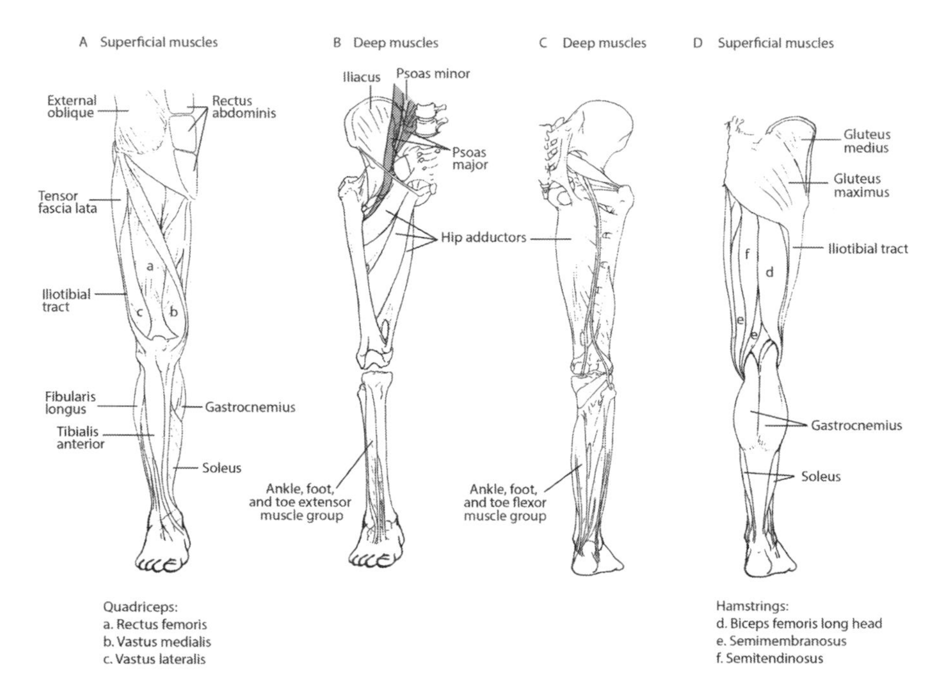


**FIGURE 5.15**
Superficial and deep muscles of the lower appendicular skeleton.

the lower torso and the leg are clearly separate, but the muscular system is quite continuous. Review Chapter 4 descriptions of the rectus abdominis, transverse abdominis, and internal and external obliques (see Figure 5.15-A). These muscles in the anterior abdomen attach on the iliac crest or pubic bone and help move the pelvis relative to the trunk while walking (Anders, Wagner, Puta, Grassme, & Scholle, 2009). The rectus abdominis, like the psoas minor, helps tilt the pelvis relative to the trunk. The obliques help elevate the lateral pelvis and rotate the pelvis relative to the trunk.

Other superficial mid-size to small muscles originate on the lateral and posterior ilium to contribute to the "hip curve" between the iliac crest and the greater trochanter: (a) *tensor fascia lata*, most anterior, with a tough broad flat tendon (the *iliotibial tract*) extending from near the greater trochanter to the lateral upper tibia below the knee (see Figure 5.15-A); (b) *gluteus medius*, on the posterior-posterolateral ilium, which inserts on the superior aspect of the greater trochanter (see Figure 5.15-D); and (c) the deeper posterior-posterolateral *gluteus minimus*, which inserts below the intertrochanteric region of the femur. Together these three muscles abduct the thigh to lift it away from the midline of the body. They can be difficult to palpate. To see them in action, steady yourself with one hand on a chair and lift one leg straight out to the side. The muscles contribute to a lesser extent to other hip motions.

In addition, the hip abductors on the weight-bearing leg activate to keep the pelvis level. Hip abductors on both sides of the body work together to help stabilize the hip joint in *single leg stance* as you balance, walk, and run (Büchler, Tannast, Siebenrock, & Schwab, 2018, p. 11; Inman, 1947). Neumann (2017) elaborates on the crucial function of hip abductors during single leg standing and gait. Depending on walking/jogging speed, the force to each hip during single leg stance is, on average, 2.8 to 5.5 times body weight. If a person stumbles, the force increases to approximately eight times their body weight (Bergmann, Graichen, & Rohlmann, 1993). Understanding these functions is particularly relevant for designers of products developed to assist in load-bearing activities: backpacks, baby carriers, lift belts, and load-bearing exoskeletons.

The large superficial and posterior *gluteus maximus* fills out the curves of the buttocks (see Figure 5.15-D). It varies in bulk person to person requiring product size and form adjustments. Its primary function is to extend the thigh at the hip joint. The gluteus maximus originates on the external surfaces of the sacrum, SI joint, and adjacent portion of the ilium. It inserts low on the posterior femoral trochanter and iliotibial tract. To locate and palpate the action of the gluteus maximus: (1) steady yourself with one hand on a chair, (2) place your other hand with your thumb on the lateral iliac crest and your fingers spanning over the top of your buttocks, just below the iliac crest, (3) lift your leg straight back, moving from the hip not the low back, (4) feel the muscle contract. Portions of the gluteus maximus also laterally rotate, abduct, and adduct the thigh. Additional small, superficial lower torso muscles act with the small muscles originating inside the pelvis to move and position the thigh. These small muscles typically do not affect product shape or function.

### 5.6.2 Muscular Components of the Thigh and Leg

A few large muscles account for most of the leg's muscle mass. Several thigh muscles originate on the pelvis and insert on the patella and/or the tibia below the knee—for example, the tensor fascia lata discussed above. Other major muscles crossing both the hip and knee joints include the quadriceps muscle of the anterior thigh (Figure 5.15-A) and the *biceps femoris* (lateral hamstring) and the *semimembranosus* and *semitendinosus* muscles (medial hamstrings) of the posterior thigh (Figure 5.15-D). These muscles are bulky, particularly in power athletes. They create complex curves around the thigh, except for the medial surface. Their pathways across the groin and the back of the knee can affect wearable product designs. The hip adductors, a group of three large muscles on the medial thigh, also originate from the pelvis but insert on the femur, above the knee (Figures 5.15-B, 5.15-C). They contribute to the bulk and shape of the medial thigh. These muscle groups—the quadriceps, hamstrings, and adductors—act through wide ranges of motion (see

Section 5.9). Consider their placement, shape, and function when designing bifurcated garments.

### *Quadriceps Muscle*

The large four-part quadriceps muscle, three superficial muscle sections and one deep section, extends the knee. The superficial divisions are: (a) the *rectus femoris,* (b) the *vastus medialis,* and (c) the *vastus lateralis.* The other division, (d) the *vastus intermedius,* lies deep to the rectus femoris. Feel the quadriceps' actions: (1) while seated, place both hands over the top of the thigh of one leg, (2) straighten your knee and lift your leg to feel the contraction of this compound muscle. Each section originates from a different place. Only the rectus femoris, which arises from the ASIS of the ilium, crosses both the hip and knee joints and therefore can be felt lateral to the iliopsoas at the groin as it helps the iliopsoas flex the hip. The v. medialis and v. lateralis originate on the proximal femur. The v. intermedius originates from a large area of the anterior femoral shaft. The four sections come together to insert on the patella, each with its own specific direction of pull. The v. medialis and v. lateralis are thought to play a role in PFP as they influence the tracking of the patella in the groove of the patellofemoral joint. The forces generated by the quadriceps reach the tibia to extend the knee, through the patellar ligament, which joins the patella to the tibial tuberosity.

### *Hamstring Muscles*

The hamstrings have two actions: (a) assisting the gluteus maximus with hip extension and (b) as the primary knee flexors. The hamstrings cover the posterior femur in two layers. The superficial layer includes the *long head* of the two-part biceps femoris (the lateral hamstring) and the two medial hamstring muscles, all which originate on the ischial tuberosity of the pelvis. The deep layer, the *short head* of the biceps femoris originates on the femur. The hamstrings all insert below the knee. The biceps femoris inserts on the fibula. The medial hamstrings insert on the tibia. To palpate the hamstrings in action: (1) stand, using one hand on a chair to balance, (2) place the free hand on the back of your thigh, (3) bend your knee and feel the muscle actions of the mid-thigh, (4) slide your hand to behind your knee, and (5) feel the cord-like hamstring tendons as you bend and straighten the knee. A product constricting the hamstring tendons at the knee may be uncomfortable.

### *Thigh Adductor Muscles*

Three large thigh adductors (*adductor magnus, adductor longus,* and *adductor brevis*) pull the thighs together and fill out the medial thigh. They originate on the pubic bone and insert on the femur. To feel their action: (1) sit in a chair, (2) place a something like a soccer ball between your knees, (3) place one hand over one medial (inner) thigh, and (4) squeeze the object to feel the muscles contract. The bulk of the inner thigh can completely fill the space between the upper thighs and influences the curve of the proximal pants inseam. The mass of the quadriceps, hamstrings, and adductors all contribute to the fullness needed in the upper pants leg.

Much like an EKG monitors heart activity (see Section 4.2.4), specialized wearable products offer an opportunity for monitoring skeletal muscle activity using *surface electromyography* (EMG) techniques. Large superficial muscles like those of the lower torso and thighs are good choices for monitoring. Technical shorts with built-in sensors which detect, measure, and transmit muscle electrical activity data (U.S. Patent No. 7,152,470, 2006) are an example. Postolache, Carvalho, Catarino, and Postolache (2017) surveyed a wide range of "smart" garments and described several garments which use EMG technology.

### *Q-Angle: A Proposed Measure for Pants Design*

Although *Q-angle* (named for its relationship to the quadriceps muscle) was developed as an objective two-dimensional anatomical measurement to try to understand PFP, it may help designers improve pants designs. For designers, the Q-angle may serve as a measure to associate the angle of the thigh between the hip and the knee in a frontal plane with pants-leg drape from the hipline. Q-angle is defined as the angle between two intersecting lines of pull at the patella (refer to Figure 5.10). The measurement relates (a) the direction of the quadriceps muscle's pull—between the ASIS and the center of the patella—to (b) the line of pull from the center of the patella to the tibial tuberosity—through the patellar ligament. Stated another way, although the knee is primarily a hinge joint, the force of the quadriceps does not follow a straight line from the hip joint to the tibial tuberosity as it pulls on the patella to extend the knee.

Three landmarks are used for the Q-angle: (a) the center of the patella, (b) the tibial tuberosity, and (c) the anterior superior iliac spine (ASIS). A goniometer can be used to measure the angle with the center of the patella as the apex of the angle. The angle is measured with the knee in full extension (Amis, 2007; Weiss, DeForest, Hammond, Schilling, & Ferreira, 2013). The normal Q-angle for women (17°) is larger than for men (14°). The gender difference relates to differences in pelvic width—a woman's pelvis is typically wider. Other factors also influence the Q-angle: (a) the angle between the femoral neck and the femoral shaft, (b) the length of each femoral component, and (c) the natural external rotation of the tibial tuberosity with full knee extension—the *"screw-home"* mechanism (Hallén & Lindahl, 1966). Q-angle may affect pants' grainline placement, but more study is needed.

### *Muscular Components of the Anatomical Leg*

The muscles of the anatomical leg are grouped in layers around the leg below the knee. Calf and anterior leg muscle tendons extend into the foot just beyond the ankle as well as to the bones next to the small joints of the foot and toes and play a key role in foot and toe motions. The largest and most superficial muscles move the ankle joint (Figures 5.15-A, 5.15-D). The smaller and deeper muscles control movements of the *arch* of the foot and the toes (Figures 5.15-B, 5.15-C). The large superficial muscles are the most relevant for designers.

The *gastrocnemius* and *soleus* muscles (on the back of the leg) *plantar flex* the ankle. They are most noticeable when someone lifts their heels off the floor to "walk on tiptoes" or wears high-heeled shoes. The gastrocnemius originates from the back of the femur just above the knee and inserts on the heel bone via the Achilles tendon, so it also flexes the knee. The soleus originates below the knee with a similar insertion; it acts only at the ankle.

The Achilles tendon is a crucial structure of the anatomical leg. It transfers the power of the gastrocnemius and soleus to the foot, pushing the body forward with each foot step. Achilles tendon inflammation or injuries, seen in up to 24% of athletes over a lifetime, can be very difficult to treat (Munteanu et al., 2015). A wearable product, an *ankle-foot orthosis* (*AFO*), has been commonly used to immobilize and protect an injured/inflamed Achilles tendon (Figure 5.16), but this clinical practice has not been proven to be beneficial. In their randomized controlled study of exercise plus a customized AFO versus exercise and a sham AFO for Achilles tendinopathy treatment, Munteanu et al. (2015) were unable to demonstrate a significant positive effect from a specific AFO style. This suggests new approaches—perhaps including different wearable products—could be tried.

Physiologically specialized muscle fibers (fast-twitch and slow-twitch) in the leg muscles combine to produce strength, speed, and endurance. Thigh and leg muscles generally contain both fiber types, but the numbers of each fiber vary person to person and fiber type proportions change with age (Evans & Lexell, 1995). Optimal lower appendicular musculoskeletal function includes posture maintenance (an endurance function) as well as motion, which combines strength and speed. The calf muscles, for example, actively maintain upright posture when standing still, but also provide power for walking.

If the calf muscles are overactive, as they are in many neurological problems, a child or adult may tend to be a "toe-walker." AFOs are often used to provide a sustained stretch to such muscles, to help them to relax, or in conjunction with specific medications designed to relax the calf muscles and restore a more normal gait pattern. The Achilles tendon can also shorten and restrict normal ankle extension due to either overactive calf muscles or wearing shoes with more than a 1-inch heel for long hours on a daily basis.

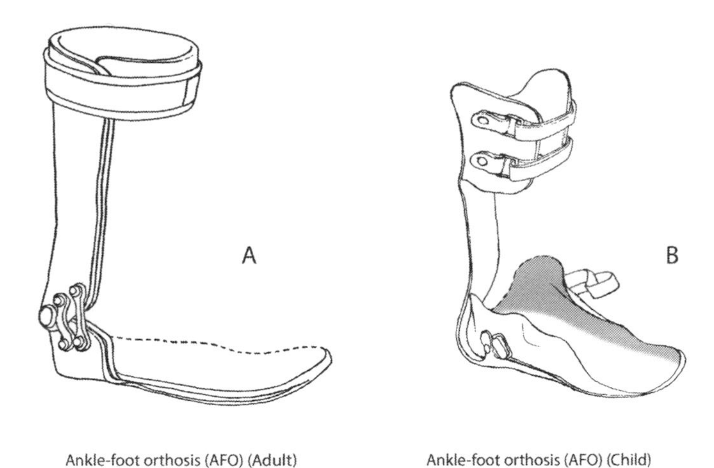


**FIGURE 5.16**
Ankle-foot orthoses, for an adult and a child.

The *tibialis anterior* is the largest muscle at the front of the leg. It extends (*dorsiflexes*) the ankle and produces subtle contours of the anterior leg. The anterior leg muscles are active if you "walk on your heels." When the tibialis anterior is weak, the foot "drops" while walking, increasing risks for tripping and falling and increasing the energy demands of walking. AFOs are frequently used for *foot drop* to keep the foot at a 90° angle with the anatomical leg and make walking safer and easier. *Fibularis longus* (a long, narrow muscle) extends from the fibular head (a prominence of the proximal fibula) around the posterior of the lateral ankle to the bottom of the foot near the big toe. Feel the fibularis longus' action: (1) place your hand over the mid-portion of the muscle on the outside of your leg, (2) raise the outside edge of your foot off the floor to feel the muscle contract. Read more about ankle, foot, and toe muscles in Chapter 8.

## 5.7 Lower Torso Fat: Gender Differences

Muscles contribute to body shape and size. Fat also adds contours and volume. Fat serves the same purposes for women and men; it stores energy and insulates the body. Review the discussion and definitions in Section 2.6 and recall that subcutaneous adipose tissue (SAT) deposit locations differ between women and men (see Figure 2.14). Men collect some SAT in the upper thighs, while women deposit SAT more broadly: (a) over the abdomen down to the pubic arch; and over the (b) lateral lower torso, (c) buttocks, and (d) thighs. These differences further affect fit and sizing of lower torso and leg products for men and women.

## 5.8 Summary: Gender and Activity Differences Influence Lower Torso and Leg Wearable Products

Female-male differences in external genitals, bone structure, fat deposition, and to a lesser degree muscularity all create needs for distinct female-male lower torso product shapes and sizes. Especially for women's lower torso apparel, the greater breadth and lesser height of the female pelvis, along with Q-angle, dictates dimensions, curved seam contours, grainline orientation, and placement of shaping devices like darts and seams. Body changes related to pregnancy often require shape and size accommodations as pregnancy advances (refer to Section 5.4.7). Male fat patterning with relative lack of low abdominal fat may require different shaping techniques with fewer contoured fitting devices. However, if a man's mid-torso fat is excessive, different fitting approaches are necessary (read more in Chapter 6).

Female-male body differences at the crotch also affect product shapes and sizes, especially in apparel. A fly front opening is usually included in men's slacks/pants to allow access to the penis for urination. Slacks/pants crotch curve seams are shaped to include ease for the penis and testicles. Read more about crotch curve shaping for men's products in Chapter 6, Section 6.9.5. Harnesses anchored at the crotch, like parachute harnesses, need to accommodate the genitals for comfort and safety. Adductor muscle bulk and any overlying fat at the medial thigh, in men and women, affect product shape and size.

Athletes of both sexes may require sport-specific accommodations. Horse-racing jockeys need adductor strength and endurance, thus, have muscular anteromedial thighs and need extra room in pants' upper thigh as well as anterior leg fullness or stretch fabric to accommodate their flexed knees while seated on their horse. Runners, with powerful gluteus maximus (hip extensor) muscles need extra fullness over the buttocks. Men, and female athletes specifically, may require accommodations for more muscular calves.

## 5.9 Lower Torso and Leg Range of Motion

The muscle actions described above move the pelvis, thigh, anatomical leg, and foot through variable ranges of motion. Pelvic girdle joints move in very small arcs, measured in single digits. Lower appendicular skeletal joint motions occur in broad arcs, with up to 135–140° of motion. The ROM discussed for each joint below incorporates any/all biomechanical actions at the joint—a summation of actions into a single ROM value. (Recall the motions are based on a 360-degree system, discussed in Section 2.4.3.)

As with the upper appendicular skeleton, the International Society of Biomechanics proposed standardized terminology for joint motions of this region to facilitate communication among researchers (Wu et al., 2002). Except as otherwise noted, the ROM values presented in the following sections are derived from the Roaas and Andersson (1982) study of lower extremity joint ROM in a normal young adult male population, including their comparisons to previously published norms. Range of motion, as an anatomical feature, and apparel function need to be considered together in wearable product design and vice versa. Park et al. (2015) studied the impact of firefighter gear on lower body motion and recommended an ergonomic redesign of firefighter boots to improve lower body mobility and safety.

### 5.9.1 Pelvic Girdle Motion

Pelvic girdle motion is essential for (a) weight-shifting—one leg to another, (b) pelvic tilt, and (c) walking (*reciprocal gait*). There are two types of pelvic girdle motions: movement of the whole pelvis and movement of the pelvic girdle components at the pelvic girdle joints.

Movements of the whole pelvis can be documented in a sagittal plane (pelvic tilt), in a transverse plane, and in a coronal plane. Measuring pelvic tilt is technically challenging and produces less precise range of motion values than measurements elsewhere in the region (Alviso, Dong, & Lentell, 1988). Best estimates, measured from a neutral standing posture (seen in Figure 5.12-A), from Alviso, Dong, and Lentell's small study (1988) are anterior pelvic tilt of 8° and posterior pelvic tilt of 6°. Measuring pelvic motions in transverse and coronal planes is best done with gait analysis tools. Inman, Ralston, and Todd (1981) in their definitive work, *Human Walking*, report pelvic rotation during normal walking—in a transverse plane—of approximately 4°to either side, or a total of 8°. This rotation involves motion between the individual lumbar vertebra as well as between L5 and the sacrum, effectively moving the entire pelvis. When you stand on one leg, the pelvis (on the side of the non-weightbearing leg) drops in the coronal plane approximately 5°, relative to the weight-bearing leg. The coronal plane motion occurs at the hip joint. These three whole pelvic motions contribute to each individual's characteristic walking pattern.

In addition to these motions of the pelvis as a unit, there is motion at the joints within the bony pelvis. The important thing for designers to know about pelvic girdle component motion is that it occurs at both sacroiliac joints and at the pubic symphysis, but the motion is very limited. At the SI joint, the sacrum and the pelvic innominate bones both move, relative to one another. Lund, Krupinski, and Brooks (1996) report normal sagittal plane (nodding) sacral motions in the range of 2–10 mm (0.08–0.39 in.). Jacob and Kissling (1995, p. 359) present the normal range of the same motion as 1° in each direction. On the other side of the SI joint, the innominate bones rotate small distances anteriorly and posteriorly (Smidt, McQuade, Wei, & Barakatt, 1995).

Avoid unnecessary restriction of these small motions. A more detailed analysis of pelvic motion at the sacroiliac joints and pubic symphysis is beyond the scope of this book; see Greenman (1996) for more in-depth information (pp. 308–310).

### 5.9.2 Hip Motion

The hip is a ball-and-socket joint, like the shoulder, with similar motions: flexion/extension in a sagittal plane, abduction/adduction in a coronal plane, and medial/lateral rotation in a transverse plane (Figure 5.17). Hip circumduction uses flexion/extension and abduction/adduction motions together, producing a conical movement, allowing a person to trace a circle with the foot. If tight hip flexor muscles create an exaggerated static pelvic tilt or the hip has arthritic changes, hip motions may be restricted. With those possibilities in mind and remembering how commonly ROM variation occurs at any joint, how might the hip motions and approximate ROMs in Figure 5.17 change?

Muscles which contribute to hip flexion (Figure 5.17-A) include the iliopsoas and, to a much lesser extent, the rectus femoris section of the quadriceps. Starting from a standing position, they lift the thigh through about 120° of motion, bringing the knee up, to a position in front of the mid-torso. If you stand on one foot and move the other leg straight behind you, you are using the large gluteus maximus muscle to extend the hip. Hip extension is normally approximately 15°. The hip flexion/extension arc of motion totals about 135°.

The tensor fascia lata, gluteus medius, and gluteus minimus (labeled abductor m. in Figure 5.17-B) move the femur laterally 40–45° in a coronal plane. Tightness in these muscles or in the iliotibial tract, on the lateral thigh, may limit the opposing motion, hip adduction. The adductor muscle group of the thigh (adductor m. in Figure 5.17-B) pulls the thigh toward and across the midline. Normal hip adduction ROM is approximately 30°. Movement in this arc is most commonly seen (in combination with some hip flexion) when you cross one thigh over the other while sitting. In standing, the supporting leg gets in the way of full adduction, preventing the non-weight-bearing leg from moving through a complete ROM.

A combination of several superficial and deep, large and small muscles rotates the femur in the hip joint. Normal hip rotation ROM totals 75–85°. The subject for the hip joint medial and lateral rotation illustration (Figure 5.17-C) is lying supine, with the thigh supported on a horizontal surface, the knee flexed, and the anatomical leg unsupported. The anatomical leg moves as the subject rotates the thigh in a transverse plane in the hip ball-and-socket joint. Although it may appear this motion arises from the knee, the anatomical leg cannot rotate about the knee because, as a hinge joint, the knee is limited to sagittal plane motions. Figure 5.17-C does not use a 360° reference frame, but rather shows only the extent of the medial and lateral rotations.

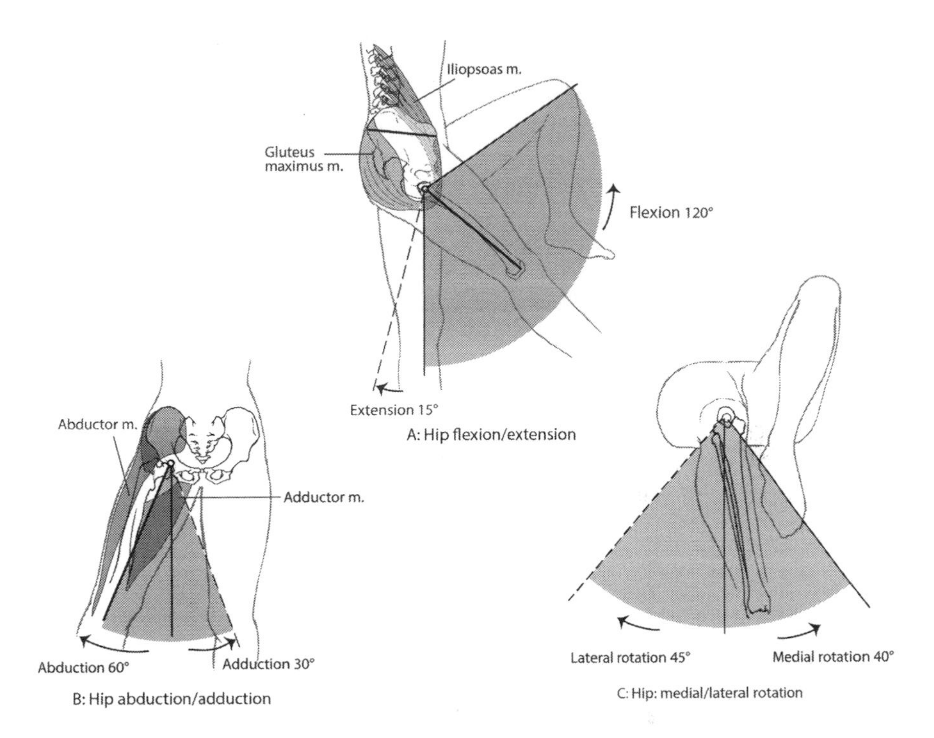


**FIGURE 5.17**
A: Hip flexion/extension B: abduction/adduction C: medial/lateral rotation.

### 5.9.3 Knee Motion

Normal motion of the leg at the knee, a synovial hinge joint, is limited to flexion/extension in a sagittal plane, with a very small external rotation of the tibia at full knee extension, the "screw home" motion—which increases joint stability. The quadriceps muscle extends the knee and the hamstring muscles flex the knee. The muscles generate three distinct actions during flexion and extension: rolling, spinning, and sliding with movement (Gustafson, Takenaga, & Debski, 2018, pp. 7–8). Unlike the elbow, which articulates with both bones of the forearm, the femur articulates with only the tibia at the knee. In the primary knee motion—rolling—the paired smooth round cartilage-covered condyles of the femur roll (rotate) on the cartilage-covered slightly concave condyles of the tibia as the knee flexes and extends. The patella, as it tracks between the femoral condyles, does not affect the hinge action of the knee joint, but rather increases the efficiency of the quadriceps' action (Kaufer, 1979).

The knee, in a standing position, is normally straight, at 180°. If the back of the knee is convex in standing, the knee is hyperextended. In Figure 5.18, the subject is seated, the proximal thigh is supported, and the distal thigh and anatomical leg are unsupported. As in the Figure 5.17-C, Figure 5.18 does not use a 360° reference frame, but rather shows only the extent of knee flexion

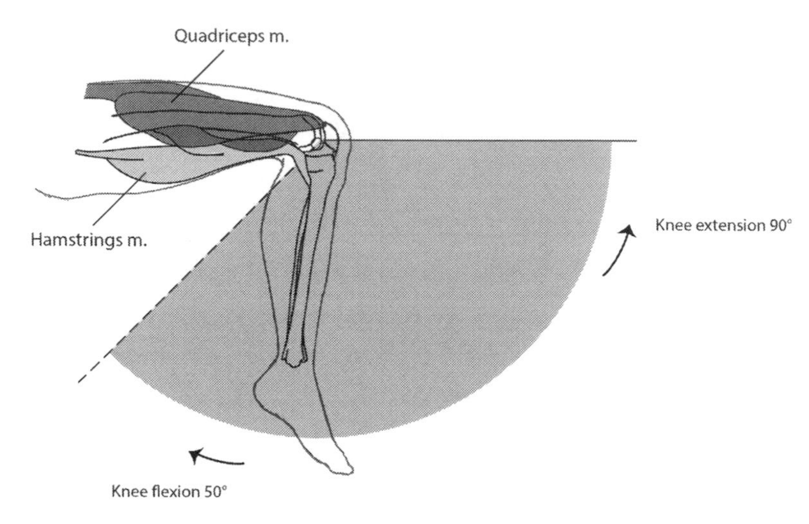


**FIGURE 5.18**
Knee flexion/extension.

(50°) and extension (90°) from a normal seated position. From a standing position, the hamstrings move the knee through 140° when fully flexing the joint.

### 5.9.4 Tibial/Fibular Motion

Designers involved in cast and brace design for treatment and prevention of ankle sprains need to be aware of the motion across the syndesmosis joint between the tibia and fibula. Although of limited extent, movement between the tibia and fibula is greater in the distal leg, near the ankle and occurs in conjunction with motion at the ankle joint between the leg and foot. Ranges and directions of this motion are not clearly established, but normal motion between the two bones is felt to be important for weight-bearing and the flexion/extension hinge action of the ankle (Huber, Schmoelz, & Bölderl, 2012). Read more about ankle joint anatomy, function, and range of motion in Section 8.1.1.

## 5.10 Circulatory, Nervous, and Lymphatic Systems in the Lower Torso and Legs

The circulatory, nervous, and lymphatic systems in the lower torso and legs function much as they do in the upper torso and arms (Sections 4.2, 4.6, and 4.11). This section highlights (a) differences between these two body regions, (b) specific regional anatomy, and (c) particular lower torso and leg wearable products.

### 5.10.1 Circulatory System in the Lower Torso and Legs, Including Wearable Product Considerations

The circulatory system responds to the large muscle heat generation in the lower torso and legs and aids the skin in heat dissipation, influencing thermal balance. Recall the importance of testicular temperature for male fertility and the influences of anatomy and apparel. Apparel creates a microclimate around the body interacting with the circulatory system. Lower body and leg surface area, tissue volume, and essential functions demand a large volume of circulating blood. Arteries carry blood to the region, veins return it to the heart. As blood is pumped back to the heart one-way *valves* close to prevent backflow. See Liu, Guo, Lao, and Little (2017, p. 1124) for an excellent illustration of venous valve function and failure. Review Figure 2.16A and B, arterial and venous pathways.

Gravity favors pooling of blood in the lower body, legs, and feet, where excess blood can cause discomfort, decrease the amount of blood available to the rest of the body, and lower blood pressure to suboptimal levels. Compression stockings and other wearable products have been used for decades to decrease gravity-influenced blood pooling in the legs. Sports compression garments covering large portions of the body and limbs, particularly in power sports, show some clear circulatory and other physiologic effects—although more research is needed (Engel & Sperlich, 2016; MacRae, Laing, & Partsch, 2016).

#### *Arteries and Veins in the Pelvis*

The descending aorta (refer to Figure 4.2) divides into two large arteries which travel through the lower torso near the psoas major muscles. These paired arteries subdivide into internal and external branches. The internal branches provide arterial blood to the organs of the digestive, urinary, and reproductive systems and muscles within the pelvis. The external branches deliver blood to arteries of the legs.

Blood returns from the legs to the heart through the pelvis in large paired veins (parallel to the two arteries) and then through the inferior vena cava, the major vein below the diaphragm. Whether wearable products (such as sports compression garments) affect blood return from the lower torso to the heart remains to be seen. Sports compression shorts and pants cover the lower torso and are available in thigh to ankle lengths. Working from a hypothesis that compression shorts improve jumping power, Kraemer et al. (1996) studied vertical jump power and jumping endurance in collegiate volleyball players wearing thigh-length compression shorts. They found better sustained jumping performance, although they did not look for increased venous return as a contributing factor.

### *Arteries and Veins in the Penis*

An erection, reliant on blood circulation in the penis, is necessary for intercourse to occur. A series of complex interrelated changes in the nervous and circulatory systems create the erection. An erection results from two physiological factors: (1) increased arterial blood flow through deep arteries to expandable tissues within the penis, while (2) the venous outflow from the penis is restricted by the expanding tissues (Smith et al., 2011, p. 32). Condoms need to fit the erect penis to provide an effective barrier to both sperm and infections.

### *Arteries and Veins in the Leg*

The *femoral artery* is the primary arterial blood supply for the leg. It crosses from the pelvis to the leg at the groin. It travels along the medial thigh, sending branches to the tissues throughout the thigh, and then wraps around the medial femur to the back of the knee where it is renamed the *popliteal artery*. Below the knee, on the posterior side of the tibia, it branches into the *posterior tibial* and *fibular arteries*—two vessels which supply blood to the anatomical leg, ankle, and foot. Arterial pulses can be found in the groin (near the iliopsoas muscle) and at the back of the knee between the medial and lateral hamstring tendons (refer to Figure 2.17). Problems with low blood pressure or with *peripheral arterial disease* (*PAD*) may make it difficult to feel pulses at these points. PAD is caused by fatty deposits in the arteries which can restrict blood flow. Restricted arterial blood flow in the lower torso and legs causes leg pain, and, in the extreme, tissue death, which can lead to amputation of the foot and/or leg. A prosthetic limb, a custom-fit wearable product to substitute for the amputated limb, may be worn to help maintain independence, mobility, and quality of life.

Blood from the leg returns to the heart through superficial and deep veins, influenced by the pumping of respiration (refer to Figure 2.16-B and Section 4.3.1). Lower extremity muscle contractions, primarily from the calf, pump up to 90% of the blood in the deep veins back to the heart (Black, 2014). Moore, Dalley, and Agur (2014, p. 620) call this action the *musculovenous pump*. An overly tight wearable product can obstruct venous flow and cause color change and discomfort below it as venous blood collects. On the other hand, the body's blood pumping mechanisms may serve as inspiration for new and innovative wearable product designs.

### *Wearable Products to Facilitate Venous Blood Flow from the Legs to the Heart*

Support pantyhose and body shapers may facilitate venous blood flow from the legs by providing mild generalized compression. Medical compression stockings require a prescription and provide specific levels of pressure (review Section 4.12.4). Conrad Jobst developed the concept of *graduated compression stockings* (*GCS*) in 1949 using a *bobbinette weave* (Bergan, 1985, pp. 532–538). Bobbinette, or bobbin net, weave is a type of hexagonal net structure. Jobst later patented the stockings as well as a measuring template for size determination (U. S. Patent No. 2,691,221, 1954; U.S. Patent No. 2,829,641, 1958). The compression in GCS is relatively stronger at the ankle compared to the knee, compared to the thigh. Thigh-high GCS or other compression devices are used (often along with medications) for hospitalized patients to prevent *deep vein thrombosis* (*DVT*)—blood clot—which may break loose and travel to the lungs (Stanton, 2017). Inflatable leg wraps with pumps are most often the product chosen for DVT prevention for hospitalized bed-bound patients. When patients start walking, thigh-high GCS are worn, but the stockings tend to slide down because of the tapered thigh. Wade, Paton, and Woolacott (2016) surveyed studies on compression stocking compliance and found that patients preferred knee-length stockings to thigh-high stockings. If a patient has a DVT, GCS are frequently used to try to prevent post-DVT leg symptoms, which are similar to *chronic venous disease* (*CVD*).

CVD includes *varicose veins*, problems of leg and foot swelling, skin color changes in the leg, and/or development of weeping sores on swollen legs (*venous ulcers*). Varicose veins are defined as subcutaneous veins larger than 3 mm (0.12 in.) in diameter when standing (Gloviczki et al., 2011). They are generally dark colored and enlarged, with irregular bulges. Primary structural or biochemical vein abnormalities are the likely causes, although blood clots, proximal venous obstruction, or inflammation may also contribute. *Spider veins*, tiny dilated leg veins, represent a very mild form of CVD (Liu, Guo, Lao, & Little, 2017, p. 1123). Varicose veins and CVD may be managed with several types of wearable product including: compression stockings, various multiple-layered bandages and wraps—some with elastic and non-elastic components and/or creams/ointments—and pneumatic compression devices (Gloviczki et al., 2011). Wearable products are a mainstay in CVD treatment although no product type is clearly superior (Amsler, Willenberg, & Blättler, 2009; Dolibog et al., 2014).

Ibegbuna, Delis, Nicolaides, and Aina (2003) developed a testing method to see whether GCS wear increased venous return from the calf before, during, and after walking. They found GCS increased return during walking. Wittens et al. (2015), in the Clinical Practice Guidelines of the European Society for Vascular Surgery (ESVS), emphasize the important role of compression in conservative treatment of CVD. However, older persons often have both PAD and CVD and compression products may exacerbate PAD (Van der Velden, Pichot, Van Den Bos, Nijsten, & De Maeseneer, 2015).

Compression stocking wear is beneficial for CVD patients who wear the product regularly for many hours a day. Medical compression stockings were the most commonly prescribed treatment for CVD in a study from Germany (Pannier, Hoffmann, Stang, Jöckel, & Rabe, 2007). The CVD patients who wore compression stockings reported improved symptoms over time. However, two-thirds of them stopped wearing the stockings due to: itching, skin rash or irritation, constriction of tissues under the stocking, and trouble keeping the stockings in place. In addition, GCS are difficult to don, especially for older people with hand arthritis, weakness, or limited truncal flexibility.

#### *Developments in New Compression Wearables: Compression Stockings*

GCS, as a standard non-surgical CVD treatment for decades, provide definite clinical improvements. To try to improve patient GCS use and compliance, researchers have reassessed (a) stocking pressure levels, (b) stocking pressure distribution, (c) compression garment coverage, and (d) alternative materials and manufacturing methods for compressive devices. Lattimer, Azzam, Kalodiki, Makris, and Geroulakos (2013) compared four compression stocking styles with combinations of two compression pressures and below knee and above knee lengths. All styles were beneficial, so they concluded the wearer could select compression level and length depending on personal preference. Mosti and Partsch (2014) changed the compression patterns on the leg—increasing the calf compression relative to the ankle. The modification improved blood return. Castilho, Dezotti, Dalio, Joviliano, and Piccinato (2018) broadened GCS user categories to study compression stocking use in amateur runners, documenting improved blood return from the leg. Technical textiles play an important role in new developments in compression products. In their review of the characteristics and mechanisms of compression textiles used to treat CVD, Liu et al. (2017) cite research from physiology, pathophysiology, biomechanics, material science, and textile engineering.

#### *Compression Garment Innovations for Other Medical Problems*

GCS and other new wearable compressive devices are used as treatment for several other health conditions. *Orthostatic hypotension* (positional low blood pressure) occurs after getting up from lying down or sitting. It can manifest as lightheadedness and is often associated with decreased blood return to the heart. Granberry, Abel, and Holschuh (2017) developed a dynamic, mobile, and active knit compression garment incorporating smart materials to treat orthostatic hypotension.

Patients with an entirely different health problem with a similar name, postural orthostatic tachycardia syndrome (POTS), experience episodic lightheadedness and fainting related to both circulatory system and nervous system changes when they stand up. Heyer (2014) showed improvement of POTS symptoms when patients wore a compression garment over the abdomen and legs. He interpreted the positive results as improved venous return to the heart. Granberry, Ciavarella, Pettys-Baker, Berglund, and Holschuh (2018) proposed a lower body compression garment for POTS patients using shape memory alloys. Duvall et al. (2017) proposed a new shape-memory alloy compression garment design suitable as treatment for a variety of conditions. Hu and Lu (2015) reviewed elastic fiber classes used in sportswear, including characteristics of elastomeric fibers which respond to body temperature changes. The elastomer variations have potential for use in compression garments.

#### *Advanced Wearable Technology Fit and Function*

New compression mechanisms, plus new materials and combinations of materials, show potential for improving compression devices and garments. Technology innovations led to a new compression garment category—advanced functional garments. Success of these garments relies on proper location of sensors and garment actuating components on the body. Granberry, Duvall, Dunne, and Holschuh (2017) developed "sizing and design strategies for an advanced functional compression garment for the lower leg" (p. 10).

### 5.10.2 Peripheral Nerve Structures in the Pelvis, Lower Torso, and Legs

Unlike the upper torso, the lower torso and legs only contain peripheral nervous system (PNS) elements—representing both somatic (motor and sensory) and autonomic (sympathetic and parasympathetic) functions (review Sections 4.6 and 2.3). The peripheral nerve structures include: (a) the *cauda equina*—a group of nerve roots within the spinal canal, (b) spinal nerves, (c) the lumbar, sacral, and coccygeal nerve plexuses (Netter, 1986, p. 122), and (d) peripheral nerves. Nerves in the lower torso and legs have some vulnerability to product compression and injury.

#### *Cauda Equina*

*Cauda equina* (Latin for "horse's tail") is the descriptive term given to the grouped lumbar, sacral, and coccygeal nerve roots within the lower spinal canal. These structures serve as the first segment of the final PNS pathway between the brain and spinal cord (CNS) and the pelvic organs, musculoskeletal structures, and skin of the lower torso and legs. Paired spinal nerves separate from the "horse's tail" and leave the spinal canal between the lower lumbar vertebrae and through openings in the sacrum. *Cauda equina* damage produces varying patterns of paraplegia and loss of sensation in the buttocks, groin, and legs, and impairment of the sympathetic and parasympathetic nervous system control of bladder, bowel, and sexual/reproductive functions.

#### *Lumbar, Sacral, and Coccygeal Plexuses: Structure and Function*

Spinal nerve fibers—motor, sensory, and autonomic nervous system fibers—from lumbar spinal nerves (2–5), plus the five sacral and one coccygeal spinal nerve come together within the lower abdomen and pelvis to form *lumbar, sacral, and coccygeal plexuses*. Figure 5.19 illustrates selected structures which are pertinent to product design. The 10 spinal nerves (L2-coccygeal 1) regroup into different fiber combinations in the plexuses and branch into peripheral nerves which serve motor and sensory functions in the anterior and posterior lower torso, thigh, anatomical leg, and foot. Like the brachial plexus fiber regroupings, this intricate fiber branching in the lower torso plexuses helps to limit effects of localized nerve root or spinal nerve injuries which could affect motor and sensory nerve function.

The important factors for wearable product designers are that the PNS nerves: (a) cause action in the lower torso and leg muscles and (b) carry sensations back to the brain.

The middle lumbar spinal nerves:

- Flex the hip, extend the knee and ankle, adduct the thigh
- Carry sensation back from the inner upper thigh and genitals, lateral and anterior thigh, and medial leg and foot

The lower lumbar and upper sacral spinal nerves:

- Extend the hip, flex the knee, abduct the hip
- Carry sensation back from the posterior leg, thigh, and buttocks regions plus the foot

The lower sacral spinal nerves:

- Innervate the external sphincter muscles and muscles in the genitals
- Carry sensation from the central "saddle" of the lower torso, around the external genitals and the anus

The lumbar, sacral, and coccygeal plexuses lie within the pelvis and are relatively protected from trauma.

#### *Lower Torso and Leg Autonomic Nervous System: Structure and Function*

The sympathetic and parasympathetic divisions of the autonomic nervous system affect many organ systems in the lower body and legs. The sympathetic chain extends into the pelvis (review Section 4.6.1). Sympathetic nerve fibers supply circulatory, digestive, urinary, and reproductive system components. They make hairs stand on end and increase sweat output. The parasympathetic nerve fibers from sacral spinal nerves 2–4 play roles in the digestive, urinary, and reproductive systems in the lower torso.

#### *Peripheral Nerve Structure and Function in the Legs: Motor Nerves, Sensory Nerves, and Nerve Compression Sites*

Leg peripheral nerves (Figures 5.19 and 5.20) emerge from the lumbar and sacral plexuses mixing motor, sensory, and autonomic fibers. They reach the skin and deep tissues including muscles, joints, blood vessels, and sensory receptors. The motor fibers connect to all the lower limb muscles. Study the schematic drawing in Figure 5.20. Look at the *femoral* and *sciatic nerves* in the thigh. Find the sciatic nerve branches—the *tibial* and *common fibular nerves*, the major peripheral nerves of the leg—and the locations of the muscles they innervate.

These nerves assist in mobility activities. They provide sensory feedback on muscle activity, joint actions, and information about body position and movement. There are many cutaneous sensory nerves in the lower torso and legs and their nerve distributions vary individual-to-individual (McCrory, Bell, & Bradshaw, 2002). Individual variations may originate from nerve fiber regrouping in the plexuses. Figure 5.20 shows one possible distribution pattern.

*Meralgia paresthetica* is a relatively common purely sensory nerve compression in the lower extremity. It involves the *lateral femoral cutaneous nerve of the thigh* (refer to Figure 5.19), a branch of the lumbar plexus. The compression site is in the groin, medial to the ASIS. Symptoms may include numbness, tingling, and discomfort over the lateral thigh, from the trochanter to the knee. Tight products: jeans, military armor, police uniforms, and seat belts have all been associated with this problem (Cheatham, Kolber, & Salamh, 2013). Pressure on the common fibular nerve at the fibular head, such as from a poorly fitting AFO, can compromise both motor and sensory components of the nerve (Rigoard, 2017). A tibial nerve compression site is located distally

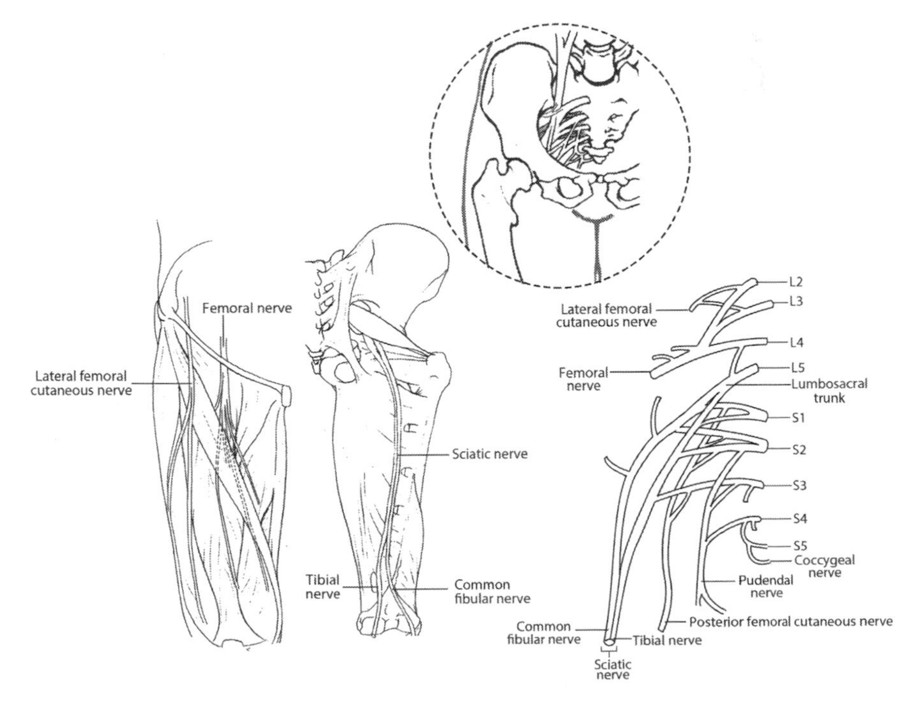


**FIGURE 5.19**
Lower torso nerve plexuses and their relationships to the pelvis and major nerves of the thigh.

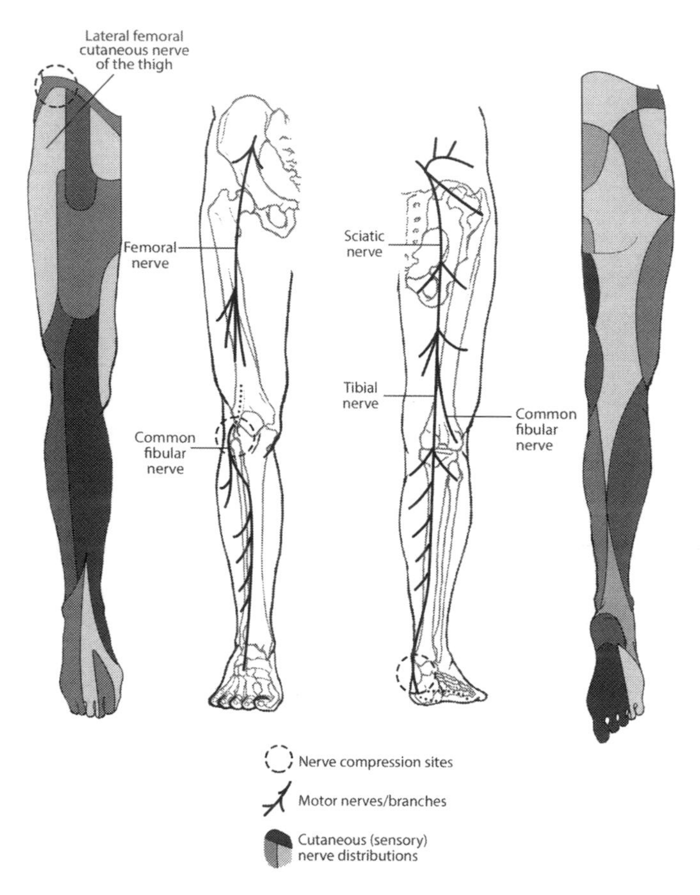


**FIGURE 5.20**
Lower extremity nerves: Anterior and posterior, motor and sensory.

behind and below the medial malleolus at the *tarsal tunnel*. Compression at the site may lead to sensory disturbances of the heel and/or sole of the foot and weakness in the foot muscles. For additional information read Section 8.5.1. The *tarsal tunnel syndrome* is rare compared to the more familiar carpal tunnel syndrome at the wrist (discussed in Chapter 7). Wearable products for carpal tunnel syndrome abound, few if any wearable products are available for the rarer lower leg problem (McSweeney & Cichero, 2015).

### 5.10.3 Lymphatic System in the Lower Torso and Leg, Including Wearable Product Considerations

Lymphatic capillaries drain spaces between cells in the lower torso and legs and return lymph fluid to the vascular system (refer to Section 2.7.3. Figure 2.16-C, Figure 2.18). The lymphatic capillaries associated with pelvic organs drain to lymph nodes along the major blood vessels in the pelvis.

Lymph from the genitals and legs drains through lymph nodes in the groin. Lymphedema (the abnormal collection of lymph in the tissues) has many causes—and research continues to expand knowledge of lymphedema's anatomical and physiological bases (Wigg & Lee, 2014). Lymph fluid differs in composition from CVD edema, however compression product treatments have been similar for the two conditions. No-stretch multi-layer bandages wrapped around the leg appear to work with the calf muscles to promote lymphatic flow (Lymphoedema Framework, 2006), much as the musculovenous pump helps empty deep leg veins. Lymphedema research suggests that new and more effective treatment options, including wearable products, are on the horizon (International Society of Lymphology, 2013; Wigg & Lee, 2014).

In developed countries, cancer treatment-related lymphedema occurs in the lower torso, genitals, and legs much as it does in the arm after breast cancer treatment (Section 4.12.4). It is treated similarly to arm lymphedema (Douglass, Graves, & Gordon, 2016). However, the most common cause of lymphedema in the legs and genitals is the tropical mosquito-borne parasitic infection, *filariasis*. Parasites collecting in lymph nodes and lymph channels result in inflammation. In 2012, 120 million people were infected with filariasis and 40 million people had significant lymphedema effects (Brady, 2014). The presence of lymphedema fluid increases the risk of secondary local infections compounding the problem by increasing inflammation and swelling. *Elephantiasis*, severe filarial lymphedema, can distend the genitals and legs to many times the normal size. Compressive multi-layer bandaging, commonly used in the legs, can also be used to reduce lymphedema of the penis and scrotum (Zuther & Norton, 2013, pp. 286–289). Compression garments for both men and women can be designed to provide genital, lower torso, and leg compression (Zuther & Norton, 2013, pp. 290–291).

*Filarial lymphedema* treatment is complicated by tropical climate and regional poverty. Bernhard, Bernhard, & Magnussen (2003) documented successful treatment in a tropical setting—bandaging followed by compression stocking wear. The method's expense is a drawback. Ryan (2017) and Ryan and Narahari (2012) suggest treatments which emphasize self-care and minimal technology. New simply-designed wearable products suitable for a tropical climate would be welcome to reduce the lifelong disabling effects of the condition. Products using economical local materials could make wearable products more widely available.

## 5.11 Lower Torso and Leg Skin, Including Wearable Product Considerations

The skin covering the lower torso and legs accounts for more than one-half of the integument—the largest body organ system. It is characterized by the structures and functions described in Section 2.8 and illustrated in

Figure 2.21: hair and eccrine and apocrine sweat glands. Skin here, like elsewhere on the body, has several protective functions.

Hair distribution and texture may be of interest to designers. Hair of varying length and coarseness is distributed over the lower torso and limbs. Garn (1951, illustration p. 502) divided the body skin surface into 11 separate regions based on hair growth patterns; there are 5 different hair regions in the lower torso and legs. He reported considerable individual variation in hair patterns and hair density.

Harkey (1993) concisely describes human hair anatomy and physiology. Three types of hair are found in the lower body and leg region: vellus hair, intermediate hair, and terminal hair. Hair changes with hormonal changes of puberty. Children have *vellus hair*: almost invisible, very fine, short, nonpigmented hair, on the body and limbs. Adult men and women have *intermediate hair*—intermediate in length and shaft size—on the arms and legs. Hair in adult pubic areas is coarse, long *terminal hair.* In addition, men have male sexual hair follicles on the abdomen which produce hair similar to pubic hair.

In general, the skin covering the lower torso and legs relates to other body systems in the same way it does in the upper torso and limbs (as discussed in Chapter 4). The following additional wearable product considerations may also be relevant to skin in other regions of the body. The skin and skin structures of the perineal region deserve special attention and are discussed in Section 5.11.2.

Although skin is water resistant, skin structures are permeable to some chemicals, which can be both a benefit and a risk. On the positive side, patches applied to the skin to deliver medication to a specific location have helped control some nerve pain (Finnerup et al., 2015). Birth control patches rely on skin permeability. On the negative side, workers exposed to pesticides require more protection than the skin alone provides. This is particularly true in the perineal region where chemical penetration of the skin is reported to be the highest in the body—42 times greater than in the foot (Holmgaard & Nielsen, 2009). Many factors influence this *dermal absorption*—will your product design need to consider this skin property?

### 5.11.1 Lower Torso and Leg Skin, Excluding the Perineal Region

Skin healing, thermal regulation, and sweat production all relate to wearable product design. Sores and injuries in the legs can be very difficult to heal for many reasons. Specialized wearable products, wound dressings, can be life-changing for patients with lower extremity skin wounds associated with chronic diseases such as diabetes or CVD. Brown, Ashley, and Koh (2018) review the mechanisms of healing and types of wound dressings. They identify wearable technology options for wound monitoring as part of a wound healing program. Research suggests eccrine sweat glands support both wound healing and the skin microbiome (Alam, Hardman, Paus, & Jimenez, 2018; Poblet et al., 2018).

Evaporation of sweat is the body's most efficient thermoregulatory activity. Watery eccrine sweat contributes to body thermal regulation in the lower torso and legs as it does elsewhere in the body (see Sections 2.8.1, 4.13.2). Designers who know (a) where people sweat, (b) when they sweat, (c) how much they sweat, and (d) what kind of sweat is produced in specific locations can design wearables to (a) help manage evaporative heat dissipation/body cooling, (b) minimize garment odor and recommend washing methods and frequency, and (c) program thermal manikins to test prototype garments.

Taylor and Machado-Moreira (2013) conducted a literature review on water and electrolyte losses through eccrine sweat glands. Total body sweat production equals at least 500 to 750 mL (17 to 25 oz) per day (Hodge & Brodell, 2018). Eccrine sweat gland density (number of glands/cm$^2$) and resulting water loss across the skin varies from one location to another (Taylor & Machado-Moreira, 2013, Table 1, Figure 4). The lower torso/buttocks and legs produce at least one-half the total body sweat volume, a factor to take into consideration when making material and garment feature choices for garments worn in this area. Li and Tokura (1996) studied seasonal heat acclimatization in two groups of young women—one group wearing knee-length skirts, one group wearing full-length trousers. They compared several factors, including sweating-related body mass loss. They concluded that participants wearing skirts improved their heat tolerance. Sweat rates influence eccrine sweat composition suggesting that sweating in extremely hot environments or during very strenuous exercise may create differences in garment sweat staining and garment cleaning needs.

Documenting the where, when, and how much of sweat glands/sweat production requires use of sensors which conform and adhere to the skin. The sensors cannot interfere with accurate data collection or the heating/exercise protocols used to generate sweat. Taylor and Machado-Moreira (2013) outline the challenges and technical considerations which go into designing and using such wearable sensors. Testing systems have used many sensor designs and technologies (Brueck, Iftekhar, Stannard, Yelamarthi, & Kaya, 2018; Morris, Coyle, Wu, Lau, Wallace, & Diamond, 2009; Wei et al., 2013).

### 5.11.2 Perineal Region Skin: Secretions, Sweat, Hair, and Wearable Product Considerations

Perineal region skin structures serve several purposes: protection, thermal regulation, and social signaling. This skin has either a dry surface (the stratum corneum–like most body skin—see Section 2.8.1) or is a moist mucous membrane. The mucous membrane structures of the perineal region are closely related to the urinary, digestive, and reproductive systems and their orifices.

### *Perineal Region Secretions*

The urethral, anal, and vaginal linings are mucous membranes; these specialized skin structures serve important roles. The urethral and anal membranes secrete mucus—thick protective fluid—which helps protect the skin cells from harsh chemicals in urine and feces and from bacterial infections. This mucus passes from the body with the urine and feces.

Vaginal secretions provide more lubrication than protection. Levin (2003) outlines several important features and functions of vaginal lubrication.

- The vaginal mucous membrane moisture is maintained by very small amounts of *cervicovaginal mucus* (*CVM*)—20 to 60 mg (.0007 to .002 oz)/day—made up of fluids from many pelvic region structures (Taherali, Varum, & Basit, 2018).
- CVM volume fluctuates according to hormonal changes and sexual stimulation. The vagina produces up to 1 gm (0.035 oz) of watery mucus near the time of *ovulation*—enough to leak from the vagina and wet an underwear crotch panel.
- CVM plays a role in conception, helping sperm reach the egg at the time of ovulation.

The metabolic products of the natural vaginal microbial community (the VMC discussed in Section 5.4.5) add to the volume of the CVM (Huggins & Preti, 1981).

CVM production, unlike a menstrual flow, occurs continuously in small but varying amounts, making vaginal secretion management a relevant topic for women's underwear designers. Some women use panty liners to absorb these vaginal secretions. Panty liners do not appear to alter perineal region health but do carry economic and environmental costs. An industry-supported literature review concluded that consistent long-term panty liner wear is safe and does not promote vaginal/perineal region yeast infection or urinary tract infections (Farage, Bramante, Otaka, & Sobel, 2007). "Breathable" panty liners, incorporating microporous instead of impermeable films as barriers to bodily fluid leak-through, aim to reduce heat and moisture build-up in the perineal region. Giraldo et al. (2011) conducted a controlled study with 53 women wearing breathable panty liners (three/day) and 54 women wearing their own underwear, for 10–12 hours daily over 75 consecutive days. They found no differences between the two groups in vaginal infections or in irritation or inflammation of the vagina or perineal skin.

Huggins and Preti (1981) described vaginal odors and secretions in detail with some references to intravaginal wearable products; they addressed vaginal odors and secretions relative to (a) chemical composition and cyclic variations, (b) contributing structures and processes, (c) secretion collection and methods to quantify, (d) analytical methods, (e) health conditions associated with malodorous vaginal discharge, and (f) early research on possible chemical communication associated with vaginal secretions. Pause (2017) relates the odors of vaginal secretions to cyclic changes in female hormones and associates vaginal secretion odors with human reproductive behavior. For more information on vaginal secretion scent see Michael, Bonsall, and Warner (1974).

#### *Perineal Region Sweat and Wearable Product Considerations*

Sweat glands are distributed throughout the body—except there are no sweat glands on the *glans penis*, clitoris, or labia minora (Hodge & Brodell, 2013). The eccrine sweat glands in the perineal region and groin function as elsewhere in the body. Apocrine sweat glands (also known as odoriferous sweat glands) like those found in the axilla (refer to Sections 4.13.2 and 4.13.3) are found in the perineal skin around the anus, in the labia majora in women and in the skin of the scrotum and *prepuce*—the fold of skin covering the glans penis (Hodge & Brodell, 2018). They are known for producing malodorous perspiration. A third gland type, the anorectal "sweat" gland reported by van der Putte (1991) and further characterized by Konstantinova et al. (2017) as a "mammary-like gland" has no defined function at this time. It is difficult to find sweat volume information for the axillary or perineal regions. Quantifying sweat in these regions, especially with sensors, is difficult due to hair and the concave/mobile anatomical configurations.

The smell of sweat in the perineal region concerns individuals, wearable product designers, and the scientific community. Bacterial interaction with sweat has been associated with clothing odor. Antimicrobial textile finishes and metallic nanofibers have been used in attempts to decrease garment odors and laundering frequency, particularly in sportswear. Antimicrobial textile treatments may alter the bacterial composition of the skin microbiome—with unclear consequences. Callewaert et al. (2014) showed that textile fiber types grow different kinds of bacteria, and therefore smell different after exposure to sweat. Klepp, Buck, Laitala, and Kjeldsberg (2016) compared residual odors (after garment use during exercise) on treated and untreated sportwear textile samples made of wool, cotton, and polyester. They found comparatively less odor on wool and cotton textiles and concluded "the *problem* that odor-control textiles seek to solve is *not* a problem" (p. 312) [for garments made from cotton or wool].

Höfer (2006) explored the principles of body odor formation from sweat and the intricate relationships between antimicrobial textiles, skin-borne flora, and odor. Sweat composition varies—influenced by sweat gland type, sweat rate, diet, hormonal status, health status, and even genetics (Havlíček, Fialová, & Roberts, 2017). Eccrine sweat is mostly water and not particularly malodorous. The volatile substances in apocrine sweat, on the other hand, appear to play a role in social communication, and what smells attractive to one person may smell bad to another (Lübke & Pause, 2015). For instance, androstadienone, a male hormone derivative from apocrine sweat glands, appears to influence how attractive women find a man (Saxton, Lyndon, Little, & Roberts, 2008). Textile odor-control treatments also relate to other issues: environmental concerns regarding textile manufacturing and laundry practices, health risks—including illness related to drug-resistant bacteria, and economic costs to the consumer. Wearable product designers need to consider and balance this complex set of variables.

Medical textiles may need antimicrobial treatments to kill bacteria and control spread of disease (Sun, 2016). As an example, in addition to avoiding occlusive textiles or textiles irritating to the skin, persons with atopic dermatitis (see Section 2.8.5) may benefit from wearing garments incorporating silver nanoparticles—to reduce problems of secondary skin infections (Mobolaji-Lawal & Nedorost, 2015). However, such interventions may have unintended consequences. Kaweeteerawat, Na Ubol, Sangmuang, Aueviriyavit, and Maniratanachote (2017) report development of resistance to multiple antibiotics in common bacteria exposed to silver nanoparticles and emphasize the need for the judicious use of silver nanoparticle technology in consumer products. Use of antimicrobial treatments beyond critical necessity raises the risk of creating epidemics due to "super bugs," microbes highly resistant to known antibiotic treatments.

#### *Perineal Region Hair*

Prior to adolescence, hair in the perineal region is fine, soft, and barely noticeable. Male and female pubic hair patterns develop with the hormonal changes of adolescence and hair becomes darker, coarser, and more curled (Tanner, 1962, pp. 32–33). At full sexual maturity, attained between the ages of 12 and 17 years, pubic hair shows a characteristic inverted triangle with a horizontal upper margin across the mons pubis and hair extending to the labia or scrotum and the upper aspects of the thighs. Adhesives used to secure IMPs and MMPs to underwear may adhere to pubic hair causing discomfort.

Pubic hair grooming practices (shaving, trimming, waxing) have a long history and have become relatively common in the U.S. (Ramsey, Sweeney, Fraser, & Oades, 2009). Rowen et al. (2016) surveyed a diverse group of women and noted variations in this cosmetic practice by age and race—young white women tended to be most likely to use pubic hair grooming methods—facts which may interest underwear and swimsuit designers.

## 5.12 Lower Torso and Leg Landmarking and Measuring

Lower torso and leg landmarking and measuring may be simple or complex. Three measurements, waist circumference, hip circumference, and length can suffice to draft a tube-shaped skirt. A more detailed and complex landmarking and measuring scheme is needed to determine body dimensions at the hip/leg/crotch intersections to apply to pants patterns. Information on pants landmarking and measuring is included in Chapter 6 because pants pattern dimensions rely on placement of the waistline. Chapter 5 landmarking and measuring focuses on the unique aspects of lower torso fit for gender-specific underwear. While reading this section, refer to landmarking and measuring foundations in Appendix A, and details of landmarking and measuring the lower torso and leg in Appendix D. Landmarking and measuring information for the perineal region is scarce in the wearable product literature, so perineal region measuring is not included in Appendix D.

### 5.12.1 Landmarks and Measurements for Lower Torso and Leg Designs

While landmarking bony structures of the pelvis is relatively straightforward and non-invasive, measurements of the surface characteristics of the reproductive structures is highly sensitive and potentially embarrassing.

#### *Basic Lower Torso and Leg Landmarking and Measuring*

Lower torso products are designed for function and/or fashion. Coverage varies for each product type. Some products, like pants and pantyhose, cover the torso and legs; and others, like a knee brace, cover a limited area of the leg. Jeans and tailored slacks constructed of minimal stretch woven fabrics require a detailed landmark and measuring process to fit curves and contours that must be accommodated with a close fit. The multi-contoured lower torso-leg intersection is particularly difficult to landmark and measure. The perineal region including external genitals in the intersection with the skeletal foundation of the upper leg, muscles, and fat of the inner thigh are difficult to measure manually. The area is occluded when using scanning technology with a standing position. In addition, motion changes the size, form, and relationships of the body parts. Individual morphological variations, especially around this area, make fitting a range of sizes difficult.

#### *Gender-Specific Considerations*

The lower torso, particularly the perineal region and genitals, is both culturally and physically sensitive to study. Although all measurement and landmarking activities require tact and consideration of the individual being measured, delicacy is even more important for lower torso/perineal landmarking and measuring. Medical personnel have guidelines for approaching female pelvic examinations which may be helpful (Bates, Carroll, & Potter, 2011). If possible choosing same gender measurers may be wise. Sometimes social customs and demands dictate the presence of a chaperone during landmarking and measuring.

Wearable products for the lower torso are usually gender specific, due to differences in male and female anatomies. In western cultures, skirts are worn by women while in other cultures skirts, kilts, and wraps are sometimes worn by men. The basic skirt is a simple tube shape requiring just a few measurements: waist circumference, *hip depth*, hip circumference, and total length.

Bifurcated garments (lower torso products with the bottom section divided into two parts called legs) are more complex and require more measurements adding: *crotch depth*, *crotch length*, upper thigh and mid-thigh circumferences, knee circumference, lower leg circumference, and ankle circumference. Lengths from the waist and/or hip to each lower limb circumference are also necessary. Products for any portion of the lower limb typically require circumference measurements which relate to the product design, and a landmark and measurement scheme to position segments of the product related to a reference landmark, like the center of the patella (knee cap). For example, baseball catcher's shin guards are articulated so that segments move to bend at the ankle and the knee. Circumferences can be adjusted to fit using straps, while positioning the flexing segments at joints is important for comfort and protection. Other products, like urine collection bags (Section 5.3.1), are difficult to fit and suspend due to the conical shape of the lower limb and weight of the bag. Compression with elastic bands as an attachment method for urine collection bags is not entirely successful. Other measuring, fitting, and suspension methods are needed.

Materials for lower torso and limb products vary from knit structures made of "power" net for body compression, to "comfort" stretch incorporating a small percentage of elastomeric fibers in a woven fabric, to wovens and non-wovens with little stretch, to rigid hard-shell materials used in protective equipment. Each material requires a different approach for best fit and size.

### 5.12.2 Focus on Fit and Sizing of Men's and Women's Underwear

Underwear is worn for many reasons. Functional purposes include protecting outer garments from sweat, oil, vaginal secretions, urine, and feces and protecting the wearer from outerwear abrasion and irritation. Some underwear is designed to protect the wearer from environmental dangers, temperature extremes, and impact trauma. As an extra layer, a second skin between a person's skin and outerwear, underwear may preserve modesty. In contrast, underwear may be designed to emphasize genitals and enhance sexual attraction.

Male/female genital differences are obvious and affect underwear pattern shapes and fit especially at the garment crotch. Clothing history demonstrates that outerwear and fashions of the time affect underwear styles. Men's and women's underwear styles have evolved from loose-fitting draped and wrapped garments to form-fitting, pieced and sewn garments. Early sewn styles relied on the bifurcated method of two identical "leg" or pants pieces sewn together with a U-shaped seam running from the front of the body, between the legs, to the back of the body. Positioning the lower U-portion of the seam close to the perineal region can be uncomfortable and does not provide body-conforming fit which then influences outerwear styles. Pattern adaptations at the crotch are necessary to more closely fit female and male anatomies—a crotch panel for women's underwear and a pouch for men's underwear.

Fiber innovations have influenced underwear styles. Rayon was introduced in the U.S. in 1924 as a modified natural fiber made from raw cellulose. Rayon could be made into lightweight knitted underwear fabrics that were less bulky than natural fiber fabrics. Introduction of nylon fibers in the 1930s, spandex in the 1950s, and knit technology improvements further influenced underwear styles with body-conforming properties (Ewing, 1978). Fabrics that could shape to the body resulted in new underwear pattern shaping strategies.

Although underwear, for both men and women, covers the perineal area including the genitals, very little research or information is available that details how to landmark, measure, or fit the anatomical features. The underwear crotch must, in some way, mimic and/or accommodate the size and contour of the perineal area, the genitals, and the relationships of the urinary, reproductive, and digestive system orifices. The shape and fit of the crotch area is likely based on a manufacturer's proprietary body measurement information and/or on pattern shapes used continuously over the years. Apparel and medical literature do not provide extensive information that is easily applied to underwear design.

### *Men's Underwear Fit and Sizing*

There are two basic underwear styles for men—boxers and briefs. Boxers use a U-shape joining device for the R and L pattern pieces and incorporate more fit ease than briefs to be comfortable. Men's briefs, so named because of their "brief" size in comparison to boxers, are sometimes called "slips" in Europe (Cole, 2010). Briefs incorporate a pouch to cradle the penis and scrotum and are made of knit fabrics to allow a close fit. Boxer-briefs combine boxer and brief features. Refer to Figures 5.1, 5.3, and 5.4 for illustrations of the male perineal region and the penis to visualize how a pouch pattern needs to contour to the body. The pouch extends from above the penis to cup under the scrotum. The pouch and the area of the brief below the anus should provide absorbency, comfort, and cleanliness.

The *codpiece,* first worn around the end of the 14th century, may be thought of as the earliest protective and concealing pouch for men's genitals, and perhaps the forerunner to today's pouch designs. The original purpose was to span the opening between the two legs of men's hose. The name codpiece is derived from an "archaic term for the scrotum and was known as 'bragetto' in Italian and 'braguette' in French" (Cole, 2010, p. 16). What started as a simple cover for the genitals evolved into a prominent feature of wealthy men's outerwear. The codpiece was often padded to accentuate power and used as a convenient storage container. Modern men's underwear designs retain some codpiece shaping and positioning.

Cole (2010, p. 81) describes the introduction of the "Jockey" brief. According to Cole, Cooper Underwear Company (current Jockey Company) created men's underwear that provided absorbency along with "masculine support" in the form of a pouch constructed of two layers of soft rib-knit fabric. The garment was based on the jock strap (Section 5.3.6) that was already on the market, thus the product name "Jockey." The elastic waistband and leg band designs were borrowed from jock strap designs. Cooper also introduced the inverted Y-shaped fly eliminating the need for button closures.

Other companies have developed fly designs with variations on access to the penis for urination—from the Jockey "Y" to horizontal or vertical slot positions, or no slot. Information on men's preferences for pouch and fly designs, or how or whether men use the fly in briefs when toileting is not available. In addition, no data are available to guide the shapes, dimensions, or placement of the pouch and the access slot in relationship to the body. An informal survey of commercial patterns for men's underwear, indicates that standard industry grading practices are used to incrementally size the pattern pieces, including the pouch. Studies of traditional grading practices applied to other apparel categories, indicate that these methods do not provide good fit for many people (Schofield & LaBat, 2005). Given the range of male genital dimensions (Section 5.4.1), it is logical that the same body-product misfit may be found with briefs.

### *Women's Panty Fit and Sizing*

Women sometimes wear boxer styles, but body-conforming panties of knit fabrics are more prevalent. Panties feature a crotch panel to shape to the perineal region. The panel extends from near the mons pubis, wraps between the legs below the perineal region, and joins to the back pattern piece (see Figure 5.12). The forerunner of modern panties was the *"petit bateau"* pants or "little boat" in French—originally designed for children (Barbier & Boucher, 2010). Visualize the crotch panel shaped somewhat like a little boat. While data on female external genitals is limited to length measurements in a roughly transverse plane, the mons pubis and labia also have depth in a sagittal plane. Although crotch panels are contoured to some extent by easing the sides of the panel to a leg elastic, further contouring the crotch panel to a concave, boat-shape may improve panty comfort. An additional advantage of the concave shape may be to support an IMP or MMP.

Information on optimum positioning of the crotch panel, anterior position to posterior position, is not available. However, you can visualize the panel position by referring to Figures 5.1 and 5.12 to see the two body perineal region plane positions, on three posture types. The female urethral meatus and the vaginal opening are located in the forward sloping anterior plane, while the anal opening is in the backward sloping posterior plane. The crotch panel should curve around the perineal region and extend over the openings in both planes. Postural differences, especially the tilt of the pelvis front to back, affect positions of the body planes and will affect the effectiveness of the crotch panel position.

The limited anthropometric data for women's perineal region dimensions and orifice positions, relative to other body features, indicates a considerable range of measurements. Increasing the crotch panel length may better serve women with longer anterior to posterior measurements.

The crotch panel should absorb or wick moisture away from the perineal region orifices: urethral meatus, vagina, and anus. There is some disagreement as to the best fiber type for the panel. Some styles are made of cotton while others use synthetic wicking fibers. Cotton has the advantage of softness and absorbency but may quickly reach absorption capacity. Wicking fibers can transport moisture from the body but require a less humid environment on the outerwear side of the panel compared to the body surface to draw moisture away from the body.

Better understanding of the perineal region and genital anatomical structures related to underwear fit and sizing is needed. Much of the discussion regarding male and female underwear shape, fit, and sizing can be applied to the design of IMPs and external MMPs discussed earlier in this chapter.

## 5.13 Designing for the Lower Torso and Legs: Conclusion

When designing wearable products for the lower torso, perineal region, and legs consider gender variations, external reproductive organ vulnerabilities, and musculoskeletal flexibility and agility. Consider the functions of internal lower torso organs. Keep the normal mobility of the pelvic organs in mind. Avoid products that either constrict, continuously displace, or otherwise limit natural organ functions. Incorporate motion into a design to the extent possible for axial and appendicular skeletal functions. Build the design on the body, for the body, choosing the details of the underlying anatomy that are appropriate for the design goals.

# 6

# *Designing for Mid-Torso Anatomy*

Designers, whether out of aesthetic or functional concerns, often focus on the body's *mid-torso*. The mid-torso overlies the abdomen. The mid-torso is an area of the body where upper torso and lower torso anatomical structures and functions overlap, so designers must consider form and function of both when designing any product that covers the mid-torso. Many designs feature a waistline that must in some way align with the waist of a person. A standard definition of *waist* is that it is the narrowest part of the body in the mid-torso. For product design purposes a more useful definition is required.

**Key points:**

- Waist circumference and product waistline definitions
- Waist circumference location variations: Product effects
- Body shape affects the waist circumference location and the product waistline placement
- Product waistline dimensions and features may affect comfort and health
- Product waistline placement affects pants crotch length and shape
- Incorporate anatomical variations into pattern shapes and product forms

Two terms that are important when discussing the mid-torso are waist circumference (WC) and waistline. Waist circumference, sometimes called *waist girth,* is the measurement taken around the body at a specific mid-torso location. The *product waistline* refers to the feature of the product that encircles the body and relates to the WC and its location in the mid-torso. The product waistline can be described as an aesthetic design element or as a functional feature for locating the product on the body. The product waistline placement in the mid-torso affects the length and width of the upper torso and/or lower torso product components that attach to the product waistline. Both aesthetic and functional approaches depend on knowing the body WC measurement and location.

The *product waistline measurement* is the WC with ease added or subtracted, depending on the product purpose. Ease can be added to the WC measurement (positive ease), for example 2.54 cm (1 in.) to allow body motion and comfort. Or an amount can be subtracted from the WC measurement (*negative ease*) to determine the product waistline measurement. Stretch-knit swimsuits, body shapers, and corsets are examples of products with negative waistline ease.

WC is clearly defined as a measurement while the anatomical position for taking the measurement varies. The WC location is more or less horizontal relative to the vertical axis of the body and falls somewhere in the mid-torso. Disciplines differ in defining WC location and in methods used to measure the WC. Health care literature frequently describes the WC measurement in relation to fat deposits and distribution. Health care methods are adopted by some design fields. A designer determines location based on the desired product aesthetic and/or function. For example, fashion designers may select a WC location to emphasize the contours of the body. The field of anthropometry, focusing on function and use, documents WC location and measurement for use in designing wearable products, furniture, automobiles, and interior environments.

## 6.1 Waist Circumference Location

Locating and measuring the WC and determining waistline placement and dimensions are necessary steps in designing a range of products, from formal dresses to tool belts. Apparel designed for the lower torso, pants and skirts, anchors to a waistline feature, a finished edge or waistband. A product waistline—with a small amount of positive ease and smaller than the lower body circumference—can be used to suspend a product from the body. A waistline with negative ease—elastic band, drawstring, or belt—compresses the body and can also be used to secure the product to the body. Upper torso products, like baby carriers and backpacks, can be designed so that some of the load is carried by a waistline feature, like a belt. Some wearable products span the upper torso and the lower torso, connecting at the waistline. The product may be shaped to fit to a WC smaller than the upper or lower body using *fitting devices* like darts, gathers, folds, and pleats.

Locating the waist and measuring the WC of one person is easier than determining WC location and dimensions for a range of people. In a study of 200 males and females in all BMI categories Daniell, Olds, and Tomkinson (2010) found that site location variation can have significant impact on girth measurements, "especially waist girths on lean females" (p. 757). Stewart, Nevill, Stephen, and Young (2010), from a study using body scanning to locate and measure the WC on 62 males and 32 females, concluded that girth exhibits significant variation according to site and sex. They found that WC differed by 4.9% in males and 11.7% in females.

Body shapes—and therefore WC and location—vary greatly, even within one size. Variations in personal preferences for waistline placement and ease amounts differ person-to-person making the task of locating and measuring the WC even more difficult. Ashdown and DeLong (1995) found that participants in a fit test were able to perceive pants waistband ease variations as small as plus or minus 0.5 cm (0.2 in.). Personal preferences are based on comfort and also aesthetics. Waist circumference locations therefore can vary according to personal waistline placement preference.

## 6.2 Waist Circumference Location Methods

It is essential that every person collecting anthropometric data for WC and waist location be trained to use consistent procedures. And, if you are using a data base of WC measurements, know how the WC was located and measured.

Look at Figure 6.1 which illustrates two different female body types, from multiple views. The two contrasting body types are a female with balanced posture and little evident body fat, and a female with significant fat distributed throughout the mid-torso. Consider how you might approach locating a waist circumference for each body type, and how body type affects methods for locating and measuring WC. The three major WC location methods are:

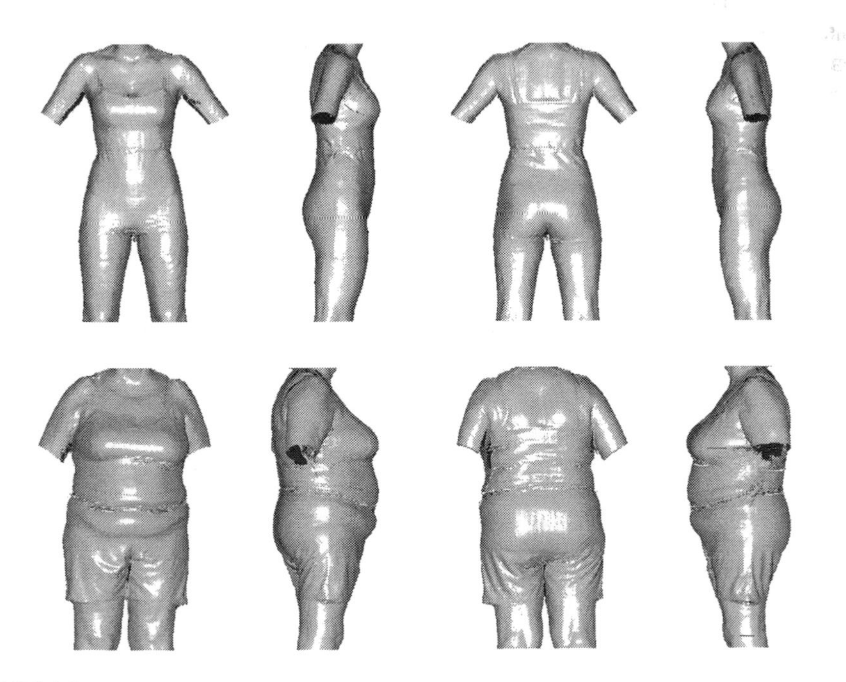

**FIGURE 6.1**
Comparison of two body types: Where would you locate the waist circumference? (Courtesy of the Human Dimensioning© Laboratory, University of Minnesota.)

(a) wearer-defined, (b) observer defined from visible body features, and (c) observer defined from palpable body features.

A product waistline is sometimes divided into four segments referencing four body landmarks: center front, center back, R *side waist point* and L side waist point. The side waist point has no distinct anatomical reference point, so is located with visual inspection after the waist circumference location is determined. Location is based on desired aesthetics and methods of joining front and back product elements. See Appendix E for more information.

### 6.2.1 Wearer-Defined Product Waistline Placement and Waist Circumference Location

Some manual and body scan landmark and measuring methods rely on asking the person "Where do you want the (product) waistline on your body?" or "Where is your waist?" According to the question, two different, but related procedures are used. One is based on the person's product waistline placement preference, and the other on where the person thinks their natural

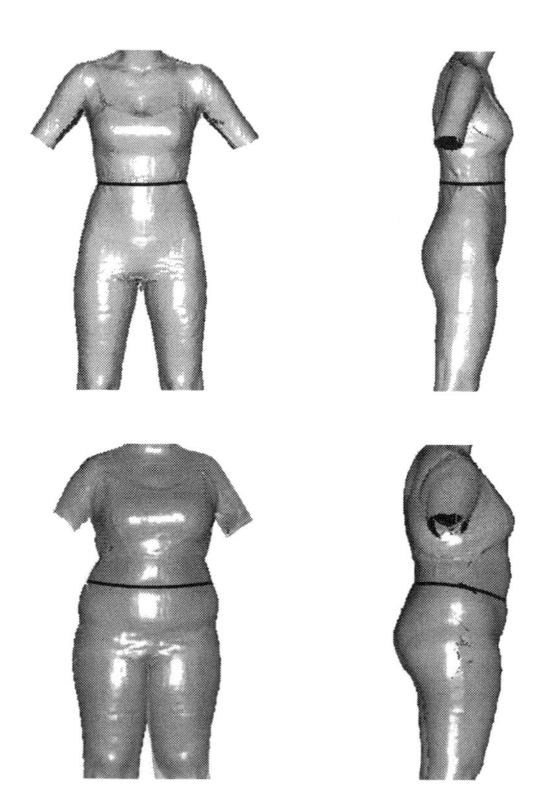

**FIGURE 6.2**
Wearer-defined product waistline placement on two body types. (Courtesy of the Human Dimensioning© Laboratory, University of Minnesota.)

waist circumference is on the body unrelated to a product. If you are collecting measurements for a specific product, you can ask the person to grasp his/her own waist where he/she would like to wear the waistline or waistband. Then landmark the person at the preferred location and measure the WC there. If the WC is measured with no specific product in mind, ask the person to bend side-to-side and locate the bend point on left and right sides of the body, then place a narrow band at the desired location and measure the circumference. Note that this individually defined location is not necessarily the smallest circumference. See Figure 6.2 showing wearer-defined waistline/waist locations. The waistline/waist on an ideal body type is usually parallel to the floor. The person with extra mid-torso fat has a wearer-defined waistline/waist that dips anteriorly.

With either method, locating the WC is based on identifying vague, individually determined landmarks. In addition, as demonstrated in the illustration, the WC may not be parallel to the floor. Although circumference measurements collected with this method vary person-to-person, the result may be a more realistic waistline placement for a product—where the person will actually wear the product. The design problem is incorporating the individually defined placement locations into a product design for mass production. If you are working with a team of designers; come to agreement on how WC relates to the product you are designing, then where the WC is located, and how it is measured. This may differ with product purpose—functional or aesthetic.

### 6.2.2 Observer-Defined Waist Circumference Location Methods from Visible Body Features

Methods more objective than personal preference, are used in attempts to produce reliable and repeatable measurements. Anatomical features give WC dimension and location cues. Features are either visible on the body surface or are internal skeletal features that can be palpated. These objective methods can still result in very different WC locations and therefore variable WC measurements.

Accurately and reliably identifying the WC location using surface visible body features can be a challenge with body scan data. Two observer defined WC location methods rely on identifying surface-visible features measuring at: (a) the level of the *omphalion* parallel to the floor and (b) the narrowest width of the torso (anterior view or lateral view). If features are not clearly visible on scans, the more time-consuming process of locating features by palpation and attaching markers to the body may be necessary.

#### *Measuring at the Omphalion, the Center of the Umbilicus*

The umbilicus is described, in common terms, as the "belly button" because it is a small round feature at the belly or anterior mid-torso. Navel is another term for umbilicus. It is an easily identified external anatomical feature common to all humans. It is the residual physical evidence of the umbilical cord, the connection between a fetus in the uterus and the mother. The *omphalion* is the middle of the umbilicus and is often used as a landmark for anthropometric studies.

The omphalion measuring method can be described with a few steps. For a detailed explanation, read Appendix E on waist circumference landmarking and measuring. The first step is to locate the omphalion at the anterior of the mid-torso. While the umbilicus and omphalion are easily identified, the precise location differs person-to-person depending on body morphology. The umbilicus may protrude from the body surface, sometimes called an "outie," or the umbilicus may indent with an indefinite small hollow area, an "innie." An omphalion that protrudes from the body surface can be detected with a body scanner, while an indented omphalion may have to be manually landmarked with a small protruding marker, before scanning.

With a manual method; the linear distance, floor-to-omphalion, is measured with an anthropometer. The anthropometer is then used to place marks the same distance from the floor at three additional locations, center back and L and R sides. The four marks establish a WC location that is parallel to the floor. The resulting WC location may not represent the best product waistline placement, especially if the subject varies from an ideal body or posture.

The omphalion method is appropriate when measuring and designing for a population of people with near-ideal body types. For example, the ANSUR anthropometric studies used omphalion as the WC location landmark, with data collected from military personnel who met fitness standards and were ages 17 to 50, with most subjects under 30 (Gordon et al., 1989, p. 54). Using this method to design products for other body types, especially people with excess mid-torso fat, will result in products with misaligned waistline placement. Look at the body scan of the person with excess body fat compared to the scan of a person with minimal fat (Figure 6.3). With excess fat the omphalion sags toward the floor. For this person, the omphalion method places the back waistline too low, well below the natural indentation of the body where a product would comfortably settle.

#### *Natural Waist: Maximum Concavity at Mid-Torso from Anterior View or Lateral View*

The maximum concavity of the mid-torso body shape, anterior or lateral view, can be used to locate the WC (Figure 6.4). For the anterior view, use the L and R points of maximum concavity as landmarks to locate the waist circumference. The anterior view works well for a body type with minimal fat; however, people with more fat in the mid-torso may have several contours that could be identified as the WC location. And a person's body silhouette may differ, lateral view compared to anterior view.

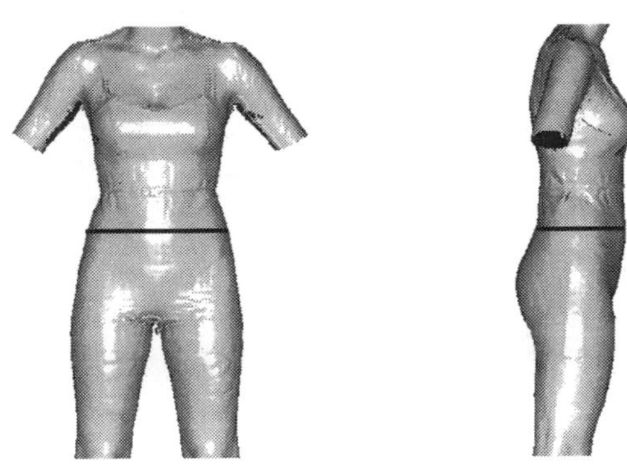

Female form with minimal mid-torso fat

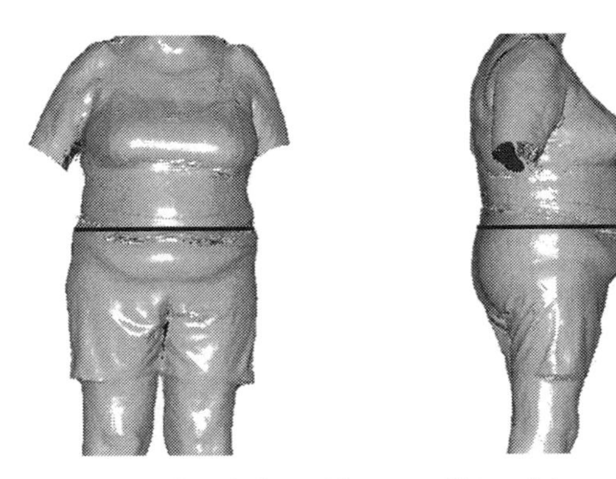

Female form with excess mid-torso fat

**FIGURE 6.3**
Omphalion waist circumference location method on two body types. (Courtesy of the Human Dimensioning© Laboratory, University of Minnesota.)

For the lateral view method, locate the maximum concavity of the lumbar spinal curve, the "small of the back." The waist circumference is measured parallel to the floor based on the maximum concavity location. The method is used with automatic measure methods in some body scan software (Gill et al., 2014). However; Han, Nam, and Shin (2010) found in developing an algorithm for automatic landmark identification, that this landmark description did not give a precise location point and varied for different body types. Posture and amount of body fat will affect this location. As shown in Figure 6.4, a body type with a straight back and little SAT will result in an indefinite location—many points may be considered the maximum concavity.

### *Natural Waist Circumference: Smallest Mid-Torso Circumference*

*Natural waist circumference* is the term used by Kouchi (2014) to describe the smallest WC. He says to locate and measure the natural WC, wrap a belt around the body where it "settles naturally" (p. 91) (see Figure 6.5 illustrating

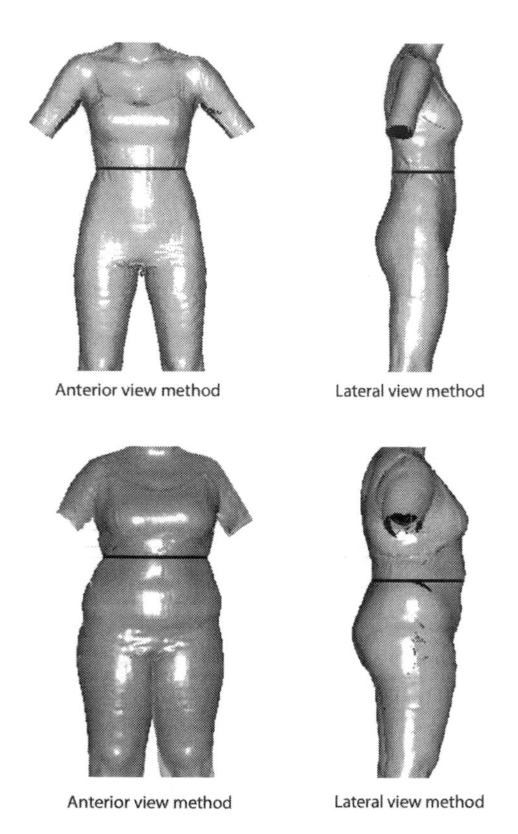


**FIGURE 6.4**
Natural waist circumference location methods using maximum concavities on two body types. (Courtesy of the Human Dimensioning© Laboratory, University of Minnesota.)

natural waist circumference). Then draw short horizontal lines (underneath and parallel to the belt) at the center front, center back, right and left sides—a total of four lines. Replace the belt with a tape measure and record the circumference. A related method is to take several circumference measurements throughout the mid-torso and select the smallest circumference. Body scanner software can quickly measure mid-torso circumferences and identify the smallest circumference.

Wang et al. (2003) say the smallest waist method is recommended frequently and is easy to locate on most people. They identify two problems: (a) some people have several areas that could be considered smallest, and (b) identifying the smallest waist is especially difficult for obese or extremely thin people. Gill et al. (2014) studied automated body scanner waist definitions on a sample of 106 females. They found that the anteriorly viewed narrowest point is misleading and is likely higher than the true waist height location. WC data collected with this method and reported in a database may not be useful for product designers because exact location varies person to person.

### 6.2.3 Using Palpable Skeletal Features to Identify Waist Circumference Location

Internal anatomy can help determine the best location for a product waistline. For example, the belt for a backpack may need to rest on the pelvis for support. With extensive practice, skeletal structures can be palpated, providing cues for placing landmarks to measure the waist. Wang et al. (2003) describe three WC locations that rely on identifying internal structures by palpating the lower rib cage and the iliac crest (see Figure 6.5). The palpation locations include: (a) immediately below the lowest anterior [10th] rib; (b) the midpoint between the lowest lateral [10th] rib and the iliac crest; and (c) immediately above the iliac crests, laterally.

#### *Waist Circumference Immediately below the Ribs (Lowest Anterior)*

Wang et al. (2003) say that a WC immediately below the lowest anterior (non-floating) ribs is easy to locate on all subjects, even the obese. Bra bands often encircle the body at the circumference immediately below the breasts (bust), relying on the rib cage structure for stability. However, this location is too high for most waistline features.

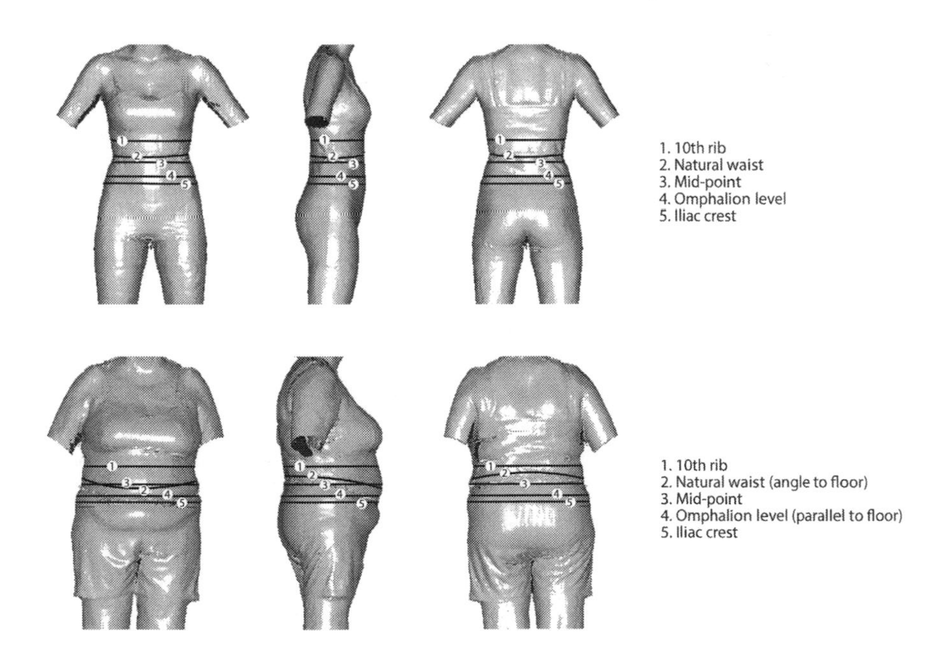


**FIGURE 6.5**
Waist circumference locations based on body features on two body types. (Courtesy of the Human Dimensioning© Laboratory, University of Minnesota.)

### *Waist Circumference Midpoint between the Lowest Rib and the Iliac Crest*

The International Standards Organization (ISO) defines waist circumference as a horizontal circumference measured midway between the lowest rib and the upper iliac crest (ISO 7250-1, 2008). Most anthropometric studies using this method say to locate and landmark structures on the right torso. This method works best for ideal, R-L, symmetrical bodies. Wang et al. (2003) give details for the torso midpoint method. The instructions are to locate the lowest point of the tenth rib and the upper edge of the iliac crest and landmark the sites. Then locate one-half the distance between the points as the WC level. Wang et al. (2003) state that this is the most time-consuming method because identifying the anatomical structures may be difficult and consistency in measuring is not assured. Variations within the mid-torso region, excess SAT for example, can make it difficult to palpate skeletal structures, therefore affecting circumference measurements.

### *Waist Circumference Immediately above the Iliac Crest*

Locating the WC immediately above the iliac crests may be useful for some products like low-rise jeans and some men's pants. Wearable products like "fanny" packs, ammunition belts, and tool belts are often positioned on or slightly above the iliac crests. Wang et al. (2003) state that locating the iliac crests is technically difficult, depending on pelvic structure and posture. Clearly defining the landmark location and developing precise directions for palpating and marking is important. Position of the pelvis, sometimes called pelvic tilt, affects iliac crest location. Read more about pelvic tilt related to product fit later in this chapter.

## 6.3 Waist Circumference Defined by Health Professionals

Health professionals share an interest in measuring the WC as an indicator of health. Health organizations, like the World Health Organization (WHO), also find that locating and measuring the WC is difficult. The WHO (2011) measurement protocol specifies measuring the WC at the "level parallel to the floor, midpoint between the top of the iliac crest and the lower margin of the last palpable rib in the mid axillary line" (p. 20). They note that other health organizations have used the umbilicus as the level of WC measurements. The U.S. National Institutes of Health (NIH) and the National Health and Nutrition Examination Survey (NHANES) conducted by the U. S. Centers for Disease Control (CDC, 2013), differ from WHO, specifying locating WC at the top of the iliac crest (WHO, 2011). As with design professionals, WC location methods vary even among major health organizations. When using health organization data for product design, consider the procedures that were used to collect the data.

## 6.4 Body Scanning and Waist Circumference Locations

When using WC measurements collected from body scans, know how the circumference location is determined and programmed into the software. If you are doing the scanning, you can manually landmark each person with markers that are visible in a scan. Such markers are placed using manual methods, visual and/or palpation. Landmark subjects in the scan position, because skin and garments stretch and move as the body changes position, displacing the markers (Kim, LaBat, Bye, Sohn, & Ryan, 2014).

Scanner company technology, including programmed landmarking procedures, are usually proprietary and methods will evolve as technology advances. Simmons and Istook (2003) compared manual anthropometric methods to methods used by scanner companies in the late 1990's. They stated that *[TC]*$^2$ located waist as the smallest circumference between the bust and hips, determined by first identifying the "small of the back" and then moving up or down a "pre-determined amount" for a "starting point" (p. 319). At that time, [TC]$^2$ did not detect and use the omphalion as a landmark. In contrast, Cyberware's WC was located in reference to the "navel." Simmons and Istook stressed that scanner companies should provide absolute and repeatable measurement definitions to assure apparel production accuracy. Gill (2015), in a comprehensive review of scanner technologies, urged standardization of landmark and measuring procedures to improve fit and sizing of manufactured wearable products.

## 6.5 Anatomical Regions in the Mid-Torso

When designing a product and determining best location for a waistline and product features in the mid-torso, consider the underlying anatomy. Saladin (2007), and McKinley and O'Laughlin (2006) section the abdomen into nine regions for medical diagnosis and description purposes. Figure 6.6 (left) illustrates nine abdominal regions with the approximate locations of the underlying anatomical features. Recall that the internal organs are pliable and constantly change position depending on many factors like breathing, digestion, and body posture—standing versus sitting versus lying down. WHO (2011), concerned about body changes related to digestion, considered measuring WC after a subject fasted overnight; but decided the requirement was not feasible for large-scale

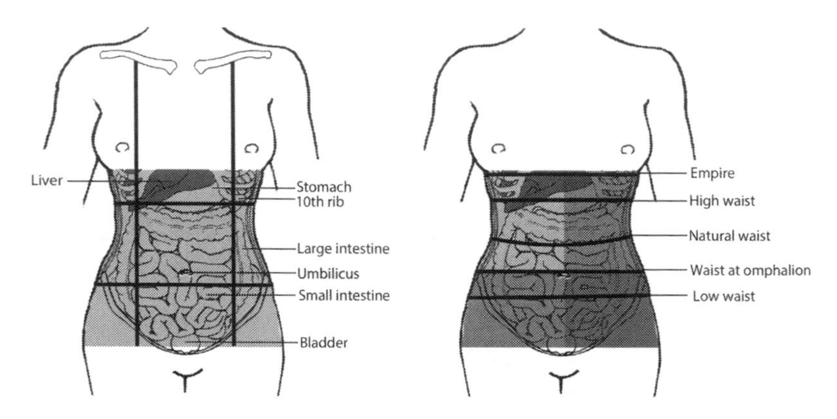


**FIGURE 6.6**
Abdominal regions and approximate waistline placement.

studies. The bony skeletal structures are more stable but can move to some extent. For example, the rib cage expands and contracts with respiration. Three horizontal regions—upper, middle, and lower—with three vertical sections—R, L, and center—help to locate internal structures. Boundaries between regions, shown in Figure 6.6, can be used to discuss relationships to products.

The upper region is bounded superiorly by the diaphragm and extends down to a horizontal plane at the lower edge of the tenth rib. This horizontal region overlies the lower rib cage, liver, and stomach. The gallbladder is posterior and inferior to the liver. The pancreas is posterior to the stomach and the spleen is posterior to the lateral stomach. The middle horizontal region extends to a plane through the iliac crest. The ascending and descending colon (large intestine) and most of the small intestine are located in this region. The lower region encompasses the bladder, ovaries, uterus, vagina, and some of the small intestine with the rectum of the colon posterior to these structures. The three regions can further be sectioned with two vertical planes bisecting the clavicles dividing each horizontal region into three sections. These parasagittal planes may correspond to contoured fitting seams in a garment, stays of a strapless gown or undergarment, or to a rigid support or pliable webbing of a functional wearable product like a parachute. While designing the product and determining where to place the product waistline on the body, consider how the waistline relates to, and may affect, internal anatomical structures.

## 6.6 Product Waistline Related to Mid-Torso Regions

A waistline can be placed anywhere in the mid-torso; and may, or may not, be parallel to the floor. Common apparel waistline placements include: *empire waistline*, high waistline, natural waistline, waistline at the omphalion, and low waistline. Figure 6.6 (right) illustrates approximate locations of these waistlines and where they cross underlying anatomy. These waistline placement descriptions relate to the static body and the approximate locations of organs within the body. Organs will shift as the body moves to different positions—standing, stretching, or supine—and organs will continuously move as the body moves. Read the sections on upper torso and lower torso ROM to consider how waistline placement interacts with the body in motion.

### 6.6.1 Empire Waistline (Underbust) and High Waistline Product Placements

The empire waistline, a womenswear style feature, encircles the rib cage directly below the bust, sometimes referred to as the *underbust circumference.* This location is also the approximate level of a bra band. A tight empire waistline can interfere with respiration. The high waistline can be a design feature for a full torso garment like a dress or jumpsuit, or the top line of a high waistline pants design. Maternity wear designs often use a high waistline above the expanding mid-torso for comfort and aesthetics. Some harnesses used to suspend the body, like parachutes and safety harnesses, fasten to a band at this location. As illustrated in Figure 6.6 (right), the high waistline lies over the lower edge of the anterior rib cage and may be uncomfortable as it bridges from the hard edge of the rib to the softer compressible organs. Harnesses need to grasp the rib cage to do their job, high-waist pants may not stay in place if they sag below the ribs, because gravity pulls them down to the next widest surface, below the natural waist.

The organs underlying these two waistline locations in the upper abdominal region include: part of the liver, gallbladder, part of the stomach, parts of the intestine, and the pancreas, and sections of the kidneys. The deep organs may not be affected by waistline placements in this region; however, pressure on a full stomach may cause discomfort. In contact sports, the kidneys in the upper and middle mid-torso are often protected with impact-absorbing padding.

### 6.6.2 Natural Waistline and Waistline Placements at the Omphalion Product Locations

Product waistline features can lie over the body's natural waist location or the body omphalion landmark location. The natural waistline corresponds to the location of the natural waist circumference which is illustrated in Figure 6.6. It is a common placement for a belt or waistband. The waistline at the omphalion is illustrated slightly caudal to the natural waist circumference described earlier. WC locations related to both of these waistline designs vary greatly person to person. These product waistline placements may lie over the middle part of the transverse colon, as well as above many parts of the small intestine, blood vessels to and from the lower limbs, the ascending and descending colon, and parts of the kidneys. Constriction at the natural waist or at the omphalion level will act on the mid-torso organs. Organs will be displaced towards the upper torso affecting respiration and digestion, and towards the lower torso affecting the position of female reproductive organs and the functioning of structures associated with elimination of urine and feces.

Corsets are perhaps the most extreme example of constricting the natural waist. Banner (1983) states that throughout much of the 19th century the goal for women was to achieve a waist circumference of 45.72 cm (18 in.). She alludes to the health effects of extreme corset lacing, saying the practice "caused headaches and fainting spells and may have been a primary cause of the uterine and spinal disorders wide-spread among nineteenth-century women" (p. 48). O'Followell and Lion (1908), recognized the problems associated with corset wear, writing in the early 1900s. They discussed and illustrated numerous changes in the internal structures of the thorax and abdomen related to corsets and discouraged the use of corsets in growing girls. Figure 6.7 is a photograph of a corset dated 1887 from the collection

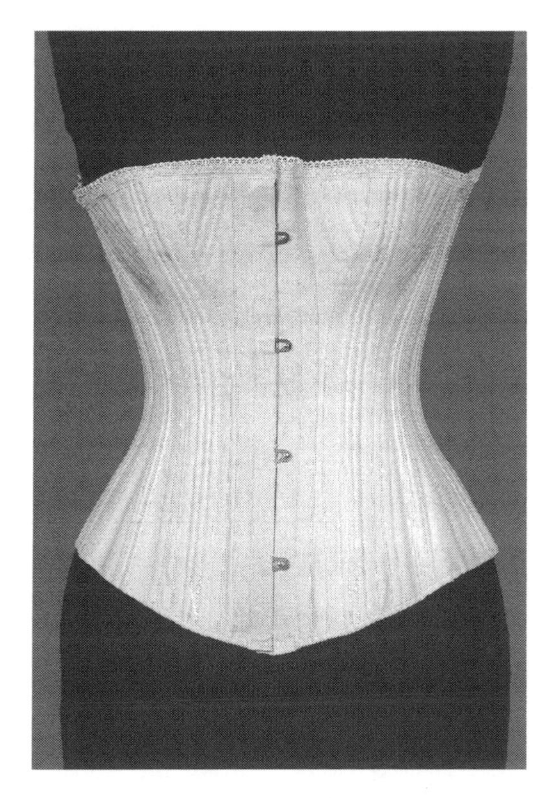

**FIGURE 6.7**
Corset with boning, 49.5 cm (19.5 in) waistline measurement, dated 1887. (Courtesy of the Goldstein Museum of Design, University of Minnesota.)

of the Goldstein Museum of Design, University of Minnesota. The waistline measurement is 49.5 cm (19.5 in). It is typical of a corset of that period with steel boning to maintain the woman's erect posture and a slim waist circumference. Gau (1998) reproduced two historic physiologic studies using modern medical equipment to determine the effects of corseting on lung capacity and comfort of historic re-enactors. Her research goal was to provide guidelines for corseting re-enactors. She found tight-lacing (three inches less than natural WC measurement) resulted in an average 9% loss of lung function, and shortness of breath was reported in varying degrees by the subjects. She recommended limiting lacing to decrease the WC no more than 10% of the natural WC, and that re-enactors should exercise regularly to avoid muscle atrophy resulting from over-reliance on corset wear.

Today, body shapers are worn by women and men to try to achieve a smooth, contoured mid-torso. Body shapers use knit fabrics and elastomeric fibers to apply pressure around the mid-torso. Focused pressure, at selected body areas, can be applied by using variations in elastomer *denier* (fiber diameter size) and density of the knit structure. Lyu (2016) studied the effects of women's body shapers on the mid-torso. Participants of all sizes described their mid-torso while wearing a body shaper as slimmer, smoother, smaller, and having less fat (Figure 6.8). The person doesn't actually have less fat; the tissues are displaced with compression from the garment for a smoother appearance.

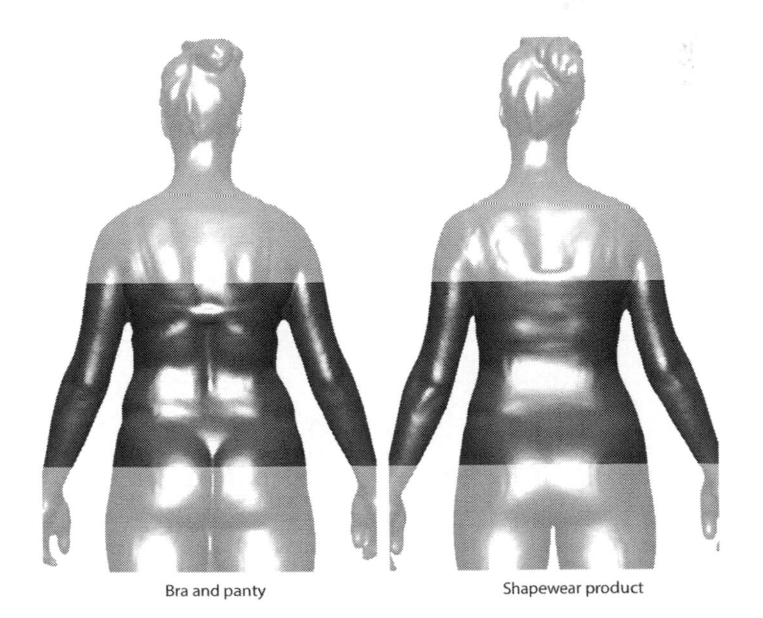


**FIGURE 6.8**
Female wearing a bra and a panty, and wearing body shaper + bra + panty. (Adapted from Lyu, S. (2016). *Posture modification effects using soft materials structures*, Doctoral Dissertation, University of Minnesota (Unpublished doctoral dissertation), University of Minnesota, Minneapolis, MN. With permission.)

### 6.6.3 Low Waistline Placement

The low waistline is located somewhere between the lower margin of the omphalion, and on/or above the iliac crests. Low-rise pants can be supported by prominent iliac crests and a drop-waist or *chemise* dress waistline may fall approximately at this level. Soft organs in the lower abdominal region are: parts of the small and large intestines, the appendix, bladder, ovaries, and uterus. Because the low waistline falls above the iliac crests, it is unlikely to interfere with organ functions in the lower region. However, if a person has body tissue (fat or muscle) that obscures the iliac crests, the product will not be supported by the pelvis and may cause downward pressure on the bladder.

The iliac bones of the pelvis are a more substantial skeletal element than the mobile rib cage. Products that encircle the pelvis are unlikely to compress tissues within the pelvis. However, substantial fat deposits—SAT and/or DAT—may cause organ compression, as the volume of the lower torso extends beyond the confines of the pelvic bones. Figure 6.9 illustrates fat distributed throughout the mid-torso which can compress soft organs, especially if external pressure from a product is applied. The *superior cluneal nerves* cross the posterior iliac crest transmitting sensations from the skin of the buttock. These nerves may be affected by pressure from a product resulting in numbness or altered sensation (Gardner & Bunge, 2005, p. 32).

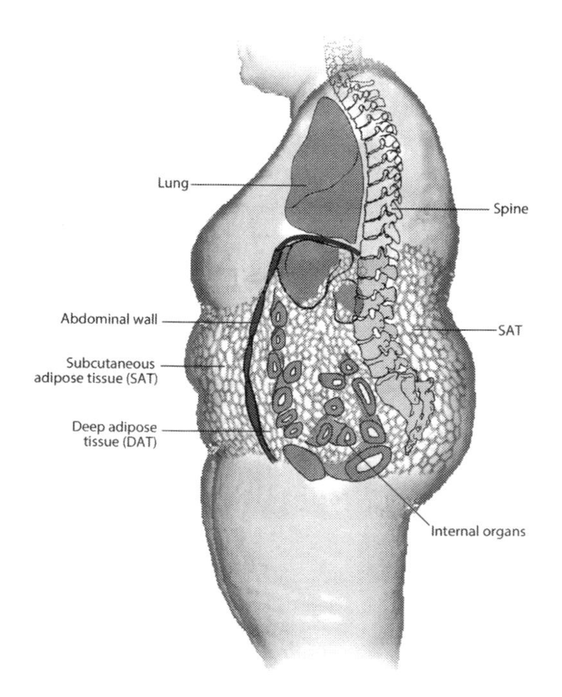


**FIGURE 6.9**
Mid-torso body fat distribution: Subcutaneous adipose tissue (SAT) and deep adipose tissue (DAT).

## 6.7 Body Type Determined by Mid-Torso Form and Shape

Human bodies are complex three-dimensional forms with multiple combinations of height, width, and depth that result in a diverse array of shapes and sizes. Because the human body is so complex, categorization methods have been developed to simplify the human body into a limited number of types. Wearable product designers and apparel manufacturers have proposed using body type to design fit into a product. Body type classifications most often depend on mid-torso form and shape.

Many art and design books define *form* as three-dimensional and *shape* as two-dimensional. Pipes (2009), in his book introducing basic principles of design, says that form refers to the apparent solidity or three-dimensionality of an object. He says that shape has width and height, but no perceived depth (p. 43). Body type is often based on viewing the shape of the body from an anterior view. Sometimes the lateral view is considered.

Body scan technology gives researchers tools to create silhouettes and to capture point, length, surface, shape and volume data (Bye, LaBat, & DeLong, 2006). A key strategy will be integrating the two-dimensional and three-dimensional scan data and developing methods of applying the data to designing products for all body types (Gill, 2015).

Why consider mid-torso body shape? Wearable products are often based on two-dimensional pattern shapes that are joined to form a three-dimensional product to conform to a body form. Researchers have tried to find a direct relationship between a human body shape and pattern shapes used to form a product. Gazzuolo (1985) stated that body shape and pattern shape show some relatedness, but no direct relationship. The following review of selected methods used to categorize body shape highlights unanswered questions.

## 6.8 Mid-Torso Body Type Categorization Methods

Sizing systems discussed in previous chapters use quantitative methods, like bust circumference for women's upper torso garments, to place people into size categories. In contrast, body type category methods have been based on qualitative assessments of body shape and form. Defined criteria determine a specific body type. Recent research explores quantitative methods to sort bodies into type categories.

Many fields have used body type categorization methods. Psychology has related psychological profiles to body form. Medicine uses body types to relate to health risks. Apparel design manufacturers have used body types to communicate product fit to consumers. For example, body type labels have been popular with jeans manufacturers in attempts to help consumers select a style and size that may best fit their body type (Gribbin, 2014).

Designing wearable products to meet the needs of real people with a range of body types is a challenge. Body forms and shapes vary based on gender, age, ethnicity, health, and nutrition. Some tactics are to: (a) custom-design and fit products, (b) use design features that adapt and shape to many body types, or (c) develop sizing systems based on the analysis of numerical measurement data. Identifying and categorizing body types is another approach to try to design and size products to better fit a range of people. Many of the categorization methods focus on the mid-torso as the key element in determining body type. An individual's mid-torso body form and shape are determined by the morphology of his or her anatomical structures including skeletal structures, internal anatomy, distribution of SAT and DAT, and layers of muscle.

### 6.8.1 Body Type Categories: Comparison to an Ideal

Body type categories, especially for apparel, often start with the concept of an ideal body as the basis for comparison (DeLong, 1998; Roach-Higgins, Eicher, & Johnson, 1995). Some writings refer to the average body as the basis for comparison. However, an average body is not equivalent to the ideal body. Average has statistical meaning as the central tendency of a set of numbers, so the average body has little meaning related to designing a wearable product for a range of body types. If you measure the WC for people of varying sizes and calculate the average, a product waistline based on that measurement is likely to fit no one.

Each culture determines an ideal male and female body, and ideals influence wearable product designs (Figure 6.10). Joseph-Armstrong (2006) states that the Western ideal female figure WC, which most apparel companies use for product development, is 25.4 to 31.75 cm (10 to 12 ½ in.) smaller than the bust and hip (p. 22). She gives the example of ideal female measurements as a 91.44 cm bust (36 in.), 66.04 cm waist (26 in.), and 91.44 cm hip (36 in.). Western fashion has emphasized an ideal with a concave, curvy waist. The desired figure balances the hip and the bust. This figure is often described as an "hourglass" shape. The wider female pelvis compared to the upper torso skeletal structures gives a distinct concave waist contour—and a good product suspension location. Ridgway, Parsons, and Sohn (2017) found that women prefer apparel that gives the illusion of an ideal figure.

The male ideal body is described as an inverted triangle with broad shoulders tapering to the waist and then the hip (Hollander, 1994). See Figure 6.10 illustrating an ideal male form. A muscular upper body, shoulders, and biceps—compared to other body segments—is preferred. Men's suits are tailored to come as close as possible to the ideal body with padding added to the shoulder and sleeve cap. However, many men exhibit an all-around uniform tube form, related to central fat distribution (Taylor, Grant, Williams, & Goulding, 2010)

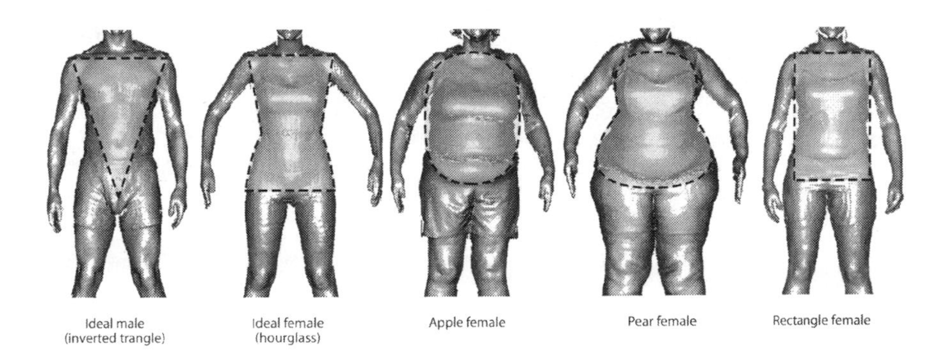


**FIGURE 6.10**
Body types: Male and female ideals; female apple, pear, and rectangle. (Courtesy of University of Human Dimensioning© Laboratory, University of Minnesota.)

## 6.8.2 Body Types: Psychology and Health

Sheldon's (1940) body type classifications are cited as an early attempt to understand the varieties of human bodies. Sheldon and colleagues at Harvard and the University of Chicago studied photographs of 400 nude males to sort the bodies into types related to what they called an "average" male body. Although Sheldon's method describes the whole body, much of the analytical focus is on the mid-torso. He describes three body types: ectomorph, mesomorph, and endomorph. Related to the mid-torso, the endomorph has "dominant digestive viscera," the ectomorph has "relative predominance of linearity and fragility," and the mesomorph is "rectangular in outline" (p. 5). The theories proposed from the study that the physical body is tied to "constitutional psychology" are now discredited, but the early attempt to categorize body types is still seen as ground-breaking.

Equating mid-torso shapes with fruit shapes is an intuitive and visual way to communicate body type (see Figure 6.10). Health and medicine fields use the images to convey a visual measure of the relationship between mid-torso shape and health. Waist/hip ratio (WHR) also is used to quantify mid-torso shapes. Abdominal obesity is defined as waist/hip ratio above 0.90 for males and above 0.85 for females. The WHO describes a body with more weight around the waist as apple-shaped, and at higher poor health risk than people with pear-shaped bodies (more weight around the hips) (Fu, Hofker, & Wijmenga, 2015).

Population trends of body shapes reported from large-scale studies, like those conducted by WHO, can be useful for designers. Ford, Maynard, and Li (2014) reviewed data on the mean WC of males and females from the National Health and Nutrition Examination Survey (NHANES) data, collected 1999–2012. They found that "the overall age-adjusted mean WC increased progressively and significantly from 95.5 cm (37.6 in.) to 98.5 cm (38.8 in.)" (p. 1151). This is the equivalent of one size increase based on many of the grading schemes used to size apparel. Designers may need to change product waistline dimensions and specifications, and develop new designs to accommodate these changing body dimensions. A challenge in fitting the apple shape—often a result of an expanding WC—is suspending a product from a mid-torso with a larger WC compared to the hip circumference. There is no waist indentation on which to suspend the product. With increasing obesity rates, designers will need to rethink how to fit and flatter non-ideal body types.

### 6.8.3 Body Types: Apparel

Many body shape classification methods for apparel focus on the female mid-torso (Minott, 1974; Simmons, 2002; Devarajan & Istook, 2004; Connell, Ulrich, Brannon, Alexander, & Presley, 2006). Minott proposed methods of using body types to draft pattern shapes to better fit each body type. Apparel design books focusing on women show style modifications to make the apple- or pear-shaped body appear more like the ideal hourglass figure. Apple and pear shapes are also used to describe men's figures in contrast to the tube shape and the ideal inverted triangle. Donnanno (2016) describes corrections to a man's suit jacket to meet the inverted triangle ideal. He describes men's body type variations as: (a) "curved spinal column" resulting in expansion of the upper back and contraction of the chest (a stooped, round-shouldered posture) and says it is the most common balance defect; (b) "forward hip tilt" (associated with increased lumbar lordosis) resulting from a posture adjustment to balance increased belly protrusion, (c) "stout" with pronounced belly fat and no posture change, and (d) "pear-shaped" with narrow shoulders and wider hip and thigh (pp. 217–220).

### 6.8.4 Body Type Classification Methods: Technical Aspects

Popular press sewing books as well as research studies have attempted to identify body types and then incorporate the types into apparel patterns.

#### *Body Type, Pattern Shape, and Product Form*

Minott (1974, 1978) and Pouliot (1980) developed methods of applying body type categories to pattern shapes before body scan technology was available. They attempted to apply body type characteristics to products. Their methods aid in understanding body-to-product relationships. Minott authored several books on analyzing body forms to draft dress, skirt, and pants patterns; with the goal of a better match between the body form and the product form. Pouliot (1980), in her master's thesis, focused on classifying mid-torso and lower torso body types to draft pants patterns for women.

Minott (1974, 1978) presented detailed instructions on qualitative analysis of body types and specific pattern adjustments to fit each type. She described six hip types and three posture types for women. Minott's hip types are based on comparisons to the WC. She gave detailed explanations for altering skirts and pants based on identifying a combination of hip and posture types. Minott's posture types for pants fit are described later in this chapter. Minott's hip types were determined by viewing the figure anteriorly. The hip types include: average, little difference, heart, semi-heart, diamond, and rounded diamond. Her average hip type is equivalent to the ideal figure. The little difference type is similar to the rectangular figure, showing little change of silhouette contour between the hip and the WC location. The heart hip type has pronounced roundness just below the waist. A slight variation of the heart type is the semi-heart with the hip fullness located midway between the waist and thigh with no pronounced thigh fullness. The diamond type is equivalent to the pear shape described earlier, with pronounced prominence at the thigh and smaller waist and upper torso silhouette. The rounded diamond exhibits fullness at the upper hip and the thigh, with a pronounced concave curve between the two. Minott acknowledged that many women have an asymmetric figure, right to left, with one hip higher than the other or one hip larger. The asymmetries relate, no doubt, to anatomical differences, muscular or skeletal; but Minott does not describe underlying anatomy. The result is a WC location not parallel to the floor. Minott gives detailed instructions on drafting pants and skirt patterns based on formulas for each type including: waist and hip ease amounts, amount of curve at the side seam, and dart sizes.

Pouliot (1980) used *somatography*, a body graphing technique developed by Douty (1968), and identified five body types to aid in drafting pants patterns. Somatometry projects a person's anterior and lateral silhouettes on a grid so that the outlined shapes can be analyzed. Pouliot identified three anterior view shapes—average, round, and triangle. Two lateral shapes were defined by finding the "midpoint of the profile waistline," drawing a line perpendicular to the floor, and then determining if the person has the most weight distributed in front of or in back of the line. The two lateral views are described as "weight in front" and "weight in back" (p. 44–45). The somatographs were used to develop a method of incorporating the body silhouettes into pants patterns. The patterns were used to construct test pants for five subjects and the pants' fit was evaluated on each subject by a panel of experts. This early study, with a limited number of subjects, found that the front crotch curve did not provide good fit, further reinforcing the difficulty of directly transferring a two-dimensional representation of a three-dimensional body form to the product. Schofield, Ashdown, Hethorn, LaBat, and Salusso (2006), in a study of pants fit on 176 women 55 years and older, found that introducing a body shape variable in test pants improved subject satisfaction with fit.

### *Identifying Body Types from Body Scan Data*

Body scanning technology adds new capabilities for analyzing and classifying body types. Many of the methods use the mid-torso as the defining area for classification. Unfortunately, these scanner-derived proposed body types have not been tested and/or reported in product applications. Studies that classify body shapes using body scanning include the frequently cited Female Figure Identification Technique (FFIT) for Apparel© and the Body Shape Assessment Scale (BSAS) ©.

Researchers at North Carolina State University developed the FFIT to analyze body scans of females to classify body shapes (Simmons, 2002; Devarajan & Istook, 2004). The objective was to "utilize software that can take 3D data and 'sort' it into congruous and related shape categories based on measurements, proportion, and shape" (Simmons, 2002, p. 97). Simmons, collaborating with [TC]$^2$, developed shape-sorting software based on criteria developed from a literature review of body shape classifications. Simmons (2002) identified nine figure types, and Devarajan and Istook (2004) validated the types by scanning and classifying a large sample of 531 women. The system is based on body form comparison to the hourglass figure. The shapes are described as: (a) hourglass with equal hip and bust, and with a defined waist; (b) bottom hourglass with larger hip circumference than bust circumference and with a defined waist; (c) top hourglass with bust circumference larger than hip circumference and with a defined waist; (d) spoon based on a large hip and bust circumference difference (compared to hourglass) and with a lower bust-to-waist ratio (compared to hourglass) with the high hip-to-waist ratio high—characterized by a "shelf" at the high hip (Simmons, p. 111); (e) rectangle with bust and hip fairly equal and bust-to-waist and hip-to-waist ratios low, and no discernible waist; (f) diamond with stomach, waist, and abdomen larger than the bust; (g) oval with the average of the person's stomach, waist, and abdomen less than the bust; (h) triangle with larger hip than bust and ratio of hip-to-waist small, with no defined waist; and (i) inverted triangle with larger bust than hip and small bust-to-waist ratio. The researchers gave no specific guidelines on using the categories to differentiate product shape or form but proposed the method as a starting point for improving product fit.

Connell, Ulrich, Brannon, Alexander, and Presley (2006) developed the Body Shape Assessment Scale (BSAS)© to categorize shape from body scans. They used scales from previous studies and writings, and experts' opinions to develop the scales. They proposed three whole body scales. One of the scales, the Body Shape Scale, identifies the mid-torso shapes of hourglass, pear, rectangle, and inverted triangle. They also developed six scales to assess component body parts. One scale is pertinent to mid-torso assessment, the Front Torso Shape Scale. It is a lateral view used to analyze shape from the base of the bust over the abdomen. The "b" shape is described as flat between the base of the bust and the waist, then rounded below the waist; "B" is rounded above and below the indented waist; and "D" is a continuous rounding from the base of the bust through the abdomen with little or no waist definition. The lateral view is often ignored for mid- torso assessment, but this scale adds information that should be helpful in developing products to fit a variety of body shapes. Alexander (2003) studied the relationship of BSAS © body shapes and apparel fit problems. She found that fit problems self-reported by study participants related to body shape, hip shape, bust shape, and back curvature.

Improved wearable product fit, which considers mid-torso as a major factor, may result from new approaches to categorizing human bodies that are less reliant on traditional methods of segmenting and analyzing body data. Methods of analyzing large data sets, including three-dimensional and four-dimensional (motion) data, are being developed that could result in body type classification methods that can be applied to fit and sizing of clothing (Bougourd & Treleaven, 2014). We present a sample of new methods that make use of technology but need to be tested on real humans with varied anatomies.

Peña, Viktor, and Paquet (2012) propose using data mining of population demographics and body scans to determine body types for marketing purposes. In a related study, Paquet and Viktor (2014) propose methods to segment and cluster body scan data. They introduce the term "human body space" as the representation which corresponds to a population they are trying to typify. They use "isometry-invariant descriptor creation algorithms" to index body regions that are made of soft tissues (p. 123), major components of the mid-torso. When tested with representative samples, their method has potential for partitioning people into body types that may be useful for marketing purposes. At this point, these large-scale data set explorations have not been tested by producing wearable products to fit a range of body types.

Liu et al. (2016) propose an approach using a mathematical technique to categorize body shapes, calling it a "fuzzy method" based on fuzzy numbers. They say that traditional classification methods are "discrete, using crisp and rather dichotomous classifications" (p. 60). Applying the method to a sample of 116 Chinese women, 20–30 years old, they identified three lower body shapes. They state that the crucial lower body classification dimensions are height, waist girth, and the difference of hip and waist. While the proposed method appears to have potential, the mathematical manipulations were conducted on a limited number of similar body types, and the body types were not tested in application to an apparel sizing system or garment fit tests. Even with sophisticated digital and computational tools, there is much work to be done to apply body types to product shape and form. Bifurcated products, commonly referred to as pants, are difficult to fit to the three-dimensional body.

## 6.9 Effects of WC Location and Body Type on Bifurcated Product Shapes

A bifurcated product is any lower torso product with the bottom section divided into two parts called legs, named after the body parts that are covered. Pants, trousers, jeans, culottes, shorts, and jumpsuits are examples of bifurcated products. For purposes of this chapter we will use the generic term "pants" to indicate a bifurcated product, unless a reference uses a different term. Female consumers are the least satisfied with pants fit compared to all other apparel items (Feather, Ford, & Herr, 1996; Goldsberry, Shim, & Reich, 1996; LaBat & DeLong, 1990; Schofield, Ashdown, Hethorn, LaBat, & Salusso, 2006). Shaping a two-dimensional pattern to fit the complex three-dimensional contours of the hip, crotch, and thigh is difficult. McKinney, Gill, Dorie, and Roth (2017) found that two current apparel pattern drafting methods for women's trousers do not effectively incorporate body shape into patterns and therefore the products. They conclude, "scholars should conduct research to understand body-to-garment relationships that result in good fit and to incorporate 3-D body shapes into pattern-drafting methods" (p. 30). 3D body scanning technologies are now making it possible to do this type of research, based on developing pattern-making theory (McKinney, 2007). The challenge in fitting men's and women's pants designs is analyzing the complex geometry of the body formed from the relationships of waist, abdomen, buttocks, legs, and the intersecting body segments at the body crotch.

Successful two-dimensional patterns for bifurcated products capture three-dimensional body form into the two-dimensional pattern. Figure 6.11 illustrates a standard pants pattern related to an ideal female form. The pattern crotch relates to the planes of the anatomical crotch described in Chapter 5. Components of the pattern are:

- *Crotch seam* is the U-shaped seam bisecting the torso and curving beneath the anatomical crotch.
- *Crotch level* is a pattern drafting line parallel to the floor at the base of the crotch seam that relates to the lowest level of the anatomical crotch.
- *Crotch depth* (women's patterns) or *rise* (men's patterns) is the two-dimensional pattern drafting measurement from the waistline to the base of the pants crotch seam, perpendicular to the crotch level.
- *Crotch length* is the distance from the center front waistline, around the crotch base, to the center back waistline measured along the curve of the crotch seam.
- *Crotch extension* is a line, one on the front pattern and one on the back pattern, along the crotch level line that extends to cover the inner thigh.
- *Crotch point* is the terminal point of the front or back crotch extension where the crotch extensions of the center front and center back crotch seams meet.
- *Outseam* is the seam extending from waistline to the desired pants leg length, joining front and back pieces of the pants.
- *Inseam* is the inner thigh seam joining front and back legs of the pants.
- *Body depth* is the space formed between the front and back pants patterns. It represents the space the body occupies in the pants.

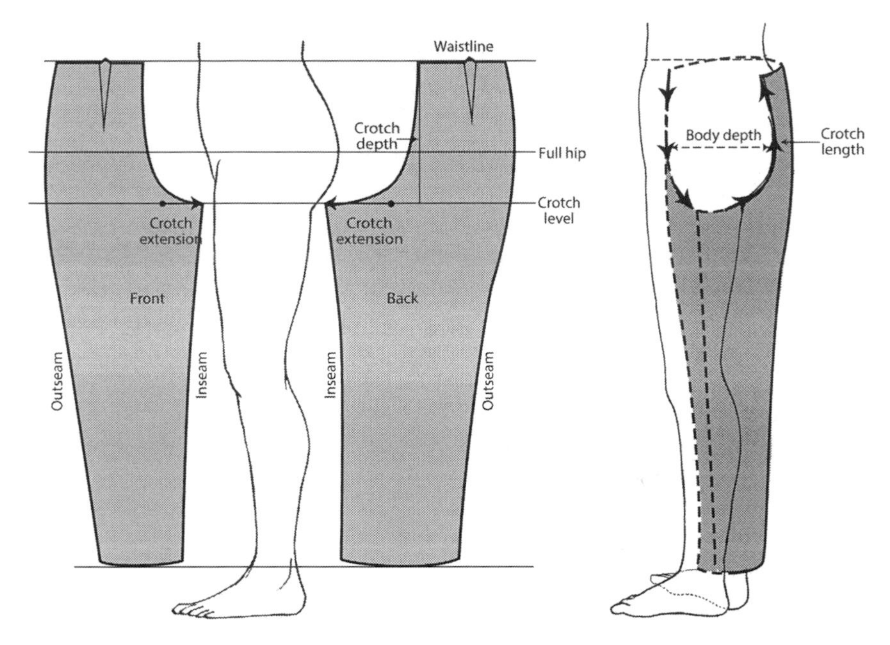


**FIGURE 6.11**
Standard pants pattern related to an ideal female body.

Bifurcated products are shaped with a cylinder that wraps around the abdomen and buttocks and is shaped (wider or narrower) to fit the WC. The torso cylinder is divided into two cylinders to wrap around each leg. As the torso cylinder is divided (bifurcated) and transitions to the two leg cylinders; a curved seam, the *crotch curve*, divides the torso cylinder. The crotch curve seam extends from the anterior and posterior WC and curves around the anatomical crotch. The length and shape of the crotch curve are dependent on the product's waistline placement, body form, and body posture.

McKinney et al. (2017) found that traditional pattern drafting methods are based on easy-to-calculate formulas that often do not relate to the human body. Pants pattern drafting methods demonstrate this disconnect. The pants crotch extension length is determined based on an untested theory that the inner thigh measurement is a percentage of the hip circumference measurement. Knowles (2005), giving instructions to draft pants, provides a table of front and back crotch extensions based on hip circumference (p. 79). For example, a 104.14 cm (41 in.) hip requires a 9.21 cm (3.625 in.) front crotch extension and a 14.61 (5.75 in.) back crotch extension. To our knowledge, there are no studies of human anatomy that prove this relationship.

An assumption of most pants pattern drafting methods is that the U-shape of the pants crotch curve will in some manner relate to the lateral silhouette of the body—center front waistline between the thighs to the center back waistline—which also forms a U-shape. This "negative space" was described by McKinney (2007), in two-dimensional terms, as "depth at abdomen level" and "depth at hip/buttock level" (p. 43). See Figure 6.11 illustrating body depth related to a pants pattern.

Studies have been conducted on the complexities of shaping a bifurcated garment to satisfactorily fit a range of body types (McKinney, 2007; McKinney, Bye, & LaBat, 2012; McKinney et al., 2017). Most studies attempt to categorize body form based on analyzing an anterior view of the hip shape, combined with analyzing a lateral view of the buttock shape to determine a crotch curve shape. With scan technology, the crotch curve can be mapped, although scanners miss data points occluded by the leg. Studies have found that simply applying the body curve shape to the pattern shape does not result in improved pants fit (McKinney, 2007; Pouliot, 1980). In a study comparing two pants pattern drafting methods, McKinney et al. (2017) identified several problems including: (a) using proportions from body measurements that do not correspond with each other, (b) using body measurements from established standards, (c) applying ease amounts based on proportions or standard amounts, (d) methods used to smooth and blend lines, and (e) disregard of body shape. The three-dimensional geometry of the body and the many materials that can be used for pants make shaping pants to fit very difficult.

### 6.9.1 Pelvis Position (Hip Tilt) Effects on Pants Crotch Curve

Minott (1974) used the hip shapes described earlier in this chapter combined with three posture types, that she describes as hip tilt types, to propose methods of drafting pants patterns to fit different body types. The posture types she described include: (a) "average," fairly straight posture with seat "neither tucked under or held out" with center front floor-to-waist measuring the same as, or up to 1.27 cm (1/2 in.) longer than, the center back floor-to-waist measurement; aligned posture results in balanced pants; (b) "tilted hip-forward" tends to stand in a slouched manner with a flat low seat and a "high roll" below the front waist or has prominent hip bones; posture causes a "baggy" pants seat; (c) "tilted hip-backward" has a "lower tummy" with the "posterior out and up" resulting in pants that may hike up over the seat (pp. 11–12).

Like Minott (1974), several patternmaking books use the terms "hip-tilt forward" and "hip-tilt backward," but the exact relationship of hip tilt to anatomy is not explicitly described. Hip tilt can be described more clearly by referring to specific movements of skeletal structures of the pelvis. See Figure 6.12 illustrating pelvic tilt (Kendall, McCreary, Provance, Rodgers, & Romani, 2005). Focus on the pelvis, following the relative positions of the coccyx and the iliac crest for each position. The *pelvis in the neutral position* is what Minott (1974) and others refer to as an average posture. The *anterior pelvic tilt posture* shifts the coccyx up and out (Minott's hip tilt backward, p. 12). The *posterior pelvic tilt posture* tucks the coccyx under, shifting the rim of the pelvis to a more parallel-to-floor orientation (Minott's hip tilt forward, p. 11).

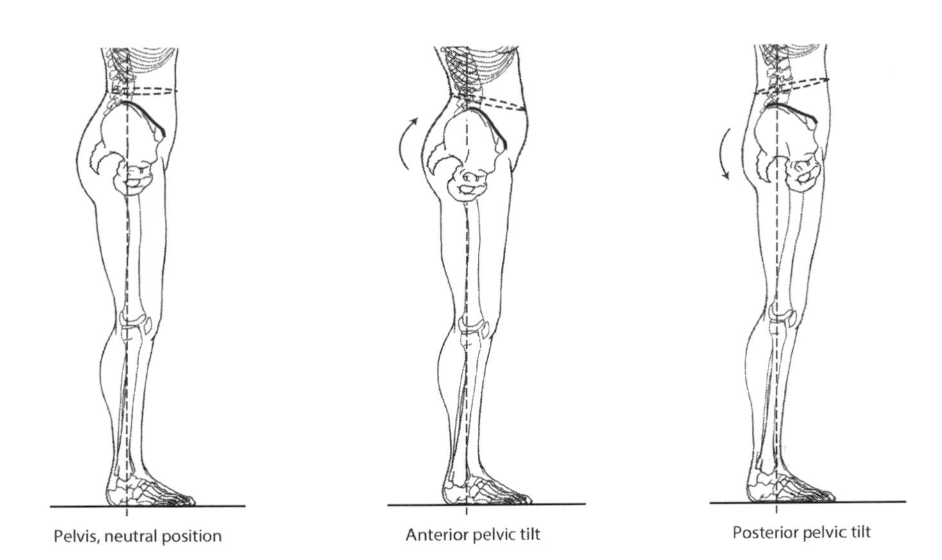


**FIGURE 6.12**
Pelvic tilt and waistline placement: Neutral position, anterior pelvic tilt, posterior pelvic tilt.

Also notice how pelvis position affects the position of the WC which, in turn, will affect the product waistline position. The neutral pelvis position WC is parallel to the floor. In the anterior pelvic tilt position, the WC slants toward the floor in the front, and in the posterior pelvic tilt position, the WC tilts in the opposite direction. As the pelvis tilts and the product waistline shifts, the pants crotch curve will change, especially the relationship of the front and back crotch lengths and curve shapes.

Song and Ashdown (2012) developed "a set of basic pants patterns optimized for three lower body shape groups" (p. 315). The shapes were determined from anterior and lateral views of scans of varying body types. Principal component analysis was used to identify five variables that grouped the three shapes: (a) "curvy shape," the curviest silhouette between the waist level and the hip level and the most prominent abdominal silhouette; (b) "hip tilt shape," the most prominent buttocks and a lower body that is tilted toward the back; (c) "straight shape," a non-curvy silhouette and less prominent buttocks. Expert judges compared pants, worn by 27 participants, made from the "body shape" patterns to pants made from a traditional pattern method. They concluded that patterns developed for specific body shapes could improve fit.

### 6.9.2 Lateral Body Shape Effects on the Pants Crotch Curve Shape

McKinney, Bye, and LaBat (2012) conducted a study to try to determine the relationship of pants crotch curve to the body. Seven subjects were scanned, body measurements extracted, and body crotch curve shapes mapped from

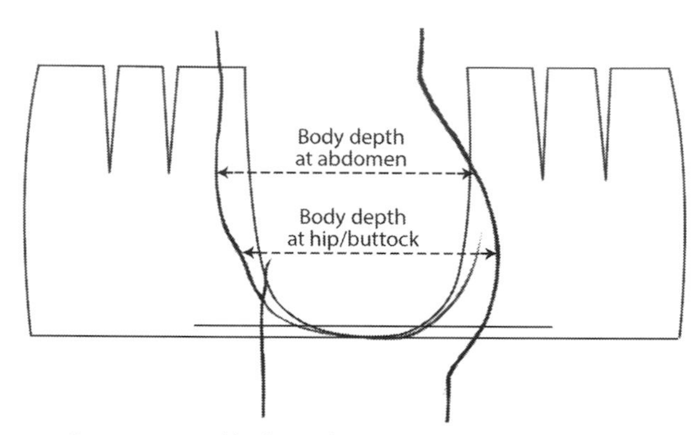


**FIGURE 6.13**
Body/manikin shape and pants crotch curve shape relationships. Pants pattern and body crotch curve superimposed, lateral view. (Adapted from McKinney, E. (2007). *Towards a three-dimensional theory of pattern drafting: relationship of body measurements and shapes to pattern measurements and shapes* (Unpublished doctoral dissertation), University of Minnesota, Minneapolis, MN. Used with permission.)

the scans. Note that the crotch curve does not measure the length of the back crotch curve between the buttocks, but is a curve that follows the shape of the buttock mimicking a pants crotch curve. Pants were custom tailored for each subject to achieve the best possible fit. The body crotch curve and the pattern crotch curve were compared. The pants crotch curve shape and the body crotch curve shape showed little relationship. The researchers state that the best fit for all participants was a pattern hip depth narrower than the body depth at the hip level. See Figure 6.13 illustrating the pants pattern to body relationship for one subject. Many variables, along with crotch curve shape, can contribute to good fit of a pattern to the body: fitting device size and placement, side seam curve, ease amounts, and fabric variables. Change in one of the variables will affect all other variables.

### 6.9.3 Anatomical Variations and Crotch Curve Shape and Length

Although studies to date have not found a direct relationship between the body and a pants pattern, it seems logical that distribution of body fat in the mid-torso will affect crotch length and shape in some way. Study the body form with extra mid-torso fat distribution and the related pattern shapes (Figure 6.14). Note how a waistline placement change affects the crotch curve. Several studies have focused on the pants crotch curve related to the U-shape formed from the body front waist, between the thighs, to the back waist. In addition, thigh shape and size related to hip shape and size should be studied for effects on the pants crotch curve.

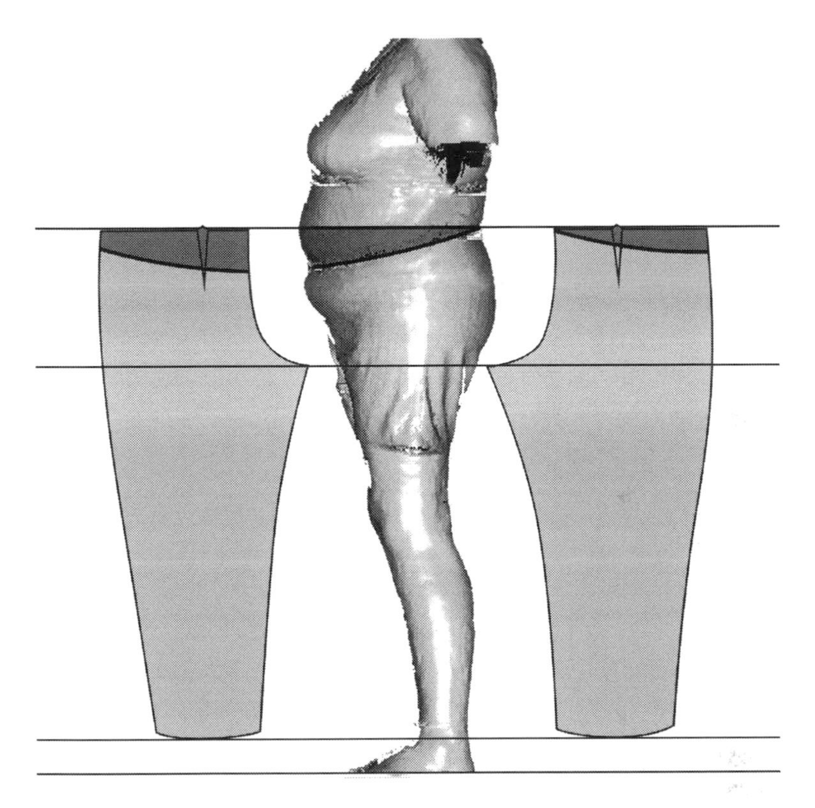

**FIGURE 6.14**
Pants crotch length variations on one non-ideal body type.

Gu, Lin, Su, and Xu (2017) proposed a method of determining optimum *crotch distance ease*, the crescent of space between the crotch curve of the pants and the crotch of the body. They scanned four manikins and used a surface modeling algorithm to interpolate occluded surfaces. Although the method has potential, the form and solidity of a manikin crotch area is much different from a human form with skeletal structure and soft tissue at that perineal region.

Another approach to designing the pants crotch curve to fit and accommodate the body, especially the body in motion, is to incorporate a gusset into the crotch seam (Figure 6.15). A gusset is a diamond-shaped piece of fabric with multi-directional stretch, either a knit fabric or a woven fabric with the bias of the fabric placed to stretch with the body. The gusset added at the pants crotch curve, can provide more crotch depth ease and can allow active movement. Although it is unlikely that the first designer to use a crotch curve gusset in a design had studied perineal anatomy, the coincidence of body area diamond shape and the shape of the gusset demonstrates that the body can provide design inspiration.

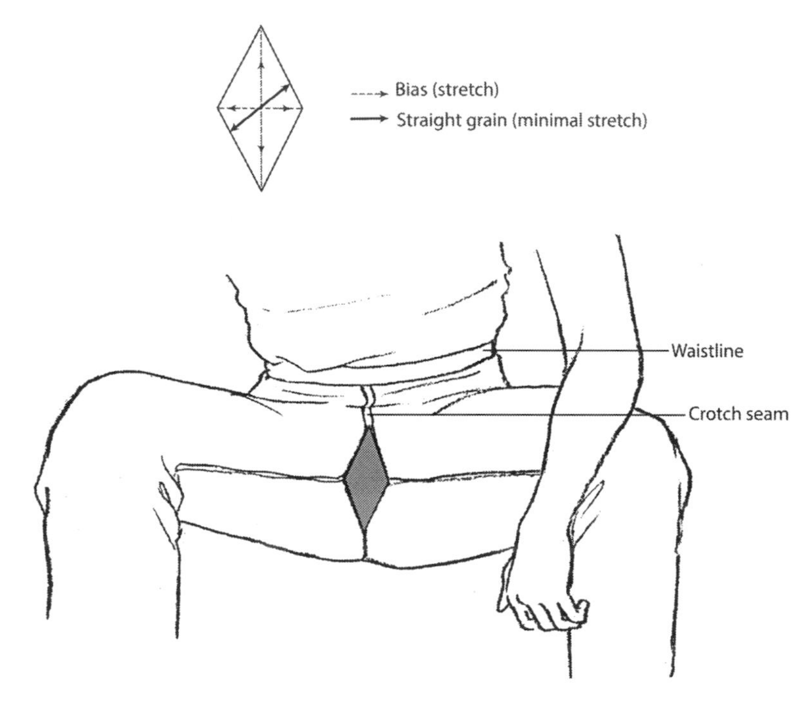


**FIGURE 6.15**
Gusset inserted into a pants crotch seam.

### 6.9.4 Pregnancy Effects on Pants Shape and Fit

Body changes due to pregnancy—growth of the fetus, expanding uterus, and related changes in the woman's shape and size—affect the crotch curve (see Chapter 5 on pregnancy). A typical strategy for designing maternity wear pants is to incorporate a stretchy panel to the front pattern pieces where the pants cover the expanding belly. Okabe and Sugimoto (2007) studied body forms of 942 women in the last trimester of pregnancy. They noted the two major changes needed to develop comfortable pants are increased total crotch length and "elevated anterior waistline height" (p. 763); a fairly standard approach. Changing posture throughout pregnancy will also affect front and back crotch length proportions. They suggest that, beyond pattern dimension changes, new materials should be developed that will change with the increasing abdomen, resulting in better fit and comfort. Materials development is no doubt needed; however, a better understanding of the changing pregnant figure in all size ranges could result in unique pattern shapes that adjust through several months of pregnancy.

### 6.9.5 Pants Crotch Curve Related to Male Genital Anatomy

Male external genitals will affect crotch shape (Review Chapter 5 on male external genital anatomy). Some research has been published on crotch curve shape for women's pants; however, very little is written about shaping the pants crotch curve for men. Knowles (2006) states that the distance of the pants crotch level from the body for menswear depends on the style. The crotch level of formal trousers, for example, is positioned a distance (not specified) below the body; while the crotch level of other styles, such as jeans, is positioned next to the body (p. 110). Knowles also explains that for styles that fit closely to the crotch (depth and length) additional distance may be added at the left front crotch extension "to accommodate a man's physique because most men 'dress' toward the left" (p. 112). By "physique" we presume Knowles is referring to the genitals.

Stone (1976) recommends incorporating additional ease to the left crotch seam (crotch extension). From illustrations in her men's tailoring book, it appears she is recommending extending the left crotch point by 1.27 cm (1/2 in.); however, she gives no reason for the pattern adjustment. Rinehart (1975) describes how to adjust the front and back crotch curve for men's pants. She based her information on talking with experienced tailors. In reference to accommodating male genitalia she says that the term "dress" refers to the "side toward which a man places the intimate parts of his anatomy when he puts on his pants" (p. 41). She says most men dress left and "a right dresser is sort of a southpaw in the tailoring trade" (p.41). In contrast to Knowles (2006) and Stone (1976), she recommends reducing the right crotch extension by 1.27 cm (1/2 in.). Some tailoring books give directions on adjusting the *pitch* or slope of the back crotch curve from crotch level to the waistline to accommodate different male body shapes and to use characteristics of the fabric grainline, especially bias, to conform to the buttocks.

Henson (1991) proposed a method for determining the crotch curve best fit shape for men's pants. She developed a measurement method and instrument she called a *flexicurve* to reproduce the crotch curve (of the body) including identifying the crotch point (Henson, 1991, p. 57). She collected data on only five subjects. She used the flexible curve to shape to the subject's crotch by locating one end of the flexicurve at the subject's self-identified front WC level and then had the subject form the flexible curve between the legs and over the genitals. The subject then located the terminal point of the flexicurve at his posterior WC. A plumb line suspended from the flexicurve was used to locate the crotch point. The flexicurve was removed from the body and the shape of the flexicurve was traced on paper to determine each subject's crotch curve. Henson compared subject crotch curves to a pants pattern she acquired from a menswear company and found several differences. One result was similar to a finding by McKinney et al. (2012) that the pattern crotch curve depth at the hip level was narrower than the body depth at the same location. She did not incorporate and test the flexicurve acquired crotch shapes in patterns drafted for the subjects, so a fit test was not conducted. Henson (1991) concluded that the method could provide valuable information, but specifying waist location, establishing ease amounts, and determining effects of fitting devices would need to be considered (p. 82).

Clearly there is more work to be done in developing methods to satisfactorily fit bifurcated garments for men and for women; recognizing the relationship of WC location to the other pants components is essential. Body scanning and virtual draping have the potential of solving the puzzle. There are technical issues to resolve, for example, when using a full-body scanner, data points at the perineal region/genitals are occluded by the legs. Current practice is to use computer patching methods to estimate and fill in missing data. Better understanding of anatomy by designers and use of small hand-held scanners to reach occluded areas will contribute to more accurate pants fit.

We do not prescribe solutions for shaping the crotch curve for body types but suggest that more research is needed to determine methods of translating three-dimensional body data to two-dimensional pattern shapes, especially for pants design and fit. A good starting point in solving dissatisfaction with pants fit is to recognize that there are a variety of body shapes and sizes based on individual anatomy. In current apparel product design, the varieties of body shapes are not considered in developing pants pattern shapes. Some apparel manufacturers use body shape categories to market products, but most are not based on in-depth knowledge of anatomy.

## 6.10 Motion at the Mid-Torso

Motion at the mid-torso, specifically at the waist, relies on motions initiated in the upper-torso and/or lower-torso spinal column. For a detailed discussion of these motions read the sections of Chapters 4, 5, and 9 on body motion. Simply stated, the basic movements related to mid-torso are: flexion or forward bending, extension or backward bending, and lateral flexion or bending to each side. Refer to Figure 4.33 showing these movements. Another mid-torso movement is rotation to the right or left around a vertical axis (Palastanga, Field, & Soames, 2002, p. 512); a twisting motion.

A key factor in designing most products that cover a limited area of the mid-torso, centered on the waist, is to avoid restricting body motion. Wide belts can limit the mid-torso motions described above and can be uncomfortable. In the case of products worn in work settings, a wide belt can be a safety hazard by preventing forward flexion. Back support belts, specifically designed to limit mid-torso movements, are purported to prevent back injuries. However, many studies of back support belt use dispute this idea. Giorcelli, Huges, Wassell, and Hsiao (2001) studied 17 males and 11 females lifting boxes weighing 9.4 kg and determined that spine movements were limited, but they did not investigate back injuries. Their final conclusion was that subjects lifted more slowly and used "squat-lift" techniques. Related to this finding, Imker (1994) states that back support belts, at best, serve as a reminder of proper lifting techniques, and that back support belts are unnecessary. He suggests that the best approach is to engineer hazards out of the job, which would include training on safe lifting.

## 6.11 Waistline Product (Belt) Fit

Belts are a major waistline product category worn almost exclusively at the WC location, so we describe sizing methods for belts. Upper and lower torso products that attach to a waistline are described in other chapters. Belts use a fairly simple direct measure system, relating the belt length to the WC measurement. Some belt manufacturers recommend measuring the person's WC and then selecting a belt size with the same measurement. Other manufacturers give directions to select a belt size one size up from the pants waist size. Most belts use a fastener system that allows adjusting the circumference of the belt. To determine the size of a standard belt with a buckle and prong on one end, and holes to secure the prong at the other end; measure the length in inches or centimeters from the prong location to the middle hole. Standard belts have five holes, so the third hole is the key to the belt size. Another fastening method is the adjustable ratchet belt, sometimes called a hole-less belt. The fastener allows a range of sizes as it uses "teeth" in the buckle that engage to grasp the belt material, cloth or leather, anywhere along the length of the belt.

Two factors to consider when designing any type of belt are: (a) the variations in WC locations based on anatomy and personal preference and (b) the increasing obesity rates which could affect size category expansion to match increasing WC dimensions.

## 6.12 Designing for the Mid-Torso: Conclusion

The body mid-torso, intersection of the upper torso and the lower torso, determines shape and form for many products. Mid-torso anatomy—relationships of skeletal components contributing to posture, movable and vulnerable soft organs, amount and distribution of fat—determines optimum location of wearable product features for this region. A key to good fit and comfort for many products is locating, measuring, and fitting the elusive waist which varies by body type and personal preference. Waist circumference location especially affects the form and fit of bifurcated products. Body type classification, mostly focused on the mid-torso, has been used to try to simplify and improve product fit. Body scan technology may hold the key to solving the body-to-product challenges of designing for the mid-torso.

# 7

# *Designing for Hand and Wrist Anatomy*

Hands, complex and dynamic anatomical structures, provide a highly specialized interface between the body and the environment. They contribute to several vital activities: grasping, manipulating, sensing, gesturing, and more. The hand and wrist may be protected and enhanced with wearable products; however, wearable products may also limit the natural capabilities of the hand and wrist.

**Key points:**

- Basic anatomy of the hand and wrist:
  - Bones, ligaments, fascia, muscles, and tendons build dexterous, yet vulnerable, hands.
  - Blood, nerve, and lymph systems reach all the way to the fingertips.
  - Hand and wrist mechanics allow movements for activities including self-care, work skills, perceiving the environment, defensive and offensive movements, emotional touch, and expression through gesture.
- Hand and wrist products: Protection, mobility, replacement
  - Environmental challenges: Thermal protection, manipulation stresses and strains; burn, abrasion, and cut vulnerabilities.
  - The "cold" hand limits manipulation capabilities and is prone to injury.
  - Handwear requirements: Mobility, protection, and comfort.
  - Prosthetics can simulate hand and wrist functions.
- Fit and sizing:
  - Function is essential in many handwear products.
  - Handwear shapes: Simple to precision fit for fingers and the thumb.

We define *handwear* as a product that covers any portion or all of the hand and/or wrist. Unless otherwise indicated, the term hand is used to include the fingers. Handwear varies in purpose and coverage, from a decorative ring on one finger, to a wristwatch covering a narrow area of the wrist, to a chemical protection glove that extends from the fingertips to the elbow. Winter mittens and gloves may be the first products that come to mind. Intricately shaped glove fragments estimated to date to 1300 B.C. have been found in Egyptian tombs of the pharaohs, including King Tutankhamen (Stall-Meadows, 2004, p. 287). Handwear includes fashion gloves, impact protective products like hockey goalie gloves, gloves to protect from pathogens in medical settings, thermal insulating handwear for extreme cold, braces and splints to protect and stabilize, and prosthetics to replace the hand. This chapter focuses mainly on mittens and gloves. Chatterjee, Jhanji, Grover, Bansal, and Bhattacharyya (2015) list numerous types of gloves in two categories: (a) commercial and industrial and (b) sport, recreational, and specific-use gloves. Among their examples: cut-resistant gloves, firemen's gauntlets, military gloves, billiards gloves, scuba-diving gloves, and wheelchair gloves—to name just a few. Handwear design decisions—selection of shape, materials, and methods of making—affect style, comfort, and the ability to perform a task or motion.

Humans depend on good hand and wrist function, and at times the right and left hand must work together to perform a desired action. Vision, in particular, and the central nervous system, in general, contribute to successful use of one or both hands. *Eye–hand coordination* describes the linked control of eye movement and hand movement. While we can gather information using touch alone, visual input triggers many hand actions and determines successful completion of tasks. Action lags a split second behind sight of an object (Scrafton, Stainer, & Tatler, 2017). The coordinated functioning of the wrist, palm of the hand, and the fingers is possible because of the framework of small bones and the muscles, tendons, and ligaments that support and move the bones. Blood, nerve, and lymph pathways running throughout the hand and wrist are essential for hand health and function.

*Handedness* is the preference and skillfulness of using one hand instead of the other, so a person is right-handed or left-handed. The preferred hand is the dominant hand. Most people are right-handed. Very few people are *ambidextrous* with equal skill using either hand (Smits, 2011). Because use of an extremity influences the growth and, therefore, the mature size of the limb, the dominant hand is usually slightly larger (Mueller & Mulaf, 2002; Turner & Pavalko, 1998). Choose ready-to-wear gloves and mittens, constructed as right-left mirror images, by fitting the product to the dominant hand. Medical splints, custom made to fit the hand, are formed to the morphology of the person's left and/or right hand. Baseball catchers' mitts fit a catcher's preferred catching hand. Some ambidextrous pitchers use a glove designed with six fingers in the mitt that can be worn on either hand.

## 7.1 Wrist and Hand Bones, Hand Joints, and the Radiocarpal Joint

The wrist and hand contain many small bones that provide flexibility and dexterity (Figure 7.1). A network of ligaments and fascia join the bones, while tendons and muscles move the bones in relationship to each other. Following distally from the styloid processes of the radius and the ulna, (bumps you can palpate on either side of your wrist), the next set of bones are the *carpal bones,* commonly called the wrist. This group of eight small irregularly-shaped bones in the base of the hand bridges between the more stable bones of the hand, the *metacarpals,* and the radius and ulna of the forearm. The narrowest circumference of the wrist/forearm, where a bracelet or fitness or health monitoring band might be worn, is located (a) near the distal ends of the radius and ulna or (b) over the proximal row of carpal bones. The metacarpal bones are the bones of the palm of the hand and the base of the thumb. The next set, and most distal bones, are the *phalanges* of the fingers and thumb.

### 7.1.1 Carpal Bones

The carpal bones are grouped into two rows, with four bones in each row (Figure 7.1). The four proximal bones, adjacent to the radius and ulna, starting from the lateral (thumb) side, are the *scaphoid, lunate, triquetrum,* and *pisiform.* When translated from Latin, these terms describe the shapes of the bones: boat-shaped, moon-shaped, triangular, and pea-shaped (Saladin, 2014, p. 187). The distal row from the lateral side are the *trapezium, trapezoid, capitate,* and *hamate.* The scaphoid, particularly susceptible to fracture, links the two rows of carpal bones (Katz, 2014, p. 5). Scaphoid fractures usually occur from a fall on the outstretched arm (McKinley & O'Loughlin, 2006, p. 23).

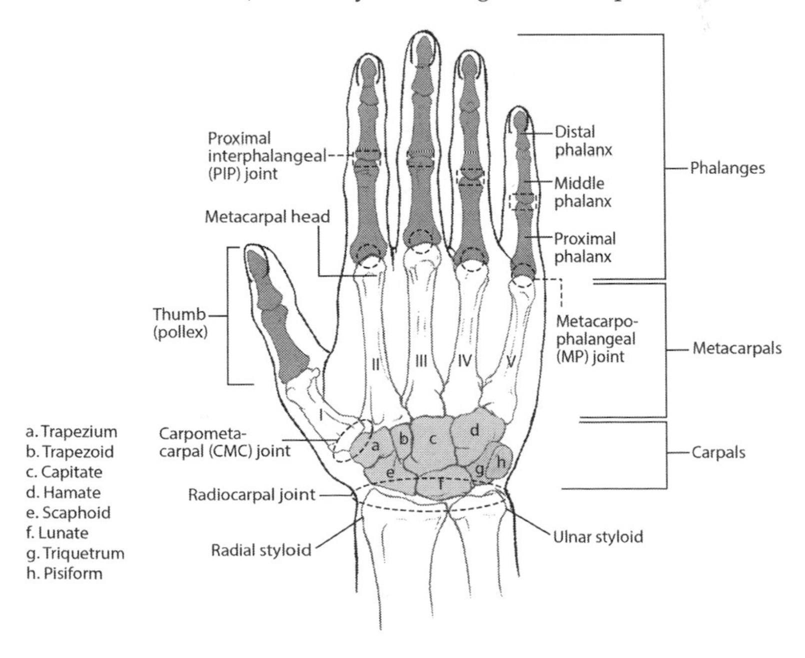


**FIGURE 7.1**
Bones of the hand and wrist: Phalanges, metacarpals, and carpal bones.

Six carpal bones—the pisiform, triquetrum, hamate, capitate, trapezoid, trapezium (with its hook-like tubercle)—make up the *carpal arch*. The arch is shaped into a tunnel with a thick connective tissue ligament, the *flexor retinaculum*, bridging the space between the medial and lateral sides of the base of the arch (Drake, Vogl, & Mitchell, 2015, p. 798). The tunnel provides a protective, yet flexible, pathway for nerves and blood vessels connecting the hand to the body (Read about carpal tunnel syndrome in Section 7.4.1).

### 7.1.2 Metacarpals

The metacarpals, in the palm of the hand, bridge between the carpals and the phalanges. Metacarpal I lies proximal to the two thumb phalanges and functions as part of the thumb. Metacarpal V is proximal to the little finger, with II, III, and IV in order between I and V. Metacarpals II-V form the palm of the hand. The shaft of each metacarpal bone is called the body; the end closest to the wrist is the base; and the end closest to the phalanges of the fingers is the head. The heads of the metacarpals form the knuckles of the distal dorsal hand, that you can see and feel when you make a fist. A direct compressive blow to the head or knuckle of the smaller fifth metacarpal may result in what is called a boxer's or a brawler's fracture. This type of injury is associated with fist fights, automobile accidents, bicycle accidents, and human or animal bites (Sueki & Brechter, 2010, p. 481).

### 7.1.3 Phalanges

Bones in the distal thumbs and in the fingers are called phalanges. *Phalanx*, the singular term, is used for one of these small bones. The thumb is also called the *pollex*. The thumb has only two phalanges, a proximal phalanx and a distal phalanx. The other four fingers have three phalanges each: proximal, middle, and distal.

*Interphalangeal (IP) joints* link a phalanx to a neighboring phalanx. The *distal interphalangeal (DIP) joints* are the end joints of the fingers and thumbs. The *proximal interphalangeal (PIP) joints* are the middle joints of the fingers. The DIPs and PIPs form the knuckles of the fingers. The *metacarpophalangeal (MP) joints* form the knuckles between the metacarpal bones in the palm and the fingers, and the second joint from the tip of the thumb. The PIP is the most commonly injured finger joint (Baugher & Graham, 2014). Finger joint injuries are often treated by stabilizing the finger with a split. Buddy taping can also be used—taping the injured finger to a neighboring finger to stabilize it while it heals.

The thumb, with its two phalanges and its metacarpal bone, participates in almost all human hand functions in cooperation with the other fingers. It is a key player in providing precision, power, and complexity. The index and middle fingers, with their related metacarpals, are essential in gripping, separately or in conjunction with the thumb. The thumb, index, and middle fingers are referred to as the *dynamic tripod* due to their precision gripping abilities (Yu, Chase, & Strauch, 2004, p. 32). Fingers can be spread wide (abducted), held close together (adducted), curled into a fist, and moved in various coordinated ways to perform a task. The little finger and ring finger cooperate with other fingers and the palm to hold objects. The little finger can also cooperate with the thumb and index finger to grip an object. Observe how your hand moves when manipulating chop sticks, knitting needles, and other small objects.

The mobility of the thumb is made possible by the first carpometacarpal (CMC) joint (Figure 7.1). The CMC is a saddle joint (see Chapter 2, Section 2.1.3) between metacarpal I and the trapezium bone. However, this mobility and ability to move in multiple directions makes the joint vulnerable. In some cases, the metacarpal or "rider" can fall off the trapezium or "saddle" increasing likelihood of arthritis and pain.

### 7.1.4 Focus on Products Protecting Bony Hand and Wrist Structures

Many athletic wearable products are designed to prevent impact injury to the wrists, hands, and fingers. For example, boxing gloves are padded to protect the boxer's hands and the opponent's body from impact. U.S.A. Boxing (2013) requires boxers in the lighter weight categories to wear 10 oz. gloves, and boxers in heavier weight categories to wear 12 oz. gloves, which assumes that heavier padding is required for larger boxers. No mention is made of padding placement or structure, new lighter weight impact protective materials, surface texture of the glove, or design of the form of the glove.

Baseball and softball gloves are designed to protect the athlete's fingers and hands from impact of the ball. Gloves are designed for the player's position: catcher's mitts, first baseman's mitts, infielder's gloves, pitcher's gloves, outfielder's gloves, and switch-thrower's gloves. The glove structure is suited to the performance requirements. For example, the catcher's mitt is much like a mitten, with no individual fingers and extra padding at the mitt's center pocket, which also makes a good target for the pitcher and extra protection for the carpal bones. In contrast, the infielder's glove is smaller with a shallow pocket so that the player can easily remove the ball to throw it quickly to a baseman (Rosciam, 2010). Hicks (2015) found that even with glove wear, catchers "commonly develop abnormal blood flow in the ulnar artery, digital ischemia in the index finger, and index finger hypertrophy (p. ii)" from continual trauma to the hand. She recommends glove designs be modified to further minimize risk of hand injury. For a thorough overview of the mechanisms of impact and impact protection methods for wearable products, read Watkins' and Dunne's (2015) chapter on impact protection.

### 7.1.5 Arthritis: Hand and Wrist Joint Damage

Arthritis affects the body's joints and causes pain and difficulty in moving. There are several types of arthritis that may affect the hand: osteoarthritis, rheumatoid arthritis, and psoriatic arthritis. Osteoarthritis is the most common, resulting from wear and tear and aging, with progressive loss of the cushioning cartilage within a joint (Clapham & Chung, 2014). The DIP joints of the fingers and the CMC joint of the thumb are most often affected. Osteoarthritis of the thumb is more common for women than men (Bernstein, 2015; Armstrong, Hunter, & Davis, 1994), which may have implications for handwear design. The joint can be stabilized with a splint, typically worn at night (Figure 7.2), allowing the joint to rest for a period of time (Egan & Brousseau, 2007). Splints are also worn during some activities to help decrease thumb pain.

Rheumatoid arthritis (RA), related to immune system damage to the synovial lining of the joints (Clapham & Chung, 2014, p. 61), commonly attacks joints of the hand. Severe finger and wrist deformities can result, as the disease stretches and destroys ligaments and cartilage. Egan et al. (2003) reviewed published studies of splint use by patients with RA and found insufficient evidence to draw conclusions about the effectiveness of splints in decreasing pain or increasing function. They found that some splints decreased grip strength and dexterity. Decreased range of motion was common while wearing a splint but was not a deterrent to splint wear. Through an iterative prototyping process, Goncu-Berk and Topcuoglu (2017) developed a smart glove for RA patients that uses electrical stimulation for pain relief, joint immobilization, and compression. Although the designers were concerned about glove appearance detracting from social interaction, participants were more interested in improved quality of life.

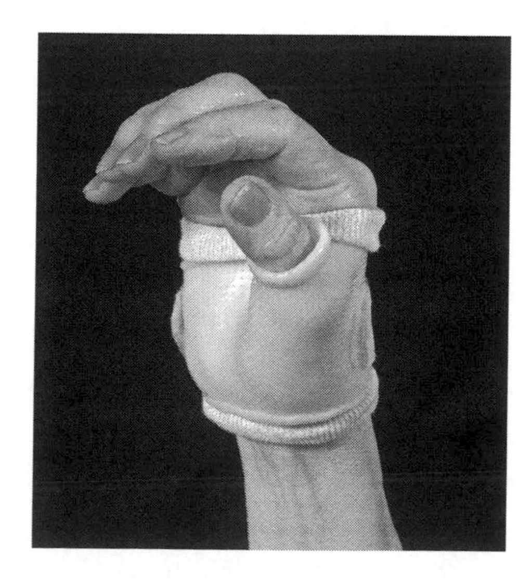

**FIGURE 7.2**
Splint to stabilize and rest the CMC joint of the thumb. (Photograph by David Bowers, University of Minnesota College of Design Imaging Lab.)

Psoriatic arthritis, characterized by inflammation of the skin (psoriasis) and the joints, also causes pain and impedes hand and wrist function. Adaptive aids and wrist splints (Mease, 2008) that look much like the molded CMC splint in Figure 7.2, but extend from the hand to the mid-forearm, can help alleviate pain.

### 7.1.6 Radiocarpal (Wrist) Joint: Connecting the Hand and Forearm

The radiocarpal (wrist) joint, the key connector between the hand and forearm, is complex in structure and function and dependent on good alignment of the adjacent forearm and carpal bones. Like at the elbow (Chapter 4, Section 4.4.4), the wrist flexes and extends and pronates and supinates. In addition, the hand and wrist deviate medially and laterally on the forearm. Grasping and fine motor activities require the wrist to be partially extended in relation to the forearm (Jenkins, 2002, p. 214), in what is called a *functional position*. In this position, the hand is said to be "cocked-up" in relation to the forearm. The term is descriptive of the action used to pull the hammer back on a gun.

Wearable products can be used to protect the bones of the lower forearm and the wrist from trauma, or to support and position broken bones as they heal. Knapp (1952) stresses that when the radius and/or ulna are immobilized in a protective wearable, the product should allow maximum motion of the elbow and fingers and that the wrist joint should be in the functional position. In the functional position, the wrist is slightly extended and deviated toward the little finger, the fingers are moderately flexed at the MP and PIP joints, and slightly flexed at the DIP joints. The volar and dorsal muscles crossing the wrist are at equal tension, so the fingers can flex with the least effort (Levangie & Norkin, 2005, p. 346). See further discussion of hand positions for activities in Section 7.3.3.

### 7.1.7 Sports Product Wearables to Prevent Wrist and Forearm Injuries

Rønning, Rønning, Gerner, and Engebretsen (2001) studied wrist injuries in snowboarders. They found that wearing a wrist splint shaped in the cocked-up position helped prevent wrist and forearm injuries while snowboarding. Schieber et al. (1996) reported similar findings for wrist splints worn by in-line skaters. These splints allow a degree of hand and finger motion, while protecting the hand and wrist bones and ligaments if the sportsperson takes a tumble.

## 7.2 Hand and Wrist Ligaments, Tendons, and Muscles

A complex network of anatomical structures stabilize and mobilize wrist and hand bones and joints. Ligaments span joints to keep them contained and positioned, strengthening the skeletal framework. Recall from Chapter 2,

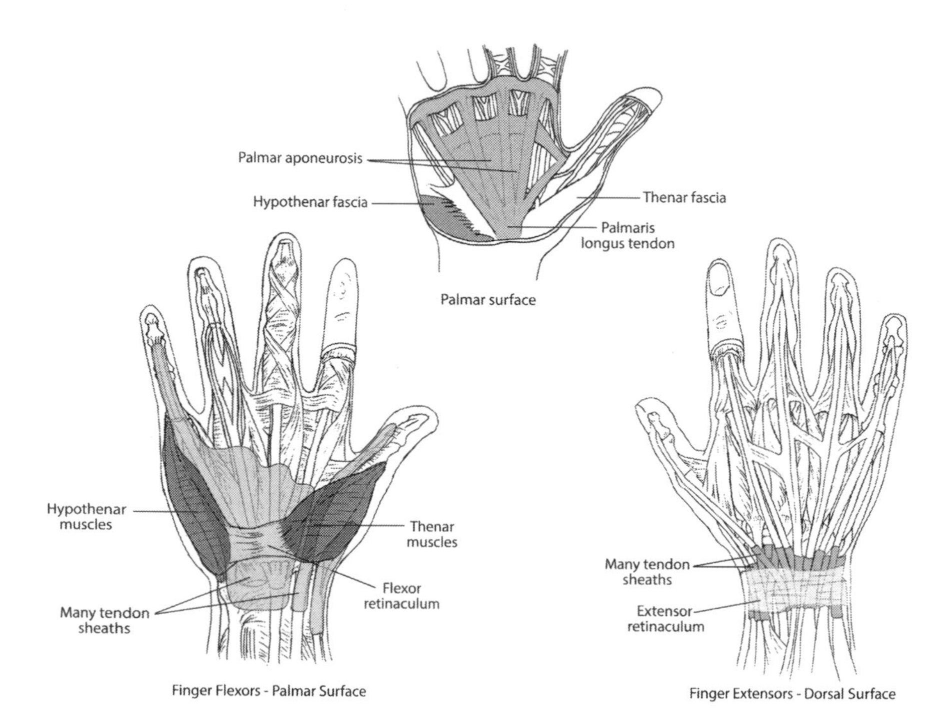


**FIGURE 7.3**
Hand tendons, fascia, and muscles.

Section 2.5.5, that a tendon is a tough, thick tissue band and an aponeurosis is a flat tendon sheet, both formed from opaque, non-contractile connective tissue. Muscles (contractile tissues) facilitate motion. Lubricant filled cushions adjacent to and encircling the linear tendons, *tendon sheaths*, promote tendon sliding in the tight connective tissue compartments within the palm and in the palmar and dorsal wrist. Bursae, along with tendon sheaths, extend from the hand to the fingers wrapping the flexor tendons to reduce friction "where the tendons rub against other structures" (Kendall, 1983, p. 25). Understanding how these structures work together in the hand and wrist is important when designing any handwear product. The complex framework of wrist and hand tendons, tendon sheaths, fascia, and aponeuroses can serve as inspiration for new handwear products. See Figure 7.3 for illustrations of some of the tendons, muscles, and fascia.

### 7.2.1 Ligaments and Other Connective Tissue Structures of the Hand and Wrist

Numerous ligaments bridge the distal radioulnar joint and the radiocarpal joint, which encompasses the distal radius and three of the eight carpal bones of the wrist. Ligaments cross the joints between the eight carpal bones.

A wrist sprain involves damage to ligaments in these locations. Rigid wrist splints designed to immobilize and support the joints help decrease swelling in traumatized tissues, promoting healing.

The palm contains a network of connective tissues that enable the cupping abilities of the hand. The palmar aponeurosis is a flexible, sturdy base that can conform to and support objects. The hypothenar and thenar fascia encase underlying muscles and serve as pliable supports for them. The palmaris longus tendon links the palmar aponeurosis to the palmaris longus muscle in the medial flexor forearm muscle group (refer to Figure 7.3).

Many muscles whose tendons attach to hand and wrist bones work in coordinated fashion to achieve a multitude of hand motions (Figure 7.4). Perhaps more than in other areas of the body, it is easy to observe muscles as "engines" and tendons as "cables" as they work together to move and position bones of the wrist and hand. Look at your forearm, wrist, and hand—both dorsal and ventral surfaces. Palpate your upper extremity from elbow to fingers with the hand and wrist at rest and in motion. Note the shapes, soft fleshy areas, and cord-like structures.

### 7.2.2 Muscles Moving the Wrist, Hand, and Fingers

*Extrinsic muscles* in the forearm/arm and *intrinsic muscles* in the hand generate hand and wrist motions. Extrinsic muscle tendons extend from forearm muscle bellies to bones in the wrist and hand. Review Chapter 4, Section 4.5.3 on the muscles and tendons of the forearm. Feel the muscles of your right forearm distal to your elbow. Then run your left hand along your forearm toward your wrist. Note that the muscles narrow into firm cords that you can see and feel on the dorsum of your hand and the palmar side of your wrist. The most superficial and central tendon belongs to the palmaris longus muscle, as it travels to connect with the palmar aponeurosis. Wave your right hand. Note how the muscles in the forearm and in the fleshy areas of the hand contract to manipulate the tendons in the hand. Learn more about hand and finger motion and range of motion in Section 7.7.

Extrinsic muscle tendons are held in place at the wrist with wide, fibrous bands—like the cuff of a shirt or a wide bracelet—called retinacula. The retinacula help to form the oval shape of the carpal tunnel that was described in Section 7.1.1. The flexor retinaculum joins the carpal bones on the palmar side and the *extensor retinaculum*, as seen in Figure 7.3, is superficial to the dorsal surface of the carpal bones (McKinley & O'Loughlin, 2006, pp. 368–369).

Intrinsic muscles occupy both superficial and deep locations within the hand (see Figure 7.3). These muscles contribute to *fine motor control*—the coordination of muscles, bones, and nerves—to produce small, exact movements. Activities like threading a needle or tying a fly-fishing knot demonstrate fine motor control. Facilitation of hand/finger dexterity is an important design

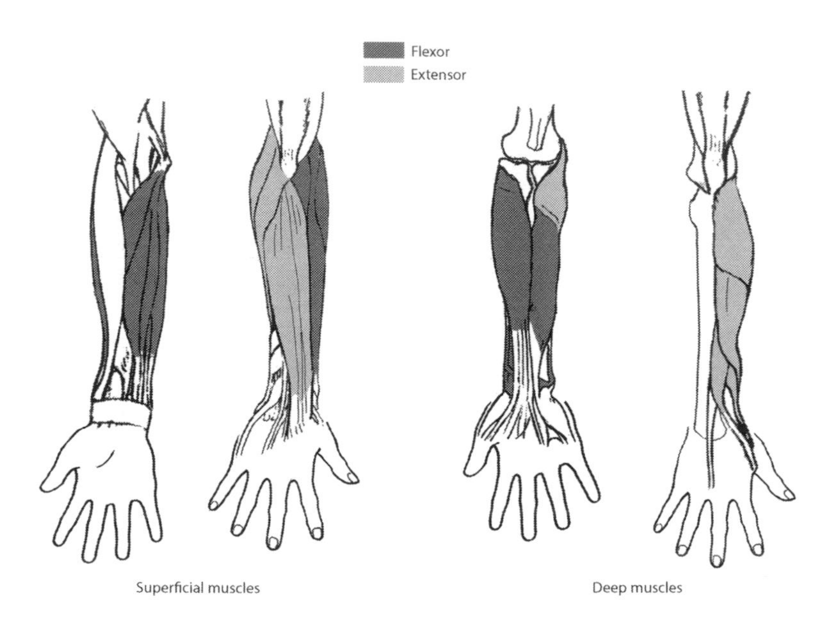


**FIGURE 7.4**
Extrinsic hand muscles.

goal for many handwear products. New materials, pattern shapes, and joining methods may advance handwear design that avoids imposing limitations on fine motor control.

*Thenar* muscles (at the base of the thumb) lie on the palmar side of the hand, along the first metacarpal. Feel the muscles by flexing your thumb to touch your index finger. Find the *hypothenar* muscles, a group of muscles on the palm, at the base of the little finger. They abduct the little finger and assist with little finger flexion. The *interossei* and *lumbrical* muscles abduct and adduct the fingers and help to stabilize their positions. These muscles are located in the deep spaces between the metacarpal bones in the palm. They are difficult to feel, except for the interosseous muscle between the thumb and the index finger. When you pinch the web space between the thumb and index finger, you can feel the 1st dorsal interosseous muscle on the dorsum of the hand as you move your thumb toward and then away from your index finger. All these muscles work together to stabilize and flex, extend, abduct, adduct, and *oppose* the digits. There are no muscle bellies in the fingers along the phalanges—only tendons arising from either the forearm's extrinsic muscles or the intrinsic muscles of the hand. As with muscles throughout the body, the names of the muscles acting on the hand describe what they do, where they attach, and occasionally some other characteristic. For example, *opponens pollicis,* in the thenar group, is a muscle that moves the thumb (pollex) to oppose other fingers and *flexor digiti minimi* flexes the little finger.

## 7.3 Forearm, Wrist, and Hand: Natural and Functional Positions

The forearm, wrist, and hand structures are highly integrated—characterized by natural forms and positions that should be considered to design functional, comfortable, and safe handwear. The evolution of the human hand with its opposable thumb is crucial in the advancement of human capacity and intelligence. Napier (1970) emphasized the importance of finger-thumb *opposition* for human emergence from an undistinguished primate background. In the human hand, compared to other primate hands, the thumb is longer than the fingers. This comparative elongation allows the thumb to oppose all four fingers, increasing the versatility of the hand and the variety of hand positions and grips possible—from the "power grip" to the "precision grip" (McGinn, 2015, pp. 21–22). Precision grip permits writing, rolling a small object between the fingertips, holding a strand of hair, and manipulating surgical tools. McGinn (2015) describes other attributes that make the human hand remarkable: it is "highly innervated for sensation and perception" and "educable" with astonishing motor memory to fluidly perform complex sequences like playing the piano (p. 24).

### 7.3.1 Hand Surface Regions

When the body is in motion surface areas can expand or contract (refer to Section 2.5.5). There are two surfaces of the hand, the *palmar surface* (includes the palm and flexor surfaces of the fingers and thumb) and the *dorsal surface* or the back of the hand (Figure 7.5). Each surface exhibits distinct regions that allow the hand to function in many ways.

#### *Regions of the Palmar Surface of the Hand*

Notice the skin crease lines at flexion points on the palmar surface. Folklore and palmistry refer to length and configuration of the lines as conferring information about an individual such as fate, intelligence, and longevity. However, no scientific evidence establishes these relationships. There is no subcutaneous fat under the flexion creases and the skin adheres directly to underlying fascia along the lines (Yu et al., 2004, p. 4). The creases do not correspond exactly to underlying joints (Mallouris et al., 2012; Sokolowski, Griffin, Carufel, & Kim, 2018), but might be useful as landmarks for some products or as inspiration for designing handwear.

Skin creases demarcate three areas of the palmar surface of the hand: *palmar region*, *finger region*, and *thumb region*. Flex your MP joints to see the *palmar digital crease* formed at the junction between the palm and the base of your four fingers. The finger region is distal to the palmar digital crease. Flex your wrist to see the creases bounding the palm and the forearm. Raise your thumb into a position perpendicular to the palm to see the *thenar crease* curving from the

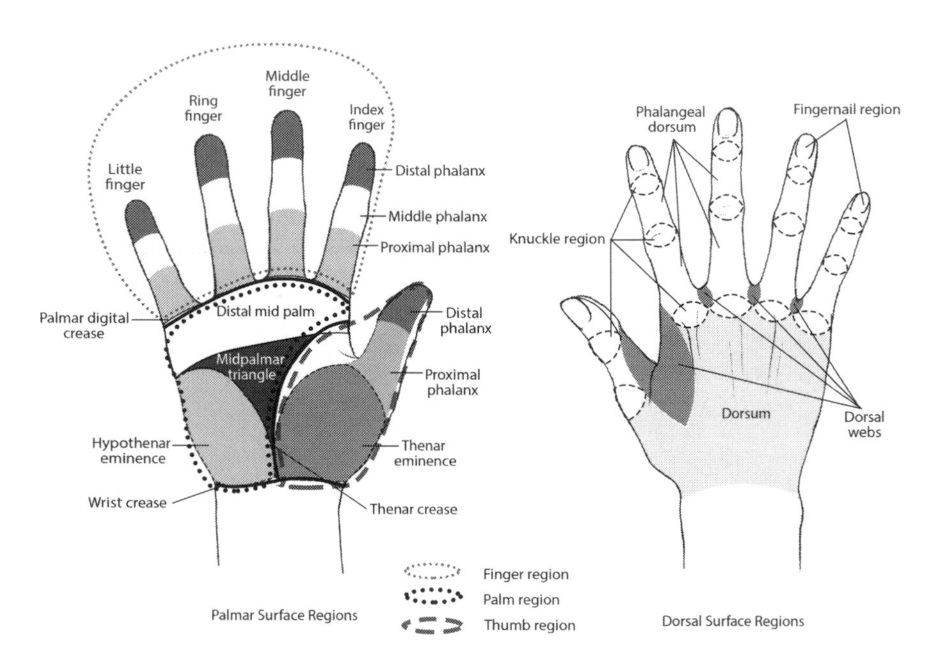


**FIGURE 7.5**
Regions of the hand, palmar surface and dorsal surface (right hand). (Modified from Figure 1–5. Anatomic regions on the palmar surface of the hand and Figure 1–6 Regions on the dorsal surface of the hand. H-L. Yu, R. A. Chase, and B. Strauch, 2004, Atlas of Hand Anatomy and Clinical Implications, pp. 7–9. Copyright 2004 by Mosby.)

*wrist crease* to approximately the midpoint between the thumb and index finger. The palmar region between the palmar digital creases and wrist crease has three areas: *midpalmar* (subdivided into the *distal midpalm* and *midpalmar triangle*), *thenar,* and *hypothenar* regions (Yu et al., 2004, p. 7). Within these areas, two "meaty" areas of the palm correspond to the thenar and hypothenar intrinsic hand muscles discussed above, and a third slightly mounded area, the distal midpalm, overlies the MP joints of the four fingers. The three areas surround the central concavity of the midpalm, the midpalmar triangle, with its underlying sheet-like central palmar aponeurosis.

The *thenar eminence* at the base of the thumb is variably plump depending on the size of the muscles, and it is demarcated by the thenar crease. The *hypothenar eminence* is the mound of muscles on the opposite side of the palm. It is smaller and often does not have a clear line of demarcation. Note the mobility of the web of the thumb—the span of soft tissue between the thumb and the index finger—as you move your thumb toward your midpalm.

### Regions of the Dorsal Surface of the Hand

Look at the opposite side, dorsal surface, of your right hand and see Figure 7.5. The regions are: *dorsum,* fingernail, *phalangeal dorsum,* knuckles, and *dorsal webs* (Yu et al., 2004, p. 8). Make a fist to see the flat, somewhat rectangular surface called the dorsum of the hand. Then extend your fingers to see the fingernail region. The knuckle region lies over the joints. When the fingers are extended, the areas over the joints exhibit an excess of wrinkled skin, especially over the middle joints of the fingers. When a finger is flexed, the excess skin glides over the length of the flexed joint. The phalangeal dorsum covers the smooth skin areas between the joints of all of the fingers. Note the dorsal web region, at the bases and between spread (abducted) fingers. The flexible web-like skin connector between the bases of the fingers allows and limits motion.

Knowing the names of the regions of the hand, their general functions, and interrelationships is useful in designing handwear. The hand is complex and multifunctional, so designing a product that covers all, or a portion of the hand and wrist, while not inhibiting multiple functions is challenging. The underlying anatomical structures work in concert to make the hand and wrist an integral part of whole body function.

### 7.3.2 Arches of the Hand

The hand's skeletal framework and its related connecting tissues naturally form the hand into a cup or dome shape. Place your hand on a flat surface in a relaxed position, then flatten your hand to the surface noting how the hand adapts from cupped to flat. The relaxed hand position—related to the functional forearm/hand position—is referred to as the *hand-at-rest position.* In this position, the palmar surface of the hand curves naturally into a slight concave shape. The dome is made of two transverse and five longitudinal arches that "provide a postural base for the hand in exerting its precise, delicate, dexterous functions, in contrast to gross power functions" (Yu et al., 2004, p. 48) (see Figure 7.6). The transverse carpal arch tends to maintain its form, while the metacarpal transverse arch, through the heads of the metacarpal bones, is movable and adapts to perform tasks. The mobile longitudinal arches extend from the fingertips to the wrist. These adaptable arches mean that the hand can assume many positions.

Many handwear products—hockey gloves, baseball catcher's mitts, ski gloves, space suit gloves—are shaped to replicate the hand-at-rest position to prevent the hand from working against the product, which could cause fatigue and pain. Figure 7.7 illustrates the hand-at-rest and a hockey glove that is shaped to the hand-at-rest position.

### 7.3.3 Hand Positions for Activities

Moore, Dalley, and Agur (2014) describe basic functional positions of the hand: *power grip, hook grip, precision grip, precision handling grip, fingertip pinch, loose grip,* and *firm grip* (p. 772). The index and middle fingers, with their related metacarpals, are essential for gripping, separately or in conjunction with the thumb (see Figure 7.8). Several variations are possible within each

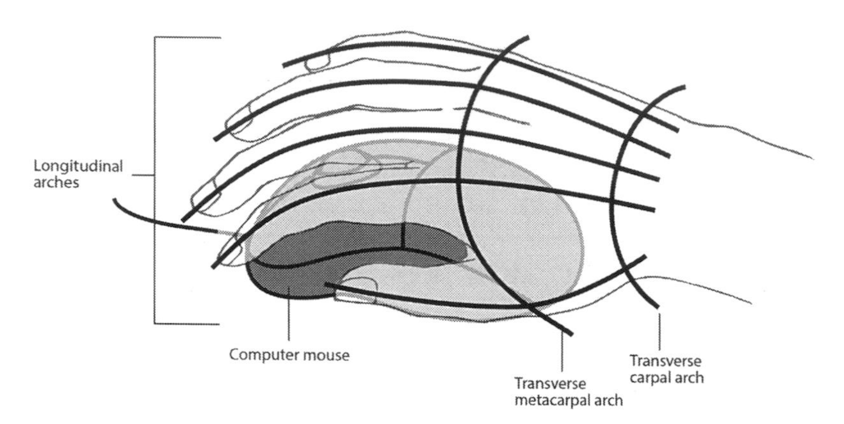


**FIGURE 7.6**
Hand transverse carpal and metacarpal arches, and longitudinal arches.

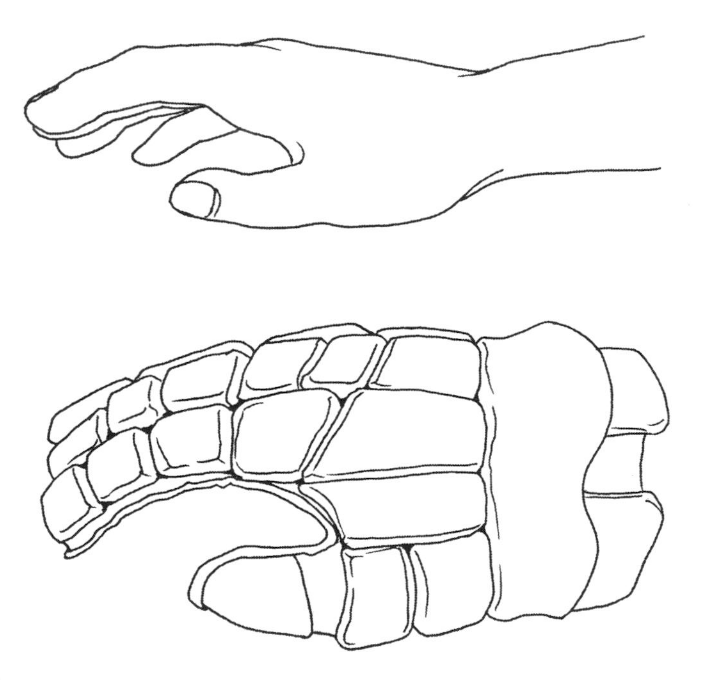

**FIGURE 7.7**
Hand at rest and hockey glove designed to accommodate hand at rest.

basic position. Fingertip pinch positions include: pulp-to-pulp pinch with pads of the thumb and the first finger touching the object, tip-to-tip pinch with the tips of the fingers in contact, or nail-to-nail pinch using the fingernails to secure a small object (Yu et al., 2004, p. 36). As you perform your daily tasks, take note of the many different hand and finger positions you use, and how regions of the hand and wrist work together.

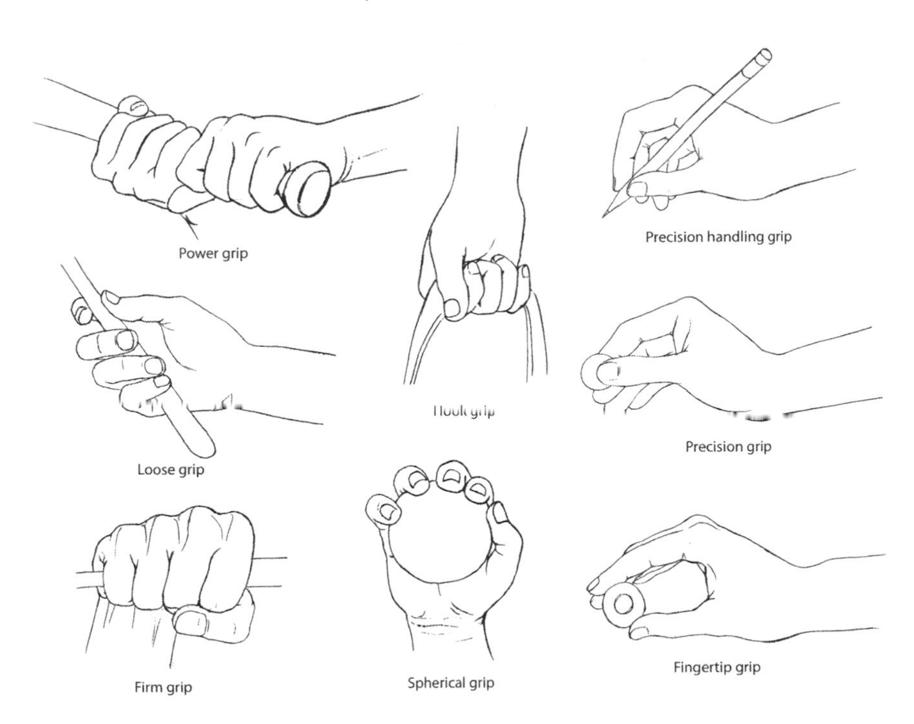


**FIGURE 7.8**
Hand positions during specific activities.

Before designing a handwear product, understand the essential and desired hand positions a person will use while wearing the product. Some tasks necessitate consideration of many of the functional positions, while others may require only a few functional positions. Each position requires that the hand adapts to use different combinations of bones, ligaments, tendons, and muscles. The power grip is used to grasp an object, for example getting a firm grasp on a hammer to strike a nail. A baseball batter holds a bat with both hands, ready to swing, with a double power grip. The hook grip uses flexion of the fingers to resist downward gravitational pull, for example when carrying a heavy handbag. A precision handling grip, used to write with a pen, requires pressure on the pen and flexibility to manipulate the pen. Surgeons perform precision tasks holding small instruments, while wearing protective gloves. The gloves must allow tactile feedback while maintaining an impermeable barrier. The precision grip is used when holding a tiny object in a stationary position. Some manufacturing assembly jobs use this hand and finger position while encased in a hand covering. The fingertip pinch, somewhat like the precision grip, uses the thumb and one or two fingers to hold a small object with opposing pressure. The loose grip and firm grip positions are used to encircle and grasp an object such as a gymnast grasping the high bar loosely or firmly. In this situation the gymnast's hands are subject to friction as they rotate around the bar.

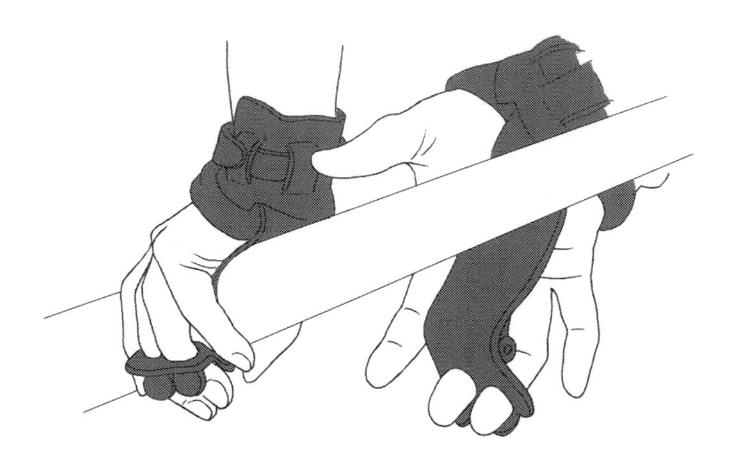

**FIGURE 7.9**
Gymnast's hand grip.

In addition to chalk dust to reduce resistance, many gymnasts wear a protective device that must adapt, along with the gymnast's hand, to a loose or firm grip while allowing smooth rotation around the apparatus bar (Figure 7.9). A well-designed wearable product for the hand and wrist allows full range of motion, while serving a primary function: fashion, protection, or improvement of performance.

## 7.4 Nerves, Blood Vessels, and Lymphatics of the Hand and Wrist

Nerves, blood vessels, and lymphatics extend through the limited confines of the fingers and must move signals and fluids along these pathways to enable good hand function. Review the basics of the related systems in Chapter 2. As the pathways travel from the core of the body, they progressively decrease in size to fit the small spaces in the fingers. The smaller structures are more vulnerable to pressure and damage. However, their pathways are also closer to the surface of the body, offering easier access for use in monitoring body functions with wearable products.

### 7.4.1 Nerves of the Hand and Wrist: Structure, Function, and Product Considerations

The three nerves (median, ulnar, and radial) that serve the hand allow a person to feel pain, temperature variations, light touch, vibration, and position (Figure 7.10). At the same time, the motor fibers of the same nerves carry

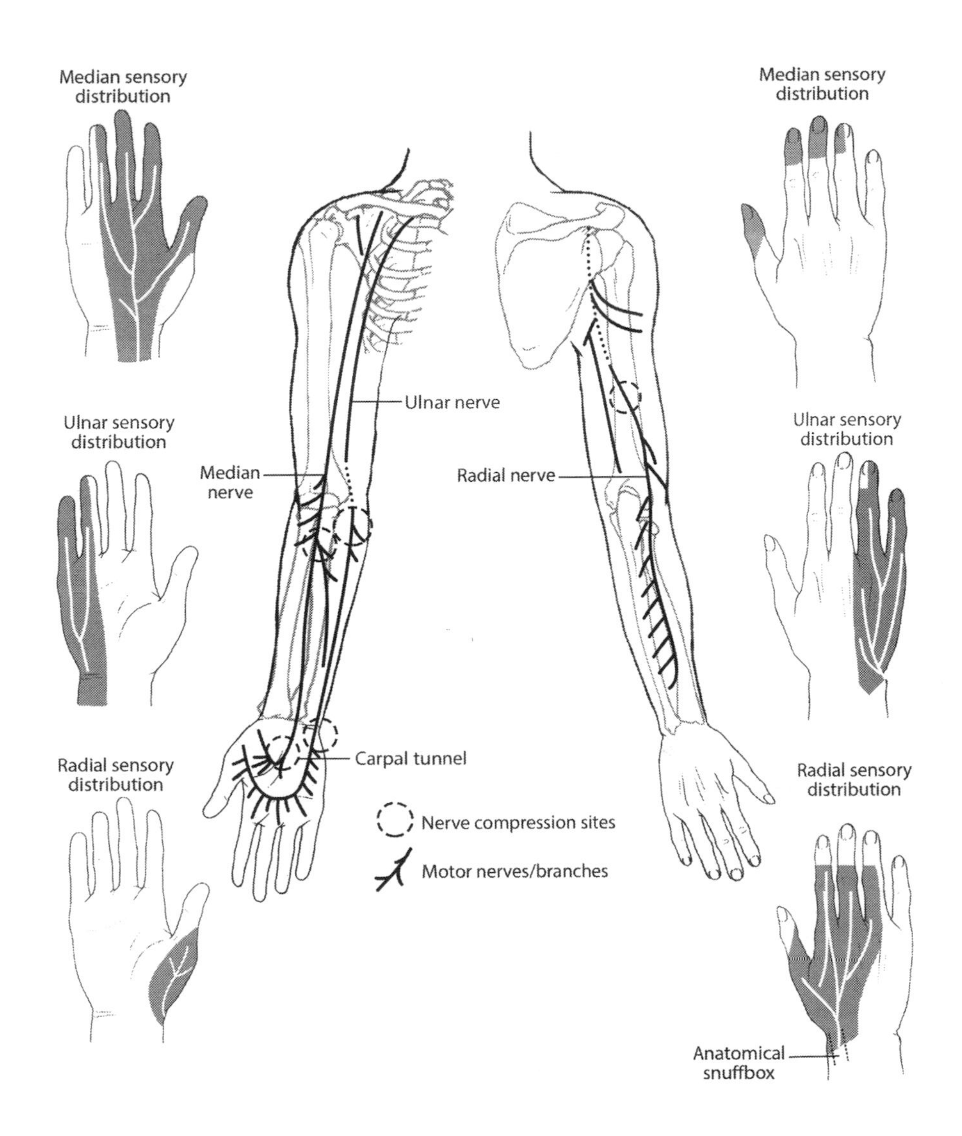


**FIGURE 7.10**
Median, radial, and ulnar nerves: Distributions to skin and muscles; anatomical snuffbox.

stimuli to the extrinsic and intrinsic hand muscles, initiating and terminating muscle contractions. Figure 7.10 schematically illustrates these motor nerves and the muscle locations they innervate. Multiple branches of all three nerves reach extrinsic hand muscles in the forearm. The muscles innervated by the median nerve primarily flex the wrist and fingers. The muscles innervated by the radial nerve primarily extend the wrist and fingers. The ulnar nerve sends branches to two extrinsic hand muscles—both are flexors.

### Median Nerve

The median nerve innervates palmar skin; over both sides of the distal phalanx of the thumb, index, and middle fingers; and the adjacent half of the ring finger. It provides motor control to most muscles of the thenar eminence, facilitating thumb opposition. As described earlier, the thumb, index, and middle finger are responsible for most vital human hand functions.

Injuries to the median nerve at any of the compression sites (shown with dotted circles in Figure 7.10) affect motor function in the hand and wrist, and may affect sensory function in the hand. Carpal tunnel syndrome, a common problem, is caused by pressure on the median nerve within the carpal tunnel. It creates altered sensation in the distribution of the median nerve and decreased thenar muscle function—both significantly impair hand use. The carpal tunnel is filled with many structures—multiple tendons from the extrinsic flexor muscles in the forearm, their tendon sheaths, and the median nerve—tucked into this small space (Figure 7.11). Untreated, carpal tunnel syndrome can lead to permanent nerve damage. Overuse, rheumatoid arthritis with associated tendon and tendon sheath swelling, and cysts at the carpal joints can cause carpal tunnel syndrome (Drake et al., 2015, p. 798). Due to inflammation, the median nerve may become tethered within the carpal tunnel, causing another variation of carpal tunnel syndrome. Treatment for carpal tunnel syndrome may include medication, surgery, nerve gliding exercises, and/or activity modification. Braces or splints that limit palmar flexion of the hand and wrist may be used to position the wrist to decrease pressure on the median nerve.

### Ulnar Nerve

Injury to the median nerve is not completely debilitating because the ulnar nerve provides motor control to some of the smaller thenar muscles as well as the many remaining intrinsic hand muscles. The ulnar nerve supplies sensation to the palmar and dorsal hand near the little finger, on to the little finger and the adjacent half of the ring finger. The ulnar nerve is susceptible to compression at the elbow (sometimes referred to as hitting the "crazy bone") and in the hypothenar palm. Press firmly, beyond the wrist creases and distal to the prominent pisiform (the most proximal carpal bone on the little finger side of the hand). You have found the ulnar nerve in that area when you sense a deep, aching pain.

### Radial Nerve

The radial nerve is the third sensory nerve of the hand, but it has no motor fibers in the hand. It emerges from the brachial plexus, wraps around the humerus, and travels along the dorsum of the forearm before it reaches the hand. It passes through the *anatomical snuffbox* on the dorsolateral (thumb) side of the wrist. See the cutaneous nerve distribution of the radial sensory

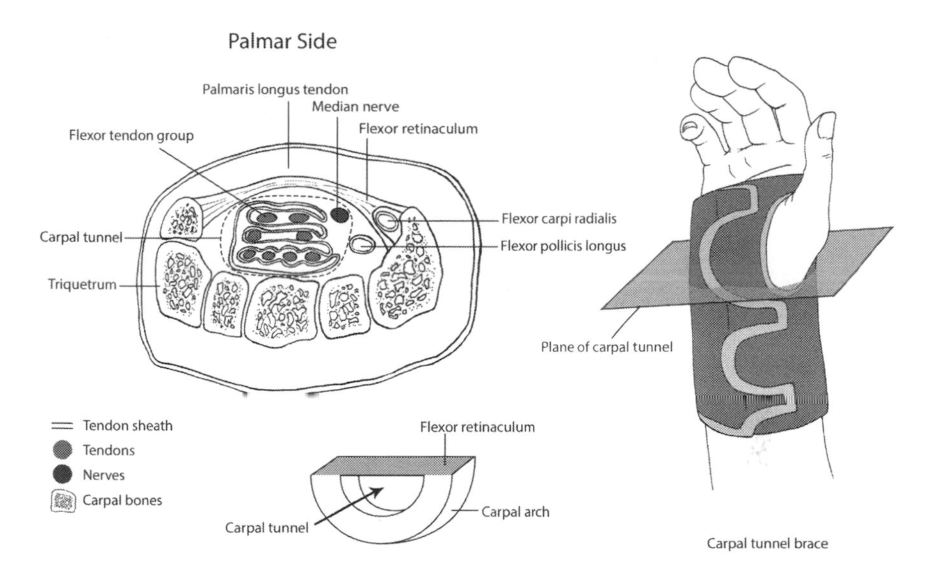


**FIGURE 7.11**
Carpal tunnel anatomy and carpal tunnel brace, carpal tunnel schematic.

nerve and the position of the anatomical snuffbox in Figure 7.10. Calling this structure a snuffbox comes from the historical practice of placing snuff (a tobacco product) in the triangular hollow over the dorsal wrist before raising the hand to the nose to inhale the snuff (Baugher & Graham, 2014, p. 51). The radial sensory nerve can be palpated as it crosses the thumb extensor tendon about two inches above the radial styloid.

## *Product Considerations*

Nerves in the hand and wrist are susceptible to localized damage that can limit function. Avoid tight product features at the vulnerable points of these nerves. Wrist restraints, hand cuffs, or any tight wearable around the distal forearm can compress the radial sensory nerve and cause sensory change in the thumb side of the dorsum of the hand, parts of the index and long fingers, and the radial side of the thumb. If the radial nerve is damaged in the upper arm, as it wraps around the humerus, or in the forearm, the person may exhibit "wrist drop," an inability to extend the wrist—often accompanied with loss of radial sensory nerve sensation. You can simulate the limitations of the injury by supporting the volar forearm with your opposite hand and letting your hand drop downward. Then with your "dropped hand" try to pick up a pencil or pen and try to write. A wrist/forearm brace to hold the hand in a position of function, wrist somewhat extended and ulnar deviated, increases grip strength and helps with fine motor hand activities.

Injuries to the ulnar nerve at either of the compression sites (shown with dotted circles in Figure 7.10) will affect motor and sensory function in the hand. Shea and McClain (1969) describe the nature of possible ulnar nerve injuries, as the nerve passes through and over the bones and tissues of the carpal region. Cyclists' gloves have extra padding over the carpal bones, especially over the pisiform and hamate bones, at the little finger side of the wrist, for protection of the ulnar nerve.

Damage to nerve pathways anywhere between the brain and the wrist and hand, whether from mechanical injury related to trauma or surgery, medical conditions, radiation, medications, chemotherapy, nutritional deficiencies, or unknown causes will affect how well messages are relayed to and from the hands. Can designers create new wearable products that compensate for lost functions, support weak hands and fingers, or relieve pain to compensate for nerve damage from these varied causes?

Handwear, as a protective barrier between the hand and the environment, can also become a nerve message disruptor to varying degrees. The earliest surgical gloves were not widely accepted by surgeons, perhaps due to the rather thick and bulky "India" rubber used as the major glove material (Randers-Pehrson, 1960). Today, surgeons' gloves are quite functional, but may have reached limits of thinness and flexibility. Medical professionals sometime double-glove (wear two pairs of surgical gloves) to decrease risk of infection from an accidental tear, cut, or puncture. Fry, Harris, Kohnke, and Twomey (2010) studied the effect of double-gloving on performance of 53 surgeons during surgical procedures. They found that double-gloving did not have substantial impact on manual dexterity or tactile sensitivity compared with no gloves or single-gloving. This study suggests that, in this application, designers have successfully met a primary goal of protective handwear design: to provide maximum protection while preserving sensory and motor function. Some handwear products, like firefighters' and astronauts' gloves, which are by necessity thick and cumbersome, restrict sensory feedback from the environment. This remains a design challenge.

### 7.4.2 Arteries and Veins of the Wrist and Hand; Product Considerations

The circulatory system of the wrist, hand and fingers exhibits the usual blood flow circular pattern to the heart and back, but also displays several important adaptive features (Figure 7.12). Arterial structures in the hand have specialized linkages with protective effects for the hand tissues and functions. If some of the veins near the wrist are blocked, branching within the superficial venous drainage pathways leaving the hand and wrist allows a degree of compensation. These pathways are visible in the dorsal hand and on both sides of the wrist.

#### *Arteries of the Wrist and Hand*

The ulnar and radial arteries emerge from the forearm to feed blood to the wrist and hand, supplying nutrients and warmth. Notice the complex

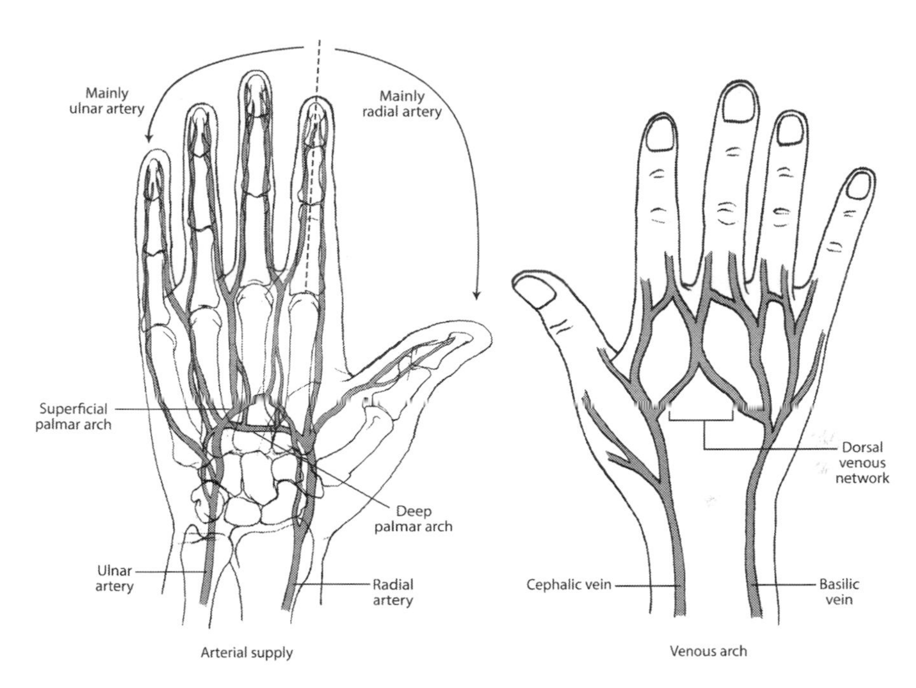


**FIGURE 7.12**
Circulatory system in the hand and wrist.

network of arterial vessels in the wrist and hand (Figure 7.12). When the radial and ulnar arteries reach the hands, they converge into loops across the palm, the deep palmar and the superficial palmar arches. Pairs of arteries to each digit of the hand arise from the palmar arches with further interconnections and branching, all the way to the fingertips. These configurations help insure the tissues of the hand and fingers will be perfused with blood even if one of the major feeding arteries is blocked or injured. Because the ratio of surface area to volume in the fingers is so high, the fingers would rapidly cool without this extensive arterial blood supply. If the fingers cool too much, the sensory nerves work improperly and overall hand function decreases. However, it is also important to remember that, because of the volume of blood traveling to the hand, blood loss from cuts in the arm/forearm can be significant. It is better to save a person's life by using a *tourniquet* to tightly encircle and constrict the arm, even if it interrupts blood flow to the hand and may cause nerve damage. Qualified medical treatment should be sought as soon as possible.

Hand and wrist arteries can be monitored to determine health status. Medical professionals routinely determine heart rate and pulse quality at an accessible wrist pulse point over the radial artery proximal to the thumb (review Chapter 2, Section 2.7.2 on locating pulse points at the wrist). A second pulse point can be felt at the ulnar artery that lies over the volar distal ulna. Arteries in the fingertips can be used to measure oxygen in the blood. Pulse oximetry was described in Chapter 2, Section 2.7.2 as a non-invasive method for monitoring oxygen levels in the blood. A finger "pulse-ox" monitoring device looks much like a kitchen "chip clip" with two opposing parts. It fastens over the dorsal and volar fingertip. Oxygen content of pulsing arterial blood is determined by measuring light absorption. The fingertip is easily accessed and small enough to allow the light from one side of the clip to penetrate through the tissues and be picked up by the other side of the clip (Severinghaus & Honda, 1987).

### Veins of the Wrist and Hand

Veins of the hand lie close to the surface (superficial) and deep within the tissues. The bulk of the oxygen-depleted blood leaves the hands through the venous system visible on the dorsum of the hand (Figure 7.12). Look at the back of your hand to see the superficial, blue-shaded veins near the skin surface. These veins drain into the major superficial forearm veins, the *cephalic* and *basilic* on the dorsomedial and dorsolateral forearm. Smaller superficial veins can be seen on the volar forearm at the wrist. Tight garments or wearable devices that encircle the forearm/arm may compress the relatively thin-walled veins. A swollen, dusky red, or blue-tinged hand warns that venous compression may be occurring.

### Hand Circulatory System: Thermal Protection

Understanding the networks of the hand's circulatory system is necessary to design products to warm or cool the body and to monitor some body functions and efficiencies. In cold conditions the *arterioles*, the final distributing arterial vessels of the hands, can reduce blood flow to the hand surface and the fingers. This automatic response helps maintain warm blood in the body core where it is needed to protect vital organs. The result can be cold hands or damaged tissues in extended periods of extreme cold. In these conditions, hands are susceptible to both *non-freezing cold injury* (peripheral cold injury without tissue freezing) and very serious, but less common, *frostbite*, with ice crystals forming in the tissues and obstructing blood vessels (Whitaker, 2016). Frostbitten distal fingers may require amputation. Hands and feet are both susceptible to frostbite, as the body uses the same tactics to preserve core heat by moving warm blood away from the farthest structures (Read Chapter 8, Section 8.5.2 on frostbite and feet).

Although cold injuries are similar in the hands and feet, design strategies for handwear and footwear differ as hand and foot functions differ. Handwear for thermal protection requires good to excellent dexterity depending on the wear situation. The relationship of the body part to the cold environment also differs. Feet lose heat through conduction when in contact with the cold ground, while hands are not in direct contact with the ground for long periods of time. The typical strategy of preventing conduction heat loss for feet is to incorporate thick, insulating barrier materials between the bottom of the foot and the ground.

Several factors specific to the hands complicate thermal balance. The extensive surface area of the fingers promotes heat loss through radiation and convection. A classic book on thermal comfort (Fourt & Hollies, 1970), discussing the difficulty of maintaining heat in the hands, notes that the small diameter and the extensively curved surface areas of the fingers limit insulation methods. The small cylinder of each finger—with two interconnected, but tiny, arteries—limits the volume of warm blood reaching the finger.

Fourt and Hollies (1970) point out that when additional insulating product materials or electrical heat sources are available in limited supply, "its application over the main body is more effective than only over the hands and feet" (p. 61). Opposite to the shift of blood toward the core in the cold, blood flow to the extremities increases when the body core is heated, pushing heat to the extremities. Working from this theory—that maintaining or increasing body core heat helps maintain extremity warmth—Koscheyev, Leon, Paul, Tranchida, and Linder (2000) heated four different body areas and measured the warmth and comfort of their subjects' fingers. After comparing protocols warming the (a) head, (b) upper torso/arm, (c) upper torso/arm/head, and (d) legs/feet, they found heating the upper torso/arm/head most effective. They recommended zonal heating for astronauts' extra-vehicular activity (EVA) suits to enhance finger flexibility and dexterity and improve work performance in outer space.

Most thermal glove designs use insulating materials that are bulky, so strategies to reduce material bulk are important. Handwear shape and form also affect thermal efficiency. Compare the warmth of gloves and mittens. Gloves allow independent finger motion and manipulation of objects, but warmth can dissipate from each finger due to heat transfer to the colder environment. Mittens enclose digits II through V in one space, so that warmth from blood flowing to each finger is shared. Mitten patterns encase the thumb separately to allow thumb/finger opposition, so the thumb is isolated from heat of the other digits.

Raynaud's is a rare disorder that affects the arteries, usually of the fingers. For brief periods of time the arteries in the fingers narrow in *vasospasm*. The fingertips turn white or blue, and pain and tingling can be felt in the fingers as the arteries return to normal size. There is no apparent cause for primary Raynaud's, while secondary Raynaud's may be caused by many factors including disease that affects arteries, repetitive actions that damage nerves, or injury (National Heart, Lung, and Blood Institute [NHLBI, n.d.]). Thermal handwear may be useful in preventing episodes of vasospasm by keeping hands warm in cold environments. The NHLBI advises Raynaud's patients to avoid tight garments or accessories around the wrist, hand, and fingers.

### 7.4.3 Lymphatics of the Hand and Wrist

Lymph fluid from the fingers drains through the arm and then through lymph nodes in the axilla (Chapter 2, Section 2.7.3). Breast cancer treatments that disrupt lymph drainage lead to lymphedema of the arm (Chapter 4, Section 4.12.4). Lymphedema-swollen fingers are uncomfortable and make everyday tasks difficult as flexion is restricted. Compression gauntlets, covering the metacarpals and wrist, or gloves that also encompass the fingers are used to try to control swelling. Wrapping the fingers and hand with compression bandages is another technique to manage edema (Burt & White, 1999, pp. 75–76). However, bulky products around the fingers lessen dexterity and obscure sensation, limiting daily activities. New design approaches are needed to address these problems (LaBat, Ryan, & Sanden-Will, 2016).

## 7.5 Skin and Fingernails of the Hand: Structure, Function, and Product Considerations

Compare skin thickness on the palm and the dorsal surfaces of your hands. Note where you can see structures under the skin and how the skin moves as you move your hands and fingers. Look at your fingernails and think of the ways you use your fingernails including scratching your skin to relieve an itch, picking up a small object, or scraping a dirt speck from a surface. Although fingernails and the skin covering the hand are resilient, they are vulnerable to abuse and environmental excesses. Because of the many ways we use our hands every day, exposing them to moisture, heat, pressure, abrasion, chemicals, and other threats to the integrity of the skin and nails, handwear may be needed for protection.

### 7.5.1 Palmar and Dorsal Skin

Palmar skin is *glabrous* (hairless) with a much thicker epidermis than the hairy skin of the dorsum (Figure 7.13). Interdigitating structures in palmar skin prevent tearing between the dermis and the epidermis (Yu et al., 2004, p. 72), a distinct advantage for this hard-working surface of the hand. Friction stimulates the palmar stratum corneum to form *calluses*, areas of thickened skin, to protect against mechanical damage (Yu et al., 2004, p. 72). Recall from Section 7.2.1 that the palmar skin attaches to connective tissues of the palm—densely interlaced networks such as the palmar aponeurosis and fascial coverings of the thenar and hypothenar muscles.

Work gloves are worn to protect the skin from abrasion, fire and heat, caustic and biological materials, extended periods of time in water, and more. Some sports products protect the hand and fingers from abrasion. Johnson (2015) explains that archers wear arm guards and finger tabs to protect their

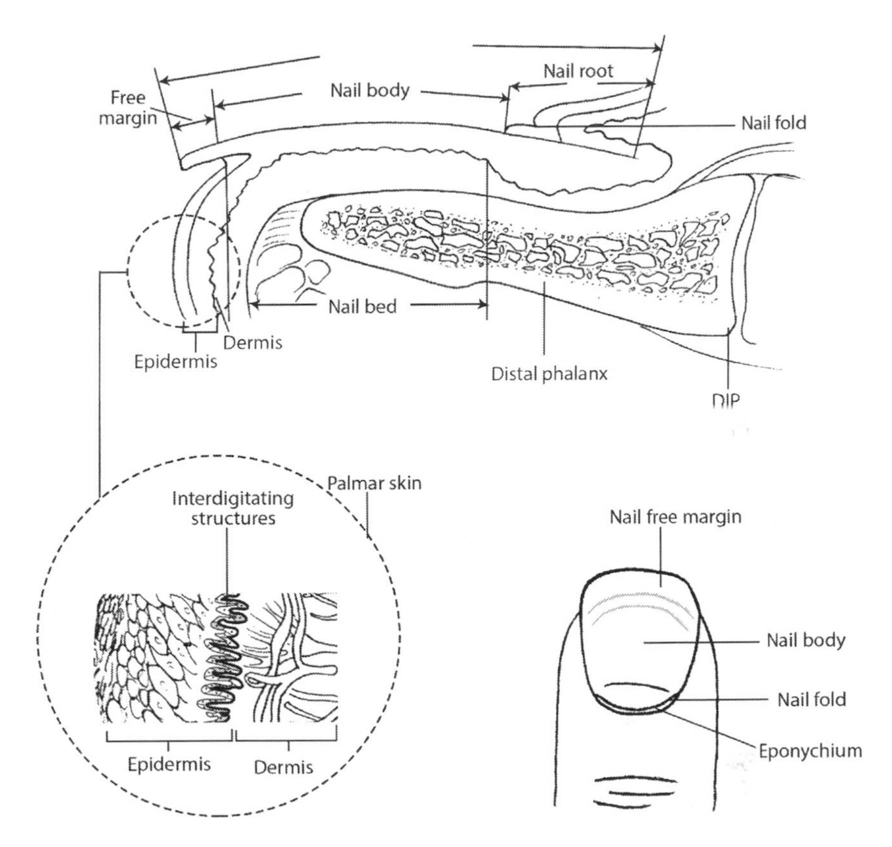


**FIGURE 7.13**
Hand skin, fingernail and nail bed.

skin from abrasion and impact. An arm guard is worn on the arm that supports the bow, keeping the bowstring from striking the forearm as it releases the arrow. A tab is worn on the hand that pulls back the string and positions the arrow, to prevent painful pressure from the string and blistering of the fingers as the arrow shaft slides by (Figure 7.14). The tab may also aid in correctly positioning the fingers to steady the arrow.

Palmar skin has no hair follicles or sebaceous glands but has many eccrine sweat glands. Skin moisture, if not excessive, can help in gripping an item. In addition, the flexibility and conformability of skin makes firm gripping possible. Handwear can interfere with this natural gripping feature. Protective gloves may impair manual performance, such as grip strength; and gloves with surfaces that do not adhere to an object may lead to discomfort, pain, and musculoskeletal disorders (ASTM, 2015c). Adhesion and grip strength are important for work and sports gloves, and tasks must be successfully performed while wearing protective gloves. ASTM (2015c) offers a test to measure gripping performance of gloves. Some athletes wear gloves that must allow secure gripping of sports equipment. For example, a quarterback

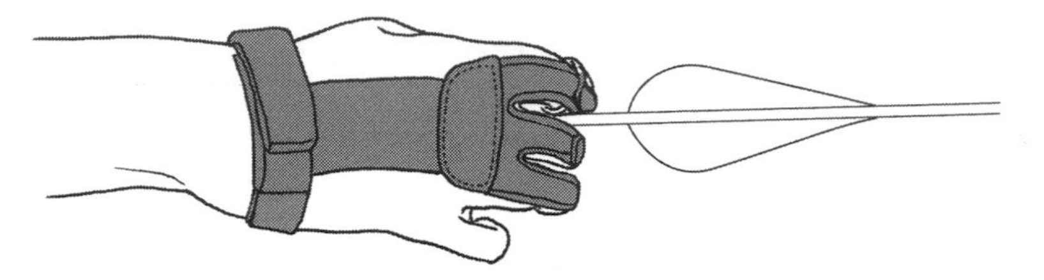

**FIGURE 7.14**
Archery finger tab.

playing in cold weather needs a glove that keeps the hands warm while maintaining dexterity to grasp the football. Soccer goalie gloves protect from abrasion and impact while providing a surface to grip the ball.

Touchscreens on phones and other devices rely on contact with a fingertip. High end touchscreens use *capacitive technology* with anything, like skin, that holds an electric charge activating the screen. Gloves or mittens block the circuit of the skin's electric charge with the screen. Some mitten designs feature a removable finger section to bare the fingertips so a touchscreen can be used. Another method incorporates conductive threads into glove fingertips so that touch with the glove can complete the circuit and activate the screen.

Some touchscreen technologies rely on fingerprint recognition. Because fingerprints are distinct for each person, access should be assured for an individual once the fingerprint pattern is established on the device. Cuts or scrapes on the fingertip can distort the fingerprint and may, therefore, limit the usefulness of this identification method. Fingerprints are formed by epidermal ridges that form distinct patterns, but there is no accepted explanation of how the ridges are formed (Kücken & Newell, 2012). The researchers propose that the patterns are created as a result of a buckling instability in the basal cell layer of the fetal epidermis. Handwear covers the ridges that form a person's fingerprint. Arora, Cao, Jain, and Paulter (2016) are developing technology to 3D print fingerprint patches that could be adhered to a glove fingertip. This technology would be an efficient method of incorporating device activation into a glove; however, it also presents identity theft and security issues.

*Dorsal skin* is thin and less wear resistant than palmar skin. It is supple and elastic with subcutaneous layers of loose connective tissue and minimal adipose tissue between the skin and deep structures. This framework accommodates marked changes in surface area as the hand and fingers flex. Observe your dorsal hand skin as you make a fist. Note how your skin moves as your hand shape changes. Place your hand flat on a table and notice the wrinkles over your PIP joints. Then curl your fingers into a fist and watch the wrinkles stretch over the joints. The slack skin allows flexion motion. As the knuckle skin smooths out with flexion, the PIP joint circumference decreases slightly, a fact that helps a ring slide over the joint. The skin from the wrist to the nail base lengthens about 3 cm (1.2 in.) with the hand fully flexed (Yu et al., 2004, p. 77).

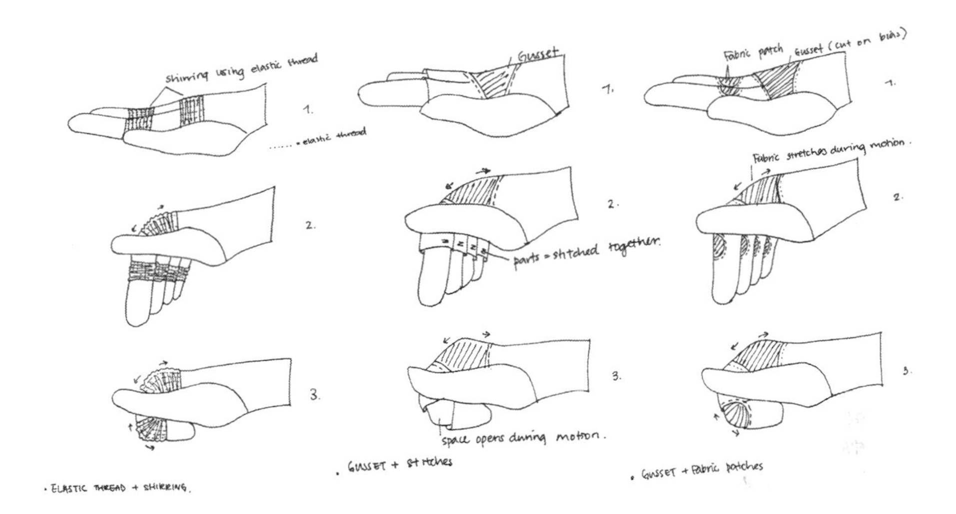


**FIGURE 7.15**
Ideation sketches for motion features in handwear.

Skin-tight gloves need to incorporate materials that have similar glide and stretch factors to allow motion. Leather has been the material of choice for tight-fitting fashion gloves because it is a natural skin product and emulates human skin motion to some extent. Handwear structuring methods can be used to accommodate hand size and shape changes. Stall-Meadows (2004) reports, "ancient gloves were manufactured with numerous pieces, as many as 150 in some instances, compared with today's gloves of fewer than ten pieces" (p. 288). Glove piecing and seaming permutations have not been thoroughly explored. Further exploration could lead to new handwear design and manufacturing methods. Figure 7.15 shows a brainstorming exercise with just a few examples of glove structure options for hand motion.

### 7.5.2 Skin Sensitivity

Protective gloves are necessary wearable products for people in many occupations. Boman, Estlander, Wahlberg, and Maibach (2005) assembled a series of articles on types of protective gloves, manufacturing methods, and glove standards from several countries. When gloves are worn for extended periods of time, with heat and perspiration build-up, skin sensitivities become a concern. The hands' expansive skin surface is susceptible to contact dermatitis from handwear (see Chapter 2, Section 2.8.5). In-depth research and innovative design strategies could lead to improvements in materials and product structures that expedite (or at least don't impede) hand function while maintaining skin health.

Latex allergies, a health hazard for both medical professionals and patients, proliferated in the 1980s and 1990s related largely to use of medical gloves of natural rubber latex (NRL), a product derived from rubber tree sap. The latex contains a protein that can cause reactions ranging from contact dermatitis to *asthma* symptoms to life-threatening *anaphylaxis*. Chowdhury and Maibach (2004) provide extensive information on the basic science and clinical management of latex intolerance. Because latex allergies can be so severe, other glove materials are now available for medical gloves. However, after Palosuo, Antoniadou, Gottrup, and Phillips (2011) reviewed the modifications made in the manufacture and use of NRL gloves in response to NRL allergies; they suggested latex gloves be reconsidered for medical use because of many highly desirable qualities. An exception is known cases of latex allergy or sensitization. Latex gloves remain widely available for non-medical use and may be a risk for allergic individuals, as "Manufacturers are not required to label home and community products which contain natural rubber" (Spina Bifida Association, 2015). Research continues to identify materials that compare more favorably with latex, to further improve medical gloves and other products made from latex substitutes.

### 7.5.3 Fingernails: Anatomy and Product Considerations

Fingernails serve functional purposes much like a small tool, or can be decorated as a fashion accessory—natural nails can be painted with fingernail polish, and fake nails can be glued to the natural nail surface.

#### *Fingernail Anatomy*

Fingernails, like hair and toenails, are comprised mostly of the protein keratin. The keratinized cells overlap "like shingles, forging a clear, shell-like casing" (Tsiaras, 2004, p. 23). Review Chapter 2, Section 2.8.1 for basics on fingernail structure and function and refer to Figure 7.13. The fingernail covers and protects the dorsal surface of each distal phalanx. Starting near the DIP joint, the living nail plate, fed by a rich capillary bed, grows continuously into the translucent fingernail. Slightly convex from side to side, the *nail body* extends distally from the nail fold to the free nail margin. The *hyponychium*, thickened skin under the free margin of the fingernail, occasionally develops painful cracks in dry and/or cold environments despite the interdigitations between the epidermis and the dermis. This often occurs at the juncture between the lateral nail fold and the hyponychium. The *cuticle*, a thin layer of epidermis (the *eponychium*), borders the proximal and lateral margins of the nail.

#### *Fingernails and Product Considerations*

Bandaging to heal cracks of the convex curvature of the fingertip adjacent to the nail margin and the hyponychium is difficult to maintain because of the hemispherical configuration of the fingertip, DIP joint motion, and the need for hand washing. A design solution that conforms to the fingertip, and withstands motion and water, would be welcome. If the cuticle tears or irregularly detaches from the fingernail, a *hangnail* develops. Bandaging for hangnails is only slightly less problematic than for cracked fingertip skin—a bandage encircling the cylinder of the finger can adhere to itself but is still subject to DIP joint motion and hand washing and drying.

Handwear can obscure fingernail function, for example preventing precision grip of small items. Some work gloves incorporate materials that assist in gripping small tools, attempting to substitute for the tool-like fingernails. Accommodating variations in fingernail length is almost impossible in mass-produced handwear. Glove finger lengths typically relate to the apex of the fleshy fingertip tissue, rather than to the end of the fingernail. However, if a glove doesn't accommodate a person's fingernail length, the glove may be uncomfortable and painful. A relatively unique problem, delaminated or detached fingernails, reported by astronauts wearing protective space gloves is particularly challenging. The enclosed area of the glove with no mechanism to dissipate heat creates a moist atmosphere which, when combined with abrasion of the fingernails against the inside of the glove, causes this painful condition (Opperman, Waldie, Natapoff, Newman, & Jones, 2010).

## 7.6 Hand Variations: Morphology and Deformities

Hand variations are common, but not often incorporated into handwear designs. Understanding and incorporating gender differences into work glove designs can make a difference in how well a person can perform a task. Hand deformities can greatly affect handwear use when a product is uncomfortable or does not fit at all. Hand variation due to aging is neither well understood nor often studied.

### 7.6.1 Gender and Age Differences

Numerous studies state that men, on average, have greater hand length and breadth than women (Gordon et al., 1988; Greiner, 1991; Pheasant & Haslegrave, 2006; Poston, 2000). However, hand dimension ranges for men and women vary widely. Understanding the gender demographics for a handwear product is important. Park, Park, Lin, and Boorady (2014) interviewed 54 firefighters and found dissatisfaction with firefighters' gloves, especially for firefighters with small hands, and for females. Hsiao, Whitestone, Kau, and Hildreth (2015) conducted an anthropometric study of 863 male and 88 female firefighters' hand dimensions and stress the importance of producing products with "at least one or two glove sizes based on female hand models" (p. 1370).

Hsiao et al. (2015) also point out the lack of information on age and hand size in the literature. They found some differences between firefighters younger than, and older than, 40 years of age. They identified significant differences in total hand length and breadth, palm length and breadth, thumb length and breadth, and finger breadths—but not finger lengths. They do not suggest how to use this information in product design. With aging populations around the world, further study of hand dimensions for older people is warranted, especially for people older than 40 years. New approaches to assistive devices for hand activities and handwear for older populations could promote a better quality of life for people as they age.

### 7.6.2 Hand and Finger Deformities

Common hand and/or finger deformities require adapted design strategies. Muscle imbalances between the finger flexors and extensors and IP or MP joint instability related to injury or to inflammation from arthritis all can cause crooked fingers (Figure 7.16). Finger deformities include: *mallet finger, swan-neck deformity,* and *boutonniere (buttonhole) deformity* (Yu et al., 2004, p. 352–354). Mallet finger, sometimes called *baseball finger,* is often caused by a sudden flexion force to the DIP joint of an extended finger. With the extensor muscle connection to the distal phalanx disrupted, the extended finger looks like a hammer or mallet. Hyperextension of a loose PIP joint with flexion at the DIP joint produces a swan-neck deformity,

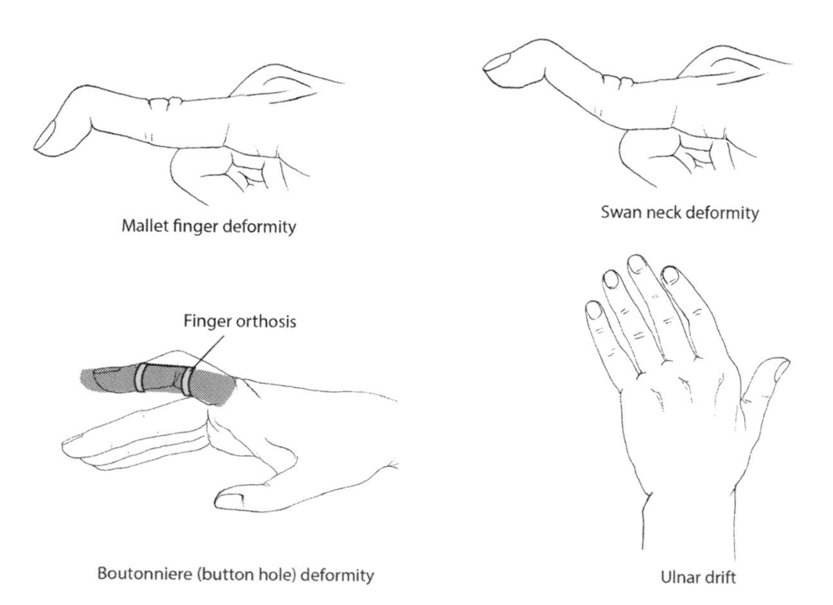


**FIGURE 7.16**
Deformities of the fingers: Mallet finger, swan-neck finger, boutonniere (buttonhole) finger, ulnar drift of the fingers; finger orthosis.

commonly seen with rheumatoid arthritis or PIP trauma. The bent finger then looks like the curve of a swan's neck. With time and no intervention, a mallet finger can progress to a swan-neck deformity. A boutonniere (buttonhole) deformity may also result from an injury or rheumatoid arthritis. The PIP joint flexes and the DIP joint is in extension. The name comes from the proximal phalanx popping through a misplaced tendon giving the appearance of a finger popping through a buttonhole. A small rigid finger splint with a dorsal connector over the PIP joint and ring segments encircling the phalanges on either side of the deformed joint can be used to stabilize or try to straighten a finger with a boutonniere deformity. Turning the splint so the connector is on the palmar side of the finger corrects a swan neck joint deformity (Bender, 1990a, p. 583). Some splints are designed to look like jewelry.

One deformity, *ulnar drift,* a shift of the fingers toward the ulnar side of the hand, occurs at the MP joints and affects digits II-V. Most often rheumatoid arthritis causes the deformity—resulting in a hand and finger shape that will not fit into standard handwear designs. Ready-to-wear handwear made of stretchy materials may conform enough to the altered bends and curves of finger and hand deformities to be serviceable, but may prove difficult to pull on. Comfortable and functional handwear for someone with a hand or finger deformity, with or without splints, may require custom fit and generous sizing.

## 7.7 The Hand in Motion: Range of Motion, Biomechanics, and Handwear Design

Wrist, hand, and finger motions, crucial to human capabilities, can be divided into motions to (a) position the hand in space and (b) facilitate manipulation of objects. Many structures contribute to hand function.

### 7.7.1 Wrist, Hand, and Finger Range of Motion

Motion at the radiocarpal joint, between the radius and the carpal bones, positions the hand in space relative to the forearm. It is influenced by the elbow motions of pronation and supination. Motions at all the remaining wrist, hand, and finger joints help provide dexterity. Motions at the five carpal-metacarpal joints—in the body of the hand and the base of the thumb—help to position the thumb and fingers relative to the body of the hand. The motions create the distal transverse palmar arch and contribute to the longitudinal arches of the hand and motions of the fingers (Levangie & Norkin, 2005, p. 321). Motions at the metacarpal-phalangeal joints further position the fingers and thumb relative to the body of the hand. Motions at the interphalangeal joints, between the proximal, middle, and distal phalanges of the thumb and fingers straighten and curl the digits.

### *Radiocarpal Joint Motion*

Radiocarpal (wrist) joint motions position the pronated or supinated hand relative to the forearm. Recall from Chapter 4, Sections 4.15.2 and 4.15.3, how the radiocarpal joint affects abilities to use the hand for grasping activities in pronation (a palm down position) and for carrying and lifting in supination (the palm up or "soup cupping" position). The four primary motions of the pronated or supinated hand at the radiocarpal joint are flexion, extension, and radial and ulnar deviation (Figure 7.17). Extrinsic hand muscles produce these motions. Recall that the extrinsic muscle bellies arise in the distal arm and in the forearm, but the tendons cross the radiocarpal joint to attach to bones in the hand. The numerous carpal ligaments help to stabilize small movements of the eight tiny carpal bones as the extrinsic hand muscles act.

Extension of the hand on the wrist is more limited than flexion, with approximately 70 degrees of motion, compared to 90 degrees of flexion. Radial and ulnar deviation motions are smaller: radial deviation, toward the thumb side of the hand, is approximately 20 degrees, while the hand moves approximately 30 degrees into ulnar deviation, toward the little finger side of the hand. Some muscles in the flexor forearm muscle group power wrist

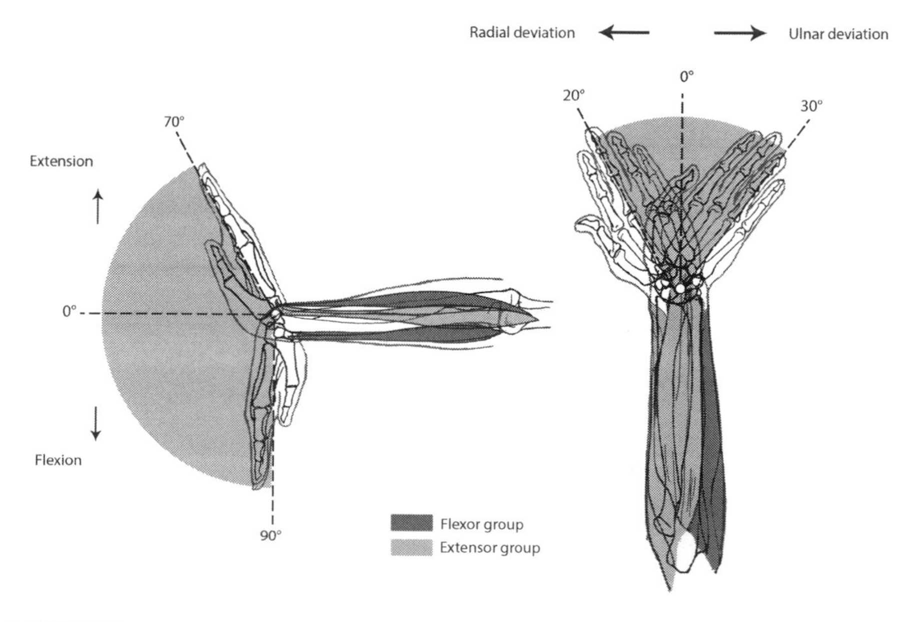


**FIGURE 7.17**
Wrist motions: Flexion, extension, radial and ulnar deviation with muscles and range of motion.

flexion, while the extensor forearm muscle group moves the hand into extension on the forearm. Specific flexor and extensor muscles, along either the radius or the ulna in the forearm, contribute to the respective radial or ulnar deviation motions at the wrist.

### *Carpal-Metacarpal and Metacarpal-Phalangeal Joint Motions*

Intrinsic hand muscles act primarily on the carpal-metacarpal and metacarpal-phalangeal joints of all five digits. The carpal-metacarpal (CMC) joint of the thumb is very mobile and consequently frequently develops osteoarthritis. Motions at this first CMC joint (Hoppenfeld, 1976, pp. 90–91) include: (a) thumb abduction of approximately 70 degrees, in a plane perpendicular to the palm, (b) adduction of the thumb back to its neutral position (Figure 7.18, A), (c) thumb extension of approximately 50 degrees at the carpal-metacarpal joint in a plane roughly parallel to the palm (Figure 7.18, B), and (d) flexion beyond neutral, moving the metacarpal bone approximately 15 degrees toward the ulnar side of the palm (Yu et al., 2004, p. 26). Occasionally the extension/flexion motion is designated as radial thumb abduction and thumb adduction.

You can feel the 1st CMC motion: (1) locate the R first CMC joint by placing your L thumb on the palmar aspect of the base of your R thumb (near the

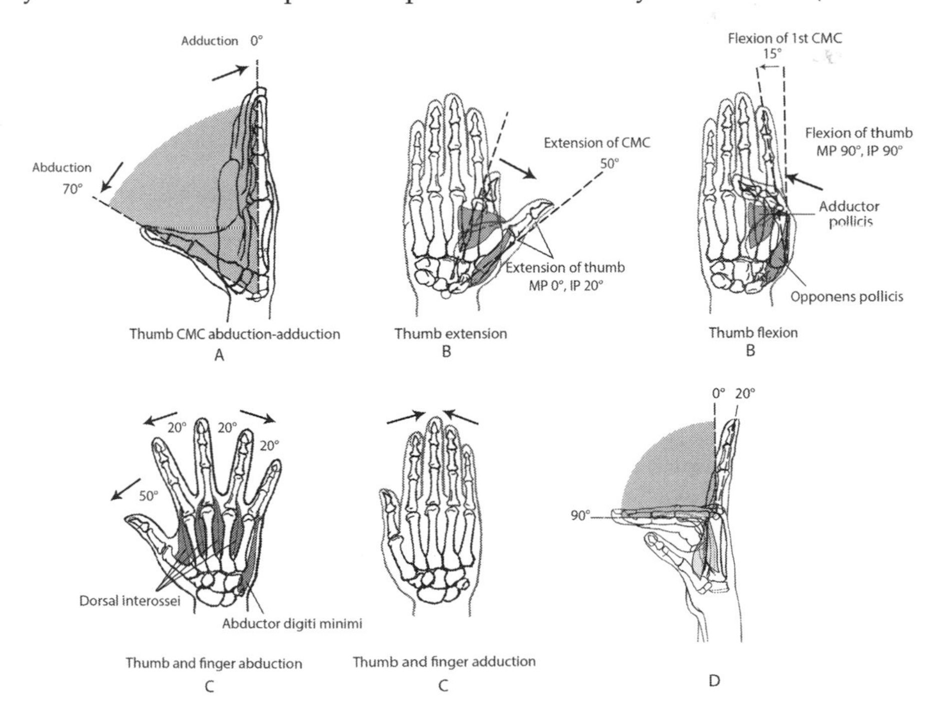


**FIGURE 7.18**
Carpal-metacarpal and metacarpal-phalangeal range of motion.

distal wrist crease), (2) place your longest L finger on the dorsal hand, opposite your thumb, (3) pinch lightly and test out the following motions. Start from a neutral position with the thumb touching the radial side of the palm. Referencing the motions in the figure, move your thumb through abduction and adduction in a plane perpendicular to the palm. Move your thumb through extension and flexion in a plane roughly parallel to the palm.

CMC joints for digits II and III move very little, but those for digits IV and V twist enough to increase the palmar concavity of the transmetacarpal arch of the hand, allowing cupping of the palm and gripping an object like a hammer handle. The CMC joints, with the metacarpal-phalangeal (MP) joints for digits II through V, also facilitate our ability to spread our fingers apart and bring them back together. Finger abduction range of motion is approximately 20 degrees for each finger and adduction returns each finger back into contact with the adjacent fingers (Figure 7.18, C). To appreciate the motions for CMC joints II-V: (1) grasp the heel of your R hand with the thumb and index finger of your L hand just distal to the carpal tunnel, over the base of the metacarpal bones, (2) cup your R hand, noting the motion at the base of metacarpal IV and V, (3) feel the CMC motions in your R palm as you fan your fingers out and close them back together. These are all small motions.

Muscle actions across the MP and IP joints of the thumb flex and extend the tip of the thumb with each of the CMC movements. Move your L thumb and longest finger distally along the R thumb to the large joint (MP joint) between the metacarpal and the proximal phalanx. Bend and straighten your R thumb—observe the MP joint range of motion—90 degrees of flexion with 0 degrees extension (to neutral). Then see that the IP joint, the most distal joint of the thumb, has a range of motion from 90 degrees of flexion to 20 degrees of extension (Figure 7.18, B). The thumb also rotates due to a combination of the CMC motions, to allow opposition of the thumb tip to fingers II-V. When the thumb is in opposition, the thumb metacarpal is abducted in the plane perpendicular to the palm—the MP joint is flexed, and the IP joint may be flexed or extended.

The MP joints of the other fingers flex 90 degrees and extend 20 degrees (Figure 7.18, D). Move your L thumb and index finger to the distal palm region, over the MP joint of any finger. Move the II-V fingers together to observe their range of motion. MP extension of these fingers may be best seen by placing your hand, palm down, on a flat surface and trying to lift your fingertips off the surface. With maximal MP and PIP joint flexion, but DIP joint extension, the fingertips can reach almost to the carpal tunnel.

### *Motions at the Interphalangeal Joints of the Fingers*

Compared to the motions of the thumb and the joints within the body of the hand, the motions of the PIP and DIP joints of digits II to V are simple (Figure 7.19, A). Each of the PIP joints flex approximately 100 degrees and extend to neutral while each DIP joint flexes to 90 degrees and extends

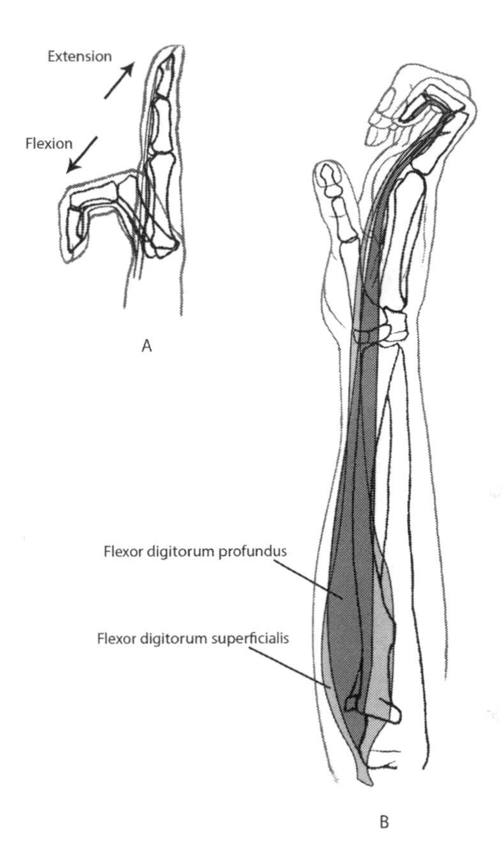


**FIGURE 7.19**
Interphalangeal joint motion.

20 degrees. While PIP motions are quite easy to examine, you may need to brace your middle phalanx against your L thumb and immobilize the PIP in extension with your L index and middle fingers to appreciate the motions at the DIP. In maximum IP joint flexion, the fingertips can touch the distal palm region over the MCP joints. The extrinsic hand muscles flex (and extend) the IP joints (Figure 7.19, B).

### 7.7.2 Hand and Wrist Biomechanics

The extrinsic hand muscles generate the greatest biomechanical forces in the hand. A series of restraints along the tendons, ranging from the tendon sheaths and retinacula at the wrist (previously discussed relative to the carpal tunnel) and small pulley-like structures in the palm and along the volar fingers guide and constrain the actions of the long flexor tendons. The IP joint extensor tendons, visible on the dorsum of the hand, are less restrained and easier to see in the hand than the flexor tendons. Comprehensive biomechanical studies of this important human feature are scarce due to the intricacy of structures and the complicated nature of hand, thumb, and finger function. Research related to development of hand prosthetics provides the greatest insight into hand biomechanics, as engineers attempt to replicate hand function for amputees.

### 7.7.3 Hand Prosthetic Purpose and Design

The hand's functional complexity makes replacing a human hand with a prosthetic a difficult design challenge. Serlin (2002) defines prosthetics as "artificial additions, appendages, or extensions of the human body" (p. 25). Ott, Serlin, and Mihm (2002) give an overview of prosthetic design and development over the years. They describe a cable and shoulder harness device to move prosthetic arm and hand joints. Developed in World War II, it is still used today. The shoulder harness, using scapulo-thoracic motions opposite the amputation side, triggers movement at a prosthetic elbow and hand through a cable mechanism. Early designs used a hook-shaped mechanical hand replacement performing only hold and release functions. *Myoelectric activation*, a more complex method of controlling prosthetic motion, uses electrical signals from the muscles in the arm above the amputation. The wearer learns to contract a specific muscle, then the muscle signal is amplified and relayed to an electronic device that activates the hook or a complex mechanical hand (Bender, 1990b, p. 1020).

Hand prosthetics are becoming more sophisticated in appearance and function. Lighter weight materials have been developed and methods of activating motion are improving. 3D printing has opened new possibilities for quick and inexpensive production of hand prosthetics.

## 7.8 Handwear Fit and Sizing

Hand shapes are relatively consistent, person-to-person, but hand sizes vary widely according to age, underlying bone and joint structure, and intrinsic hand muscle bulk. Handwear fit and sizing also depend on the ways individuals use their hands.

### 7.8.1 Translating the Hand and Wrist to Glove/Mitten Patterns

Handwear uses are numerous and varied, therefore handwear shapes and forms vary from simple mitten shapes to complex, multi-component gloves. Methods of translating the human hand to the product shape and size include drafting and *dipping*. Drafting and draping were described in Chapter 1. Dipping is similar to draping. It relies on the accuracy of a three-dimensional form to represent some part of the body, with the product materials shaped to the form. For gloves, a hand-like form is dipped into a thermoplastic solution that is then cooled and solidified to replicate the surface of the hand (Boman, Estlander, Wahlberg, & Maibach, 2005). Protective seamless gloves are manufactured with the dipping method. Most mitten patterns are drafted using simple length and circumference measurements, and are made in one-size-fits-all to simple multiple sizes—small, medium, and large. Gloves made of many components stitched or fused together use more complex measuring schemes, with drafted patterns.

Mitten and glove patterns are developed and shaped to meet the use and purpose of the design (Figure 7.20). A simple mitten pattern is a traced outline of the hand with the thumb extended. A more complex mitten pattern moves the thumb piece to a position to more easily accommodate the thumb in opposition to the other digits.

Ready-to-wear gloves are usually composed of ten parts or fewer (Genova, 2012). For close fit allowing hand and wrist motion, Stall-Meadows (2004) suggests, "Usually the more pieces in a glove, the closer the fit. However, the addition of stretch fibers, such as spandex and knitted yarns, can create a close fit with minimal glove parts" (p. 291). There are five basic glove components to shape to hand anatomy. A *trank* forms the glove front and back covering the palm and dorsum. *Fourchettes* are long narrow pieces of material that help shape the cylinder of each finger by forming the sides of the glove's finger components. Gussets or *quirks* are triangular or diamond-shaped pieces placed at the base of a finger or thumb piece to allow movement. The thumb piece or pieces shape to the thumb. A cuff may be added to the glove to extend coverage onto the forearm. The dorsum trank and fingers of high-quality leather gloves are slightly longer than the palm-side trank and fingers. The longer edges are eased into the fourchettes so that the glove

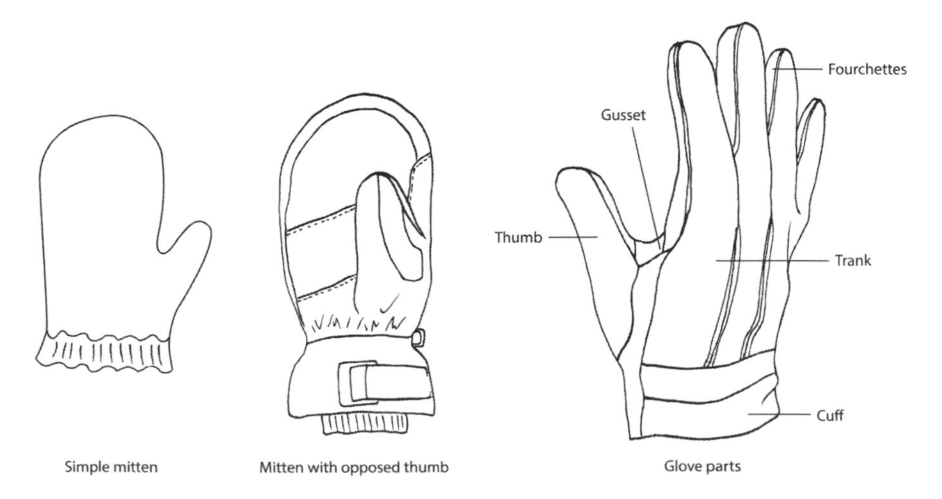


**FIGURE 7.20**
Mitten shapes: Extended thumb and opposed thumb, and glove components.

is shaped to the hand-at-rest position (Redwood, 2016). Glove component key points should match to the hand key points and a measuring system for the hand should correspond to the handwear components. See Appendix F for details on hand and wrist landmarking and measuring instructions and for illustrations of hand key points.

### 7.8.2 Handwear Fit, Sizes, and Size Systems

Rosenblad-Wallin (1987) states that handwear size and fit have a "strong influence both on dexterity and comfort;" and that length measurements are based on skeletal dimensions, while circumference measurements are influenced by muscular volume. Xavier Jouvin, a French medical doctor, is credited with developing the first glove sizing system in 1834, making mechanized production of fashion gloves possible (Wilcox-Levine, 2014). He identified 32 hand sizes and five width profiles. He proposed 320 hand sizes based on the dimensions he collected, combined with what he described as a few other factors (Redwood, 2016). Translating Jouvin's many hand sizes into S.K.U. glove sizes is not feasible for mass-produced handwear, but his findings reinforce that hand sizes and shapes vary greatly.

Modern size systems vary from one-size-fits-all, to variations of extra-small sizes to extra-large sizes, to numbering systems. Most companies size-selection directions specify to measure hand length (from wrist to the longest finger) and the largest circumference around the palm and dorsum. The number used to select the glove size is most often related to the circumference measurement, for example a seven-inch circumference is a size 7 woman's glove. However, there are no standards for hand dimensions that correlate to glove numbering systems used for ready-to-wear gloves.

Fashion gloves have traditionally been offered in five lengths from wrist-length to a glove length that reaches approximately half-way between the elbow and the shoulder point. The glove lengths are sometimes sized by "button length" with one button equaling about 1–1/2 in. For example, a two-button glove is wrist length, and a six-button glove is half-way to the elbow (Genova, 2012). This convention appears to arise from tradition with no studies correlating human dimensions with button size and spacing on the glove.

Hand anthropometric databases are available from sources focused on handwear design for special populations, for example military and protective services like firefighters and police including the technical report, *Hand Anthropometry of U.S. Army Personnel* by Greiner (1991). Hsiao et al. (2015) provide an overview of approaches to firefighter hand anthropometry and structural glove sizing. General population data for ready-to-wear and fashion handwear are not available because manufacturers tend to develop proprietary anthropometric data for their target markets. Griffin et al. (2018) used scanning technology to collect data on dynamic hand positions of 15 men and 15 women. They concluded that a large-scale anthropometric study is needed to improve designs of functional gloves.

Kwon, Jung, You, and Kim (2009) developed a glove sizing system after reviewing similar work of other researchers (Robinette & Annis, 1986; Rosenblad-Wallin, 1987; Roebuck, 1995). Kwon et al. (2009) followed a systematic sequential process. They: (1) selected size interval lengths, (2) established size categories, (3) analyzed size distribution, and (4) determined a size system. They applied the method to data collected from 1774 male and 2208 female soldiers (Gordon et al., 1988). Their process has potential for application to other target markets. Some of the major findings and suggestions include: (a) hand length and hand circumference are the key dimensions to use for glove sizing, (b) 1.3 cm (0.51 in.) is an appropriate size interval length, (c) glove sizing for men and women should be different—their system identified eight male sizes and six female sizes with only one overlap in size dimensions, and (d) size designations should be based on anthropometric data and not glove dimensions. Developments in 3D hand scanning will add to the understanding of hand anthropometry, both static and dynamic, and facilitate efficient collection of hand data.

## 7.9 Hand and Wrist Landmarking and Measuring

Hand and wrist landmarking and measuring systems may be simple or complex (Appendix F). Wristbands for fashion watches and health monitoring wristbands, like the Fitbit™, are typically adjustable. One circumference measurement with an estimate of the range of circumference measurements for the target market is adequate to determine band length. Mittens use a bi-dimensional measuring system: (a) one length measurement from a location at the level of the ulnar stylion and radial stylion to the tip of the longest finger, and (b) a measurement for palm circumference typically at the largest circumference around the palm and dorsum—which may vary in location person-to-person. Some systems add hand breadth to indicate how circumference is distributed to the palm and dorsum. Gloves require a more detailed measuring scheme to properly fit the palm of the hand, the thumb, and each finger. If the glove is articulated, requiring movement features at joints, then measurements for each digit segment are also necessary.

### 7.9.1 Product Purpose

Choice of landmarks and measurements depend on the product purpose. As with other wearable products, hand and wrist product purposes fall into broad categories: aesthetic or functional, with some overlap. Most of the handwear described in this chapter are mittens or gloves. There are no standard classification methods for these products, but a variety of categories have been proposed. Boman et al. (2005) describe "special gloves" related to occupations: firefighters, butchers, divers, electricians, and many more. They also place gloves worn with total encapsulating chemical protective suits in this category. Mellström and Boman (2005) classify protective gloves by use and thickness: disposable gloves (0.0007 to 0.25 mm), household gloves (0.20 to 0.40 mm), industrial gloves (0.36 to 0.85 mm), and special gloves (no thickness specified).

A carefully developed landmarking and measuring scheme helps to produce products to meet work requirements, protect, and prevent injury to workers' hands. Fashion product landmarking and measuring systems are often based on tradition or proprietary data developed by a company.

### 7.9.2 Product Materials

Product materials influence landmarking and measuring systems and production methods. Gloves can be stitched, knitted, or made of non-stretchy or stretchy fused components or a combination of the two. Knitted gloves can be fashioned in the shape of the hand. Sewn gloves of non-stretchy, stable materials are structured with numerous pieces to allow close fit. Incorporating stretchy materials, woven textiles with elastomeric fibers, or knit materials can reduce the numbers of pieces and the numbers of landmarks and measurements needed.

Traditional fashion gloves made of leather can conform and shape to the hand, depending on the quality and pliability of the leather. Many leather types are used in gloves, from very lightweight lambskin to heavy-duty cowhide used for work gloves (Redwood, 2016). The measuring scheme can incorporate the "stretch" of the leather as a type of negative ease (see Chapter 1 on ease in products).

Many work and special use gloves are manufactured with the dipping technique described in Section 7.8.1. The forms used in dipped gloves are the basis for the glove sizing system. The pliability of the thermoplastic materials also affects fit of gloves. Surgical and medical gloves of natural latex and gloves of alternate low allergenic materials fit closely and are very pliable. Household gloves are less form-fitting and usually made of natural rubber, polyvinyl chloride (PVC), or a plastic impregnated textile. Industrial gloves are heavier and made of natural or synthetic rubber, leather, textiles, or a combination of these materials. Bulkier materials typically result in handwear that does not fit closely to the hand, uses a limited number of landmarks and measurements, but require an understanding of how the stand-off distance (hand to glove interior) will affect hand function.

### 7.9.3 Product Coverage

Gloves may cover the entire hand and fingers and extend to the wrist or beyond. Fingerless gloves cover the hand over the palm and dorsum. Fingerless gloves are often used to improve grip. For example, motorcycle and cycling gloves designed to grip handlebars may pad the palm of the hand while relying on the gripping ability of bare skin of the fingers to grasp the handlebars.

Handwear length dimensions are typically determined by measuring from proximal arm or hand landmarks, like the level of the ulnar stylion and radial stylion, to the most distal end of the fleshy fingertip—not to the tip of the fingernail. Incorporating fingernail length into a handwear product is impossible unless the product is custom fit, because there are so many variations in personal preferences for fingernail length. Detailed measuring plans are required for gloves with numerous pieces that cover extensive areas of the hand and fingers (Appendix F).

## 7.10 Designing for the Hand and Wrist: Conclusion

Handwear products are varied and serve many purposes, from everyday use to specialized applications. Because good hand function is essential to human activities, design requirements are demanding even for common activities. Koo, Teel, and Han (2016) studied home gardeners' preferences for protective gloves. Gardeners who were questioned demanded many glove characteristics: skin cut and puncture protection, dirt and insect protection, UV ray protection, water resistance/waterproofness, dexterity, movement comfort, breathability, ease of donning and doffing, durability, and ease of care. Although this example is specific to gardening gloves—other handwear may have similarly detailed design requirements. Gloves for protection in dangerous environments, for example biohazard and radiation protective gloves, must meet stringent requirements. Gloves worn by astronauts during extra-vehicular activities (EVA) are perhaps the most complex handwear products to design, requiring support of most body systems, while allowing the hand to complete complex activities in an extreme environment. With the knowledge you have gained in this chapter, try repeating the hand observation exercise in Chapter 1, Section 1.1.2. Have your design horizons expanded?

# 8

# *Designing for Foot and Ankle Anatomy*

Feet are an interface between the body and the environment. They are complex, dynamic structures with specific mechanical functions. Footwear may change the mechanics of the foot. It also may protect feet from environmental hazards.

**Key points:**

- Basic anatomy of the foot/ankle:
  - Bones, ligaments, fascia, muscles, and tendons build strong and resilient feet.
  - Blood, nerve, and lymph systems reach all the way to the toe tips.
  - Ankle and foot mechanics create sturdy but vulnerable structures.
  - *Foot morphology* differences and deformities present design challenges.
  - Foot and ankle structures need protection.
- Footwear: Protection and risks
  - Environmental challenges: Thermal, impact, and foot/ground contact.
  - The "wet" foot is prone to injury.
  - Cultural and social demands: Fashion and/or function trade-offs.
  - Footwear: Risks to normal feet.
  - The compressible, deformable foot: Fight it or fit it.
  - Footwear: Design for stability and mobility.
- Fit and sizing for diverse markets:
  - Multiple factors affect foot shapes.
  - The "last" as basis for shoe shape and size.
  - Non-uniform sizing systems.

Feet act as a foundation for the body. The ankle connects the foot to the lower leg. We define *footwear* as a product that covers any portion, or all, of the foot and/or ankle. Therefore, footwear may vary from "barefoot" shoes to fly-fishing waders that integrate a boot with coverage to the waist. Everyday shoes and boots are perhaps the first products to come to mind, along with socks and hosiery. Walford (2005) states that in 10,000 years of history, footwear has taken many forms to "service and compliment human bodies, influenced by environment, morality, practicality, economy, and beauty" (p. 168). McDowell, in *Shoes: Fashion and Fantasy* (1989), draws from art, literature, cultural history, museum finds, and architectural artifacts to present a history of shoes which shows that no shoe style is completely new. The book, with a preface by Manolo Blahnik, describes men's and women's footwear from ancient Egyptian sandals to modern high-heeled fashion shoes. Social and cultural customs influence footwear design, with footwear often coordinated to the styles worn on the body above the feet. Function is the primary purpose of footwear; however, fashion trends, and even foot fetishes, impact footwear design.

Design decisions—selection of shape, materials, and methods of making—affect not only style, but also comfort and health. The foot and ankle—the joint between the lower leg and the foot—are extremely sensitive to both environmental factors and to footwear. Both footwear and the environment can eventually, if not immediately, affect a person's whole body. Conversely, a person's health affects foot health. Feet are vulnerable to injury and often require some form of protection. Functional/protective footwear meets a variety of needs. Medical footwear includes orthopedic shoes, casts, ankle-foot orthoses and other orthotics, and soft boots. Sports and athletic footwear include walking and running shoes, hiking boots, and other sport-specific shoes. Military and other work boots are designed to protect feet in extreme conditions. Clean-room booties, worn over everyday shoes, protect a sterile/dust-free environment from contamination.

Feet, as the foundation for posture—defined as the bearing or carriage of the body—strongly impact biomechanics at the ankles, knees, hips, and pelvis; and also influence mechanics in the entire back, neck, and head. Musculoskeletal health, posture, and gait can all depend on functional feet and ankles, so what you wear on your feet is important. Gait, the way we move on foot, depends on coordinated movements of the body, starting at the foot and ankle.

Focus on the dynamic and complex foot and ankle anatomy, as well as on the function of the body systems in the feet. Bones, the ankle joint, fascia, and ligaments provide the foot strength and stability but also allow flexibility. Muscles and tendons lend the foot form and dynamic support while facilitating motion. The lymph system extends to the toe tips and must work efficiently against gravity to move waste products that would otherwise collect in the lower extremities. An intricate network of nerves sends and receives messages from the brain to initiate and stop motion and to signal pain, light touch, temperature, and pressure. Blood vessels extend to the farthest toe, nourishing the tissues, while warming and cooling the feet. Skin covering the surface of the foot is susceptible to pressure and abrasion from the environment and footwear. Toenails are a distinct feature of the foot as fingernails are in the hand.

A person's overall health or a specific health problem, like diabetes, greatly affects foot function. Normal foot variations and deformities affect footwear design. Sizing and fit strategies, affected by foot variations, are essential in designing footwear. The foot and ankle are key structures in human body movement. Footwear designs can either promote healthy foot function or inhibit it. We challenge designers to design footwear that promotes good foot function and health, while meeting the users' demands for style and fashion.

## 8.1 Foot Bones and Ankle Joint

With a structure parallel to the wrist and hand, the ankle and foot are made up of many small bones that provide flexibility. The foot and ankle must also bear the weight of the body. According to Moore, Dalley, and Agur (2014, p. 610), the bones and joints distribute weight to maintain balance when standing, conform to uneven terrain, and absorb shock. In addition, the bones transfer weight from the heel to the forefoot and toes as we walk.

The foot is made up of 26 bones and 33 joints (Xiao, Luximon, & Luximon, 2013). The structures of the foot and ankle can be described with two different

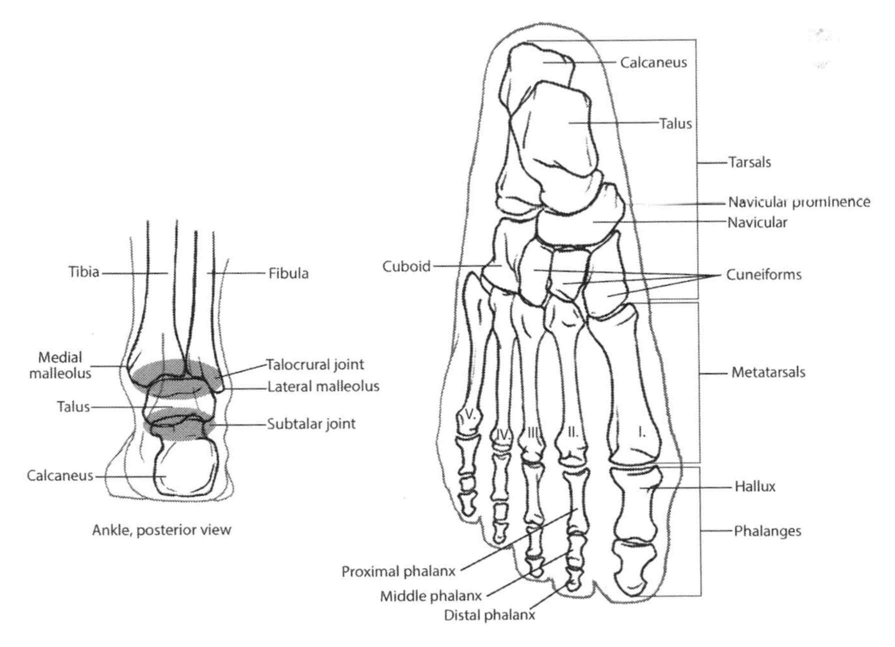


**FIGURE 8.1**
Bones and features of the ankle and foot.

organizing principles: (a) by bone groupings and (b) as functional regions, related to foot motions. Study Figure 8.1 illustrating the bones of the foot. The posterior view of the ankle illustrates the calcaneus, the largest bone of the foot, the talus above, and the *malleoli* (plural of malleolus), the protruding lower ends of the tibia and fibula of the leg. The dorsal view of the foot illustrates the three bone groups in the foot; the *tarsals*, the *metatarsals*, and the phalanges.

### 8.1.1 Tarsals

The tarsal bones form the rear of the foot and can be subdivided into proximal bones and distal bones. The proximal bones include the calcaneus and the talus. The calcaneus is the largest tarsal bone. You can feel its solid structure at the back and bottom of your foot. The calcaneal (Achilles) tendon attaches to the calcaneus and extends to the strong muscles on the posterior of the leg. The talus, as the uppermost foot bone, interfaces with the bones of the leg above it, the tibia and the fibula, to form the ankle (*talocrural joint*). Find the joint by locating the intersection of the leg and the foot. Refer to the illustration of the lower limb in Chapter 5 if needed. Run your hands down the inside and outside length of your leg from about mid-calf until you feel two protrusions. On the medial side of your leg you will feel the medial malleolus which is the distal end of the tibia. Along the outside feel the lateral malleolus which is the most distal portion of the fibula. The medial malleolus is slightly higher. Because these protrusions of the ankle do not lie on the same transverse plane (parallel to the ground), the total circumference for footwear, like high-top, pull-on boots, should be designed to comfortably encircle both malleoli. The tibia transmits the weight of the body to the talus. The fibula, with the ankle ligaments, acts as a strut to stabilize the talus beneath the tibia. The talus is unusual, as no muscles attach to it (Hamill & Knutzen, 2003). The talus rests on the calcaneus below, forming the *subtalar joint*. The ankle structure, while amazingly sturdy, is very flexible and also susceptible to sprains as described later in this chapter.

The *distal tarsal group* includes the navicular bone, three cuneiform bones and the cuboid bone. A portion of the navicular bone, the *navicular prominence*, projects on the medial side of the middle of the foot, inferior and anterior to the medial malleolus. Distal to the navicular, the three cuneiform bones—medial cuneiform, intermediate (middle) cuneiform, and lateral cuneiform—line up across the foot. These bones, plus the more lateral cuboid bone, connect the calcaneus and talus to the metatarsal bones.

### 8.1.2 Metatarsals

The metatarsals, similar in name to the metacarpal bones of the hand, are five long bones lying between the distal tarsal bones and the toes, above a large portion of the sole of the foot. The metatarsal bones are identified by

Roman numerals with the medial named metatarsal I, leading to the "big" toe, and the lateral labeled metatarsal V, leading to the "little" toe. The distal ends of the metatarsals are called the *metatarsal heads.*

### 8.1.3 Phalanges

Like bones of the fingers, the bones of the toes are called phalanges. The "big toe" is also called the *hallux* and has two phalanges, while the other toes each have three phalanges; proximal (closest to the metatarsals), middle, and distal. Toe lengths and the relationships of one toe length to another can vary (Figure 8.2). The three common foot types based on toe length are the "Egyptian type" with the hallux the longest toe, the "Greek type" with the second toe the longest, and the "square type" with similar lengths for all toes (Xiong, Rodrigo, & Goonetilleke, 2013). The "Egyptian type" occurs most often. Consider foot type differences when designing footwear.

### 8.1.4 Foot Bones and Ankle Joint Related to Footwear

According to Torrens, Campbell, and Tutton (2012) when the foot is placed into footwear, the shape of the foot changes through "movement of the bones and deformation of the flesh" (p. 158). Figure 8.3 demonstrates how various footwear designs and heel heights affect bone alignment. As the calcaneus is elevated, the bone relationships change. Note in the illustration inset that the talus is rotated so much beneath the tibia that there is no cartilage to cushion weight at part of the bone interface. Activities and culture influence affect how we "shape" the foot. Foot binding, historically practiced in China for several centuries, was an extreme method of deforming and reshaping the pliable tissues and bones of the foot to the societal ideal of a very small, lotus

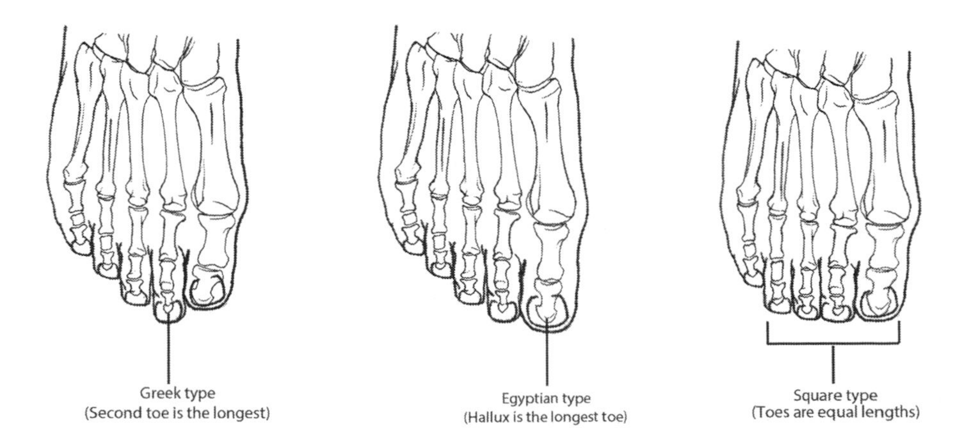


**FIGURE 8.2**
Foot types (Greek, Egyptian, Square).

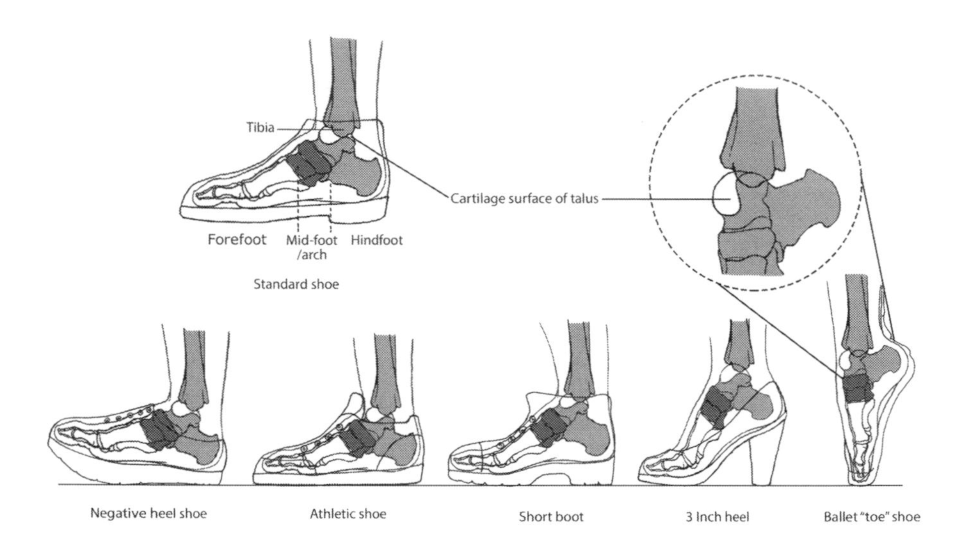


**FIGURE 8.3**
Shoe styles and bone alignment (right foot).

flower-shaped, foot. The reshaped feet severely limited walking (Howard & Pillinger, 2010). Understanding the relationship of foot bones and the ankle joint to footwear designs and footwear components is essential in making design decisions.

Organizing the foot bones by mechanical function separates the proximal and distal tarsal bones (Hamill & Knutzen, 2003, p. 209). This results in three bone groups. The calcaneus and talus comprise the *hindfoot*, commonly called the heel. The hindfoot takes weight. The *midfoot* includes the navicular, three cuneiforms, and cuboid. This cluster of distal tarsal bones and joints between the hindfoot and the metatarsals translates the energy of the foot hitting the ground into motion. And, the *forefoot* or anterior region extends from the proximal end of the metatarsals to the tips of the toes, providing push-off to propel the body forward. Study the relationship of the foot bones and ankle joint to the components of two shoe types, a lace-up dress shoe and an athletic shoe (Figure 8.4). Shoe terminology may be less consistently defined than the anatomical and functional regions.

The typical lace-up man's shoe is worn for business and dress, while athletic shoes are worn for many different athletic activities, from running to soccer (football) and are often designed for a specific activity. For purposes of this chapter, a basic athletic shoe is illustrated. Look at the illustrations and note common terms for both shoes. The *upper* covers the dorsum of the foot, the *counter* wraps around the calcaneus, the *sole* is the interface between the *plantar* aspect of the foot and the ground, the *toe box* encompasses the toes, and the *tongue* underlays the shoe lacings. Athletic shoes typically have many more components than dress shoes, some of them with specific functions described in the next sections (S. Sokolowski, personal

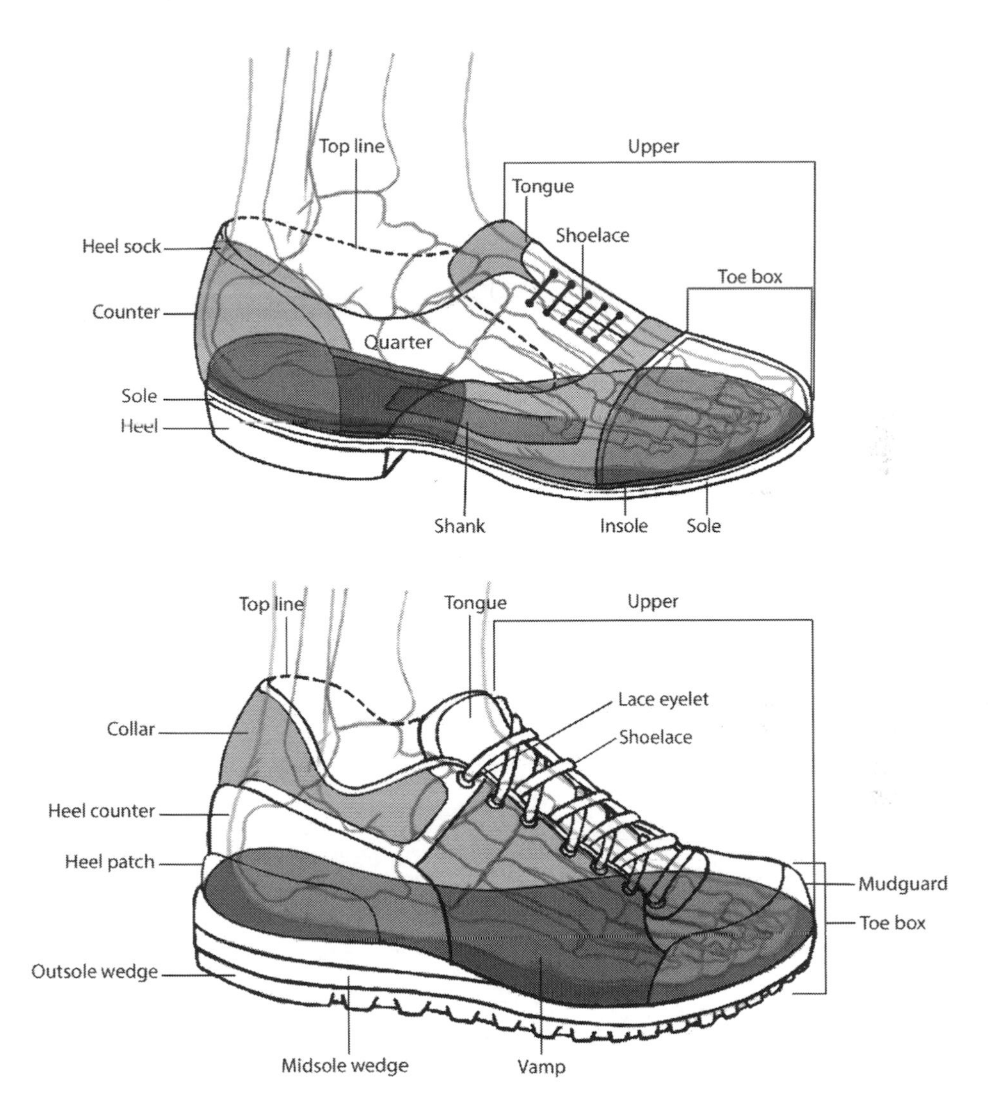


**FIGURE 8.4**
Shoe components and bones of the foot; dress shoe and athletic shoe.

communication, October 19, 2016). For details on the many athletic shoe styles, design approaches and methods of making, read *How Shoes Are Made* (2015) by Motawi.

### *Ankle Joint Footwear Coverage*

A dress shoe *top line*, the upper edge of the shoe opening surrounding the foot, clears the ankle. But some footwear styles encompass the ankle joint. In boots and high-top basketball shoes parts of the upper may extend beyond the top of the foot. The designer has to reach a balance between covering and protecting the ankle and allowing the motion needed for an activity.

### Rear Shoe Components

Look at the hindfoot in the illustrations and note the shoe components. In most cases the components are the same or similar for the two shoe types. The *heel counter*, located inside and/or outside of the shoe, is usually inflexible and firm, intended to hold the foot in place (Kirby, 2010; McPoil, 1988). It acts as a reinforcement against collapse of the back portion of the shoe. Eliminating the heel counter can be a cost-saving or fashion measure. A padded *heel sock* inside the shoe is designed to prevent blisters. The *collar* of the athletic shoe can be shaped in many ways with some styles extending up and along the Achilles tendon, while other designs are curved to avoid resting on the tendon. A *heel patch* can be used to add strength and padding to the heel counter. The dress shoe heel clearly is a separate component. There are low-heeled styles and high-heeled shoes. Alternately, the athletic shoe may incorporate some heel elevation in the *midsole* and/or *midsole wedge*. *Negative heel shoes* (Figure 8.3) reverse the incline of the foot compared to other heel types. Heel height is an individual preference based on style choice and desired comfort.

### Mid-Shoe Components

The mid-shoe region is not precisely defined. The midfoot may be supported in the mid-shoe with arch supports and/or shoe lacings. In the dress shoe, two upper shoe components, the *quarters*, cover the *instep* from superior to inferior, anterior to the ankle joint. The instep is the arched middle portion of the foot which includes both the midfoot bones and proximal forefoot bones. The quarters hold lacings and can add some size adaptability. The *tongue* lies under the lacings and provides cushioning. A *shank*, made of rigid materials and embedded in the sole of the dress shoe, is included in some shoe styles to prevent the sole from collapsing on itself. For example, a shank is often needed to support the midsole of a high-heeled style. A shank is not used in athletic shoes that must flex throughout the sole. In athletic shoes a wedge, sometimes called a medial post, is often included at the medial arch to attempt to prevent the foot from rolling inward (Kirby, 2010).

### Toe Box/Vamp Shoe Components

Shoe components that encase the forefoot should allow comfortable space for the toes. Recall the variation in foot types and how a shoe design might fit, or not fit, all types. The *vamp* covers the distal metatarsal bones and may merge with the toe box which covers the phalanges. The toe box is the domed area of the shoe above the toes. Women's fashion shoes with a pointed toe box squeeze toes to an unnatural shape. Compressing the bones with a narrow, low, or short toe box can result in immediate discomfort and/or permanent damage. (Read Section 8.7 on foot deformities.) The *toe puff*, an extra layer in the most distal and topmost portion of the toe box, helps keep the toe box from collapsing on the toes. Depending on materials and structure the toe box may prevent impact injury. For example, work and military boots are often constructed with a rigid, reinforced toe box and are called steel-toed boots. Athletic shoes typically have a pliable toe box, but they must be shaped to adequately accommodate toe height and length as well as foot motion during activity. As the name indicates, the *mudguard* is an addition to the toe box protecting the shoe from soil and abrasion.

### *Sole of Foot and Sole of Shoe*

The *sole* of footwear, the bottommost layer, may be divided into three parts: *insole, midsole,* and *outsole.* Outsole materials in both dress and athletic shoes must be sturdy enough to resist wear and tear from ground contact. The three layers of the sole give the athletic shoe designer the ability to address different consumer needs and preferences. Athletic shoe outsoles of pliable materials are designed to cushion the foot and provide traction. Various tread formations have been developed in attempts to give traction for a specific sport. The story of Bill Bowerman, co-founder of Nike, pouring urethane into his wife's waffle iron to create a textured outsole to add traction is legendary (Knight, 2016). Look at a variety of athletic shoes and notice the great variation in textures and placement of outsole components.

The midsole of the athletic shoe has become another area of design experimentation and commercial competition with material variations, injection of air in different configurations, many methods of shaping and compartmentalizing, and color variation. Some alternatives are purely aesthetic, while others are intended to cushion or stabilize the foot.

The insole may conform to the plantar contours of the foot. Insole materials, amount and placement of cushioning can vary. Most athletic shoe insoles can be removed and replaced with a custom orthotic, a custom shaped or molded shoe insert designed to enable some joint motions, while limiting others. Read more about custom-made orthotics, insoles that are constructed for a wearer's specific needs, in Section 8.8.

Understanding the relationship of the shoe components to the bones is important to designers. The anatomical structures that attach to and move the bones are also important to an understanding of foot and ankle anatomy. The next sections describe ligaments, fascia, tendons, and muscles; the structures that stabilize and mobilize the foot and ankle.

## 8.2 Ankle and Foot Joints, Ligaments, and Fascia

Each point of contact between bones in the foot and ankle is a joint. Given the number of bones and the varied bone shapes in the ankle and foot, many of the joint types described in Chapter 2 can be found in the foot. The joints are key in allowing the bare foot to accommodate to irregular walking surfaces. When *arthritis* (inflammation of the joints) occurs in the foot, joint motion can become painful and/or restricted and the foot may become more vulnerable to injury. Inflammation is a combination of pain, swelling, and warmth in the tissues. Footwear specifically designed to help limit joint motion may be necessary.

Ligaments are essential components of the foot and ankle, stabilizing the bones at the joints while also allowing motion and flexibility. They are frequently named for the bones where they attach. The ligaments at the ankle are the largest and strongest in the region, but they are also most often injured (Figure 8.5). Acute ligamentous ankle sprains (stretching or tearing of fibers in a ligament) occur in many sports as well as in everyday life (Hootman, Dick, & Agel, 2007). Normal motion at the ankle includes inversion and eversion (Figure 8.6-A and Figure 8.6-B). Inversion ankle sprains tend to happen when the planted foot quickly *inverts* relative to the leg, although they can also occur when the foot is "searching for ground," such as when unexpectedly stepping off a curb. The tibia shifts on the talus while the foot turns onto the lateral side (Hamill & Knutzen, 2003). Refer to Figure 8.6-C. The injury most often involves the *anterior talofibular ligament*, and possibly also the *calcaneofibular ligament*, and/or the *posterior talofibular ligament*. The medial, *deltoid*, ligament is stronger and less likely damaged unless the foot *everts* very forcefully with the tibia shifting laterally on the talus while the foot turns to the medial side (Figure 8.6-D). The lateral malleolus, the distal end of the fibula, tends to limit the extent of eversion motions. Torn ligaments may heal with scarring, but are never the same after a sprain, so some athletic footwear is designed to try to prevent injury.

Braces worn after an injury are designed to firmly support the ankle and promote tissue healing. They are typically constructed from stiff and/or rigid materials. Some braces are also designed to try to prevent sprains. Leppänen, Aaltonen, Parkkari, Heinonen, and Kujala (2014) noted generally positive effects for such devices in several randomized controlled trials. Taking a different approach; Fong, Chan, and Mok (2013) reported their work to design an "intelligent anti-sprain sport shoe" (p. 584). Their method relies on (a) the sport shoe device detecting foot and ankle motion and (b) calculating ankle inversion velocity and (c) then making the lateral leg muscle, which everts the foot, respond sooner than its natural reaction time. The goal is to protect the ankle joint from an inversion sprain.

The superficial and deep fascia, in sheets of tissue, envelop the muscles and support tendons, blood vessels, and nerves. Fascia is a living tissue

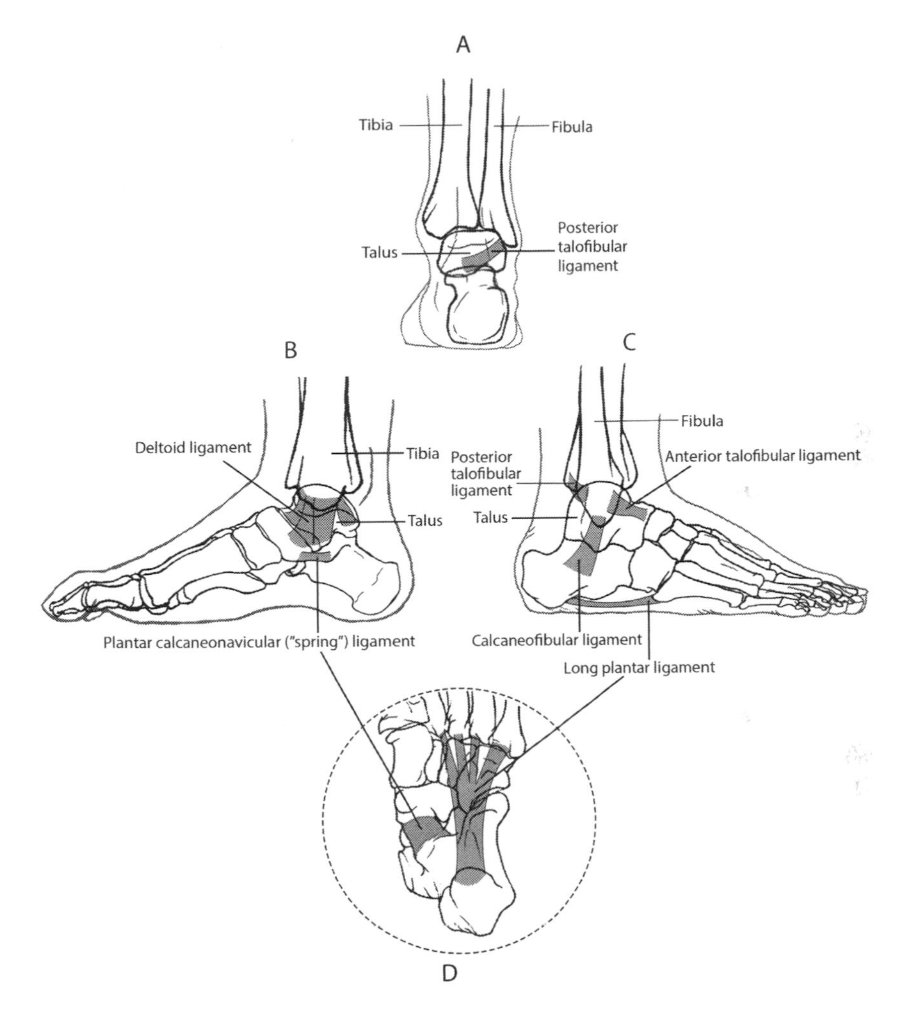


**FIGURE 8.5**
Key ligaments of the ankle and foot (right). Views: A. Posterior, B. Medial, C. Lateral, D. Plantar.

which constantly regenerates itself. This ongoing remodeling process helps to keep the foot flexible unless foot motion is restricted in some way. The *plantar fascia*—a thin sheet of tissue toward the bottom of the foot, with its thicker central portion, the *plantar aponeurosis*—holds the parts of the foot together, protects the sole from injury and supports the foot (Moore et al., 2014, p. 610). Tightness in the plantar aponeurosis creates a mechanical *windlass effect* (Hicks, 1954), raising and shortening the arch when the big toe is passively extended (Fuller, 2000).

*Plantar fasciitis*, inflammation of the plantar fascia, causes pain at and in front of the heel, which often is severe with the first steps taken in the morning. McKinley and O'Loughlin (2006) point to several possible causes including weight-bearing activities (lifting heavy objects, running, or walking),

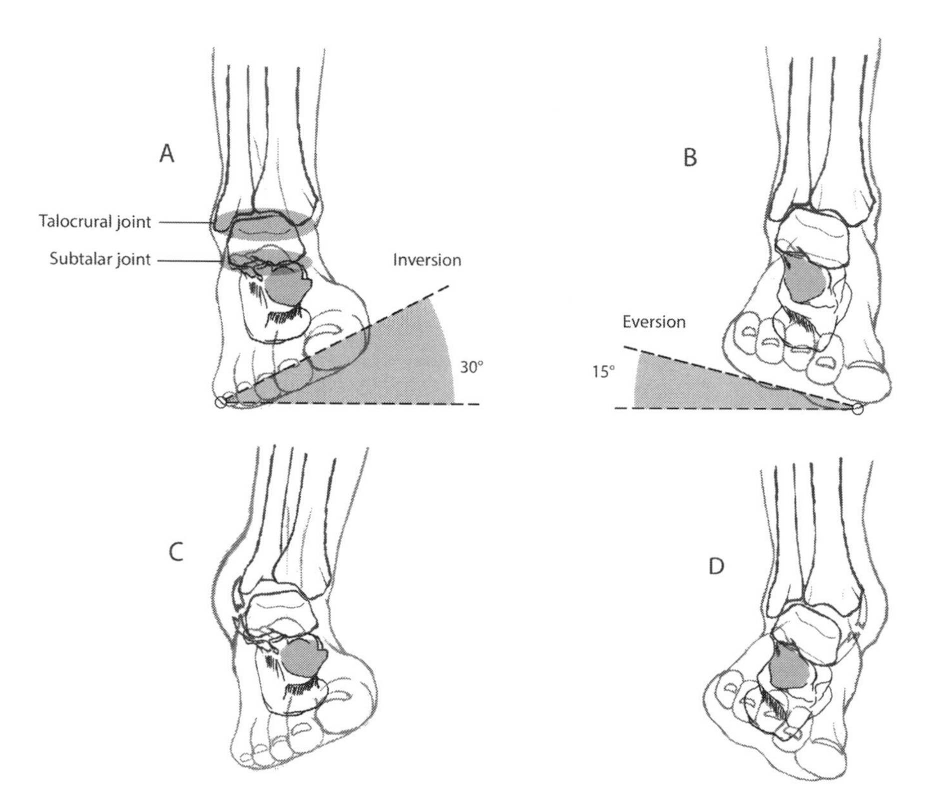


**FIGURE 8.6**
Ankle inversion and eversion (right foot). A & B Normal range of motion; C. mechanism of inversion sprain; D. mechanism of eversion sprain.

excessive body weight, improperly fitting shoes, and poor biomechanics due to wearing high-heeled shoes or having flat feet. An orthotic, when combined with an adjustable soft splint worn at night, has been found to provide effective treatment of plantar fasciitis for some people (Lee, Wong, Kung, & Leung, 2012).

## 8.3 Foot "Arches" and Foot "Dome"

The foot is often conceptualized as having two to four separate arches that provide support for the weight of the body. The elevations of the arches keep sensitive nerves and blood vessels from contact with the ground when standing. Look at the illustrations of foot arches in Figure 8.7 for views of transverse arches (posterior transverse arch viewed from posterior, 8.7-A, anterior transverse arch from anterior, 8.7-D) and longitudinal arches (8.7-B from medial view, and 8.7-C from lateral). The *medial longitudinal arch* runs

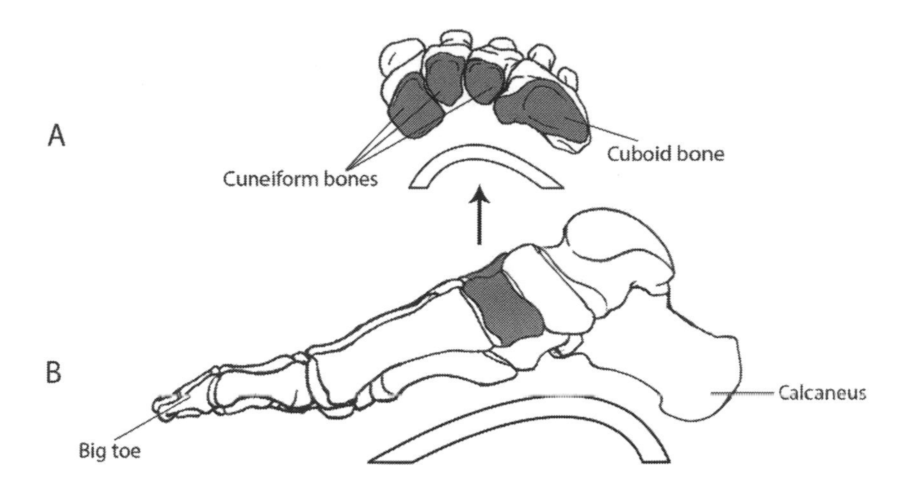


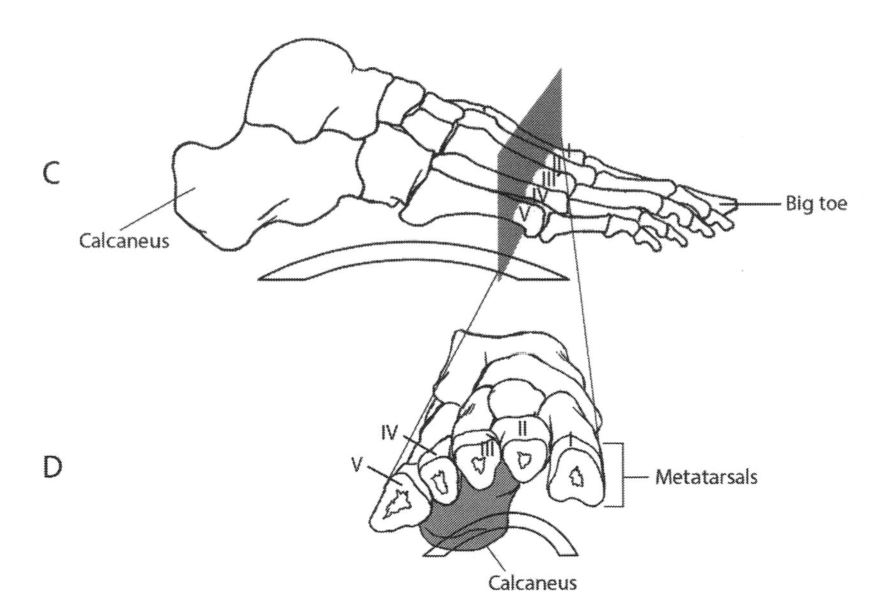


**FIGURE 8.7**
Foot arches (right foot); medial, lateral, and transverse. A. Posterior transverse arch-posterior view, B. medial longitudinal arch, C. lateral longitudinal arch, D. anterior transverse arch-anterior view.

from the calcaneus to the hallux and includes the calcaneus, talus, navicular, and cuneiform bones, and metatarsals I-III. It is the highest of the arches and typically keeps the medial side of the foot from touching the ground. It prevents compression of the soft structures on the bottom of the foot. The lateral longitudinal arch extends from the little toe to the heel and is formed by the calcaneus and cuboid bones, and metatarsals IV and V. This arch is only slightly raised. It is evident as the narrowed lateral edge of a person's footprint, although it may be elevated and not evident. Ligaments throughout the hindfoot, midfoot, and proximal forefoot establish primary arch stability (Hollinshead & Jenkins, 1981, p. 335) and stability in static standing (Basmajian & De Luca, 1985, p. 259, p. 351). As seen in Figure 8.5, the *plantar calcaneonavicular ligament* supports the highest portion of the medial longitudinal arch. It is credited for allowing some elasticity, or "spring" in this arch (Jenkins, 2002, pp. 357–358). The *long plantar ligament,* extending along most of the lateral longitudinal arch, connects the calcaneus and cuboid to the proximal ends of the three lateral metatarsals. The transverse arch or arches run perpendicular to the longitudinal arches (Gray, 1966, p. 372). A posterior transverse arch is formed from the most distal row of tarsals (cuboid and cuneiforms) and an anterior transverse arch from the distal ends of all five metatarsals. The foot bones are not just arranged in medial-lateral and anterior-posterior relationships, but also in superior-inferior positions.

McKenzie (1955) first proposed that the separate foot arches could be integrated and compared to a "half dome" (p. 1068). See Figure 8.8 for an illustration of the arches forming a dome. McKeon, Hertel, Bramble, and Davis (2015) also describe the arches of the foot forming a "functional half dome" (p. 3). When standing with the medial malleoli next to each other and touching, the two half domes meet to form a complete dome (Gray, 1966, p. 372). In this unified structure the bones, ligaments, tendons, and muscles of the foot all work together to form an arch. The origin, or top of each half dome, is the talus (Figure 8.8). When the foot bears weight, the "half-arches, reaching anteriorly and posteriorly [flatten] . . . under the strain" (McKenzie, 1955, p. 1069). When a walker bears weight in a heel-toe pattern, each of the arching elements radiating from the talus, from the heel to the toes, via the lateral foot, takes weight in turn. The plantar arches are also superimposed on Figure 8.8, designated A-D as they are in Figure 8.7.

Bones form the scaffold for the arches while ligaments, tendons, and muscles hold the pieces together, integrating the separate arches and providing strength and flexibility. Understanding arch characteristics can be useful in designing any footwear as foot "architecture" varies even between normal individuals. A person's footprint gives an indication of arch height and shape. Footprints show that some people have "high arches" and others have "flat feet" (Figure 8.9). The footprint displays the front and back of the foot and the lateral outline of the middle of the foot while the medial foot is not apparent. You can look at your footprint and footprints of others using a simple technique. Wet your feet and step on dry concrete to get a quick image of your footprint. If you want a more permanent record use bath powder or corn starch to imprint the foot on black paper. Try this with several people; if possible people who wear the same size shoe. Note differences in footprints. Variation in arch height is reflected in the width of the proximal forefoot. If there is only a narrow rim of lateral foot showing, the medial longitudinal

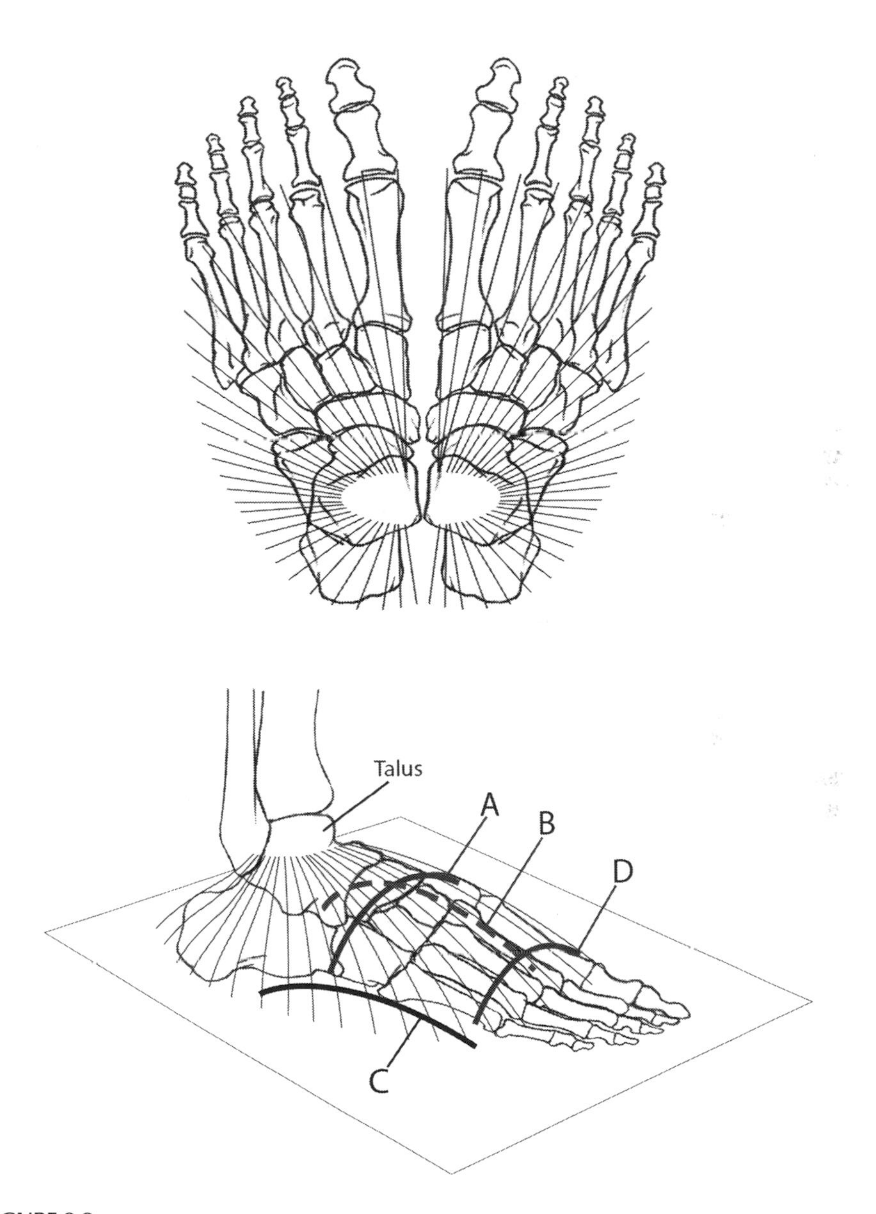


**FIGURE 8.8**
Foot arches as half dome: Top view and quarter view with planes. (Modified and combined from "The foot as half-dome," J. McKenzie, 1955. *British Medical Journal*, 1, p. 1068. Copyright 1955 British Medical Association and "The foot core system: A new paradigm for understanding intrinsic foot muscle function," P. McKeon, J. Hertel, D. Bramble, and I. Davis, 2014. *British Journal of Sports Medicine*, 0, p. 3. doi:10.1136/bjsports-2013-092690. Copyright 2014 British Association of Sports and Medicine.)

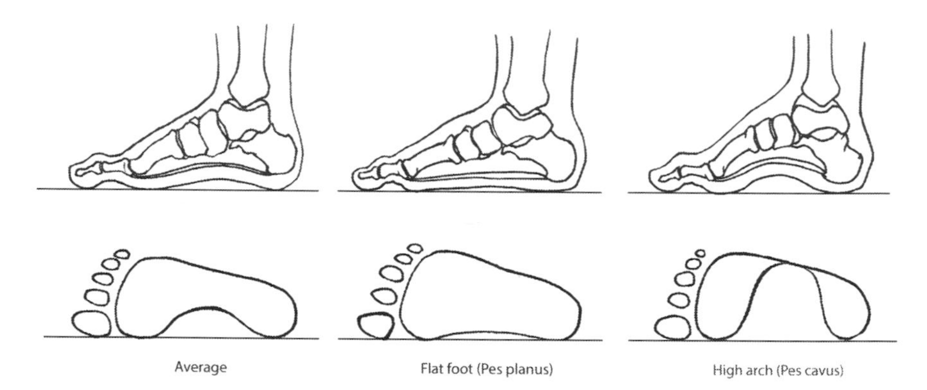


**FIGURE 8.9**
Arch heights and footprints.

arch is "high." When the footprint is wide in the proximal forefoot, it implies a "flat foot." In addition, some people have a broader forefoot with a narrow heel, while others have a narrow forefoot and wider heel. We think of footwear as a direct interface to the ground surface, and may harbor the misconception that the sole of the shoe replicates the sole of the foot. Look at the footprint shapes and then look at the sole of any shoe you own. It is highly unlikely that you will find a match.

## 8.4 Ankle and Foot Tendons and Muscles

Muscles and tendons also help stabilize the ankle and foot while they facilitate motion. The tendons of nearly all the muscles which originate in the leg (below the knee) insert and act on bones in the foot. You can see and feel some tendons of the foot and ankle when you activate leg muscles. Place your bare foot on the ground, keep your heel on the ground, and raise your toes and the sole of the foot up (away from the ground). Feel the strong extensor tendons which dorsiflex/extend the foot and toes up, much like puppet strings, on the top (dorsum) of the foot. You can see the large *tibialis anterior tendon* as it runs on the dorsum of the foot from the anterior ankle (slightly medial) towards the big toe. Feel the calf muscles contract when you point your toes down, plantar flexing your foot with the Achilles tendon. See Figure 8.10 for the illustration of these two motions. Tight or poorly fitted footwear over any of these tendons can inhibit function and cause pain.

The natural rest position of the toes (Figure 8.11-D) gives no indication of the underlying complexity. The 20 intrinsic muscles of the foot originate and insert within the foot and are grouped into dorsal muscles and plantar muscles. As in other parts of the body, foot muscles closest to the surface are generally the largest and, from fit and vulnerability concerns, are of most

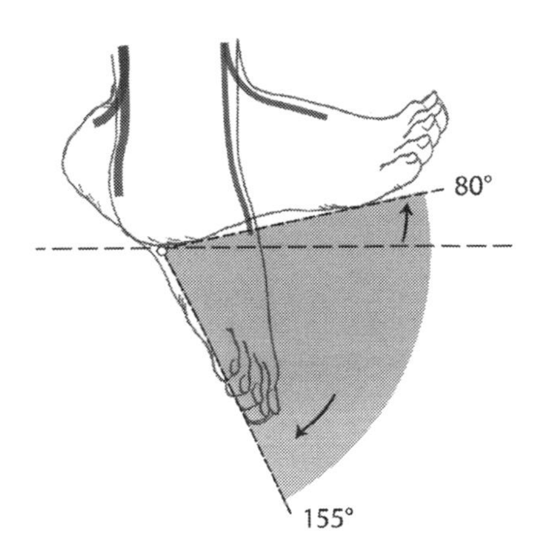


**FIGURE 8.10**
Ankle extension and flexion range of motion.

interest to designers. Two dorsal muscles—the *extensor hallucis brevis* and the *extensor digitorum brevis*—lie above the cuboid, the lateral cuneiform and the intermediate cuneiform bones. Keeping the sole of your bare foot on the floor, raise your toes up (away from the floor) and you can feel these muscles contract over the dorsal forefoot. Refer to Figure 8.11-A for this toe extension motion.

The 14 muscles located on the plantar aspect (nearest the sole) act as a group to provide support and balance while standing and help to maintain the arches of the foot particularly during walking and running (Moore et al., 2014, p. 610). The plantar muscles are grouped into four layers and generally act as follows: (a) flex the toes—curl the toes downwards (Figure 8.11-B), (b) spread the toes apart—abduct the toes (Figure 8.11-E), and (c) move the toes back together—adduct the toes (Figure 8.11-F). The superficial layer consists of three relatively bulky muscles: the *flexor digitorum brevis*, which attaches to both the medial and lateral sides of the middle phalanges, the *abductor hallucis*, which abducts the hallux, and the abductor digiti minimi, which abducts the small toe. The second, slightly deeper layer, working with the toe extensors in the leg, facilitates flexion and extension of the lateral four toes. The third deeper layer helps stabilize the transverse arch of the foot and the hallux and the other phalanges. The deepest layer lies between the metatarsals and abducts and adducts the phalanges (Figure 8.11-E, Figure 8.11-F). These muscles also help the foot adjust to uneven ground and play a major role in moving the body in all directions. A combined action of flexing the proximal interphalangeal (PIP) joints of the toes and extending the distal interphalangeal (DIP) joints produces a gripping action of the toes, helpful when trying to balance. See Figure 8.11-C. Like the bones, some of the foot muscle names are similar to the names of the hand muscles, but the intrinsic foot muscles generally do not perform the precise movements that the hand muscles must. Even so, the toes need adequate space to move

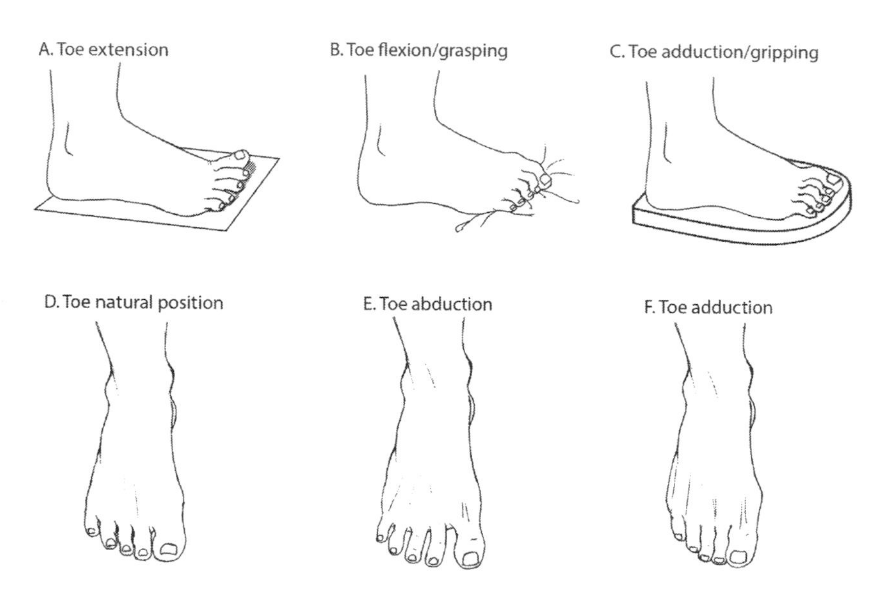


**FIGURE 8.11**
Toe motions.

naturally inside the toe box, particularly for the PIP/DIP flexion/extension gripping action of the toes. See Figure 8.12 for an illustration of foot muscles to consider in footwear design.

The intrinsic muscles stabilize the foot and the extrinsic muscles originating in the leg are the "global movers to generate foot motion via their long tendons" (McKeon et al., 2015, p. 3). Together, the intrinsic and extrinsic muscles control movement and stability of the arch. McKeon et al. (2015), expanding on McKenzie's work, describe the "foot core" or "dome" as the essential structure for all foot function. They also suggest that footwear which isolates the foot from interacting with the ground surface is responsible for many foot problems. They recommend foot strengthening exercises and minimal footwear, or no footwear, to promote natural foot strength and resiliency.

Larson and Katovsky (2012, p. 242) advocate that runners use barefoot-style shoes, but caution that runners need to gradually increase periods of wear to adapt to new forces on the feet. Altering the interaction of the foot with the environment may be desirable in specific instances. Slipping is a risk for a gymnast mounting and dismounting an apparatus. Sokolowski, Hansen, and Roether (2011) developed a minimum-coverage gymnast's shoe to provide traction on a variety of surfaces. The major portion of the footwear covers the plantar aspect of the forefoot with a traction element. The footwear, secured to the foot with minimal strapping on the dorsal and heel aspects, leaves the foot as "bare" as possible.

On the other hand, in studies of soccer boots (cleats), researchers concluded that footwear stabilizes the foot arch and ankle (Fukano, 2015) and significantly improves postural stability compared to a barefoot condition

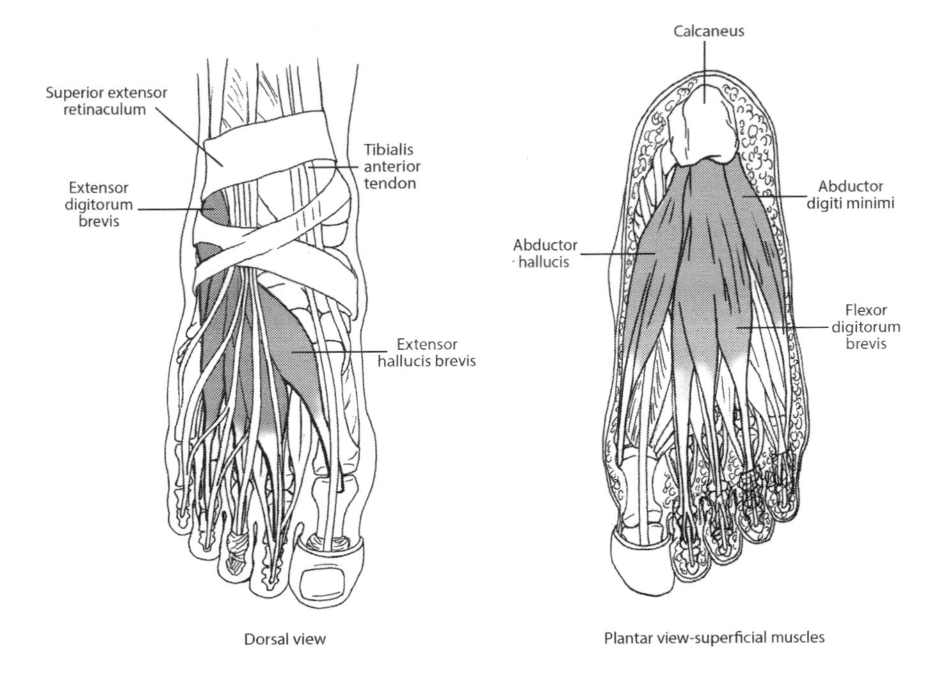


**FIGURE 8.12**
Superficial muscles and tendons of the foot and ankle.

(Notarnicola et al., 2015). Presumably, both these studies were done with subjects who routinely wear shoes. Would the results have been the same if the subjects routinely went barefoot? The level of physical fitness of the athlete will also affect foot and ankle stability.

Protecting tissues from the ground with footwear can be particularly important for individuals with foot problems. Neurological disease may produce reduced sensation, weakness, or abnormal muscle function leading to altered anatomical alignments. Foot changes may also arise from injury, arthritis in the joints, impaired circulation, lymphedema, or congenital foot malformations.

Some footwear does not provide solid foot contact with the ground, putting added stresses on the bones, ligaments, fascia, tendons, and muscles. Consider high-heeled footwear that elevates the hindfoot off the ground. Csapo, Maganaris, Seynnes, and Narici (2010) compared lower limb and foot muscles and tendons of women who wore high heels with a minimum heel height of 5 cm (2 in.) for an average of 60 hours a week to women who wore flat shoes. They determined that the high-heeled shoe wearers had shorter calf muscle fibers and less springiness in the Achilles tendon with resulting reduced ankle range of motion. Age and physical condition of the individual also influence these body/footwear interactions.

Ice skates and in-line skates challenge the foot-to-ground relationship. Skaters learn new, whole-body balance skills as they interpret ground information from a single centered sagittal blade instead of from the full breadth of the foot. The boot of an ice skate needs to a "hug" the skater's foot with just the right amount of compression: not too tight and not too loose, because figure skaters depend on "responding, in the moment, to what their ice skates are telling their body" (Gravestock, 2013, p. 65).

Painters and construction workers sometimes use stilts to work on tasks above their heads, to avoid having to move ladders around. This is an extreme test of the foot-to-ground relationship. Pan, Chiou, Kau, Bhattacharya, and Ammons (2009) found that the height of the stilts, 46 cm (18 in.) to 102 cm (40 in.), combined with narrow or offset foot placements, significantly increases postural instability and risk of injuries from falls. The use of stilts has been discouraged and even prohibited in some locations. Workers using stilts reported a sense of postural instability related to foot placement. The researchers suggest that if the wearer feels unsafe wearing the stilts, they probably are unsafe. Use caution when designing footwear that limits solid foot contact with the ground.

## 8.5 Nerves, Blood Vessels, and Lymphatics of the Ankle and Foot

Networks of nerves, blood vessels, and lymph vessels serve the ankle and foot. See Figures 2.9 (nervous system) and 2.16 (blood vessels and lymphatics) in Chapter 2 for basics on these systems. Feet, as the most distal parts of the body, require that all of these systems successfully travel the length of the body and operate efficiently. Nerve messages travel through a series of branches from the brain to the ankle and foot and back again. Oxygenated blood follows a circuit from the lungs to the heart and through the branching arteries to reach the tips of the toes and then is pushed back to the heart and lungs to release waste products and be renewed to begin the circuit again. Foot and ankle lymphatics follow pathways similar to the blood circulation system.

### 8.5.1 Nerves of the Ankle and Foot

Similar to the hand and wrist, multiple peripheral nerves pass by the ankle to innervate the foot. Nerves are distributed to the plantar aspect and the dorsum, and have sensory and/or motor functions. Muscles (described in Section 8.4) contract in response to the stimuli of motor nerves. End branches of the fibular nerve of the leg supply both the dorsal intrinsic muscles of the foot and the anterior and lateral muscles of the leg which act on the foot and ankle. Branches of the tibial nerve connect to the plantar intrinsic muscles and the posterior muscles of the leg which act on the foot and ankle.

The distribution of sensory nerves in the foot is more complex than that of the motor nerves. There are more sensory nerve endings in the plantar foot than on the dorsum. Sensation from the sole of the foot travels back to the spinal cord via five different sensory branches of the three major nerves of the lower extremity. Three of these nerves leave the sole of the foot at the medial border of the calcaneus, the fourth in front of the medial malleolus, and the fifth nerve wraps behind the lateral malleolus after collecting sensory impulses from both lateral sole and lateral dorsum of the foot. On the dorsal foot, four different sensory branches of the same three major nerves depart the foot at various points on the anterior and medial ankle. The result of these motor and sensory patterns is that if one of the major leg or thigh nerves is damaged, a substantial nerve supply remains.

### Nerve Message Disruptors

Intuitively, one might expect that footwear disrupts the sensory input from the environment and the ground to the foot. In addition, a growing body of literature on fall prevention in the elderly indicates that plantar sensation is critical to gait and balance (Hatton, Rome, Dixon, Martin, & McKeon, 2013; Kavounoudias, Roll, & Roll, 1998; McKeon, & Hertel, 2007). Limited studies comparing use of shoes with not wearing shoes are contradictory. In a report comparing dynamic balance in healthy barefooted subjects to subjects in both standard and minimal running shoes, Rose et al. (2011) found better balance in barefoot conditions. Designers may wish to consider the many factors that may alter nerve messages in the foot: the environment, disease and treatments for disease, injury from mechanical forces, and foot-to-ground contact. When the foot contacts a cold surface, heat conduction from the foot chills the foot and nerves, decreasing pressure sensations from the sole of the foot, and increasing body sway (Magnusson, Enbom, Johansson, & Pyykkö, 1990). This factor reinforces the importance of designing well-insulated footwear.

*Peripheral neuropathy* results from diseased or damaged peripheral nerves causing weakness, numbness, and/or pain. Feet are often the first body part to demonstrate nerve disease or damage because peripheral nerves in the legs and feet travel a long distance from the end of the spinal cord to the feet (refer to the nervous system information in Chapter 2). Diabetes frequently leads to peripheral neuropathy and foot care becomes essential, including wearing socks that absorb moisture and shoes that cushion yet support the feet (Tyrrell & Carter, 2009). Treatment for disease, like chemotherapy drugs to treat cancer, can also cause nerve damage (Windebank & Grisold, 2008).

Outside mechanical forces can cause nerve damage. Fortunately, the locations of both sensory and motor nerves in the foot and ankle are relatively protected from footwear pressure. However, one type of *compressive neuropathy* (nerve damage caused by localized pressure) is a painful condition known as *Morton's neuroma* (Young, Niedfeldt, Morris, & Eerkes, 2005).

Seen predominately in women, it is attributed to damage of small sensory branches by the metatarsal heads and/or ligaments located in the interspaces between the toes. Although the precise cause is unclear, it is believed to stem from hyperextension of the toes in high-heeled shoes and/or from altered bony architecture from a *bunion* (Kimura, 2013, p. 778). Read more about bunions and other foot deformities later in this chapter. Katirji and Wilbourn (2005) state, "Typically, the [Morton's neuroma] symptoms are relieved by non-weight bearing and by shoe removal" (p. 1507). They also state that nerve compression may be relieved with footwear modifications, including larger widths, low-heeled shoes, and limitation of toe extension.

### 8.5.2 Blood Vessels of the Ankle and Foot

The network of leg arteries ends in the ankle and foot (Figure 2.16). The *anterior tibial artery* of the leg feeds the *dorsalis pedis artery* of the foot. In a healthy person, the pulse at this artery can be found quite easily over the central portion of the navicular bone, presenting an accessible site to monitor heart rate. As the dorsalis pedis reaches the top of the forefoot it branches across to feed into each toe. The *posterior tibial artery*, coursing behind the medial malleolus, carries rich amounts of blood to the sole of the foot and toes. Blood from the foot drains through superficial veins into deep veins proximal to the ankle joint, to take advantage of the musculovenous pump of the calf (see Section 5.10.1). Research on the influence of high-heeled shoes on venous function in young women (Tedeschi Filho, Dezzotti, Joviliano, Moriya, & Piccinato, 2012) demonstrated impairment of the function of the musculovenous pump. There was decreased return of venous blood, proportional to heel height, from the lower leg and foot.

Blood vessels in the ankle and foot play a significant role in controlling skin and surface tissue temperature. Because the body constricts peripheral circulation in normal response to cold stress, it is easy to understand that by the time blood reaches the very distant toes, it may have cooled considerably—especially in cold environments. Determining the extent of cooling is not simple, but Khanday and Hussain (2015) developed a mathematical model to estimate peripheral temperature loss with severe cold exposure. As with hands, when exposed to extreme cold, feet are susceptible to both non-freezing cold injury and very serious, but less common, frostbite (see Chapter 7, Section 7.5.2). *Frostnip* can occur before frostbite and is defined as transient formation of ice crystals in the tissues. Frostbite is a much more significant injury (Whitaker, 2016). Severe frostbite may require amputation of the affected area.

Prevention of cold injury is the best strategy. Footwear can play a major role in cold protection (Whitaker, 2016). Footwear fitted and designed to avoid constriction of arteries and veins helps maintain thermal equilibrium in peripheral tissues. Blocking foot-to-ground contact with insulating materials is an often-used strategy. Hamlet (2000) describes footwear features which predispose to frostbite: a narrow or too flat toecap not allowing space for thick, insulating socks and the toes; toes compressed from a shrunken, worn midsole that has moved up to crowd the toes; or boots fitted to the length of the hallux, but too short for a longer second toe producing frostbite of the second toe tip. Once tissues have been damaged by frostbite, orthotics such as a walking cast/boot may be helpful in the healing process.

The opposite environmental condition is extreme heat. Warm environments increase the blood flow to the periphery and combined with sweating of the soles of the foot, an increased volume of blood in the feet can help to cool the body, especially if the feet are lightly covered or bare. But increased blood flow may have another, less desirable, impact—swollen feet.

Socks worn inside other footwear contribute to foot health and comfort, affecting thermal balance and moisture transfer. Van Amber, Wilson, Laing, Lowe, and Niven (2015) found that fiber content plays an important role, but the effect is small in comparison to effects of fabric structure. From laboratory tests they concluded that terry fabrics were "the most thermal and water vapor resistant, least permeable to water vapor, most absorbent, and most conductive" (p. 1274). While considering footwear design variables, consider the contribution of socks to the footwear ensemble.

### 8.5.3 Lymphatics of the Foot and Ankle

Lymph from the medial foot follows the great saphenous vein and drains directly to superficial inguinal lymph nodes (refer to Chapter 2, Figure 2.16-C). Lymph from the lateral foot follows the small saphenous vein and drains initially to the popliteal lymph nodes and then by deep lymphatic vessels to the deep inguinal nodes (Moore et al., 2014, p. 620). With increased blood flow to the feet, plus the force of gravity, more fluids leak out of the circulation, causing edema (swelling). It must be recycled through the lymphatics. Throughout the lymphatic system, the natural flow is from the periphery to the center of the body. The distance of the foot from the thoracic duct in the upper chest increases the risk of lymph pooling in the foot and ankle. Natural lymph flow from the feet and ankles may also be limited by diseased or damaged lymph nodes creating obstructions along the pathway. Cancer, radiotherapy that damages the vessels, trauma, and chronic infections may cause lymphedema (Frowen, O'Donnell, & Gordon, 2010).

As in the legs, the most common cause of lymphedema in the feet and ankles is the tropical mosquito-borne parasitic infection, filariasis. It causes inflammation in the lymph nodes and channels. Foot infections, a problem in climates and cultures where shoes are less commonly worn (Moore et al., 2014, p. 624), exacerbate the problem. Elephantiasis, when the affected foot and/or leg swells to a gigantic size, often after repeated secondary infections, is especially disabling. The opportunity to design unique and affordable footwear to help reduce the effects of lymphedema for these populations is evident.

*Podoconiosis,* a non-communicable tropical disease found in economically disadvantaged populations, is another cause of lymphedema. It occurs when silica particles penetrate into bare feet of workers exposed to volcanic soils. Particles collect in the lymph nodes of the leg and groin, blocking lymph flow. Davey and Newport (2007) state that prevention includes wearing "robust shoes and socks" (p. 889), but the challenge is cost and availability of appropriate footwear.

## 8.6 Skin and Toenails

Skin serves as the protective covering for the ankle and foot. Skin thickness varies by area depending on the degree of protection needed. The skin of the sole is thicker than the skin of the dorsum, especially in the major weight-bearing areas—the heel, lateral edge, and ball of the foot (Moore et al., 2014, p. 610). A subcutaneous plantar fat pad of specially organized adipose tissue cushions the muscles, ligaments, and bones above (Hills, Hennig, Byrne, & Steele, 2002). Toenails, essentially hardened dead skin cells, protect the toes. The skin of the sole is hairless and infused with many sweat glands (but no oil glands), all of which create design challenges.

Footwear designs must vary according to the environment. Temperature conditions and the amount of environmental moisture must all be considered. Hot, wet environments, paired with the physical protection needed during military operations, present the most arduous demands. One of the body's major cooling mechanisms is sweat production. When boots limit sweat evaporation, the skin may become locally saturated and excessive moisture contributes to a loss of mechanical integrity of the skin of the foot (Torrens et al., 2012).

Strategies to manage sweat from the sole of the foot include frequent changes of socks or using wicking material in the socks and the boot liner. Cold and wet environments as well as hot/wet environments present foot perspiration management issues, largely related to using waterproofing materials for the sole and upper, which restrict perspiration evaporation. Hofer, Hasler, Fauland, Bechtold, and Nachbauer's (2014) study of the microclimate in ski boots, found that socks and liners absorb as much as 45.5 g. (1.6 oz.) of liquid in a short period of time. Footwear design strategies can influence the enclosed environment by venting perspiration, trapping moisture below the sock liner, and/or using newly developed and more efficient absorbent materials.

Contact and abrasion from footwear can cause blisters, calluses, and *corns* on the skin of the foot. Blisters are the result of shearing, compression, and scraping forces leading to mechanical separation between the epidermis and dermis. The separation can then fill with fluid forming a blister

(Van Tiggelen, Wickes, Coorevits, Dumalin, & Witvrouw, 2009). Blisters may form very quickly. Mailler and Adams (2004) state that blisters are the most common complaint of runners. It has been suggested that socks offer a means to reduce blister formation. However, the effects of sock friction on foot skin are unclear. Tasron, Thurston, and Carré (2015) tested sock/skin friction using human subjects. They measured the friction generated by the dry plantar surface of the foot sliding against five running sock materials and found no consistent friction differences. They concluded that the natural variation of a person's skin has more effect on friction levels than the knit pattern or fiber types. The numerous possible combinations of individual foot variations, shoe/boot options, and sock choices further complicate the understanding of effects of friction on foot skin.

Calluses and corns usually arise from footwear friction or increased loading over a length of time. Calluses, generalized areas of thickened skin, tend to develop on the sole and plantar aspects of the toes (Young et al., 2005), although a *pinch callus* may be found on the medial side of the hallux (Tiberio, 1988). Corns are isolated localized areas of thickened skin and may be either hard or soft. Hard corns tend to form on dorsal toe surfaces, while soft corns are frequently located between toes. Preventing footwear friction is the best way to avoid skin injuries. Strategies include controlling foot moisture, choosing non-irritating footwear materials, careful construction of the footwear interior to avoid friction spots, size selection to match foot size, and designs to accommodate natural foot shape.

Toenails, like fingernails, arise from the stratum corneum of the skin and are made of thin, dead, scaly cells that are packed together and filled with parallel fibers of hard keratin (Saladin, 2014). These hard, clear shells offer some protection to the ends of the toes but are less functional than fingernails. Review nail anatomy shown in Chapter 7, Figure 7.13. When a portion of the nail plate punctures the soft tissue of the toe, it is called an *ingrown toenail*. Often seen on the hallux, the condition can be quite painful with redness, swelling, and sometimes infection. Risk factors include a genetic predisposition and family history, *hyperhidrosis* (excessive sweating), and poor foot hygiene (Heidelbaugh & Lee, 2009). Constricting footwear and footwear that prevents evaporation or absorption of moisture can worsen the condition.

## 8.7 Foot Variations: Morphology and Deformities

Foot morphology refers to form and structure, going beyond foot length and width dimensions and arch types to include three-dimensional form. Accommodating variations of foot size, shape, and structure in footwear is a challenge. Some shoe styles are available in varying widths, but other factors besides length and width of the foot will affect foot comfort. Lace-up shoes and adjustable straps on sandals, splints, and braces give some mid-shoe size adaptability. Materials that stretch and mold to individual foot shapes can be used, being mindful of support and comfort needs. Other foot variations have been explored related to body mass, gender, and age.

### 8.7.1 Body Mass Effects

As the population of obese people continues to increase, the effects of body mass index (BMI) on footwear size and shape must be recognized. Hills, Hennig, McDonald, and Bar-Or (2001) found that with increased body weight plantar pressure increased, while standing and walking, under the heel, under the arch at the midfoot, and at the metatarsal heads. The highest pressures were found underneath the arch and at the metatarsal heads. Gravante, Russo, Pomara, and Ridola (2003) found that plantar pressure and contact area were greater for obese young adults. Park (2013) used 3D foot scanning to determine effects of obesity on foot morphology. She determined that obesity is positively related to foot width, a finding consistent with the Gravante et al. (2003) study. In addition, the scanning process revealed that obesity not only widens feet (from a two-dimensional perspective), but also increases the height of the ball girth on both feet (i.e., voluminous forefeet). The conclusion is that obesity positively affects both the width and volume of feet, which could affect footwear shaping for obese people.

### 8.7.2 Gender Differences

Gender plays a significant role in foot morphology. Fessler, Haley, and Lal (2005) found that for a given *stature* (total body height), male foot lengths were longer than women's. Using foot scanning with 847 subjects; Krauss, Grau, Mauch, Maiwald, and Horstmann (2008), found some basic differences in foot shapes of men and women. For feet of a particular length, men's feet were both wider and thicker. For both genders, shorter feet are more voluminous (wider and thicker) and longer feet are typically narrow and flat. Additional design considerations arise from physiological gender differences. Smith et al. (2013) found that men perspired more than women, making a case for using different materials in footwear based on gender.

### 8.7.3 Foot Deformities

Foot deformities may be genetic, neurologic, arthritic, traumatic, or footwear-induced. Foot abnormalities require careful choice of footwear to accommodate foot structures that often are not considered in off-the-shelf products. Footwear can be adapted with the use of orthotics. Another, more costly, option is custom-made footwear. The most common foot abnormalities that can affect design and fit of footwear are: bunions, *hammertoes, claw*

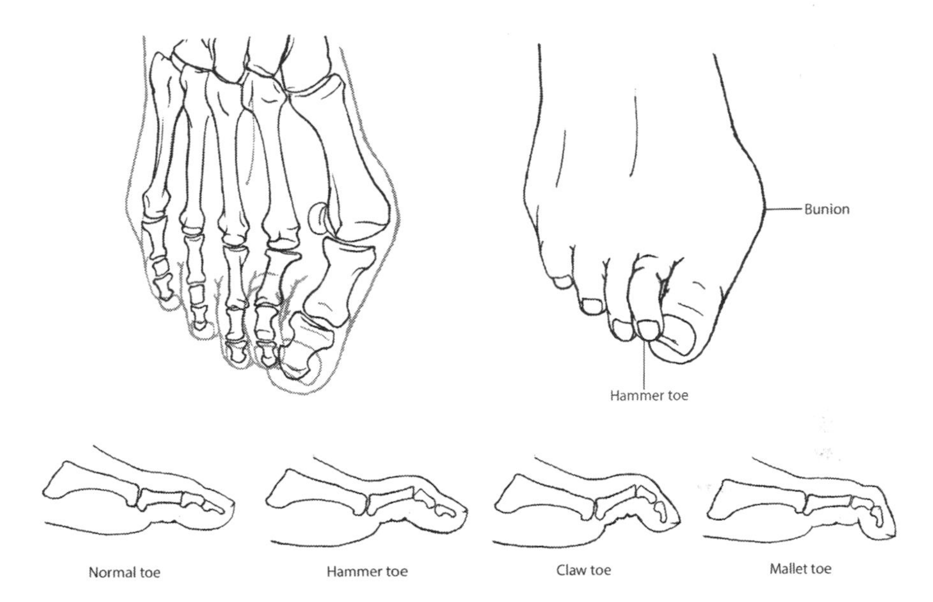


**FIGURE 8.13**
Deformities of the foot and toes.

*toes, mallet toes, pes planus,* and *pes cavus.* See Figure 8.13 for illustrations of foot and toe deformities and Figure 8.9 for an illustration of pes planus and pes cavus.

### Bunions

*Hallux valgus deformity* (bunion) is a common foot problem. Experts disagree if footwear is a cause or a contributing factor to a genetic predisposition. O'Connor, Bragdon, and Baumhauer (2006) state that women are two to four times more likely to develop bunions. Ferrari, Hopkinson, and Linney (2004) measured the hallux angle with the 1st metatarsal and found that women are predisposed to bunions due to natural hallux adduction. However, Mafart (2007) has studied the changes in the human foot over time and states that the presence of a bunion suggests the influence of footwear, especially higher heeled shoes and boots and that women's footwear has had a deleterious effect.

A bunion is characterized by lateral deviation of the big toe at the first metatarsophalangeal joint (joint at the base of the hallux). The result is a noticeable bulge at the medial side of the foot. In many ways, this condition mimics the shape of a pointed-toe shoe with the toe box forming a point instead of a curve to match natural toe alignments and lengths. Bunions may be accompanied with *bone spurs* (projections on the surface of the bone) and bursitis (an inflammation of the bursa). Various splints have been designed to abduct the hallux in attempts to slow progression of the deformity.

#### *Toe Deformities*

A hammertoe has a claw-like appearance as the proximal interphalangeal joint (PIP) linking the distal segments of the toe is constantly flexed, most commonly in the second toe. Similar conditions with descriptive names are claw toe and mallet toe. These conditions develop over time or due to arthritis as the muscles that flex and extend the joints of the toes weaken. The curved and cramped bend of the toe may also be due to wearing footwear that does not fit the length of the toe. Quick childhood growth spurts in children may lead to children wearing shoes that are too small, cramping the toes and resulting in deformities. Some wearable products are used to try to alleviate pain associated with these conditions, including metatarsal pads inside the shoe and cushioning sleeves worn on affected toes.

### 8.7.4 Arch Extremes

Refer to Figure 8.9 to see the arch height extremes of pes planus and pes cavus. Pes planus (flat foot) is a deformity of the medial longitudinal arch with the arch touching the ground. The condition may be *congenital* (develops when the baby is in the uterus), or related to obesity or weak supporting tissues. If there are major structural changes in the ankle joint and tarsal alignment, the foot flattens, resulting in a lower than normal medial malleolus as the ankle shifts inward. The altered alignment may cause boot fit problems. Custom-designed arch supports are often prescribed for pes planus. People who spend hours standing may have temporarily flattened arches. Shoes with good arch and ankle support can help compensate for the problem, or shoes can cause the problem by reducing exercise and normal foot movement. Pes cavus is characterized by a noticeably high arch that does not flatten when bearing weight. Pes cavus is either congenital or idiopathic (unknown cause) and often treated with custom shoes or shoe inserts that support the arch (Tyrrell & Carter, 2009).

## 8.8 The Foot in Motion: Range of Motion, Biomechanics, and Footwear Design

Motion in the ankle and foot incorporates how far joints move and how they move. Footwear affects both factors.

### 8.8.1 Foot and Ankle Range of Motion

Ankle and foot joint motions combine to allow accommodation to the environment and to gait alterations. Although the foot has many bones and many joints, there are two chief motions of the ankle and foot: (a) a simple up-and-down at the ankle and (b) a side-to-side motion in the hindfoot. The talocrural (ankle) joint is a hinge-type synovial joint allowing only a simple up-and-down flexion (plantar flexion, point foot and toes down) and extension (dorsiflexion, point foot and toes up) (Kapit & Elson, 2014, p. 39). Total motion is approximately 75 degrees, with about 10 degrees of extension and the remaining 65 degrees as flexion. See Figure 8.10. The foot inverts about 30 degrees and everts around 15 degrees at the hindfoot, with the motion (illustrated in Figure 8.6) primarily at the subtalar joint (the joint between the talus and the calcaneus) and the other joints between the tarsal bones.

Motion at the subtalar joint is much more complex than the motion at the talocrural joint. The subtalar motion includes movement in sagittal, frontal, and transverse planes. Subtalar joint motion is a major part of foot eversion, moving the foot into a "flatfoot" position; and inversion, moving the foot in the opposite direction to rest on the lateral margin. Inversion looks like the sole of the inverted foot is facing toward the other foot. Eversion/inversion motions help the foot accommodate to rough ground as well as allowing sudden sharp turns and lateral movements.

Extension of the metatarsal-phalangeal joint of the hallux is the one important isolated movement within the forefoot. The motion, when restricted, limits toe-off in walking and severely limits forward propulsion in running. As Fuller (2000) elaborates in his discussion of the windlass mechanism in the foot, tightness in the plantar aponeurosis may contribute to this restriction as well as to bunions. The natural toe motions (see Figure 8.11) rely on coordinated function of the intrinsic and extrinsic muscles of the foot. The ability to perform these motions, along with the range of motion for them all, varies from one person to the next. Actively using the foot muscles results in greater flexibility and ability. Individuals born with upper limb deficiencies can show remarkable foot dexterity, using their feet and toes for activities usually done with the hands.

### 8.8.2 Foot and Ankle Biomechanics

Footwear affects foot biomechanics, so the dynamic biomechanics of the foot should be considered in footwear design. This section presents a starting point for understanding the actions and interactions of anatomical foot structures. Running or walking both start with actions of the feet and ankles. The sequence for walking begins with *heel strike*, landing the body weight on the heel of one foot on the ground. Heel strike transitions to *foot flat/mid-stance*, with full foot-ground contact and full weight-bearing on the foot. The final motion of the gait cycle is *toe-off* with the hallux and other toes bending and pushing the body forward. The sequence overlaps and repeats with the other foot.

Walking motions and running motions are different. While running, both feet are off the ground at the same time for a period of the time. When running and barefoot, or wearing barefoot-style shoes, heel strike is often eliminated or at least minimized and replaced with a different foot strike pattern, with initial ground contact landing more of the foot (Lieberman, 2012). Variations to heel strike in running are midfoot strike and forefoot strike with the person's natural gait landing first on one of these areas of the foot. The propulsive force of toe-off becomes more important, particularly as speed increases. In athletic shoes, runners tend to land with a heel strike more often than with a midfoot or forefoot strike. Designers often strive to facilitate and protect motions of heel strike and toe-off through midsole and outsole shaping and materials.

### 8.8.3 Athletic Shoes, Minimalist Shoes, or No Shoes, Which Is Best?

Which is best? In recent years, a great deal of research has focused on this question. The whole body is involved in and impacted by gait (see Chapter 9), so footwear design choices are of tremendous importance (Schrödter, Brüggemann, Hamill, Rohr, & Willwacher, 2016).

Footwear designed for specific activities, such as running, suggests the possibility of adapting to functional biomechanical variations in the foot. "Arguably, the purpose of the running shoe or orthosis is to control the timing, rate, and excursion of the subtalar joint. . . [and of the leg joints and muscles] for optimal shock absorption and force transmission" (Yamashita, 2005, p. 802). Yamashita provides a detailed discussion of indications for and characteristics of three categories of running shoes: cushion shoes, stability shoes, and motion-control shoes. Issues as diverse as the amount of dorsiflexion of the hallux, variations in the angulation of the subtalar joint as well as arch height and angulation of the forefoot relative to the rest of the foot need consideration, both in static and dynamic activities, for optimum design.

In contrast, Lieberman (2012) challenges the utility and use of running shoes, from an evolutionary perspective. He states, "Running injuries are highly multifactorial, and no single factor, such as shoe design, will explain more than a fraction of the injuries" (p. 71). He advocates greater attention to running form than to footwear. Richards, Magin, and Callister (2009), reviewing 32 publications—clinical trials and systematic reviews—from 1981 to 2009, were unable to locate evidence-based research to support use of running shoes with elevated cushioned heels and pronation control to promote health and performance for distance running. (Pronation is a combination of hindfoot/midfoot eversion and forefoot abduction, while the opposite motion is supination; a combination of hindfoot/midfoot inversion and forefoot adduction.) They conclude: ". . . the true effects of PCECH [pronation control, elevated cushioned heel] running shoes on the health and performance of distance runners remain unknown" (p. 160).

Parallel to these concerns, footwear manufacturers have developed many variations of a *minimalist shoe*. Research on the effectiveness of such designs requires a common language and standard for evaluation. Working with international experts, Esculier, Dubois, Dionne, Leblond, and Roy (2015) developed a definition of minimalist shoes; "Footwear providing minimal interference with the natural movement of the foot due to its high flexibility, low heel to toe drop, weight and stack height, and the absence of motion control and stability devices" (p.1). They also developed and tested a rating scale, the *Minimalist Index*. Chen, Sze, Davis, and Cheung (2016) found significant foot and leg muscle volume increases in runners who made a gradual (six-month) transition from training with traditional running shoes to use of minimalist shoes with a Minimalist Index score of 92%. Squadroni, Rodano, Hamill, and Preatoni (2015) found differences between brands when they compared foot strike pattern and kinematics during running in a number of minimalist shoes.

### 8.8.4 Orthotic (Orthosis) Purpose and Design

Orthosis is a word from Greek, meaning to make straight. Torrens et al. (2012) define an orthotic as "the mechanical support, or orthosis, that enables the range of motion of a given joint or joints, but limiting movement beyond the norm" (p. 145). Tyrrell and Carter (2009) say that foot orthotics are custom-made mechanical supports fashioned to alter the interaction between the foot and the ground and/or to alter the alignment of the anatomy of the foot. Foot orthotics typically require precise construction to meet therapy goals. The book *Therapeutic Footwear* (Tyrrell & Carter, 2009) gives detailed directions on measuring and fitting footwear for special needs. See Figure 8.14 for an illustration of some orthotic designs. The simple orthotic insert illustrated in 8.14-A replaces the right shoe insole. Note the contour of the medial arch. Another orthotic design of pliable textiles with foot and ankle coverage is illustrated in 8.14-B. The shoe must coordinate with any orthotic; the orthosis and shoe are considered as a single therapeutic unit (Tyrrell & Carter, 2009).

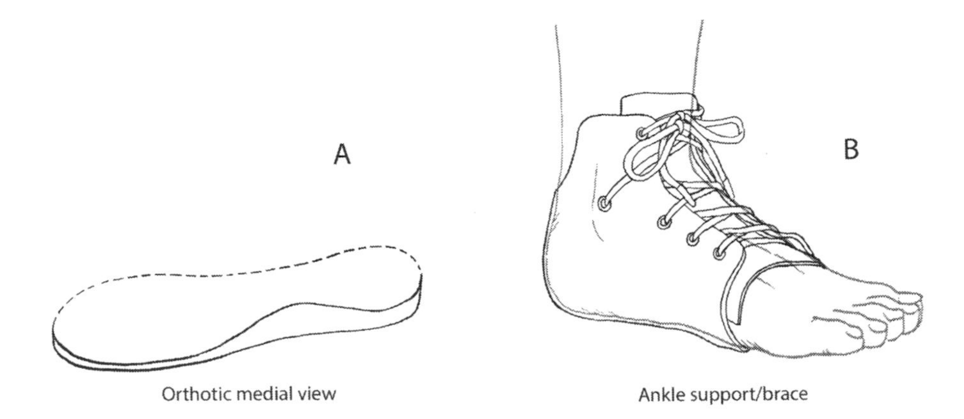


**FIGURE 8.14**
Ankle and foot orthotics for the right foot.

Yamashita's (2005) introduction to construction of orthotics for runners presents a sound basis for designers, but also illustrates the importance of detailed biomechanical analysis, skill, and experience in assessing foot problems. As with shoe styles, clear evidence regarding the efficacy of use of orthotics and insoles is difficult to find. Leppänen, Aaltonen, Parkkari, Heinonen, and Kujala (2014) reviewed randomized, controlled studies of insoles/orthotics worn for sports activities and suggested the inserts are helpful. However, they suggested caution in interpreting the studies because of a perceived risk of bias. Bonanno, Landorf, Munteanu, Murley, and Menz (2017) studying clinical trials of foot orthoses and shock-absorbing insoles for injury prevention found benefits in some areas of injury prevention for foot orthoses, but none for shock-absorbing insoles. They also advised interpreting the results with caution due to concerns about the quality of the methodology in the research studies they reviewed.

Attesting to the importance of foot function throughout the lower extremity kinetic chain, Barton, Menz, and Crossley (2011) identified patient criteria that predicted successful treatment of patellofemoral pain in subjects with pronated feet. They used a prefabricated, non-customized foot orthosis with arch support and rear foot wedging. Foot orthosis effects also occur further up the kinetic chain, above the knee in the muscles of the thigh and pelvic girdle (Hertel, Sloss, & Earl, 2005) and even into the paraspinal muscles in the low back (Bird, Bendrups, & Payne, 2003).

## 8.9 Footwear Fit and Sizing

Because feet are important to whole body function and comfort, good footwear fit is crucial. Selecting the optimal size is important. Methods to size protective footwear products are similar to methods used for general use footwear. Footwear fit and sizing methods are different from methods used to size other apparel items. Most footwear fit and sizing depends on the form and dimensions of a footwear model called a *last*.

### 8.9.1 Translating the Human Foot to a Last

Last methods are based on hundreds of years of tradition and craftsmanship that have evolved with changing technologies. Rather than a three-dimensional representation of the foot, the last is "the model form or shape around which most footwear is made" (Tyrrel and Carter, 2009, p. 28). The last incorporates fit and style ease. Fit ease can also be incorporated into adjustable fasteners like shoe lacings. Luximon and Luximon (2013) give detailed step-by-step information on how lasts are made. See Figure 8.15 for illustrations of shoe lasts (from left to right): athletic shoe, flat shoe, cowboy boot, and high-heeled shoe. The last provides locations to secure the shoe to the foot and reflects the form and style of the shoe, while allowing a degree of fit and comfort for the intended wearer. Figure 8.15 also illustrates the relationship of a high-heeled shoe last to anatomically positioned foot bones, demonstrating that the human foot and such a shoe last are not identical. The heel cross-section (A) displays the last symmetry and how the hindfoot outline and foot bones do not match the last outline. At the instep cross-section (B) at the midfoot-forefoot junction, there are notable differences. The last is not only narrower than the foot, but also shallower. The last shows little elevation under the arch, however, an insole may be added to the shoe to provide some contour and support to the arch. The cross-section at the phalanges (C) reveals compression of the hallux, medially and superiorly, while the smaller phalanges have some space allowed in the last above the toes which will form the shoe toe box. Note that the forefoot at this level is wider than the last.

The athletic footwear industry is contributing to changed models of the foot with sophisticated assessment tools like 3D foot scanning (Saunders & Chang, 2013; Rajulu & Corner, 2013). Yamashita (2005) illustrates last variations developed for running shoes intended to compensate for specific biomechanical/motion patterns in the foot. New methods of measuring and assessing human foot morphology are changing industry concepts of foot shapes, sizes, and shoe construction. Nácher et al. (2006) developed a model for predicting footwear fit based on 3D scan and shoe fit preferences

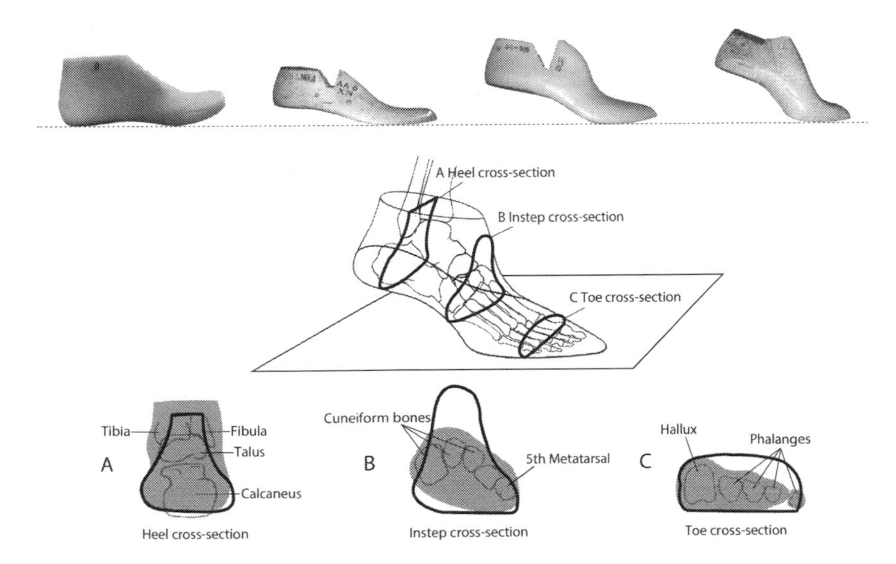


**FIGURE 8.15**
Shoe lasts and last to foot relationships. Shoe lasts from the collection of Patrick Kerrigan, Lone Oak Leather, Andover, MN. (Shoe last and foot outlines with permission from: "Shoe-last design templates," by Y. Luximon and A. Luximon, 2013, *Handbook of Footwear Design and Manufacture*, A. Luximon, Ed. p. 223. Copyright 2013 Woodhead Publishing Limited.)

of 316 females. Large-scale studies of foot shapes and sizes are possible with fast, accurate scanning methods.

As in apparel, footwear is graded to produce the range of sizes for the target market. Footwear grading has been based on custom and tradition. Sokolowski (1999) used a method similar to methods used for apparel pattern development and grading to study women's foot proportion changes through a range of sizes. Five women were selected from 161 volunteers to represent a sizing system (XS, S, M, L, XL) that had been developed from statistical analysis of a large anthropometric data base. Plaster foot models, made by casting the dominant foot, were partitioned into 24 sections, with the sections based on traditional foot landmarks. The plaster models were then draped, and the draped pattern sections were compared through the five sizes. A major finding was that, unlike in standard foot sizing practice, the eight sections of the plantar aspect (sole of the foot) had no linear relationship between sizes. Some sections did not change in shape or dimensions, while others changed a great deal. This exploratory study illustrates that foot proportions are far more complex than current footwear sizing methods represent. 3D scanning offers the possibility of new options for grading. Mochimaru, Kouchi, and Dohi (2000), applying allometric principles and computer graphic deformation techniques, manipulated foot scan proportions to produce a 3D last pattern for a wide foot from a 3D scan of a narrow foot. Krauss, Valiant, Horstmann, and Grau (2010) compared women's feet to last shapes used to produce running shoes for women. The lasts did not reflect female foot shapes, bringing the practice of "down-grading" men's lasts for women's shoes into question.

### 8.9.2 Footwear Fit, Sizes, and Size Systems

Developing a sizing system and communicating the system to consumers is an integral part of designing footwear. Footwear is widely available in retail outlets and online, so providing consumers with methods of measuring their feet to select the best size is necessary. Customers can select a size by using a measuring method like the Brannock™ device (Appendix A), size charts, or selecting a size based on past experience with a brand that fits well. Online retailers have developed simple measuring tools such as printable foot templates for the customer to determine foot length and width related to that retailer's sizing scheme.

Size selection may also be based on intended use. Athletes may have different requirements when selecting a size compared to the average consumer. McHenry, Arnold, Wang, and Abboud (2015) found, in a study of 56 rock climbers, that 55 of the climbers wore ill-fitting, excessively tight footwear and that 91% of the participants experienced foot pain but tolerated it due to the sport's necessity of securely gripping rock without the shoe slipping on the foot. The climbers' acceptance of very tight footwear carried over to street shoes. They wore street shoes much smaller than their measured foot would indicate. Other researchers found that rock climbing leads to foot deformities and injuries including, among other maladies, infected nails, cuts, bruises and bunions (Killian, Nishimoto, & Page, 1998). Suffering for a sport may be part of its customs.

Sizing systems vary by country and comparing meanings and numbering schemes from one part of the world to another is confusing. In general, sizing systems can be grouped into two broad categories with origins in either Europe or Asia. The European (sometimes called Paris Point), U.S., and UK systems are based on the last length, which incorporate an ease component, indicated by a size number, for example European 42. The English system is the oldest. Some systems use it as a starting point, but then revise size labeling. The Japanese or Chinese system is based on actual foot length. Most of the systems are based on foot length and girth. For example, the American system relates length to a numerical label and width to a letter label (AAAA is the narrowest and EEE is the widest). Most American over-the-counter shoes are offered in limited widths, often only average, with wide offered for some styles.

The International Standards Organization (ISO) is attempting to clarify the process of size development and selection by issuing standards so that size systems can be "translated" from one system to another. The ISO recently released two standards (2015): *Footwear sizing: Conversion of sizing systems*, ISO/TS 19407:2015; and *Footwear sizing: Vocabulary and terminology*, ISO/TS 19408: 2015. Footwear designers are advised to keep up-to-date on footwear sizing systems and how each system relates to another. Questions to consider: Can your footwear product fit and function with a simplified, S, M, L, XL, system without compromising on function? Where will you market your product, regionally or worldwide; and what sizing system will you use to convey your footwear sizes?

Databases for feet are not as readily available as databases for other body areas. Some three-dimensional measurement data bases are included in large-scale anthropometric studies, like ANSUR I and ANSUR II. As discussed in this chapter, the last—a three-dimensional representation of the foot—is used to shape most footwear. More work compiling three-dimensional databases that incorporate the complex interactions of two-dimensional measurements with morphology are needed. Arch type categories, beyond the simple approach of footprint outlines, are needed to truly change how footwear is designed, shaped, and sized. With quickly advancing technologies this information and new footwear modeling are on the horizon.

## 8.10 Foot and Ankle Landmarking and Measuring

Beyond using available databases and commercially available lasts, understanding foot and ankle landmarks and measurements is necessary for any type of footwear design. Landmarks and measurements for footwear should be based on purpose or use, materials used, product coverage, and—because sizing systems are different around the world—a determination of the market for the product. Review basic instructions for landmarking and measuring in Chapter 1. See Appendix F for details on foot and ankle landmarking and measuring instructions and for illustrations.

Footwear types and purposes vary greatly, so the choice of landmarks and the measurement methods will also vary. Locating and defining the midfoot with landmarks and measurements will assure optimum location of materials at the arch. As discussed earlier in this chapter, influencing the design and shaping of the last that will be used as the base for a design is as important as understanding basic foot and ankle landmarks and measurements. Footwear shaping and sizing is being revolutionized by new measuring methods, especially 3D scanning that is now being used to collect two-dimensional measurements and three-dimensional form information. 4D scanning adds a fourth dimension, time, to 3D scan data. This technology, which incorporates motion, will benefit athletic footwear design. As a more sophisticated understanding of foot morphology evolves, footwear design and fit can improve.

### 8.10.1 Product Purpose

Choice of landmarks and measurements depend on the product purposes. As with other wearable products, footwear purposes and products fall into broad categories: aesthetic or functional, with some overlap.

Footwear with an aesthetic focus often follows social and cultural dictates; some reflecting fashion trends (Kern, 1975; Riello & McNeil, 2006; Rossi, 1977; Wright, 1922). As demonstrated in the Figure 8.15 illustration, fashion shoe lasts may bear little relationship to real feet, particularly in cross-section. Park and Curwen (2013) conducted in-depth interviews with women ranging in age from 18 to 85 and found that they were always compromising between the ideal and reality when selecting fashion footwear; sacrificing comfort for what they thought of as footwear that meets fashion expectations.

In the functional/protective footwear category, thermal, impact, abrasion, puncture, and contamination protection all have different design and specific shaping requirements. For example, steel-toed work boots are intended to protect the distal forefoot from impact for those in construction trades. Sports or athletic shoes are designed, shaped, and sized to meet the motion and support needs of the athlete. For feet with anatomic or physiologic variations, or specific functional limitations, therapeutic footwear incorporates support, protection, and special accommodations for the total foot or for selected areas of the foot.

### 8.10.2 Product Materials

Soft materials shape to the foot and ankle, decreasing the number of measurements needed, while rigid materials require a greater understanding of foot morphology to avoid constricting the foot, or risking abrasion at contact points. Knitted socks and hosiery are typically offered in limited sizes and simple shapes, from tube socks to stockings shaped to the calcaneus contour. Incorporation of the stretchy fiber spandex generally makes limited size offerings adaptable to a range of foot sizes. Compression and support stockings for the feet and legs also use elastic materials, but of varying degrees of stretch, sometimes within one item, thus requiring more detailed and extensive landmarking and measuring procedures. Pressure, measured in millimeters of mercury (mmHg) in compression garments, may decrease from relatively high pressure in the foot, to lesser values in the leg and thigh to accommodate the mechanics of the circulatory and the lymphatic systems. Cold weather boots, in addition to requiring insulation and waterproofing materials, also require attention to fit, to help prevent cold injury (Hamlet, 2000).

Leather, the traditional shoe material, has natural characteristics of human skin. It is protein based, pliable, shapeable, and able to absorb and transport perspiration (O'Keefe, 2012, p. 18). The leather upper of a shoe is typically smoothed over the last, relying on the natural capability of the leather to adapt to a shape (O'Keefe, 1996, pp. 17–19). When a person wears the shoe, the expectation is that the leather will, to a degree, shape to variations within a size. A leather shoe can, to a degree and over time, mold to the foot, according to the leather's thickness and pliability. Synthetic leathers are also used for footwear. These products are often less expensive than leather and can be engineered to shape to the foot using the thermoplastic characteristics of the material. Synthetic materials are moisture resistant, but do not allow foot perspiration to escape.

Protective footwear uses slightly different materials from those for general use footwear. Hiking boots that need to protect from irregular rocks and spiny plants are made of thicker, but less flexible leather than street shoes. Astronauts' boots must be both pressurized and heavily insulated to withstand the vacuum and temperature extremes of space.

Materials used for an insole with some type of formed arch will affect stance, gait, and comfort. Witana, Goonetilleke, Xiong, and Au (2009) found that people are very sensitive to materials that underlie the midfoot. Subjects in the study were able to identify changes in both midfoot shape and cushioning material. The instep also appears to be an area of discomfort for alpine skiers (Pinter, Eckelt, & Schretter, 2010)

### 8.10.3 Product Coverage

Footwear product coverage varies from minimalist shoes to thigh-high boots. Minimal coverage footwear includes "flip-flops," with a cushioning sole held in place by a "thong" device held between the hallux and the second toe. Minimal coverage footwear must, with some method, adhere to the foot while providing the desired product purpose. Fasteners need to avoid sensitive anatomical structures, should secure the product to the foot throughout foot motion, and should not restrict foot function. Look for anatomical features that can anchor the product without causing undue stress to the foot.

Maximum coverage footwear includes products such as military boots, winter boots, and, in the extreme, astronaut boots or deep-sea diving boots. Design of maximum coverage footwear demands detailed measuring schemes, a comprehensive understanding of the three-dimensional form of the foot and ankle, and of the chain reactions footwear can cause throughout the body. Footwear of rigid materials covering naturally mobile foot and ankle areas may cause injury in the short or long term.

### 8.10.4 Combined Product Purpose, Product Coverage, Product Materials

Alpine ski boots illustrate design decision interactions concerning product purpose, materials, and product coverage. Alpine ski boots are made of an outside hard shell and an inner pliable liner (also called an inner boot). The rigid shell is needed because downhill skiing puts significant mechanical stress on the foot and ankle. The liner allows some fit adjustability, especially if thermoformable, while insulating the foot and ankle. Colonna, Nicotra, and Moncalero (2013) conducted an in-depth review of alpine ski boot materials, designs, and standards. They describe the challenges of encasing the foot in a hard plastic shell with the design goals of efficient transfer of loads from the skier to the ski edge, quick connection of the boot with the binding along with quick release in case of a fall, shock absorption, protection from injury related to overloads during falls, and comfort including uniform pressure and optimum temperature and humidity inside the boot. If a boot is too loose the skier can't control the ski edges; if a boot is too tight it is uncomfortable. Freestyle and mogul skiing require some flexibility of specific areas of the boot, so some styles use a bellows-type tongue that can flex, yet spring back to its original shape. Several studies have shown that the cuff height of the rigid boot (the coverage) affects the location of leg injuries. A high boot can cause knee injuries, while lower boot heights are known to cause "boot-top" fractures (Höflin & van der Linden, 1976; Karpf, Mang, & Hoerterer, 1982; Lyle & Hubbard, 1985; Shealy & Ettlinger, 1987).

## 8.11 Designing for Foot and Ankle Anatomy: Conclusion

Feet and ankles are complex in form and function. They are, perhaps, more vulnerable than other body regions due to our upright mobile nature and the wide variety of environments in which we move. Foot health is likely to improve when individuals vary the footwear they use in the course of a day and week.

Designers using this overview of anatomy, structural variations, and foot and ankle function can expand the range of products available to consumers. Collaboration among footwear designers, health care providers, and materials and structural engineers enhances both the design process and the resulting end-products. Some excellent resources for footwear designers include: *The Science of Footwear* (Goonetilleke, 2013), *Handbook of Footwear Design and Manufacture* (Luximon, 2013), *Neale's Disorders of the Foot: Clinical Companion* (Frowen, O'Donnell & Gordon, 2010), *How Shoes Are Made* (Motawi, 2015), and *Therapeutic Footwear* (Tyrrell & Carter, 2009). Although the last two give guidelines for selecting and designing footwear for specific foot problems, the guidelines are also useful in designing footwear for the normal, healthy foot.

# 9

# *Designing for Whole-Body Anatomy and Function*

Designing wearable products which provide a successful interface between the wearer and an environment requires integration of information about the full spectrum of human anatomy and function and detailed knowledge of the environment. Chapter 9 takes a macro approach, looking at the whole body, and describes some factors to consider in whole-body wearable product design. Zoom out from the regionally-focused discussions of previous chapters to think about how to apply your knowledge and to develop design approaches for the whole body.

**Key points:**

- Design for the whole body involves consideration of all or most body systems.
- The static body is the foundation for wearable product design.
- Static whole-body variations—height, posture, and age—influence wearable product design strategies.
- Designing for the dynamic body, the body in motion—from subtle position changes to continuous motions of walking and running—requires accommodation of range of motion (ROM) in varying degrees.
- Extended body positions—bending and stretching—require accommodation of ROM maximums for all involved body regions.
- Simulated body forms, manikins and digital body models, are useful tools in wearable product design when based on human body anatomy and function.
- The Extra-Vehicular Activity (EVA) space suit is the ultimate microenvironment and serves as an example of the designer's need to understand and incorporate human body structure and systems into wearable products.

Total body (whole-body) landmarking and measuring methods are described in Appendix H.

## 9.1 Whole-Body Wearable Products

A whole-body wearable product may be an all-in-one product covering most of the body, or several products worn together. In many cases, the whole-body wearable is intended to protect the body, and in some cases the environment is protected from the wearer.

### 9.1.1 Extent of Coverage

Some wearables, especially protective wearables, cover the torso, limbs, head, feet, and hands. Several specific terms apply to whole-body wearable products—for fashion and function. Fashion terms *unitard, body suit,* and *"cat" suit* imply one-piece products with upper torso, lower torso, arm, and leg components combined. Performers may wear whole-body coverage; consider sports mascots' costumes, which enclose the entire person yet require full ROM. Functional whole-body product terms include *coverall* and personal protective equipment (PPE). A coverall, as the term implies, is an all-in-one garment with attached hood and perhaps foot and hand coverings.

PPE is worn for many work situations and combines a coverall with additional equipment, like a mask or respirator. PPE level of protection varies relative to the hazards in the environment, with Level A as the highest and Level D the lowest. ASTM Committee F23 has jurisdiction over PPE standards published by ASTM. The committee web site categorizes standards as: physical, chemical, biological, radiological, and flame and thermal hazards (https://www.astm.org/COMMITTEE/F23.htm). The Occupational Safety and Health Administration (OSHA) and National Institute for Occupational Safety and Health (NIOSH) provide PPE selection guidelines.

An *ensemble* of products, like retail winter wear, is also considered whole-body coverage. Assembling separate winter wear products meets a goal—staying warm (Figure 9.1). The products—boots, pants, parka, mittens, face mask, and hat are assembled to protect the person from the cold environment. If one piece in the ensemble lacks thermal protection or the layers severely limit motion, the total ensemble fails to protect the wearer.

### 9.1.2 Whole-Body Products for Protection from the Environment

The body needs protection in extreme environments. Significant environmental threats—temperature extremes, chemical and biological hazards, solar and radiation effects—require whole-body protection. One design strategy, to eliminate as many points of body exposure as possible, encases the body in a protective envelope. The thermal protection *immersion suit* designed for first responder cold water rescue is a good example (Figure 9.2). The full-body insulating suit, as a barrier between the body and cold water, prevents

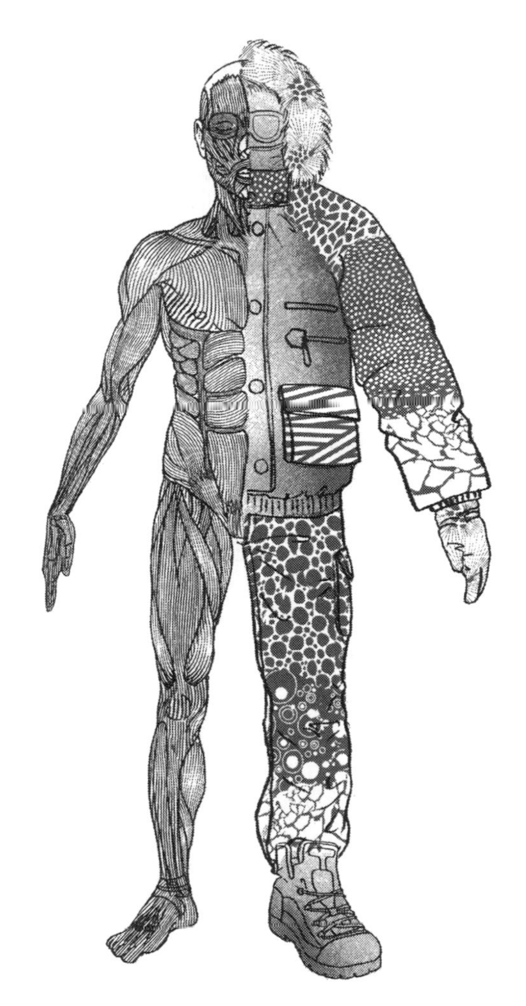

**FIGURE 9.1**
Ensemble of products for cold environment protection.

heat loss by conduction, a universal heat transfer mechanism. Without a protective insulating layer, a person immersed in cold water quickly loses body heat to the surrounding water and hypothermia sets in. Cold-water immersion suits reduce the rate of temperature loss by limiting the body surface area in contact with the water (Tipton, 1989). The suit must maintain a dry layer of insulation next to the skin to meet the protection goal. Sweeney and Taber (2014) describe the trade-offs of immersion suit design: donning simplicity for quick response to an emergency while assuring a water-tight seal at the closure and any other openings. The suit must also fit a variety of body types and sizes while maintaining its protective qualities.

Other whole-body specialized wearable products include asbestos abatement suits, *blast furnace suits, explosive ordnance (bomb) disposal suits,* and

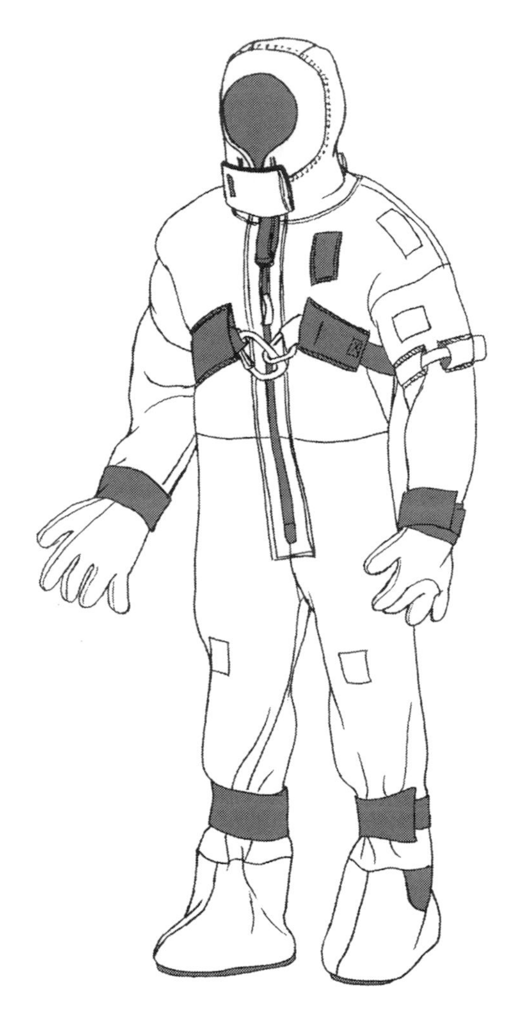

**FIGURE 9.2**
Total body immersion suit for cold water rescue.

hazardous materials *(hazmat)* suits. Trade-offs in whole-body wearable product design are typical. The major design objective must be met; other objectives are secondary.

### 9.1.3 Whole-Body Products to Protect the Environment

Some protective wearable products are designed to protect the environment, or others in the environment, from human contamination. Cleanroom suits prevent human contaminates—hair, dead skin cells, and clothing fibers—from infiltrating delicate mechanisms as they are manufactured. A surgeon's total-body ensemble protects the patient from contamination by the surgeon. And, the surgeon may need protection from a patient's blood-borne disease. In this wear situation, meeting protection goals for both the surgeon and the patient is essential.

### 9.1.4 Physiological Effects of Wearing Whole-Body Products

Protective wearable products are essential in many settings; however, full-gear protective ensembles can pose a physiological strain on the wearer (ASTM, 2018, 2016c). Encasing the body increases thermal stress in most environments. In high-temperature environments, heat stress associated with protective gear may become life-threatening. A full-body protection wearable that limits motion adds to physiological strain. Designers who understand anatomical structures and functions can reduce negative product effects by carefully selecting materials and designing product forms to work with the body. An essential part of this design approach is understanding the static body as the foundation for a design and incorporating dynamic body dimensions into the design.

## 9.2 Designing for Total Body Anatomy: The Static Body

A wearable product must fit the static body and move with the dynamic body for aesthetics, comfort, and safety. The static, upright body, is typically the starting point for designing wearable products. The systems that shape the static body are the skeleton as the framework, muscles as the connectors and stabilizers, and fat distribution filling out the body form. Refer to Chapter 1, Figure 1.5, to see an illustration of a person in the static anatomical/anthropometric pose. Observable total body characteristics are height, width/depth (girth), and posture. As a first step, the designer thinks of a product, for example a coverall, on a person in this pose. The person is pictured in general terms, not as an individual—especially when designing for a target market. Market segments are often based on height categories plus assumed height-to-girth proportions. Individual posture variations are not factored into the design and fit of the product. However, posture influences how a product shapes to the body.

### 9.2.1 Static Body Characteristics: Height and Width/Depth Indicators

Static body characteristics—height and width/depth indicators—influence product fit, function, and aesthetics. Individuals can be characterized by their height to width and depth relationships. For example, people may be described as tall and thin, tall and muscular, or short and stocky.

### Height

*Height* is the measure of the total standing body, from the floor to the top of the head. Height reflects the sum of the "vertical" skeletal components as modified by habitual posture, primarily of the spine—but pelvic tilt and muscle tightness, especially at the iliopsoas, play a role. Height is determined by a person's anatomy, particularly the lengths of the long bones of the legs, and total skeletal structure. Bone grows in response to mechanical stresses put on it by the pull of muscles. Exercise helps bones stay strong and plays an essential role in preventing osteoporosis with associated vertebral compression fractures and height loss. Genetic heritage, nutrition, health status, and age all influence height.

Stature is often used interchangeably with height. For example, the CAESAR® study uses the term stature for total body height. Other body height dimensions include acromial height, axilla height, calf height, cervicale height, chest/bust height, and neck height. See Appendix H for the landmarking and measurement techniques used to determine these dimensions. These height dimensions divide total body height into segments that relate to product positioning. An over-the-body measure, vertical trunk circumference, is used to develop one-piece products like coveralls which encompass the upper and lower torso. The measurement is taken with a tape measure from one shoulder down to the groin, between the legs, and around the body returning to the same shoulder. When compared to height the measurement suggests a person's height to width/depth distribution. See Appendix H for definitions and anthropometric data collection techniques.

### Width/Depth Indicators: Weight, Circumference(s), Girth

Weight and circumference(s)—including girth(s)—reflect body fat quantity and fat patterning and indirectly convey information about body width and depth. BMI or body mass index, the weight-to-height relationship used in health and medical fields, gives a rough approximation of total body fat in adults, but not fat distribution. Most individuals know—exactly, or approximately—what they weigh.

Circumference measurements are typical width/depth indicators used for wearable products. Circumference measurements—bust/chest, waist, and hip—help capture the three-dimensional nature of a person, but do not give an idea of distribution of the circumference measurement, front-to-back or laterally. The skeleton is the primary contributor to shoulder and hip widths. Muscle bulk, more so in men, and fat, particularly at the hip in women, add to these body widths. Two-dimensional width comparisons between the bust/chest, waist, and hip levels, measured from an anterior view, gives an idea of torso fat distribution. The primary contributor to body depth is fat, followed by muscle and the skeleton. Depth, measured from a lateral view, at the level of a circumference measurement combined with the position of the coronal orienting plane, shows front to back body mass distribution. *Girth* is a measure around the middle of something. Waist circumference is typically the body's girth measurement. However, as discussed in Chapter 6, locating a person's waist is difficult, so girth alone is insufficient as an indicator of a person's width-to-height relationship. Visual analysis of a body's width/depth to height relationship is often used to develop body type categories. Chapter 6, Section 6.8 describes body type categorization methods based on mid-torso assessment including body girth. The categorization methods span from Sheldon's (1940) categories of ectomorph, mesomorph, and endomorph to recent 3D body scanning methods developed specifically for wearable product design. Ashdown (2014) describes the difficult task of categorizing the numerous combinations of body shapes and sizes during development of wearable products and size selection methods, especially when trying to limit numbers of body measurements. Pei, Park, Ashdown, and Vuruskan (2017) propose a new sizing methodology with applications for whole-body wearable products which may solve some of the problems.

#### *The Challenge of Fitting Whole-Body Products Using Static Body Characteristics*

Fitting whole-body products requires an indication of weight related to height. Although people usually know their height and weight, and it seems logical that a person could choose the right size for a whole-body product based on height and weight, whole-body product sizing isn't simple. At one time, men's suit categories were based on a height-to-depth relationship, with suit proportions described as average, stout, and slim. Using height and weight as a sizing method may suffice for coveralls and other whole-body products which don't conform to the body. Pantyhose also use height/weight bi-dimensional size charts, but pantyhose four-way stretch materials allow fit to a variety of height and weight combinations. A new whole-body sizing system incorporating relationships between height, muscle bulk, and fat quantity and fat patterning is needed.

### 9.2.2 Posture: A Static and Dynamic Body Characteristic

Each individual has a habitual posture, a distinct carriage, characterized by how regions of the body align relative to one another. Posture, the body's position in space, is normally a dynamic state. It is inherently three-dimensional and implies balance about a center of gravity. Spinal alignment is central to posture. Posture variations add to the challenge of designing whole-body wearable products. Ready-to-wear companies develop clothing designs using fit models and/or manikins as the basic form for designs. The model and/or manikin determines fit and form of all products. Models and manikins usually have erect and aligned postures, while customer postures vary greatly. Figure 1.12 in Chapter 1 illustrates a Misses size 12 manikin used to develop womenswear designs, and the outlines of scans of three women, all Misses size 12. Compare the postures of the manikin and the women. Visualize a product shaped to each posture type and imagine how the product will drape differently on each. Mass-produced products have not incorporated posture variations, but body scanning technologies offer the possibility of developing posture variation categories, along with simple height and width/depth categories. Adding posture characteristics to customary anthropometric measurements could result in better product fit.

### 9.2.3 Aging Effects on Body Height and Posture

Normal development and aging change human body height and posture through a lifetime. Posture is a whole-body characteristic that modifies body height and affects wearable product placement, particularly for whole-body wearable products. Typical female body changes can be seen in Figures 9.3A and 9.3B. The greatest differences are evident at the ends of the age scale.

Notice that the infant's spine is C-shaped (left side of Figure 9.3A), so a one-piece baby garment should shape to the convex posterior body curve. In a baby's first year, significant changes occur in the spine. As the baby starts to crawl and lift his/her head, a slight curve forms in the cervical region, and when a child starts to walk, another curve forms in the lumbar region. The resulting slight S-curve of the spine makes sustained bipedal mobility possible (Saladin, 2007, pp. 260–263).

Relatively greater brain development before birth makes an infant's head proportionally larger than the torso and legs. This relationship makes it important to obtain age-related anthropometric data when designing infant and toddler garments that are donned and doffed over the head (see also Chapter 3, Section 3.10). Throughout infancy and childhood, the figure is characterized by an anteriorly protruding mid-torso. There is no defined waist indentation for suspending lower torso products, so most children's whole-body wearable products are suspended from the shoulders. These distinctive proportions gradually change, until they become more adult-like by middle childhood.

Moving to the right, the fourth set of body forms in Figure 9.3A demonstrates a female body at puberty. The mature figure starts to emerge, and secondary sexual characteristics become evident (refer to Chapters 4 and 5). The female starts to develop breasts and the waist curve appears as the pelvic girdle widens. Differences at this stage are apparent person-to-person, depending on fat distribution and muscular development. The fifth set of body forms in Figure 9.3A illustrates the adult female with fully attained height and developed body form.

Figure 9.3B, left side, continues the age progression with anterior and lateral views of a pregnant female. The sequential breast and torso changes of pregnancy, as seen in Figure 5.9 and discussed in Chapters 4 and 5, lead to the late-pregnancy postures seen here. Notice the exaggerated curve

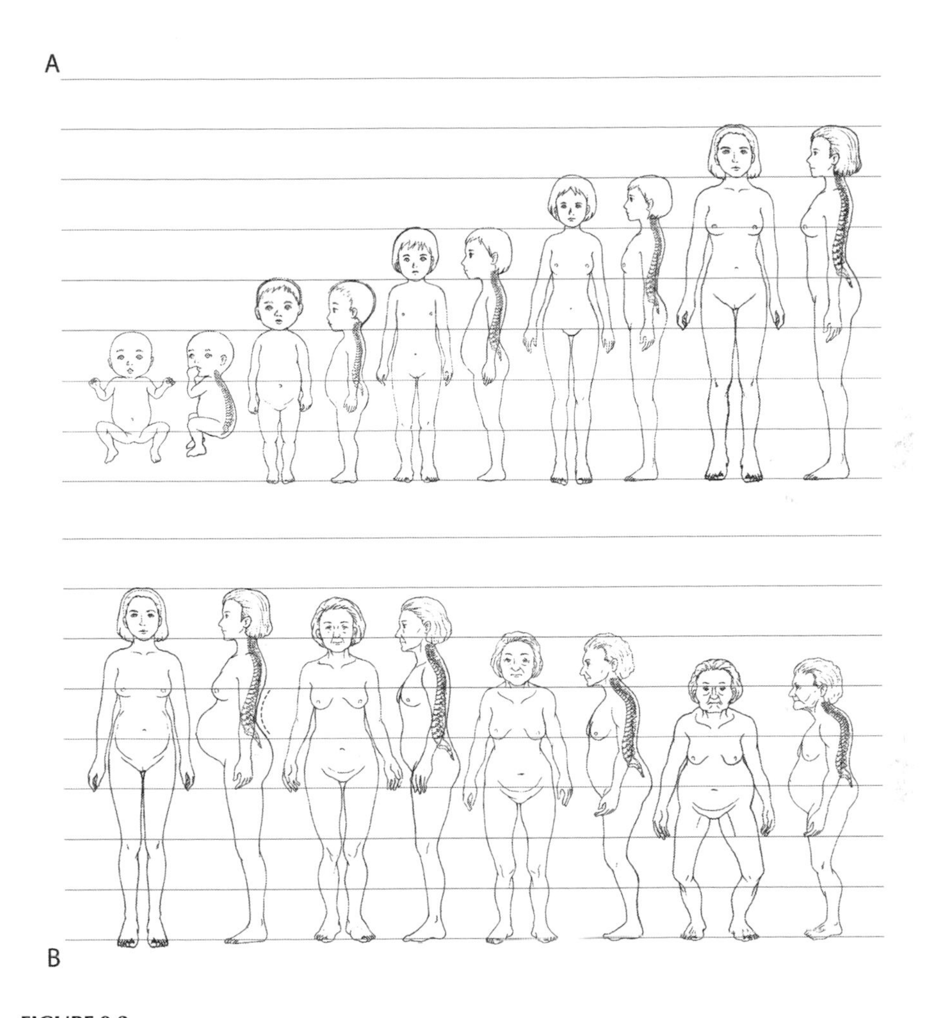


**FIGURE 9.3**
(A) Height and posture through life stages: Infant to adult woman. (B) Height and posture through life stages: Pregnant female to elderly woman.

in the lumbar spine and the change in the center of gravity as the woman counterbalances the weight at the anterior of the body. A pregnant woman's figure and posture affect wearable product form, balance, and positioning on the body.

The second set of body forms in Figure 9.3B illustrates a female after menopause with significant height loss and posture change due to spinal changes. Spine changes in such women in their 50s include normal intervertebral disc degeneration, possible compression fractures in the vertebrae due to osteoporosis (refer to Chapter 4), and probable changes in pelvic tilt related to the spinal changes. A compression fracture in an osteoporotic vertebra distorts the vertebral form unpredictably. A vertebra may simply collapse and lose height. The vertebra may become symmetrically or asymmetrically wedge-shaped (or not) in sagittal and/or coronal planes. The thoracic spine seems particularly vulnerable to osteoporotic compression fractures, but they do occur in all spinal regions. Unfortunately, multiple osteoporotic vertebral compression fractures can occur over time and each fracture tends to alter spinal length and posture.

Spinal structural changes lead to changes of function in the paraspinal muscles. In the healthy spine, they stabilize the axial skeleton and support upright posture (refer to Chapter 4, Section 4.5.2). The paraspinal muscles do not function as effectively when compression fractures change the distance between their muscle origins and insertions. The change in paraspinal muscle function leads to further posture changes. Still other posture adjustments develop to allow the eyes to keep looking ahead and maintain the body's center of gravity over the feet. In addition to the pain a compression fracture causes, secondary posture changes often cause neck and back pain.

Spinal structural and/or secondary posture changes can bring on scoliosis and/or kyphosis. They, in turn, impact other body functions, including breathing, and widen the circle of negative effects from osteoporotic vertebral fractures. Osteoporosis-related postural alterations can eventually progress to the extreme seen in the fourth set of body forms in Figure 9.3B. This degree of height and posture change begins to affect standing and walking balance. Note the elderly woman figure's wider base of support in the anterior view in the same set of body forms, a sign of poor balance.

As head position shifts forward due to kyphosis, the anterior torso length decreases and back length increases. Posture changes raise garment aesthetic and fit considerations, and too much product weight can exacerbate posture changes. Coats constructed of lighter weight materials may help reduce posture-related neck and back pain as well as help prevent further postural deterioration. Figure 9.4 shows a body scan (lateral view) of a woman with osteoporotic spinal posture change and the custom-fit bodice pattern developed for her (Ryan, 2006). Her posture change demanded adjustments to both the front and back of the pattern. The anterior neck-to-waistline length is decreased and the posterior neck-to-waistline length is increased. Armscye shapes changed because of the center front and back length changes and the need to join the pieces with a vertical side seam, but also because her head and shoulders have shifted anteriorly relative to her waist. Her kyphosis and change in shoulder position influenced shoulder seam placement, garment suspension from the shoulder, sleeve cap shape, and sleeve drape.

Aging-related posture and height changes are visually evident; however, all body systems are affected throughout the aging process, some more dramatically than others. Most of the changes will affect product design decisions, especially whole-body product decisions. Be aware of the age range of your wearer. Start with foundational knowledge in anatomy, and then research factors specific to the age demographic that will affect a product design.

**FIGURE 9.4**
Bodice pattern for osteoporotic figure, with lateral body scan view. (From K. Ryan (2006). *Aesthetically unique, specially sized clothing for women with osteoporotic posture changes* (Unpublished master's thesis). University of Minnesota, St. Paul, MN with permission.)

## 9.3 Designing for Total Body Anatomy: The Dynamic Body

The human body is never totally static. Try standing "still" for 15 minutes and notice how frequently you shift and adjust your body, sometimes very subtly. Although slight position changes do not often affect product design, imagine how product requirements change as a person transitions from standing "still" to walking, and then to running. Products worn during specific work activities should allow ease of the motion necessary to accomplish the work. How do product requirements change when work environments (a) demand superhuman feats, (b) are incredibly small, or (c) call for maximum reach? Although an initial whole-body design idea may be envisioned on a body in a static upright posture, whole-body wearables must move with the body into positions at or near ROM extremes.

### 9.3.1 Dynamic Body: Gait

Gait, the term used to describe an individual's manner of moving on foot, is a complex activity. Terms used to describe gait—stroll, saunter, cruise along, skip, walk, run, gallop, race—each bring a characteristic movement pattern to mind. Product weight, bulk, and aerodynamics change gait patterns and efficiency. Gait involves the total body, although we often think of it as involving only the lower body. Numerous sets of muscles work on the bones and joints of the spine, pelvis, thigh, leg, and foot to facilitate gait. Muscles also act on the arms and shoulders during gait to assist with balance and to increase gait efficiency. These total body motions increase with increased speed (and distance covered). Think about your arm swing as you stroll along—there isn't much movement. If you hurry to catch a bus, you likely find your arms pumping as you twist your pelvis and stretch your legs to cover more ground at a faster pace.

### Walking

The action of walking can affect product designs, and wearable products can influence walking. Inman, Ralston, and Todd (1981) define walking as "a process of locomotion in which the erect, moving body is supported by first one leg and then the other . . . [with] a brief period when both feet are on the ground [in *double support*]" (p. 2). Almost any wearable product will affect how a person walks, from footwear to headwear. Chapter 8, Section 8.8, gives in-depth information on the foot in motion—foot motion effects on footwear design and footwear effects on the body. Weight suspended from the upper body, such as a backpack or baby carrier, changes posture and how a person walks.

Protecting the body in cold environments without restricting mobility is a challenge. Insulation in cold protective wearable products often adds bulk which restricts body motion. O'Hearn, Bensel, and Polcyn (2005) used motion capture to study the mobility of soldiers with and without cold weather gear. They concluded bulky clothing worn for thermal protection constrains movement including arm movement, alters the walking pattern, and even affects resting posture. Any of these changes in usual body function could result in reduced responses in sports, work, and combat situations. There is a trade-off between protection and mobility—with a need to decide which capability is most important for a given situation.

Prosthetic and orthotic design requires an understanding of the walking process (Inman, Ralston, & Todd, 1981) as well as knowledge of anatomy and the body/product interactions throughout the course of wear (Staker, Ryan, & LaBat, 2009). Successful prosthesis design applies biomechanical knowledge of *ground reaction forces* (Everett & Kell, 2010, p. 129) and an understanding of how the body generates the periodic movement of each foot from one position of support to the next (Inman et al., 1981, p. 2). Prosthesis design teams often include surgical, medical, and rehabilitation specialists.

### Running

Running propels the body forward more rapidly than walking. Both feet are off the ground simultaneously at one point in the running sequence. Jogging and sprinting, which use different whole-body positions and motions, are variations of running. Athletic wear designers want to assure products facilitate, rather than impair, running motions and performance. There are numerous studies on running styles and efficiency, many of which have been conducted by major athletic wear companies. Oggiano et al. (2013) reviewed state-of-the-art skin suit technologies and the related sport aerodynamic functions. They determined that many factors affect running including: the athlete's body (shape, size, and posture), the specific running motion, the garment (stretch, air permeability, finishes on materials, construction, surface structure, yarn type, seam position), and total garment fit. Designers must meet the challenge of selecting from multiple variables when designing a product for runners.

### 9.3.2 Mechanized Walking and Gait: Exoskeleton Products

Exoskeleton design requires in-depth knowledge of the skeleton, nerve and muscle control functions, and natural gait patterns. An exoskeleton, as the term implies, is a skeleton-like structure that surrounds the exterior of the body. In some ways the product mimics the internal skeleton of the wearer. Exoskeletons may fit head to foot (or cover only the lower limbs) and serve to extend a human's natural load capacity in military and industrial settings (Yang, Gu, Zhang, & Gui, 2017, p. 4). Or they may function as a rehabilitation aid (Moreno, Figueiredo, & Pons, 2018, p. 89). Even when covering just the lower limbs, the wearer's whole body is involved in moving with the exoskeleton. Moreno et al. (2018) reviewed developments in lower limb exoskeletons and describe several major research directions. Yang et al. (2017) give an overview of exoskeleton development in the U. S., China, Japan, and South Korea.

In research on lower limb functions which contribute to total body gait, Shorter, Kogler, Loth, Durfee, and Hsiao-Wecksler (2011) developed a portable powered ankle-foot orthosis for daily use. The product attaches to the lower limb, but its effects are on the total body—enhancing walking function, assisting in gait training in physical therapy, and helping to improve strength and ROM.

### 9.3.3 Dynamic Body: Extended Positions

Understanding walking and running patterns, and using the patterned motions when designing is important. However, other motions affect comfort and function. Notice throughout the course of a day the many different positions you assume—not just sitting, standing, and walking—but bending, reaching, stretching, and more. Every job requires specific motions, from office worker to construction worker. Each athlete—baseball pitcher, tennis player, golfer—exhibits characteristic motions.

Assuring that a wearable product moves with the body for basic function, comfort, and aesthetics is important. Product position that remains stable with the dynamic body is essential for many wearable technologies. If a health monitoring sensor moves out of place as the body moves, the sensor will lose the body signal or the signal may be distorted. Continued innovation in wearable sensing, with many new applications especially in health and medical settings, requires further understanding of how garments move with the body, so that sensors are optimally positioned at all times (Griffin, Compton, & Dunne, 2016; Zeagler, 2017). Dunne, Gioberto, and Koo (2011) proposed a method for measuring garment movement for wearable sensing, concluding that movement is dependent on garment properties, like textile extensibility and wearing [fit] ease.

Designing motion features for a whole-body wearable product is complex. New methods are being developed to record, analyze, and use motion data in product design. A typical approach is to break down a continuous motion into a series of positions; and then design product motion features, like stretch panels and pleats, to accommodate the motion.

### *Documenting, Analyzing, and Translating Extended Positions for Product Design*

Motion can be recorded with many methods: gesture sketches, photographs, and motion capture technology. Motion analysis requires knowledge of the skeleton, typical joint movements, and ROM for each joint. ROM descriptions and limits of joint motions described in Chapters 3 through 8 serve as a starting point for understanding body motion and extent of motion. Direct transfer of the observed motions and ROM into a product is more difficult, requiring knowledge of product variables: materials, methods of joining product segments, and how a total product will move with the body. Several approaches for measuring and applying extended motion are described in this section.

Ashdown and Watkins (1992) used a "slash-and-spread" technique to visualize and measure where a protective coverall must adjust as the wearer moves from a static pose to extended positions. A non-woven coverall was slashed with a honeycomb pattern throughout its surface. The tiny slashes remain closed when the wearer is standing upright. With body motion, the slashes in stressed areas open indicating where more garment movement ease is needed.

Griffin, Lastovich, Bye, and LaBat (2016) were tasked with redesigning a protective coverall for a major company to improve function and comfort. They scanned workers wearing the coverall in a series of extended motion positions (Figure 9.5). A panel of expert judges viewed the scans to analyze stress areas and marked "hot spots" on a sketch of the coverall. The research/design team used the information to develop prototypes with strategically placed stretch features. A series of prototypes were developed and tested. The redesigned coverall improved motion and comfort (Figure 9.6). Along with motion accommodation, the stretch features allow fitting for a range of sizes.

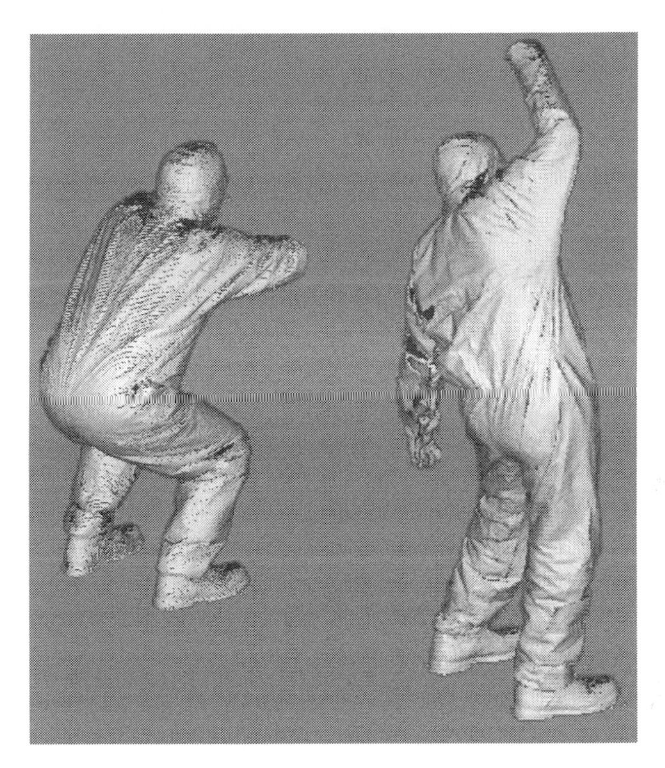

**FIGURE 9.5**
Extended motion positions while wearing a protective coverall. (Courtesy of the Human Dimensioning© Laboratory, University of Minnesota.)

In a similar process, Simoes (2013) viewed the mobile body and used the observations to develop pattern shapes. After in-depth observation and documentation of body positions of six women wearing a standard coverall, she identified several in-motion positions: (a) forward inclination of the torso and backward projection of the buttock, (b) forward projection of the elbows and wrists and increased space between flexed arms and torso, (c) forward/outward projection of the knees and increased space between the flexed knees. She used observations to develop a series of what she calls coverall "mobile block patterns." The low-tech but lengthy process has potential for independent designers who do not have access to sophisticated motion capture technology.

### *Motion Analysis Tools*

Motion capture as a tool for assessing product fit and designing products for motion is in its early stages. A motion capture set-up using reflective markers is pictured in Figure 9.7. Markers are placed at key motion sites, typically joints. A thorough understanding of the skeleton and how it moves is necessary when placing markers for a motion capture study. At this time, motion

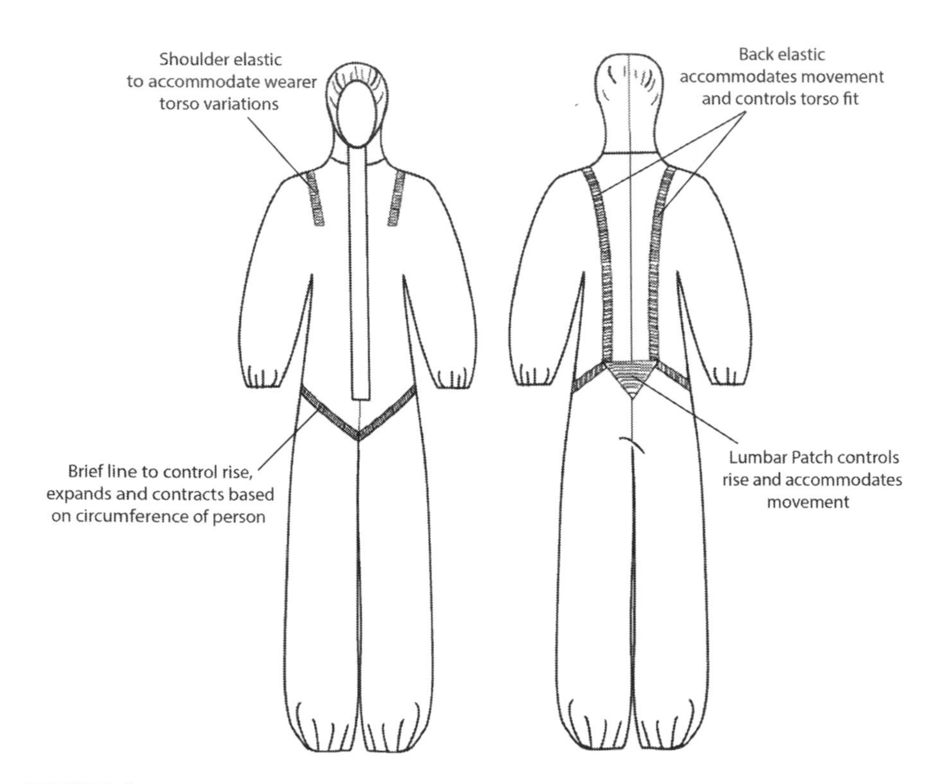


**FIGURE 9.6**
Protective coverall with stretch features. Design by L. Griffin, T. Lastovich, E. Bye and K. LaBat (2016). (Courtesy of the Human Dimensioning© Laboratory, University of Minnesota.)

capture for product design has focused on limited regions of the body, possibly due to the complex nature of the whole body in motion.

Sohn and Bye (2014) studied motions at the posterior upper torso. Although limited to one area of the body, they suggested ways of applying the data to apparel design. Several studies assessed effects of wearing a product by recording and analyzing motion of a person in minimal clothing and then in the wearable product. Prototypes can be analyzed and revised to assure maximum ROM. Song, Beard, and Ustinova (2016) used motion capture to study the effects of an upper body compression garment on golf swing. Compression's positive effects, as discussed in Chapter 5, were desired without restricting peak athletic performance. Park et al. (2015) used motion capture to study the impact of firefighter gear on lower body ROM. Firefighter design goals are maximum protection and maximum function. Studying whole-body motion and whole- body wearables is a more complex problem, but methods of quickly analyzing and applying the vast amount of motion capture data are improving.

Zong and Lee (2011) suggest combining 3D body scanning and motion capture to: (a) build motion simulation models, (b) examine mobility and function of apparel, and (c) design and develop custom-made products.

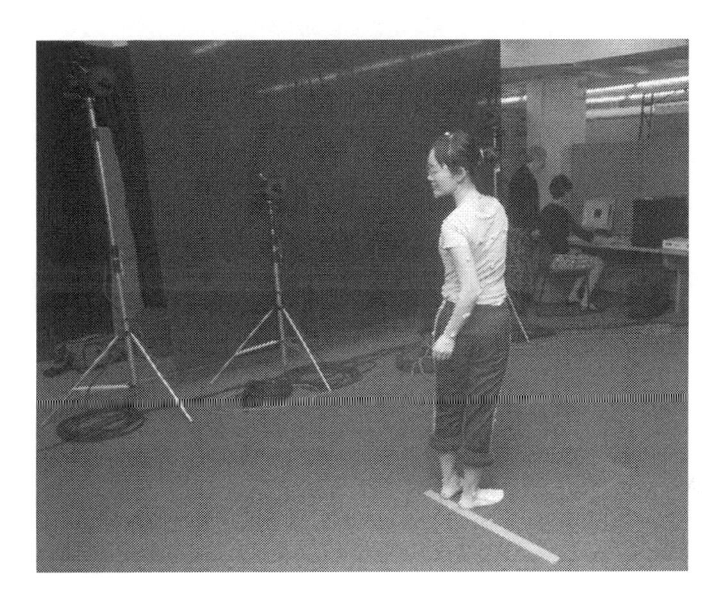

**FIGURE 9.7**
Motion capture system. (Courtesy of the Human Dimensioning© Laboratory, University of Minnesota.)

Several resources are available for use in analyzing how body motion affects a wearable product. ASTM Standard F3031-17 (2017) describes a standard practice for ROM evaluation of first responder's protective ensembles. The standard is proposed for use by fire and rescue departments to evaluate and select protective clothing and equipment. Although the standard does not specify acceptable ROM ranges, it can be used to compare ease of motion between prototypes.

### *Skin Stretch as Indicator of Body Motion*

Along with ROM, how skin stretches over joints has been used in designing product stretch features. You can observe skin stretch on your index finger. Refer to Appendix F.1, dorsal view, to see the joint prominence landmark locations for the index finger. With your finger extended, on the dorsal side of the hand, mark the junction of the finger and the hand. Draw another line over the joint prominence of the DIP, the most distal finger joint. Measure from the first line to the second. Flex all your finger joints and measure the length again to determine skin stretch. The change in length gives an indication of how much motion accommodation is needed in a product. Watkins and Dunne (2015) review several methods for analyzing skin stretch with applications to product design (pp. 54–57). Andrews, Verburg, Cooper, and Frick (1977) reviewed research on skin stretch and determined extent of skin stretch in several body areas with possible applications to product design: 14% across the back during elbow bending, 45% for buttocks in bending from a standing position, and 50% for the elbow fully bent (p. 55). Andrews et al. (1977) advise that other factors come into play when designing a product, such as the product's "slip" between skin and fabric, so direct use of skin stretch in a wearable does not always result in best fit and comfort.

The human body is in constant motion. New methods of recording, analyzing, and applying motion data to wearable products are needed. New materials, product structures, and methods of manufacture will contribute to the success of wearable products with maximum function.

## 9.4 Wearable Product Testing: Manikins, Digital Human Models, and Humans

Manikins and digital human models can substitute for real humans in developing wearable products. A life-size manikin approximates the human body form. A manikin and/or digital human model can "stand-in" for a human, with the advantage that the manikin never changes and is always available. For some products, testing on human subjects is the best method to determine product success.

### 9.4.1 Manikin Testing

Manikins are also called dress forms, body forms, and stands. Manikins are typically made of solid, non-compressible materials from wood to papier-mâché to dense foam. The characteristics and responsiveness of compressible human flesh are difficult to replicate. Some companies have developed compressible forms for bra design using breast prosthesis materials to simulate the human breast. When compared to humans, most manikins reflect an idealized body shape and form. Body scanning technology can be used to produce manikins based on scans of real people. Scan-derived manikins are more realistic in proportion, form, and posture.

Instrumented manikins can test specific product functions. Manikin testing is especially useful in the early stages of prototype development allowing quick evaluation of many variations.

Manikins can be used to sort through various prototypes for best performance before producing a design. Holmér (2004) gives a history of manikin development, how they are used in testing, and describes a newly developed breathing manikin. Walking and running manikins can test product prototypes in motion. Testing the effectiveness of clothing's thermal protection on a thermal manikin or sweating thermal manikin before a design is implemented and marketed is useful and often necessary. A thermal manikin measures heat loss, and a sweating thermal manikin measures evaporative heat loss, simulating human perspiration production.

McCullough (2009) describes a process for using manikins to evaluate cold weather protective clothing. Refer to ASTM F1291-16 (2016b) for standards on thermal manikin testing.

Manikins can be used to test products in situations when human testing is not possible. For example, fire protective clothing can be tested on a manikin with sensors which can predict skin damage from flame (Wang, Wang, Zhang, Wang, & Li, 2015). McQuerry, Barker, and DenHartog (2018) developed four *firefighter turnout gear* prototypes that combined variables for body heat venting, stretch, and *modularity*—modules that can be interchanged. The prototypes and a control (a currently available total body turnout garment) were tested on a sweating thermal manikin. The results indicated that adding ventilation openings and modularity improved heat release from the turnout gear.

Starr, Cao, Peksoz, and Branson (2015) tested body armor designs and materials on a sweating thermal manikin in an environmental chamber. Although they could quickly sort through materials and design options to select one optimal design with this method, they encourage testing body armor systems on real persons—manikin testing cannot replicate physiological responses of a real person, especially core body temperature fluctuations that occur during different activities and in different environments.

### 9.4.2 Digital Human Models

*Digital human model* developers are moving toward more realistic replication of human body tissue compressibility and motion. At this time, digital human models are available to test product forms and shapes, but not physiological responses to products. Cheng and Robinette (2009) describe developments in the field of static and dynamic human shape modeling. A static shape model provides a digital human form in a particular pose, and a dynamic shape model addresses shape variations due to pose changes or while the subject is in motion. The technology is developing quickly, moving beyond the early *avatars* which often looked unrealistic and moved in mechanical ways. Digital technology opens the possibility of forms with life-like skin and underlying tissue characteristics, which will change how prototypes are developed. Technologies for digitally characterizing and replicating product materials are also improving. Even with the most sophisticated digital human models, if the materials do not conform to the body in manners that predict real product performance, the design will not be successful. The future of these technologies is promising as both human and materials models become more realistic.

### 9.4.3 Human Testing: Laboratory and Wear Tests

*Human testing*, sometimes called *in vivo testing*, uses responses of a person to evaluate a product or ensemble of products. Human testing can be conducted in a controlled environmental chamber in a laboratory or in a natural setting. Human testing can measure physiological responses to a product, and/or collect information concerning a person's opinions of the product.

ASTM F2668-16 (2016) presents a standard practice for chamber testing, including specifications for the chamber, a treadmill, and how to measure physiological responses. The standard describes collecting rectal temperature (a body core temperature indicator), skin temperature, whole-body sweating, and heart rate. A test using human participants in a natural setting is called a *wear test*. Physiological measures and/or subjective measures can be used. Physiological tests are similar to those used in a laboratory setting. Subjective measures include having test participants complete questionnaires or, in long-term tests, keep journals about their experiences while wearing the product(s).

## 9.5 Extra-Vehicular Activity (EVA) Space Suit

The *Extra-Vehicular Activity (EVA)* space suit acts as the ultimate mediating variable between the body and the environment (refer to Chapter 1, Section 1.1). A space suit serves total body survival needs in an extremely hostile environment. The astronaut's need for healthy function of the body systems is the same in space as on Earth. The space environment, however, is very different from Earth's environment. There are physiological hazards—no oxygen for respiration, no atmospheric pressure in the vacuum of space, dramatically reduced gravitational forces, extreme exposure to UV radiation from the sun, and vast temperature variations (Watkins, 2005). Space exploration requires Earth-like environments inside the spacecraft and inside the EVA space suit to sustain human body systems. Technical reports from the early days of space suit design describe developers' predictions of how the human body would function in space, and detailed information on the strategies they proposed to prevent negative effects of space flight (Annis & Webb, 1971; Webb & Annis, 1967).

Watkins (2005) defines a space suit as, "the total life-support system for Extra-Vehicular Activity (EVA) that takes place outside the shelter of the spacecraft" (p. 200). The EVA space suit creates a microenvironment capable of supporting human body systems. See Figure 9.8, a photo of the NASA XPS space suit. Its purpose is to protect and support the complex functioning of the vulnerable human body. Watkins (2005) lists the functions of the space suit's layers and accessories as: (a) pressure application, (b) environmental thermal and impact protection, (c) astronaut thermal and general comfort, (d) provision of breathable oxygen, (e) abrasion protection, (f) eye protection from solar radiation, (g) a structure to allow mobility, and (h) communication. All these demands must be met while allowing the astronaut to move and perform tasks while confined in the suit.

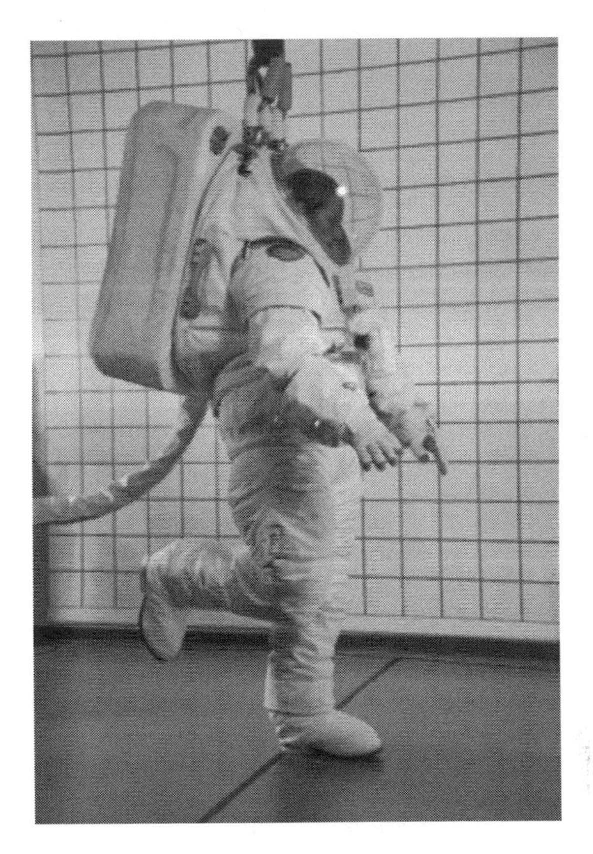

**FIGURE 9.8**
XPS prototype space suit. Credit: NASA/Bill Stafford, Aug. 6, 2017.

The space suit ensemble is made up of many layers; starting with the layer nearest the body (C. Compton, personal communication, September 13, 2018):

- The *thermal comfort undergarment* (TCU) is an optional first layer worn closest to the body. It is a form-fitting, long-sleeved, easily laundered upper torso garment.
- The maximum absorbency garment (MAG) is the first lower torso layer (see Sections 5.2.2 and 9.5.2).
- The *liquid cooling and ventilation garment (LCVG)* is a separate underwear-like suit worn close to the body surface. It is designed to substitute for the body's normal on-earth thermal regulation (see Section 9.5.1).
- A pressurized "bladder" made of rubber or polyurethane to substitute for Earth's atmospheric pressure (see Section 9.5.7).
- A restraining layer of polyester or nylon that keeps the bladder from blowing up and bursting like a balloon when it is pressurized.

- An outer suit made of multiple layers of materials: (a) inner nylon layer, (b) five to seven layers of aluminized Mylar for thermal protection, and (c) outer layer of ortho-fabric with a Kevlar™ aramid rip-stop grid fabric for mechanical protection (micrometeoroid deflection).

Other components help complete the space suit microenvironment: space helmet, boots, gloves, and the EVA *PLSS* (portable life support system). The PLSS looks like a large backpack. It handles many of the input/output functions that, in Earth's atmosphere, transpire between a human and the environment. One important function is supporting the respiratory cycle—providing oxygen and removing carbon dioxide. The PLSS contains three "loops" that feed to and from the astronaut providing these vital functions: (a) ventilation for oxygen supply, oxygen circulation, $CO_2$ removal, and air cooling; (b) liquid cooling to circulate water to and from the LCVG, and (c) a thermal heat rejection loop (Skoog, 2013, p. 230). The PLSS may also include a drinking water supply, a urine collection method, and a snack administration mechanism.

A space suit must work with many of the body systems. The following sections discuss selected systems and their functions within the space suit microenvironment: circulatory and integumentary systems, digestive and urinary systems, nervous system, reproductive system, and skeletal and muscular systems.

### 9.5.1 Wearable Products for Space: Circulatory and Integumentary Systems

The circulatory system and integumentary system (skin surface) play important roles in maintaining thermal balance in Earth's atmosphere and within the space suit. In space "heat is transferred through radiant energy in the form of direct or reflected sunlight" (Thomas & McMann, 2012, p. 21). Temperature range is extreme, exceeding 250°F (121°C) in direct sun to −397°F (−233°C) in total shade. Every space exploration destination has different temperature ranges, but all are life-threatening.

Two thermal threats are present in space: radiant energy from the sun and heat generated within the enclosed space suit. Radiant energy can be deflected with the suit's reflective outer layer. The heat generated by the body in the suit is a more complex problem. The enclosed suit poses a thermal balance issue—heat builds up inside the suit and perspiration cannot be dissipated by evaporation. When an astronaut works in space, her/his muscular exercise can increase heat production 10–20 times over resting levels, a serious problem (Buckey, 2006, p. 103). Excess heat must be removed.

Koscheyev and Leon (2014) present an overview of space suit thermal comfort strategies. The LCVG, briefly described above, is the current technology. It resembles one-piece knit winter underwear interlaced with tubing that is held in close contact with the body. Cool liquid circulates through the tubing and body heat is transferred to the liquid via conduction. The warmed liquid is carried to a refrigeration unit in the PLSS where

> [t]he water runs across a porous metal plate that is exposed to the vacuum of outer space on the other side. Small amounts of water pass through the pores where it freezes on the outside of the plate. As additional heated water runs across the plate, the heat is absorbed by the aluminum [plate] and is conducted to the exposed side. There the ice begins to sublimate (turns into water vapor) and disperses in space. Sublimation is a cooling process. Additional water passes through the pores and freezes as before. Consequently, the water flowing across the plate has been cooled again and is used to recirculate through the suit to absorb more heat (NASA, 2012, p. 2).

Researchers have studied the most effective and efficient tubing patterns to remove heat. Blood vessels close to the body's surface are the most effective in transferring heat and therefore skin areas with good circulation are the best sites for tubing placement (Kim & LaBat, 2010; Koscheyev, Coca, & Leon, 2007; Leon et al., 2009; Koscheyev & Leon, 2014).

An air circulation system works in conjunction with the water circulation system. As air circulates through the suit it draws perspiration-laden air into a water separator in the PLSS. The extracted moisture is collected in a reservoir to add to the liquid circulating through the LCVG, and the drier air is circulated through the suit. The astronaut controls the cooling systems to meet individual comfort needs (NASA, 2012).

### 9.5.2 Wearable Products for Space: Digestive and Urinary Systems

Digestive system input and output—nutritional intake and waste disposal—continue in space. The space helmet may incorporate a drink valve and a food-stick (Thomas & McMann, 2012, p. 20). However, the astronaut cannot manually manipulate features inside the helmet. The design challenge is positioning the nutrition devices for easy access by the mouth, while not obstructing sight-lines through the helmet's visor. Toileting in the space suit environment, especially during long periods of EVA, raises a need to develop management strategies for digestive waste and perineal region hygiene.

Extended EVA periods, anywhere from six to nine hours, may require digestive and urinary waste collection methods. Male and female astronauts wear maximum absorbency garments (MAGs), adult diapers with an inner super-absorbent polymer layer, to collect urine and fecal matter. In the early years of space flight, condom catheters (refer to Chapter 5, Figure 5.5) were worn by male astronauts to collect urine. Female urine collection devices are more difficult to design (refer to Chapter 5, Section 5.3.1). New waste management methods for space suits are being developed, but the MAG is the method of choice at this time (Cofield, 2018).

### 9.5.3 Wearable Products for Space: Nervous System

Neurological functions are affected in space; some issues have been addressed with space suit designs. Clément and Reschke (2008) describe negative space flight effects on sensory functions including eyesight, hearing, taste, and sense of smell. Eyesight effects, reported over many years of spaceflight, include decreased visual acuity and color perception, and changes in intraocular pressure (Clément & Reschke, 2008, p. 104). Further exploration of physical change causes and development of wearable products that might alleviate the problems are warranted. The astronaut's eyes need protection while on EVA because space lacks the Earth's atmosphere, which filters harmful UV rays of the sun. Helmet visors coated with a very thin layer of a metallic alloy, typically gold, filter the rays.

The astronaut must be able to clearly hear and have control of a speech transmission method to communicate with colleagues in the spacecraft. A closely fitted "Snoopy" cap, named after the cartoon character, is worn as the first layer inside the space helmet. The cap contains input and output messaging devices that must be optimally aligned with the individual astronaut's ears and mouth. Hearing changes detected during and after space flight are attributed to constant noise in the spacecraft (Clément & Reschke, 2008, p. 108). Space suit design may not prevent hearing problems, but neither should space suit designs worsen the problems.

Other sensory effects have been noted, for example changes in taste and smell. Theories to explain changed sensory perceptions include effects of nasal congestion due to fluid shift to the head, response to space motion sickness, and stress from workload, little sleep, and spacecraft noise (Clément & Reschke, 2008, p. 108). Spacesuit designs may not solve these long-term effects of time in space, but products worn in the spacecraft may alleviate uncomfortable effects.

EVA task performance, including tool manipulation, requires preserved tactile sense in the hands (refer to Chapter 7, Section 7.4). Skoog (2013) states, "Tactility is of major importance for the design of pressurized [space] gloves, and the smallest object that should be able to be grasped is 8-12 mm" (p. 216). Space glove design for protection, mobility, and tactile sense is a design challenge. Applying pressure within the glove and promoting tactility is especially difficult (Thompson, Mesloh, England, Benson, & Rajulu, 2010). Thomas and McMann (2012) review approaches to space glove design. While the trend is to make space suit sizing modular, custom fit of space gloves may be the only way to provide the best mobility, while preventing injury from the glove's internal structures (refer to Chapter 7, Section 7.5.3, delaminated fingernails).

Weightlessness is an unnatural state for a person accustomed to Earth's atmosphere. Space motion sickness (SMS) is a negative space travel effect attributed to altered nervous system perceptions of pressure on the body. Solutions include products worn while in the spacecraft. Clément and

Reschke (2008) describe products worn by Russian cosmonauts to relieve weightlessness-induced SMS: (a) pressurized insoles to add a feeling of weight to the feet; (b) load suits with adjustable elastic bands to add pressure to the chest, back, abdomen, side, and leg; (c) a pneumatic occlusion cuff worn on the hip to try to reduce or prevent shifting of body fluid toward the head; and (d) a neck pneumatic shock absorber (NPSA), a cap that provides a load to the cervical vertebrae and neck muscles to try to reduce SMS. However, reports of effectiveness for these products are anecdotal.

### 9.5.4 Wearable Products for Space: Reproductive System

Reproductive function in space, especially on long space flights, are a concern. Buckey (2006) stresses the importance of contraception methods on long space flights, because little is known about influences of weightlessness and in-space radiation exposure on a developing fetus (pp. 215–216). Some information on menstruation while on a space mission is available through interviews with female astronauts (Wright, 2002). Early concerns were due to the effects of zero gravity—could menstrual blood pool in the abdomen?—a condition called *retrograde menstrual flow* (Buckey, 2006, p. 215; Jennings & Baker, 2000). After several women traveled in space, it became clear that menstruation in zero-gravity does not cause flow abnormalities and earth-bound options for menstrual management work just as well in space (refer to Chapter 5, Section 5.4.2) Suppressing the monthly menstrual cycle with hormone pills or a hormonal intrauterine device (IUD) (Teitel, 2016) could serve both contraceptive and menstrual management needs.

### 9.5.5 Wearable Products for Space: Skeletal and Muscular Systems

Weightlessness has negative effects on bone and muscle, promoting loss of bone integrity and muscle mass. The *Gravity Loading Countermeasure Skinsuit* (GLCS) was developed to counteract those effects (Waldie & Newman, 2011). The theory is that the body will work against the suit much like it works against gravity on Earth to maintain bone and muscle mass. The sleeveless full-body skinsuit is designed to be worn in the spacecraft, not with the EVA suit. A non-stretch canvas yoke stabilizes the suit over the shoulders. The body and legs of the GLCS are made of a bi-directional elastic weave. Stirrups anchor the suit under the feet. Waldie and Newman (2011) tested the comfort and mobility of the suit with positive results. Testing the GLCS during space flight will prove its efficacy. The Russian space program developed a wearable device with similar goals called the *Pingvin* or "Penguin" suit that has been worn by cosmonauts for several years. However, no studies are available on effectiveness (Waldie & Newman, 2011).

The spinal intervertebral discs, in contrast to the bones and muscles, appear to benefit in space, as the spine elongates under weightless conditions.

Changes in body size and form as results of bone, spine, and muscle changes will affect space suit fit.

### 9.5.6 Wearable Products for Space: EVA Suit Size and Fit

For best space suit function, adequate, if not perfect, fit is necessary. Early space suits were custom fit to each astronaut. Traditional space suit measurements for custom-fit suits included: stature (total height), vertical trunk circumference, chest breadth, chest depth, hip breadth, chest circumference, bicep circumference, knee height, crotch height, and thigh circumference (refer to Appendices on landmarking and measuring).

Each person's unique skeletal structure, posture, and muscle and fat distribution contribute to body form and size but when weightless, further fit complications arise. In space, many body dimensions change due to short-term (fluid shift) and long-term (muscle atrophy) processes, so it's especially difficult to determine best fit ahead of time (B. Holschuh, personal communication, August 24, 2018).

Even with a full set of individual astronaut body dimensions, custom space suits do not fit perfectly. Digital human modeling may allow estimates of weightlessness effects for individual astronauts. Modeling technology, combined with 3D scanning and 3D printing currently being used to compare three-dimensional digital astronaut body forms with space suit prototypes, may help resolve space suit fit issues.

#### *New Space Suit Sizing Strategies*

New space suits currently under development by NASA—including the Z series (Z-1, Z-2, Z-3) and XPS suits (refer to Figure 9.8)—may solve some fit and sizing issues. The completion target for the Z-2 is a Mars surface mission in the mid-2030s (McFarland, 2016). Reports indicate that the Z-2 is being developed specifically with a smaller body size in mind and incorporates a *hard upper torso* (HUT) of *S-glass* and carbon fiber (Ross et al., 2014). The new generation space suits use *modular sizing* (Skoog, 2013, p. 216). For example, a long suit arm could replace a short suit arm to fit an astronaut with a longer body arm length. Modularity means that attach/detach junctures must match, for example where a suit arm attaches to the torso at the shoulder, however, "one of the prices of modularity is a compromised ability to optimally fit smaller sizes because the standard [i.e. torso] components must accept the larger sizes" (Ross et al., 2014). Another adaptable sizing approach is to use adjustable straps on the inside and outside areas of the suit, which can be lengthened or shortened to fit an individual astronaut. Benson and Rajulu (2009) describe some approaches to space suit sizing and fit and suggest that objective fit tests are necessary to assure best fit for each astronaut.

Space suit materials also affect fit and sizing methods. *Semi-rigid suits*, combining soft and hard components, have been the prevailing concept for over 30 years (Skoog, 2013, p. 222). Using soft materials can allow good mobility, but with higher suit pressures, suit volume and stiffness become a problem. Suit pressures must also be coordinated with the gas composition of the air/oxygen circulating inside the space suit and accommodate respiratory system function (Skoog, 2013, pp. 216–218). Suits made entirely of hard materials are pressure independent—the solid shell contains and limits expansion of the bladder layer. The problem is that the hard-shell suits are heavier and designing bearings (at joints) for mobility is difficult. Individual hard components cannot be adjusted for body size differences.

#### *Body Structure Changes in Space*

Woolford and Mount (2006) found that body changes are evident even after short periods of time in space. Microgravity fluid shift changes body proportions—especially in hands, legs, torso, and the face—and the spine elongates, increasing body height by as much as 3% (p. 930). Measurements of the astronaut taken on Earth to determine space suit dimensions may not reflect body shape and size in space. Future space suit components may be 3D printed in space, allowing custom-printed parts to meet the changing anthropometric needs of each astronaut (Thryft, 2018). This could be an advantage for long space flights.

Body dimension changes in space, over the short term, are mostly attributed to fluid shift. On Earth, gravity pulls blood toward the legs, with circulatory system functions pushing the blood back toward the heart. In space, zero-gravity means blood is more equally distributed throughout the body, evident in the puffy-faced look of astronauts while in space. The biggest changes will appear in circumference measurements. Muscle mass loss will result in decreased cross-sectional areas affecting circumference measurements (Hinghofer-Szalkay, 1996). Buckey (2006) notes calf circumference decreases significantly in space, and no doubt other muscular areas will change. Bone density changes on long space flights will likely affect posture, possibly causing space suit misfit.

### 9.5.7 Wearable Products for Space: EVA Suit Mobility

An astronaut must be able to move fairly freely to accomplish work assignments. Roach (2010) states, "A pressurized space suit is a heavy-duty body-shaped balloon—almost more of a tiny inflated room than an article of clothing. Fully pressurized, it's all but unbendable without some sort of joints" (p. 139). The astronaut may have to work against the suit, disrupting natural biomechanics of the body, including restricting or altering joint motion which can have short-term and long-term effects on the body (Cullinane, Rhodes, & Stirling, 2017).

***Space Suit Range of Motion Considerations***

Some information is available on NASA and Zvezda (Russian space agency) range of motion (ROM) guidelines for space suits (England, Benson, & Rajulu, 2010; Skoog, 2013, p. 217). ROM while wearing a space suit is restricted, especially when compared to ROM of a minimally clothed person (refer to ROM sections in Chapters 3 through 8). For example, normal ROM for elbow extension and flexion is 150 degrees (refer to Chapter 4, Figure 4.40), while NASA accepts 130 degrees (England et al., 2010). The bulky materials, the stiffening "balloon" effect of the suit bladder, and shapes of space suits don't allow maximum ROM. An alternative to the gas-pressurized "bladder" suit is the *mechanical counter-pressure (MCP) suit* introduced in the early days of space exploration (Annis & Webb, 1971; Webb & Annis, 1967). An MCP suit, because it eliminates the "balloon" effect, improves mobility. Newer tight-fitting MCP suits apply pressure by mechanically squeezing the wearer inside a skin-tight suit (Holschuh, Obropta, Buechley, & Newman, 2012).

***Modular Space Suit Joint Mobility Issues and Solutions***

Fitting any wearable product to allow needed mobility through transition regions between the limbs and torso, for example at the shoulder, is a challenge (refer to Chapter 1, Figure 1.8). The problem is magnified in spacesuits. In space, the goal is to allow natural motion at body joints where the components attach, but the *reach envelope* and shoulder joint position are different when a person reaches laterally versus anteriorly. Review scapulothoracic and glenohumeral joint motions in Chapter 4, Section 4.4.4. Unlike traditional garments, space suit materials are variably flexible; rigid space suit components do not wrinkle to allow reach. Modular attachment points must remain tightly sealed so they don't have the give of an armscye seam. Therefore, it isn't possible to design and join a space suit arm that effectively accommodates both reach positions. The current space suit design trade-off decision optimizes one arm position according to the specific tasks the astronaut needs to perform. For example, if job tasks require anterior reach to manipulate tools in front of the body, the suit limits lateral reach and the astronaut must deal with the limitations, each in her/his own way (B. Holschuh, personal communication, August 24, 2018). The major space suit design challenge remains—allow maximum body performance while insuring health and safety in the space environment.

### 9.5.8 Wearable Products for Space: Continuing Research

As space exploration expands beyond the moon, and time in space increases, space suit design needs to evolve to address astronaut needs during and after space travel. NASA (2014) identifies muscle atrophy, reduced orthostatic tolerance, reduced aerobic capacity, sensorimotor dysfunction (including gait instability and altered dynamic visual acuity), and decreased bone density and changed bone architecture as post-flight health effects (page 4–211). Can wearable in-flight countermeasures mitigate these effects in full or in part? Will wearable products be part of the solution in preventing these negative effects?

Thomas and McMann (2012) reviewed new approaches to designing space suits from government agencies like NASA, private industries, and entrepreneurs. University faculty and students contribute talent and resources to solving space wearable technology challenges (Simon, Dunne, Zeagler, & Pailes-Friedman, 2014). The varied approaches, uses of materials, and structures can provide an inspiration for designers tackling complex body/product/environment problems.

## 9.6 Designing for the Human Body: Conclusion

Designing any wearable product for the human body presents a challenge and an opportunity. Whole-body wearable products, more than products limited to an isolated body part or region, demand understanding of anatomy, physiology, and biomechanics. Wearable products that fit the static and dynamic body enhance product appearance. A product which facilitates body functions promotes health and comfort and may result in better performance. The need for precise whole-body wearable product fit and function increases as environmental hazards intensify.

Whole-body wearable design projects—especially for protective products—often involve collaborative efforts of a team of people. Teamwork demands good communication skills and a shared understanding of terminology and methods. Design team members who develop knowledge and understanding of human body form and function can work effectively with other team members to meet any number of difficult design challenges.

Marvel at the complexity of the human body while respecting its limitations to surmount design challenges in the ever more complex world we live in. When wearable product designers go beyond surface observations of the human body to fully understand anatomy and body function—safe, fully functional, and innovative products will result.

# *Appendix A: Landmarks and Measurements Foundations*

Use this guide as a reference when designing products to fit the human body and to help understand research that references landmarks and measurements. The definitions and directions are based on human anatomy. The landmarking and measuring information given in the Appendices is intended as background information, not a comprehensive manual, for wearable product designers. To conduct a full-scale anthropometric study, seek training from experienced anthropometrists.

Landmarking and measuring information coordinates with the book chapters:

- Table A.1 lists Appendices B–H, Landmarks and Measurements by Body Region.
- Appendix B: Head and Neck Landmarks and Measurements relates to Chapter 3, Designing for Head and Neck Anatomy.
- Appendix C: Upper Torso and Arm Landmarks and Measurements relates to Chapter 4, Designing for Upper Torso and Arm Anatomy.
- Appendix D: Lower Torso and Leg Landmarks and Measurements relates to Chapter 5, Designing for Lower Torso and Leg Anatomy.
- Appendix E: Waist Landmarks and Measurements relates to Chapter 6, Designing for Mid-Torso Anatomy.
- Appendix F: Hand and Wrist Landmarks and Measurements relates to Chapter 7, Designing for Hand and Wrist Anatomy.
- Appendix G: Foot and Ankle Landmarks and Measurements relates to Chapter 8, Designing for Foot and Ankle Anatomy.
- Appendix H: Total Body Landmarks and Measurements relates to Chapter 9, Designing for Whole-Body Anatomy.

Each appendix consists of two parts, the first on landmarking and the second on measuring. Look over the accompanying illustrations to visualize the procedures, recalling what you have learned about the anatomical features that are the basis for the procedures. Depending on the product you are designing, you may use all of the procedures or just a few. You may decide to adapt some of the procedures. Use or develop a precise and consistent process for all people who are being landmarked and measured.

**TABLE A.1**

Landmarks and Measurements for Body Regions and Total Body

| Appendix B. Head and Neck Landmarks and Measurements | |
|---|---|
| **B.1 Head and Neck Landmarks** | **B.2 Head and Neck Measurements** |
| Alare (R and L) | Bitragion chin arc |
| Cervicale | Bitragion crinion arc |
| Chelion (R and L) | Bitragion coronal arc |
| Crinion | Bitragion frontal arc |
| Ear lobe bottom (R and L) | Bitragion submandibular arc |
| Ear point (R and L) | Bitragion subnasale arc |
| Ear top (R and L) | Bizygomatic breadth |
| Ectocanthus (R and L) | Crinion inion arc |
| Ectoorbitale, posterior (R and L) | Ear breadth |
| Frontotemporale (R and L) | Ear length (total) |
| Glabella | Ear length above tragion |
| Gonion (R and L) | Ear protrusion |
| Infrathyroid | Head circumference |
| Inion (external occipital protuberance) | Head depth |
| Menton | Menton-sellion length |
| Neck (anterior, R lateral, and L lateral) | Neck circumference, base |
| Opisthocranion | Neck circumference, infrathyroid |
| Orbitale | |
| Otobasion superior (R and L) | |
| Promenton | |
| Pronasale | |
| Sellion | |
| Stomion | |
| Submandibular | |
| Subnasale | |
| Suprasternale | |
| Tragion (R and L) | |
| Trapezius point (R and L) | |
| Zygion | |
| Zygofrontale (R and L) | |

(*Continued*)

**TABLE A.1 (CONTINUED)**

Landmarks and Measurements for Body Regions and Total Body

| Appendix C. Upper Torso and Arm Landmarks and Measurements | |
|---|---|
| **C.1 Upper Torso and Arm Landmarks** | **C.2 Upper Torso and Arm Measurements** |
| Acromial point (R and L) | Acromion to radiale length |
| Biceps point, lateral flexed | Axillary arm circumference |
| Breast mound (female) boundaries (R) | Biacromial breadth |
| Bustpoint (R and L) | Biceps circumference, flexed |
| Clavicle point | Biceps circumference, not flexed |
| Deltoid point (R and L) | Breast mound (nude) |
| Midshoulder | Bustpoint/thelion, breadth |
| Midspine | Center back length at scye depth level (cervicale to armscye depth mark) |
| Neck, anterior, lateral (R and L) | Chest breadth |
| Olecranon (center) | Chest/bust circumference at the thelion (male), bustpoint (female) |
| Radiale | Chest circumference at scye |
| Scye or armscye | Chest circumference below breast |
| Styloid process of the radius | Chest/bust depth |
| Styloid process of the ulna | Elbow circumference |
| Suprasternale—in Head and Neck | Forearm circumference (flexed) |
| Tenth rib (bottom of the rib cage) | Forearm to hand length |
| Thelion | Interscye or cross-back length at axillary fold level |
| Trapezius point (R and L) | Interscye or cross-back length at the midscye |
| Wrist (dorsal) | Radiale (proximal) to stylion the stylion process of the radius, length |
| | Scye (armscye) circumference |
| | Scye (armscye) arc (anterior and posterior) |
| | Shoulder circumference |
| | Shoulder (acromion) to elbow (olecranon) length (two-dimensional) |
| | Shoulder length |
| | Shoulder neckpoint (trapezius point) to bustpoint/thelion length |
| | Sleeve length from C7 to olecranon |
| | Sleeve length from C7 to ulnar styloid process |
| | Strap length |
| | Waist back length (back waist length) |
| | Waist front length (front waist length) |

(Continued)

**TABLE A.1 (CONTINUED)**

Landmarks and Measurements for Body Regions and Total Body

|  |  |
|---|---|
| **Appendix D. Lower Torso and Leg Landmarks and Measurements** | |
| **D.1 Lower Torso and Leg Landmarks** | **D.2 Lower Torso and Leg Measurements** |
| Anterior-superior iliac spine of the pelvis | Buttock circumference |
| Buttock point, lateral, (R and L) | Buttock depth |
| Buttock point, posterior | Buttock height |
| Calf | Calf circumference |
| Dorsal juncture of the foot and leg | Crotch height |
| Gluteal furrow point | Hip breadth |
| Iliocristale | Knee height at midpatella |
| Inner thigh (estimation method) | Lower thigh circumference |
| Inner thigh (plumb line method) | Thigh circumference |
| Lateral femoral epicondyle (standing and sitting) | |
| Midpatella | |
| Suprapatella | |
| Tibiale | |
| Trochanter (seated) | |
| Trochanterion (standing) | |
| Lower Torso Landmarks (sited, not landmarked) | |
| Dorsal juncture of calf and thigh | |
| Posterior superior iliac spine, posterior point of the crest of R iliac | |
| **Appendix E. Waist/Waistline Landmarks and Measurements** | |
| **E.1 Waist/Waistline Landmarks** | **E.2 Waist/Waistline and Product Related Measurements** |
| Waist (omphalion location) (R, L, anterior, posterior) | Crotch depth (over the hip curve arc) |
| Side waist point (R and L) | Crotch length (omphalion level) |
| | Crotch length, posterior (omphalion level) |
| | Waist circumference (omphalion level) |
| | Waist breadth (omphalion level) |
| | Waist depth (omphalion depth) |

*(Continued)*

**TABLE A.1 (CONTINUED)**

Landmarks and Measurements for Body Regions and Total Body

|  |  |
|---|---|
| **Appendix F. Hand and Wrist Landmarks and Measurements** | |
| **F. 1 Hand and Wrist Landmarks** | **F.2 Hand and Wrist Measurements** |
| Finger base I, II, III, IV, V | Finger breadth at the joint, dorsal and/or ventral |
| Finger crotch I, II, III, IV | Finger length to base, I, II, III, IV, V |
| Finger flexion creases on ventral side of hand | Finger length to crotch base, I, II, III, IV, V |
| Finger joint prominence marked on dorsal side, I, II, III, IV, V | Finger segment length, joint crease to joint crease |
| Fingertip, I, II, III, IV, V | Hand breadth |
| Metacarpal II | Hand circumference |
| Metacarpal V | Hand length |
| Stylion, dorsal stylion and ventral stylion | Palm length |
| Ulnar edge of the distal wrist crease | Wrist circumference |
| **Appendix G. Foot and Ankle Landmarks and Measurements** | |
| **G.1 Foot and Ankle Landmarks** | **G.2 Foot and Ankle Measurements** |
| Fifth metatarsophalangeal protrusion | Ankle circumference |
| First metatarsophalangeal protrusion | Ball of foot circumference |
| Lateral malleolus | Ball of the foot length (from calcaneus) |
| Medial malleolus | Bimalleolar breadth |
| | Foot breadth (horizontal) |
| | Foot length |
| | Heel breadth |
| | Heel-ankle circumference |
| | Lateral malleolus height |
| | Medial malleolus height |
| **Appendix H. Total Body Landmarks and Measurements** | |
| **H.1 Total Body Landmarks** | **H.2 Total Body Measurements** |
| See landmark directions in the related body section Appendices | Acromial height (standing) |
| | Axilla height (standing) |
| | Calf height |
| | Cervicale height (standing with head in Frankfort plane) |
| | Chest/bust height (standing) |
| | Neck height (standing with head in Frankfort plane) |
| | Stature/total body height (standing with head in Frankfort plane) |
| | Vertical trunk circumference |

## A.1 Body Position for Landmarking and Measuring

Body position, or how the person sits or stands when being measured, can affect the measurements that are collected. Directions indicate which body position to use, standing or sitting. Refer to Chapter 1 for information on the anatomical position. You may use the anatomical position as the basic standing position, or an alternative. Decide the best standing position for the product you are designing. Seated positions should be on a solid, flat surface—chair or platform—with no cushion or contoured surfaces. The chair, stool, or platform should position the person so that the flexed knee is at about 90 degrees. Some measurements require that the person stand or sit with the head in the Frankfort plane (see Chapter 1) to assure the head is in proper alignment for a specific measurement. Positioning a person in a "dynamic pose" (squatting, flexing, bending, etc.) may be necessary when measuring for some products like one-piece protective clothing. For people with special needs, you may need to adapt positions. Body position for manual measurement methods and body scan methods often differ. For scanning methods, a person may have to stand with legs farther apart and arms held a distance from the body so that areas of the body are not obscured. Consistent body positions for everyone being measured is important. If you are using measurements from a data base, try to find out which positions were used to collect the data.

The point in time of the respiratory cycle when the measurement is recorded may affect some positions and therefore measurements. For example, chest circumferences vary from full exhalation to full inhalation, possibly affecting product dimensions. Callaway et al. (1991, p. 44) state that various anthropometric studies have used different breath cycle points from maximum inhalation and exhalation, to normal or quiet inhalation or expiration, to mid-respiration. Establish a protocol for taking measurements during a breathing cycle or try to determine which cycle point was used for a data base you are using, especially if breath cycle will affect your product design. Unless otherwise noted, "maximum point of quiet respiration" (ANSUR, 1988) is used for the directions in the Appendices. Observe the person as they go through several natural breathing cycles. Do not ask them to "breathe naturally" as this request may alter their normal breathing. Note the point between exhalation and the beginning of the next inhalation. Record the measurement at that point in time.

What the person wears can also affect landmarking and measuring accuracy and applicability to a product. Scan suits that have been used for some studies, can compress body tissue and do not represent what people typically wear under a product (Kim, LaBat, Bye, Sohn, & Ryan, 2014). For most products, people are measured wearing their own underwear or underwear that is supplied for a study.

## A.2 Landmarking

Landmarks are points on the body established to consistently locate measurements, and are defined relative to anatomical features. Procedures for locating and marking the landmarks are described with individual exceptions and variations noted. If one side of the body is landmarked and measured, the convention is to measure the right side of the body. There may be some exceptions based on an individual's anatomy. If an exception is made, note that the left side of the body was measured for a specific person. A product design may necessitate measuring the left side of the body or both sides of the body.

After locating a landmark on a person, use a removable adhesive marker or a marking pen or pencil to draw the landmark on the body. Draw a small mark: a dot, an X, or line, as indicated in the procedures. Some landmark locations have distinctive physical characteristics and need no additional markings placed on the body. Avoid drawing or placing an adhesive marker on an area that is easily irritated. For example, the lateral corner of the eye is usually not physically marked. Landmark procedures are also appropriate for placing reflective or transmitting markers used with motion capture systems.

## A.3 Measuring

Measurements are either two-dimensional: lengths, depths, breadths, and heights; or three-dimensional over-the-body surface: circumferences and arcs. Two-dimensional measurements are taken with an anthropometer or caliper. A standing anthropometer can be separated into smaller sections to measure a body segment, like leg length. Specialized anthropometers and calipers are available to measure specific body areas. The generic terms anthropometer and caliper are used in these Appendices to indicate two-dimensional measurement techniques. A tape measure (steel or plastic-coated) is used for three-dimensional, over-the-body measurements. Consult a detailed anthropometric manual for descriptions and uses of these instruments (see resources listed below). If you are conducting a study for designing your product, all equipment should be disinfected after each person is measured, for example, wipe down the tape measure with alcohol.

## A.4 Measurements Related to Product

It can be challenging to relate measurements to the product you are designing. First, understand the structure of the product and how the product should relate to the body. Baggy sweatpants need a minimum of measurements and a loosely fitted crotch curve, while slim-fit jeans require a precise relationship of the crotch curve measurements, waist, and hip measurement locations. Medical products require an understanding of underlying anatomical features and how to position the product on the body surface. Think how a product will be used, if it will be moved or repositioned during use, and how repositioning can affect product dimensions.

## A.5 Appendices Illustrations

The illustrations in the Appendices show the male and female figures with outlines of generic, closely fitted underwear. The skeleton is included as a frame of reference for placing landmarks. Locating the skeletal structures can be fairly easy if the person is slim, but rather difficult if the skeletal frame is covered with extra adipose tissue and/or muscle. For those persons, visualize the underlying skeletal structures and approximate the locations.

## A.6 Sources for Appendices Landmarking and Measuring Directions

Landmark and measuring directions were adapted from several sources including:

- *Anthropometry and biomechanics* (1995). Man-Systems Integration Standards. Volume I, Section 3. National Aeronautics and Space Administration. This standard is no longer maintained by NASA, but is available to download from NASA. https://msis.jsc.nasa.gov/downloads.htm
- Callaway, C. W., Chumlea, W. C., Bouchard, C., Himes, J. H., Lohman, T. G., Martin, A. D., . . . Seefeldt, V. D. (1991). Circumferences. In T. G. Lohman, A. F. Roche, & R. Martorell, (Eds.). *Anthropometric standardization reference manual* (pp. 39–54). Champaign, IL: Human Kinetics Books.
- Clauser, C., Tebbetts, I., Bradtmiller, B., McConville, J., & Gordon, C. (1988). *Measurer's Handbook: U.S. Anthropometric survey, 1987–1988* (Technical report NATICK/TR-88/043, unlimited distribution). Natick, MA: United States Army Natick Research, Development and Engineering Center. http://www.dtic.mil/dtic/tr/fulltest/u2/a202721.pdf

- Greiner, T. M. (1991). *Hand anthropometry of U.S. Army personnel* (Technical report NATICK/TR-92/011, unlimited distribution). Natick, MA: United States Army Natick Research, Development and Engineering Center. http://www.dtic.mil/get-tr-doc/pdf?AD=ADA244533
- Holtzman, J., Gordon, C., Bradtmiller, B. Corner, B.D., Mucher, M., Kristensen, H., Paquette, S. & Blackwell, C. L. (2011). *Measurer's handbook: US Army and Marine Corps Anthropometric surveys, 2010–2011* (Technical report NATICK/TR11/017, unlimited distribution). Natick, MA: United States Army Natick Research, Development and Engineering Center. http://www.dtic.mil/dtic/tr/fulltext/u2/a548497.pdf
- Human Systems Information Analysis Center (1994). *Anthropometric data analysis sets manual.* Ohio: Wright-Patterson Air Force Base. http://mreed.umtri.umich.edu/mreed/downloads/anthro/ansur/ADAS-Dimension_Definitions.pdf
- Kim, D-E, LaBat, K., Bye, E., Sohn, M-H., Ryan, K. (2014). A Study of Scan Garment Accuracy and Reliability. *Journal of the Textile Institute, 106* (8), 853–861. doi: 10.1080/00405000.2014.949502
- Kouchi, M. (2014). Anthropometric methods for apparel design: Body measurement devices and techniques. In D. Gupta, & N. Zakaria (Eds.), *Anthropometry, apparel sizing and design* (pp. 67–94). Cambridge: Woodhead.
- Lee, H. Y., Hong, K., & Kim, E. A. (2004). Measurement protocol of women's nude breasts using a 3D scanning technique. *Applied Ergonomics, 35*(4), 353-359.
- Lohman, G., Roche, A. F., & Martorell, R. (Eds.). (1991). *Anthropometric standardization reference manual.* Champaign, Illinois: Human Kinetics Books.
- Minott, J. (1978). *Fitting commercial patterns: The Minott method.* Minneapolis, MN: Burgess Publishing Company.
- Mallouris, A., Yiacoumettis, A., Thomaidis, V., Karayiannaki, A., Simopoulos, C., Dakagia, D., & Tsaroucha, A. (2012). A record of skin creases and folds. *European Journal of Plastic Surgery, 35,* pp. 847–854. doi: 10.1007/s00238-012-0774-3
- NASA. (1978). *Anthropometry for designers. NASA Reference Publication* 1024. https://ntrs.nasa.gov/search.jsp?R=19790003563
- Roebuck, J.A. (1995). *Anthropometric methods: Designing to fit the human body.* Santa Monica, CA: Human Factors and Ergonomics Society.
- Simmons, K. P. & Istook, C. L. (2003). Body measurement techniques: Comparing 3D body-scanning and anthropometric methods for apparel applications. *Journal of Fashion Marketing and Management, 7*(3), 306–332. doi: 10.1108/13612020310484852
- Singer Sewing Reference Library. (1989). *Sewing pants that fit.* Minnetonka, MN: Cy DeCosse Incorporated.

## A.7 Equipment and Devices

- Flexible steel tape measure or flexible plastic-coated tape measure.
- Brannock™ device for foot width and depth measurements (see Figure G.2).
- Caliper: Typically a beam caliper, a small caliper used to measure length segments, for example olecranon to radiale. Several other caliper types are used for anthropometric studies.
- Anthropometers: Standing anthropometer and modified height gauge (small anthropometer, sometimes described as a sliding anthropometer) (see Figure A.1).
- Hand block: Small block of wood approximately 12.5 cm (5 in.) × 5 cm (2 in.) × 2.5 cm (1 in.) for hand measurements. The hand block can be constructed using supplies from a lumber yard or craft company.

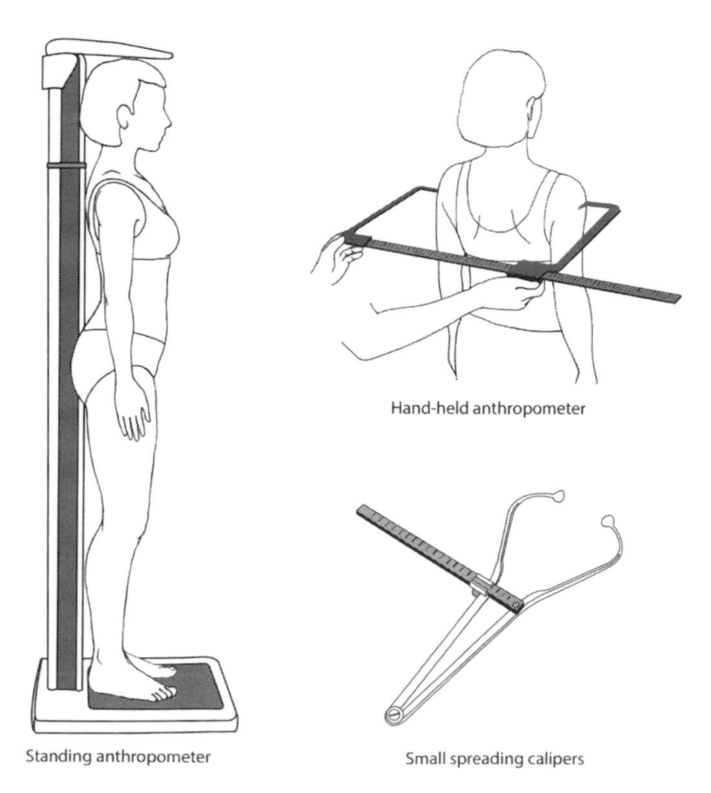


**FIGURE A.1**
Anthropometers and caliper used to measure lengths, widths, and depths.

- The anthropometrist/designer may need to create specialized tools when measuring for a specific product. The goal when collecting measurements from several, to many people is to assure that measurements are replicable and reliable. Tools that consistently position each person can aid in developing a consistent landmark and measurement plan.

# *Appendix B: Head and Neck Landmarks and Measurements*

Head and neck landmarks and measurements are described in this appendix, landmark information in Section B.1 and measuring procedures in Section B.2. Refer to the related anatomical structures in Chapter 3. Look over the accompanying illustrations to visualize the procedures, recalling what you have learned about anatomical features of the head and neck that are the basis for procedures. Landmarks are illustrated in Figures B.1 and B.2 and measurements are illustrated in Figures B.3 and B.4.

## B.1 Head and Neck Landmarks (Alphabetical Order)

***Alare (R and L):*** The most lateral point on the wing of the nose.

*Procedure:* View person from the front and visually locate the outer flare of the wing of the nose. Place small mark.

***Cervicale:*** The superior palpable point of the posterior spine of the seventh cervical vertebra. The spine of the 7th cervical vertebra is typically the most prominent vertebral spine of the back of the neck.

*Procedure:* The person is standing. Have the person bend his/her head forward. Stand behind the person and palpate with the pad of the index finger. Have the person slowly bring the head up into the Frankfort plane (ear canals and inferior orbit parallel to the floor). Locate the superior point of the spine of the 7th cervical vertebra (cervicale). Place a small mark.

*Exceptions and variations:* Some people will have no discernible prominence at the 7th cervical vertebrae. In those cases, estimate the point. Some people will have several vertebrae that are equally prominent. In that case, mark the spine closest to the posterior base of the neck as established by a lightweight fine-chain necklace draped around the neck. Be sure that the person is standing erect, looking straight ahead.

***Chelion (R and L):*** The most lateral point of the juncture of the fleshy (mucosal) tissue of the lips with the facial skin at the corner of the mouth.

*Procedure:* View person from the front and visually locate the outermost corner of the lips (R and L). This landmark may not need a marker.

***Crinion:*** The lowest point of the hairline on the forehead in the midsagittal plane.

*Procedure:* The person is standing. Stand in front of the person and locate, by inspection, the lowest point of the hairline on the forehead in the midsagittal plane. Some people will have a definite point often referred to as a "widow's peak." Place a small mark.

*Exceptions and variations:* For bald people or people with receding hairlines, this landmark is often not marked (if the product relates to hairline location, estimate that location).

***Ear lobe bottom (R and L):*** The lowest point of the ear on its long axis.

*Procedure:* View the person from the side and front, and visually locate the lowest point of the earlobe. A marker is usually not needed for this landmark.

***Ear point (R and L):*** The lateral point of the outer ridge of the ear farthest from the head.

*Procedure:* View the person from the front, and visually locate the outer ridge of the ear. A marker is usually not needed for this landmark.

***Ear top (R and L):*** The highest point of the ear on its long axis.

*Procedure:* View the person from the side, and visually locate the highest point of the ear. A marker is usually not needed.

***Ectocanthus (R and L):*** The outside corner of the eye formed by the meeting of the upper and lower eyelids.

*Procedure:* View the person from the front and side to locate this point. A marker is usually not needed.

***Ectoorbitale, posterior (R and L):*** The posterior point on the frontal process of the zygomatic bone at the level of the outer corner of the eye (ectocanthus).

*Procedure:* With the person facing forward, stand in front of the person and palpate the zygomatic bone following posteriorly from the outer corner of the eye. Place a small mark.

***Frontotemporale (R and L):*** The point of deepest indentation of the temporal crest of the frontal bone above the brow ridges.

*Procedure:* View the person from the front and palpate the brow ridges above the eye along the eyebrow locating the arch of the ridge. Place a small mark.

***Glabella:*** The slight indent on the frontal bone midway between the bony brow ridges.

*Procedure:* View the person from the front and visually inspect for the center point between the eyebrows. Place a small mark.

*Exceptions and variations:* The prominence of the ridge of the brow varies person to person.

***Gonion (R and L):*** The most lateral point on the posterior angle of the mandible.

*Procedure:* View the person from the right side inspecting for the most lateral point. Palpate to locate the angle of the mandible. Place a small mark.

*Exceptions and variations:* The angle of the mandible varies person-to-person and amount of muscles and fat may obscure the angle.

***Infrathyroid:*** The inferior point in the midsagittal plane of the thyroid cartilage (Adam's Apple).

*Procedure:* With the person standing and head in the Frankfort plane, stand in front of the person and gently palpate the smooth lateral bulge over the body of the thyroid cartilage moving downwards until you feel the space between the thyroid cartilage and cricoid cartilage just below it. Locate the bottom point of the thyroid cartilage in the midsagittal plane. Draw a short horizontal line at this point.

*Exceptions and variations:* Be sure that the person's head is in the Frankfort plane.

***Inion (external occipital protuberance):*** Most prominent point of the external occipital protuberance, at the center of the posterior cranium.

*Procedure:* With the person seated, stand behind the person and with one hand place a finger on each trapezius muscle. Run fingers upward along the trapezius muscle to find approximate insertion of the muscle to the skull. The inion is located between the two insertions.

*Exceptions and variations:* This location is often within the hair, so landmarking with a pen or adhesive marker may not be possible. Technician may have to palpate for this location as a measurement is being taken.

***Menton:*** The inferior point of the mandible in the midsagittal plane (bottom of the chin).

*Procedure:* With person standing and head in the Frankfort plane, with teeth together (not clenched), stand in front of the person. Palpate for a point at the lower center of the mandible (chin) and place a small mark.

*Exceptions and variations:* Be careful not to distort the soft tissue under the chin if drawing a landmark.

***Neck (anterior, R lateral, and L lateral):*** Anterior and lateral points of the base of the neck. These landmarks are used to locate the curve of a fitted neckline that encircles the base of the neck.

*Procedure:* To facilitate location of these marks, use a lightweight fine-chain necklace approximately 50.8 cm (20 in.) in length. With the person standing, place the midpoint of the chain at the cervicale and drape the ends of the chain around the neck so that the ends hang parallel to each other over the chest (see Figures A.1 and A.2). Draw a short horizontal line along the chain at the posterior neck and an intersecting perpendicular line at the location of the cervicale. Place a similar short line along the chain at the lateral neckpoints where the chain intersects with the crest of the trapezius muscle. To mark center front, hold a straight edge or ruler at the base of the neck and parallel to the floor. This is an estimation based on the product position (may correspond to the top of the manubrium or to the clavicle protuberances at both sides of the manubrium). The ruler should be perpendicular to the chain at both sides of the neck. Draw a short line along the ruler. Mark the center front on the line with an X or short line.

*Exceptions and variations:* The four landmarks help describe the circular shape of a basic neckline. Variations in muscle and adipose tissue will affect landmark placement. Try to visualize the product on the body in relationship to the base of the neck.

***Opisthocranion:*** The posterior point on the back of the head.

*Procedure:* With the person standing, head in the Frankfort plane. Stand at the right side of the person. Locate the landmark position by palpating along the back of the head. Place a small landmark.

*Caution:* Hair covering the cranium will affect how the landmark is marked. Unless the person has very little hair, a wig cap or lightweight skullcap should be used to hold hair to the outline of the cranium.

***Orbital:*** Also called infraorbitale. The lowest point on the anterior border of the bony eye socket.

*Procedure:* With the person standing facing forward, stand in front of the person and gently palpate the bony eye socket under the eye to locate its lowest point. Draw a small dot at the location (an adhesive marker may be uncomfortable).

*Exceptions and variations:* Persons may be uncomfortable with the measurer palpating so closely to the eye. Use a delicate touch and tell the person exactly what you are doing. Also be sure to thoroughly wash your hands before palpating and marking this landmark.

***Otobasion superior (R and L):*** The anterior superior point of the juncture between the ear and the head.

*Procedure:* View the person from the side and the front and visually locate where the ear intersects with the skull. A marker is unnecessary.

***Promenton:*** The most anterior projection of the soft tissue of the chin in the midsagittal plane. Relates to the underlying center of the mandible and is superior to the menton.

*Procedure:* View the person from the right side and locate the most anterior projection that will also be visible when viewing the person from the front. Place a small mark.

*Exceptions and variations:* A cleft chin (natural indentation at the prominence of the chin) variation may be used as the landmark location.

***Pronasale:*** The point of the most anterior projection of the soft tissue of the tip of the nose in the midsagittal plane.

*Procedure:* View the person from the right side and visually locate the most anterior projection of the tip of the nose. Place a small mark.

*Exceptions and variations:* A deviated septum, broken, or crooked nose (variations from average) may require that the designer consider type of product and locate the landmark in the best position.

***Sellion:*** The point of the deepest depression of the nasal bones at the top of the nose.

*Procedure:* With the person standing, stand at the right of the person and palpate the point of deepest depression of the bridge of the nose in the midsagittal plane. Place a small mark.

*Exceptions and variations:* Some persons will not have a distinct depression or deep point at this location. Use your judgment on how you will use this landmark in relationship to the product.

***Stomion:*** The point of intersection of the upper and lower lip in the midsagittal plane when the mouth is closed.

*Procedure:* View the person from the front and locate this landmark. A marker is usually not needed.

***Submandibular:*** The juncture, in the midsagittal plane, of the lower jaw and the neck.

*Procedure:* The person stands with head in the Frankfort plane. Use a pencil to find the location. Gently roll the pencil along the underside of the chin toward the neck until the pencil stops. Draw a short horizontal line at this location.

***Subnasale:*** The intersection point of the philtrum (vertical groove above the upper lip) with the base of the nose.

*Procedure:* View the person from the front to locate this intersection. May be landmarked with a horizontal line.

***Suprasternale:*** The inferior point of the sternal notch of the manubrium (top of the sternum).

*Procedure:* With the person standing, stand in front of the person and palpate to locate the bottom of the sternal notch. Place a marker or draw a short horizontal line at this point.

*Exception:* May or may not correspond with anterior neck point.

***Tragion (R and L):*** The superior point on the juncture of the cartilaginous flap (tragus) of the ear with the head.

*Procedure:* Palpate to locate the tragus of each ear. Find the superior point of the attachment to the head. Place a small mark.

*Exceptions and variations:* Avoid distorting the soft tissue while drawing this landmark.

***Trapezius point (side neck point) (R and L):*** The point at which the anterior border of the trapezius muscle crosses the lateral neck landmark.

*Procedure:* With the person standing, ask the person to place their R hand on his/her L shoulder. This position should help to outline the trapezius muscle of the right shoulder. Stand at the right side of the person, move your fingertips along the top edge of the shoulder toward the neckline, palpating the mass of the trapezius muscle to locate its anterior border. Draw a short line from the neck toward the shoulder at the point where the anterior border of the muscle crosses the lateral neck landmark and place the landmark at the crossing point of the two lines. Repeat for the L side.

***Zygion (R and L):*** The most lateral point on the zygomatic arch.

*Procedure:* The person stands, looking straight ahead. Ask the person to inhale and exhale twice to relax their facial muscles. Stand in front of the person and palpate the most lateral point on each side of the zygomatic arch. Place a small mark.

***Zygofrontale (R and L):*** The most lateral point of the frontal bone where it forms the upper margin of the bony eye socket.

*Procedure:* View the person from the front and palpate along the ridge of the frontal bone locating the most lateral point. Place a small mark.

## B.2 Head and Neck Measurements (Alphabetical Order)

Head and neck measurements are either two-dimensional or over-the-body. For example, breadth of the head is measured using a caliper or sliding anthropometer and head circumference is measured with a flexible tape measure. Refer to Figures A.3 and A.4 for measurement illustrations.

***Bitragion chin arc:*** The over-the-body distance between the R and L tragion across the anterior point (promenton) of the chin.

*Landmarks:* tragion (R and L) and promenton.

*Procedure:* The person is seated with mouth closed, but relaxed. Stand in front of the person. Use a tape measure to measure the over-the-body distance from the R tragion landmark over the point of the chin (promenton) to the L tragion landmark.

*Cautions, exceptions, variations:* Exert just enough tension on the tape to maintain contact with the skin. The chin will be slightly compressed. Be sure that the 0-point of the tape is on the R tragion landmark.

***Bitragion crinion arc:*** The over-the-body distance between the R and L tragion across the top of the forehead at the crinion.

*Landmarks:* tragion (R and L) and crinion.

*Procedure:* The person is seated. Stand in front of the person and use a tape measure to measure the distance between the R and L tragion landmarks across the top of the forehead. Ask the person to assist by placing their index finger on the tape measure at the crinion location.

*Cautions, exceptions, variations:* Be sure the 0-point of the tape is on the tragion point. The hairline locations of persons will vary. If a person is bald, the tragion location may be estimated or this measurement may be excluded. Record these exceptions and the reason for not taking the measurement.

***Bitragion coronal arc:*** The over-the-body distance between the R and L tragions across the top of the head in a coronal plane.

*Landmarks:* tragion (R and L).

*Procedure:* The person is seated with head in the Frankfort plane. Stand in front of the person and use a tape measure to measure over-the-body distance between the R and L tragion landmarks across the top of the head in the coronal plane.

*Cautions, exceptions, variations:* Exert enough pressure to compress the hair and be sure to hold the 0-point of the tape on the R tragion landmark.

***Bitragion frontal arc:*** The over-the-body distance between the R tragion and the L tragion across the mid-forehead.

*Landmarks:* tragion (R and L).

*Procedure:* The person is seated. Stand in front of the person. Use a tape measure to measure the over-the-body distance between the R tragion landmark and the L tragion landmark with the tape passing across the forehead just above the ridges of the eyebrows (supraorbital ridges).

*Cautions, exceptions, variations:* Be sure that the 0-point of the tape is on the R tragion landmark and exert just enough pressure on the tape to maintain contact with the skin.

***Bitragion submandibular arc:*** The over-the-body distance between the R and L tragions across the submandibular landmark at the juncture of the jaw and the neck.

*Landmarks:* tragion (R and L) and submandibular.

*Procedure:* The person is seated. Stand in front of the person and use a tape measure to measure the over-the-body distance from the R tragion landmark to the L tragion landmark across the submandibular landmark.

*Cautions, exceptions, variations:* Be sure the 0-point of the tape measure is on the R tragion landmark and exert just enough tension on the tape to maintain light contact with the skin.

***Bitragion subnasale arc:*** The over-the-body distance between the R and L tragion across the bottom of the nose (subnasale).

*Landmarks:* tragion (R and L).

*Anatomical feature (not landmarked)*: subnasale.

*Procedure:* The person is seated. Stand in front of the person and use a tape measure to measure the over-the-body distance between the R and L tragion landmarks. The top edge of the tape should cross the lowest point of the bottom of the nose (subnasale).

*Cautions, exceptions, variations*: Do not compress the soft tissue under the nose.

***Bizygomatic breadth:*** The maximum horizontal breadth of the face between the zygomatic arches.

*Anatomical features (not landmarked):* Zygomatic arches.

*Procedure:* The person is seated. Stand in front of the person and use a caliper or sliding anthropometer to measure the maximum breadth of the face between the cheekbones (zygomatic arches).

*Cautions, exceptions, variations:* Exert only enough pressure to make sure the tips of the measuring device are resting on the zygomatic arches. Do not compress the soft tissues over the zygomatic arches.

**Crinion inion arc:** The over-the-body distance between the crinion and the inion along a midsagittal plane of the cranium.

*Landmarks:* crinion and inion.

*Procedure:* The person is seated. Stand at the right side of the person and use a tape measure to measure the over-the-body distance between the crinion landmark and the inion landmark with the tape as close to the scalp as possible.

*Cautions, exceptions, variations:* The crinion and inion may be inexact as the amount of hair and hairline can affect locations. For bald people or people with receding hairlines, this landmark is often not marked (if the product relates to hairline location, estimate the location).

**Ear breadth:** The greatest breadth of the ear between the superior otobasion (the point where the ear intersects with the skull) and a point on the outside edge of the ear. The ear breadth is perpendicular to the long axis of the ear (not necessarily parallel to the floor).

*Anatomical features (not landmarked):* superior otobasion.

*Procedure:* The person is seated. Stand at the person's right and use a caliper or a sliding anthropometer to measure the maximum breadth of the ear perpendicular to its long axis. The tip of the measuring device is placed in front of the ear at the juncture of the top of the ear with the head (otobasion superior) and oriented in a line perpendicular to the long axis of the ear. The other device blade reaches to the outer edge of the ear.

*Cautions, exceptions, variations:* Exert only enough pressure to maintain contact with the skin. Be careful not to distort the soft tissue of the ear. Angle of the ear breadth, R to L, will vary by individual.

**Ear length (total):** The length of the ear from the top of the ear to the bottom of lobe.

*Antatomical features (not landmarked):* Most superior edge of upper ear, most inferior extension of lower ear (ear lobe).

*Procedure:* The person is seated looking straight ahead. Stand at the right of the person and use a caliper or sliding anthropometer to measure the length of the ear from its highest to lowest points on a line parallel to the long axis of the ear. Hold the fixed blade of the device at the bottom of the ear and move the slide bar to reach the top of the ear.

*Cautions, exceptions, variations:* Exert just enough pressure to maintain contact with the skin. Do not distort the ear with the measuring device.

**Ear length above tragion:** The length of the anterior portion of the ear.

*Landmarks:* tragion.

*Anatomical features (not landmarked):* Most superior tip of the top of the ear above the tragion.

*Procedure:* The person is seated. Stand at the right of the person and use a caliper or sliding anthropometer to measure the distance from the tragion landmark to the top of the ear on a line parallel to the long axis of the ear. The tip of the measuring device is placed at the tragion with only enough pressure to hold the device in place.

*Cautions, exceptions, variations:* The tilt of the long axis of the ear will vary person to person. Do not compress the soft tissue of the ear.

***Ear protrusion:*** The distance the wing of the ear protrudes from the skull.

*Anatomical features (not landmarked)*: mastoid process (bony ridge behind the ear), most lateral point (ear point) of the wing of the ear.

*Procedure:* The person is seated. Stand behind the person and use a caliper or sliding anthropometer. Measure the horizontal distance between the mastoid process and the outside edge of the ear at its most lateral point.

*Cautions, exceptions, variations:* Keep the horizontal bar of the measuring device perpendicular to the head. Do not distort the soft tissue of the ear.

***Head circumference:*** The maximum circumference of the head above the supraorbital ridges and ears.

*Anatomical features (not landmarked):* supraorbital ridges and otobasion (superior, R and L). These features typically are not physically landmarked but used as points of reference when measuring.

*Procedure:* The person is seated. Stand to the right of the person. Use a tape measure to measure the maximum circumference of the head above the attachment of the ears to the head (otobasion superior). The bottom of the tape measure should pass just above the ridges of the eyebrows (supraorbital ridges) and around the back of the head.

*Cautions, exceptions, variations:* Use just enough pressure to compress the hair. Typically the tape measure will be higher in the front than in the back. Observe the person to determine where the maximum circumference location is. The location may vary person to person. You may also redefine this measurement to match the requirements of the product you are designing, but be sure that you define the measurement and maintain consistency in location of measurements for all people being measured.

***Head depth:*** The two-dimensional measurement of the cranium from glabella to farthest protuberance of the back of the head. This measurement may be horizontal or angled relative to the floor.

*Landmark:* glabella.

*Anatomical features (not landmarked):* Opisthocranium, posterior protuberance of the cranium (back of head).

*Procedure:* The person is seated. Stand at the right of the person. Use a caliper or sliding anthropometer to measure in the midsagittal plane, the distance between the browridges (glabella) and the posterior point on the back of the head. Place one tip of the measuring device on the glabella and move the other tip along the back of the head to determine the maximum protuberance.

*Cautions, exceptions, variations:* Hold the device with light pressure on the glabella and enough pressure at the back of the head to compress the hair.

***Menton-sellion length:*** A two-dimensional measurement taken from the bridge of the nose to the base of the chin.

*Landmarks*: Menton and sellion.

*Procedure:* The person is seated with mouth closed, teeth together, but not clenched. Stand to the right of the person and use a sliding anthropometer or caliper to measure in the midsagittal plane, the distance between the menton landmark at the bottom of the chin and the sellion landmark at the deepest point of the nasal root depression. Place the fixed extension of the caliper on the sellion. Exert light pressure to maintain contact between the caliper and the skin.

*Cautions, exceptions, variations:* Be sure that the person's mouth is closed, but teeth are not clenched.

***Neck circumference, base:*** The base of the neck. Used as the measurement for a fitted neckline or collar. Typically angled with cervicale higher than the center front landmark.

*Landmarks:* lateral neck markings (R and L), center front neck landmark and cervicale (7th vertebrae) back neck landmark.

*Procedure:* The person is standing with the head in the Frankfort plane. Stand behind the person and use a tape measure to measure the circumference of the base of the neck. The bottom of the tape should match each landmark and form a smooth circle. Exert only enough tension to maintain skin contact. Do not constrict the neck. Ask the person to assist by lightly placing a finger at the center front of the tape.

*Cautions, exceptions, variations:* Be sure that the person's head is in the Frankfort plane.

***Neck circumference, infrathyroid:*** Circumference of the neck at the infrathryroid level, parallel to the floor.

*Landmark:* Infrathyroid (Adam's apple).

*Procedure:* The person is standing with the head in the Frankfort plane. Stand at the right of the person and use a tape measure to measure the circumference of the neck at the level of the infrathryoid landmark. Keep the tape measure perpendicular to the long axis of the neck. Exert only enough tension to maintain skin contact, do not constrict the neck.

*Cautions, exceptions, variations:* Recall that the Adam's apple is more prominent in men, so locating this position for women is more difficult. An alternate location may be mid-neck, approximately one-half distance between under the chin and the base of neck.

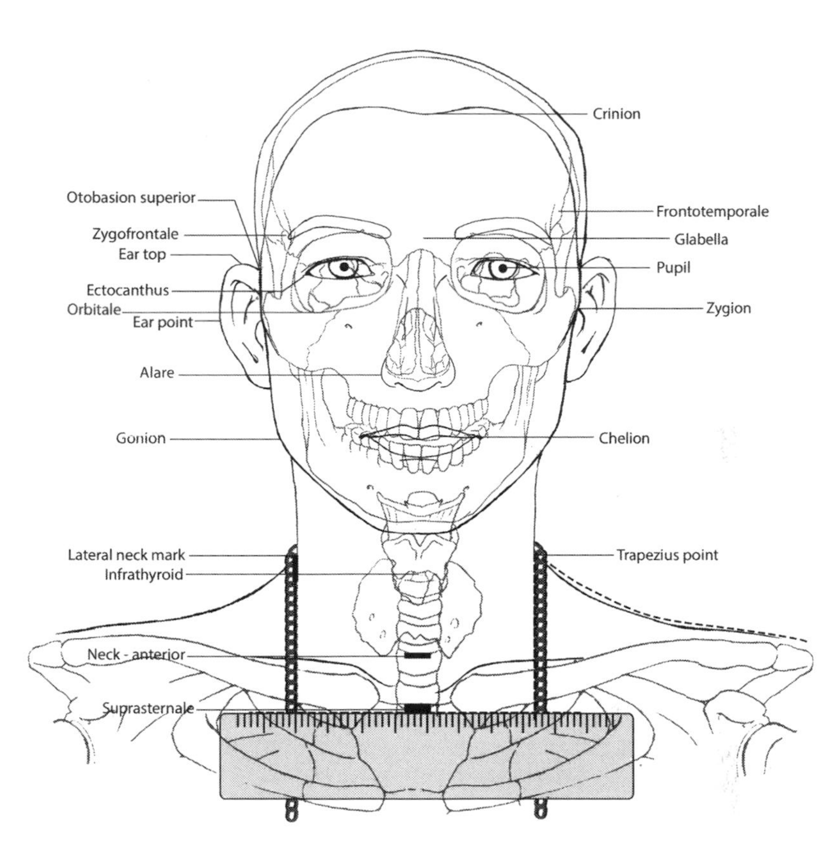


**FIGURE B.1**
Head and neck landmarks, anterior view.

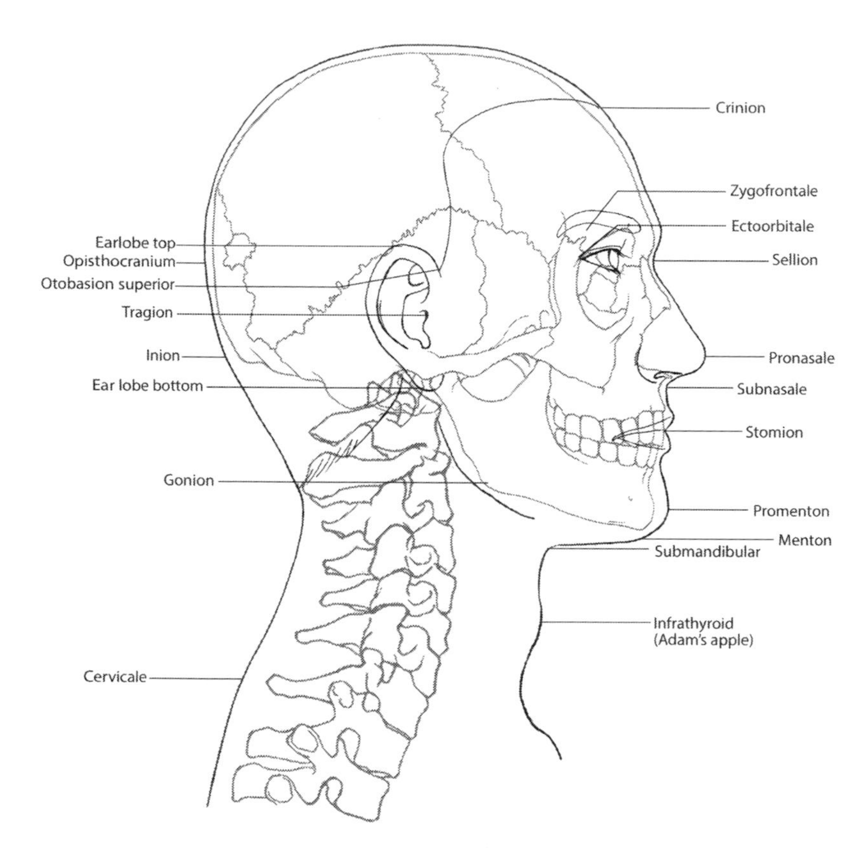


**FIGURE B.2**
Head and neck landmarks, lateral view.

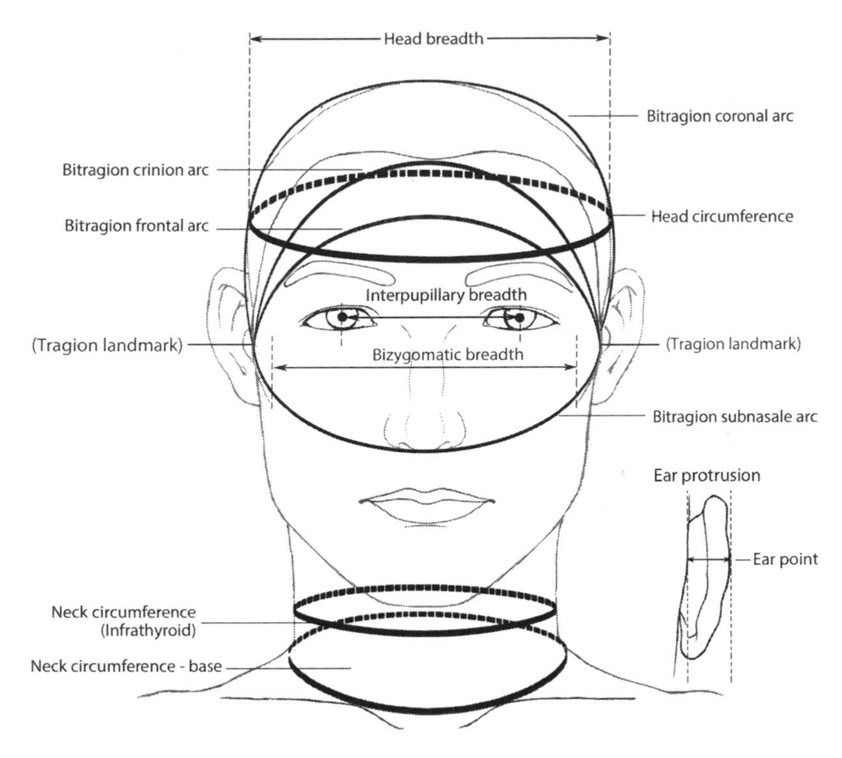


**FIGURE B.3**
Head and neck measurements, anterior view.

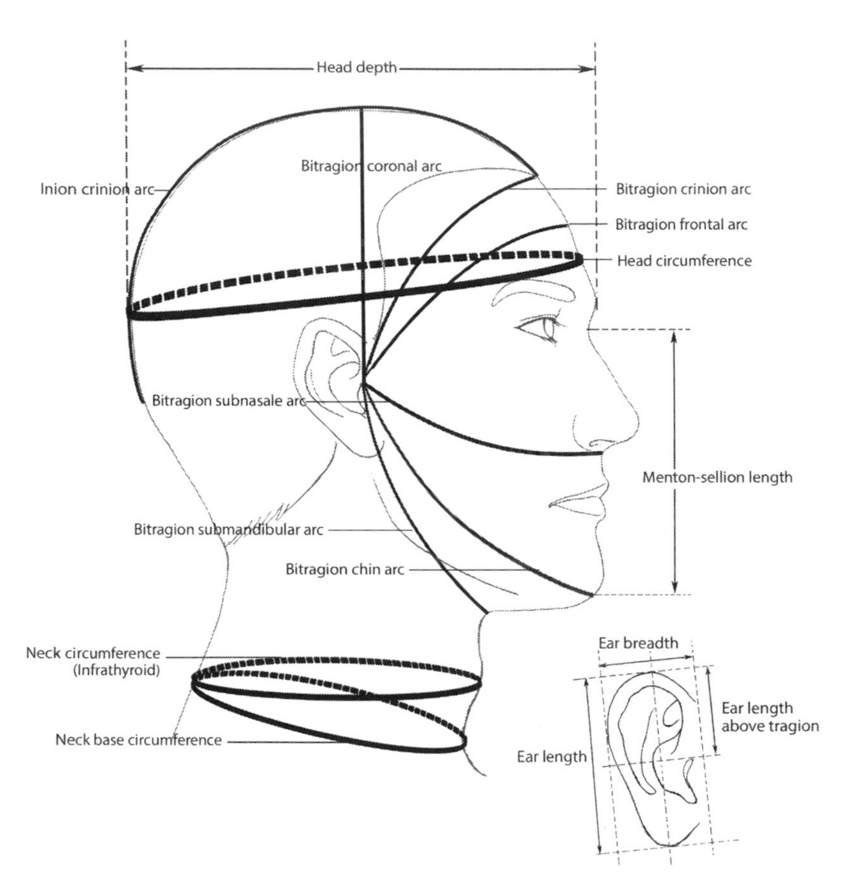


**FIGURE B.4**
Head and neck measurements, lateral view.

# Appendix C: Upper Torso and Arm Landmarks and Measurements

Upper torso and arm landmarks and measurements are described in this appendix. Some landmarks and measurements of head/neck and neck/upper torso overlap at the neck/upper torso depending on the product. Some structures of the upper torso and arm are difficult to landmark and measure, because the interlinking parts move in relationship to each other in complex ways. The products designed for upper torso and arm, especially if covering the whole area must not only fit, but other requirements must be considered: aesthetics, comfort, and/or protection. Some products also cover the hand, so understanding the connection of arm-to-wrist-to-hand may be necessary.

A tailored jacket requires more detailed measurements for a close fit in a non-stretchy woven textile, compared to an oversized sweater of a stretchy knit that requires a few basic measurements. Medical monitoring devices require knowledge of placement and functions of underlying anatomical structures so that the device is placed optimally. Anthropometric methods for landmarking and measuring the breast are not described in most anthropometry guides. This appendix includes some basics on landmarking the boundaries of the breast and some suggestions for measuring the breast mound. The landmark and measuring scheme for the breast typically has to be integrated with a landmarking and measuring scheme for the entire upper torso, for example bra design requires measurements for cup size and for band and strap dimensions.

See Chapter 4 for upper torso and arm anatomical details. Landmarks are illustrated in Figures C.1, C.2, C.3, and C.4. Measurements are illustrated in Figures C.5, C.6, C.7, C.8, and C.9.

## C.1 Upper Torso and Arm Landmarks (Alphabetical Order)

*Acromial point (R and L):* The intersection point of the lateral border of the acromial process and a line visualized running along the crest of the trapezius muscle, from the side neckpoint to the intersection with the point of the acromion (see details on locating the acromion and acromial point in Chapter 4). This landmark relates to the typical location of the intersection of a shoulder seam and the top of the sleeve cap of a basic fitted torso pattern for men or women.

*Procedure:* The person is standing. Stand behind the person and palpate the tips of both shoulders. Draw a line along the lateral bony border of each shoulder. Then stand at the right of the person and lay a tape measure along the shoulder originating at the trapezius point (refer to instructions on locating the base of the neck), so that the front edge of the tape lies over the clavicle point, and crosses the acromial border line at the tip of the shoulder. Draw a short line along the front edge of the tape where it crosses the acromial border. Repeat the process for the L shoulder. See Figure C.2.

*Exceptions and variations:* Muscles and adipose tissue overlaying the bony acromion may make location of this landmark difficult. Visualize the product in relationship to the person's body and use the visualization to approximately locate the landmark. Assure that all people who are landmarking and measuring use the same methods.

***Biceps point (lateral flexed):*** The highest point of the R flexed biceps as viewed from the person's right side.

*Procedure:* Person stands with the R upper arm extended forward horizontally and the elbow flexed about 90 degrees. The hand is tightly clenched forming a fist and held facing the head. Stand to the right of the person and locate the highest point of the flexed biceps. Draw a short line perpendicular to the long axis of the upper arm passing through the landmark.

*Exceptions and variations:* This "point" is ill defined as related to individual variations in muscle and adipose tissue. Consider the relationship to the product and the purpose for the landmark and related measurements.

***Breast mound (female) boundaries, R:*** The boundaries of the nude breast as reference for scanning (adapted from Lee, Hong, & Kim, 2004). The natural boundaries of breast tissue are difficult to define with visual inspection only. The technique for locating boundaries is sometimes called the "folding technique" and is used for cosmetic surgery, but may be adopted for products requiring definition of the breast mound. The technique should be explained to the participant and consent acquired, as the technique may be seen as intrusive. An alternative is to have the person being measured position their own breast according to your instructions. A female technician should landmark and measure the person, accompanied by an attendant. See Figure C.4.

*Procedure:* The person is nude from the waist up and in the standing position. 1) Stand in front of the person and place your left hand over the person's right breast, covering the nipple with your fingers extended toward the axilla. Push the breast mound firmly, but gently, against the rib cage. Notice the skin fold that is formed around the top of the breast mound. Place a mark at the fold and the intersection of the arm (FAP=front axilla point). Place a mark at the top point of the fold (UBP=upper breast point). 2) Stand to the left of the person and place your right hand over the breast with your fingers extended toward the sternum. Push firmly, but gently, toward the midline of the sternum and note the fold that is formed. Place a marker at the fold formed at the center of breast (IBP=inner breast point). Without breast compression, mark the lowest point or inframammary fold (BBP=bottom breast point). The procedure for exception can be used if the bottom breast point is covered by a ptotic breast.

*Exception:* Ptotic breast obscuring the BBP. Gently lift the breast to locate and mark the lowest point. See measurement methods for body scanning ptotic breasts.

*Additional landmarks* **for bra design may include:** front neck point, center shoulder point, upper breast point on a line from R axilla to L axilla, outer breast point (on a line from the FAP to the side waist point) that intersects with the IBP level, side waist point, nipple apex.

***Bustpoint (R and L):*** The anterior points of the bra cups. The landmark relates to the nipple point (see description below) of the nude female figure, although for most product purposes the person wears a bra and the bustpoint is located on the surface of the bra as the apex of vertical and lateral curves of the bra cup.

*Procedure:* The person is standing. Stand at the right of the person and site the most protruding point of the bra (or nipple) over each breast. Place a small mark. See Figure C.1.

*Exceptions and variations:* The person landmarking must use knowledge of anatomy and product in locating this landmark, especially with large variations in breast shape and size. Measurements for each individual will vary depending on the bra that is worn. Ask people to wear the bra they would wear with the product that is being designed. If the person does not wear a bra with the product, then measure the nude figure.

***Cervicale:*** See Appendix A.

***Clavicle point (R and L):*** The superior points of the lateral ends of the clavicles.

*Procedure:* The person is standing. Stand behind the person and palpate the top of the lateral ends of the clavicles (collar bones) near the tips of the shoulders until you locate the most superior points. Place a small mark on each point. See Figure C.1.

*Exceptions and variations:* Consider variations in muscle and adipose tissue overlying this skeletal structure.

***Deltoid point (R and L):*** The lateral point of the R deltoid muscle, and the lateral point of the L deltoid muscle at the level of the R deltoid point.

*Procedure:* The person is standing. Stand at the right of the person and locate the most protruding point of the R upper arm overlying the deltoid muscle. A standing anthropometer may be used to mark the position (height from floor) of the R deltoid point. Then move the anthropometer to the L arm and mark the position at the same height as the R deltoid. See Figure C.1.

*Exceptions and variations:* Decide how this landmark and related measurements will be used. Individual variations will affect location of the landmark, especially for persons with one shoulder higher than the other.

***Midshoulder:*** The point on top of the right shoulder midway between the neck (R trapezius point or visualization of the base of the neck) and the tip of the shoulder (acromial point, R).

*Procedure:* The person is standing. Stand behind the person and lay a tape measure along the top of the shoulder from the trapezius point, at the juncture of the neck and shoulder, to the acromion landmark at the tip of the shoulder. Divide the full-length measurement by two. Then apply that length from the acromion along the length of the shoulder. Place a short landmark line, approximately 2 cm (1 in.) perpendicular to the shoulder line. Typically the landmark crosses over the top of the trapezius muscle.

*Exceptions and variations:* Adipose tissue may obscure the trapezius muscle.

***Midspine:*** A line down the center of the back (Figure C.2).

*Procedure:* The person is standing. Stand behind the person. Using a flexible plastic ruler as a guide, draw a vertical line about 10 cm (3 in.) long, in the midsagittal plane with the line beginning at a point about 4 cm (1.5 in.) below the cervicale (base of neck). The mark correlates with a garment center back seam position.

*Exceptions and variations:* The position may be further visualized by having the person squeeze their scapulae (shoulder blades) together to locate a position midway between the scapulae.

***Neck, anterior, lateral, R and L:*** See Appendix A.

***Olecranon (center):*** A point on the center of the curvature of the right olecranon process (elbow point) with the elbow flexed approximately 115 degrees (Figure C.1).

*Procedure:* With the person standing, have her/him clench both hands, raising the arms to bring the clenched fists together so that the metacarpophalangeal and proximal interphalangeal knuckles are touching. With the dorsal (outer) surface of the hands facing outwards and the palm (volar or inner) facing inwards, the person raises the arms until they are approximately parallel to the floor. The forearms and fists are in a straight line. Stand to the right of the person and by inspection locate the center of the curvature of the elbow. Draw a short vertical line through the point. Repeat for the L arm.

***Radiale:*** The highest point on the outside edge of the radius.

*Procedure:* The person is standing with arms relaxed. Stand to the right of the person and palpate the hollow on the lateral side of the elbow. Try to locate the upper end of the radius. Without distorting the skin over the area, draw a small mark (Figure C.1).

*Exceptions and variations:* The position may be difficult to locate. Try grasping the wrist of the person and gently rotating the arm (radiating the radius bone around the ulna) while palpating the bone to find the position.

***Scye or armscye:*** Points at the intersection of the upper torso and the arm that relate to the armhole (armscye) of a fitted basic garment. Several points encircle the shoulder/arm joint: anterior scye on the torso, anterior scye on the upper arm, anterior horizontal scye (R and L), posterior horizontal scye (R and L), midscye (R and L), scye level at midspine.

*Procedure:* The person is standing. Have the person place their right hand on their hip. Move around the person to visualize and mark each position. See Figures C.9, C.2, and C.3.

- *Locate the lower U-shape curve for a fitted armscye:* Use a short plastic ruler to locate a plane parallel to the floor for the lower armscye. The position relates to the axilla but allows for movement and comfort in a product so is placed approximately 2.5 cm (1 in.) below the axilla (armpit) depression. Have the person carefully lower his/her arm so that the ruler is grasped between torso and arm (see Figure C.9). Make sure the ruler is parallel to the floor. Instead of using a ruler, you may make a device described by Minott (1978).
- *Anterior (horizontal) scye of the torso/arm (R and L):* Stand in front of the person. Draw a short line where torso and upper arm meet at the ruler, R and L. Mark lower armscye limits for both sides of the body, as with posterior scye.
- *Posterior (horizontal) scye of the torso/arm (R and L):* Stand behind the person. Draw a short line where torso and upper arm meet at the ruler, R and L. Mark lower armscye limits for both sides of the body.
- Remove the ruler after both sides have been marked.
- *Posterior scye seamline (R and L):* Use a flexible-curve ruler to shape a curve from the acromial point to the line marked as posterior horizontal scye. Draw a curve along the ruler around the back of the torso/arm intersection and corresponding to a visualized armscye. Draw R and L armscye seams (Figures C.2 and C.3).
- *Midscye (pattern notch point):* Using a flexible tape measure, measure the distance from the acromial point to the line marked as posterior scye limit. Draw a short line perpendicular to the scye seamline at one-half the noted distance. This is the typical placement for notches, marks on the sleeve and bodice/torso patterns used to match the sleeve to the bodice when the two garment parts are sewn together.

- *Scye level at midspine:* Place a ruler (parallel to the floor) connecting the R and L posterior horizontal scye marks. Draw a short line parallel to the floor and perpendicular to the center back or midspine (Figure C.2).

*Exceptions and variations:* The person measuring must be familiar with construction and fit of a basic garment (fitted bodice for women, fitted suit jacket for men) and be able to visualize the armhole (armscye) placement on each person. Various methods are used to locate these difficult to define landmarks. Alternate methods are described in some apparel pattern making and drafting books.

***Styloid process of the radius:*** The lowest point of the bottom of the radius at the wrist.

*Procedure:* With the person standing, arms relaxed and palms toward the hips. Stand in front of the person and lightly grasp the person's right hand. Gently palpate from the thumb upwards to locate the end of the radius. Place a small mark.

*Exceptions and variations:* This area is crossed by tendons so it may be necessary to flex the hand up and down at the wrist to find the position.

***Styloid process of the ulna:*** The lowest point of the bottom of the ulna, typically characterized by a noticeable bump on the outside of the wrist.

*Procedure:* The person is standing as for locating the styloid process of the radius. Stand to the side of the person and palpate along the length of the ulna to locate the protrusion of the distal end of the ulna. Place a small mark.

***Suprasternale:*** See Appendix A.

***Tenth rib (bottom of the rib cage):*** The inferior point of the tenth rib.

*Procedure:* The person is standing. Stand in front of the person and gently palpate along the bottom of the tenth rib to locate the lowest point. Draw a short horizontal line along the rib. Repeat for L side if needed for the product.

*Exceptions and variations:* The person may be sensitive to touch in this area, so avoid prolonged palpation. Use firm pressure to locate the position. Variations in muscle and adipose tissue may make landmark location more difficult.

***Thelion:*** See bustpoint.

***Trapezius point (R and L):*** The point where the anterior border of the trapezius muscle crosses the neck landmark (top or crest of the shoulder seam).

*Procedure:* The person is standing. Ask the person to place the right hand on his/her left shoulder. Stand to the right of the person and palpate the trapezius muscle to locate the border, typically the crest of the front and back shoulder. Draw a short line intersecting with the base of the neck. Mark the left side using the same procedure.

*Exceptions and variations:* Amount of fat and muscle may obscure locating the trapezius muscle. If so, estimate location of a shoulder line seam at the top of the shoulder extending from the neck to the acromion.

***Waist:*** See Appendix E.

***Wrist (dorsal):*** A line across the back of the wrist originating at the radial styloid process landmark and perpendicular to the long axis of the arm.

*Procedure:* The person is standing. Ask her/him to form fists with both hands, raising the arms to bring the clenched fists together so that the metacarpophalangeal and proximal interphalangeal knuckles are touching. With the dorsal (outer) surface of the hands facing outwards and the palm (volar or inner) facing inwards, the person raises the arms until they are in a position roughly parallel to the floor. The forearms and fists are in a straight line. Stand in front of the person looking at the stylion landmark. Draw an approximately 5 cm (2 in.) line across the dorsal surface of the wrist beginning below the stylion.

*Exceptions and variations:* The line should be perpendicular to the floor when the arms are held in the position described. The origin should be the styloid process of the radius and continue toward the styloid process of the ulna. The line may be thought of as the location of the bottom of a sleeve.

## C.2 Upper Torso and Arm Measurements (Alphabetical Order)

Upper torso and limbs measurements are either two-dimensional, such as chest breadth, or three-dimensional, such as chest circumference. The two-dimensional measurements are taken with a caliper or small sliding anthropometer; and the over-the-body measurements with a flexible tape measure. Measurements described in this chapter are listed in alphabetical order, but the measuring procedure is more efficient if you start at the neck and proceed down the torso of the body, typically doing two-dimensional measurements first, followed by circumference measurements. If the measurement scheme includes the lower body, begin with upper body measurements and continue the length of the lower body.

***Acromial height, standing:*** See Appendix H on landmarking and measuring the total body.

***Acromial height, sitting:*** See Appendix H landmarking and measuring the total body.

***Acromion to radiale length:*** The two-dimensional length from the acromion level to the level of the radiale landmark on the R elbow.

*Process:* The person is standing. Stand to the right of the person and use a caliper to measure the distance between the acromial landmark on the tip of the shoulder and the radiale landmark on the elbow. Both blades of the caliper barely touch the body without compressing the body or indenting the skin (Figure C.6).

*Exceptions and variations:* This measurement is typically a shorter length than the length taken over-the-body with a tape measure.

***Axilla height:*** See Appendix H on measuring the total body.

***Axillary arm circumference:*** Circumference of the arm at a level as close as possible to the juncture of the arm and the chest at the axilla. See Figure C.5.

*Procedure:* Tell the person to move positions as directed. First position: Person stands with both fists resting on the hips, top of pelvis. Second position: Person drops both arms to the sides. With the person in the first position, place the tape measure as close as possible under the axilla. With the person in position two, encircle the arm with the tape measure and note the measurement. The tape measure should be held perpendicular to the length of the arm. The tape measure should correspond to the level of the anterior scye on the upper arm landmark and be held with enough pressure to maintain contact with the skin while not exerting pressure that compresses the tissues.

*Exceptions and limitations:* Variations in muscle and adipose tissue may affect how high the tape can be positioned in the axilla.

***Biacromial breadth:*** A two-dimensional planar distance between the R and L acromion landmarks. See Figure C.6.

*Procedure:* The person is seated. Stand behind the person. A beam caliper or sliding anthropometer is used to measure the distance between the R and L acromial landmarks at the tips of the shoulders. The measurement is taken at the maximum point of quiet respiration. The tips of the caliper should rest on the body.

*Exceptions and limitations:* If the landmarks cannot be seen when standing behind the person, stand in front of the person.

***Biceps circumference, flexed:*** Circumference of the largest dimension of the biceps in the flexed position.

*Procedure:* The person stands and extends the R arm forward, flexing the elbow to about 90 degrees with the hand in a fist, and fingers of the fist held facing the head. Tell the person to exert maximum effort in "making a muscle." Stand to the side of the person and place a tape measure at the position of the biceps point landmark. Hold the tape measure perpendicular to the long axis of the arm. Measure the circumference using enough pressure on the tape measure to maintain contact with the body without compressing the tissue.

***Biceps circumference, not flexed:*** The biceps measurement may also be taken with the arm in a relaxed, non-flexed position at the side.

*Procedure:* Location of the tape is the same as the flexed position passing over the biceps point landmark. This location is also described in some texts as measuring the maximum circumference of the upper arm approximately midway between the acromion and the olecranon (elbow point). Coordinate the measurement location with the product design requirements. See Figure C.5.

***Bustpoint/thelion breadth:*** Distance between bustpoint/thelion, L to R.

*Procedure:* The person is standing. Stand in front of the person and use a caliper to measure the distance between the R and L bustpoints on women and the center of the thelion on men. On women, the landmark is typically placed (drawn or adhesive mark) on the bra. The landmark is not drawn on men, but the apex of the nipple (thelion) is used as the landmark. The measurement is taken at the maximum point of quiet respiration.

*Exceptions and limitations:* See information on placing the landmark on women.

***Breast mound (nude):*** The breast mound may be viewed as a hemisphere. It is defined with breast boundary landmarks: front axilla point (FAP), upper breast point (UBP), inner breast point (IBP), bottom breast point (BBP), the breast point (nipple apex; BP), and the supersternale (sternal notch). See Figure C.4.

*Procedure*: Measurements can be collected with a body scanner or manually with a flexible tape measure. The person may be standing, prone with breast suspended, or supine. Typical measurements, proceeding clockwise around the breast are: 1) UBP to BP, 2) Supersternale to BP, 3) IBP to BP, 4) BBP to BP, 5) OBP to BP, and 6) FAP to BP. See Figure C.4.

***Center back length cerivicale to scye level at midspine:*** The over-the-body length extending from the cervicale to the scye level at midspine.

*Procedure:* The person is standing with the head in the Frankfort plane. Stand behind the person. Position the 0-end of the tape measure at the cervicale landmark and measure to the landmark of the armscye depth.

*Caution:* The posture of the person influences this measurement. Assure that the person is standing with the head in the Frankfort plane.

***Cervicale height (standing):*** See Appendix H on landmarking and measuring the total body.

***Cervicale height (sitting):*** See Appendix H on landmarking and measuring the total body.

***Chest breadth:*** The maximum horizontal breadth of the chest at the level of the R bustpoint/thelion landmark (Figure C.6).

*Procedure:* The person is standing with weight distributed equally on both feet, and arms held slightly away from the body, so that the caliper or anthropometer can be placed at the sides of the chest. Stand in front of the person and place one end of the caliper or anthropometer at the side of the body, level with the R bustpoint for women or R thelion for men. Extend the other end of the device to the L side of the body holding it horizontal to the floor exerting only enough pressure to maintain contact between the device and the body (skin or bra). The measurement is taken at the maximum point of quiet respiration.

*Exceptions and limitations:* Breast tissue (female or male) should not be included in this measurement. Avoid this when necessary, by tilting the blades of the caliper upwards and approaching the bony sides of the chest from below the breasts. On some males the latissimus dorsi (heavy muscles at the back of the axilla) bulge beyond the bony sides of the chest. These muscles are typically not included in the measurement.

***Chest/bust circumference at the thelion (male) or bustpoint (female):*** The circumference measurement is taken parallel to the floor at the level of the bustpoint for women/thelion for men. For women, this measurement is sometimes called full bust circumference.

*Procedure:* The person is standing. Stand in front of the person. Use a tape measure to encircle the body. The goal is to measure the circumference over the bust/thelion points with the tape parallel to the floor. If the bustpoints/thelion points are not parallel to the floor, use knowledge of anatomy and the related product to determine best placement of the tape (Figure C.5).

*Exceptions and limitations:* It is difficult for one measurer to see if the tape is parallel to the floor around the body. An assistant may help to position and hold the tape in place, or have the person stand with his/her back to a mirror, siting the tape position in the mirror. The tape should span any hollow space over the spine between the scapulae. Persons will vary greatly in amount of soft tissue at this location. The tape should be held snugly to the body while avoiding excessive tissue compression. This measurement can vary greatly with inhalation and exhalation, so the measurement is taken at the maximum point of quiet respiration; however, if the product must accommodate full inhalation, also take a measurement at maximum inhalation instructing the person how to breathe for each step of the measuring procedure.

***Chest circumference at scye:*** Circumference measurement of the upper chest. See Figure C.5.

*Procedure:* The person is standing. Stand in front of the person. There are two positions for this measurement. First position: person places both fists on top of the hips (lateral pelvis). Place a tape measure around the torso, passing over the scye level at the midspine landmark. Second position: after the tape is in place have the person drop his/her arms with palms facing the thighs and arms relaxed. Hold the tape snugly without compressing the flesh. Take the measurement.

*Exceptions and limitations:* Depending on a person's body form, this measurement may not be parallel to the floor. See exceptions and limitations for chest/bust circumference at thelion/bustpoint.

***Chest circumference below breast:*** Chest circumference below the fullest breast or chest. See Figure C.5.

*Procedure:* The person is standing. Stand in front of the person. The measurement is taken at the inferior juncture of the lowest breast with the rib cage, on women at the inframammary fold. For women with ptotic breasts, the tape may be positioned under the breast tissue. The tape may follow the bra band.

*Exceptions and limitations:* See exceptions and limitations for chest/bust circumference at thelion or bustpoint.

***Chest/bust depth:*** The depth of the body viewed laterally.

*Procedure:* The person is standing. Stand at the right side of the person. A caliper held parallel to the floor is used to measure the distance between the chest/bust at the level of the R bustpoint (women)/R thelion (men) and the torso back. This measurement can vary with inhalation and expiration so is taken at the maximum point of quiet respiration (Figure C.7).

*Exceptions and limitations:* Take care to exert just enough pressure to maintain contact between the caliper and the body without compressing the tissue.

***Chest/bust height:*** See Appendix H on landmarking and measuring the total body.

***Elbow circumference:*** Circumference of the elbow at the level of the olecranon. See Figure C.5.

*Procedure:* The person is standing with the arms straight and held slightly away from the body. Stand at the right of the person and use a tape measure to measure the circumference of the elbow in a plane perpendicular to the long axis of the arm. The tape is positioned at the olecranon.

*Exceptions and limitations:* Use only enough pressure to maintain contact between the tape and the body. Assure that the person does not hyperextend their elbow.

***Forearm circumference (flexed):*** Circumference of the forearm distal to the flexed elbow.

*Procedure:* The person is standing with the right upper arm extended forward anteriorly and parallel to the floor with the elbow flexed 90 degrees and the hand clenched in a fist. Stand in front of the person. Measure the forearm at the position just distal to the crease between the upper arm and the forearm (elbow crease). The measurement is perpendicular to the long axis of the forearm. Exert just enough pressure to maintain contact between the tape and the body.

***Forearm to hand length:*** A two-dimensional measurement from the olecranon point to the farthest extension of the hand with the forearm parallel to the floor. Distance seen in Figure C.6.

*Procedure:* The person is standing with the upper arm at the side and the R elbow flexed 90 degrees and hand open and extended. Kneel at the right side of the person. Use a caliper to measure the distance between the olecranon (back of the tip of the elbow) to the tip of the middle finger (dactylion). The fixed blade is placed at the back of the olecranon and the caliper arm extended to reach the end of the middle finger. Exert only enough pressure to maintain contact between the caliper and the skin.

*Exceptions and limitations:* Fingernail length, which can vary greatly person to person and time to time, can affect the length of this measurement, so the measurement is taken to the end of the fingertip, not the end of the fingernail.

***Interscye or cross-back length at axillary fold landmark:*** A measurement across the back parallel to the floor at the level of the axillary fold.

*Procedure:* The person is standing. Stand behind the person and measure the distance over-the-body between the top of the R posterior axillary fold (where the arm meets the torso) and the top of the L posterior axillary fold (where the arm meets the torso). Hold the tape against the skin over-the-body except where it spans the hollow of the back at the spine.

*Exceptions and limitations:* Be sure the tape lies smoothly across the back parallel to the floor and does not arch up over the scapulae (shoulder blades). Take the measurement at the maximum point of quiet respiration.

*Exceptions and limitations:* If the person shifts shoulder position, the measurement will change.

***Interscye or cross-back length at the midscye:*** A measurement across the back, parallel to the floor at the level one-half the distance between the acromion and the lower limits of the axilla (Figure C.5).

*Procedure:* The person is standing. Stand behind the person and use a tape measure to measure the distance between the L and R armscyes at a position approximately one-half the distance between the acromion (shoulder point) and the axilla lower level. Hold the tape on the skin except where it spans the hollow of the back. The tape should not arch up over the scapulae (shoulder blades). The measurement is taken at the maximum point of quiet respiration.

*Exceptions and limitations:* If the person shifts shoulder position, the measurement will change.

***Neck circumference-infrathyroid and base of neck:*** See Appendix A on landmarking and measuring the head and neck.

***Neck height, lateral:*** See Appendix H on landmarking and measuring the total body.

***Radiale (proximal) to the stylion process of the radius, length:*** The length of the radius (as close as possible to the total length) from the radiale landmark on the elbow to the stylion landmark on the wrist (Figure C.1).

*Procedure:* The person is standing with arms at their sides and the right palm facing anteriorly. Stand at the right of the person and use a caliper to measure the distance between the radiale and the stylion process of the radius.

*Caution:* Place the caliper parallel to the long axis of the radius. Maintain light pressure at the points of the caliper.

***Scye (armscye) circumference:*** An over-the-body measurement using the acromial point as the major landmark.

*Procedure:* The person stands in two different positions. First position: the person stands with the right fist resting on the lateral pelvis. Stand at the person's right side and place the end of the tape measure on the acromial point, then pass the tape under the axilla to encircle the armscye bringing the tape to meet the start point. Second position: the person lowers his/her arm and the measurement is read at the acromial point. Posterior view of this measure in Figure C.2.

*Caution:* This is an imprecise measurement as the amount of pressure is difficult to determine, and muscle and fat greatly influence the shape and contours of the armscye area. Train everyone taking the measurement to try to assure consistency in measuring technique.

***Scye (armscye) arc, anterior and posterior:*** An over-the-body measurement using the acromial point as the major landmark dividing anterior and posterior armscye lengths, with the lower level at the intersection of the torso and the arm.

*Procedure:* The person is standing. Stand at the person's right side and place the end of the tape measure on the acromial point. Measure from the acromion to the armscye depth landmark (or top of the armscye mark ruler) (Figure C.9).

***Shoulder circumference:*** Circumference measurement encircling the R and L deltoid landmarks. See Figure C.5.

*Procedure:* The person is standing in front of a mirror, so that you can see anterior and posterior of the upper body, or work with an assistant who can help site the tape. The 0-end of the tape is placed at the level of the maximum protrusion of the deltoid muscle on the right side. The tape is passed around the person keeping the tape parallel to the floor. Encircle the body touching the L deltoid landmark and bring the tape back to the R landmark. Take the measurement at the maximum point of quiet respiration.

***Shoulder (acromion) to elbow (olecranon) length (two-dimensional):*** The length of the upper arm (Figure C.6).

*Procedure:* The person is standing with the R arm flexed at the elbow to 90 degrees with the hand straight and palm facing medially. Stand at the R of the person and use a caliper to measure the distance from the acromial point landmark to the bottom of the olecranon (elbow point). Place the stationary end of the caliper at the acromial point and use as little pressure as possible to meet the bottom of the olecranon with the movable end of the caliper.

*Caution:* Do not distort tissue around the acromial point, but maintain body contact with the caliper.

***Shoulder length:*** Shoulder length from the trapezius point (side neck point) to the acromial point (approximates the length of the top of the shoulder seam of a fitted garment). See Figure C.3.

*Procedure:* The person is standing. Stand to the right of the person, place the tape measure at the trapezius point landmark at the base of the neck. Read the measurement at the acromial point.

*Caution:* Be aware if the person has tensed or "hunched" the shoulders which will affect the measurement. Coach the person through several relaxing breaths, inhaling and exhaling, before taking the measurement.

***Side neckpoint (trapezius point) to bustpoint/thelion length:*** Measurement from the side neckpoint (intersection of the base of neck and shoulder line) to the bustpoint in women/thelion in men.

*Procedure:* The person is standing. Stand in front of the person and measure the distance between the trapezius point (side neckpoint) to the bustpoint on women/thelion on men. Place the 0-end of the tape at the trapezius point (side neckpoint). Exert only enough pressure on the tape to prevent a slack tape. Take the measurement at the maximum point of quiet respiration as seen in Figure C.5.

*Exceptions and limitations:* Muscle, adipose and breast tissue vary greatly person to person; especially for women. If breast and muscle tissue protrude creating a hollow between the side neckpoint at the bustpoint/thelion, position the tape to span the hollow—unless the product design requires measuring over-the-body.

***Sleeve length from C7 to olecranon:*** The length from the center back neck to the olecranon (point of elbow). See Figure C.8.

*Procedure:* The person is standing with R arm elbow flexed at approximately 45 degrees. Alternate arm positions are used depending on the product being designed. Stand to the R of the person and place the 0-end of the tape measure at the C7 landmark, extend the tape over the acromion to the olecranon.

*Caution*: This measurement may require help of an assistant to position the tape over the contours of the body.

***Sleeve length from C7 to ulnar styloid process:*** This measurement is taken continuously with the C7 to olecranon measurement. Record the C7 to olecranon measurement and then immediately continue to the ulnar styloid. The measurement is often used as the sleeve length measurement for men's shirts (Figure C.8).

*Procedure:* The person stands with the right arm flexed at approximately 45 degrees. Stand to the right of the person and place the 0-end of tape at the C7 landmark, extend the tape over the acromion to the olecranon and on to the styloid process of the ulna.

*Caution:* This measurement may require help of an assistant to position the tape over the contours of the body.

*Caution:* Be aware that the measurement will be affected if the person has tensed or "hunched" the shoulders. Coach the person through several relaxing breaths, inhaling and exhaling, before taking the measurement.

***Strap length:*** Distance from the bustpoint/thelion on the right side, around the back of the neck to the bustpoint/thelion on the L side.

*Procedure:* The person is standing with hands at the sides of the body and the head in the Frankfort plane. Stand in front of the person and place the 0-end of the tape at the R bustpoint/thelion (or ask the person to hold the tape in place). Extend the tape around the posterior base of the neck. Bring the tape to the L side bustpoint/thelion. Record the measurement at maximum quiet respiration.

*Caution:* Assure that the person is holding the head in the Frankfort plane. Avoid compressing tissues.

***Waist back length (back waist length):*** The posterior over-the-body distance between the cervicale landmark and level of the omphalion (or alternate waist location), (Figures C.5 and C.7).

*Procedure:* The person is standing with the head in the Frankfort plane. Stand behind the person and place the 0-end of the tape measure at the cervicale landmark. Extend the tape over the contours of the back to the level of the omphalion landmark. See Appendix E for alternate locations of a waistline related to a product. Exert only enough pressure to prevent a slack tape. Record the measurement at maximum quiet respiration.

***Waist front length (front waist length):*** The over-the-body distance between the suprasternale landmark at the sternal notch (manubrium) to the omphalion (or alternate waist location—see Chapter 6), anterior, (Figures C.5 and C.7).

*Procedure:* Stand in front of the person and place the 0-end of the tape measure at the sternal notch (manubrium) landmark. Measure to the omphalion (or alternate waist location). See Appendix E for alternate measurement locations of the waist related to omphalion and a product. Exert only enough pressure to prevent a slack tape. Record the measurement at maximum quiet respiration. For women, the tape may span over the bra or lay over the sternum in the hollow between the breasts, depending on the product being designed and how it will be worn. For products that will drape over the breasts and not shape to the concavity between the breasts, make a bridge or span from one bustpoint to the other, using a piece of masking tape or strip of paper to "bridge" between the breasts.

***Waist measurements including waist circumference, waist breadth, and waist depth:*** See Appendix E on landmarking and measuring the waist.

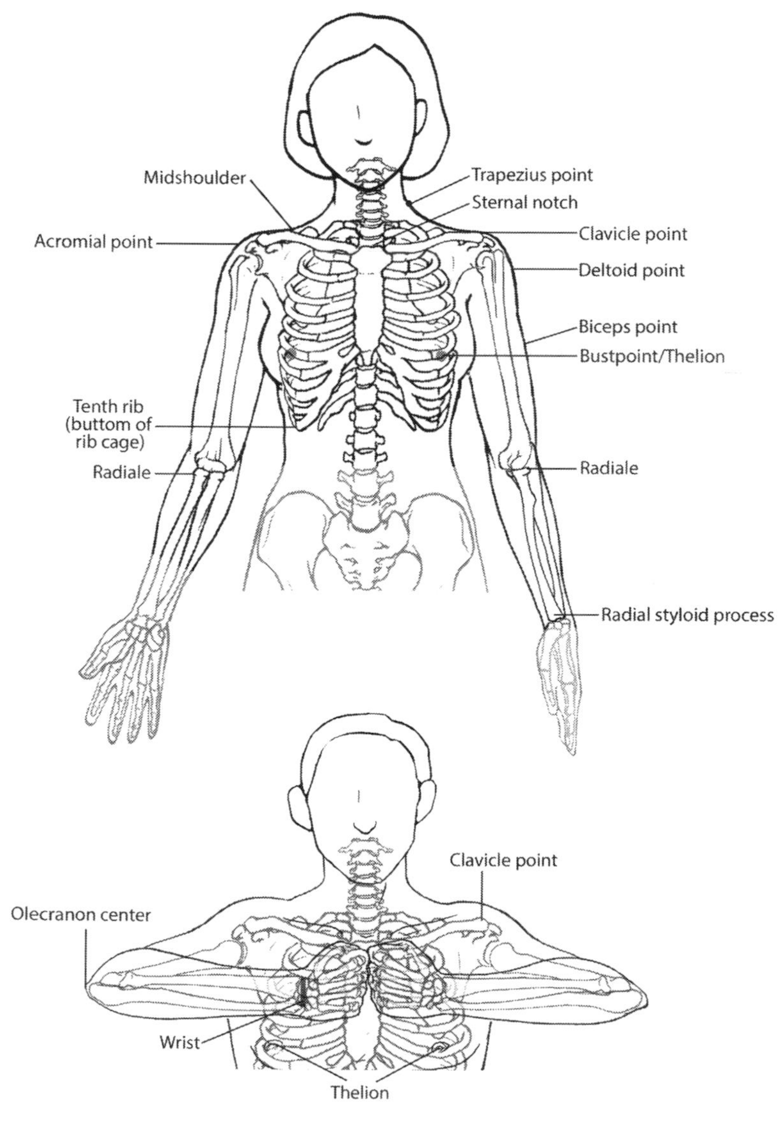


**FIGURE C.1**
Upper torso and arm landmarks, anterior view.

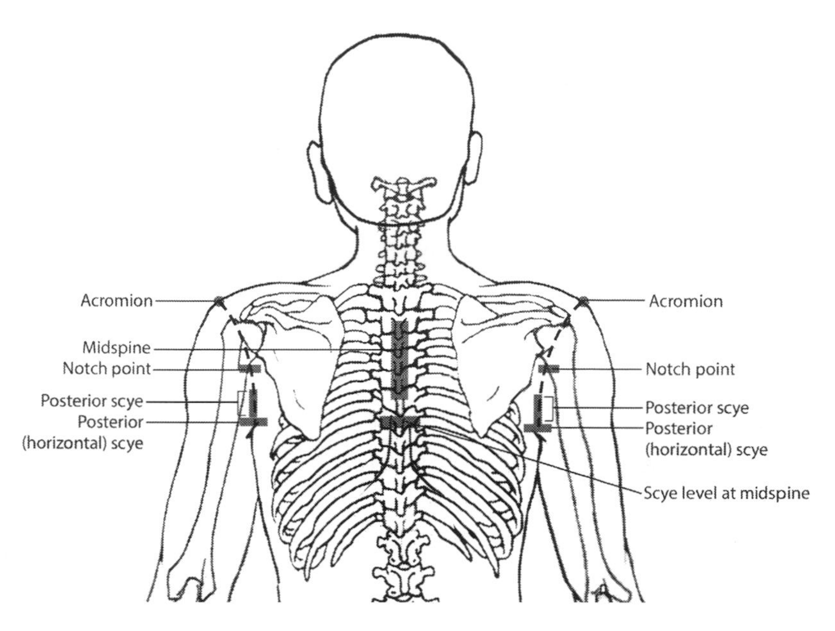


**FIGURE C.2**
Upper torso and arm landmarks, posterior view.

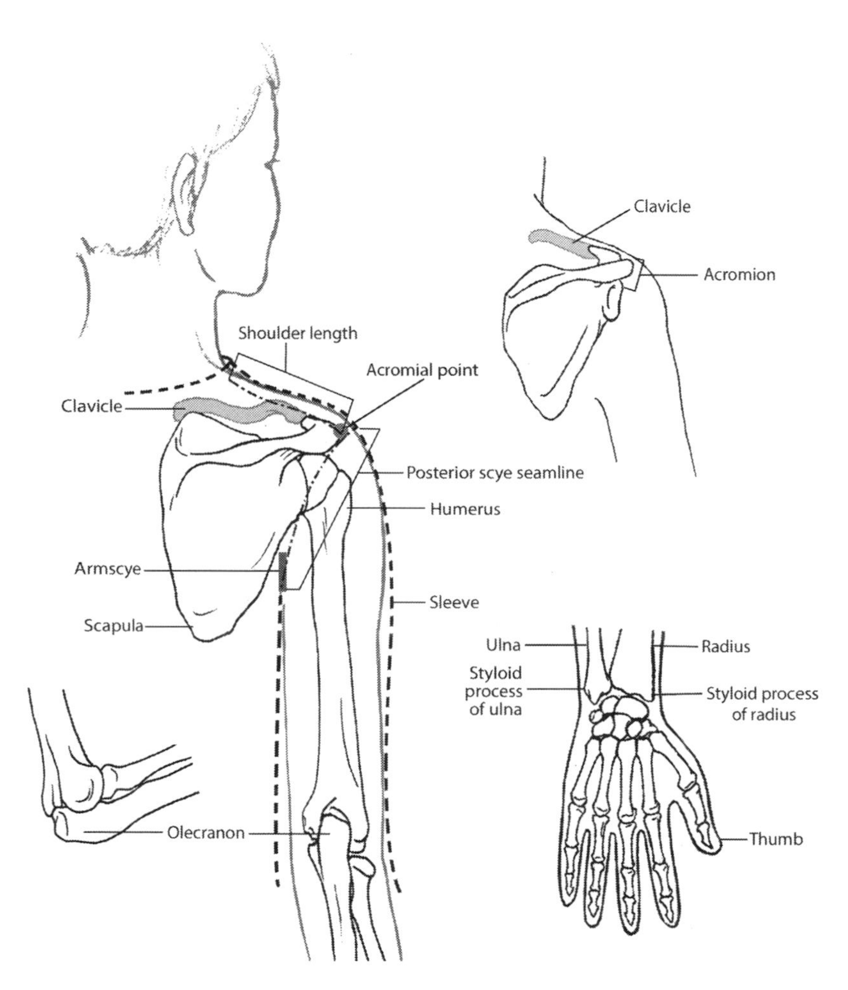


**FIGURE C.3**
Armscye and arm landmarks.

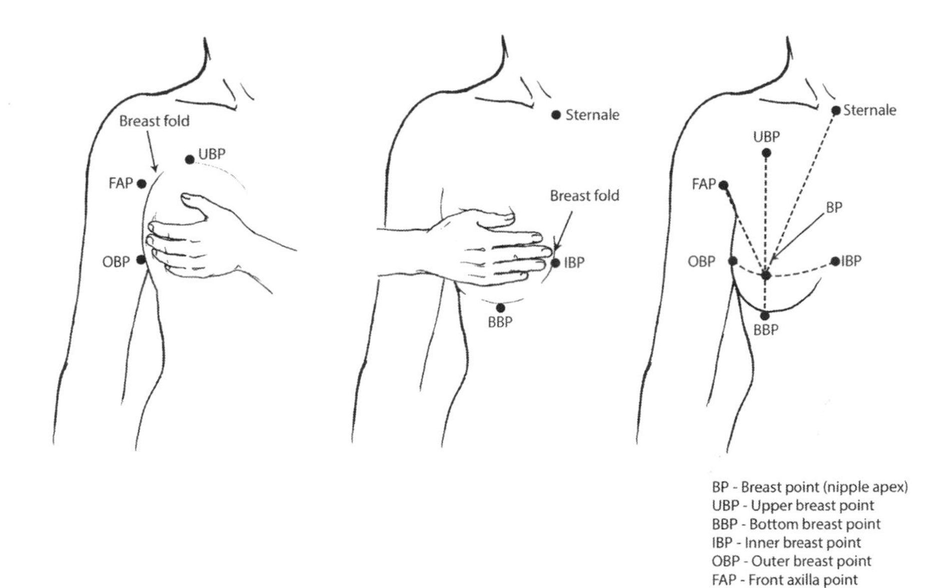


**FIGURE C.4**
Defining the breast mound boundary landmarks, and measuring the breast mound hemisphere.

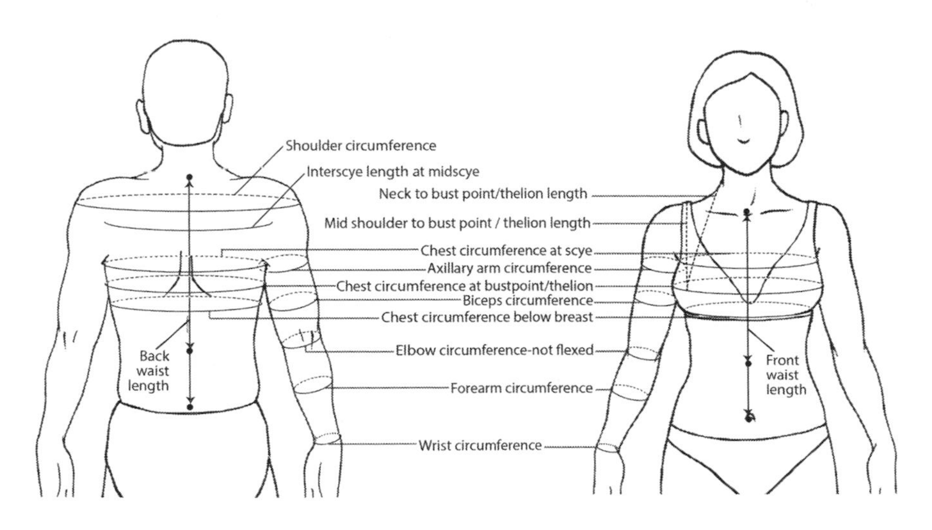


**FIGURE C.5**
Upper torso and arm circumferences and lengths, anterior and posterior views.

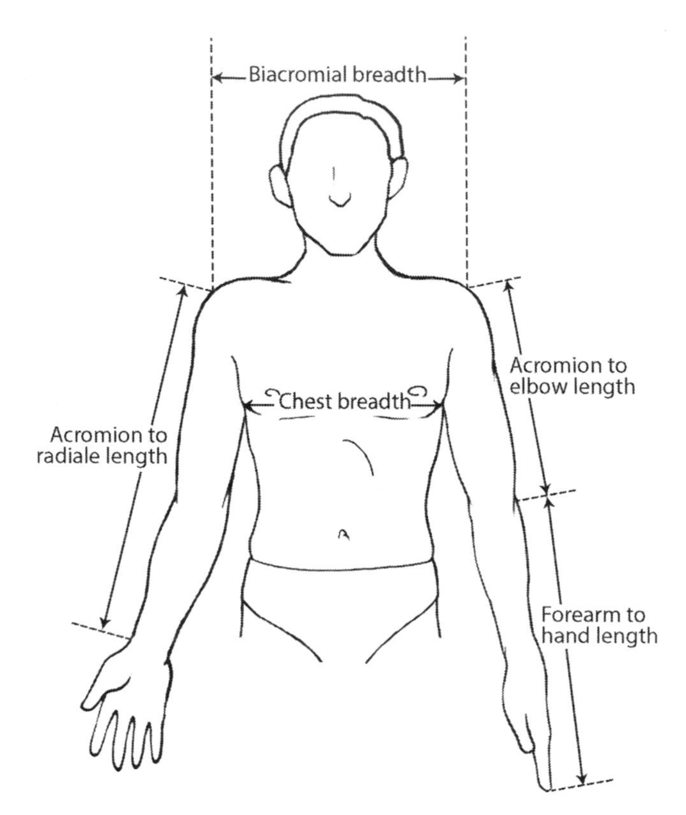


**FIGURE C.6**
Upper torso and arm, lengths, breadths, and depths.

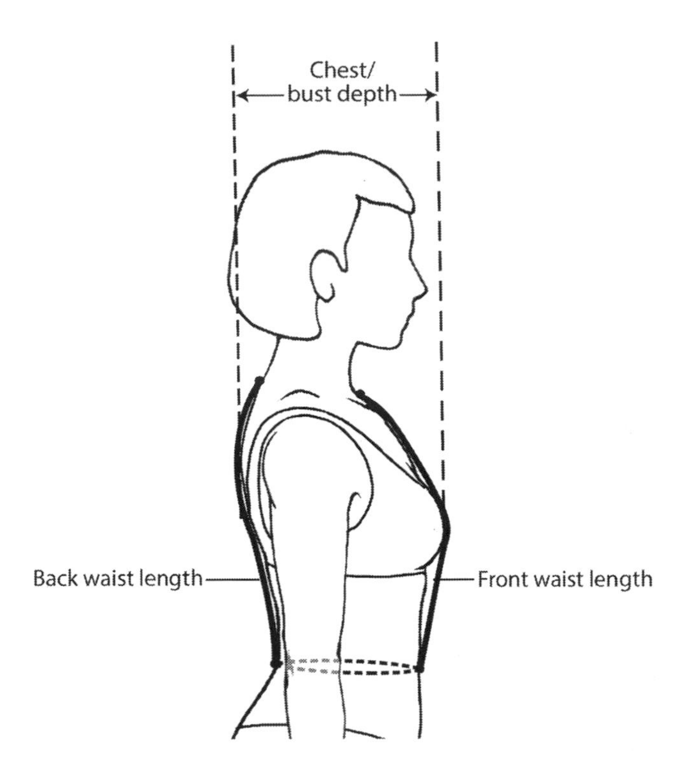


**FIGURE C.7**
Upper torso lengths and breadth, lateral view.

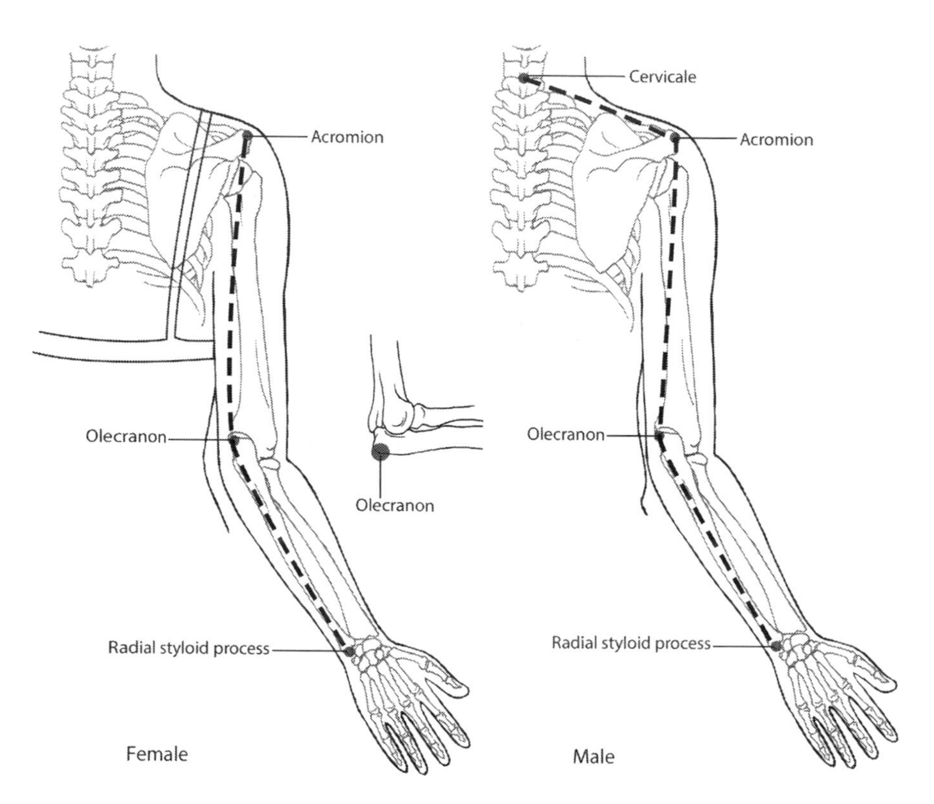


**FIGURE C.8**
Arm length measurements.

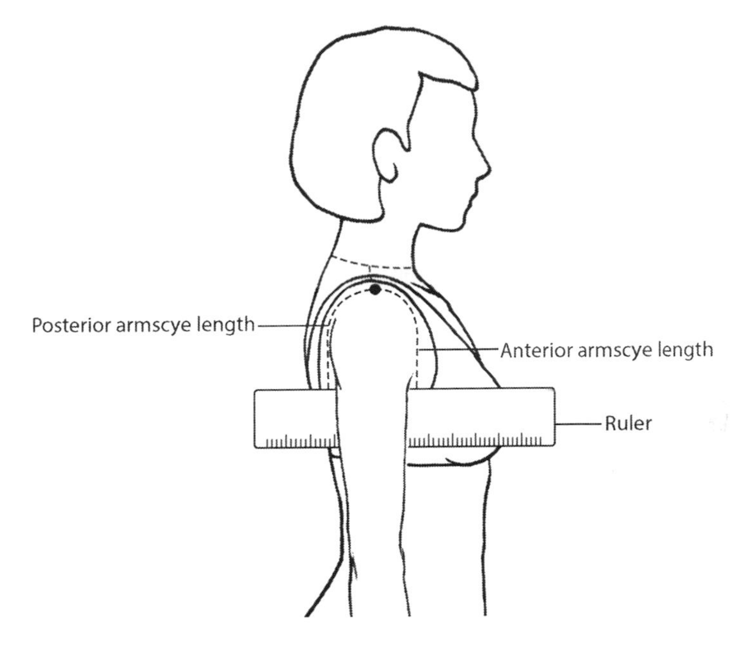


**FIGURE C.9**
Armscye landmarks, and posterior and anterior lengths (over-the-body).

# *Appendix D: Lower Torso and Leg Landmarks and Measurements*

The information on landmarking and measuring the lower torso and leg is inseparably related to landmarking and measuring the waist/waistline, which is discussed in Appendix E on landmarking and measuring the waist/waistline. Other lower torso and limb landmarks and measurements are described in this appendix.

Landmarks are illustrated in Figures D.1 and D.2 and measurements are illustrated in Figures D.3 and D.4.

## D.1 Lower Torso and Leg Landmarks (Alphabetical Order)

***Anterior-superior iliac spine of the pelvis (R and L):*** The anterior protrusions of the R and L iliac crests of the pelvis (Figure D.2).

*Procedure:* The person is standing. Stand in front of the person, and explain what you will do to locate this landmark. Palpate the top of the pelvis to locate the top (crest) of the pelvis (see the description in Chapter 6 of pelvis positions/hip tilt variations). Place a small mark at the point on both sides of the body. If this is a key landmark for the product you are designing, provide appropriate garments or undergarments for the person so that the points are easily located.

*Exceptions and variations:* Fat and muscle may obscure the bony structures of the pelvis. You may have to approximate the locations.

***Buttock point (R lateral and L lateral):*** The fullest part of the buttocks marked at the R and L hip, based on the level of the buttock point, posterior. See Figure D.2.

*Procedure:* The person is standing. Kneel or sit to the right of the person. Site the fullest protrusion of the buttock and place a landmark (straight short line parallel to the floor) at that level above the trochanter on the R hip. Use a standing anthropometer to measure this height from the floor. Move the anthropometer to the L side and mark the R side distance on the L hip.

*Exceptions and variations:* Fat and muscle will vary one person to another, so locating an exact point may be difficult.

***Buttock point, posterior:*** The fullest part of the buttocks as sited from a lateral view, used to locate the full seat (sometimes called hip) circumference measurement. This landmark is sited and not landmarked, although the position may be marked with a vertical line or sticker (Figure D.2).

*Procedure:* The person is standing. Kneel or sit to the right of the person. Site the fullest protrusion of the buttock.

*Exceptions and variations:* Fat and muscle will vary one person to another, so locating an exact point may be difficult.

***Calf:*** The maximum posterior protrusion of the calf. Used as the location for measuring the maximum calf circumference.

*Procedure:* The person stands on a platform. Kneel or sit to the right of the person to site the fullest part of the calf. Place a landmark (short line parallel to the floor) on the lateral side of the calf along the length of the fibula. Use a standing anthropometer to measure this height from the floor. Then use this height measurement to place a landmark on the person's L calf. See Figure D.2.

*Exceptions and variations:* The maximum protrusion or fullness of the calf may cover a length vertically more than 2.5 cm (1 in.); if so, place the mark at the lowest part of the protrusion.

***Dorsal juncture of the foot and leg:*** The approximate location of the juncture of the lower limb with the top of the foot, often the location of a skin crease (Figure D.2).

*Procedure:* The person is standing. Ask the person to flex their knees to about a 30-degree angle so that a skinfold forms at the juncture of the foot and the leg. Place your index finger at this juncture and have the person stand erect again. At this location, place a short landmark line perpendicular to the length of the lower limb.

***Gluteal furrow point:*** The point of the lowest crease of the buttock, at the juncture of the buttock and the thigh (Figure D.1).

*Procedure:* The person is standing. Kneel or sit behind the person and draw a short line at the lowest point of the crease at the intersection of the R buttock and upper thigh.

*Exceptions and variations:* Variations in muscle and fat may affect location of this landmark. Briefs or panties may accentuate this landmark location.

***Iliocristale:*** The highest palpable point of the R pelvis. Often defined as one-half the distance between the anterior superior iliac spine and the posterior superior iliac spine (Figure D.1).

*Procedure:* The person is standing. Stand in front of the person and use both hands positioning them to encircle the person's mid-section to locate the top rim of the pelvis. Locate the anterior and posterior points of the iliac crests and place a landmark approximately one-half the distance between the crests.

*Exceptions and variations:* Variations in muscle and fat may affect the location of this landmark.

***Inner thigh (estimation method):*** A vertical line at the inner thigh dividing anterior and posterior thigh, approximately equally. Typical placement for a pants inseam (see Chapter 6 on standard pants pattern parts and seams) (Figure D.2).

*Procedure:* The person stands with the L foot placed on a platform so that the L knee is flexed approximately 90 degrees. Stand at the left of the person and draw a 7–10 cm (3–4 in.) line bisecting the thigh (the landmark), placing the upper limit at about the level of the gluteal fold and extending downwards.

*Cautions:* This is an approximate location. Train all measurers so that they are using the same visual criteria to locate the landmark. Variations in muscle, fat, and posture will affect location of the landmark.

***Inner thigh (plumb line method):*** A vertical line at the inner thigh using a plumb line to locate the center of gravity. This alternate method, to the estimation of equal thigh division method described above, is based on the observation that anterior/posterior hip tilt can affect location of a pants inseam. Typical placement for a pants inseam (see Chapter 6 on standard pants patterns).

*Device:* Make a "plumb line" device. Cut two pieces of string. String A to measure approximately 25 cm (10 in.). String B to measure approximately 50 cm (20 in.) longer than the person's crotch length (or longer than the largest person being measured). Attach a weight (small fishing line weight or similar) to one end of String A. Tie a small ring or loop on the opposite end. Thread the ring unto String B so that it slides freely (Figure D.2).

*Procedure***:** Place a narrow belt or string around the person's waist (at the desired placement of the pants waistline). Attach one end of String B to the belt/string at center front waist. Pass String A between the person's legs and attach the string to the belt/string at the center back waist. Have the person place their L foot on a platform with L knee at approximately 90 degrees. See that the weight suspends freely and adjust it with the help of the person so that it hangs perpendicular to the floor. Draw a line approximately 7–10 cm (3–4 in.) along the length of the suspended string.

***Lateral femoral epicondyle (standing and sitting):*** Lateral point of the R epicondyle of the femur. This is the approximate knee pivot point (Figure D.2).

*Procedure:* This landmark may be located in both the standing and seated positions and location may vary depending on the person's position. Distinguishing landmarks (X for standing and 0 for seated) should be used to note the differences, especially if the landmark locations will affect the design of a product.

*Standing landmark:* The person stands with weight distributed equally. Palpate the knee to locate the lateral femoral epicondyle. You may ask the person to lift the knee to help in locating the position. Place the landmark at the designated point.

*Seated position:* With the person seated, use the procedure for the standing landmark location to locate the landmark.

*Exceptions and variations:* Muscle and fat may obscure location of this landmark. View the knee as a motion pivot point to assist in locating the point.

***Midpatella:*** The midpoint of the patella (knee cap) as viewed anteriorly (Figure D.2).

*Procedure:* The person stands without "locking" the knees. Kneel in front of the person. Gently grasp the top and bottom ridges of the patella between your thumb and index finger. Locate a point approximately one-half the distance between the upper and lower edges and place a small mark.

*Caution:* Some persons may find it difficult to relax the knee without bending it.

***Suprapatella:*** The upper edge of the patella (knee cap) (Figure D.2).

*Procedure:* The person is standing. Palpate the patella to locate the upper most edge of the patella.

*Caution:* Same as for midpatella.

***Tibiale:*** The superior palpable point on the lateral condyle of the R tibia.

*Procedure:* The person stands on a raised platform or sturdy table with weight equally distributed. Stand in front of the person and with one hand grasp the condyles (bony prominences) just below the patella with your thumb and forefinger. Ask the person to flex their knee so that you can more easily feel the structures. Ask the person to return to the standing position and place a small X at the lateral condyle.

*Variations:* Amount of muscle and adipose tissue varies from person-to-person. If the condyle is not easily palpated, estimate the best location.

***Trochanter (seated):*** The center point of the trochanter of the femur on the seated person.

*Procedure:* The person is seated with knees flexed to approximately 90 degrees. Stand to the right of the person and palpate the lateral surface of the greater trochanter near the hip joint to estimate the center. Place a small mark at this position (Figure D.2).

*Exceptions and variations:* Muscle and fat will obscure the location of this landmark. View the intersection of the femur and the pelvis as a pivot point for motion to locate the point.

***Trochanterion (standing):*** The upper edge of the greater trochanter of the femur with the person standing (Figure D.1).

*Procedure:* The person is standing. Stand in front of the person. Palpate the upper edge of the trochanter on both limbs. Ask the person to move the limb anteriorly and posteriorly to help locate the position. Mark this point with a short line parallel to the floor.

*Exceptions and variations:* Muscle and fat will obscure location of this landmark. View the hip as a pivot point for motion to locate the point.

*Caution:* This is a difficult landmark to locate and best approximation often has to suffice. If there are several measurers conducting the study, consensus and training on location is helpful.

**Lower Torso Landmarks (Sited, Not Landmarked):**

***Dorsal juncture of calf and thigh:*** The person is seated. Site the juncture between the R calf and the thigh behind the knee.

***Posterior superior iliac spine, posterior point of the crest of the R iliac:*** The person is standing. Stand behind the person and site the top edge of the iliac spine of the pelvis, the person may have a dimple (skin indentation) at this point (Figure D.1).

## D.2 Lower Torso and Limb Measurements (Alphabetical Order)

***Buttock circumference:*** The circumference of the lower torso at the level of the maximum protrusion of the R buttock, parallel to the floor (Figure D.3).

*Procedure*: The person stands with heels together on a platform. Stand to the right of the person and ask him/her to flex their arms at the elbow to expose the side hip. Locate the buttock point (R and L lateral) landmarks and site the maximum protrusion of the posterior buttock. Encircle the buttock with a tape measure, the tape is parallel to the floor. Have an assistant help with tape placement if necessary.

***Buttock depth:*** The horizontal depth of the torso at the level of the maximum protrusion of the R buttock. See Chapter 6 for an explanation of the pants pattern hip depth space (Figure D.3).

*Procedure:* The person stands with heels together on a platform. Stand to the right of the person and ask him/her to flex the arms at the elbow to expose the side hip. Locate the buttock point (R and L lateral) landmarks and site the maximum protrusion of the posterior buttock. Use a caliper to measure the depth of the lower torso at the level of the maximum protrusion of the R buttock. Place the fixed blade at the posterior level and extend the movable blade. Maintain the blades in the parallel to floor position. To avoid interfering with the caliper blade, it may be necessary to ask male participants to adjust the genitalia.

***Buttock height:*** The vertical distance from the floor to the height of the maximum protrusion of the R buttock (Figure D.3).

*Procedure:* The person stands on a platform. Stand to the right of the person and ask him/her to flex the arms at the elbow to expose the side hip. Locate the buttock point (R lateral) landmark. Use a standing anthropometer to measure the distance from the floor to the buttock point landmark.

***Calf circumference:*** The horizontal circumference (maximum) of the R calf (Figures D.3 and D.4).

*Procedure:* The person stands on a platform with heels approximately 10 cm (4 in.) apart and weight equally distributed over both feet. Stand to the right of the person and observe the person's R calf to determine the approximate location of the largest circumference. Encircle the calf at the location with a tape measure. Record the circumference. Depending on the product shape and desired fit, measurements slightly above and below the first measurement may be recorded.

***Crotch height:*** The vertical distance from the floor to the crotch (an unmarked landmark) (Figure D.3).

*Procedure:* The person stands on a platform. Stand in front of the person. Ask the person to spread their legs enough to place the standing anthropometer between their legs. Ask the person to move back to the original position and have the person raise the upper blade of the anthropometer to place firm, but comfortable pressure at the crotch level. For males, the blade is typically placed to the right of the genitalia. The measurer may exert further firm pressure to assure that the blade is in position.

*Variations and cautions:* The placement level of the anthropometer blade will depend on how the measurement will be used and if fit ease will be added to a wearable product that encompasses the perineum and genitals. Respect the person's modesty and comfort level.

***Crotch Length (Omphalion):*** See Appendix E on waist landmarking and measuring, refer to Chapter 6 on waist location and waistline placement variations.

***Hip breadth:*** The distance between the lateral buttock landmarks (at the sides of the hips) (Figure D.4).

*Procedure:* The person stands with arms held slightly away from the sides of the body. Stand in front of the person. Use a caliper to measure the distance between the lateral buttock landmarks. Hold the caliper parallel to the floor. If underwear covers the landmarks, ask the person to hold the garment out of the way (be sure that the person does not distort soft tissue).

***Knee height at midpatella:*** The vertical distance from the floor to the midpatella landmark (Figure D.3).

*Procedure:* The person stands on a platform with the heels together and weight distributed equally. Stand at the right of the person. Place the base of a standing anthropometer adjacent to the R leg. Move the sliding blade to touch the midpatella landmark. Assure that the anthropometer is perpendicular to the floor. Record the distance.

*Caution:* The person should not lock their knees. You can help the person relax the knee joint, by firmly grasping the leg just above the knee for a few seconds. Before touching the person, tell them you will help them position the leg by grasping the knee.

***Lower thigh circumference:*** The circumference of the R thigh at the level of the suprapatellar landmark (Figures D.3 and D.4).

*Procedure:* The person stands with feet approximately 10 cm (4 in.) apart. Stand at the right of the person and encircle the thigh at the level of the suprapatellar landmark (just above the top of the patella) with the tape measure. Assure that the tape is parallel to the floor.

*Caution:* The person should not lock their knees. You can help the person relax the knee joint, by firmly grasping the leg just above the knee for a few seconds. Before touching the person, tell them you will help them position the leg by grasping the knee.

***Thigh circumference:*** The circumference of the thigh at the juncture with the buttock (below the gluteal fold) (Figures D.3 and D.4).

*Procedure:* The person stands on a platform with feet apart just enough so that the thighs do not touch. Ask the person to place their R hand on their chest. Stand at the right of the person and use a tape to measure the circumference of the thigh at the juncture with the buttock (gluteal furrow point). Hold the tape perpendicular to the long axis of the thigh.

*Caution:* Assure that the person does not tense the thigh muscles. Avoid placing the tape in a furrow of the buttock. When measuring overweight or obese persons, avoid placing the tape in the fold and develop a standard procedure, determining the tape distance from the juncture of the buttock and the thigh. Ask the person to place the tape between the legs, high around the thigh. Respect the person's modesty and comfort level.

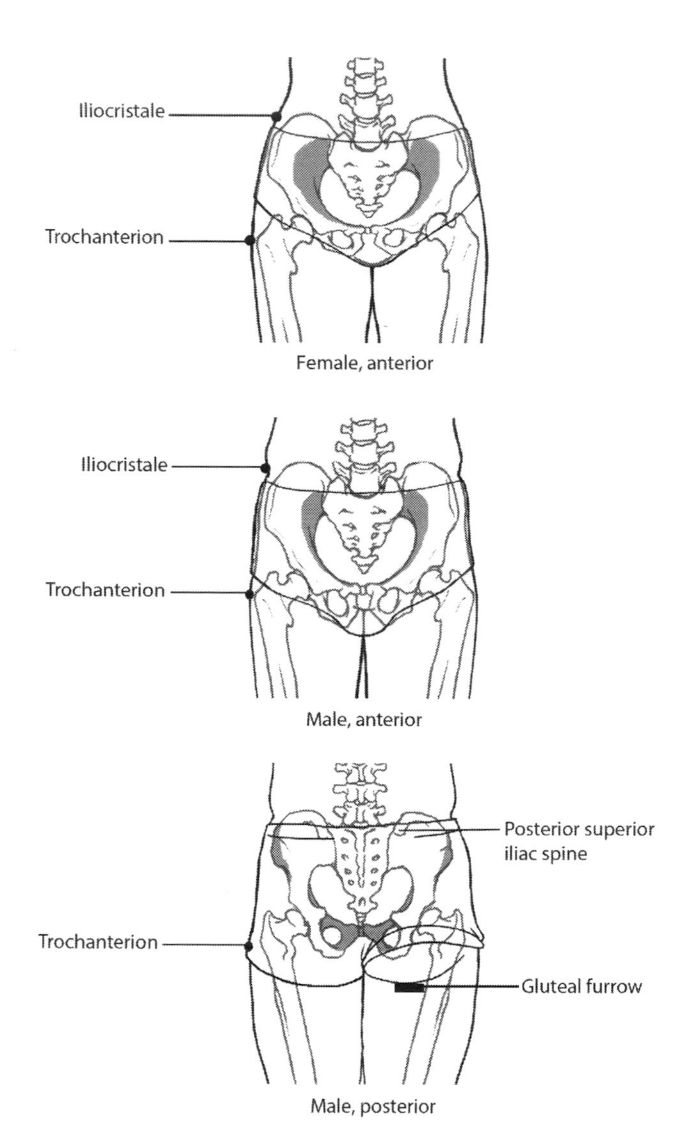


**FIGURE D.1**
Lower torso landmarks, anterior and posterior views.

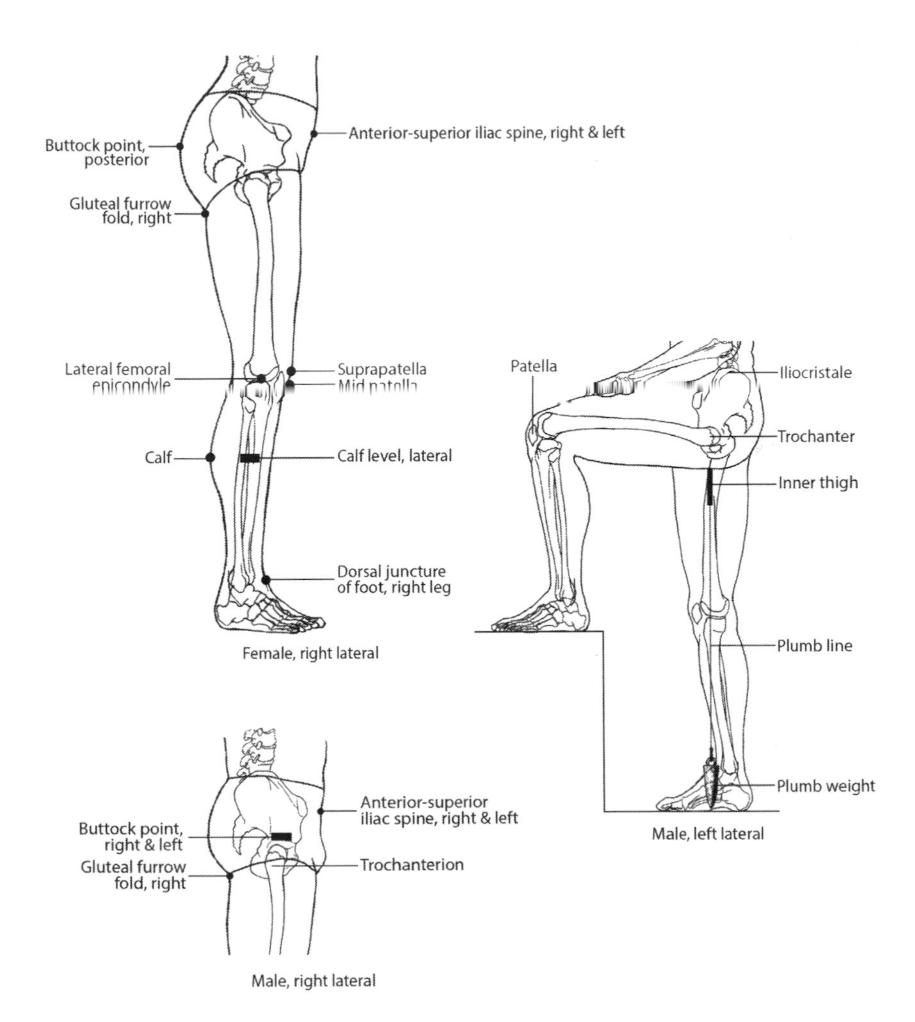


**FIGURE D.2**
Lower torso and leg landmarks, lateral views.

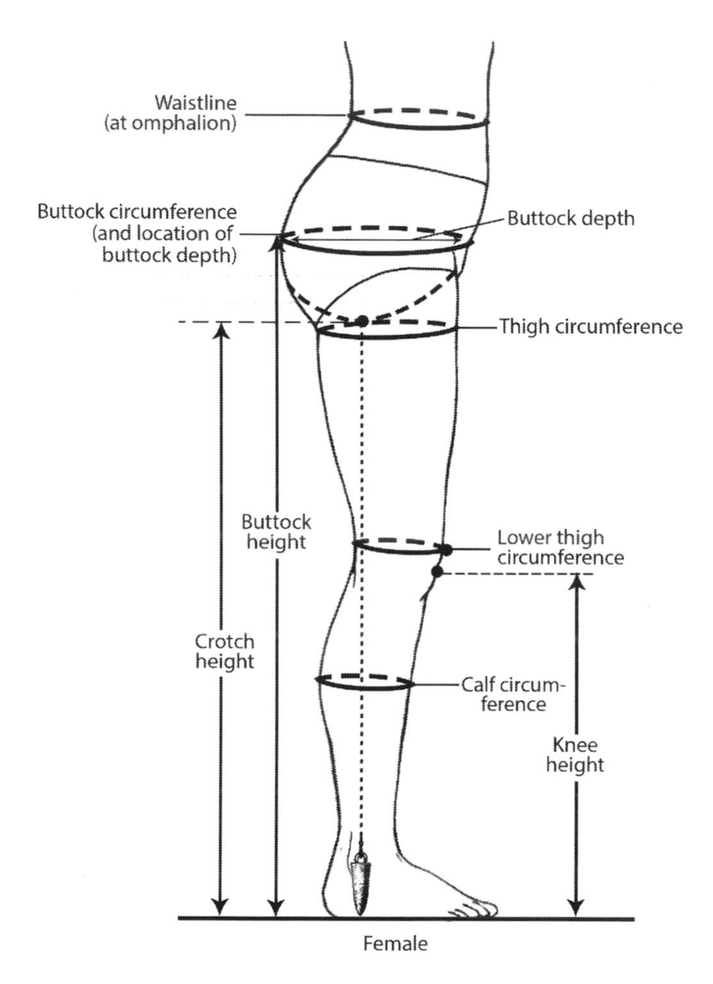


**FIGURE D.3**
Lower torso and leg measurements, lateral view.

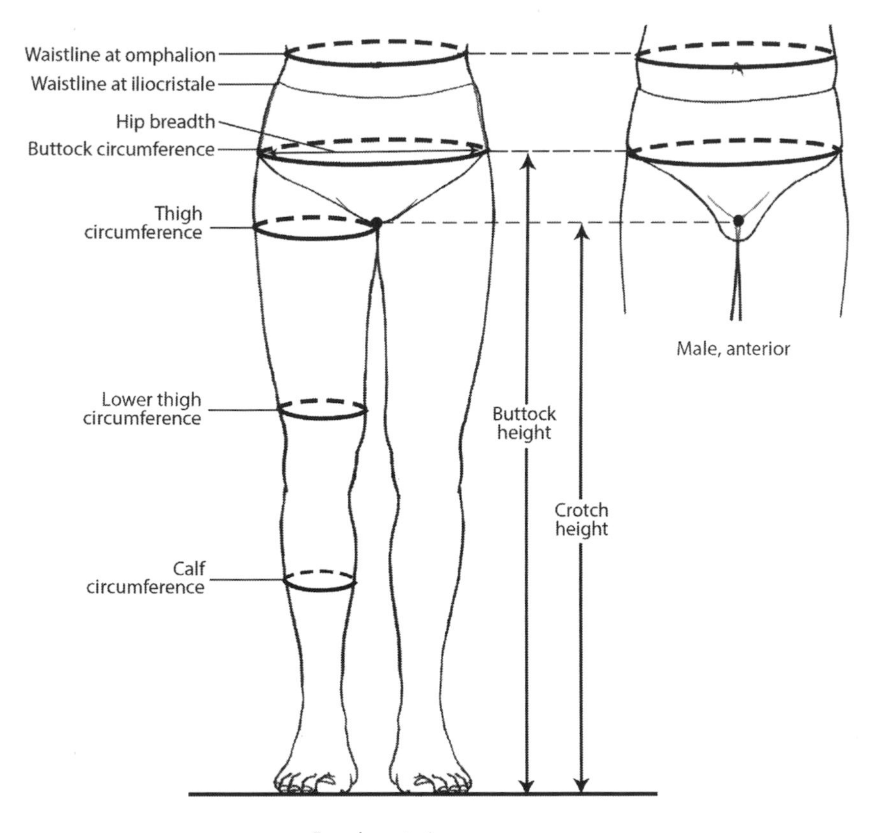


**FIGURE D.4**
Lower torso and leg measurements, anterior view.

# *Appendix E: Waist/Waistline Landmarks and Measurements (Alphabetical Order)*

Waist/waistline landmarks and measurements that are related to the iliocristale and the omphalion (navel) are described in this appendix. These locations are readily observed anatomical features; however, these measurements are limited in usefulness for establishing waistline landmarks and measures related to wearable products. Modify these directions or develop unique waist/waistline landmarks and measurements for the product you are designing. Review Chapter 6 on defining and locating the waist/waistline related to wearable products. Landmarks and measurements applied to drafting the crotch seam of bifurcated lower body apparel patterns (pants, slacks, trousers, jeans, etc.) are included in this appendix. The crotch seam shape and length depend on location of the waist/waistline, along with other lower body dimensions and a person's posture.

Landmarks are illustrated in Figure E.1 and measurements are illustrated in Figures E.2 and E.3. The landmark for the iliocristale is illustrated in Figure D.1, as related to the lower torso. Waist/waistline circumference is also illustrated in Figures D.3 and D.4, as related to the lower torso.

## E.1 Waist/Waistline Landmarks (Alphabetical Order)

***Waist, omphalion location method (R, L, anterior, and posterior):*** The omphalion (center of the navel) and its related posterior, and lateral landmarks at the same location parallel to the floor.

*Instrument*: Standing anthropometer. Read Chapter 6 on alternative waist/waistline location methods, and adapt the omphalion method as needed.

*Procedure:* The person is standing. Stand in front of the person and site the omphalion. Draw a 4 cm (2 in.) horizontal line to the right of the omphalion. Place a standing anthropometer on the floor and move the blade point to lightly touch this landmark height.

***Side waist point (R & L):*** Move the anthropometer to the R side of the person and draw a 4 cm (2 in.) horizontal line at this level; located approximately at the bisection of the anterior and posterior of the body (relates to approximate side seam location of classic pants). Place marker or draw a line. For a non-ideal figure, the designer/measurer determines best location based on aesthetics or product structure. Repeat on the L side. Draw a 4 cm (2 in.) line at this height at the center spine.

*Variations:* This method locates the waist parallel to the floor. The person's iliac crests may not lay parallel to the floor, reflecting an asymmetrical pelvis laterally. The asymmetry will affect the waist/waistline level. If the product is based on locations of the iliac crests, revise the landmark procedure accordingly, using an alternate waist/waistline method described in Chapter 6.

## E.2 Waist/Waistline and Product-Related Measurements (Alphabetical Order)

***Crotch depth:*** The over-the-body measurement at the R hip, seated position, over the curve of the body; waist/waistline to the sitting surface.

*Procedure:* The person sits on a flat, solid surface with knees flexed at a 90-degree angle and the R arm extended anteriorly and flexed. Kneel at the right side of the person. Place the 0-end of the tape measure at the approximate location of a side-seam of pants (midpoint on the waist). Shape the tape along the curve of the hip to the full hip level, then extend the tape in a straight line to the sitting surface. Record the measurement.

*Variations:* For some products and some pants pattern drafting methods, a two-dimensional measurement is taken with an anthropometer. One blade of the anthropometer is placed at the sitting surface. The movable blade is placed at the level of the waist/waistline.

***Crotch length (omphalion level):*** The over-the-body length from the omphalion, between the legs, to the center back omphalion level landmark.

*Procedure:* The person is standing with legs spread just slightly to allow passing the tape between the legs. Ask the person to move their legs together after the tape is placed. Stand to the left of the person and place the 0-end of the tape at the back omphalion landmark level. Pass the tape between the legs and bring to the omphalion mark. For males, the tape is usually placed to the R of the scrotum and held securely without compressing the flesh. Also for males, the tape is typically slightly to the R of the omphalion.

*Variations:* Crotch length for pants/trousers/jeans depends on the style of the garment, with waist line level varying from hip level to above the omphalion. The described measurement is based on the anatomical position of the omphalion. Read Chapter 6 on waist/waistline variations and adjust the landmarks for the crotch length measurement based on the product waistline location.

***Crotch length, posterior (omphalion level):*** The total crotch length measurement is often divided into two sections, anterior and posterior. The crotch length measurement described above is taken in sequence with the posterior crotch length measurement.

*Procedure:* Prepare to take this measurement by positioning a step stool slightly in front of the person's L foot. Take the crotch length measurement as described above. Record the total measurement. Ask the person to place his/her L foot on the step stool. Record the measurement at the inner thigh landmark.

***Waist circumference (omphalion level):*** The circumference of the waist at the level of the omphalion encompassing the waist (omphalion) landmarks.

*Procedure:* The person is standing. Stand in front of the person. Place a tape measure at the omphalion landmark and pass the tape around the body so that the tape lays over all of the landmarks. Have the person stand in front of a mirror so that you can site the tape around the body, or have an assistant help to site the tape. Record the measurement at the person's maximum point of quiet respiration, applying just enough pressure to maintain contact between the tape and the body.

***Waist breadth (omphalion level):*** The distance between the lateral omphalion landmarks.

*Procedure:* The person is standing. Stand in front of the person. Use a caliper to measure the distance between the L and R omphalion landmarks. Hold the caliper parallel to the floor. Record the measurement at the person's maximum point of quiet respiration, applying just enough pressure to maintain contact between the caliper and the body.

*Caution:* Do not compress the soft tissues.

***Waist depth (omphalion level):*** The distance between the front and the back of the waist at the level of the omphalion.

*Procedure:* The person stands with the R hand on the chest. Stand at the right of the person. Use a caliper to measure the horizontal distance between the anterior and posterior omphalion landmarks. Hold the caliper parallel to the floor. Record the measurement at the person's maximum point of quiet respiration, applying just enough pressure to maintain contact between the caliper and the body.

*Caution:* Do not compress the soft tissues.

***Note on depth and breadth measurements based on omphalion level:*** Read Chapter 6 on effects of using the omphalion as the key landmark for waist location. This method does not consider asymmetry—one hip higher than the other, or hip/pelvic tilt that varies from the ideal body type. Use an alternative method for locating waist/waistline, if asymmetry is prevalent in the people you are measuring. Alternative methods are described in Chapter 6.

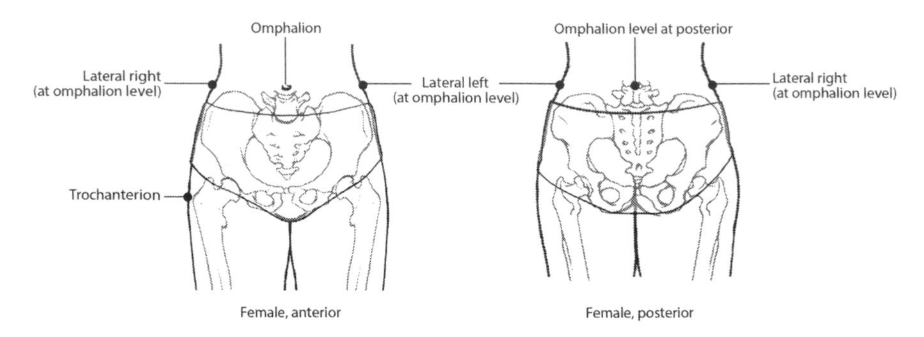


**FIGURE E.1**
Waistline landmarks, anterior and posterior views.

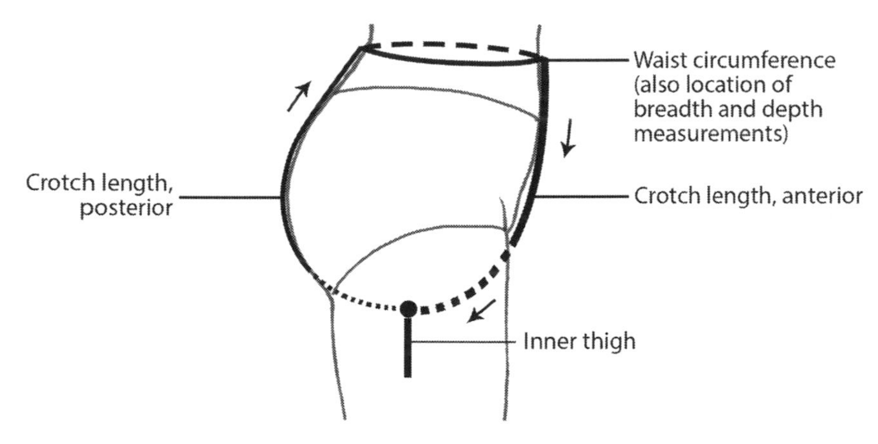


**FIGURE E.2**
Crotch length measurements, total and posterior.

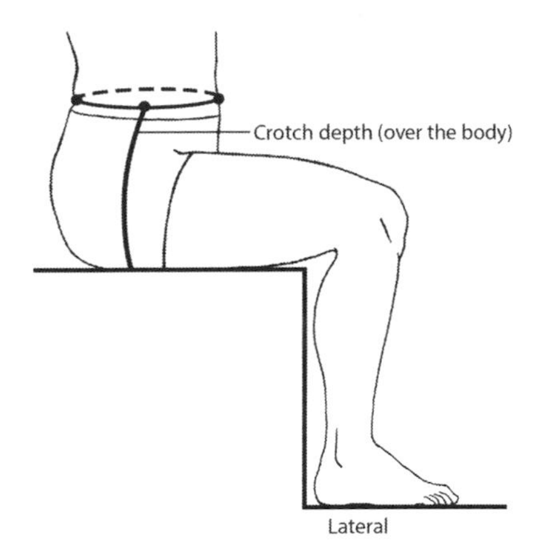


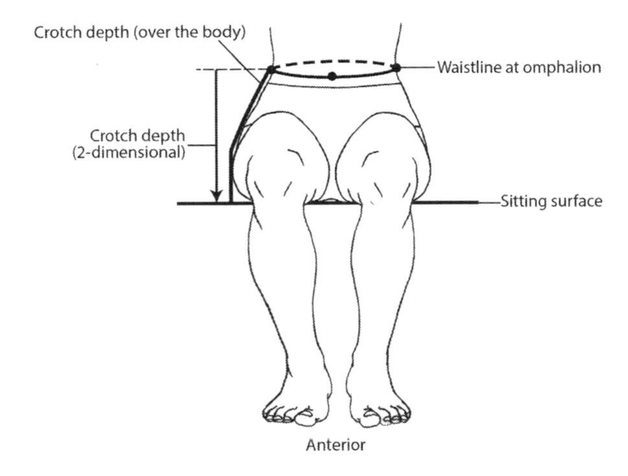


**FIGURE E.3**
Crotch depth (seated), over-the-body and two-dimensional.

# *Appendix F: Hand and Wrist Landmarks and Measurements*

Hand and wrist landmarks are described in this appendix. Markings are placed on the R wrist and hand unless otherwise noted. Directions in this appendix use the anatomical terms ventral and dorsal to distinguish the hand orientation for landmarking and measuring. Persons being measured may not be familiar with the terms, so when instructing them on hand position, use the common terms palm (ventral) and top of hand (dorsal). Ask the person to remove their wristwatch and rings, if possible. For security purposes, have the person keep all valuables with him/her during the process.

For some studies, the dominant hand is landmarked and measured. The dominant hand is the person's preferred hand for fine motor skill tasks such as handwriting. Record the hand that was measured along with the measurements. Depending on the product and its function, hand and wrist landmarking and measuring may be very detailed. If the hand must perform intricate movement while wearing a product, each joint segment should be measured. Several hand measurements refer to skin creases, fixed and permanent lines related to underlying anatomical structures, which are useful topical features for measuring the hand and wrist (Mallouris et al., 2012).

These directions specify measuring finger length to the fingertip and to exclude fingernail length. Consider how to incorporate long fingernails into some products; for example, fashion gloves.

Basic landmarks and measurements are described. For a detailed, comprehensive procedure for the wrist and hand, consult comprehensive instruction manuals; for example, *Hand Anthropometry of U.S. Army Personnel* (Greiner, 1991); a technical report distributed by the NATICK Research, Development and Engineering Center.

Some landmark locations are difficult to landmark with an adhesive marker, for example the "crotch" between fingers. A marking pen is the best tool to use for these small landmarks. The directions may also be used to place reflective or transmitting markers used for motion capture systems, however the markers must be small in proportion to the hand.

Landmarks are illustrated in Figure F.1 and measurements are illustrated in Figure F.2.

## F.1 Hand and Wrist Landmarks (Alphabetical Order)

***Finger I base:*** The proximal flexion crease of the metacarpophalangeal joint of the thumb.

***Finger II base:*** The proximal flexion crease of the metacarpophalangeal joint of the index finger.

***Finger III base:*** The proximal flexion crease of the metacarpophalangeal joint of the middle finger.

***Finger IV base:*** The proximal flexion crease of the metacarpophalangeal joint of the ring finger.

***Finger V base:*** The proximal flexion crease of the metacarpophalangeal joint of the little finger.

*Procedure for location of finger base, all fingers:* Have the person flex their hand and note where the wrinkles form at the base of the fingers. The location is typically not landmarked although a line may be drawn at the base of each finger on the palmar side of the hand.

***Finger crotch I:*** The deepest indentation of the space between the first and second fingers.

***Finger crotch II:*** The deepest indentation of the space between the second and third fingers.

***Finger crotch III:*** The deepest indentation of the space between the third and fourth fingers.

***Finger crotch IV:*** The deepest indentation of the space between the fourth and fifth fingers.

*Procedure for location of finger crotch, all fingers:* Ask the person to spread their fingers as far as comfortably possible. Look at the "web" between each finger and mark, with a short line perpendicular to the length of the finger, the deepest indentation of the web. Extend the line approximately 1 cm (.5 in.) to the dorsal and ventral sides of the hand. Finger crotch landmarks are visible from both the ventral and dorsal sides of the hand.

***Finger flexion creases on ventral side of hand:*** Each segment of each finger may be landmarked and measured. Use the flexion crease between each segment as viewed on the ventral side of the hand as the delineating mark between each segment. The crease may be used as the landmark or a line may be drawn along the crease. If there is more than one crease between segments draw a line along the most proximal crease.

***Finger joint prominence marked on dorsal side, I, II, III, IV, V:*** Each joint prominence (apex of knuckle) of each finger may be landmarked on the dorsal side of the hand. They are marked with solid lines in Figure F.1. These landmarks are necessary when designing and sizing a handcovering that must flex with the hand. When the hand is flexed, dorsal lengths increase substantially while ventral lengths decrease.

*Procedure:* Ask the person to form a fist with the R hand and place the hand on a table surface with the elbow supported by the table. The thumb should wrap around the flexed fingers. With the hand dorsal side up, place a small mark at the center of the most protruding point of each metacarpophalangeal joint. Then have the person rotate the fist so the ventral side is exposed and mark the remaining interphalangeal joint projections.

***Fingertip, I, II, III, IV, V:*** The most distal point of each finger (not including the fingernail).

*Procedure:* The person is seated with hand placed ventral side up, fingers extended. Locate the point of each fingertip. Do not include the fingernail length. This landmark is typically not physically marked.

***Metacarpal II:*** The most lateral point of the R metacarpophalangeal joint II (at the base of the index finger).

*Procedure:* The person is standing, elbow bent at 90 degrees, with R hand extended, ventral side up. Stand in front of the person. Support the person's hand with your hand and palpate, with your other hand, the metacarpophalangeal joint II. Locate the most laterally protruding point of the joint. The landmark is located at the base of the index finger. Draw a short line at the protuberance perpendicular to the long axis of the finger.

***Metacarpal V:*** The most medial point of the R metacarpophalangeal joint V (at the base of the little finger).

*Procedure:* The person is standing, elbow bent at 90 degrees, with R hand extended, and ventral side up. Stand in front of the person. Support the person's hand with your hand and palpate, with your other hand, on the medial side of the hand from the wrist distally to the first joint, the metacarpophalangeal joint V, to locate the most protruding point. Draw a short line at the protuberance, perpendicular to the long axis of the finger.

***Stylion, dorsal stylion and ventral stylion:*** The inferior point of the bottom of the radius, which may be extended with a line to delineate the base of the hand at the intersection of the hand and the lower arm.

*Procedure:* The person is standing, elbow bent at 90 degrees, with R hand extended, ventral side up. Stand in front of the person. Support the person's hand with your L hand. With your R hand, palpate the length of the person's thumb to the intersection of the hand and the lower arm until you feel the edge of the radius. Draw a small X at this point. See further directions for intersection of hand and wrist with ulnar edge procedures.

***Ulnar edge of the distal wrist crease:*** The projection of the distal wrist crease to the ulnar edge of the wrist. Along with the stylion, this location approximates the axis of rotation for the wrist.

*Procedure:* This landmark is typically located immediately after marking the stylion. Positions are as described for locating the stylion. Palpate the length of the person's hand along the outer edge of the little finger to locate the edge of the ulna. Draw a small X at this point. To mark the intersection of the hand and the lower arm, you may draw a line from the X along the dorsal (back) surface of the wrist and along the ventral (front) surface of the wrist. To facilitate location of the lines, ask the person to flex their wrist to locate the skin wrinkles that can be used to visualize locations. Visualize a product that must flex at the hand-forearm intersection.

## F.2 Wrist and Hand Measurements (Alphabetical Order)

Wrist and hand measurements are two-dimensional, like hand length; and over-the-body measurements, like wrist circumference.

***Finger breadth at the joint, dorsal and/or ventral:*** The breadth of each joint (knuckle) of each finger from the dorsal or ventral side depending on the product. The joint is typically the largest breadth of the finger. The midpoint of each finger segment could also be measured.

*Procedure:* This measurement is difficult to accurately measure with manual methods. A small caliper may be used to measure the midpoint width of each finger segment. The person places their R hand on the table, with elbow supported. Ask the person to spread their fingers as far as comfortably possible. Place the small caliper at the knuckle landmark, with the caliper blades as close to the finger as possible without compressing the flesh. Digitizing methods also have been used to document the two-dimensional shape of the handprint. Simple, low-tech, methods are to draw the outline of the hand, or use a photocopy machine to capture the outline of the hand.

***Finger length to base, I, II, III, IV, V:*** The length of each finger measured from the tip of the finger (not including the fingernail) to the crease at the base of each finger.

***Finger length to crotch base, I, II, III, IV, V:*** The length of each interior side of each finger measured from the tip of the finger (not including the fingernail) to the crotch base landmark. Alternately the total length from one fingertip to the crotch of the adjoining fingertip.

***Finger segment length, joint crease to joint crease:*** The segment length of a finger measured from the center of each joint crease to the next.

***Hand breadth:*** The breadth of the hand between the landmarks at metacarpal II and V.

*Procedure:* The person sits at a table and places the R hand on a hand block. Have the person hold the fingers together with the thumb abducted at approximately a 45-degree angle so that the caliper can be placed between the medial side of the hand and the thumb. Press the person's hand firmly, but gently into the block and ask the person to maintain this pressure during the measurement procedure. The middle finger is aligned with the long axis of the forearm. Stand in front of the person and use a caliper to measure the breadth of the hand between metacarpal II and metacarpal V landmarks.

*Cautions:* The amount of pressure the person exerts on the surface can affect this measurement; too little pressure will result in a smaller measurement and excess pressure will result in a larger measurement. The caliper should not compress or indent the skin.

***Hand circumference:*** The circumference of the hand encompassing the metacarpal II and V landmarks.

*Procedure:* The person sits at a table and places the fingertips of the hand on the hand block in order to slightly elevate the palm of the hand. Place the tape measure under the palm of the hand before the next step. Ask the person to hold the fingers together with the thumb abducted at approximately 45 degrees, so that the tape measure can be placed between the medial side of the hand and the thumb. The middle finger is aligned with the long axis of the forearm. Stand in front of the person and encircle the metacarpal II and metacarpal V landmarks with the tape measure.

*Cautions:* Exert only enough pressure on the tape to maintain contact with the hand.

***Hand Length:*** The length of the hand between the stylion landmark on the wrist and the tip of the middle finger (dactylion III).

*Procedure:* The person sits at a table with the palm of the hand on the table surface and the distal phalanges placed on the hand block. Ask the person to hold the fingers together with the thumb abducted at approximately 45 degrees. The middle finger is held positioned parallel to the long axis of the forearm. Press the person's hand firmly but gently to the surface of the table and ask the person to the hold the position while you are taking the measurement. Stand to the left of the person and position a sliding caliper so that the beam of the caliper is parallel to the long axis of the arm and the fixed blade point is at the stylion. Carefully slide the movable bar to the tip of the middle finger.

*Caution:* Do not let the person flex or hyperextend their hand. Take the measurement at the tip of the finger and do not include fingernail length.

***Palm length:*** The distance between the center of the crease at the base of the middle finger (finger III, base) and the ventral stylion landmark on the wrist.

*Procedure:* The person sits and places the hand on the table (no hand block) with the ventral side up. Ask the person to hold the hand in a straight line aligned with the forearm without flexing or hyperextending the hand. The fingers are held together with the thumb abducted approximately 45 degrees. Stand to the left of the person and position the sliding caliper to measure the distance between the center of the crease at the base of the finger (finger III, base) and the ventral stylion landmark on the wrist. The fixed blade of the caliper is placed at the base of the finger and is parallel to the long axis of the person's arm. Extend the movable arm to the ventral stylion landmark.

*Caution:* Do not let the person flex or hyperextend their hand. If there are multiple creases at the base of finger III, choose the most proximal crease.

***Wrist circumference:*** The circumference of the wrist at the level of the styion and perpendicular to the long axis of the forearm.

*Procedure:* The person stands with the upper arm relaxed and the elbow flexed at 90 degrees with the palm up. Stand in front of the person. Place a tape measure with the upper edge of the tape just below the bony prominence of the stylion, and the lower edge of the tape just above the pisiform bone in line with the little finger. Encircle the wrist with the tape and read the measurement at ventral mid-wrist.

*Caution:* Exert only enough pressure on the tape to maintain contact between the tape and the skin, without indenting the flesh.

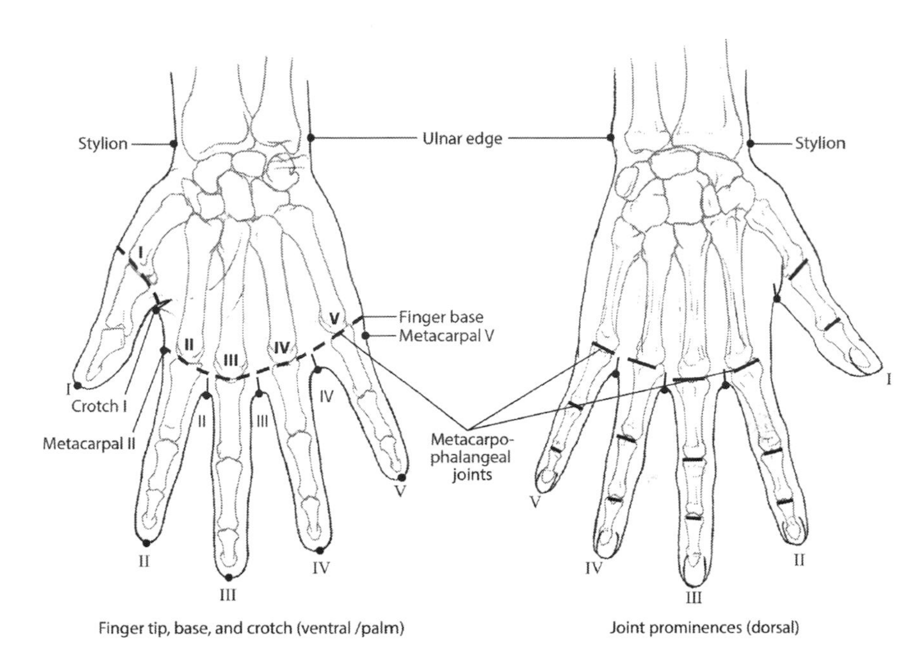


**FIGURE F.1**
Hand and wrist landmarks, ventral and dorsal surfaces.

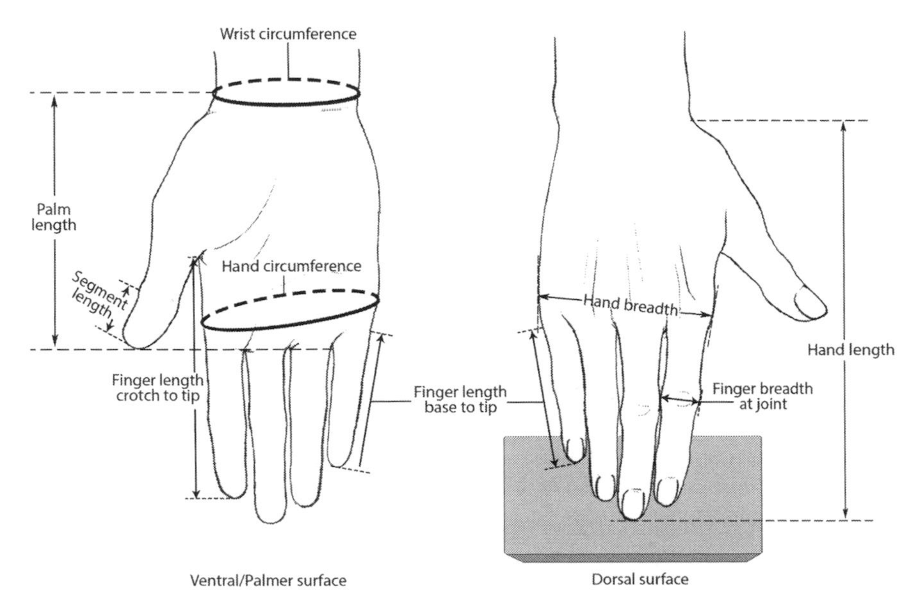


**FIGURE F.2**
Hand and wrist measurements, ventral and dorsal surfaces.

# Appendix G: Foot and Ankle Landmarks and Measurements

Ankle and foot landmarks are described in this appendix. Markings are placed on the R foot unless otherwise noted. For some studies, the dominant foot is landmarked and measured. In that case, record which foot and ankle were measured. Unless otherwise noted, have the person stand on a stable platform or low table approximately 46 cm (18 in.) high. A Brannock™ device is specified for length and breadth measurements. The device is used widely by footwear retailers to measure the foot, and is available from the manufacturer (https://www.brannock.com). Alternatively, a foot box can be constructed. Directions for making a foot box are included in the *NATICK measurers handbook* (Clauser, Tebbetts, Bradtmiller, McConville, & Gordon, 1988, Appendix C, p. 259).

See Figure G.1 for foot and ankle landmark illustrations and Figures G.2 and G.3 for measurement illustrations.

## G.1 Foot and Ankle Landmarks (Alphabetical Order)

***Fifth metatarsophalangeal protrusion:*** The most lateral protrusion of the foot at the fifth metatarsophalangeal joint.

*Procedure:* The person stands with weight equally distributed with the medial edge of the R foot parallel to a line perpendicular to the front edge of the platform. Stand in front of the person and locate the maximum protrusion on the lateral edge (little toe side) of the foot. Draw a short line at the center of the protrusion.

***First metatarsophalangeal protrusion:*** The most medial protrusion of the foot at the first metatarsophalangeal protrusion.

*Procedure:* The person stands with weight equally distributed with the medial edge of the R foot perpendicular to the front edge of the platform. Stand in front of the person and locate the maximum protrusion on the medial edge (hallux/big toe side) of the foot. Draw a short line at the center of the protrusion. The tip of the big toe of some people may adduct beyond the landmark location; ignore the splayed out toe and focus on the first joint protrusion.

***Lateral malleolus:*** The point of the lateral malleolus (lateral ankle bone).

*Procedure:* The person stands with their weight equally distributed. Stand at the person's right to locate the farthest protrusion of the malleolus. Place a marker or an X at this location.

***Medial malleolus:*** The point of the medial malleolus (medial ankle bone).

*Procedure*: While locating the lateral malleolus, locate the medial malleolus by palpating for the location of the tip of the tibial bone. Place a marker or X at this location.

## G.2 Foot and Ankle Measurements (Alphabetical Order)

***Ankle circumference:*** Minimum circumference of the ankle.

*Procedure:* The person stands with weight equally distributed and the feet approximately 10 cm (4 in.) apart. Stand to the right of the person. Use a tape measure to measure the circumference above the lateral and medial malleoli, parallel to the floor.

***Ball of foot circumference:*** The circumference encompassing the first metatarsophalangeal protrusion and the fifth metatarsophalangeal protrusion.

*Procedure:* The person stands with weight equally distributed and feet approximately 10 cm (4 in.) apart. Stand in front of the person. Place the tape measure under the ball of the foot, bring the tape around the foot to encompass the first and fifth metatarsophalangeal landmarks. Record the circumference.

***Ball of the foot length (from calcaneus):*** The length from the furthest protrusion of the calcaneus to the first metatarsophalangeal protrusion mark.

*Procedure:* The person stands with the R foot on the Brannock device and the L foot on a board of equal height to the device to balance the person's stance. Ask the person to distribute weight equally on both feet. Stand to the right of the person and make sure that the back of the heel is lightly touching the back of the device and that the long axis of the foot is in line with the long axis of the device. When the foot is correctly positioned, move to the front of the person. Measure the distance from the back of the heel to the ball of the foot by moving the pointer of the vertical slide to the first metatarsophalangeal protrusion landmark. Read the measurement from the device scale.

***Bimalleolar breadth:*** The breadth from the lateral malleolus to the medial malleolus.

*Procedure:* Note that the protrusions of the malleoli are typically not parallel to the ground. This measurement is taken parallel to the floor, using a small anthropometer. The person stands with weight equally distributed and feet about 10 cm (4 in.) apart with the long axis of the foot parallel to the side edge of the podium. Stand behind the person and position the caliper with one blade touching the farthest protrusion of the lateral malleolus. Spread the caliper blades so that the opposite blade is in line with the medial malleolus and the caliper is parallel to the platform.

***Foot breadth (horizontal):*** The maximum breadth of the foot.

*Procedure:* The person stands with the R foot on the Brannock device and the L foot on a board of equal height to the device to balance the person's stance. Ask the person to distribute weight equally on both feet. Stand to the right of the person and make sure that the back of the heel is lightly touching the back of the device and that the long axis of the foot is in line with the long axis of the device. Move the point of the vertical slide to the first metatarsophalangeal protrusion landmark. After assuring that the foot is aligned, measure the maximum breadth of the foot by moving the horizontal slide of the device until it is just touching the protrusion. Read the measurement from the device scale.

***Foot length:*** The maximum length of the foot.

*Procedure:* The person stands with the R foot on the Brannock device and the L foot on a board of equal height to the device to balance the person's stance. Ask the person to distribute weight equally on both feet. Stand to the right of the person and make sure that the back of the heel is lightly touching the back of the device and that the long axis of the foot is in line with the long axis of the device. After assuring that the foot is aligned, locate the extent of the foot length by placing a small block of wood against the tip of the longest toe (typically the hallux, but may be the second toe, refer to Chapter 8 for variations). Use only enough pressure to touch the tip of the toe. The objective is to measure the total length of the foot that should be accommodated by footwear. Read the length from the device scale.

***Heel breadth:*** The maximum distance between the medial and the lateral points on the inside and outside of the heel, at or posterior to the lateral malleolus landmark.

*Procedure:* The person stands on the raised platform with feet spread apart approximately 10 cm (4 in.) and their weight equally distributed. Stand behind the person. Use a caliper to measure the maximum distance between the medial and lateral points on the inside and outside of the heel. Hold the caliper parallel to the podium surface. Take the measurement just above the level of the platform at the most protruding points of the curve of the heel and parallel to the surface of the platform. Exert just enough pressure to touch the sides of the foot.

***Heel-ankle circumference:*** The circumference of the foot encompassing the ankle and the base of the heel.

*Procedure:* The person stands with feet spread apart approximately 10 cm (4 in.) and weight equally distributed. Stand in front of the person. Place the tape at the base of the heel where it first contacts the platform and wrap the tape around the foot at an angle to touch the dorsal juncture of the foot and the leg. The tape should span any hollows of the foot. Exert only enough pressure to maintain contact between the tape and the foot.

*Caution:* Have an assistant help by holding the tape at the base of the heel so that it does not slip out of place.

***Lateral malleolus height:*** The vertical distance between the floor and the lateral malleolus landmark.

*Procedure:* The person stands with heels together and weight distributed equally. Stand to the right of the person. Use a small anthropometer or short ruler to measure from the platform to the lateral malleolus landmark.

*Caution:* Your sight line should be level with the blade of the gauge. If using a short ruler (A), place another short ruler (B) or straight edge at the malleolus perpendicular to ruler A to record the measurement from ruler A.

***Medial malleolus height:*** The vertical distance between the floor and the medial malleolus landmark.

*Procedure:* The person stands with heels together and weight distributed equally. You may have to ask the person to move their feet apart to position the measuring device. Stand in front of the person. Use a small anthropometer or ruler, as described in the lateral malleolus height directions, to measure from the platform to the medial malleolus landmark.

*Caution:* Your sight line should be level with the blade of the gauge to avoid taking a measurement that is not perpendicular to the platform.

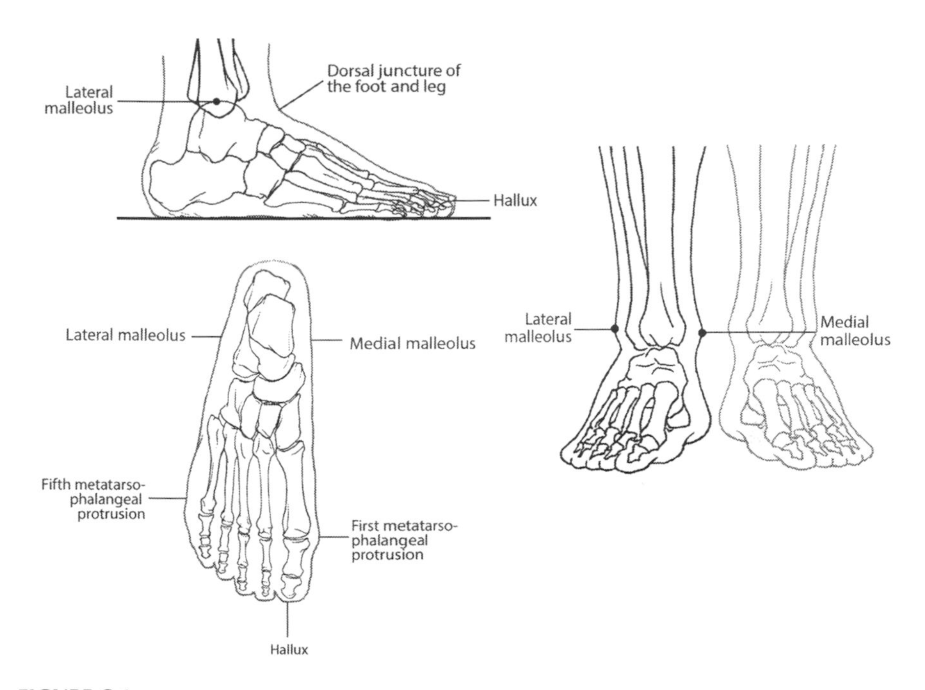


**FIGURE G.1**
Foot and ankle landmarks.

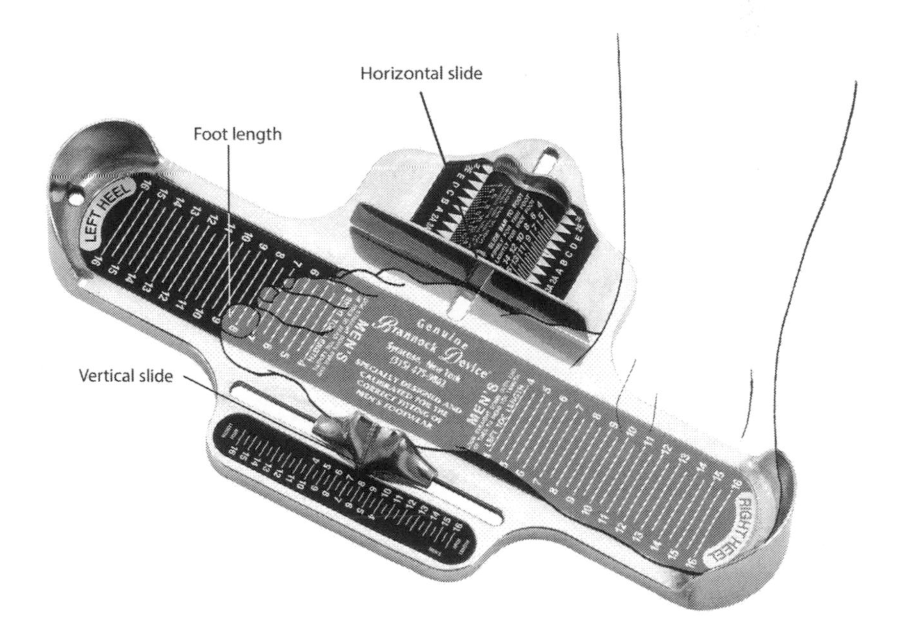


**FIGURE G.2**
Brannock™ device to measure foot length and breadth. Brannock™ device photograph with permission by the Brannock Company. Foot illustration by L. Wen.

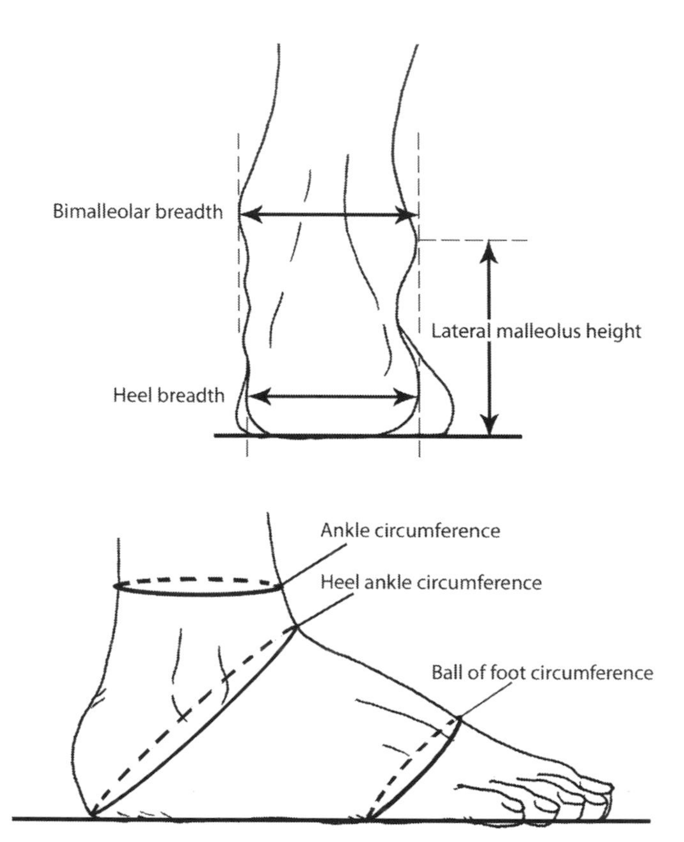


**FIGURE G.3**
Foot and ankle circumferences, breadths, and height.

# *Appendix H: Total Body Landmarks and Measurements*

Total body measurements span two or more body segments described in Appendices A-G. Most of the measurements are two-dimensional and are taken from a specified landmark to the floor. Some measurements are over-the-body using a tape measure. For example, a protective coverall may require a measurement of the torso from center front neck (or alternately the suprasternale) over-the-body between the legs and continuing to the center back neck (7th cervical vertebra, cervicale).

Landmark descriptions for the total body are described throughout Appendices A–G. Measurements are illustrated in Figures H.1 and H.2.

## H.1 Total Body Landmarks

Landmark descriptions for total body are described throughout Appendices A–G.

## H.2 Total Body, Two-Dimensional Height Measurements

*Procedure for all measurements:* The person stands with palms facing the body. Stand to the right of the person. An anthropometer is used to measure the vertical distance from the floor to the landmark.

*Caution:* The person's respiration may affect these measurements as the person may slightly raise their shoulders or other areas of the body when inhaling, so the technician should be aware to take the measurement at the maximum point of quiet respiration. The person's position, and therefore measurement, may vary as the person shifts body weight.

***Acromial height (standing):*** The distance from floor to the height of the acromion.

***Axilla height (standing):*** The distance from the floor to the level of the lower armscye.

***Calf height:*** The distance from the floor to the fullest part of the calf.

***Cervicale height (standing with head in Frankfort plane):*** Distance from floor to the cervicale.

***Chest/bust height (standing):*** The distance from the floor to the bustpoint for women/thelion for men.

***Neck height (standing with head in Frankfort plane):*** The distance from the floor to the trapezius point.

***Stature/total body height (standing with the head in Frankfort plane):*** The distance from the floor to the top of head.

## H.3 Total Body (Over-the-Body) Measurements

***Vertical trunk circumference:*** The circumference of the trunk from the R midshoulder over the fullest chest/bust, between the legs encompassing the perineum/genitals, and continuing over the buttock point, returning to the midshoulder.

*Procedure:* The person is standing with feet approximately 10 cm (4 in.) apart and weight equally distributed. Ask the person to assist in passing the tape measure between their legs and positioning the tape (for men, the tape passes to right of the scrotum). Stand to the right of the person. The 0-end of the tape measure is placed at the midshoulder landmark and passed over the fullest part of the chest/bust, between the legs and encompassing the perineum/genitals, then continues over the buttock point (not landmarked) and on to meet the original start point at midshoulder. Assure that the person returns to an upright standing position after the tape is placed. Exert only enough pressure to maintain contact with the body. The tape will span hollow areas of the body. Record the measurement at the maximum point of quiet respiration.

*Variations:* An alternate measurement is with start point at the center front neck (or alternately the suprasternale) over-the-body between the legs and continuing to the center back neck (7th cervical vertebra, cervicale).

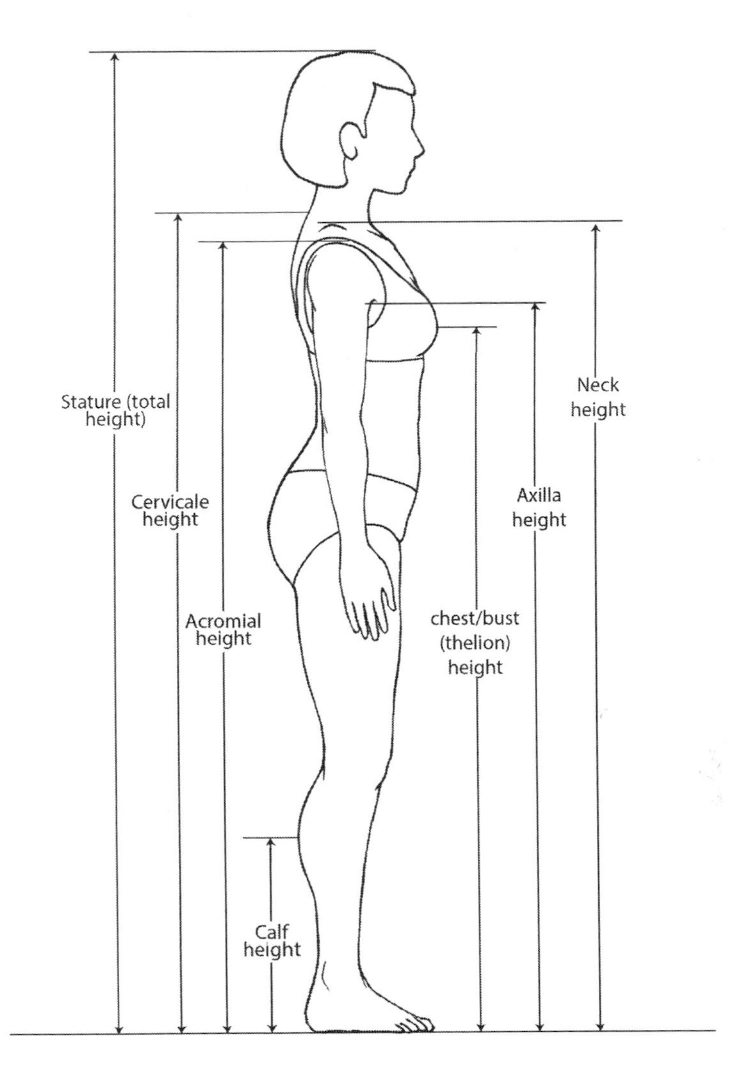


**FIGURE H.1**
Total body height measurements, lateral view.

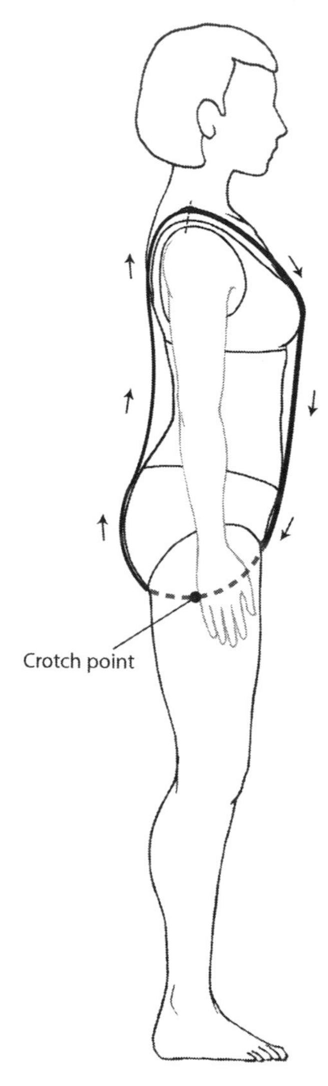


**FIGURE H.2**
Vertical trunk circumference, midshoulder (R) to midshoulder (R), lateral view.

# *Glossary*

**3D scanners:** equipment to capture data points from a three-dimensional object, e.g. human body
**abduction:** movement of a body part away from the midline of the body
**abductor digiti minimi:** the small hand or foot muscle which moves the little (*minimi*) finger or toe (*digiti*) away from the central axis of the hand or foot (*abducts*)
**abductor hallucis:** muscle on the medial foot which abducts the big toe (*hallux*)
**accessories:** items worn with the main wearable product to enhance aesthetics or function
**acetabulum (pl. acetabula):** hip socket
**Achilles tendon:** strong, flat *connective tissue* band (*tendon*) which attaches the calf muscle to the heel bone (*calcaneus*)
**acoustic:** pertaining to hearing or sound
**acromioclavicular joint:** joint at the front of the shoulder where the collar bone (*clavicle*) and the shoulder blade (*scapula*) join
**acromion:** the most lateral bony protuberance of the shoulder blade, or *scapula*
**actin:** specialized protein threads (*contractile protein myofilaments*) in muscle capable of contracting; also see *myosin*
**active marker system:** motion capture method of *landmarking* the body, uses electronic signals transmitted from the marker to a receiver
**Adam's apple:** common name for the anterior bulge of voice-box portion of the windpipe (*trachea*) in the mid-neck
**adduction:** movement of a body part toward the midline of the body
**adductor brevis:** one of three muscles on the *medial* thigh which moves the thigh bone (*femur*) toward the midline of the body
**adductor longus:** one of three muscles on the *medial* thigh which moves the thigh bone (*femur*) toward the midline of the body
**adductor magnus:** one of three muscles on the medial thigh which moves the thigh bone (*femur*) toward the midline of the body
**adipocyte:** fat cell
**adipose or adipose tissue:** anatomical terms for what is commonly called fat
**aerodynamic:** allowing smooth airflow around something
**aesthetics:** perception of an object, e.g. apparel, within a cultural context based on a set of principles related to beauty or art
**agonist:** a contracting muscle which moves a body part, with a mutually resisting muscle—an *antagonist*, which helps to control, modify, or balance the motion
**alveoli (pl. of alveolus):** tiny air sacs of the *lungs*, where the body exchanges carbon dioxide from the blood for oxygen from the air

**ambidextrous:** with equal skill using either hand

**ambulatory blood pressure monitoring:** continuous blood pressure monitoring during daily activities over a time period, see *blood pressure monitor*

**ambulatory electrocardiography:** measurement of heart electrical activity over a time period, during daily activities

**amputee:** an individual who has lost one or more limbs or limb segments to trauma or surgery

**anal fissure:** a crack in the moist lining (*mucous membrane*) of the *digestive system*'s exit from the body (*anus*)

**anal sphincter:** the circular muscle (*sphincter*) which serves to open and close the *anus*, releasing or retaining solid body waste (*feces*)

**anaphylaxis:** a serious, sometimes life-threatening, allergic response characterized by hives, soft tissue swelling, trouble breathing and a drop in blood pressure

**anatomical crotch:** the angle between the lower torso and the medial proximal thighs, extending from the pubic bone anteriorly to the buttocks crease posteriorly

**anatomical position:** standing with arms at the sides and palms facing forward, a standard body position used to describe relationships between body parts

**anatomical snuffbox:** the triangular hollow over the *dorsal* wrist between *extensor* tendons to the thumb and index finger

**anatomy:** the study of body structure, from Greek term meaning cut up or cut open

**angular vein:** a superficial *vein* lying between the eye and nose, extending toward the *jaw line*

**ankle:** the joint between the lower leg and the foot

**ankle-foot orthosis (AFO):** a mechanical support (*brace*) to maintain ankle joint position, stability, or to assist ankle movement

**annular ligament:** a key part of the proximal *radioulnar joint* which allows the forearm bones to move relative to one another, and the hand to turn palm-down and palm-up

**ANSUR I and ANSUR II:** (U.S. Army Anthropometry Surveys) anthropometric study of U.S. Marine Corps and Army men and women

**antagonist:** see *agonist*

**antecubital fossa:** the depression in front of the elbow joint

**anterior:** refers to a structure being more in front toward the chest and belly than another structure; *ventral or volar*

**anterior and posterior cruciate ligaments:** strong ligaments within the knee joint which help provide knee stability

**anterior chamber:** the space between the eye surface (*cornea*) and the colored central ring of the eye (*iris*)

**anterior superior iliac spine (ASIS):** the most anterior bony prominence of the upper rim (*iliac crest*) of the pelvis

**anterior talofibular ligament:** lateral ankle ligament most vulnerable to an ankle sprain; also see *lateral ankle ligaments*

**anterior tibial artery:** a named artery of the *anterior* lower leg

**anthropometer:** anthropometry tool used to measure body widths, depths, and heights

**anthropometric:** referring to anthropometry

**anthropometry:** study of the body using surface measurements

**antibodies:** infection fighting proteins in the blood

**anus:** opening for passage of solid waste (*feces*) at the lower end of the *digestive system*

**aorta:** primary *artery* of the *systemic circulation*; arises from the heart's L *ventricle*

**aortic arch:** curved portion of the aorta just after it leaves the heart

**apocrine glands:** multicellular *glands* producing odorous secretions including *pheromones*; i.e. sweat *glands* found in the armpits and groin; apocrine secretions include some products of glandular cell breakdown

**aponeurosis:** a flat sheet of tough *connective tissue* (*tendon*) which anchors a muscle to a bone or another muscle

**apparel:** personal attire, often used in reference to a ready-to-wear clothing category, e.g. women's apparel, athletic apparel, fashion apparel

**apparel drafting methods:** formulas applied to body measurements to draw two-dimensional patterns intended to shape to the form of the body

**apparel sizing systems:** methods of grouping body measurement data from a range of people into segments for apparel production and/or consumer size selection

**appendicular skeleton:** the bones of the upper and lower limbs and the bones connecting them to the spine (the *pectoral* and *pelvic girdles*)

**appendix:** a skinny sac-like *digestive system* structure pouching out from the beginning of the *ascending colon* on the R side of the body, also called *vermiform appendix*

**aqueous humor:** the fluid filling the *anterior* and *posterior chambers* of the eye

**arch (of a spinal vertebra):** the curved *posterior* bony section of a *vertebra* covering the cavity housing the *spinal cord*

**arch (of the foot):** the curved structure of the inferior aspect of the foot, made up of several foot bones, supported by ligaments and tendons, which supports the weight of the body above it

**areola:** the circular, more densely pigmented zone of the breast around the *nipple*

**arm:** the anatomical term for the upper limb segment between the shoulder and the elbow

**armscye (garment armhole):** edge of an upper torso garment that encircles the arm/shoulder intersection, seam to attach a sleeve to the garment torso

**arrector pili:** small involuntary muscles attached to the hair in the hair follicle

**arteriole:** smallest arterial blood vessel, the connecting structure between an *artery* and *capillary*

**artery (pl. arteries):** relatively thick-walled blood vessels carrying blood away from the heart to all body *tissues*

**arthritis:** *inflammation* of the joints; a disease process causing joint inflammation

**articular processes:** *vertebral* bony projections which bridge between the vertebral arch and the joint between adjacent vertebrae

**articulated:** interconnected

**asbestos abatement suits:** personal protective equipment to guard against breathing or skin contact with asbestos

**ASIS:** see *anterior superior iliac spine*

**aspiration:** the movement of food, liquid, or a foreign body from the upper airway into the lungs

**asthma:** a chronic breathing disorder characterized by wheezing and difficulty taking a breath

**athletic apparel:** clothing designed for wear during athletic (sports) activities

**athletic cup:** anatomically-shaped rigid protective shell, to cover both the penis and scrotum

**athletic supporters:** commonly called jock straps; undergarment consisting of a textile pouch and straps worn to support the penis and scrotum during vigorous physical activity

**atlas:** the first cervical vertebra; C1

**atopic dermatitis:** a chronic relapsing skin disease which involves an itch-scratch cycle; can be mild to severe

**atrium (pl. atria):** upper chamber(s) of the heart

**atrophy:** to shrink from disuse; commonly used to refer to muscle wasting or deterioration

**auricle:** the visible portion of the external ear, includes the *helix*, *scapha*, and *tragus*

**auriculotemporal nerve:** a *sensory* nerve behind the *TMJ* and traveling upward in front of the ear

**autoimmune disease:** a condition in which the body's infection/foreign body fighting capacity is misdirected toward normal body *tissues*

**autonomic:** referring to specific *motor* nerves which we do not consciously control. See *autonomic nervous system*

**autonomic nervous system:** portion of the nervous system which is not under voluntary control: nerves to *smooth muscle*, cardiac muscle, and glands (including sweat glands), where we cannot consciously decide on the muscle actions, includes *sympathetic* and *parasympathetic* divisions

**auxetic material:** somewhat flexible textile or other structure that reacts to high-speed impact by changing properties from soft and flexible to stiff and protective; used for body armor, sports pads

**avatar:** a figure that represents a person

**axial skeleton:** skeleton of the head and trunk; includes the skull, plus the spine and rib cage in the torso
**axilla:** arm pit
**axillary tail:** the portion of the breast which extends from the breast body into the armpit
**axis of rotation:** center of a joint, location for *goniometer* pivot point placement
**ball-and-socket joint:** a geometric description for the *glenohumeral joint* at the shoulder or for the hip joint
**baseball finger:** a finger with a mallet-like appearance as the *DIP* linking the most distal phalanges is constantly flexed, often to 90°; often caused by a sudden flexion force
**basement membrane:** a thin, ridged structure that firmly attaches the *epidermis* to the *dermis*, increasing skin durability
**basic straight skirt:** a textile cylinder that wraps around the hips and is contoured to the waist circumference
**basilic vein:** major superficial forearm vein draining the dorsal *ulnar* side of the hand
**beam caliper:** anthropometric measuring tool used to measure body widths, depths
**bespoke:** custom made
**biceps brachii or biceps:** the major muscle on the front of the upper arm, acts on shoulder and elbow joints
**biceps femoris:** lateral hamstring, a major muscle on the back of the thigh that acts on the hip and knee joints; comprised of two "heads" or sections: (a) a *long head*, spanning from the *ischial tuberosity* of the *pelvis* to below the knee (bending the knee and straightening the hip) and (b) a *short head*, originating on the *femur* and continuing to below the knee, which bends the knee
**bi-dimensional measuring scheme:** two measurements used to draft a simply shaped product, e.g. stocking cap, also see next entry
**bi-dimensional sizing system:** two body measurements used to communicate the dimensions (size) of a product
**bifurcated product; bifurcated garment:** lower torso product with the bottom section divided into two parts called legs
**bile:** a mix of chemicals produced by the liver to aid *digestion* of fatty foods
**bio-hazard suits:** personal protective equipment to guard against inhaling or skin contact with infectious agents and/or bodily fluids which may carry infectious agents
**biomechanics:** the study of the actions of natural mechanical forces and energy in and on the body
**bipedal mobility:** standing and/or getting around on two feet; getting around on two feet
**birth control diaphragm:** device inserted in the vagina to cover the cervix of the uterus, intended to block *sperm* from entering the uterus

**bladder:** a hollow *organ* serving as the body's reservoir for the *urinary system* product, *urine*

**bladder capacity:** the total amount of *urine* a person can tolerate holding in the *bladder*

**bladder suit:** a space suit containing a gas-pressurized bladder to simulate Earth's atmospheric pressure

**blast furnace suit:** garment worn to protect from extreme heat in manufacturing situations

**blister:** a pocket of fluid between the epidermis and dermis resulting from shearing, compression, and/or scraping forces

**blood:** a complex fluid consisting of *plasma* (the liquid portion of blood) and *formed elements* (blood cells)

**blood pressure:** the pressure of circulating blood on the walls of the arteries

**blood pressure cuff:** a band placed circumferentially around a limb and attached to a *blood pressure monitor* to record *blood pressure*

**blood pressure monitor:** An electronic or mechanical device, often with a recording capacity; most often attached to a pliable textile *blood pressure cuff* to document *blood pressure*

**blood vessels:** anatomical structures which carry blood within the body

**bobbinette weave:** a type of hexagonal net structure

**bodice:** the portion of a garment which covers a woman's upper torso

**body (of a rib):** the *shaft*, the main/central portion of a rib

**body (of a vertebra):** see *vertebral body*

**body core:** the torso's internal space where the essential organs of life are contained and protected

**body form methods:** anthropometric techniques to record information about the surface, shape, and volume of the body

**body systems:** inter-related anatomical structures of the body which combine into a functional whole and perform specific physiologic functions

**bony joint:** a joint which does not move; although it began as a fibrous or cartilaginous joint at an early stage of development

**bony pelvis:** bony ring formed by the *sacrum* and *coccyx* of the spine with the R and L *innominate* bones, including two *sacroiliac* (SI) joints and the *pubic symphysis*

**boutonniere (buttonhole) deformity:** joint alignment changes seen in the *IP* joints of a finger with flexion of the PIP joint and extension of the *DIP* joint due to tendon defects, commonly seen in people with rheumatoid arthritis

**bowel movement:** the evacuation of *digestive* waste from the body

**boxer brief:** men's undergarment combining features of boxer and brief styles

**bra (brassiere):** woman's undergarment worn to support the breasts

**bra band size designator:** bra size based on an under the bust circumference measurement, typically paired with cup size designator

**bra cup size designator:** bra size based on breast volume, typically paired with a band designator

**brace:** a mechanical support, most often used to stabilize a joint
**brachial artery:** the major *artery* of the arm
**brachial plexus:** an interlacing network of *motor, sensory,* and *autonomic nervous system* fibers from cervical spinal nerves 5 through 8, plus the first thoracic spinal nerve
**brachioradialis:** a superficial *forearm* muscle which helps *flex* the elbow joint
**brainstem:** the *CNS* center for head and face motor control, facial sensation; special senses of hearing, equilibrium, and taste; and regulation of automatic body functions; a vulnerable connector to the spinal cord, acting as a physical anchor for the "floating brain"
**Brannock™ device:** anthropometric device used to measure foot length and breadth
**brassiere:** see *bra*
**breathing cycle:** includes a breath in, followed by a breath out, a *respiratory cycle*
**bridge (of eyeglass frame):** connecting section of eyeglass frame above the *bridge of the nose*
**bridge of nose:** upper section of the nose, between the eyes, formed by the paired *nasal* bones
**bronchial tree:** tubular *respiratory* structures that branch repeatedly in the thorax
**bronchioles:** final branches of the *bronchial tree,* the *respiratory* equivalent of *capillaries*
**bronchus (pl. bronchi):** one of two main branches of the *trachea,* further branching into additional smaller tubular structures, *bronchi*
**bunion:** a *lateral* deviation of the big toe at the first *metatarsophalangeal* joint resulting in a noticeable bulge at the *medial* side of the foot
**burn, first-degree:** minor burn which affects the epidermal layer of the skin
**burn, second-degree:** a deeper burn (compared to first degree), involving both the epidermal and dermal skin layers
**burn, third-degree:** a burn through the two top skin layers, which reaches into the third, deep, *subcutaneous layer*
**bursa (pl. bursae):** small fluid-filled pouch, often between mobile *tissues; bursae* allow smooth motion of body parts as they work together
**bursitis:** inflammation of a *bursa;* often seen adjacent to joints or other structures which are moved repetitively (between adjacent muscles, between bone and skin, or where a tendon slides over a bone)
**bust:** clothing term, undefined area of the breast mound covered by a wearable product
**bust point:** apex of a dart pointing to the fullest projection of the bust of a woman's bodice
**CAESAR®:** Civilian American and European Surface Anthropometry Resource Project; anthropometric study of 2400 U.S. and Canadian and 2000 European civilians, conducted 1998 to early 2000
**calcaneal tendon:** see *Achilles tendon*
**calcaneofibular ligament:** see *lateral ankle ligaments*

**calcaneus:** heel bone, the largest bone of the foot

**caliper:** *anthropometric* tool used to measure body widths, depths, and heights

**callus:** a generalized area of thickened skin caused by friction or pressure

**capacitive technology:** electronic devices, like touch screens, which respond to skin or anything else that holds an electric charge

**capillary (pl. capillaries):** the tiniest blood vessel of the *circulatory system;* microscopic structures which bridge between the smallest arteries (*arterioles*) and tiny veins

**capitate:** one of the distal *carpal bones* of the wrist, part of the *carpal tunnel*

**carbohydrates:** the group of sugar-like organic substances used by the body, after the *digestion* of food

**carbon dioxide:** a waste product of the body's oxygen and nutrient use

**cardiac:** referring to the heart

**cardiac muscle:** involuntary muscle which keeps the heart pumping without conscious effort

**carotid arteries:** large arteries running through the neck which carry blood to the head and brain

**carpal arch:** a curved structure, similar to the *arch* of the foot, comprised of six of the eight carpal bones—the *pisiform, triquetrum, hamate, capitate, trapezoid, trapezium,* and the ligaments which stabilize them; a structure essential to preserve and facilitate hand function

**carpal bones:** the bones of the carpal arch plus the lunate and scaphoid bones, residing in the proximal hand; commonly called the wrist; from the Latin *carpus*

**carpal tunnel syndrome:** a condition variably characterized by hand pain, numbness, and *thumb/index finger* weakness, which results from pressure on, or excessive motion of, a major nerve at the wrist, the *median nerve*

**carrying angle:** a medical term referring to how the forearm is positioned relative to the arm, in the *anatomical position*

**cartilage:** a resilient, firm white *connective tissue* found in *synovial* joints as well as elsewhere in the body, e.g., the tip of the nose

**cartilaginous joints:** immobile or partially mobile joints; adjacent bones bound together by cartilage

**cauda equina:** the group of *spinal nerve roots* within the spinal canal, *distal* to the tip of the *spinal cord*

**caudal:** refers to a structure closer to the feet

**cellulite:** the puckering skin changes often seen on women's buttocks and thighs

**center of gravity:** balance point of the body, sometimes called center of mass; changes with body position; in the *anatomical position,* its approximate location is *anterior* to the second *sacral* vertebra

**central nervous system (CNS):** *nervous system* components found inside the skull and *vertebral column* (brain, *brainstem,* and *spinal cord*)

**cephalic vein:** major superficial forearm vein draining the dorsal radial/thumb side of the hand

**cerebellum:** the brain structure serving as the center for coordination; lies below the *cerebral hemispheres* toward the back of the head; occupies 10% of the volume of the brain, but holds over 50% of the *neurons*

**cerebral hemispheres:** large paired R and L brain structures serving *cognitive functions*, making up the major portion of the human brain

**cerebrospinal fluid (CSF):** the colorless liquid, which fills the spaces around and within the central nervous system inside the skull and spine

**cervical:** pertaining to the neck

**cervical plexus:** interlacing *nerve fibers* from the first four cervical *spinal nerves*

**cervical vertebra:** one of the seven bony segments of the spine (*vertebral column*) in the neck

**cervicale:** tip of the *spinous process* of the seventh cervical vertebra, visible and easily palpated when the person bends their head forward

**cervicovaginal mucus (CVM):** variably thick liquid found in the *cervix* and lubricating the *vagina*; secreted by the *vaginal mucous membrane* and other regional pelvic structures

**cervix:** the opening of the uterus into the *vagina*

**CF:** see *cystic fibrosis*

**chondrosternal joint:** junction between the *cartilage* of a rib and the *sternum*

**chronic venous disease (CVD):** includes *varicose veins*, problems of leg and foot swelling, skin color changes in the leg, and/or development of weeping sores (*venous ulcers*) on swollen legs

**circulatory system:** inter-related structures to circulate blood throughout the body, including the heart and blood vessels: *arteries, arterioles, veins*, and *capillaries*

**circumduction:** a movement combining flexion, abduction, extension, and adduction; resulting in a continuous circular movement of a limb

**clavicle:** collar bone

**claw toe:** a toe abnormality similar to a *hammertoe* in which the toe develops a claw-like appearance as the *proximal interphalangeal joint* (PIP), linking proximal segments of the toe, is constantly flexed, with lesser flexion in the *distal interphalangeal joint* (DIP)

**clitoris:** the sensitive structure which lies *anterior* to the vaginal opening and inferior to the pubic symphysis

**clot:** a semi-solid mass of blood

**CMC:** referring to a joint between a *carpal* and *metacarpal* bone, see *first carpometacarpal joint* as CMC most commonly refers to the *first carpometacarpal joint*

**CNS:** see *central nervous system*

**coccygeal:** referring to or involving the *coccyx*

**coccyx:** the tailbone, varies from person to person in form; three to five very small vertebrae which may or may not be fused

**codpiece:** pouch concealing or accentuating the male genitals; element of 14th century clothing worn by wealthy men

**cognitive functions:** thoughts, emotions, personalities, memories, dreams, and plans for the future that reside in the human brain

**coitus:** sexual intercourse

**collagen:** a specific connective tissue protein; in the *dermis*, it gives the skin both a structural framework and strength

**collar (of an athletic shoe):** a shaped piece in some athletic shoe styles extending up and along the Achilles tendon, some designs are curved to avoid resting on the tendon

**comfort:** physical state of ease, pain-free

**common fibular nerve:** one of two major peripheral nerves serving *motor*, *sensory*, and *autonomic* functions below the knee; previously called the *common peroneal nerve*

**commotio cordis:** cardiac arrest caused by a blow to the chest

**compression force (compressive force):** pushing inward, causing an object to become compacted

**compression fracture:** a bone break in which one surface of the bone is pushed inward toward the opposite bone surface, most often used to describe a *vertebral body* fracture

**compression garment:** item of clothing that compresses the body and can change the body form from the original, for aesthetic, functional, or therapeutic applications

**conception:** merger of *sperm* and egg (*ovum*)

**concussion:** a disturbance of brain function caused by the brain shaking in the skull

**condom:** a sheath worn over the erect penis to collect ejaculated semen and act as a barrier between sexual partners

**condom catheter:** a pliable *condom*-like sheath worn over the penis and connected to a wearable *urine* collection bag

**condylar:** associated with a *condyle*

**condylar joint:** a synovial joint with oval convex surface(s) which articulate with geometrically elliptical surface(s)

**condyle:** an oval convex surface of one bone that articulates with an elliptical depression of another; a knuckle-like prominence on a bone

**cones:** light-sensitive cells concentrated toward the center of the *retina*, responsible for vision in a lighted environment, and for color perception

**conjunctiva:** thin *mucous membrane* which is the outermost layer of the exposed eyeball and contiguous with the inner layer of the eyelid

**connective tissues:** specialized variably soft/hard supporting and connecting *tissues* of the body

**constipation:** decreased frequency of fecal elimination; feces may be dry, hard, and/or difficult to eliminate

**contact lens:** thin lens (product) that rests directly on the surface of the eye to correct vision or change pupil color

**contraception:** prevention of conception
**contraceptive:** product or device to prevent conception
**contractile protein myofilaments:** threads of protein capable of contracting; microscopic muscle components
**control (key) measurements:** the basic measurements used to develop a sizing system
**Cooper's suspensory ligaments:** specialized supportive connective tissue structures of the breast
**copulation:** sexual intercourse
**coracoid process:** an *anterior* beak-shaped projection of the *scapula*
**core temperature:** temperature measurement recorded within a body cavity (mouth, ear, rectum) which reflects internal body temperature, as opposed to a measurement recorded from the skin's surface
**cornea:** the clear *anterior* portion of the eye
**corns:** isolated localized areas of thickened skin; may be either hard or soft
**coronal (frontal) planes:** planes running through the body from side to side, perpendicular to sagittal planes, dividing the body into unequal front and back parts; a coronal plane through the shoulders, or, alternatively, through the geometric center of the body is commonly used as a reference plane
**coronary arteries:** first arteries to branch from the aorta; carry blood to the heart muscle
**cortical bone:** bone which is structurally strong, dense, and found near bone surfaces
**costochondral joint:** junction between the sternal end of each bony rib section and the cartilage connecting the rib directly or indirectly to the sternum
**costovertebral joint:** junction between the *vertebral* end of each bony rib section and adjacent *thoracic vertebrae*
**costume:** clothing depicting a character, historical period, or country culture
**counter:** segment of a shoe that wraps around the calcaneus
**coverall:** all-in-one garment with attached hood and perhaps foot and hand coverings
**cranial:** refers to a structure being closer to the head
**cranial bones:** bones which form the *cranium*, the portion of the skull which encases and protects the brain and sense organs
**cranial cavity:** largest internal hollow area of the skull; acts as a container for the brain and *CSF*
**cranial nerve VII:** see *facial nerve*
**cranial nerves:** *PNS nerves* that extend from the *brainstem* to structures in the head and specific parts of the torso
**cranium:** see *cranial bones*
**cremaster muscle:** muscle surrounding the *testes* which contracts to elevate them within the *scrotum*

**cricoid cartilage:** a ring-shaped cartilage at the inferior edge of the voice box (*larynx*) in the superior section of the *trachea*
**cricothyroid membrane:** the thin tissue connecting the *thyroid cartilage* of the voice box and the *cricoid cartilage*
**crotch, anatomical:** angle between the lower torso and the medial proximal thighs, extending from pubic bone anteriorly to buttocks crease posteriorly
**crotch, garment:** product section covering the anatomical *crotch*
**crotch curve seam, crotch seam:** U-shaped seam that extends from center front (*anterior*) waist to center back (*posterior*) waist, bisects the right and left sides of the body; joins the two sections of a bifurcated garment
**crotch depth:** two-dimensional pattern drafting measurement from the waistline to the base of the pants crotch seam, perpendicular to the crotch level; term used with women's patterns
**crotch distance ease:** crescent of space between the crotch curve of pants and the crotch of the body (anatomical *crotch*)
**crotch extension:** a line, one on the front pattern and one on the back pattern, along the *crotch level* line that extends the pattern to cover the inner thigh
**crotch length:** distance from the center front waistline, around the crotch base, to the center back waistline, measured along the curve of the *crotch seam*
**crotch level:** pattern drafting line parallel to the floor at the base of the *crotch seam*; relates to the lowest level of the anatomical *crotch*
**crotch point:** terminal point of the front or back *crotch extension* where the crotch extensions of the center front and center back crotch seams meet
**crystalline lens:** lens of the eye; bends light to focus images on the *retina*; located behind the colored *iris* of the eye
**CSF:** see *cerebrospinal fluid*
**cutaneous:** pertaining to the skin
**cuticle:** see *eponychium*
**CVD:** see *chronic venous disease*
**CVM:** see *cervicovaginal mucus*
**cystic fibrosis (CF):** a genetic disorder that affects the *respiratory* and *digestive* tracts, with resulting very thick, sticky *mucus* which accumulates in the respiratory and digestive structures
**danger triangle of the face:** triangular region of the central face including the nose and upper lip; an infection inside the triangle can transmit bacteria to, and infect, the brain if the *angular vein* drainage is blocked
**dart:** a triangular-shaped fold of fabric in a garment, stitched in place, to shape a tube of fabric to the curved body; reduces excess fabric at a smaller circumference
**DAT:** see *deep adipose tissue*

**deep adipose tissue (DAT):** fat tissue occupying internal body spaces, especially the *abdomen*
**deep vein thrombosis (DVT):** a blood clot found in a deep vein in the leg
**defecation:** process of voluntarily eliminating *digestive system* waste (*feces*) from the body at a socially appropriate time and place
**dehydrated:** lacking sufficient fluids in the body tissues to promote healthy body function
**delaminated:** referring to finger/toenails splitting into layers and/or detaching from the nail bed—the pink tissue visible beneath the nail
**deltoid ligament:** primary stabilizing ankle ligament at the medial ankle; stronger ligament less likely damaged in an ankle sprain than lateral ligaments
**deltoid m.:** major superficial muscle bridging between the shoulder area and the upper arm, produces the contour distal to the anterior, medial, and *posterior* aspects of the *acromion*; increased deltoid muscle volume broadens the shoulders
**denier:** fiber diameter size
**dens:** a tooth-like structure which projects upward from the body of the second *cervical vertebra* (C2) to articulate with the *anterior* arch of the *atlas* (C1)
**depression:** primary jaw motion for opening the mouth; lowering the *mandible*
**dermal absorption:** penetration of the skin by chemical toxins or medications
**dermatitis:** general term for a red, itchy skin irritation from a specific cause; e.g. (a) contact dermatitis—including irritant and allergic varieties—and (b) atopic dermatitis, also called eczema
**dermatome:** an area of skin innervated by a single *spinal nerve*
**dermis:** the middle layer of the skin
**descending aorta:** the aorta as it turns downward from the heart
**design process:** step-wise (iterative) method to solve a design problem: problem identification, research, problem definition, ideation, design development, evaluation
**diabetes:** disease caused by a deficit of natural *insulin* or abnormal insulin function
**diarrhea:** frequent liquid bowel movements
**diastolic:** referring to the pressure obtained when the heart muscle rests and the heart refills with blood (between beats); the second number of a *blood pressure* reading
**digestion:** the process of breaking food down into nutrients which the body can use
**digestive system:** inter-related hollow anatomical structures (mouth, *esophagus*, stomach, small and large intestines, *rectum*, *anus*) and internal *organs* (*liver*, *gall bladder*, *pancreas*) which work together to complete *digestion*
**digestive:** referring to *digestion*

**digital human model:** replicates the surface architecture of a human body, manipulated in a computer using digital data
**DIP:** see *distal interphalangeal joint*
**dipping:** method to make a seamless hollow product that replicates the surface of a three-dimensional form, the form is dipped into a thermoplastic solution, then cooled and solidified to produce a three-dimensional "skin"
**dirndl:** a gathered skirt
**dissect:** to cut apart for study
**distal:** used in reference to the limbs (arms, legs) and refers to a structure being further away from the median plane or attachment of the limb than another structure
**distal interphalangeal joint (DIP):** joint between the most distal two *phalanges* of a finger or toe
**distal midpalm:** subdivision of the *midpalmar area* of the hand; a slightly mounded area overlying the *MP joints* of the four fingers
**distal radioulnar joint:** articulation of *radius* and *ulna* at the wrist
**distal tarsal group:** row of bones found between the *calcaneus* and the *metatarsal bones* of the foot; the cuboid bone, three cuneiform bones, and the navicular bone which connect the calcaneus and *talus* to the metatarsal bones
**distance between lenses:** common eyeglass frame measurement
**diuretic:** a food or medication which increases the flow of urine
**dorsal:** refers to a structure being more in back of, toward the body back, than another structure
**dorsal surface:** in general, the back side of the body (*posterior*), in the hand and foot the surface opposite the *palm* of the hand and *sole* of the foot
**dorsal webs:** fleshy web-like tissue between the fingers or toes, just distal to the body of the hand or foot
**dorsalis pedis artery:** artery on the top of the foot; a *pulse point*
**dorsiflexion:** to bend the ankle, raising the *dorsum* of the foot to the front of the anatomical *leg*
**dorsum:** top of the foot or hand
**double support:** the period during walking when both feet are on the ground
**draft:** (verb) draw a pattern outline based on body measurements or (noun) the product of that action
**drafting methods:** see *apparel drafting methods*
**DVT:** see *deep vein thrombosis*
**dynamic:** in motion
**dynamic tripod:** the thumb, index, and middle fingers; term used when referring to their precision gripping abilities
**dysmenorrhea:** cramping and pain associated with *menstruation*
**ear canals:** superficial skull cavities connecting the *auricle* (external ear) with the structures of the *middle ear*
**ear lobes:** fleshy lower portion of the *pinna*

**eccrine sweat glands:** *sweat glands* producing watery non-scented sweat; found throughout almost all of the body
**eczema:** see *atopic dermatitis*
**edema:** any collection of fluid in soft tissues of the skin and the tissue spaces beneath the skin, which causes swelling
**ejaculation:** the rhythmic forcible expulsion of *semen* (*sperm* plus other reproductive system fluids) from the *penis*
**elastin fibers:** part of the *dermis* of the skin; provide pliability and elasticity
**electrocardiography:** measurement of heart electrical activity
**electrodes:** electrical conductors used to detect muscle's electrical energy changes; includes devices used to detect such signals from both heart and *skeletal muscle*
**elephantiasis:** severe *filarial lymphedema* in the *scrotum*, foot, and/or leg
**elevation:** primary jaw motion for closing the mouth; upward motion of the *mandible*
**empire waistline:** womenswear style feature, encircles the rib cage directly below the bust
**endocrine system:** a collection of *glands* located throughout the body which secrete chemical products (*hormones*) that affect *organs* or *tissues* in the body
**endometrium:** mucous membrane with a rich blood supply, which lines the uterus
**enzymes:** specialized proteins, produced by the body, which promote specific chemical changes, such as digestive processes
**epidermis:** the outermost skin layer
**epiglottis:** a *laryngeal cartilage* just above the *larynx* which helps to protect *respiration* by directing food away from the airway into the esophagus
**eponychium:** a thin layer of *epidermis* bordering the proximal and lateral margins of the nail, commonly called the *cuticle*
**erect:** rigidly upright
**erection:** enlarged, firm state of the *penis* or *clitoris*
**ergonomic:** pertaining to human factors
**ergonomists:** human factors specialists
**esophageal hiatus:** a relatively small opening in the diaphragm for the esophagus
**esophagus:** a flat flexible *fibromuscular* tube linking the mouth and the stomach
**eustachian tube:** an anatomical structure which runs from the ear to the *nasopharynx* allowing equalization of air pressure between the *middle ear* and the *external acoustic meatus*
**EVA:** see *Extra-Vehicular Activity*
**evaporative cooling:** cooling the body by evaporating perspiration (sweat) from the body surface
**eversion:** descriptor of the body motion of turning outward, used only with foot/ankle motions

**evert:** to turn outward
**exhalation:** release of the waste products of *respiration* from the lungs when letting a breath out, lung volume decreases
**exocrine glands:** *glands* which deliver their secretions to the skin surface; e.g. *sweat glands* and the *mammary* (milk-producing) *glands* of the breast
**exoskeleton:** external skeleton-like frame that surrounds the exterior of the body for rehabilitation or extending human capabilities
**explosive ordnance (bomb) disposal suits:** garment worn for protection from explosion while disabling an explosive device
**extension:** straightening a joint; the opposite of flexion
**extensor digitorum brevis m.:** small, superficial *intrinsic* foot muscle located on the *dorsal lateral* foot; extends the lateral four toes
**extensor hallucis brevis m.:** small, superficial *intrinsic* foot muscle located *medial* to *extensor digitorum brevis* on the *dorsal* foot; extends the big toe
**extensor retinaculum:** a wide, fibrous band superficial to the dorsal surface of the carpal bones which holds the dorsal extrinsic muscle tendons in place at the wrist
**external acoustic meatus:** the external opening of the *ear canal* of the skull
**external ear:** ear structures distal to the *tympanic membrane;* includes the *auricle* and *external auditory meatus* (ear canal)
**external genitalia (genitals):** the male and female external organs of reproduction; the *penis* and *scrotum* with *testes* (male); vulva (*labia* and *vaginal opening*) and *clitoris* (female)
**external jugular veins:** moderately sized superficial neck veins returning blood toward the heart from structures on the exterior of the skull
**external oblique:** the most superficial lateral abdominal wall muscle
**external occipital protuberance**, or **inion:** a palpable prominence on the back of the occipital bone which lies at the junction of the skull and the neck; used as a *landmark*
**external rotation:** turning a limb away from the midline of the body— e.g. seen at the arm at the shoulder and the leg at the hip
**Extra-Vehicular Activity (EVA):** astronaut activity outside the shelter of the spacecraft
**extrinsic muscles:** muscles which act on a specific body part but originate and are found elsewhere; e.g. extrinsic muscles in the forearm generate hand and wrist motions, extrinsic muscles in the anatomical *leg* generate ankle and foot motions
**eye sockets:** major superficial skull cavities which contain the eyes
**eye-hand coordination:** linked control of eye movement and hand movement
**facet (of rib):** a small flat surface on the sternal end of a rib
**facet joints:** spinal joints between the *articular processes* of adjacent vertebrae
**facial bones:** multiple skull bones, some very small, which provide each person with much of their characteristic facial appearance

**facial nerve (cranial nerve VII):** a *PNS motor* nerve extending from the *brainstem* and leaving the skull in front of the ear near the *TMJ*; controls the muscles of facial expression, among others
**false ribs:** ribs 8–12 which do not attach directly to the sternum
**fascia:** a loose and then denser connective tissue, which acts much like "glue" to hold different tissues together
**fashion apparel:** clothing designed to meet mass market demands, based on changing aesthetic ideals
**fat:** specialized connective tissue distributed throughout the body, also called *adipose* or *adipose tissue*
**fat patterning:** *SAT* distribution; distinguished from the accumulation of *DAT*
**feces:** non-digestible food, *digestive* waste products, and bacteria eliminated from the body during *defecation*
**female condom:** a female-controlled barrier product, an internal wearable product useable for both pregnancy and *STD* prevention
**female urination device:** collection device to allow women to stand to *void*
**femoral artery:** primary arterial blood supply for the leg
**femoral head:** a hemispheric structure on the proximal end of the *femur* which fits into the *acetabulum* of the *innominate bone* of the *pelvis* in a *ball-and-socket joint*
**femoral neck:** section of the femur between the *femoral head* and the *trochanter*
**femoral nerve:** major nerve of the *anterior* thigh; carrying *motor, sensory,* and *autonomic nerve* fibers
**femoral shaft:** the long middle section of the *femur*
**femur:** the thigh bone; largest bone of the leg
**fertility:** the ability to conceive children
**fertilization:** see *conception*
**fetus:** developing infant
**fibrocartilage:** a cartilage with many fibers
**fibromuscular:** made of muscle and *connective tissue*
**fibrous joint:** adjacent bones bound by *collagen* fibers
**fibula:** smaller and more lateral of the two bones in the anatomical *leg*
**fibular artery:** continuation of the *popliteal artery* in the *posterior lateral* anatomical *leg,* running along the *fibula*
**fibular head:** the bony prominence of the proximal *fibula* on the outside of the leg, toward the back of the knee
**fibularis longus:** a long, narrow muscle which extends from the *fibular head,* a prominence of the proximal *fibula,* around the *posterior* of the lateral ankle to the bottom of the foot near the big toe
**filarial lymphedema:** *lymphedema* caused by the common tropical infection, *filariasis*
**filariasis:** the most common cause of *lymphedema* in the lower torso, feet, and ankles; a tropical mosquito-borne parasitic infection;

parasites collecting in *lymph nodes* and *lymph channels* result in *inflammation*

**fine motor control:** the coordination of muscles, bones, and nerves to produce small, exact movements

**finger region:** palmar hand region including the fingers distal to the palmar digital crease

**fingertip pinch:** one of the basic functional positions of the hand

**firefighter turnout gear:** firefighters' clothing and accessories, e.g. helmet, gloves, boots

**firm grip:** one of the basic functional positions of the hand

**first carpometacarpal joint:** joint between the wrist and thumb that allows you to move your thumb toward your little finger

**first metatarsophalangeal joint:** joint at the base of the big toe (*hallux*)

**fit:** how a product/garment conforms to the form and size of a body, characterized as poor fit or good fit

**fit ease:** length or width added to a body measurement and applied to a garment or product measurement; for comfort, ease of motion, and/or smooth appearance

**fit model:** a person who represents a company's ideal body type

**fitted sleeve:** tube-shaped garment segment conforming closely to the arm of the body

**fitting devices:** darts, gathers, folds, and pleats to reduce fabric dimensions from a large body area to a smaller area, e.g. skirt fullness gathered into a waistband

**flaccid:** soft

**flexicurve:** a flexible measuring device used to reproduce curves, e.g. the crotch curve (of the body)

**flexion:** bending a limb segment at a joint

**flexor digiti minimi m.:** the small *hypothenar* muscle which flexes the little finger toward the palm

**flexor digitorum brevis m.:** the small intrinsic superficial plantar foot muscle which flexes the toes

**flexor retinaculum:** a thick connective tissue ligament that bridges the space between the medial and lateral sides of the base of the carpal arch

**flexor retinaculum of the foot:** thick connective tissue ligament that bridges the space between the medial *malleolus* and the *calcaneus*

**floating ribs:** ribs 11 and 12, two shorter ribs which do not attach to the sternum

**follicle:** root of a hair, part of the *dermis* of the skin

**foot drop:** a *gait* problem associated with weakness of the *tibialis anterior m.* and characterized by an inability to fully *dorsiflex* the foot at the ankle, leading to problems of stubbing the toes and/or tripping and falling while walking

**foot flat/mid-stance:** a phase of *gait* characterized by full foot-ground contact and full weight-bearing on the foot

**foot morphology:** foot form and structure, going beyond foot length and width dimensions and arch types to include three-dimensional form
**footwear:** a product that covers any portion, or all, of the foot and/or ankle
**foramen magnum:** large opening at the base of the skull through which the spinal cord passes into the *spinal canal*
**forearm:** upper limb segment between the elbow and wrist
**forefoot:** *anterior* region extending from the proximal end of the *metatarsals* to the tips of the toes
**foreign body:** an object, large or small, which is not normally found or wanted in the body
**form:** a three-dimensional body or representation of a body
**formed elements (of the blood):** red blood cells, white blood cells, and platelets
**fourchettes:** narrow pieces of material shaping the cylinder of each finger by forming the sides of a glove's finger components
**fovea:** the center of the retina; the retinal area with the clearest vision
**Frankfort (sometimes spelled Frankfurt) plane:** the plane passing through the lower edge of the left bony eye socket and the upper margin of each ear canal; parallel to the floor or close to parallel to the floor in the normal head position in the standing person; sometimes referred to as the auriculo-orbital (ear-eye) plane
**frontal bone:** portion of the *cranium* just above the eyebrows; forms the forehead and the top front section of the skull
**frontal lobe:** one of the four anatomical regions of the *cerebral hemisphere* of the brain which contains neurons with specialized functions; lies under the *frontal bone*
**frontal plane:** see *coronal plane*
**frostbite:** formation of ice crystals in the *tissues* and *blood vessels*
**frostnip:** the transient formation of ice crystals in the *tissues*, can occur before *frostbite*
**full breath:** peak *lung* volume achieved during *inhalation* (breathing in)
**full inhalation:** see full breath
**functional finish (of textile):** chemical or physical substance applied to a textile to change or enhance function, e.g. water-resistant finish
**functional position:** partial extension of the wrist in relation to the forearm; facilitates fine motor activities in the hand
**gag reflex:** an involuntary contraction of muscles in the pharynx which occurs when the *posterior* part of the tongue or another structure in the back of the mouth or throat is touched
**gait:** an individual's manner of moving on foot; includes walking, running, skipping, etc.
**ganglia:** clusters of nerve cells located outside the CNS
**gap uniformity:** space between the head and the helmet throughout the helmet
**gastrocnemius m.:** large muscle on the back of the leg, the calf muscle, which *plantar flexes* the ankle

**gastroesophageal reflux:** backflow of stomach contents into the distal esophagus; commonly known as heartburn
**gathers:** see *fitting devices*
**GCS:** see *graduated compression stockings*
**gender:** male or female sex as determined at birth
**genitals:** see external genitalia (genitals)
**girdle:** see compression garment
**girth:** measure around the middle of something
**glabella:** the slightly indented center point between the eyebrows on the forehead, just above the bridge of the nose
**glabrous:** hairless skin as found on the palm and sole
**glands:** groups of cells or *organs* producing secretions
**glans penis:** distal tip of the penis
**GLCS:** see *Gravity Loading Countermeasure Skinsuit*
**glenohumeral joint:** joint connecting the arm to the *scapula*; key components include the saucer-shaped *glenoid fossa* of the *scapula* and the hemispheric *head* of the humerus
**glenoid fossa:** see *glenohumeral joint*
**gluteus maximus:** the large superficial muscle of the buttocks region; extends the thigh at the hip
**gluteus medius:** a smaller muscle, deep to the *gluteus maximus*, which acts to abduct the thigh at the hip
**gluteus minimus m.:** the deepest of the three glutei; another relatively small muscle which acts to abduct the thigh at the hip
**gonads:** *organs* which produce genetic material for *reproduction*
**goniometer:** a simple tool with two arms and a friction pivot point, with a 360-degree protractor; commonly used by medical professionals, along with very specific protocols, to document *range of motion* (*ROM*)
**grade rules, grading table:** table with set of incremental measurements based on differences in body size in a range of sizes, used to grade a pattern
**grading:** method of systematically increasing and decreasing measurements from the sample size pattern to fit the company's range of sizes
**graduated compression stockings (GCS):** medical compression stockings which require a prescription and provide limb compression which gradually decreases from the ankle to the knee to the thigh
**Gravity Loading Countermeasure Skinsuit (GLCS):** wearable product for astronauts developed to counteract negative effects of weightlessness in space
**great auricular nerve:** sensory nerve located below the ear and the *jaw line*
**greater occipital nerve:** sensory nerve at the back of the head which travels from the skin into and through the *posterior* neck muscles; may be compressed by eyewear elements
**greater trochanter:** the bony prominence on the *lateral* side of the upper thigh

**greater tuberosity:** a "knob" adjacent to the *humeral head* where *rotator cuff* muscles insert
**groin:** the junction between the leg and the lower abdomen; the fold between the abdomen and the thigh
**ground reaction force:** force of the floor acting on the body to oppose the effects of gravity
**gusset:** diamond-shaped piece of fabric with multi-directional stretch, a knit or woven fabric with the bias placed to stretch with the body
**gut microbiome:** all of the microorganisms living inside the lower *digestive* tract
**hallux:** the "big toe"
**hamate:** one of the distal *carpal bones* of the wrist, part of the *carpal tunnel*
**hammertoe:** a toe abnormality with *extension* of the *metatarsophalangeal joint* (MTP) and constant *flexion* of the *proximal interphalangeal joints* (PIPs) linking the distal segments of the toe, most commonly seen in the second toe
**hand-at-rest position:** relaxed hand position
**handedness:** the preference and skillfulness of using one hand instead of the other, so a person is right-handed or left-handed
**handwear:** a product that covers any portion or all of the hand and/or wrist
**hangnail:** a torn or irregularly detached *nail* cuticle
**hard palate:** firm *anterior* part of the roof of the mouth
**hazardous environment:** environment that may endanger human health and safety
**hazmat suits:** garment worn for protection from hazardous materials, chemicals, biological materials, etc.
**HDL:** University of Minnesota Human Dimensioning© Laboratory
**head (of rib):** bony prominence on a rib which articulates with two adjacent vertebral bodies
**head (of the humerus):** see *humeral head*
**headform:** manikin of the head
**hearing aid:** device, most often worn inside the ear, designed to improve sub-optimal hearing
**heart:** primary *circulatory system organ* which pumps *blood* into *arteries* and receives blood from *veins*
**heart event monitor:** a recording device worn to record heart rate and rhythms over a period of hours to days, also called a Holter monitor
**heart monitor:** see heart event monitor, heart rate monitor, heart rhythm monitor
**heart rate monitor:** a recording device worn to record number of heart beats per minute
**heart rhythm monitor:** a recording device worn to record the regularity of the heart beat
**heat dissipation:** release or loss of heat from the body via one or more *heat transfer mechanisms*

**heat transfer mechanisms:** (a) convection, (b) radiation, (c) evaporation of sweat from the skin, (d) conduction, (e) counter-current heat exchange from the arteries to veins, and (f) sublimation

**heel counter:** internal or external shoe structure, usually inflexible and firm, intended to hold the foot in place

**heel patch:** addition to heel counter for strength and cushioning

**heel sock:** internal shoe component to prevent blisters

**heel strike:** first phase of *gait;* when the body weight lands on the heel of one foot on the ground

**height:** the measure of the total standing body, from the floor to the top of the head

**helix:** the firm outer rim of the *auricle*

**helmet fit index:** a formula to determine space between the head and the inside dimensions of the helmet, to provide optimum head protection

**hemoglobin:** red blood cell component which binds to oxygen

**hemorrhoids:** dilated *veins* within the *anus* and/or in the canal above it

**high ankle sprain:** a tear in the distal segment of the *syndesmotic ligament* between the *fibula* and *tibia*

**hindfoot:** commonly called the heel

**hinge joint:** a geometric joint form, such as found at the elbow which allows *flexion* and *extension* motions

**hip depth:** distance measurement of the space formed between the front and back pants patterns which represents the space the body occupies in the pants

**hip tilt:** term commonly used in apparel design to discuss fit issues related body posture variations related to *pelvic tilt*

**Holter monitor:** a wearable heart monitoring device, also see *heart event monitor*

**hook grip:** one of the basic functional positions of the hand

**horizontal center line (of an eyeglass frame):** a line used as a reference for vertical measurements of the eyeglass frame itself

**horizontal reference line (of the face):** a horizontal line tangential to the upper margin of the lower eyelid

**hormone:** a chemical produced by cells in a *gland* or *organ* which regulates distant specific *organs* or *tissues* in the body

**human factors:** field of study, how humans react physically and/or psychologically to a product or other elements in the environment

**human testing (in vivo testing):** evaluation of a product or ensemble using real persons, as opposed to manikin testing

**humeral head:** a hemispheric structure on the proximal end of the *humerus* which fits into the *glenoid fossa* of the *scapula*

**humeroradial joint:** joint between the humerus and radius (one of the two bones of the forearm) at the elbow; acts with the *humeroulnar joint* as a hinge joint for elbow flexion and extension

**humeroulnar joint:** joint between the humerus and ulna (one of the two bones of the forearm) at the elbow; limits elbow joint extension

**humerus:** the bone of the upper arm; the longest and largest arm bone
**HUT:** hard upper torso component of a space suit
**hyaline articular cartilage:** see *hyaline cartilage*
**hyaline cartilage:** a translucent cartilage, with few fibers, commonly found in synovial joints
**hyoid bone:** bone at the base of the tongue muscle which props the entrance to the airway open
**hyperextension:** joint extension beyond the natural extended position
**hypertrophy:** enlargement in an *organ* or *tissue* at a cellular level
**hyponychium:** thickened skin under the free margin of the fingernail
**hypothenar:** referring to the fleshy mound on the ulnar side of the palm
**hypothenar eminence:** a group of *intrinsic* hand muscles on the palmar hand, located between the base of the little finger and the wrist, which act on the little finger
**hypothenar muscles:** see *hypothenar eminence*
**hypothenar region:** subdivision of the palmar region of the hand
**ideal body:** the desired body form and appearance, culture specific
**idiopathic:** of unknown cause
**idiopathic scoliosis:** a lateral spinal curvature to either side which occurs without known cause
**iliac:** referring to the *ilium,* the broad upper section of the *innominate* bone
**iliac crest:** the upper rim of the *innominate* bone of the pelvis, extending from near the spine to the outer abdomen
**iliacus m.:** a deep lower torso muscle originating on the interior aspect of the *innominate* bone, which joins with the psoas major to form the *iliopsoas m.*
**iliopsoas m.:** a deep muscle formed from the *psoas major* and *iliacus muscles,* in the lower torso which acts to bend the hip
**iliotibial tract:** a broad, flat, but tough *connective tissue* band (*tendon*) which connects the muscle belly of the *tensor fascia lata m.* to the lateral side of the tibia, below the knee; facilitates hip *abduction*
**ilium:** one of three bones which form separately in each half of the pelvic girdle of the developing fetus but which fuse into the *innominate bone* by adulthood
**immersion suit:** thermal protection garment worn for cold water rescue
**IMP:** see *incontinence management product*
**imperial system (of bra sizing):** a formula calculated in inches
**in vivo testing:** from Latin, "within the living"; testing with people
**incontinence, incontinent:** an inability to control the body's natural evacuations
**incontinence management product (IMP):** a padded underwear-like garment to collect and absorb feces and/or urine
**incontinence products:** absorbent pads, e.g. diapers to collect urine and/or feces
**incus:** the "anvil" bone of the middle ear
**index finger:** second digit of the hand; commonly the pointer finger

**indwelling catheter:** a *urinary catheter* left in the bladder for a period of time
**infectious agents:** organisms capable of causing infections; bacteria, viruses, or parasites
**inferior:** places a part lower than another structure
**inferior vena cava:** large major vein returning blood to the heart from the lower body and legs
**inflammation:** a combination of pain, swelling, and warmth in the tissues
**inframammary fold:** the crease below the breast
**infraorbital nerve:** sensory nerve below the eye between the nose and the center of the eye; may be compressed by eyewear elements
**infratrochlear nerve:** sensory nerve on the side of the nose at the bridge of the nose
**ingrown toenail:** a portion of the nail plate that punctures the soft tissue of the toe
**inhalation:** the act of breathing in; lung volume expands
**inion:** see *external occipital protuberance*
**inner ear:** portion of the ear internal to the *tympanic membrane* and *middle ear;* contains *sensory* cells to convert the mechanical energy of sound into electrical impulses for transmission to the brain; contains three *semicircular canals,* structures essential to the maintenance of balance
**innervated:** supplied with nerves
**innominate bone:** the "hip" bone of the physically mature body; comprised of three bones present in infancy and childhood (the *ilium, ischium,* and *pubis*) which mature and fuse into a single structure in an adult; a major component of the *bony pelvis*
**inseam:** inner thigh seam joining front and back legs of pants
**insertion (of a muscle):** the muscle end which is more movable during muscle contraction
**insole:** inner layer of the 3-layered sole of a shoe
**instep:** the arched middle portion of the foot which includes both the mid-foot bones and proximal forefoot bones
**insulin:** a hormone secreted by specialized cells of the pancreas
**integumentary system:** layers of skin cells plus other specialized cells (hair and sweat glands) which cover the body surface
**intercostal muscles:** three layers of muscles spanning the spaces between adjacent ribs
**intermediate hair:** hair found on the arms and legs which is intermediate in length and shaft size
**intermuscular septa:** *connective tissue* divisions between functional groups of muscles
**internal condom:** see *female condom*
**internal jugular veins:** large deep neck veins returning blood toward the heart from the head and brain
**internal oblique m.:** abdominal wall muscle in the middle of three muscle layers

**internal rotation:** turning a limb toward the midline of the body at the shoulder or at the hip

**interosseous m. (pl. interossei):** small deep muscles between the *metacarpal bones* (hand) and *metatarsal bones* (foot) which *abduct, adduct,* and help to stabilize the fingers and toes, along with the *lumbricals* in the hand

**interosseous membrane:** a sheet of tough, dense connective tissue between bone shafts; seen between the *radius* and *ulna* in the forearm and between the *tibia* and *fibula* in the leg

**interphalangeal (IP) joint:** joint linking a phalanx to a neighboring phalanx

**interpupillary distance:** distance between the centers of the *pupils* of the eyes

**intervertebral discs:** soft fibrocartilage structures which lie between the vertebral bodies of all the vertebrae from the upper neck to the pelvis

**intrinsic muscles:** muscles which act on a specific body part and are located within that body part; intrinsic hand and foot muscles are found in the hand and foot respectively and generate hand/finger motions and foot/toe motions

**inversion:** descriptor of the body motion of turning inward, used only with foot/ankle motions

**invert:** to turn inward

**IP joint:** see *interphalangeal joint*

**iris:** colored circular disk in the center of the eye

**ischial tuberosities:** the "sit bones," the bony contact points between the lower torso and a horizontal surface

**ischium:** one of three bones which form separately in each half of the pelvic girdle of the developing fetus, but which fuse into the *innominate bone* by adulthood

**jaw line:** an arbitrary division between the face and the neck at the jaw

**jock strap:** see athletic supporters

**joint:** where two or more bones make contact

**joint capsule:** a fibrous bubble around a *synovial joint,* containing *synovial fluid*

**keratin:** a tough fibrous protein made by skin cells

**keratinocytes:** skin cells which make the protein keratin

**key (control) measurements:** the basic measurements used to develop a sizing system

**key points:** points on a pattern or product which relate to *landmarks* on the body

**kidney:** an organ which filters waste products, including excess water, from the blood to make *urine*; the primary *organ* of the *urinary system*

**kinematic chain:** overlapping limb segments in series secured by joints, where motion at one joint results in predictable motion at the others

**kinetic chain:** see *kinematic chain*

**knuckles:** the joints between the small bones of the fingers and toes
**kyphosis:** an unusual or increased *posterior* convexity of the spine
**labia (minora and majora):** female external genital structures located along each side of the vaginal opening
**lacrimal (tear) bones:** the smallest bones of the skull, which form the medial part of the eye socket
**lacrimal apparatus:** the drainage system for the tears
**lacrimal sac:** a dilated section of the *nasolacrimal duct* at the medial corner of the eye
**lactation:** the formation and secretion of milk
**lactation mastitis:** a painful breast condition related to milk retention when a woman starts breast-feeding
**lactiferous ducts:** milk ducts originating in the *mammary gland* and opening in the nipple
**landmarking process:** sequential process to accurately and consistently locate body *landmarks* for an anthropometric procedure: assess the body for visible key characteristics, palpate for key locations, approximately locate other points
**landmarks:** reference points on the body which relate to product key points, based on locating features on the body surface which relate to anatomical structures
**Langerhans cells:** a cell type located in the *epidermis* which acts to fight off microbes such as bacteria
**laryngeal:** referring to the *larynx*
**laryngeal prominence:** portion of the *trachea* that projects anteriorly, also called Adam's apple
**laryngopharynx:** the lowest region of the *pharynx*, found behind the voice box
**larynx:** voice box, a specialized portion of the windpipe (*trachea*) that projects anteriorly in the mid-neck; the structure that allows us to produce sound
**last:** a three-dimensional form which serves as the basis for developing footwear size and form
**lateral:** refers to a structure being further away from the median plane than another structure, may also refer to a side view of the person
**lateral (fibular) collateral ligament (at the knee):** a ligament bridging the *lateral* aspect of the knee, between the *femur* and *fibula*
**lateral ankle ligaments:** stabilizing connective tissue structures at the lateral aspect of the ankle; three relatively separate ligamentous structures: *anterior* and *posterior talofibular ligaments*, and the *calcaneofibular ligament*; variably susceptible to *sprain* injuries (anterior talofibular ligament most often injured)
**lateral deviation:** side-to-side movement of the *mandible*
**lateral epicondyle:** with the body in the *anatomical position*, the protuberance at the lateral (thumb side) and distal (elbow) end of the *humerus*

**lateral femoral cutaneous nerve of the thigh:** sensory nerve innervating the proximal lateral and anterolateral aspect of thigh; also see *meralgia paresthetica*

**lateral malleolus:** protrusion of the distal *fibula* at the lateral ankle

**lateral meniscus (pl. menisci):** a partially movable crescent-shaped cartilage spacer between the *femur* and *tibia* closest to the *lateral (fibular) collateral ligament*

**lateral rotation:** rotating a long bone away from the midline of the body at the shoulder or hip joint

**latissimus dorsi:** a large superficial back muscle running between the arm near the *glenohumeral joint* and the middle and lower spine and *iliac crest;* executes many arm motions at the *glenohumeral joint*

**LCVG:** see *liquid cooling and ventilation garment*

**leg:** in anatomical terms, the portion of the lower limb between the knee and the ankle

**lens height:** common eyeglass frame measurement

**lens width:** common eyeglass frame measurement

**lesser occipital nerve:** sensory nerve in the upper neck between the ear and the back of the head; may be compressed by eyewear elements

**lesser trochanter:** prominence on the medial section of bone at the top of the *femoral shaft;* attachment site for *iliopsoas m.*, which flexes the hip

**ligament:** an important *tissue* of the *skeletal system;* a strong band of *connective tissue* which links and stabilizes bones across a joint

**limbs:** arms and legs

**linea alba:** a strong linear connective tissue structure in the *median* line of the abdomen, seen between the muscle bellies of the *rectus abdominis m.*

**linear measurement:** over-body-contour distance measurement

**liquid cooling and ventilation garment (LCVG):** close-fitting garment which removes excess body heat from the enclosed space suit

**lobe (of lung):** a division of the R or L lung, served by a *bronchus*

**long head (of the lateral hamstring):** see *biceps femoris*

**long plantar ligament:** connects the *calcaneus* and *cuboid* to the proximal ends of the three lateral *metatarsals;* extends along most of the lateral longitudinal arch

**long-waisted:** The upper torso is long in proportion to the lower body

**loose grip:** one of the basic functional positions of the hand

**lordosis:** an unusual or increased *anterior* convexity of the spine, most often in the neck or low back

**lower torso, lower torso region:** portion of the torso from the top of the *bony pelvis* to and including the external genitals (but excluding any part of the legs)

**lumbar:** referring to the section of the spine and back between the lowest rib and the *iliac crest* of the pelvis

**lumbricals:** small deep muscles between the *metacarpal bones* (hand) which help to stabilize hand and finger movements

**lumpectomy:** removal of a mass from breast tissue including a tumor or cyst plus a margin of breast tissue around the mass
**lunate:** one of the proximal *carpal bones* of the wrist; moon-shaped
**lung:** the primary *organ* of the *respiratory system*
**lymph:** a collection of excess tissue fluids, leaked plasma proteins, and remnants of cellular decomposition and infection
**lymph capillaries:** the first tiny, thin-walled, segments of the lymphatic drainage vessels
**lymph node:** collection of organized lymphatic *tissue* which acts as a filter for *lymph* and also produces *lymphocytes*
**lymphadenitis:** *lymph node* inflammation
**lymphatic:** pertaining to *lymph* or vessels carrying lymph
**lymphatic system:** inter-related anatomical structures of the body which helps drain surplus tissue fluid into the bloodstream and fight infections; major components include *lymph capillaries, lymph channels, lymph nodes,* and the *spleen*; structurally, a "blunt-ended" linear system
**lymphedema/lymphoedema:** chronic and often progressive swelling of one or more parts of the body if *lymph* drainage pathways are blocked or interrupted
**lymphocytes:** specialized white blood cells active in the immune system; found in *lymph nodes*, the *spleen*, *blood*, and *lymph*
**magnetic resonance imaging (MRI):** a medical body imaging technique which shows details of internal anatomical structures and functions; does not use x-rays but rather magnetic energy to capture the images
**MAGs:** see *maximum absorbency garments*
**malignant melanoma:** dangerous skin cancer caused by transformed *melanocytes*
**malleoli (pl. of malleolus):** the protruding lower ends of the *tibia* and *fibula* of the leg
**mallet finger:** see *baseball finger*
**mallet toe:** a toe with a mallet-like appearance as the DIP linking the most distal phalanges of the toe are constantly flexed, often to 90°
**malleus:** the "hammer" bone of the middle ear
**mammary gland:** milk producing gland of the breast, a modified *apocrine gland*
**mandible:** U-shaped lower jawbone; a separate bone which articulates with the skull at the *TMJ*; holds the lower teeth
**manikin:** simulated body form
**manubrium:** the upper section of the *sternum*
**masseter m.:** powerful muscle beneath the lateral surface of the face which, along with the *temporalis* muscle moves the *mandible*
**mastectomy:** removal of the breast and a variable amount of adjacent tissue
**mastication:** chewing
**mastoid process:** bony prominence of the skull located just behind the earlobe on each side of the head

**maternity:** related to wearable products used during pregnancy and shortly after childbirth
**maxilla (pl. maxillae):** one of the paired stationary upper jawbones, which together form the U-shaped structure holding the upper teeth
**maximum absorbency garments (MAGs):** wearable products used in space travel to collect urine and feces
**maximum bladder volume:** the total amount of *urine* a person can tolerate holding in the *bladder*
**mechanical counter-pressure (MCP) suit:** space suit design, tight-fitting, applies pressure by mechanically squeezing the wearer inside a skin-tight suit
**mechanics:** the science of the actions of forces and energy in the physical environment
**mechanotransduction:** the process by which mechanical signals affect the form and actions of all parts of the body
**medial:** refers to a structure closer to the median plane than another structure
**medial (tibial) collateral ligament (at the knee):** a ligament bridging the *medial* aspect of the knee, between the *femur* and *tibia*
**medial epicondyle:** with the body in the *anatomical position*, the protuberance at the medial (little finger side) and distal (elbow) end of the *humerus*
**medial longitudinal arch:** the primary *arch* of the foot; located on the *medial inferior* aspect of the foot between the *calcaneus* and the big toe
**medial malleolus:** protrusion of the distal *tibia* at the medial ankle
**medial meniscus (pl. menisci):** a partially movable crescent-shaped cartilage spacer between the *femur* and *tibia* closest to the *medial (tibial) collateral ligament*
**medial rotation:** rotating a long bone toward the midline of the body at the shoulder or hip joint
**median nerve:** major nerve of the forearm and hand; carrying *motor, sensory,* and *autonomic nerve* fibers; travels through the arm to the forearm; vulnerable to compression at the *carpal tunnel* and from wearable products on the *anterior* forearm, just below the elbow crease
**median plane, or midsagittal plane:** the midline plane through the long axis of the body dividing the body into equal left and right halves
**medical device:** product for diagnostic and/or therapeutic purposes, regulated by the U.S. Food and Drug Administration (FDA)
**medical product, medical wearable product:** product for diagnostic and/or therapeutic purposes, may or may not be regulated by the U. S. Food and Drug Administration (FDA)
**melanin:** skin pigment in the epidermis, produced by *melanocytes*
**melanocyte:** a cell type in the epidermis which produces the dark skin pigment melanin and, thus, determines skin color
**menopause:** permanent cessation of menstruation
**menstrual:** referring to *menstruation*
**menstrual cup:** reusable conical cup of silicone, rubber, or thermoplastic elastomer inserted in the vagina to collect menstrual flow

**menstrual cycle:** the periodic relatively consistent changes in the female reproductive structures associated with *menstruation*

**menstrual flow:** blood and endometrial tissue sloughed through the cervix

**menstrual management products (MMPs):** products designed to absorb or collect menstrual flow including *sanitary pads* (also called *napkins* or *towels*) and *panty liners* (or *shields*); and products inserted into the vagina—*tampons* and *menstrual cups*

**menstruation:** the monthly shedding of the uterus' lining in non-pregnant women

**mental nerve:** sensory nerve just below the corner of the mouth

**meralgia paresthetica:** sensory disturbance with numbness/pain caused by compression of the *lateral femoral cutaneous nerve of the thigh*, most often in the *groin*

**Merkel cells:** specialized light touch receptors of the *sensory nervous system*; found in high touch body areas (fingertips) at the junction of the *epidermis* and the *dermis* of the skin

**mesentery:** a continuous connective tissue structure which acts to support the intestine in the abdomen

**metacarpals:** the bones of the palm of the hand and the base of the thumb

**metacarpophalangeal joints (MCPs, sometimes MPs):** joints between the *metacarpal bones* and *phalanges* in the hand and the fingers, and, in the thumb, the second joint from the thumb tip

**metatarsal heads:** the distal ends of the *metatarsals*

**metatarsals:** five long bones of the foot, lying between the distal *tarsal* bones and the toes, above a large portion of the sole of the foot

**metatarsophalangeal joints (MTPs):** joints between the metatarsal bones and *phalanges* in the foot and the toes, and, in the big toe, the second joint from the toe tip

**metric system (of bra sizing):** direct use of the underbust measurement in centimeters as the bra band size designator

**middle ear:** the skull cavity just beyond the ear drum (*tympanic membrane*) containing the opening of the *eustachian tube* and three tiny bones—*malleus, incus, stapes*—which transmit sound energy to the nerve center in the *inner ear*

**middle finger:** the long finger, third digit of the hand

**midfoot:** region between the hindfoot and the metatarsals which translates the energy of the foot hitting the ground into motion

**midpalmar area:** subdivision of the palmar region, further subdivided into the *distal midpalm* and *midpalmar triangle*

**midpalmar triangle:** central triangular subdivision of the *midpalmar area* of the hand

**midsagittal plane:** see *median plane*

**midsole wedge:** midsole variation of a shoe, thinner in the front and thicker at the heel, typically die cut

**midsole:** middle layer of the three-layered sole of a shoe

**mid-torso:** middle region of the body encompassing some of the upper and lower torso
**milk let-down:** the milk ejection reflex
**Minimalist Index:** rating scale developed to evaluate effects of wearing minimalist shoes
**minimalist shoe:** type of athletic shoe with minimal interference with the natural movement of the foot, highly flexible with flat sole, and absence of motion control and stability features
**Misses sizes:** ready-to-wear size label category for women, originally developed in the U.S. in the 1930's based on age and bust size, typical range is Misses 6, 8, 10, 12, 16, 18
**mixed urinary incontinence (mixed UI):** combined *urgency* and *stress incontinence*
**MMP:** see *menstrual management products*
**model size:** prototype to fit to one body representing all product users, other sizes are developed by decreasing or increasing dimensions from the model size
**modularity, modular sizing:** parts of a wearable that can be interchanged
**mons pubis:** the mound of fatty tissue overlying the pubic symphysis in women
**morphology:** study of form (structure)
**motion capture system:** equipment to record movements of a person (or animal or mechanized object), recorded in digital format; used for animation, analysis, and application of motion; also mo-cap or mocap
**motor function:** movement generated by muscles
**MRI:** see *magnetic resonance imaging*
**mucous membranes:** specialized moist skin tissues lining body cavities in contact with air
**mucus:** thick protective fluid which helps to prevent infection of moist skin tissues
**mudguard:** addition to a shoe toe box to protect the shoe from soil and abrasion
**multi-probe method:** method to measure a distance along a plane; widths, depths, and heights
**muscle:** a specialized *tissue* with contractile fibers or cells supported by *connective tissue*, capable of moving *skeletal* components or generating flow through tubular structures in various body systems; structurally, groups of *muscle fascicles* bound together by more connective tissue
**muscle atrophy:** muscle wasting or deterioration from any cause
**muscle belly:** a "meaty" section of a muscle made of muscle fibers, which is relatively thicker than the areas of origin or insertion and often located at the muscle center
**muscle fascicle:** a level of structural *skeletal muscle* organization; groups of *muscle fibers* surrounded by an outer layer of fine connective tissue strands

**muscle fiber:** a level of structural muscle organization; a collection of *myofibrils* with associated non-contractile supporting tissues

**muscle power:** the rate at which *muscle work* is done; the product of the force of muscle contraction and the velocity of muscle contraction; e.g. running 100 m (328 ft) takes more *muscle power* than walking 100 m

**muscle tone (muscle tonus):** term describing a normal responsiveness of muscles to nervous system input, not to muscle size or shape

**muscle work:** the amount of effort to complete a motor task; the product of the force produced by a muscle's action and the distance the muscle shortens or lengthens while producing the force; e.g. lifting a full water bottle takes more *muscle work* than lifting an empty water bottle

**muscular:** referring to muscles

**muscular dystrophy:** an inherited chronic disease of *muscle* which causes significant muscle weakness

**muscular system:** inter-related muscle cells, tendons, and other *connective tissue* structures which act to move the *skeletal* structures of the body

**musculovenous pump:** powerful leg and thigh muscles which empty deep leg veins, pushing blood back to the heart

**myoelectric activation:** a technologically advanced method of controlling *prosthetic* motion; amplified electrical signals from muscles above an amputation activate an electronic device to move a distal prosthetic limb element

**myofibril:** structural level of muscle organization; a collection of *myofilaments*

**myofilaments:** threads of protein capable of contracting; unit of *muscle* structural organization

**myosin:** specialized protein threads (*contractile protein myofilaments*) in muscle capable of contracting; see also *actin*

**nail body:** the section of a fingernail or toenail which extends distally from the *nail fold* to the free margin

**nail fold:** a skin fold at the proximal end of a nail

**nail plate:** the living, growing, proximal end of a nail; fits into a skin fold named the *nail fold*

**nails (toenails and fingernails):** horny epidermal *keratin*-containing tissues composed of plates of compacted *keratinocytes*

**nasal:** pertaining to the nose

**nasal cavities:** paired spaces where air flows when breathing through the nose; the spaces extend internally from the openings at the end of the nose

**nasolacrimal duct:** commonly the tear duct, drains excess tears from the eye into the nose

**nasopharynx:** cavity in front of the upper spine behind the nasal cavity; one section of the *pharynx*

**natural waist circumference:** smallest mid-torso circumference as viewed either anteriorly or laterally, or where a belt settles naturally

**navel (umbilicus):** belly button—so called because it is a small round feature at the belly or *anterior* mid-torso—the residual physical evidence of the umbilical cord, the connection between a fetus in the uterus and the mother

**navicular prominence:** portion of the navicular bone which projects on the medial side of the middle of the foot

**neck base circumference:** measured from the center back neck *landmark*, to the right side neck point landmark, to the center front base of neck landmark, to the left side neck point landmark, and returning the tape measure to the center back neck landmark; thus, joining four landmarks

**negative ease:** amount subtracted from the *waist circumference* measurement to determine the product waistline measurement

**negative heel shoes:** shoes in which the heel is closer to the ground than the toes, reversing the incline of the foot compared to shoes with other heel types

**nerve:** a bundle of nerve fibers held together by *connective tissue*, found outside of the skull and spinal column

**nerve fiber:** a long branch from a *neuron*, a nerve cell

**nerve signal:** electrical impulse generated by chemical changes and transmitted through structures of the *nervous system*

**nervous system:** the inter-related anatomical structures (CNS structures—brain, *brainstem*, and *spinal cord*; PNS structures—*spinal nerves, spinal ganglia, plexuses, peripheral motor* and *sensory nerves, sensory receptors*; and *autonomic* (motor) *nerves*) which direct and coordinate functions in other body *systems*

**neural plasticity:** capacity to adapt and use alternate connections to restore lost functions

**neuron:** cell type found in the brain as well as in other nervous system structures which transmits chemical and electrical messages to control a wide range of body functions

**neurophysiological:** pertaining to the *physiology* (function) of the *nervous system*

**nipple:** most prominent apex of the breast; has a visible coloration difference on the nude female or male body

**non-freezing cold injury:** peripheral cold injury without tissue freezing

**nostrils:** external openings of the nose; entry/exit orifices of the *respiratory system*

**nucleus (pl. nuclei):** the control center of a cell

**numbness:** loss of sensation

**OAB:** see *overactive bladder*

**obstructive sleep apnea (OSA):** collapse of the airway in the oropharynx during sleep

**occipital bone:** *cranial* bone inferior to the *parietal bones*; forming the back and much of the bottom of the skull

**occipital lobe:** one of the four anatomical regions of the *cerebral hemisphere* of the brain which contains neurons with specialized functions including vision; lies next to the *occipital bone*

**occipitofrontalis m.:** contracts to raise the eyebrows and horizontally wrinkle the skin of the forehead

**odontoid process:** see *dens*

**olecranon fossa:** an indentation at the distal end of the *humerus*

**olecranon process:** part of the *ulna* that fits into the *olecranon fossa* of the *humerus* when the arm is straightened, and which can be felt as the tip of the elbow when the arm is bent

**olfactory cells:** nerve cells responsible for the sense of smell

**olfactory nerves:** nerves conveying information from the *olfactory cells* to the brain

**opponens pollicis m.:** muscle of the *thenar eminence* of the palm which moves the thumb across the palm to *oppose* the other fingers, facilitating pinching activities

**oppose or opposition (of the thumb):** the action which allows the thumb to move to contact the tips of other fingers for pinching and grasping activities

**oral cavity:** the mouth, a major superficial skull cavity

**orbicularis oculi m.:** the squinting muscle threaded through not only the eyelids, but also the other skin overlying the *orbits*

**orbicularis oris m.:** muscle encircling the mouth which provides the ability to pucker the lips or form an O-shape

**orbits:** see *eye sockets*

**organ:** a separate, single structure of the body which performs a specific function

**origin (of a muscle):** the fixed or mostly stationary end of a muscle

**oropharynx:** the mouth and the space directly *posterior* to the mouth; the section of the *pharynx* below the *soft palate* and above the *larynx*

**Ortho-fabric:** Kevlar™ aramid rip-stop grid fabric developed for the outer layer of space suits, for mechanical protection (deflecting micrometeoroids)

**orthosis, orthotic:** mechanical support designed to assist with body functions; i.e. a *splint* or *brace* to support or assist a weak or deformed body part; from Greek, meaning to set straight or make straight

**orthostatic hypotension:** positional low blood pressure

**OSA:** see *obstructive sleep apnea*

**osteoarthritis:** degeneration of bone and cartilage in a joint; commonly called "wear and tear" arthritis

**osteoporosis:** loss of bone mass with associated bone structural changes, decreased bone strength, and increased fracture risk; more commonly seen in post-menopausal women

**ostomy:** an artificial opening in the body wall serving as a new exit for elimination of waste

**outseam:** seam extending from waistline to the desired pants leg length, joining front and back pieces of the pants
**outsole:** bottom layer of the three-layered sole of a shoe
**ova (pl. of ovum):** the structural vehicle for the female's genetic material in reproduction; eggs
**ovaries:** paired female *gonads;* site of *ova* production; found deep in the pelvis
**overactive bladder (OAB):** an urgent need to urinate
**overbite:** the amount of overlap of the upper teeth in front of the lower teeth when the mouth is in an unstrained closed position
**ovulation:** maturation and release of an *ovum* from an *ovary*
**palm (of the hand):** the hairless, smooth, wrinkled, *volar* surface of the hand
**palmar digital crease:** crease of the palm formed at the junction between the palm and the base of the four fingers with MP joint flexion
**palmar region:** hand region between the *palmar digital creases* and the *wrist crease*
**palmar surface:** the palm and *flexor* surfaces of the fingers and thumb
**palpation:** a medical term used to describe the method of using the sense of touch to identify or examine internal structures
**pancreas:** a major *organ* of the *digestive system* located near the *posterior* body wall of the abdomen which secretes enzymes important in the digestion of fats, proteins, and carbohydrates
**pantoscopic angle or tilt:** the amount of tilt of the front of the eyewear frame
**panty liner (panty shield):** thinner, lighter form of sanitary pad, worn to absorb light menstrual flow, vaginal secretions, or small amounts of urine
**pantyhose:** a product covering lower torso and legs; typically uses weight and height as the bases for size selection
**paraplegia:** paralysis of the torso and legs (loss of muscle control) frequently accompanied by loss of sensation in the torso and legs and impairment of *sympathetic* and *parasympathetic nervous system* components of the *autonomic nervous system* including bladder, bowel and sexual/reproductive functions
**parasagittal:** parallel and lateral to the *median* plane of the body
**paraspinal muscles:** muscles running along the *spinous processes* of the spine, stabilizing the axial skeleton and supporting upright posture
**parasympathetic:** referring to a division of the *autonomic nervous system* which facilitates calming/restorative body processes, like slowing the heart rate, contracting intestinal and urinary *smooth muscle* and relaxing the sphincters promoting/allowing *urination* and *defecation*
**parietal bones:** paired *cranial* bones forming the *posterior* top of the skull, posterior to the *frontal bone*, superior to the *occipital bone*
**parietal lobe:** one of the four anatomical regions of the *cerebral hemisphere* of the brain; contains neurons with specialized functions; lies beneath the *parietal bone*

**passive marker system:** motion capture system using reflective markers placed at the joint intersections, cameras record the motion

**patella:** knee cap

**patellar ligament:** a tough ligament joining the *distal* aspect of the *patella* to the *tibial tuberosity;* a continuation of the *tendon* of the *quadriceps m.* which attaches to the *proximal* aspect of the *patella*

**patellar tendon reflex:** an automatic response elicited by tapping on the *patellar* ligament distal to the *patella;* the knee jerk elicited in routine medical exams

**patellofemoral joint:** a part of the knee joint; the gliding joint where the *patella,* suspended between the *tendon* and *ligament* connecting the *quadriceps* (the large muscle on the *anterior* thigh) to the *tibia,* tracks over the anterior distal *femur*

**patellofemoral pain (PFP):** *anterior* knee pain, around the *patella;* often more noticeable with activities requiring knee *flexion* or *extension;* specific cause(s) and treatment(s) are unclear at this time

**pattern:** two-dimensional interpretation of the three-dimensional body surface

**pectoral:** referring to a structure in or on the chest

**pectoral (shoulder) girdle:** the bones (*clavicle* and *scapula*) supporting the upper limb and connecting the arm to the *rib cage* (and, therefore, indirectly to the spine) at the upper end of the *sternum*

**pectoralis m., pectoralis major m., pectoralis minor m.:** *anterior* chest muscle(s) which *protract* the shoulder—move the arm/shoulder mechanism on the *torso; pectoralis major*—a large, superficial, fanlike muscle which covers much of the upper anterior *rib cage;* forms the *anterior* wall of the arm pit and *inserts* on the *humerus* (upper arm bone), *pectoralis minor*—a deep, small muscle between the *pectoralis major* and the *rib cage*

**pedometers:** step counting devices

**pelvic diaphragm:** sling of muscles spanning between the pelvic bones, above the *perineal region,* and supporting the internal organs

**pelvic girdle:** the R and L hip bones (*innominate bones*) joined at the *pubic symphysis* in the front of the body, the lower torso equivalent of the *pectoral girdle* (*clavicle* and *scapula*)

**pelvic organ prolapse:** varying degrees of loss of *connective tissue* support for the *bladder, rectum,* or *uterus* allowing any or all of them to sink down into and/or out of the *vagina*

**pelvic outlet:** opening in the *bony pelvis* allowing the *urinary, digestive,* and *reproductive* systems to pass to the body's exterior for expulsion of waste products or birth of a child

**pelvic tilt:** deviation of the pelvis from a neutral standing posture as measured in a sagittal plane

**pelvis:** bowl-shaped cavity of bone, muscle, and soft tissues to hold and cradle internal organs of the *urinary, digestive,* and *reproductive* systems, bounded by the *anterior* abdominal wall

**penis:** the external male reproductive organ
**perception:** recognition of a physical sensation; awareness of pressure, touch, pain, etc.
**performance characteristics (of materials):** flexibility or stability, heat retention or dissipation, resistance to impact, and more
**perineal region:** skin and soft tissue structures between the legs including external *genitals* and *anus*; lying in a more or less transverse plane
**periodontal ligaments:** strong *connective tissue* structures which anchor teeth in their sockets
**PAD:** see *peripheral arterial disease*
**peripheral arterial disease (PAD):** a chronic, often times progressive condition, caused by fatty deposits in the *arteries* which can restrict blood flow
**peripheral nerve:** a bundle of nerve fibers held together by *connective tissue*, found outside of the skull and spinal column
**peripheral nervous system:** nerve tissues connecting the brain and spinal cord to the rest of the body, outside of the skull and spinal column
**peristalsis:** intermittent *smooth muscle* contractions and relaxations
**personal protective equipment (PPE):** body covering to protect from natural or man-made environmental risks; level of protection varies relative to the hazards in the environment
**perspire:** to sweat
**pes cavus:** foot deformity with a noticeably high arch that does not flatten when bearing weight; may be congenital or *idiopathic*
**pes planus:** flat foot, a deformity of the *medial longitudinal arch*—the arch touches the ground, frequently treated with an *orthotic*
**pessary (pl. pessaries):** wearable *medical device* inserted into the *vagina* to help stabilize and support the *bladder*, *uterus*, and/or *rectum*
**PFP:** see *patellofemoral pain*
**phalangeal dorsum:** top of a finger or toe
**phalanges:** the bones of the thumb, fingers, and toes
**pharyngotympanic tube:** see *eustachian tube*
**pharynx:** the space lying behind the nasal and oral cavities and behind the larynx in the neck; commonly called the throat
**pheromones:** body chemicals produced by *apocrine glands* which communicate between persons and modify behavior
**phrenic nerve:** a *PNS* structure essential for *respiratory* function; arises from middle *cervical spinal nerve roots* and *innervates* the *diaphragm*
**physiology:** a branch of biology that deals with the functions and activities of life
**pilot study:** small scale preliminary study, to test methods, materials, cost, time, participant understanding before conducting the final study
**pinch callus:** an area of thickened skin caused by friction or pressure found on the medial side of the *hallux*

**Pingvin:** sometimes called the "Penguin" suit worn by Russian cosmonauts to counteract negative effects of weightlessness, see *mechanical counter-pressure (MCP) suit*

**pinky finger:** the little finger, the smallest finger of the hand; 5th finger

**pinna:** another name for the ear, from a word referring to its shape meaning "wing"

**PIP:** see *proximal interphalangeal joint*

**pisiform:** one of the proximal *carpal bones* of the wrist; part of the *carpal tunnel*; pea-shaped

**pitch:** slope of the back crotch curve

**pivot joint:** a specialized geometric joint such as the joint found at the elbow between the *humerus* and *radius* and between the radius and *ulna* which allows the radius to rotate around its long axis

**placenta:** a specialized structure of the *endometrium* in the *uterus* which nurtures the fetus and connects it to the mother

**planar:** along a plane; two-dimensional

**planar measurements:** distances measured along a plane of the body; width, depth, or height between two or more *landmarks*

**plane joint:** a geometric joint form in which complementary slightly convex/concave joint surfaces slide across each other; e.g. the joints between the *articular processes* of the spine

**plantar:** relating to the sole of the foot

**plantar aponeurosis:** thicker central portion of the *plantar fascia*

**plantar calcaneonavicular ligament:** sometimes called the spring ligament, it supports the highest portion of the *medial longitudinal arch*

**plantar fascia:** a thin sheet of tissue toward the bottom of the foot

**plantar fasciitis:** inflammation of the *plantar fascia*

**plantar flexion:** flexion toward the *plantar* aspect of the foot (toward the floor)

**plasma:** liquid portion of blood

**platysma m.:** broad thin sheet of muscle, originating at the *jaw line* and running down to and over the collar bone

**pleats:** see *fitting devices*

**plexus:** an interlacing network of *motor, sensory,* and *autonomic nervous system* fibers from spinal nerves giving rise to specifically named peripheral nerves

**PLSS (portable life support system):** life support backpack worn with the space suit; contains oxygen and water supplies, communication equipment, etc.

**PNS:** see *peripheral nervous system*

**podoconiosis:** a non-communicable tropical disease found in economically disadvantaged populations, which occurs when silica particles penetrate into bare feet of workers exposed to volcanic soils; another cause of lymphedema

**pointer finger:** *index finger,* second digit of the hand

**pollex:** thumb

**polyhedral:** referring to a polyhedron
**polyhedron:** a complex three-dimensional solid, whose faces are polygons
**polymers:** materials composed of repeating chemical units
**POP:** see *pelvic organ prolapse*
**popliteal artery:** continuation of the *femoral artery,* name changes at the back of the knee
**posterior:** see *dorsal*
**posterior auricular artery:** artery of the lateral scalp, lying next to the *posterior auricular nerve*
**posterior auricular nerve:** motor nerve running to the muscles of the scalp behind the ear
**posterior chamber:** the space between the colored central ring of the eye (*iris*) and the eye's lens
**posterior talofibular ligament:** see *lateral ankle ligaments*
**posterior tibial artery:** continuation of the *popliteal artery* in the *posterior medial* anatomical *leg*
**postural orthostatic tachycardia syndrome (POTS):** episodic lightheadedness and fainting related to both circulatory system and nervous system changes
**posture:** the bearing or carriage of the body
**POTS:** see *postural orthostatic tachycardia syndrome*
**power grip:** one of the basic functional positions of the hand
**PPE:** see *personal protective equipment*
**precision grip:** one of the basic functional positions of the hand
**precision handling grip:** one of the basic functional positions of the hand
**prepuce:** the fold of skin covering the uncircumcised tip of the penis (*glans penis*)
**procerus:** muscle which runs over the bridge of the nose, between the eyebrows, and produces horizontal wrinkles on the nose; activated when you frown
**product characteristics:** degree of fit, level of comfort, coverage of the body, use
**product components:** see *wearable product components*
**product crotch:** the product section covering the *anatomical crotch*
**product fit:** see *fit*
**product structure:** relationship of parts or elements in a product
**product waistline measurement:** *waist circumference* with *ease* added or subtracted, depending on the product purpose
**product waistline:** see *waistline*
**pronation:** specialized movement of the hands and/or feet; positions the hand palm-down or raises the lateral edge of the sole of the foot
**prone:** horizontal body position, face-down
**proprioception:** the sense combining position in space, balance, and/or movement in the muscular and skeletal systems
**prostate gland:** an internal male reproductive structure

**prosthesis or prosthetics:** wearable product(s) to replace body parts
**prototype:** a test product
**prototyping:** process of developing a prototype
**protraction:** forward motion of a specific body part; action occurs at the articulations of the jaw to the skull, the skull to the neck, and with arm/shoulder movement on the *torso*
**proximal:** used in reference to the limbs (arms, legs); refers to a structure being closer to the body's median plane or to the attachment of the limb to the torso than another
**proximal interphalangeal joint (PIP):** joint between the first two *phalanges* of a finger or toe; digit joint closest to the hand or foot
**proximal radioulnar joint:** joint between the *radius* and *ulna* at the elbow
**proximal tarsal group (of foot bones):** *calcaneus* and *talus*
**psoas major m.:** a large muscle found deep inside the lower torso which joins with the *iliacus* muscle to form the *iliopsoas* muscle
**psoas minor m.:** a small muscle deep inside the lower torso which helps tilt the *bony pelvis*
**ptotic:** abnormally sagging—in reference to body structure; e.g. drooping breasts
**puberty:** the human developmental stage characterized by sexual maturation; evident in males with voice change and beard growth, in females with the onset of *menstruation*
**pubic bone:** one of three bones which form separately in each half of the pelvic girdle of the developing fetus, but which fuse into the *innominate bone* by adulthood
**pubic symphysis:** the *fibrocartilaginous* joint between the R and L *innominate* bones at the *median anterior inferior* point of the *abdomen*/lower torso above the *external genitals*
**pulmonary arteries:** *arteries* branching off the *pulmonary trunk* which carry oxygen-depleted blood to the *lungs*
**pulmonary circulation:** *arteries* and *veins* which convey oxygen-depleted blood from the right side of the heart to the *lungs* to take in oxygen and release carbon dioxide and then return the oxygen-replenished blood back to the left side of the heart
**pulmonary trunk:** large *artery* which leaves the R *ventricle* to carry blood to the *lungs*
**pulmonary veins:** four *veins* which return blood from the *lungs* to the L *atrium*
**pulse:** the regular recurrent effect of the beating heart, felt throughout the body as the heart pumps blood through the arteries
**pulse oximetry:** a non-invasive method for monitoring oxygen in the blood
**pulse points:** locations on the body where the "pump" action or pulse of blood from the heart, through an artery, can be detected
**pulse qualities:** regularity, strength, and variability of the *pulse* which serve as an indicator of *circulatory system* function

**pupil:** dark central area of the front of the eye; the opening at the center of the *iris*

**Q-angle:** an objective two-dimensional anatomical measurement named for its relationship to the *quadriceps* muscle; defined as the angle between two intersecting lines of pull at the *patella;* relates (a) the direction of the quadriceps muscle's pull—between the *ASIS* and the center of the patella—to (b) the line of pull from the center of the patella to the *tibial tuberosity*—through the *patellar ligament*

**quadriceps m.:** the large four-part compound muscle of the thigh which straightens the knee; comprised of four sections: *rectus femoris, vastus medialis, vastus lateralis* and *vastus intermedius*

**quarters:** the upper shoe components which cover the *instep* from superior to inferior, *anterior* to the ankle joint

**quirk:** triangular or diamond-shaped piece placed at the base of a glove finger or thumb piece to allow movement

**radial nerve:** major nerve of the arm, forearm, and hand; carrying *motor, sensory,* and *autonomic nerve* fibers to the anatomical arm and forearm and *sensory* and *autonomic nerve* fibers to the hand; vulnerable to compression in the *posterior* anatomical *arm*

**radial styloid process:** a prominent bony protrusion of the distal forearm, on the thumb side

**radiocarpal joint:** wrist joint; joint between *radius* of the arm and three of the eight *carpal* bones of the wrist; motion at this joint positions the hand in space relative to the forearm

**radioulnar joints:** proximal and distal *articulations* between the two forearm bones which allow the hand to move palm up to palm down positions and back

**radius:** larger of the two bones in the *forearm* of the upper limb

**range of motion; range of movement (ROM):** in general, an arc of the functional motion of a specific joint as measured with a *goniometer,* in degrees, in a specified plane (most often in a *sagittal, coronal,* or *transverse* plane); provides an approximation of the extent of motion and degree of flexibility around the joint's *axis of rotation*

**receptors:** specialized *PNS* structures in skin or other peripheral *tissues* which collect *sensory* information

**reciprocal gait:** a pattern of moving on foot and moving/swinging the opposite arm and leg reciprocally; e.g. while walking, the left shoulder and arm move forward as the right heel strikes the ground, and *vice versa*

**rectum:** final section of the large intestine of the *digestive system*

**rectus abdominis m.:** the "abs;" the paired segmented muscles seen as the abdominal "six-pack"

**rectus femoris m.:** a superficial section of the *quadriceps m.* which functions to *flex* the hip and *extend* the knee

**reflex:** a predictable involuntary body movement in response to a particular *sensory* stimulus; e.g. a knee jerk in response to a *patellar ligament* tap in the doctor's office

**refraction:** the bending of light by the *crystalline lens* of the eye to focus an image on the *retina*

**refractive correction:** use of corrective lenses in eyeglasses to bend light before it reaches the lens of the eye; to compensate for poor eyesight

**regenerate:** skin's ability to restore/repair itself after an injury

**rehabilitation:** the restoration of satisfactory function for patients with physical impairments, to reduce disability in daily life

**reproduction:** the physical, hormonal, and biological processes by which humans produce human infants

**reproductive:** referring to the process of *reproduction*

**reproductive system:** *organs* and anatomical structures to support *reproduction*

**resilience:** skin's ability to recover from shape and/or length change, elasticity

**respiration:** the physical and chemical processes of breathing and taking air, especially oxygen, into the body

**respiratory:** referring to the process of *respiration*

**respiratory cycle:** a breath in, followed by a breath out, also *breathing cycle*

**respiratory diaphragm:** a dome-shaped muscular structure which acts to move air into and expel air from the lungs

**respiratory system:** *organs* and anatomical structures to support *respiration*

**resting heart rate:** *pulse* rate while the body is physically at rest; normal adult resting heat rate = 60 to 100 beats per minute

**retina:** layer with millions of light-sensitive cells in the back of the eye; connected to the *occipital lobe* of the brain by the optic nerve

**retraction:** backward motion of a specific body part; action occurs at the articulations of the jaw to the skull, the skull to the neck, and with arm/shoulder movement on the *torso*

**retrograde menstrual flow:** menstrual blood pooling in the abdomen

**rib cage:** the bony framework of the chest; includes the ribs, *sternum*, and thoracic vertebrae

**right lymphatic duct:** one of the final common pathways of the *lymphatic system* passing through the thorax; drains the lymphatic system in R upper quadrant of the body and R half of the head

**ring finger:** 4th digit of the hand, next to the small, 5th (little) finger

**rise:** men's pattern term equivalent to *crotch depth* on women's patterns; see *crotch depth*

**rods:** very light-sensitive cells located outside of the center of the retina, which do not detect color; responsible for low light or night vision

**ROM:** see *range of motion*

**rotator cuff:** group of four deep muscles originating from the *scapula* and inserting on the *humerus* which interact to rotate the arm in the *glenohumeral joint*

**sacral:** referring to the *sacrum* or region of the *sacrum*
**sacroiliac joint:** joint between the lateral aspects of the *sacrum* and the two *iliac* bones of the pelvis, joining the *axial* and *appendicular skeletons* in the pelvis
**sacrum:** the five sacral vertebrae fused into a bony shield-shaped structure
**saddle joint:** a geometric joint type in which two bone surfaces that are saddle-shaped (concave on one axis and convex on the perpendicular axis) interconnect to allow motion, e.g. the *first carpometacarpal joint* (*CMC*) between the wrist and the thumb that allows you to move your thumb toward your little finger
**sagittal planes:** orienting planes passed through the body from front to back, parallel to the median plane
**saliva:** the watery product of the salivary glands of the mouth which helps begin *digestion*
**sample size:** prototype to fit one body representing all product users, other sizes are developed by decreasing or increasing dimensions from the sample size
**sandwich board:** wearable advertisement consisting of a flat panel (board) suspended from the shoulders
**sanitary napkin:** absorbent product (pad) worn below the perineum/perineal region to absorb menstrual flow, also called *sanitary napkins, sanitary towels*, and *panty liners* (or shields)
**sanitary pad:** see sanitary napkin
**sanitary towel:** see sanitary napkin
**SAT:** see *subcutaneous adipose tissue*
**scalp:** the skin surface over the *cranium*
**scapha:** flatter inner portion of the *auricle* next to the *helix* of the ear
**scaphoid:** one of the proximal *carpal bones* of the wrist; boat-shaped
**scapula:** the shoulder blade
**scapular spine:** the bony projection on the *posterior* side of the shoulder blade
**scapulohumeral joint:** see *glenohumeral joint*
**scapulothoracic joint:** network of bones, muscles, and the bursa between the scapula and the ribs; without a bony articulation, not a true joint
**scar:** a lasting visible mark on the skin due to damage to the skin
**sciatic nerve:** the large major nerve in the lower limb carrying *motor, sensory*, and *autonomic nerve* fibers; travels through the buttocks and *posterior* thigh; divides and continues into the anatomical *leg* as the *tibial nerve* and *common fibular nerve*
**sclera:** the white portion of the eyeball
**scoliosis:** a lateral deviation of the spine to either side
**screw-home:** term to describe the mechanism of the final few degrees of knee extension which includes some rotation
**scrotum:** a muscular skin pouch containing the *testes*
**sebaceous (oil) glands:** part of the dermis; only found in hair follicles

**sebum:** oily, waxy *sebaceous gland* secretion; flows along the hair shaft to the surface, to lubricate the hair and skin; contains fatty compounds and cellular debris

**segment (of a lobe of a lung):** a portion of the lung, smaller than a lobe, but still served by a *bronchus*

**semicircular canals:** tubular structures of the *inner ear,* situated at right angles to one another

**semimembranosus:** one of two medial hamstring muscles; a major muscle on the back of the thigh, acts on the hip and knee joints

**semi-rigid (space) suit:** combines soft and hard components

**semitendinosus:** one of two medial hamstring muscles; a major muscle on the back of the thigh, acts on the hip and knee joints

**sensorial comfort:** how satisfactory a product feels to the wearer; *tactile comfort*

**sensory:** referring to *PNS* structures related to the senses including cells, fibers, and nerves; pertaining to the senses

**sensory nervous system:** portion of the nervous system including (a) special senses (vision, hearing, smell, and taste); (b) *somatic* senses (touch, pain, pressure, vibration, temperature, and *proprioception*—the sense combining position in space, balance, and/or movement in the muscular and skeletal systems); and (c) *visceral* senses of *organ* pain and stretch of *organ* walls

**septum:** a *cartilage* and bony structure which separates the left and right *nostrils* and *nasal cavities*

**serratus anterior m.:** a muscle attached to the *scapula* which also covers the bones of the rib cage

**sexual intercourse:** the act of sexual union

**sexually transmitted disease (STD):** venereal disease; any infection transmitted during vaginal, oral, or anal sex

**shank:** component in the sole of a dress shoe made of rigid materials, included in some shoe styles to prevent the sole from collapsing on itself

**shape:** a two-dimensional representation

**shapewear:** see compression garment

**short head (of the lateral hamstring):** see *biceps femoris*

**short-waisted:** upper torso of the body is short in proportion to the lower body

**shoulder:** the body region connecting the arm to the *torso*; including both the *scapulothoracic and glenohumeral* joints and the overlying muscle, fat, and skin

**shoulder point:** garment reference point to match the top of the sleeve cap to the shoulder seam

**shoulder seam:** seam joining garment front and back at the shoulder

**side seam:** seam joining garment front and back at the side of the body

**single leg stance:** a phase of *gait* with full weight-bearing on one foot while the other foot is off the ground, between *toe-off* and *heel strike*
**sinuses:** *mucous membrane*-lined cavities in the skull located above and below the eyes as well as deep inside the skull near the *respiratory* passages and *cranial* cavity which drain into the *nasopharynx*
**size designators:** numbers (Misses 10, 12, 14, etc.) or letters (S, M, L) to communicate the dimensions of a garment; vary by manufacturer
**size:** reflects an individual person's body dimensions related to a product
**sizing system:** method of organizing body measurement data for application to wearable product to determine number of sizes, size scales smallest to largest, and size intervals
**skeletal:** referring to the *skeleton*
**skeletal muscle:** muscle attached to *skeletal* structures which generates body movements; *voluntary muscle*
**skeletal system:** the inter-related bones and other anatomical structures of the *skeleton* and their interactions to support and protect the body
**skeleton:** the bony framework inside the body, made up of all the body's bones plus their joints and connecting structures
**skin:** largest organ of the body; composed of three layers: *epidermis, dermis,* and *subcutaneous* layers; with important qualities—it can *regenerate* and has *resilience*
**skin microbiome:** microbes that live on the skin without causing harm but rather with beneficial effects
**skull:** the bones of the head
**sleeve cap:** upper sleeve from *armscye* level to top of sleeve where it attaches to the shoulder seam
**sliding filament theory of muscle contraction:** explains the interaction of *actin* and *myosin myofilaments* to generate a muscle contraction
**smart materials:** materials that respond in a controlled way to a stimulus, physical and/or chemical; used for sensors and actuators
**smart textiles:** textiles that sense human body signals and/or the surrounding environment to react in a specific way
**smooth muscle:** involuntary muscle found in the stomach and intestines, *urinary system, blood vessels,* and *reproductive organs*
**social/psychological environment:** a person's surroundings beyond the physical to include societal expectations and the individual's perceptions of self in that setting
**soft palate:** soft part of the roof of the mouth
**sole (of foot):** interface between the *plantar* aspect of the foot and the ground
**sole (of footwear):** the bottommost layer, may be divided into three parts: *insole, midsole,* and *outsole*
**soleus muscle:** muscle located on the back of the leg, *plantar flexes* the ankle
**somatic:** related to structures of the body wall; also used to describe nerves associated with the *voluntary* muscles; opposite of *visceral*

**somatography:** a body graphing technique developed by Helen Douty
**space suit:** a total life-supporting mini-environment worn by an astronaut when outside of the spacecraft
**spectacles (eyeglasses):** a pair of optical lenses positioned in front of the eyes with a surrounding frame, held on the face with frame components including the frame bridge over the nose and bows (*temples*) over the ears
**sperm:** the structural vehicle for the male's genetic material in reproduction
**sphincter:** circular band of muscles
**sphygmomanometer:** a device that detects the systolic pressure and then the diastolic pressure of the arterial blood
**spider veins:** dilated tiny superficial leg veins
**spinal accessory nerve:** nerve which controls the trapezius muscle
**spinal canal:** the ovoid cylindrical space between the bony vertebral arches and the *vertebral bodies*
**spinal cord:** *CNS* structure contained within the spinal canal: connects the brain to *PNS* structures outside the *vertebral column*
**spinal nerve ganglion:** an enlargement along the more *posterior* nerve root which contains *sensory* nerve cells, and lies in the opening between the *vertebrae*
**spinal nerve roots:** separate bundles of nerve fibers from the *motor* and *sensory* divisions of the central nervous system
**spinal nerves:** *PNS* nerve structures extending from the *spinal cord* to *plexuses* in the torso and limbs
**spine:** backbone; comprised of *vertebrae*
**spine of the scapula:** a prominent ridge of the *scapula* palpable from the body surface
**spinous process:** the most *posterior* element of a *vertebra*; the portions of the spine which can be *palpated* in the middle of the back
**spleen:** a solid, blood-cell forming organ in the upper left abdomen, largest *lymphatic organ*
**splenius capitis m.:** a smaller muscle of the back of the neck, deep to the large *trapezius m.*, which acts along with the trapezius and several smaller and deeper muscles to tip the head backward
**splint:** thin piece of rigid material to maintain a fixed body position
**sprain:** stretching or tearing of fibers in a ligament
**spreading calipers:** adjustable anthropometric device to measure widths and depths
**stance:** the way a person stands
**standing anthropometer:** a device with a vertical measure and a sliding bar that can be adjusted to touch the top of the head
**standoff distance (SOD):** the measurement of the distance between the head and the inside surface of the helmet
**stapes:** the "stirrup" bone of the *middle ear*
**static:** body at rest

**stature:** total body height; sometimes used interchangeably with height
**STD:** see *sexually transmitted disease*
**sternal notch:** a large, visible dip above the *manubrium* of the *sternum,* at the base of the neck, bounded by the *sternocleidomastoid* muscles and the *sternoclavicular joints*
**sternoclavicular joint:** articulation point of the *clavicle* and the *manubrium* of the *sternum*
**sternocleidomastoid muscle:** a strap-like muscle on the side of the neck, running from behind the ear to the collarbone
**sternocostal joint:** see *chondrosternal joint*
**sternum:** breast bone
**stool:** see *feces*
**stratum corneum:** topmost layer of the *epidermis* of the skin
**stress urinary incontinence (stress UI):** involuntary loss of *urine* due to increased intra-abdominal pressure (from coughing, sneezing, or laughing) which negates the *urethral sphincters'* and *pelvic diaphragm*'s actions—forcing urine from the bladder
**stretch marks:** shiny white or grayish marks of variable width along skin tension lines on the breasts, abdomen, hips, buttocks, and/or thighs often related to changes in body weight
**striated:** term referring to the microscopic appearance of *skeletal muscle*; literally "striped"
**striated muscle fiber:** see *muscle fiber*
**stroke:** a medical condition causing damage to structures in the *CNS*, related to either blockage of blood flow to or bleeding in an area of the brain or spinal cord; associated with changes in *nervous system* function
**style ease:** length or width added to the minimum *fit ease* to produce a desired style or garment shape
**subcutaneous:** referring to a structure beneath the skin
**subcutaneous adipose tissue (SAT):** fat tissue lying just under the skin
**subcutaneous layer (of skin):** the deepest skin layer, the base of the skin
**sublimate:** a cooling process, turning ice into vapor
**sublimation:** see heat transfer mechanisms
**subluxation:** incomplete dislocation of the joint
**subluxing:** slippage in joint leading to *subluxation*
**subscapular bursa:** a fluid-filled sac that lies between the *subscapularis* and *serratus anterior* muscles in the *scapulothoracic joint*; facilitates smooth movement of the *scapula* on the *torso*
**subscapularis m.:** a *rotator cuff* muscle; rotates the *humerus* medially; originates on the concave interior surface of the *scapula*
**subtalar joint:** joint between the *talus* and *calcaneus;* participates in many foot motions
**sucking reflex:** involuntary movements of the lips, tongue, and jaw in response to light touch on the lips; facilitates an infant's breast-feeding (*suckling*)

**suckling:** breast-feeding or nursing

**sudomotor nerves:** sympathetic nerves (*PNS* motor structures) controlling the sweat glands

**superciliary ridges:** brow arches, part of the *frontal* bone

**superficial temporal artery:** arterial structure with a *pulse point*, found on the lateral aspect of the head, palpable as the artery passes over the *zygomatic process* of the *temporal bone*, in front of the ear

**superior cluneal nerves:** *sensory* nerves which cross the *posterior iliac crest* and transmit sensations from the skin of the buttocks to the *CNS*

**superior ophthalmic vein:** a deep vein draining the central face, may serve as a conduit for infection between the *danger triangle of the face* and the brain

**superior vena cava:** large major vein returning blood to the heart from the upper torso, arms, neck, and head

**superior:** places a part higher than another structure

**supination:** specialized movement of the hands and/or feet; positions the hand palm-up and raises the medial edge of the sole of the foot

**supine:** horizontal body position, face-up

**suppositories:** waxy bullet-shaped cylinders containing bowel stimulants and lubricants

**supraorbital nerve:** *sensory* nerve located just above the eye, between the nose and the center of the eye, may be compressed by eyewear elements

**supratrochlear nerve:** *sensory* nerve running upward between the *infratrochlear* and *supraorbital* nerves; may be compressed by eyewear elements

**surface electromyography (EMG):** non-invasive technique used to record muscle electrical activity from the body surface

**suspensory ligaments (of the lower torso):** tough *connective tissue* structures which help maintain the position of the *bladder* and *uterus* within the *pelvis*

**sutures:** relatively immobile *fibrous joints* between the bones of the *cranium* and of the face

**swan-neck deformity:** finger alignment abnormality associated with *hyperextension* of a loose *PIP* joint with flexion at the *DIP* joint

**sweat glands:** *exocrine glands* in the dermis which exit to the skin with two subtypes: *apocrine* and *eccrine*

**sweating thermal manikin:** instrumented manikin; measures evaporative heat loss; simulates human perspiration production

**sympathetic:** referring to a division of the *autonomic nervous system* which helps us in situations of fright, flight, or fight by altering body processes, e.g. speeds up the heart rate, shifts blood flow to the muscles, dilates the *bronchi* to allow deeper breaths, and slows intestinal and urinary *smooth muscle* contractions and tightens the *sphincters*—decreasing the need for *urination* or *defecation*

**sympathetic chain or sympathetic trunk:** the connected *sympathetic ganglia* which lie in a relatively protected location, close to the *vertebral bodies* of the spine

**sympathetic ganglia:** clusters of *sympathetic nervous system* cells found near the *spinal nerves* whose fibers connect to other sympathetic nervous system cells, *blood vessels, sweat glands* of the skin, and internal *organs*

**symphysis joint:** slightly movable *fibrocartilaginous* joint; e.g. the *pubic symphysis*

**synchondrosis joint:** a joint held together by *hyaline cartilage* which is immobile; e.g. the joint between the first rib and the *manubrium*

**syndesmosis joint:** a specialized *fibrous joint* where two bones are joined by densely organized collagen structures, *interosseous membranes;* e.g. the *radioulnar joint*

**syndesmotic ligament:** the strong fibrous *connective tissue* structure between the distal portions of the *tibia* and *fibula*, an extension of the *interosseous membrane* between the two bones

**synovial fluid:** a viscous clear lubricating fluid found in *synovial joints, bursae,* and *tendon sheaths*

**synovial joint:** most common and most mobile joint; bony joint surfaces are covered with *hyaline articular cartilage*, joint is enclosed in a *joint capsule,* and the joint is lubricated by *synovial fluid*

**synovium, synovial membrane:** the soft tissue lining of a *synovial joint* which produces *synovial fluid*

**system, systems:** see *body systems*

**systemic circulation:** *arteries* and *veins* which carry oxygen-rich blood from the left side of the heart to the body and return oxygen-poor blood to the right side of the heart

**systolic:** referring to the peak *blood pressure* produced when the heart beats; the first number of a blood pressure reading

**tactile comfort:** see *sensorial comfort*

**tactile point:** anatomical feature of the body that can be palpated (felt) to place a *landmark*

**tailored jacket:** upper torso garment with simple lines, fitted to the body form; traditional tailored jackets include support materials in the collar, lapels, front edges, and upper back; may refer to a jacket that is custom fit to a person

**talocrural joint:** ankle joint

**talus:** uppermost foot bone; interfaces with the bones of the leg above it, the *tibia* and the *fibula*, to form the ankle (*talocrural joint*)

**tampons:** absorbent cotton/rayon plugs inserted into the *vagina* with a finger or removable applicator to absorb *menstrual flow*

**target market:** segment of a population matching a company's customer descriptions including age, gender, socio-economic group, educational background, fashion adopter profile, etc.

**tarsal bones:** bones forming the rear of the foot; subdivided into *proximal* and *distal* bone groups

**tarsal tunnel:** region *posterior* to and *distal* to the *medial malleolus* at the ankle (between the malleolus and *calcaneus*); beneath the *flexor retinaculum* and deep *fascia* at the medial ankle; passageway for the medial plantar artery and the most *distal* branch of the *tibial* nerve

**tarsal tunnel syndrome:** pain and numbness along the *medial* aspect of the heel and the medial sole of the foot, plus foot and big toe muscle weakness resulting from compression of the *distal* branch of the *tibial* nerve (*motor* and *sensory*) at the *tarsal tunnel*

**taste buds:** sensory receptors for the sense of taste on the tongue

**[TC]$^2$:** company offering 3D body scanner equipment, software, and consulting services

**tears:** secretions produced to lubricate and clean the surface of the eye

**temple length:** common eyeglass frame measurement

**temples:** side pieces of eyeglass frames, also known as bows

**temporal bones:** paired *cranial* bones on the side of the head at the level of the eye socket; location of the *ear canal*

**temporal lobe:** one of the four anatomical regions of the *cerebral hemisphere* of the brain which contains neurons with specialized functions; located adjacent to the temporal bone

**temporalis m:** powerful paired superficial muscles located on the side of the head, in front of and extending up from the ear, which move the *mandible* during chewing activities

**temporomandibular joint (TMJ):** articulation between *mandible* and *temporal bone*

**tendon sheaths:** lubricant filled cushions adjacent to and encircling the linear *tendons*

**tendon:** strong, flat connective tissue band

**tendonitis:** an *inflammation* or irritation of a *tendon* caused by repetitive motions or damage from minor impacts

**TENS:** see *transcutaneous electrical nerve stimulator*

**tensile force (tension):** pulling, stretching; state of being stretched

**tensor fascia lata m.:** superficial and relatively small muscle originating from the *iliac crest* near the *ASIS* of the *bony pelvis;* acts to flex, medially rotate and abduct the thigh at the hip

**terminal hair:** coarse, long pubic hair of adults

**testes (pl. of testis), testicles:** male *gonads,* a part of the male *external genitalia* which produce *sperm* after *puberty*

**testosterone:** the primary male *hormone* secreted by the *testis*

**textile characteristics:** appearance and performance of the textile variables (fiber, yarn, structure, appearance finish, functional finish)

**thenar crease:** the crease created when the thumb is raised to a position perpendicular to the palm; the thenar crease curves from the *wrist crease* to the approximate midpoint between the thumb and index finger

**thenar eminence:** the mound of fleshy tissue at the base of the thumb containing *intrinsic* hand muscles which act on the thumb

**thenar muscles:** several intrinsic hand muscles at the base of the thumb which control motions of the thumb including: *abduction, adduction, flexion, extension,* and *opposition*

**thenar region:** a subdivision of the *palmar region* of the hand; also see *thumb region*

**thermal manikin:** instrumented manikin measures heat loss, simulates human heat loss

**thermal protection:** products and product components designed to maintain a healthy body *core temperature*

**thermoregulation:** body processes to maintain healthy *core temperature*, a form of homeostasis

**thigh:** proximal section of the lower limb between the *pelvis* and knee

**third occipital nerve:** a *sensory* nerve located in the *posterior* lateral side of the head; passes through the upper *trapezius* near the hair line; may be compressed by straps to secure eyewear

**thoracic:** referring to the thorax

**thoracic duct:** one of the final common pathways of the lymphatic system passing through the thorax

**thoraco-lumbo-sacral-orthosis (TLSO):** a spinal *orthosis* spanning the spine from the base of the neck to the pelvis; used to correct an abnormal spinal curvature or to immobilize an unstable or post-operative spine

**thorax:** the region of the body between the neck and abdomen, commonly called the chest; includes the *rib cage*, and elements of many body systems, most prominently the heart and *lungs*

**thumb region:** palmar hand region including the thumb; defined by the *thenar crease*

**thyroid cartilage:** prominent structure in the *anterior* neck; the major cartilage structure of the *larynx*; moves up and down with a swallow

**thyroid gland:** major component of the *endocrine system*; helps regulate body metabolism; found in the anterior neck *caudal* to and somewhat *lateral* to the *thyroid cartilage* and *trachea*

**tibia:** larger of the two bones in the anatomical *leg*

**tibial nerve:** one of two major peripheral nerves serving *motor, sensory*, and *autonomic* functions below the knee

**tibial tuberosity:** *anterior* prominence of the *proximal* tibia below the *patella*

**tibialis anterior m.:** the largest muscle of the anterior leg; *dorsiflexes* (*extends*) the ankle joint

**tibialis anterior tendon:** most prominent cord-like structure at the dorsal-medial aspect of the ankle joint; palpable and visible with ankle *dorsiflexion*

**tibiofibular joint:** a *syndesmosis joint* in which the *tibia* and *fibula* are connected by an *interosseous membrane*, and distally by a *syndesmotic ligament*

**tissue:** an organized collection of similar cells within the body and the substances and structures between them

**TMJ:** see *temporomandibular joint*

**toe box:** the domed area of the shoe above the toes

**toe puff:** an extra layer of materials in the most distal and topmost portion of the toe box which helps keep the toe box from collapsing on the toes

**toe-off:** phase of *gait* characterized by the *hallux* and other toes bending (*plantar flexing*) and pushing the body forward

**tongue (of a shoe):** underlays the shoe lacings

**tonsils:** specialized *lymphatic structures* in the mouth and *pharynx*

**top line:** the upper edge of the shoe opening that surrounds the foot

**torso:** trunk of the body

**total width:** common eyeglass frame measurement

**toxic shock syndrome (TSS):** a rare but serious disease related to a toxin given off by some *Staphylococcus aureus* (*S. aureus*) bacteria which can grow in the vagina

**toxins:** poisonous substances which can be absorbed and damage vital body structures and functions

**trabecular bone:** the bone type with a spongy lattice, containing blood forming elements; found in an interior bone location or near joints

**trachea:** the semi-rigid cylindrical airway running the length of the *anterior* neck

**tragus:** the small springy protrusion toward the center of the ear, near the opening of the *ear canal*

**training pants:** self-managed underwear for toddlers making toileting easier; transition garment from diapers to child's underwear

**trank:** glove front and back covering the palm and dorsum

**transcutaneous electrical nerve stimulator (TENS):** a non-invasive pain treatment which places conductive pads over innervated skin and uses low intensity electrical stimulation of the sensory receptors of the skin in an attempt to block pain signals from the *PNS*

**transit time:** the amount of time it takes food to move through the entire *digestive* tract

**transverse (cross or horizontal) plane:** plane perpendicular to the long axis of the body; the horizontal plane dividing the body at the mid-section into upper (cranial), and lower (caudal) parts; used to create a body cross-section

**transverse abdominal m.:** deep abdominal wall muscle with horizontal muscle fibers which supports and squeezes the abdominal contents

**transverse process:** a bony projection outward from a *vertebra*, arising between the *vertebral body* and the *vertebral arch*; transverse process geometry varies from one spinal region to another

**trapezium:** one of the distal *carpal bones* of the wrist; part of the *carpal tunnel*

**trapezius:** major superficial fanlike upper back muscle which moves the arm/shoulder mechanism on the *torso*

**trapezoid:** one of the distal *carpal bones* of the wrist, part of the *carpal tunnel*

**triceps, triceps brachii:** large three-part muscle on the *posterior* upper arm; straightens the elbow

**triquetrum:** one of the proximal *carpal bones* of the wrist, part of the *carpal tunnel*; triangular
**trochanter:** see *greater trochanter*
**trochanteric belt:** a relatively narrow belt which encircles the *pelvis* between the *iliac crests/anterior superior iliac spines* (*ASIS*) and the femoral *trochanters* and is tightened to push the *innominate bones* and *sacrum* together
**TSS:** see *toxic shock syndrome*
**tubercle (of rib):** prominence on *posterior* rib which *articulates* with the *vertebral transverse process*
**two-dimensional planar materials:** textiles, films, and foams used to construct wearable products
**tympanic membrane:** the first *acoustic* structure of the *external ear*; sometimes called the ear drum because it literally vibrates like the stretched head of a drum
**ulna:** the longer of the two *forearm* bones
**ulnar drift:** shift of the alignment of the finger(s) relative to the body of the hand (from the *MCP* joint level) toward the *ulnar* side of the hand
**ulnar nerve:** major nerve of the forearm and hand; carrying *motor, sensory,* and *autonomic nerve* fibers; travels through the anatomical *arm* and a groove at the elbow (the "crazy bone") where it is vulnerable to compression
**ulnar styloid process:** a prominent bony protrusion of the distal *forearm*, on the little finger side
**ultraviolet light (UV light):** radiant energy emitted by the sun, invisible to the human eye; made up of *UVA, UVB,* and *UVC* rays; only UVA and UVB rays reach the earth's surface—producing long-lasting irreparable effects on unprotected human skin
**umbilicus (belly button):** see *navel*
**underbust circumference:** encircles the rib cage directly below the *bust*
**unitard, body suit, "cat" suit:** one-piece body-conforming garment with upper torso, lower torso, arm, and leg components combined
**universal design:** product, interior, or environmental design that accommodates all users safely and comfortably
**upper:** shoe component that covers the dorsum of the foot
**ureter:** a tubular structure connecting the *kidney* to the *bladder*
**urethra:** a tubular structure with a *sphincter* to control *urine* flow from the *bladder* to the outside of the body
**urethral meatus:** the external opening of the *urethra*, the tube connecting the *bladder* to the outside of the body
**urgency urinary incontinence (urgency UI):** leakage of small or large amounts of *urine*; associated with an urgent desire to empty the *bladder*
**urinary:** referring to structures of the *urinary system*
**urinary catheter:** a tube to drain *urine* from the *bladder*
**urinary retention:** the inability to fully empty one's *urine* from the *bladder*

**urinary system:** inter-related anatomical structures of the body which produce, transport, and eliminate liquid waste; major organs and structures include: two *kidneys* and *ureters*, the *bladder*, and the *urethra*
**urination:** process of discharging *urine* from the *bladder*
**urine:** liquid body waste
**usability:** a product's ease of use
**user group:** people who will wear or currently wear a product
**UTI:** *urinary* tract infection
**UVB:** *UV* light which stimulates vitamin D production in the skin, a positive effect of sunlight
**vagina:** internal female reproductive canal spanning from the *vaginal opening* to the *cervix*
**vaginal opening:** part of the female external *genitals*; entrance to the *vagina*
**vagus nerve:** *cranial nerve X*, a *PNS* structure originating from the *brainstem*, carrying *motor*, *sensory*, and *parasympathetic* nerve fibers; it exits the skull and passes through the neck to reach and control vital *organs* of the *thorax* and abdomen, moderating *heart rate* and *digestive* actions as well as serving muscles in the head and neck and transmitting *sensory* information from the *external ear*, airway structures, and internal organs of the thorax and abdomen
**valves (of the veins or lymphatic vessels):** anatomical structures within vessels which prevent backflow of *blood or lymph*
**vamp:** shoe component that covers the *distal* metatarsal bones, may merge with the toe box
**varicose veins:** *subcutaneous veins* larger than 3 mm (0.12 in.) in diameter when standing
**vastus intermedius m.:** a deep section of the *quadriceps m.* which functions to *extend* the knee; exerts a pull along the line of the shaft of the *femur* on the *patella*
**vastus lateralis m.:** a superficial section of the *quadriceps m.* which functions to *extend* the knee; exerts a *lateral* pull on the *patella*
**vastus medialis m.:** a superficial section of the *quadriceps m.* which functions to *extend* the knee; exerts a *medial* pull on the *patella*
**vein:** relatively thin-walled blood vessel returning blood to the heart
**vellus hair:** very fine, short, non-pigmented hair, on the body and limbs which is almost invisible
**ventral:** see *anterior*
**ventricles:** lower chambers of the heart which pump blood out of the heart
**ventricles (of the brain):** four communicating cavities within the brain, filled with *CSF*
**vermiform appendix:** see *appendix*
**vertebra (pl. vertebrae):** bony segment of the spine or *vertebral column*
**vertebral arteries:** *blood vessels* running through the *posterior* neck carrying blood to the head and brain

**vertebral body:** the columnar *anterior* portion of a *vertebra*; a spinal structure serving to bear weight
**vertebral canal:** a canal through the *vertebral column*, containing and serving to protect the *spinal cord*
**vertebral column:** the spinal column, the flexible weight-bearing column of the body
**vertex distance:** measurement along the line of sight between the *cornea* and the *posterior* surface of the eyeglass lens
**vertigo:** sensation that the environment around you is spinning with you at the center
**vibrissae:** hairs inside the *nostrils*
**virtual draping:** shaping two-dimensional patterns to the three-dimensional scanned body of an individual or a digital human model
**visceral:** related to the internal *organs*
**viscoelastic properties (of breasts):** breast tissue's response to *compression* or *tension*; varies according to the proportions of *mammary gland* tissue to *fat*, cyclic changes in *hormones*, and mammary gland tissue activity related to pregnancy and breast-feeding
**vitreous humor:** the fluid filling the eye between the *crystalline lens* and the retina
**void:** a single episode of *bladder* emptying
**volar:** opposite of *dorsal*; refers to a structure being more in front of, toward the body front, in the *anatomical position*, than another structure; in the hand—the *palmar* surface
**voluntary:** generally used relative to an action or motion; to move a body part in a deliberate manner
**voluntary muscle:** see *skeletal muscle*
**waist:** the narrowest part of the body in the mid-torso
**waist circumference (WC):** measurement around the *waist* (narrowest part) of the body
**waist girth:** waist circumference; girth is a measurement around the middle of something
**waistband:** design feature attached at a skirt or pants waistline
**waistline:** a garment feature, determined by aesthetic choice and wearer preference
**watersheds (of lymphatic system):** theoretical divisions visualized on the body surface to represent the body's regional *lymphatic* drainage patterns
**WC:** see *waist circumference*
**wear test:** human participants evaluate a product in a natural setting
**wearable product:** anything that surrounds, is suspended from, attached to, or temporarily inserted into the body
**wearable product components:** product structure, materials, and fit
**wearable technology:** products adding technology of some type, e.g., health monitoring, into a wearable product

**wicking fibers:** product fibers that transport moisture from the body to a less humid surrounding environment to promote comfort
**windlass effect:** a raising and shortening of the *arch* of foot when the big toe is passively extended
**wings:** rounded skin-covered cartilage structures at the sides of the nose
**wrap angle:** the curve of the lens section of the frame around the contour of the face
**wrist crease:** narrow skinfold which develops between the forearm and palm with wrist *flexion*
**xiphoid process:** the elongated bony *caudal* end of the *sternum*
**zygoma, zygomatic bone:** cheekbone, the prominent *facial bone* at the outer lower aspect of each eye socket
**zygomatic process:** a part of the *temporal bone* which articulates with the *zygoma* in front of the ear
**zygomaticofacial nerve:** *sensory* nerve lateral to the eye which travels downward; may be compressed by eyewear elements
**zygomaticotemporal nerve:** *sensory* nerve lateral to the eye which travels upward; may be compressed by eyewear elements

# *References*


Abd. Latiff, N. S., & Yusof, N. (2016). A methodology for facial measurement towards the establishment of ready-to-wear hijab sizing. *International Journal of Clothing Science and Technology, 28*(6), 841–853. doi:10.1108/IJCST-12-2015-0140

Abel, P. (1939). Fitting of spectacles with special consideration of the anatomy of the head. *The Australasian Journal of Optometry, 22*(2), 72–83. doi:10.1111/j.1444-0938.1939.tb01565.x

Adamson, J. P., Lewis, L., & Stein, J. D. (1959). Application of abdominal pressure for artificial respiration. *Journal of the American Medical Association, 169*(14), 1613–1617. doi:10.1001/jama.1959.03000310065014

Akeson, W. H., Amiel, D., Abel, M. F., Garfin, S. R., & Woo, S. L. (1987). Effects of immobilization on joints. *Clinical Orthopaedics and Related Research*, 219, 28–37.

Alam, M., Hardman, J., Paus, R., & Jimenez, F. (2018). 1350 Human scalp skin as an abundant and accessible source for eccrine sweat gland isolation and organ culture. *Journal of Investigative Dermatology, 138*(5), S229. doi:10.1016/j.jid.2018.03.1367

Aldrich, W. (2009). *Metric pattern cutting for children's wear and babywear* (4th ed.). West Sussex, U.K.: Wiley-Blackwell, John Wiley & Sons, Ltd.

Aldrich, N. D., Reicks, M. M., Sibley, S. D., Redmon, J. B., Thomas, W., & Raatz, S. K. (2011). Varying protein source and quantity do not significantly improve weight loss, fat loss, or satiety in reduced energy diets among midlife adults. *Nutrition Research, 31*(2), 104–112. doi:10.1016/j.nutres.2011.01.004

Alexander, M. (2003). *Applying three-dimensional body scanning technologies to body shapes analysis* (Unpublished doctoral dissertation). Auburn University, Auburn, AL.

Al-Khabbaz, Y. S., Shimada, T., & Hasegawa, M. (2008). The effect of backpack heaviness on trunk-lower extremity muscle activities and trunk posture. *Gait & Posture, 28*(2), 297–302. doi:10.1016/j.gaitpost.2008.01.002

Altman, R., Asch, E., Bloch, D., Bole, G., Borenstein, D., Brandt, K., . . . & Howell, D. (1986). Development of criteria for the classification and reporting of osteoarthritis: Classification of osteoarthritis of the knee. *Arthritis & Rheumatism, 29*(8), 1039–1049. doi:10.1002/art.1780290816

Alvarez, S. E., Peterson, M., & Lunsford, B. R. (1981). Respiratory treatment of the adult patient with spinal cord injury. *Physical Therapy, 61*(12), 1737–1745. doi:10.1093/ptj/61.12.1737

Alviso, D. J., Dong, G. T., & Lentell, G. L. (1988). Intertester reliability for measuring pelvic tilt in standing. *Physical Therapy, 68*(9), 1347–1351. doi:10.1093/ptj/68.9.1347

Amador, G. J., Mao, W., DeMercurio, P., Montero, C., Clewis, J., Alexeev, A., & Hu, D. L. (2015). Eyelashes divert airflow to protect the eye. *Journal of the Royal Society Interface, 12*(105). doi:10.1098/rsif.2014.1294

Amen, D.G., Newburg, A., Thatcher, R., Jin, Y., Wu, J., Keator, D., & Willeumier, K. (2011). Impact of playing American professional football on long-term brain function. *The Journal of Neuropsychiatry and Clinical Neurosciences, 23*(1), 98–106. doi:10.1176/jnp.23.1.jnp98

American Association of Textile Chemists and Colorists (AATCC) (2013). *AATCC TM15-2013, Colorfastness to Perspiration*, AATCC Technical Manual. Research Triangle Park: American Association of Textile Chemists and Colorists. Retrieved from https://www.aatcc.org

American Heart Association. (2016) Understanding Blood Pressure Readings. Retrieved from http://www.heart.org/HEARTORG/Conditions/HighBloodPressure/AboutHighBloodPressure/Understanding-Blood-Pressure-Readings_UCM_301764_Article.jsp#.WKZd0xiZPFw

Aminsharifi, A., Hekmati, P., Noorafshan, A., Karbalay-Doost, S., Nadimi, E., Aryafar, A., . . . & ZarePoor, M. (2016). Scrotal cooling to protect against Cisplatin-induced spermatogenesis toxicity: Preliminary outcome of an experimental controlled trial. *Urology, 91*, 90–98. doi:10.1016/j.urology.2015.12.062

Amis, A. A. (2007). Current concepts on anatomy and biomechanics of patellar stability. *Sports Medicine and Arthroscopy Review, 15*(2), 48–56. doi:10.1097/JSA.0b013e318053eb74

Amsler, F., Willenberg, T., & Blättler, W. (2009). In search of optimal compression therapy for venous leg ulcers: A meta-analysis of studies comparing divers bandages with specifically designed stockings. *Journal of Vascular Surgery, 50*(3), 668–674. doi.10.1016/j.jvs.2009.05.018

Amundsen, C. L., Parsons, M., Tissot, B., Cardozo, L., Diokno, A., & Coats, A. C. (2007). Bladder diary measurements in asymptomatic females: Functional bladder capacity, frequency, and 24 hr volume. *Neurourology and Urodynamics, 26*(3), 341–349. doi:10.1002/nau.20241

Anders, C., Wagner, H., Puta, C., Grassme, R., & Scholle, H. C. (2009). Healthy humans use sex-specific co-ordination patterns of trunk muscles during gait. *European Journal of Applied Physiology, 105*(4), 585. doi:10.1007/s00421-008-0938-9

Andersen, K., Jensen, P. Ø., & Lauritzen, J. (1987). Treatment of clavicular fractures: Figure-of-eight bandage versus a simple sling. *Acta Orthopaedica Scandinavica, 58*(1), 71–74. doi:10.3109/17453678709146346

Anderson, B. E. (2012). *The Netter collection of medical illustrations. Volume 4, integumentary system* (2nd ed.). Philadelphia, PA: Elsevier Saunders.

Andrews, K. B. A., Verburg, G. B., Cooper, A. B., & Frick, J. G. (1977). Increased recovery in knitted cotton and cotton blends with comfort stretch. In N. R. S. Hollies, & R. F. Goldman (Eds.), *Clothing comfort: Interaction of thermal, ventilation, construction and assessment factors* (pp. 55–69). Ann Arbor, MI: Ann Arbor Science Publishers.

Annis, J. F., & Webb, P. (1971). *Development of a space activity suit.* NASA-CR-1892. Prepared for NASA by Webb Associates, Inc., Yellow Springs, OH. Retrieved from https://ntrs.nasa.gov/search.jsp?R=19720005428

Appelbaum, L. G., Schroeder, J. E., Cain, M. S., & Mitroff, S. R. (2011). Improved visual cognition through stroboscopic training. *Frontiers in Psychology, 2*, 276. doi:10.3389/fpsyg.2011.00276

Armstrong, A. L., Hunter, J. B., & Davis, T. R. C. (1994). The prevalence of degenerative arthritis of the base of the thumb in postmenopausal women. *Journal of Hand Surgery: British and European Volume, 19*(3), 340–341. doi:10.1016/0266-7681(94)90085-X

Arora, S. S., Cao, K., Jain, A. K., & Paulter, N. G. (2016). Design and fabrication of 3D fingerprint targets. *IEEE Transactions on Information Forensics and Security, 11*(10), 2284–2297. doi:10.1109/TIFS.2016.2581306

Artz, L., Macaluso, M., Brill, I., Kelaghan, J., Austin, H., Fleenor, M., . . . Hook, E. W., 3rd. (2000). Effectiveness of an intervention promoting the female condom to patients at sexually transmitted disease clinics. *American Journal of Public Health, 90*(2), 237–244.

Ashdown, S. P. (2014). Creation of ready-made clothing: The development and future of sizing systems. In M.-E. Faust, & S. Carrier (Eds.), *Designing apparel for consumers: The impact of body shape and size* (pp. 17–34). Cambridge, UK: Woodhead Publishing. doi:10.1533/9781782422150.1.17

Ashdown, S. P. (Ed.). (2007). *Sizing in clothing: Developing effective sizing systems for ready-to-wear clothing*, Cambridge, England: Woodhead Publishing.

Ashdown, S. P., & DeLong, M. (1995). Perception testing of apparel ease variation. *Applied Ergonomics, 26*(1), 47–54. doi:10.1016/0003-6870(95)95750-T

Ashdown, S. P. & Watkins, S. M. (1992). Movement analysis as the basis for the development and evaluation of a protective coverall design for asbestos abatement. In J. P. McBriarty, & N. W. Henry (Eds.), *Performance of protective clothing: 4th. Vol. ASTM STP 1133* (pp. 660–674) Philadelphia, PA: ASTM. doi:10.1520/STP19195S

ASTM International (1997). *Standard safety specification for drawstrings on children's upper outerwear* (ASTM F1816-97). West Conshohocken, PA: ASTM International. doi:10.1520/F1816-97R09

ASTM International. (2015a). *Standard test methods for equipment and procedures used in evaluating the performance characteristics of protective headgear* (ASTM F1446-2015-15b). West Conschohocken, PA: ASTM International. doi:10.1520/F1446-15B

ASTM International. (2015b). *Standard consumer safety specification for sling carriers* (ASTM F2907-15). West Conschohocken, PA: ASTM International. doi:10.1520/F2907-15

ASTM International. (2015c). *Standard test method for characterizing gripping performance of gloves using a torque meter* (ASTM F2961). West Conshohocken, PA: ASTM International. doi:10.1520/F2961-15

ASTM International. (2016a). *Standard consumer safety specifications for toy safety* (ASTM F963-16). West Conschohocken, PA: ASTM International. doi:10.1520/F0963-16

ASTM International. (2016b). *Standard test method for measuring the thermal insulation of clothing using a heated manikin* (ASTM F1291-16). West Conschohocken, PA: ASTM International. doi:10.1520/F1291-16

ASTM International. (2016c). *Standard practice for determining physiological responses of the wearer to protective clothing ensembles* (ASTM F2668-16). West Conschohocken, PA: ASTM International. doi:10.1520/F2668-16

ASTM International. (2017). *Standard practice for range of motion evaluation of first responder's protective ensembles* (ASTM F3031-17). West Conschohocken, PA: ASTM International. doi:10.1520/F3031-17

ASTM International. (2018). *Standard practices for evaluating the comfort, fit, function, and durability of protective ensembles, ensemble elements, and other components* (ASTM F1154-18). West Conschohocken, PA: ASTM International. doi:10.1520/F1154-18

Atkins, T., & Escudier, M. (2013). Force. In *A Dictionary of mechanical engineering*. Oxford: Oxford University Press.

Bach-y-Rita, P., & Kercel, S. W. (2003). Sensory substitution and the human–machine interface. *Trends in Cognitive Sciences*, 7(12), 541–546. doi:10.1016/j.tics.2003.10.013

Ball, R., Shu, C., Xi, P., Rioux, M. Luximon, Y., & Molenbroek, J. (2010). A comparison between Chinese and Caucasian head shapes. *Applied Ergonomics, 41*(6), 832–830. doi:10.1016/j.apergo.2010.02.002

Banner, L. W. (1983). *American beauty*. New York, NY: Alfred A. Knopf, Inc.

Barbier, M., & Boucher, S. (2010). *The story of women's underwear* (Vol. II). New York, NY: Parkstone Press International.

Barclay, A. E. (1932). The mobility of the abdominal viscera. *QJM: An International Journal of Medicine, 1*(2), 257–276. doi:10.1093/oxfordjournals.qjmed.a066585

Barclay, A. E. (1936). *The digestive tract* (2nd ed.). Cambridge, UK: Cambridge University Press.

Barnhart, K. T., Izquierdo, A., Pretorius, E. S., Shera, D. M., Shabbout, M., & Shaunik, A. (2006). Baseline dimensions of the human vagina. *Human Reproduction, 21*(6), 1618–1622. doi:10.1093/humrep/del022

Barrett, P. M., Komatireddy, R., Haaser, S., Topol, S., Sheard, J., Encinas, J., . . . Topol, E. J. (2014). Comparison of 24-hour Holter monitoring with 14-day novel adhesive patch electrocardiographic monitoring. *The American Journal of Medicine, 127*(1), 95-e11. doi:10.1016/j.amjmed.2013.10.003

Barrow, M. M. & Barrow, J. (2005). *Sun protection for life*. Oakland, CA: New Harbinger Publications, Inc.

Basmajian, J. V., & De Luca, C. J. (1985). *Muscles alive: Their functions revealed by electromyography* (5th ed.). Baltimore, MD: Williams & Wilkins.

Bates, C. K., Carroll, N., & Potter, J. (2011). The challenging pelvic examination. *Journal of General Internal Medicine: JGIM, 26*(6), 651–657. doi:10.1007/s11606-010-1610-8

Baugher, W. H. & Graham, T. J. (2014). It's only a game: The athlete's hand. In E. F. S. Wilgis (Ed.), *The wonder of the human hand* (pp. 39–53). Baltimore, MD: John Hopkins University Press.

Baumgartner, R. N. (2005). Age. In Heymsfield, S. B., Lohman, T. G., Wang, Z.-M., & Going, S. B. (Eds.), *Human body composition* (pp. 259–260). Champaign, IL: Human Kinetics.

Baumgartner, R. N., Heymsfield, S. B., & Roche, A. F. (1995). Human body composition and the epidemiology of chronic disease. *Obesity, 3*(1), 73–95. doi:10.1002/j.1550-8528.1995.tb00124.x

Bellemare, F., Jeanneret, A., & Couture, J. (2003). Sex differences in thoracic dimensions and configuration. *American Journal of Respiratory and Critical Care Medicine, 168*(3), 305–312. doi:10.1164/rccm.200208-876OC

Bender, L. F. (1990a). Upper extremity orthotics. In F. J. Kottke, & J. F. Lehmann (Eds.), *Krusen's handbook of physical medicine and rehabilitation* (4th ed., pp. 580–601). Philadelphia, PA: W. B. Saunders Company.

Bender, L. F. (1990b). Upper extremity prosthetics. In F. J. Kottke, & J. F. Lehmann (Eds.), *Krusen's handbook of physical medicine and rehabilitation* (4th ed., pp. 1009–1023). Philadelphia, PA: W. B. Saunders Company.

Ben-Menachem, E., Revesz, D., Simon, B. J., & Silberstein, S. (2015). Surgically implanted and non invasive vagus nerve stimulation: A review of efficacy, safety and tolerability. *European Journal of Neurology, 22*(9), 1260–1268. doi:10.1111/ene.12629

Benson, E., & Rajulu, S. (2009) Complexity of sizing for space suit applications. In V. G. Duffy (Ed.), *Digital human modeling: ICDHM 2009 lecture notes in computer science,* Volume 5620 (pp. 599–607). Berlin, Germany: Springer. doi:10.1007/978-3-642-02809-0_63

Bergan, J. J. (1985). Conrad Jobst and the development of pressure gradient therapy for venous disease. In J. Bergan, & J. Yao (Eds.), *Surgery of the veins* (pp. 529–540). Orlando, FL: Grune & Stratton.

Bergmann, G., Graichen, F., & Rohlmann, A. (1993). Hip joint loading during walking and running, measured in two patients. *Journal of Biomechanics, 26*(8), 969–990. doi:10.1016/0021-9290(93)90058-M

Bernhard, L., Bernhard, P., & Magnussen, P. (2003). Management of patients with lymphoedema caused by filariasis in North-eastern Tanzania: Alternative approaches. *Physiotherapy, 89*(12), 743–749. doi:10.1016/S0031-9406(05)60500-7

Bernstein, N. A. (1967). *The co-ordination and regulation of movements* (1st English ed.). New York, NY: Pergamon Press.

Bernstein, R. A. (2015). Arthritis of the thumb and digits: Current concepts. *AAOS Instructional Course Lectures, 64*, pp. 281–294.

Bird, A. R., Bendrups, A. P., & Payne, C. B. (2003). The effect of foot wedging on electromyographic activity in the erector spinae and gluteus medius muscles during walking. *Gait & Posture, 18*(2), 81–91. doi:10.1016/S0966-6362(02)00199-6

Black, C. M. (2014). Anatomy and physiology of the lower-extremity deep and superficial veins. *Techniques in Vascular and Interventional Radiology, 17*(2), 68–73. doi:10.1053/j.tvir.2014.02.002

Blount, W. P., Schmidt, A. C., Keever, E. D., & Leonard, E. T. (1958). The Milwaukee brace in the operative treatment of scoliosis. *JBJS, 40*(3), 511–525.

Bluestone, C. D. (1996). Pathogenesis of otitis media: Role of eustachian tube. *The Pediatric Infectious Disease Journal, 15*(4), 281–291.

Blunt-Vinti, H. D., Thompson, E. L., & Griner, S. B. (2018). Contraceptive use effectiveness and pregnancy prevention information preferences among heterosexual and sexual minority college women. *Women's Health Issues, 28*(4), 342–349. doi:10.1016/j.whi.2018.03.005

Boateng, J., & Catanzano, O. (2015). Advanced therapeutic dressings for effective wound healing—A review. *Journal of Pharmaceutical Sciences, 104*(11), 3653–3680. doi:10.1002/jps.24610

Boden, B. P., Dean, G. S., Feagin, J. A., & Garrett, W. E. (2000). Mechanisms of anterior cruciate ligament injury. *Orthopedics, 23*(6), 573–578.

Bogduk, N., & Mercer, S. (2000). Biomechanics of the cervical spine. I: Normal kinematics. *Clinical Biomechanics, 15*(9), 633–648. doi:10.1016/S0268-0033(00)00034-6

Boman, A., Estlander, T., Wahlberg, J. E., & Maibach, H. I., (Eds.). (2005). *Protective gloves for occupational use* (2nd ed.). Boca Raton, FL: CRC Press.

Bonanno, D. R., Landorf, K. B., Munteanu, S. E., Murley, G. S., & Menz, H. B. (2017). Effectiveness of foot orthoses and shock-absorbing insoles for the prevention of injury: A systematic review and meta-analysis. *British Journal of Sports Medicine, 51*(2), 86-96. doi:10.1136/bjsports-2016-096671

Bortole, M., Venkatakrishnan, A., Zhu, F., Moreno, J. C., Francisco, G. E., Pons, J. L., & Contreras-Vidal, J. L. (2015). The H2 robotic exoskeleton for gait rehabilitation after stroke: Early findings from a clinical study. *Journal of Neuroengineering and Rehabilitation, 12*(1), 54. doi:10.1186/s12984-015-0048-y

Bouchard, C., & Johnston, F. E. (1988). *Fat distribution during growth and later health outcomes.* New York, NY: Liss.

Bougourd, J., & Treleaven, P. (2014). National size and shape surveys for apparel design. In D. Gupta, & N. Zakaria (Eds.), *Anthropometry, apparel sizing and design* (pp. 141–166). London, UK: Woodhead Publishing. doi:10.1533/9780857096890.2.141

Boulais, N. & Misery, L. (2007). Merkel cells. *Journal of the American Academy of Dermatology, 57*(1), 147–165. doi:10.1016/j.jaad.2007.02.009

Boussu, F., & Bruniaux, P. (2012). Customization of a lightweight bullet-proof vest for the female form. In E. Sparks (Ed.), *Advances in military textiles and personal equipment* (pp. 167–195). Oxford, UK: Woodhead Publishing. doi:10.1533/9780857095572.2.167

Bowman, K. (2014). *Move your DNA: Restore your health through natural movement,* Carlsborg, WA: Propriometrics Press.

Brackley, H. M., & Stevenson, J. M. (2004). Are children's backpack weight limits enough?: A critical review of the relevant literature. *Spine, 29*(19), 2184–2190.

Brady, M., & Global Alliance to Eliminate Lymphatic Filariasis (2014). Seventh meeting of the Global Alliance to Eliminate Lymphatic Filariasis: Reaching the vision by scaling up, scaling down, and reaching out. *Parasites & Vectors 7*(46), 1–19. doi:10.1186/1756-3305-7-46

Brander, M. (1942). Tampons in menstruation. *British Medical Journal, 1*(4239), 452. doi:10.1136/bmj.1.4239.452-a

Branson, D. H. & Nam, J. (2007). Materials and sizing. In S. P. Ashdown (Ed.), *Sizing in clothing: Developing effective sizing systems for ready-to-wear clothing* (pp. 264–276). Cambridge, England: Woodhead Publishing.

Brauer, C. A., Coca-Perraillon, M., Cutler, D. M., & Rosen, A. B. (2009). Incidence and mortality of hip fractures in the United States. *JAMA, 302*(14), 1573–1579. doi:10.1001/jama.2009.1462

Braun, B. L., & Amundson, L. R. (1989). Quantitative assessment of head and shoulder posture. *Archives of Physical Medicine and Rehabilitation, 70*(4), 322–329.

Britton, C. J. (1996, November). Learning about "the curse": An anthropological perspective on experiences of menstruation. *Women's Studies International Forum, 19*(6), 645–653. doi:10.1016/S0277-5395(96)00085-4

Brouwer, T. A., Eindhoven, B. G., Epema, A. H., & Henning, R. H. (1999). Validation of an ultrasound scanner for determining urinary volumes in surgical patients and volunteers. *Journal of Clinical Monitoring and Computing, 15*(6), 379–385. doi:10.1023/A:1009939530626

Brown, M. S., Ashley, B., & Koh, A. (2018). Wearable technology for chronic wound monitoring: Current dressings, advancements, and future prospects. *Frontiers in Bioengineering and Biotechnology, 6,* 47. doi:10.3389/fbioe.2018.00047

Brueck, A., Iftekhar, T., Stannard, A. B., Yelamarthi, K., & Kaya, T. (2018). A real-time wireless sweat rate measurement system for physical activity monitoring. *Sensors, 18*(2), 533. doi:10.3390/s18020533

Brzezinski, W. (1990). Blood pressure. In H. K. Walker, W. D. Hall, & J. W. Hurst (Eds.), *Clinical methods: The history, physical, and laboratory examinations* (3rd ed., pp. 95–97). Boston, MA: Butterworths.

Büchler, L., Tannast, M., Siebenrock, K. A., & Schwab, J. M. (2018). Biomechanics of the hip. In K. A. Egol, & P. Leucht (Eds.), *Proximal femur fractures* (pp. 9–15). Cham, Switzerland: Springer. doi:10.1007/978-3-319-64904-7_2

Buckey, J. C., Jr. (2006). *Space physiology*. Oxford, U.K.: Oxford University Press.

Buckley, T., & Gottlieb, A. (1988). *Blood magic: The anthropology of menstruation.* Berkeley, CA: University of California Press.

Bullough, V. L. (1981). A brief note on rubber technology and contraception: The diaphragm and the condom. *Technology and Culture, 22*(1), 104–111. doi:10.2307/3104294

Bullough, V. L. (1985). Merchandising the sanitary napkin: Lillian Gilbreth's 1927 survey. *Signs: Journal of Women in Culture and Society, 10*(3), 615–627. doi:10.1086/494174

Burgess, L-A. (2000). *Making beautiful bras.* Wagga Wagga, Australia: R. O. Burgess.

Burt, J., & White, G. (1999). *Lymphedema.* Alameda, CA: Hunter House Publishers.

Bye, E., & Griffin, L. (2015). Testing a model for wearable product materials research. *International Journal of Fashion Design, Technology and Education, 8*(2), 139–150. doi:10.1080/17543266.2015.1018959

Bye, E., LaBat, K. L., & DeLong, M. R. (2006). Analysis of body measurement systems for apparel. *Clothing and Textiles Research Journal, 24*(2): 66–79. doi:10.1177/0887302X0602400202

Bye, E., LaBat, K., McKinney, E., & Kim, D.-E. (2008). Optimized pattern grading. *International Journal of Clothing Science and Technology, 20*(2), 79–92. doi:10.1108/09556220810850469

Cai, Y., Yu, W., & Chen, L. (2016). Finite element modeling of bra fitting. In W. Yu (Ed.), *Advances in women's intimate apparel technology.* Cambridge, UK: Woodhead Publishing in association with The Textile Institute (pp. 147–168). doi:10.1016/B978-1-78242-369-0

Callaghan, M. J., Parkes, M. J., Hutchinson, C. E., Gait, A. D., Forsythe, L. M., Marjanovic, E. J., . . . & Felson, D. T. (2015). A randomised trial of a brace for patellofemoral osteoarthritis targeting knee pain and bone marrow lesions. *Annals of the Rheumatic Diseases, 74*(6), 1164–1170. doi:10.1136/annrheumdis-2014-206376

Callard, G. V., Litofsky, F. S., & DeMerre, L. J. (1966). Menstruation in women with normal or artificially controlled cycles. *Fertility and Sterility, 17*(5), 684–688.

Callewaert, C., De Maeseneire, E., Kerckhof, F.-M., Verliefde, A., Van de Wiele, T., & Boon, N. (2014). Microbial odor profile of polyester and cotton clothes after a fitness session. *Applied and Environmental Microbiology, 80*(21), 6611–6619. doi:10.1128/AEM.01422-14

Cameron, I. D., Robinovitch, S., Birge, S., Kannus, P., Khan, K., Lauritzen, J., ... & Kiel, D. P. (2010). Hip protectors: Recommendations for conducting clinical trials—an international consensus statement (part II). *Osteoporosis International, 21*(1), 1–10. doi:10.1007/s00198-009-1055-2

Carneiro, V. H., Meireles, J., & Puga, H. (2013). Auxetic materials—A review. *Materials Science-Poland, 31*(4), 561–571. doi:10.2478/s13536-013-0140-6

Carr, D. J. (2016). Applications of blast injury research: Solving clinical problems and providing mitigation. In A. M. J. Bull, J. Clasper & P. F. Mahoney (Eds.), *Blast injury science and engineering: A guide for clinicians and researchers* (pp. 261–264). Cham, Switzerland: Springer International. doi:10.1007/978-3-319-21867-0_24

Carrilero, L. (2016). Rib and sternum injuries in an athlete. In J. A. Neumann, D. T. Kirkendall, & C. T. Moorman, III (Eds.), *Sports medicine for the orthopedic resident* (pp. 51–59). Singapore: World Scientific.

Castilho, O. T., Jr., Dezotti, N. R. A., Dalio, M. B., Joviliano, E. E., & Piccinato, C. E. (2018). Effect of graduated compression stockings on venous lower limb hemodynamics in healthy amateur runners. *Journal of Vascular Surgery: Venous and Lymphatic Disorders, 6*(1), 83–89. doi:10.1016/j.jvsv.2017.08.011

CDC (n.d.) Condoms and STDs: Fact sheet for public health personnel. Retrieved from https://www.cdc.gov/condomeffectiveness/docs/Condoms_and_STDS.pdf

Centers for Disease Control (CDC). (2013). *National Health and Nutrition Examination Survey (NHANES).*Retrieved from http://www.cdc.gov/nchs/nhane.htm

Chahla J., O'Brien L., Godin J. A., & LaPrade R. F. (2018) Return to play after multiple knee ligament injuries. In V. Musahl, J. Karlsson, W. Krutsch, B. Mandelbaum, J. Espregueira-Mendes, & P. d'Hooghe (Eds.), *Return to play in football.* Berlin, Germany: Springer. doi:10.1007/978-3-662-55713-6_47

Chalmers, L. (1937). *U.S. Patent No. 2,089,113.* Washington, DC: U.S. Patent and Trademark Office.

Chan, A. C. K. (2014). The development of apparel sizing systems from anthropometric data. In D. Gupta & N. Zakaria (Eds.), *Anthropometry, apparel sizing and design* (pp. 167–196). London, UK: Woodhead Publishing. doi:10/1533/9780857096890.2.167

Chanet, S., & Martin, A. C. (2014). Mechanical force sensing in tissues. *Progress in Molecular Biology and Translational Science, 126,* 317–352. doi:10.1016/B978-0-12-394624-9.00013-0

Chase, D. J., Schenkel, B. P., Fahr, A. M., Eigner, U., & Tampon Study Group. (2010). Randomized, double-blind crossover study of vaginal microflora and epithelium in women using a tampon with a "winged" apertured film cover and a commercial tampon with a nonwoven fleece cover. *Journal of Clinical Microbiology, 48*(4), 1317–1322. doi:10.1128/JCM.00359-09

Chasin, M. (2009). *Hearing loss in musicians: Prevention and management.* San Diego, CA: Plural Publishing.

Chatterjee, K. N., Jhanji, Y., Grover, T., Bansal, N., & Bhattacharyya, S. (2015). Selecting garment accessories, trims, and closures. In R. Nayak, & R. Padhye (Eds.), *Garment manufacturing technology* (pp. 129–183). Cambridge, UK: Woodhead Publishing. doi:10.1016/B978-1-78242-232-7.00006-0

Cheatham, R. H. (1908). *U.S. Patent No. 890,842A.* Washington, DC: U.S. Patent and Trademark Office.

Cheatham, S. W., Kolber, M. J., & Salamh, P. A. (2013). Meralgia paresthetica: A review of the literature. *International Journal of Sports Physical Therapy, 8*(6), 883–893.

Chen, C.-M., LaBat, K., & Bye, E. (2010). Physical characteristics related to bra fit. *Ergonomics, 53*(4), 514–524. doi:10.1080/00140130903490684

Chen,C-M.,LaBat,K.,&Bye,E.(2011).Bustprominencerelatedtobrafitproblems.*International Journal of Consumer Studies, 35,* 695–701. doi:10.1111/j.1470-6431.2010.00984x

Chen, T. L.-W., Sze, L. K., Davis, I. S., & Cheung, R. T. (2016). Effects of training in minimalist shoes on the intrinsic and extrinsic foot muscle volume. *Clinical Biomechanics, 36,* 8–13. doi:10.1016/j.clinbiomech.2016.05.010

Chen, X., & Wang, J. (2015). Breast volume measurement by mesh projection method based on 3D point cloud data. *International Journal of Clothing Science and Technology, 27*(2), 221–236. doi:10.1108/IJCST-11-2013-0124

Cheng, Z., & Robinette, K. (2009). Static and dynamic human shape modeling. In V.G. Duffy (Ed.), *Digital human modeling: ICDHM 2009 Lecture notes in computer science,* Volume5620(pp.3–12).Berlin,Germany:Springer.doi:10.1007/978-3-642-02809-0_1

Cho, H. S. M., Paek, J. U., Davis, G., & Fedric, T. (2008). Expanding the comfort of postmastectomy patients using the Papilla gown. *Journal of Nursing Scholarship, 40*(1), 26–31. doi:10.1111/j.1547-5069.2007.00202.x

Choi, I., Lee, S., & Hong, Y. (2012). The new era of the lymphatic system: No longer secondary to the blood vascular system. *Cold Spring Harbor Perspectives in Medicine, 2*(4), a006445. doi:10.1101/cshperspect.a006445

Choung, W., & Heinrich, S. D. (1995). Acute annular ligament interposition into the radiocapitellar joint in children (nursemaid's elbow). *Journal of Pediatric Orthopedics, 15*(4), 454–456.

Chowdhury, M. M. U., & Maibach, H. I. (Eds.). (2004). *Latex intolerance: Basic science, epidemiology, and clinical management.* Boca Raton, FL: CRC Press.

Cianchetti, C., Serci, M. C., Pisano, T., & Ledda, M. G. (2010). Compression of superficial temporal arteries by a handmade device: A simple way to block or attenuate migraine attacks in children and adolescents. *Journal of Child Neurology, 25*(1), 67–70. doi:10.1177/0883073809333534

Cianferotti, L., Fossi, C., & Brandi, M. (2015). Hip protectors: Are they worth it? *Calcified Tissue International, 97*(1), 1–11. doi:10.1007/s00223-015-0002-9

Clapham, P., & Chung, K. C. (2014). Worn down, but not out: The arthritic hand. In E. F. S. Wilgis (Ed.), *The wonder of the human hand* (pp. 54–65). Baltimore, MD: John Hopkins University Press.

Clauser, C., Tebbetts, I., Bradtmiller, B., McConville, J., & Gordon, C. (1988). *Measurer's handbook: U.S. Army anthropometric survey 1987-1988.* (Technical Report Natick/TR-88043.) Natick, MA: U.S. Army Natick Soldier Research, Development and Engineering Center.

Clément, G., & Reschke, M. F. (2008). *Neuroscience in space.* New York, NY: Springer. doi:10.1007/978-0-387-78950-7

Coffey, J. C., & O'Leary, D. P. (2016). The mesentery: Structure, function, and role in disease. *The Lancet Gastroenterology & Hepatology, 1*(3), 238–247. doi:10.1016/S2468-1253(16)300267

Cofield, C. (2018, February 20). NASA's new spacesuit has a built-in toilet. *Space.com.* Retrieved from https://www.space.com/39710-orion-spacesuit-waste-disposal-system.html

Cole, S. (2010). *The story of men's underwear* (Vol. I). New York, NY: Parkstone Press International.

Cole, T. M., & Tobis, J. S. (1990a). Measurement of musculoskeletal function. In F. J. Kottke, & J. F. Lehmann (Eds.), *Krusen's handbook of physical medicine and rehabilitation* (4th ed., pp. 20–71). Philadelphia, PA: W. B. Saunders Company.

Cole, T. M., & Tobis, J. S. (1990b). Measurement of musculoskeletal function. In F. J. Kottke, & J. F. Lehmann (Eds.), *Krusen's handbook of physical medicine and rehabilitation* (4th ed., p. 25). Philadelphia, PA: W. B. Saunders Company.

Collins, C. C., & Bach-y-Rita, P. (1973) Transmission of pictorial information through the skin. In J. H. Lawrence, & J. W. Gofman, (Eds.), *Advances in biological and medical physics: Vol. 14* (pp. 285–315). New York, NY: Academic Press.

Colonna, M., Nicotra, M., & Moncalero, M. (2013). Materials, designs and standards used in ski-boots for alpine skiing. *Sports, 1*(4), 78–113. doi:10.3390/sports1040078

Coltman, C. E., McGhee, D. E., & Steele, J. R. (2017). Three-dimensional scanning in women with large, ptotic breasts: Implications for bra cup sizing and design. *Ergonomics, 60*(3), 439–445. doi:10.1080/00140139.2016.1176258

Coltman, C. E., Steele, J. R., & McGhee, D. E. (2017). Effect of aging on breast skin thickness and elasticity: Implications for breast support. *Skin Research and Technology, 23*(3), 303–311. doi:10.1111/srt.12335

Conley, M. S., Meyer, R. A., Feeback, D. L., & Dudley, G. A. (1995). Noninvasive analysis of human neck muscle function. *Spine, 20*(23), 2505–2512.

Connell, L. J., Ulrich, P. V., Brannon, E. L., Alexander, M., & Presley, A. B. (2006). Body shape assessment scale: Instrument development for analyzing female figures. *Clothing and Textile Research Journal, 24*(2), 80–95. doi:10.1177/08873 02X0602400203

Cooklin, G. (1991). *Pattern grading for children's clothes: The technology of clothing.* Oxford, England: BSP Professional Books.

Couldrick, C. A. (2004). *A systems approach to the design of personal armour for explosive ordnance disposal.* (Doctoral engineering dissertation). Cranfield University, United Kingdom. Retrieved from https://dspace.lib.cranfield.ac.uk

Crawford, M. H., Bernstein, S. J., Deedwania, P. C., DiMarco, J. P., Ferrick, K. J., Garson, A., . . . & Smith, S., Jr. (1999). ACC/AHA guidelines for ambulatory electrocardiography: A report of the American College of Cardiology/American Heart Association task force on practice guidelines. *Journal of the American College of Cardiology, 34*(3), 912–948. doi:10.1016/S0735-1097(99)00354-X

Crisco, J. J., Wilcox, B. J., Beckwith, J. G., Chu, J. J., Duhaime, A., Rowson, S., . . . & Greenwald, R. M. (2011). Head impact exposure in collegiate football players. *Journal of Biomechanics, 44*(15), 2673–2678. doi:10.1016/j.jbiomech.2011.08.003

Croney, J. (1971). *Anthropometrics for designers.* London, U.K.: Batsford, Ltd.

Csapo, R., Maganaris, C. N., Seynnes, O. R., & Narici, M. V. (2010). On muscle, tendon and high heels. *Journal of Experimental Biology 213*(15), 2582–2588. doi:10.1242/jeb.044271

Cudejko, T., van der Esch, M., van der Leeden, M., Roorda, L. D., Pallari, J., Bennell, K. L., . . . & Dekker, J. (2018). Effect of soft braces on pain and physical function in patients with knee osteoarthritis: Systematic review with meta-analyses. *Archives of Physical Medicine and Rehabilitation, 99*(1), 153–163. doi:10.1016/j.apmr.2017.04.029

Cullinane, C. R., Rhodes, R. A., & Stirling, L. A. (2017). Mobility and agility during locomotion in the Mark III space suit. *Aerospace Medicine and Human Performance, 88*(6), 589–596. doi:10.3357/AMHP.4650.2017

Cusick, J. F., & Yoganandan, N. (2002). Biomechanics of the cervical spine 4: Major injuries. *Clinical Biomechanics, 17*(1), 1–20. doi:10.1016/S0268-0033(01)00101-2

Cuta, K., Dujovski, C., Froissard, L., Henderson, C., Rust, D., & Viner, A. (2016). *U.S. Patent No. 9,507,175 B2.* Washington, DC: U.S. Patent and Trademark Office.

Daban, C., Martinez-Aran, A., Cruz, N., & Vieta, E. (2008). Safety and efficacy of vagus nerve stimulation in treatment-resistant depression: A systematic review. *Journal of Affective Disorders, 110*(1), 1–15. doi:10.1016/j.jad.2008.02.012

Dahlqvist, A. (2018). *It's only blood: Shattering the taboo of menstruation.* Translated by Alice Olsson (translator). London, UK: Zed Books.

Daniell, N., Olds, T., & Tomkinson, G. (2010). The importance of site location for girth measurements. *Journal of Sports Sciences, 28*(7), 751–757. doi:10:1080/02640411003645703

Das, D. (2014). Composite nonwovens in absorbent hygiene products. In D. Das, & B. Pourdeyhimi (Eds.), *Composite nonwoven materials: Structure, properties and applications* (pp. 74–88). Cambridge, UK: Elsevier Science. doi:10.1533/9780857097750.74

Davey, G., & Newport, M. (2007). Podoconiosis: The most neglected tropical disease? *The Lancet, 369*(9565), 888–889. doi:10.1016/S0140-6736(07)60425-5

Del Mar, B. S. (2005). The history of clinical Holter monitoring. *Annals of Noninvasive Electrocardiology, 10*(2), 226–230. doi:10.1111/j.1542-474X.2005.10202.x

DeLong, M. R. (1998). *The way we look: Dress and aesthetics* (2nd ed.). New York, NY: Fairchild Publications.

DeLong, M., & Daly, C. (2013) Eyewear, fashion, design, and health. *Fashion Practice, 5*(1), 117–127. doi:10.2752/175693813X13559997788880

DeMarinis, M., Kaschak, T. R., & Newman, D. K. (2018). Absorbent products for incontinence. In D. K. Newman, E. S. Rovner, & A. J. Wein (Eds.), *Clinical application of urologic catheters, devices and products* (pp. 149–172). Cham, Switzerland: Springer. doi:10.1007/978-3-319-14821-2_6

Derakhshan, M. H., Robertson, E. V., Fletcher, J., Jones, G. R., Lee, Y. Y., Wirz, A. A., & McColl, K. E. (2012). Mechanism of association between BMI and dysfunction of the gastro-oesophageal barrier in patients with normal endoscopy. *Gut, 61*(3), 337–343. doi:10.1136/gutjnl-2011-300633

Devarajan, P., & Istook, C. (2004). Validation of "female figure identification technique (FFIT) for apparel©" software. *Journal of Textile and Apparel, Technology and Management, 4*(1), 1–23.

Dewey, F. E., Rosenthal, D., Murphy Jr, D. J., Froelicher, V. F., & Ashley, E. A. (2008). Does size matter? Clinical applications of scaling cardiac size and function for body size. *Circulation, 117*(17), 2279–2287. doi:10.1161/CIRCULATIONAHA.107.736785

Dodds, W. J., Hogan, W. J., Stewart, E. T., Stef, J. J., & Arndorfer, R. C. (1974). Effects of increased intra-abdominal pressure on esophageal peristalsis. *Journal of Applied Physiology, 37*(3), 378–383. doi:10.1152/jappl.1974.37.3.378

Dolibog, P., Franek, A., Taradaj, J., Dolibog, P., Blaszczak, E., Polak, A., . . . & Kolanko, M. (2014). A comparative clinical study on five types of compression therapy in patients with venous leg ulcers. *International Journal of Medical Sciences, 11*(1), 34–43. doi:10.7150/ijms.7548

Donnanno, A. (2016). *Fashion patternmaking techniques (Vol. 2): How to make shirts, undergarments, dresses and suits, waistcoats and jackets for women and men*. Barcelona, Spain: Promopress.

Douglass, J., Graves, P., & Gordon, S. (2016). Self-care for management of secondary lymphedema: A systematic review. *PLoS Neglected Tropical Diseases, 10*(6), e0004740. doi:10.1371/journal.pntd.0004740

Douty, H. I. (1968). Visual somatometry in health related research. *Journal of Alabama Academy of Science, 39*, 21–34.

Drake, R. L., Vogl, A.W., & Mitchell, A. W. M. (2015). *Gray's anatomy for students* (3rd ed.). Philadelphia, PA: Churchill Livingstone.

Drucker, A. M., Wang, A. R., Li, W. Q., Sevetson, E., Block, J. K., & Qureshi, A. A. (2017). The burden of atopic dermatitis: Summary of a report for the National Eczema Association. *Journal of Investigative Dermatology, 137*(1), 26–30. doi:10.1016/j.jid.2016.07.012

Dunne, L. E., Gioberto, G., & Koo. H. (2011). A method of measuring garment movement for wearable sensing. *Proceedings of the 15th Annual International Symposium on Wearable Computers*, San Francisco, CA: IEEE Computer Society, pp. 11–14. doi:10.1109/ISWC.2011.19

Dunne, L. E., Walsh, P., Hermann, S., Smyth, B., & Caulfield, B. (2008). Wearable monitoring of seated spinal posture. *IEEE Transactions on Biomedical Circuits and Systems, 2*(2), 97–105. doi:10.1109/TBCAS.2008.927246

Durairajanayagam, D., Sharma, R. K., Du Plessis, S. S., & Agarwal, A. (2014). Testicular heat stress and sperm quality. In S. S. du Plessis, A. Agarwal, & E. S. Sabanegh, Jr. (Eds.), *Male infertility: A complete guide to lifestyle and environmental factors* (pp. 105–125). New York, NY: Springer. doi:10.1007/978-1-4939-1040-3_8

Durfee, W., & Iaizzo, P. (Eds.). (2015). *Medical device innovation handbook*. Minneapolis, MN: Medical Devices Center, University of Minnesota. Retrieved from http://www.me.umn.edu/~wkdurfee/projects/InnovationWorkshop/InnovationHandbook/innovation-handbook.pdf

Duvall, J. C., Dunne, L. E., Schleif, N., & Holschuh, B. (2016, September). Active hugging vest for deep touch pressure therapy. In *Proceedings of the 2016 ACM International Joint Conference on Pervasive and Ubiquitous Computing: Adjunct UbiComp* (pp. 458–463). September 12–16, 2016, Heidelberg, Germany: ACM. doi:10.1145/2968219.2971344

Duvall, J., Granberry, R., Dunne, L. E., Holschuh, B., Johnson, C., Kelly, K., . . . & Joyner, M. (2017). The design and development of active compression garments for orthostatic intolerance. In ASME *Proceedings of Design of Medical Devices Conference, Frontiers in Biomedical Devices*. April 10–13, 2017, Minneapolis, MN (p. V001T01A013). doi:10.1115/DMD2017-3480

Dwyer, P. L., Lee, E. T. C., & Hay, D. M. (1988). Obesity and urinary incontinence in women. *BJOG: An International Journal of Obstetrics & Gynaecology, 95*(1), 91–96. doi:10.1111/j.1471-0528.1988.tb06486.x

Ebling, F. J. G. (1987). The biology of hair. *Dermatologic Clinics, 5*(3), 467–481.

Egan, M., & Brousseau, L. (2007). Splinting for osteoarthritis of the carpometacarpal joint: Review of the evidence. *American Journal of Occupational Therapy, 61*(1), 70–78. doi:10.5014/ajot.61.1.70

Egan, M., Brosseau, L., Farmer, M., Ouimet, M. A., Rees, S., Tugwell, P., & Wells, G. (2003). Splints/orthoses in the treatment of rheumatoid arthritis. *Cochran Database of Systematic Reviews (1)*, CD004018. doi:10.1002/14651858.CD004018

Eicher, J. B., Evenson, S. L., & Lutz, H. A. (2008). The visible self: *Global perspectives on dress, culture, and society* (3rd ed.). New York, NY: Fairchild Publications.

Eldridge, G. D., St. Lawrence, J. S., Little, C. E., Shelby, M. C., & Brasfield, T. L. (1995). Barriers to condom use and barrier method preferences among low-income African-American women. *Women & Health, 23*(1), 73–89. doi:10.1300/J013v23n01_05

Ellena, T., Subic, A., Mustafa, H., & Pang, T. Y. (2016). The helmet fit index: An intelligent tool for fit assessment and design customization. *Applied Ergonomics, 55*, 194–207. doi:10.1016/j.apergo.2016.02.008

Elsner, P., Hatch, K., & Wigger-Alberti, W. (Eds.). (2003). In G. Burg (Series Ed.), *Current problems in dermatology: Vol. 31. Textiles and the skin*. Basel, Switzerland: S. Karger AG.

Emslie, M., & Wilkie, C. (1938). New menstruation toilet. *British Medical Journal*, 2(4067), 1282. Retrieved from http://www.jstor.org.ezp1.lib.umn.edu/stable/20301768

Engel, F., & Sperlich, B. (Eds.). (2016). *Compression garments in sports: Athletic performance and recovery*. Cham, Switzerland: Springer International.

England, S. A., Benson, E. A., & Rajulu, S. L. (2010, May). *Functional mobility testing: Quantification of functionally utilized mobility among unsuited and suited subjects (NASA/TP-1010-216122)*. Houston, TX: Johnson Space Center. Retrieved from https://ston.jsc.nasa.gov/collections.trs/

Engrav, L. H., Heimbach, D. M., Rivara, F. P., Moore, M. L., Wang, J., Carrougher, G. J., . . . Gibran, N. S. (2010). 12-year within-wound study of the effectiveness of custom pressure garment therapy. *Burns, 36*(7), 975–983. doi:10.1016/j.burns.2010.04.014

Erickson, K., McMahon, M., Dunne, L. E., Larsen, C., Olmstead, B., & Hipp, J. (2016). Design and analysis of a sensor-enabled in-ear device for physiological monitoring. *Journal of Medical Devices, 10*(2), 020966. doi:10.1115/1.4033200

Esculier, J. F., Dubois, B., Dionne, C. E., Leblond, J., & Roy, J. S. (2015). A consensus definition and rating scale for minimalist shoes. *Journal of Foot and Ankle Research, 8*(1), 42. doi:10.1186/s13047-015-0094-5

Evans, W. J., & Lexell, J. (1995). Human aging, muscle mass, and fiber type composition. *The Journals of Gerontology: Series A, 50A* (Special Issue), 11-16. doi:10.1093/gerona/50A.Special_Issue.11

Everett, T., & Kell, C. (Eds.). (2010). *Human movement: An introductory text* (6th ed.). Edinburgh, UK: Elsevier, Churchill Livingstone.

Ewing, E. (1978). *Dress and undress: A history of women's underwear.* New York, NY: Drama Book Specialists.

Eze, B. I., Uche, J. N., Shiweobi, J. O., & Mba, C. N. (2013). Oculopalpebral dimensions of adult Nigerians: Report from the Enugu normative ocular anthropometry study. *Medical Principles and Practice: International Journal of the Kuwait University, Health Science Centre, 22*(1), 75–79. doi:10.1159/000339800

Fairbank, J. C., Pynsent, P. B., van Poortvliet, J. A., & Phillips, H. (1984). Mechanical factors in the incidence of knee pain in adolescents and young adults. *Journal of Bone & Joint Surgery, British volume, 66*(5), 685–693. doi:10.1302/0301-620X.66B5.6501361

Farage, M., Bramante, M., Otaka, Y., & Sobel, J. (2007). Do panty liners promote vulvovaginal candidiasis or urinary tract infections?: A review of the scientific evidence. *European Journal of Obstetrics & Gynecology and Reproductive Biology, 132*(1), 8–19. doi:10.1016/j.ejogrb.2006.11.015

Farrell-Beck, J., & Gau, C. (2002). *Uplift.* Philadelphia, PA: University of Pennsylvania Press.

Faust, M.-E. (2014). Apparel size designations and labeling. In D. Gupta, & N. Zakaria (Eds.), *Anthropometry, apparel sizing and design* (pp. 255–273). London, UK: Woodhead Publishing. doi:10.1533/9780857096890.2.255

Feather, B. L., Ford, S., & Herr, F. B. (1996). Female collegiate basketball players' perceptions about their bodies, garment fit and uniform design preferences. *Clothing and Textile Research Journal, 14*, 22–29. doi:10.1177/0887302X9601400104

Federative Committee on Anatomical Terminology. (1998). *Terminologia anatomica: International anatomical terminology.* New York, NY: Thieme. Accessed online at http://terminologia-anatomica.org/en/Terms

Feger, M. A., Donovan, L., Hart, J. M., & Hertel, J. (2014). Effect of ankle braces on lower extremity muscle activation during functional exercises in participants with chronic ankle instability. *International Journal of Sports Physical Therapy, 9*(4), 476–487.

Felner, J. M. (1990). An overview of the cardiovascular system. In H. K. Walker, W. D. Hall & J. W. Hurst (Eds.), *Clinical methods: The history, physical, and laboratory examinations* (3rd ed., pp. 57–68). Boston, MA: Butterworths.

Ferrari, J., Hopkinson, D. A., & Linney, A. D. (2004). Size and shape differences between male and female foot bones: Is the female foot predisposed to hallux abducto valgus deformity? *Journal of American Podiatric Medical Association, 94*(5), 434–452. doi:10.7547/0940434

Fessler, D. M., Haley, K. J., & Lal, R. D. (2005). Sexual dimorphism in foot length proportionate to stature. *Annals of Human Biology, 32*(1), 44–59. doi:10.1080/03014460400027581

Fetherston, C. (1998). Risk factors for lactation mastitis. *Journal of Human Lactation, 14*(2), 101–109. doi:10.1177/089033449801400209

Finch, S. R. (2003). *Mathematical constants* (Encyclopedia of mathematics and its applications, vol. 94) pp. 5–10. Cambridge, UK: Cambridge University Press.

Finnerup, N. B., Attal, N., Haroutounian, S., McNicol, E., Baron, R., Dworkin, R. H., . . . & Kamerman, P. R. (2015). Pharmacotherapy for neuropathic pain in adults: A systematic review and meta-analysis. *The Lancet Neurology, 14*(2), 162–173. doi:10.1016/S1474-4422(14)70251-0

Fong, D. T.-P., Chan, K.-M., & Mok, K.-M. (2013). Footwear for preventing acute sport-related ankle ligamentous sprain injury. In R. S. Goonetilleke (Ed.), *The science of footwear* (pp. 577–594). Boca Raton, FL: CRC Press. doi:10.1201/b13021-34

Forbes, G. (1938). Lymphatics of the skin, with a note on lymphatic watershed areas. *Journal of Anatomy, 72*(Pt. 3), 399–410.

Ford, E. S., Maynard, L. M., & Li, C. (2014). Research letter: Trends in mean waist circumference and abdominal obesity among U.S. adults, 1999-2012. *Journal of the American Medical Association, 312*(11), 1151–1153. doi:10.1001/jama.2014.8362

Forrester, J. V., Dick, A. D., McMenamin, P. G., Roberts, F., & Pearlman, E. (2016a). Chapter 1 - anatomy of the eye and orbit. In J. V. Forrester, A. D. Dick, P. G. McMenamin, F. Roberts, & E. Pearlman (Eds.), *The eye: Basic sciences in practice* (4th ed., pp. 1–102.e2). Edinburgh, UK: W.B. Saunders.

Forrester, J. V., Dick, A. D., McMenamin, P. G., Roberts, F., & Pearlman, E. (2016b). Chapter 5 - physiology of vision and the visual system. In J. V. Forrester, A. D. Dick, P. G. McMenamin, F. Roberts, & E. Pearlman (Eds.), *The eye: Basic sciences in practice* (4th ed., pp. 269–337.e2). Philadelphia, PA: W.B. Saunders.

Fourt, L., & Hollies, N. R. S. (1970). *Clothing: Comfort and function*. New York, NY: Marcel Dekker.

Franke, B. A., Jr. (2003). Formative dynamics: The pelvic girdle. *Journal of Manual & Manipulative Therapy, 11*(1), 12–40. doi:10.1179/106698103790818977

Frisbie, J. H. (2016). Normal diaphragmatic and rib cage breathing: Effects on venous return patterns in monitored human subjects. *Angiology, 4*(1), 165. doi:10.4172/2329-9495.1000165

Frowen, P., O'Donnell, M., Lorimer, D., & Gordon, B. (2010). *Neale's Disorders of the Foot: Clinical Companion*. Edinburgh, Scotland: Churchill Livingstone, Elsevier LTD.

Fry, D. E., Harris, W. E., Kohnke, E. N., & Twomey, C. L. (2010). Influence of double-gloving on manual dexterity and tactile sensation of surgeons. *Journal of the American College of Surgeons, 210*(3), 325–330. doi:10.1016/j.jamcollsurg.2009.11.001

Fu, J., Hofker, M., & Wijmenga, C. (2015, April 7). Apple or pear: Size and shape matter. *Cell Metabolism, 21*, 507–508. doi:10.1016/j.cmet.2015.03.016

Fukano, M. (2015). Biomechanical analysis of the effects of footwear. In K. Kanosue, T. Ogawa, M. Fukano, & T. Fukubayashi (Eds.), *Sports injuries and prevention* (pp. 347–353). Tokyo, Japan: Springer. doi:10.1007/978-4-431-55318-2_29

Fuller, E. A. (2000). The windlass mechanism of the foot. A mechanical model to explain pathology. *Journal of the American Podiatric Medical Association, 90*(1), 35–46. doi:10.7547/87507315-90-1-35

Furer, B. S. (1967). *Maternity dress designs with selected expandable features* (Unpublished master's thesis). Oregon State University, Corvallis, OR. Retrieved from http://ir.library.oregonstate.edu/concern/graduate_thesis_or_dissertations/xs55m

Gallagher, P., Buckmaster, A., O'Carroll, S., Kiernan, G., & Geraghty, J. (2009). Experiences in the provision, fitting and supply of external breast prostheses: Findings from a national survey. *European Journal of Cancer Care, 18*(6), 556–568. doi:10.1111/j.1365-2354.2007.00898.x

Gao, Z. G., Sun, S. Q., Goonetilleke, R. S., & Chow, D. H. K. (2016). Effect of an on-hip load-carrying belt on physiological and perceptual responses during bimanual anterior load carriage. *Applied Ergonomics, 55*, 133–137. doi:10.1016/j.apergo.2016.02.005

Gardner, E., & Bunge, R. P. (2005). Gross anatomy of the peripheral nervous system. In P. J. Dyck, & P. K. Thomas (Eds.), *Peripheral neuropathy* (4th ed., pp. 11–33). Philadelphia, PA: Elsevier Saunders.

Garn, S. M. (1951). Types and distribution of the hair in man. *Annals of the New York Academy of Sciences, 53*(3), 498–507. doi:10.1111/j.1749-6632.1951.tb31952.x

Garn, S. M. (1954). Fat patterning and fat intercorrelations in the adult male. *Human Biology, 26*(1), 59–69.

Garn, S. M. (1955). Relative fat patterning: An individual characteristic. *Human Biology, 27*(2), 75–89.

Gau, C. R. (1998). *Historic medical perspectives of corseting and two physiologic studies with reenactors* (Unpublished doctoral dissertation). Iowa State University, Ames, Iowa.

Gazzuolo, E. B. (1985). *A theoretical framework for describing body form variation relative to pattern shape* (Unpublished master's thesis), University of Minnesota, Minneapolis, Minnesota.

Gemperle, F., Kasabach, C., Stivoric, J., Bauer, M., & Martin, R. (1998). Design for wearability. In *Digest of papers. Second International Symposium on Wearable Computers*, October 19–20, 1998. doi:10.1109/ISWC.1998.729537

Genova, A. (2012). *Accessory design*. New York, NY: Fairchild Books.

Geraghty, L. N., & Pomeranz, M. K. (2011). Physiologic changes and dermatoses of pregnancy. *International Journal of Dermatology, 50*(7), 771–782. doi:10.1111/j.1365-4632.2010.04869.x

Gersak, J. (2014). Wearing comfort using body motion analysis. In D. Gupta, & N. Zakaria (Eds.), *Anthropometry, apparel sizing and design* (pp. 320–333). London, UK: Woodhead Publishing. doi:10.1533/9780857096890.2.320

Gill, S. (2015). A review of research and innovation in garment sizing, prototyping and fitting. *Textile Progress, 47*(1), 1–85. doi:10.1080/00405167.2015.1023512

Gill, S., Parker, C. J., Hayes, S., Brownbridge, K., Wren, P., & Panchenko, A. (2014). The true height of the waist: Explorations of automated body scanner waist definitions of the TC2 scanner. *Proceedings of the 5th International Conference on Body Scanning Technologies*, Lugano, Switzerland, October 21–22, 2014. doi.org/10.15221/14.055

Giorcelli, R. J., Hughes, R. E., Wassell, J. T., & Hsiao, H. (2001). Effects of wearing a back belt on spine kinematics asymmetric lifting of large and small boxes. *Spine, 26*(16), 1794–1798.

Giraldo, P. C., Amaral, R. L., Juliato, C., Eleutério, J., Brolazo, E., & Gonçalves, A. K. S. (2011). The effect of "breathable" panty liners on the female lower genital tract. *International Journal of Gynecology & Obstetrics, 115*(1), 61–64. doi:10.1016/j.ijgo.2011.04.016

Givler, V., & Mohr, P. (2005). *U.S. Patent No. 6,945,945 B2*. Washington, DC: U.S. Patent and Trademark Office.

Glenn, W. W., Holcomb, W. G., Gee, J. B., & Rath, R. (1970). Central hypoventilation; long-term ventilatory assistance by radiofrequency electrophrenic respiration. *Annals of Surgery, 172*(4), 755.

Gloviczki, P., Comerota, A. J., Dalsing, M. C., Eklof, B. G., Gillespie, D. L., Gloviczki, M. L., . . . & Wakefield, T. W. (2011). The care of patients with varicose veins and associated chronic venous diseases: Clinical practice guidelines of the Society for Vascular Surgery and the American Venous Forum. *Journal of Vascular Surgery, 53*(5), 2S–48S. doi:10.1016/j.jvs.2011.01.079

Godha, K., Tucker, K. M., Biehl, C., Archer, D. F., & Mirkin, S. (2018). Human vaginal pH and microbiota: An update. *Gynecological Endocrinology, 34*(6), 451–455. doi: 10.1080/09513590.2017.1407753

Goldsberry, E., Shim, S., & Reich, N. (1996). Women 55 years and older, Part II: Overall satisfaction and dissatisfaction with the fit of ready-to-wear. *Clothing and Textiles Research Journal, 14,* 121–132. doi:10.1177/0887302X9601400203

Goncu-Berk, G., & Topcuoglu, N. (2017). A healthcare wearable for chronic pain management: Design of a smart glove for rheumatoid arthritis. *The Design Journal: Design for the Next: Proceedings of the 12th. European Academy of Design Conference, 12–14 April 2017,* L. diLucchio, L. Imbesi, & P. Atkinson (Eds.), *20,* suppl, S1978–S1988. doi:10.1080/14606925.2017.1352717

Goonetilleke, R. S. (Ed.). (2013). *The science of footwear.* Boca Raton, FL: CRC Press.

Gordon, L. (2015). *The development of design requirements for breastfeeding apparel: A user-oriented product development approach.* Master's thesis. Retrieved from the University of Minnesota Digital Conservancy. Retrieved from http://hdl.handle.net/11299/177041

Gordon, C. C., Churchill, T., Clauser, C. E., Bradtmiller, B., McConville, J. T., Tebbetts, I., & Walker, R. A., Anthropology Research Project, Inc. Yellow Springs OH (1989). *1987-1988 anthropometric survey of U.S. Army personnel: Methods and summary statistics 1988* (Tech. Report TR-89-044). Natick, MA: U. S. Army Natick Research, Development and Engineering Center. Retrieved from http://www.dtic.mil/dtic/tr/fulltext/u2/a225094.pdf

Gorton III, G. E., Young, M. L., & Masso, P. D. (2012). Accuracy, reliability, and validity of a 3-dimensional scanner for assessing torso shape in idiopathic scoliosis. *Spine, 37*(11), 957–965.

Gould, T. E., Piland, S. G., Kzeminski, D. E., & Rawlins, J. W. (2015). Protective headgear for sports. In R. Shishoo (Ed.), *Textiles for sportswear* (pp. 213–244). Cambridge, UK: Woodhead Publishing Limited.

Granberry, R., Abel, J., & Holschuh, B. (2017). Active knit compression stockings for the treatment of orthostatic hypotension. In *Proceedings of the 2017 ACM International Symposium on Wearable Computers, Sept. 11-15, 2017, Maui HI* (pp. 186–191). doi:10.1145/3123021.3123065

Granberry, R., Ciavarella, N., Pettys-Baker, R., Berglund, M. E., & Holschuh, B. (2018, April). No-power-required, touch-activated compression garments for the treatment of POTS. In ASME *2018 Design of Medical Devices Conference, April 9–12, Minneapolis MN* (p. V001T10A006, 5 pages). American Society of Mechanical Engineers. doi:10.1115/DMD2018-6886

Gravante, G., Russo, G., Pomara, F., & Ridola, C. (2003). Comparison of ground reaction forces between obese and control young adults during quiet standing on a baropodometric platform. *Clinical Biomechanics, 18*(8), 780–782. doi:10.1016/S0268-0033(03)00123-2

Graves, G. O., & Edwards, L. F. (1944). The eustachian tube: A review of its descriptive, microscopic, topographic and clinical anatomy. *Archives of Otolaryngology--Head & Neck Surgery, 39*(5), 359–397. doi:10.1001/archotol.1944.00680010374001

Gravestock, H. (2013). Drawing on ice: Learning to create performance with and through the blade and boot of a skate. *Performance Research, 18*(6), 64–70. doi:10.1080/13528165.2013.908058

Gray, H. (1966). *Anatomy of the human body* (28th ed.). C. M. Goss (Ed.), Philadelphia, PA: Lea & Febiger.

Greenman, P. E. (1996). *Principles of manual medicine* (2nd ed.). Baltimore, MD: Williams & Wilkins.

Greiner, T. M. (1991). *Hand anthropometry of U.S. army personnel* (Tech Report No. TR-92/011). Natick, MA: U.S. Army Natick Research Development Laboratories. Retrieved from http://www.dtic.mil/dtic/tr/fulltext/u2/a244533.pdf

Gribbin, E. A. (2014). Body shape and its influence on apparel size and consumer choices. In M.-E. Faust & S. Carrier (Eds.), *Designing apparel for consumers: The impact of body shape and size*. Cambridge, UK: Woodhead Publishing Limited.

Grice, E. A., & Segre, J. A. (2011). The skin microbiome. *Nature Reviews Microbiology, 9*(4), 244–253. doi:10.1038/nrmicro2537

Griffin, L. A., & Dunne, L. E. (2016). Effects of ready-to-wear sizing conventions on sensor placement for medical wearable sensing. *Journal of Medical Devices, 10*(2), 020939. doi:10.1115/1.4033170

Griffin, L. A., Compton, C., & Dunne, L. E. (2016, September 12–16). An analysis of the variability of anatomical body references within ready-to-wear garment sizes. In *Proceedings of the ISWC 2016 Conference UbiComp 2016*, Heidelberg, Germany. doi:10.1145/2971763.2971800

Griffin, L. A., Kim, N., Carufel, R., Sokolowski, S., Lee, H., & Seifert, E. (2018, July). Dimensions of the dynamic hand: Implications for glove design, fit, and sizing. In *Proceedings of the 9th International Conference on Applied Human Factors and Ergonomics*, Orlando, Florida.

Griffin, L. A., Lastovich, T., Bye, E., LaBat, K. (2016, November 8–11). Protective coverall design development and testing. *Proceedings of the ITAA 2016 Conference*, Vancouver, B.C., Canada. Retrieved from https://lib.dr.iastate.edu/cgi/viewcontent.cgi?article=1653&context=itaa_proceedings

Gu, B, Lin, W., Su, J., & Xu, B. (2017). Predicting distance ease distributions on crotch curves of customized female pants. *International Journal of Clothing Science and Technology, 29*(1), 47–59.

Guan, J., & Bisson, E. F. (2017). Treatment of odontoid fractures in the aging population. *Neurosurgery Clinics of North America, 28*(1), 115–123. doi:10.1016/j.nec.2016.07.001

Gustafson J. A., Takenaga T., & Debski R. E. (2018). Basic Concepts in Functional Biomechanics. In V. Musahl, J. Karlsson, W. Krutsch, B. Mandelbaum, J. Espregueira-Mendes, & P. d'Hooghe (Eds.), *Return to play in football.* (pp. 3–15). Berlin, Germany: Springer. doi:10.1007/978-3-662-55713-6_1

Haas, E. (1933). *U.S. Patent No. 1,926,900.* Washington, DC: U.S. Patent and Trademark Office.

Hadjipavlou, A. G., Tzermiadianos, M. N., Bogduk, N., & Zindrick, M. R. (2008). The pathophysiology of disc degeneration: A critical review. *The Bone and Joint Journal, 90*(10), 1261–1270. doi:10.1302/0301-620X.90B10.20910

Hakestad, K. A., Torstveit, M. K., Nordsletten, L., Axelsson, Å. C., & Risberg, M. A. (2015). Exercises including weight vests and a patient education program for women with osteopenia: A feasibility study of the OsteoACTIVE rehabilitation program. *Journal of Orthopaedic & Sports Physical Therapy, 45*(2), 97–105. doi:10.2519/jospt.2015.4842

Hallén, L. G., & Lindahl, O. (1966). The "screw-home" movement in the knee-joint. *Acta Orthopaedica Scandinavica, 37*(1), 97–106. doi:10.3109/17453676608989407

Hamill, J., & Knutzen, K. M. (2003). *Biomechanical basis of human movement* (2nd ed.). Philadelphia, PA: Lippincott Williams & Wilkins.

Hamlet, M. P. (2000). Prevention and treatment of cold injury. *International Journal of Circumpolar Health, 59*(2), 108–113. PMID: 10998827

Han, H., Nam, Y., & Shin, S.-J. H. (2010). Algorithms of the automatic landmark identification for various torso shapes. *International Journal of Clothing Science and Technology, 22*(5), 343–357.

Han, S. C., Graham, A. D., & Lin, M. C. (2011). Clinical assessment of a customized free-form progressive add lens spectacle. *Optometry and Vision Science, 88*(2), 234–243. doi:10.1097/OPX.0b013e31820846ac

Hanten, W. P., Lucio, R. M., Russell, J. L., & Brunt, D. (1991). Assessment of total head excursion and resting head posture. *Archives of Physical Medicine and Rehabilitation, 72*(11), 877–880. doi:10.1016/0003-9993(91)90003-2

Harkey, M. R. (1993). Anatomy and physiology of hair. *Forensic Science International, 63*(1–3), 9–18. doi:10.1016/0379-0738(93)90255-9

Hartman, V. (2017). End the bloody taxation: Seeing red on the unconstitutional tax on tampons. *Northwestern University Law Review, 112*, 313–353.

Hatton, A. L., Rome, K., Dixon, J., Martin, D. J., & McKeon, P. O. (2013). Footwear interventions: A review of their sensorimotor and mechanical effects on balance performance and gait in older adults. *Journal of the American Podiatric Medical Association, 103*(6), 516–533. doi:10.7547/1030516

Havlíček, J., Fialová, J., & Roberts, S. C. (2017). Individual Variation in Body Odor. In: A. Buettner (Ed.), *Springer handbook of odor* (pp. 125–126). Cham, Switzerland: Springer. doi:10.1007/978-3-319-26932-0_50

Haycock, C. E., Shierman, G., & Gillette, J. (1978). The female athlete: Does her anatomy pose problems? In *Proceedings of the 19th conference on the medical aspects of sports*, Chicago, IL: American Medical Association (pp. 1–8).

Heckman, J. D., & Sassard, R. (1994). Current concepts review. Musculoskeletal considerations in pregnancy. *JBJS, 76*(11), 1720–1730.

Heidelbaugh, J., & Lee, H. (2009). Management of the ingrown toenail. *American Family Physician, 79*(4), 303–308, 311–312.

Henderson, D. (2002). *Hat talk: Conversations with hat makers about their hats—the fedora, homburg, straw, and cap.* Yellow Springs, OH: Wild Goose Press.

Henriksen, T. E., Skrede, S., Fasmer, O. B., Schoeyen, H., Leskauskaite, I., Bjørke-Bertheussen, J., . . . & Lund, A. (2016). Blue-blocking glasses as additive treatment for mania: A randomized placebo-controlled trial. *Bipolar Disorders, 18*(3), 221–232. doi:10.1111/bdi.12390

Henson, S. K. (1991). *The development of a method for determining the best-fit shape for the crotch seam of men's pants* (Unpublished master's thesis). Virginia Polytechnic Institute and State University, Blacksburg, VA.

Herbenick, D., Reece, M., Schick, V., & Sanders, S. A. (2014). Erect penile length and circumference dimensions of 1,661 sexually active men in the United States. *Journal of Sexual Medicine, 11*(1), 93–101. doi:10.1111/jsm.12244

Hertel, J., Sloss, B. R., & Earl, J. E. (2005). Effect of foot orthotics on quadriceps and gluteus medius electromyographic activity during selected exercises. *Archives of Physical Medicine and Rehabilitation, 86*(1), 26–30. doi:10.1016/j.apmr.2004.03.029

Heyer, G. L. (2014). Abdominal and lower-extremity compression decreases symptoms of postural tachycardia syndrome in youth during tilt table testing. *Journal of Pediatrics, 165*(2), 395–397. doi:/10.1016/j.jpeds.2014.04.014

Hicks, B. (2015). *A comparative analysis of softball gloves: Catcher's glove yields highest peak hand pressure.* (Master's thesis). London, Ontario, Canada: University of Western Ontario. Available from Electronic Thesis and Dissertation Repository. 3353. Accessed from https://ir.lib.uwo.ca/etd/3353

Hicks, J. H. (1954). The mechanics of the foot. II. The plantar aponeurosis and the arch. *Journal of Anatomy, 88*(Pt 1), 25–30.1 PMCID: PMC1244640

Hills, A. P., Hennig, E. M., Byrne, N. M., & Steele, J. R. (2002). The biomechanics of adiposity–structural and functional limitations of obesity and implications for movement. *Obesity Reviews, 3*(1), 35–43. doi:10.1046/j.1467-789X.2002.00054.x

Hills, A. P., Hennig, E. M., McDonald, M., & Bar-Or, O. (2001). Plantar pressure differences between obese and non-obese adults: A biomechanical analysis. *International Journal of Obesity and Related Metabolic Disorders, 25*(11), 1674–1679. doi:10.1038/sj.ijo.0801785 PMID: 11753590

Hinghofer-Szalkay, H. C. (1996). Physiology of cardiovascular, respiratory, interstitial, endocrine, immune, and muscular systems. In D. Moore, P. Bie, & H. Oser (Eds.), *Biological and medical research in space: An overview of life sciences research in microgravity* (pp. 107–153). Berlin, Germany: Springer. doi:10.1007/978-3-642-61099-8_2

Ho, S. S., Yu, W. W., Lao, T. T., Chow, D. H., Chung, J. W., & Li, Y. (2008). Comfort evaluation of maternity support garments in a wear trial. *Ergonomics, 51*(9), 1376–1393. doi:10.1080/00140130802116489

Ho, S. S., Yu, W. W., Lao, T. T., Chow, D. II., Chung, J. W., & Li, Y. (2009b). Garment needs of pregnant women based on content analysis of in-depth interviews. *Journal of Clinical Nursing, 18*(17), 2426–2435. doi:10.1111/j.1365-2702.2009.02786.x

Ho, S. S., Yu, W. W., Lao, T. T., Chow, D. H., Chung, J. W., & Li, Y. (2009a). Effectiveness of maternity support belts in reducing low back pain during pregnancy: A review. *Journal of Clinical Nursing, 18*(11), 1523–1532. doi:10.1111/j.1365-2702.2008.02749.x

Hochwalt, A. E., Jones, M. B., & Meyer, S. J. (2010). Clinical safety assessment of an ultra-absorbency menstrual tampon. *Journal of Women's Health, 19*(2), 273–278. doi:10.1089/jwh.2009.1423

Hodge, B. D., & Brodell, R. T. (2018). Anatomy, Integument, Sweat Glands: Updated 2018 Jan 7. In StatPearls (Internet). Treasure Island, FL: StatPearls Publishing; NCBI bookshelf. A service of the National Library of Medicine, National Institutes of Health. Retrieved from https://www.ncbi.nlm.nih.gov/books/NBK482278/

Höfer, D. (2006). Antimicrobial textiles, skin-borne flora and odour. In U.-C. Hipler, & P. Elsner (Eds.), *Current problems in dermatology: Biofunctional textiles and the skin* (Vol. 33, pp. 67–77). Basel, Switzerland: S. Karger AG. doi:10.1159/000093937

Hofer, P., Hasler, M., Fauland, G., Bechtold, T., & Nachbauer, W. (2014). Microclimate in ski boots—Temperature, relative humidity, and water absorption. *Applied Ergonomics, 45*(3), 515–520. doi:10.1016/j.apergo.2013.07.007

Höflin, F., & van der Linden, W. (1976). Boot top fractures. *The Orthopedic Clinics of North America, 7*(1), 205–213. PMID: 1256788

Hollander, A. (1994). *Sex and suits*. New York, NY: Alfred A. Knopf, Inc.

Hollinshead, W. H., & Jenkins, D. (1981) *Functional anatomy of the limbs and back* (5th ed.). Philadelphia, PA: W. B. Saunders.

Holmér, I. (2004). Thermal manikin history and applications. *European Journal of Applied Physiology, 92*(6), 614–618. doi:10.1007/s00421-004-1135-0

Holmgaard, R., & Nielsen, J. (2009) Dermal absorption of pesticides: Evaluation of variability and prevention. *Pesticides Research No. 124*. Environmental Medicine, Institute of Public Health, University of Southern Denmark: Danish Environmental Protection Agency. Retrieved from https://www2.mst.dk/udgiv/publications/2009/978-87-7052-980-8/pdf/978-87-7052-981-5.pdf

Holschuh, B., Obropta, E., Buechley, L., & Newman, D. (2012, September). Materials and textile architecture analyses for mechanical counter-pressure space suits using active materials. *AIAA Space 2012 Conference & Exposition Proceedings*, Pasadena, CA, September 11–13, 2012, (p. 5206). doi:10.2514/6.2012-5206

Holter, N., & Glassrock, W. (1965). *U.S. Patent No. 3,215,136*. Washington, DC: U.S. Patent and Trademark Office.

Hootman, J. M., Dick, R., & Agel, J. (2007). Epidemiology of collegiate injuries for 15 sports: Summary and recommendations for injury prevention initiatives. *Journal of Athletic Training, 42*(2), 311–319.

Hoppenfeld, S. (1976). (With collaborator Hutton, R., & illustrator Thomas, H.) *Physical examination of the spine and extremities*. New York, NY: Appleton-Century-Crofts.

Horváth, I., Hunt, J., & Barnes, P. J. (2005). Exhaled breath condensate: Methodological recommendations and unresolved questions. *The European Respiratory Journal, 26*(3), 523–548. doi:10.1183/09031936.05.00029705

Howard, R., & Pillinger, M. H. (2010). Consequences of Chinese foot binding. *Journal of Clinical Rheumatology, 16*(8), 408. doi:10.1097/RHU.0b013e3182005c7c

Hsiao, H., Whitestone, J., Kau, T-Y., & Hildreth, B. (2015). Firefighter hand anthropometry and structural glove sizing: A new perspective. *Human Factors 57*(8), 1359–1377. doi:10.1177/0018720815594933

Hu, J., & Lu, J. (2015). Recent developments in elastic fibers and yarns for sportswear. In R. Shishoo (Ed.), *Textiles for sportswear* (pp. 53–76). Cambridge, England: Woodhead Publishing Elsevier. doi:10.1016/B978-1-78242-229-7.00003-5

Hu, Y., Converse, C., Lyons, M. C., & Hsu, W. H. (2018). Neural control of sweat secretion: A review. *British Journal of Dermatology, 178*(6), 1246–1256. doi:10.1111/bjd.15808

Huang, Y. C., Chang, K. H., Liou, T. H., Cheng, C. W., Lin, L. F., & Huang, S. W. (2017). Effects of kinesio taping for stroke patients with hemiplegic shoulder pain: A double-blind, randomized, placebo-controlled study. *Journal of Rehabilitation Medicine, 49*(3), 208–215. doi:10.2340/16501977-2197

Huber, T., Schmoelz, W., & Bölderl, A. (2012). Motion of the fibula relative to the tibia and its alterations with syndesmosis screws: A cadaver study. *Foot and Ankle Surgery, 18*(3), 203–209. doi:10.1016/j.fas.2011.11.003

Huggins, G. R., & Preti, G. (1981). Vaginal odors and secretions. *Clinical Obstetrics and Gynecology, 24*(2), 355–377.

Huxley, H., & Hanson, J. (1954). Changes in the cross-striations of muscle during contraction and stretch and their structural interpretation. *Nature, 173*(4412), 973–976. doi:10.1038/173973a0

Ibegbuna, V., Delis, K. T., Nicolaides, A. N., & Aina, O. (2003). Effect of elastic compression stockings on venous hemodynamics during walking. *Journal of Vascular Surgery, 37*(2), 420–425. doi:10.1067/mva.2003.104

Imker, F. W. (1994). The back support myth. *Ergonomics in Design, 2*(2), 8–12.

Impiö, J., Karinsalo, T., Reho, A., Remes, A., Tolvanen, P., & Välimäki, E. (2006). U.S. Patent No. 7,152,470. Washington, DC: U.S. Patent and Trademark Office.

Inman, V. T. (1947). Functional aspects of the abductor muscles of the hip. *Journal of Bone & Joint Surgery, 29*(3), 607–619.

Inman, V. T., Ralston, H. J., & Todd, F. (1981). *Human walking*. Baltimore, MD: Williams & Wilkins.

International Lymphoedema Framework. (2006). *Best practice for the management of lymphoedema. ILF International consensus,* 3–52. London, UK: Medical Education Partnership Ltd.

International Society of Lymphology (2013). The diagnosis and treatment of peripheral lymphedema: 2013 consensus document of the International Society of Lymphology. *Lymphology, 46*(1), 1–11.

International Standards Organisation (ISO) (2008). *ISO 7250-1:2008: Basic Human Body Measurements for Technological Design, Part I: Body measurement definitions and landmarks.* Geneva: International Standards Organisation: https://www.ISO.org

International Standards Organization (2012). ISO 13666:2012(en): *Ophthalmic optics, spectacle lenses, vocabulary*. Geneva, Switzerland: https://www.ISO.org

International Standards Organization (2015a). *Footwear sizing: Conversion of sizing systems* (ISO/TS 19407: 2015). Geneva, Switzerland: https://www.ISO.org

International Standards Organization (2015b). *Footwear sizing: Vocabulary and terminology* (ISO/TS 19408: 2015). Geneva, Switzerland: https://www.ISO.org

Irwin, D. E., Kopp, Z. S., Agatep, B., Milsom, I., & Abrams, P. (2011). Worldwide prevalence estimates of lower urinary tract symptoms, overactive bladder, urinary incontinence and bladder outlet obstruction. *BJU International, 108*(7), 1132–1138. doi:10.1111/j.1464-410X.2010.09993.x

Isermann, P., & Lammerding, J. (2013). Nuclear mechanics and mechanotransduction in health and disease. *Current Biology, 23*(24), R1113–R1121. doi:10.1016/j.cub.2013.11.009

Jablonski, N. G. (2006). *Skin: A natural history*. Berkeley, CA: University of California Press.

Jacob, H. A. C., & Kissling, R. O. (1995). The mobility of the sacroiliac joints in healthy volunteers between 20 and 50 years of age. *Clinical Biomechanics, 10*(7), 352–361. doi:10.1016/0268-0033(95)00003-4

Janssen, K. W., van Mechelen, W., & Verhagen, E. A. (2014). Bracing superior to neuromuscular training for the prevention of self-reported recurrent ankle sprains: A three-arm randomised controlled trial. *British Journal of Sports Medicine, 48*(16), 1235–1239. doi:10.1136/bjsports-2013-092947

Jenkins, D. B. (2002). *Hollinshead's functional anatomy of the limbs and back* (8th ed.). Philadelphia, PA: W. B. Saunders.

Jennings, R. T., & Baker, E. S. (2000). Gynecological and reproductive issues for women in space: A review. *Obstetrical and Gynecological Survey, 55*(2), 109–116.

Jobst, C. (1954). *U.S. Patent No. 2,691,221A*. Washington, DC: U.S. Patent and Trademark Office.

Jobst, C. (1958). *U.S. Patent No. 2,829,641*. Washington, DC: U.S. Patent and Trademark Office.

Johnson, T. (2015). *Archery fundamentals* (2nd ed.). Champaign, IL: Human Kinetics.

Johnston, J. T., Mandelbaum, B. R., Schub, D., Rodeo, S. A., Matava, M. J., Silvers-Granelli, H. J., . . . & Brophy, R. H. (2018). Video analysis of anterior cruciate ligament tears in professional American football athletes. *American Journal of Sports Medicine, 46*(4), 862–868. doi:10.1177/0363546518756328

Jones, I., & Johnson, M. I. (2009). Transcutaneous electrical nerve stimulation. *Continuing Education in Anaesthesia, Critical Care & Pain, 9*(4), 130–135. doi:10.1093/bjaceaccp/mkp021

Joseph-Armstrong, H. (2006). *Patternmaking for fashion design* (4th ed.). Upper Saddle River, NJ: Pearson Education, Inc.

Jung, A., Schill, W., & Schuppe, H. (2005). Improvement of semen quality by nocturnal scrotal cooling in oligozoospermic men with a history of testicular maldescent. *International Journal of Andrology, 28*(2), 93–98. doi:10.1111/j.1365-2605.2004.00517.x

Kaefer, M., Zurakowski, D., Bauer, S. B., Retik, A. B., Peters, C. A., Atala, A., & Treves, T. S. (1997). Estimating normal bladder capacity in children. *The Journal of Urology, 158*(6), 2261–2264. doi:10.1016/S0022-5347(01)68230-2

Kakalios, J. (2005). *The physics of superheroes.* New York, NY: Gotham Books.

Kapit, W., & Elson, L. M. (2014). *The anatomy coloring book* (4th ed.). San Francisco, CA: Pearson Education, Inc.

Karimi, M. T., Ebrahimi, M. H., Mohammadi, A., & McGarry, A. (2017). Evaluation of the influences of various force magnitudes and configurations on scoliotic curve correction using finite element analysis. *Australasian Physical & Engineering Sciences in Medicine, 40*(1), 231–236. doi:10.1007/s13246-016-0501-7

Karpf, M. (1990). Lymphadenopathy. In H. K. Walker, W. D. Hall, & J. W. Hurst (Eds.), *Clinical methods: The history, physical, and laboratory examinations* (3rd ed., pp. 711–714). Boston, MA: Butterworths.

Karpf, P. M., Mang, W., & Hoerterer, H. (1982). Biomechanical investigation of the distal tibia fracture and its relation to the length of ski boots. In *Proceedings of Skiing Trauma and Safety, 4th International Symposium*, Munich, Germany, pp. 93–97.

Katirji, B., & Wilbourn, A. (2005). Mononeuropathies of the lower limb. In P. J. Dyck, & P. K. Thomas (Eds.), *Peripheral neuropathy* (4th ed., pp. 1487–1510). Philadelphia, PA: Elsevier Saunders.

Katz, R. D. (2014). Form follows function: The anatomy of the hand. In E. F. S. Wilgis (Ed.), *The wonder of the human hand* (pp. 5–15). Baltimore, MD: John Hopkins University Press.

Kaufer, H. (1979). Patellar biomechanics. *Clinical Orthopaedics and Related Research,* 144, 51–54.

Kavounoudias, A., Roll, R., & Roll, J. P. (1998). The plantar sole is a 'dynamometric map' for human balance control. *Neuroreport, 9*(14), 3247–3252.

Kaweeteerawat, C., Na Ubol, P., Sangmuang, S., Aueviriyavit, S., & Maniratanachote, R. (2017). Mechanisms of antibiotic resistance in bacteria mediated by silver nanoparticles. *Journal of Toxicology and Environmental Health, Part A, 80*(23–24), 1276–1289. doi:10.1080/15287394.2017.1376727

Kendall, J. (1983). *Functional anatomy and anthropometry of the hand.* (Master's Thesis). Urbana-Champaign, IL: University of Illinois.

Kendall, F. P., McCreary, E., Provance, P., Rodgers, M., & Romani, W. (2005). *Muscles: Testing and function with posture and pain* (5th ed.). Baltimore, MD: Lippincott Williams & Wilkins.

Kern, S. (1975). *Anatomy and destiny: A cultural history of the human body*. Indianapolis, IN: Bobbs-Merrill.

Khan, M. H., Victor, F., Rao, B., & Sadick, N. S. (2010). Treatment of cellulite: Part I. Pathophysiology. *Journal of the American Academy of Dermatology, 62*(3), 361–370. doi:10.1016/j.jaad.2009.10.042

Khanday, M. A, & Hussain, F. (2015). Explicit formula of finite difference method to estimate human peripheral tissue temperatures during exposure to severe cold stress. *Journal of Thermal Biology, 48*, 51–55. doi:10.1016/j.jtherbio.2014.12.010

Kılıç, A. Ş., Tama, D., & Öndoğan, Z. (2014). Clothing problems with maternity garments. From *XIIIth International Izmir Textile and Apparel Symposium (IITAS* 2014) *April 2–5, 2014*, Antalya, Turkey, pp. 63–66.

Killian, R. B., Nishimoto, G. S., & Page, J. C. (1998). Foot and ankle injuries related to rock climbing. The role of footwear. *Journal of American Podiatric Medical Association, 88*(8), 365–374. doi:10.7547/87507315-88-8-365

Kim, D.-E., & LaBat, K. (2010). Design process for developing a liquid cooling garment hood. *Ergonomics, 53*(6), 818–828. doi:10.1080/00140131003734229

Kim, D.-E., LaBat, K., Bye, E., Sohn, M.-H., & Ryan, K. (2014). A study of scan garment accuracy and reliability. *The Journal of the Textile Institute, 106*(8), 853–861. doi:10 .1080/00405000.2014.949502

Kimura, J. (2013a). Chapter 12: Anatomy and physiology of the skeletal muscle. In J. Kimura (Ed.), *Electrodiagnosis in diseases of nerve and muscle: Principles and practice* (4th ed., pp. 315–332). New York, NY: Oxford University Press.

Kimura, J. (2013b). Chapter 25: Mononeuropathies and entrapment syndromes, 9: Tibial Nerve. In J. Kimura (Ed.), *Electrodiagnosis in diseases of nerve and muscle: Principles and practice* (4th ed., pp. 777–778). New York, NY: Oxford University Press.

Kirby, K. A. (2010). Evolution of foot orthoses in sports. In M. B. Werd, & E. L. Knight, (Eds.), *Athletic footwear and orthoses in sports medicine* (pp. 19–35). New York, NY: Springer Science + Business Media, LLC. doi:10.1007/978-0-387-76416-0_2

Klepp, I. G., Buck, M., Laitala, K., & Kjeldsberg, M. (2016). What's the problem? Odorcontrol and the smell of sweat in sportswear. *Fashion Practice, 8*(2), 296–317. doi: 10.1080/17569370.2016.1215117

Knapp, M. E. (1952). Treatment of some complications of Colles' fracture. *Journal of the American Medical Association, 148*(10), 825–827. doi:10.1001/jama.1952.02930100043009

Knight, C. (1991). *Blood relations: Menstruation and the origins of culture*. New Haven, CT: Yale University Press.

Knight, P. (2016). *Shoe Dog: A memoir by the creator of Nike*. New York, NY: Scribner.

Knowles, L. A. (2005). *The practical guide to patternmaking for fashion designers: Juniors, misses, and women*. New York, NY: Fairchild Publications, Inc.

Knowles, L. A. (2006). *The practical guide to patternmaking for fashion designers: Menswear*. New York, NY: Fairchild Publications, Inc.

Knudson, D. (2007). *Fundamentals of Biomechanics* (2nd ed.). New York, NY: Springer Science+Business.

Kohl, J. V., Atzmueller, M., Fink, B., & Grammer, K. (2001). Human pheromones: Integrating neuroendocrinology and ethology. *Neuroendocrinology Letters, 22*(5), 309–321.

Konitzer, L. N., Fargo, M. V., Brininger, T. L., & Reed, M. L. (2008). Association between back, neck, and upper extremity musculoskeletal pain and the individual body armor. *Journal of Hand Therapy, 21*(2), 143–149. doi:10.1197/j.jht.2007.10.017

Konstantinova, A. M., Stewart, C. J., Kyrpychova, L., Belousova, I. E., Michal, M., & Kazakov, D. V. (2017). An immunohistochemical study of anogenital mammary-like glands. *American Journal of Dermatopathology, 39*(8), 599–605. doi:10.1097/DAD.0000000000000724

Koo, H. S., Teel, K. P., & Han, S. (2016). Explorations of design factors for developments of protective gardening gloves. *Clothing and Textiles Research Journal, 34*(4), 257–271. doi:10.1177/0887302X16653671

Koscheyev, V. S., & Leon, G. R. (2014). Spacesuits: Development and design for thermal comfort. In F. Wang, & C. Gao (Eds.), *Protective clothing: Managing thermal stress* (pp. 171–191). Amsterdam: Woodhead Publishing. doi:10.1533/9781782420408.1.171

Koscheyev, V. S., Coca, A., & Leon, G. R. (2007). Overview of physiological principles to support thermal balance and comfort of astronauts in open space and on planetary surfaces. *Acta Astronautica, 60*(4–7), 479–487. doi:10.1016/j.actaastro.2006.09.028

Koscheyev, V. S., Leon, G. R., Paul, S., Tranchida, D., & Linder, K. V. (2000). Augmentation of blood circulation to the fingers by warming distant body areas. *European Journal of Applied Physiology, 82*(1–2), 103–111. doi:10.1007/s004210050658

Kottke, F. J., & Lehmann, J. F. (Eds.). (1990). *Krusen's handbook of physical medicine and rehabilitation* (4th ed.). Philadelphia, PA: Saunders.

Kouchi, M. (2014). Anthropometric methods for apparel design: Body measurement devices and techniques. In D. Gupta & N. Zakaria (Eds.), *Anthropometry, apparel sizing and design* (pp. 67–94). London, UK: Woodhead Publishing. doi:10.1533/9780857096890.1.67

Kraemer, W. J., Bush, J. A., Bauer, J. A., Triplett-McBride, N. T., Paxton, N. J., Clemson, A., . . . & Newton, R. U. (1996). Influence of compression garments on vertical jump performance in NCAA Division I volleyball players. *Journal of Strength and Conditioning Research, 10*, 180–183.

Krauss, I., Grau, S., Mauch, M., Maiwald, C., & Horstmann, T. (2008). Sex-related differences in foot shape. *Ergonomics, 51*(11), 1693–1709. doi:10.1080/00140130802376026

Krauss, I., Valiant, G., Horstmann, T., & Grau, S. (2010). Comparison of female foot morphology and last design in athletic footwear—Are men's lasts appropriate for women? *Research in Sports Medicine, 18*(2), 140–156. doi:10:1080/15438621003627216

Krissi, H., Ben-Shitrit, G., Aviram, A., Weintraub, A. Y., From, A., Wiznitzer, A., & Peled, Y. (2016). Anatomical diversity of the female external genitalia and its association to sexual function. *European Journal of Obstetrics & Gynecology and Reproductive Biology, 196*, 44–47. doi:10.1016/j.ejogrb.2015.11.016

Krosshaug, T., Nakamae, A., Boden, B. P., Engebretsen, L., Smith, G., Slauterbeck, J. R., . . . & Bahr, R. (2007). Mechanisms of anterior cruciate ligament injury in basketball: Video analysis of 39 cases. *American Journal of Sports Medicine, 35*(3), 359–367. doi:10.1177/0363546506293899

Kücken, M., & Newell, A. C. (2004). A model for fingerprint formation. *Europhysics Letters, 68*(1), 141–146. doi:10.1209/epl/i2004-10161-2

Kuhlmann, A. S., Henry, K., & Wall, L. L. (2017). Menstrual hygiene management in resource-poor countries. *Obstetrical & Gynecological Survey, 72*(6), 356–376. doi:10.1097/OGX.0000000000000443

Kumar, K., Mandleywala, S. N., Gannon, M. P., Estes, N. A. M., III, Weinstock, J., & Link, M. S. (2017). Development of a chest wall protector effective in preventing sudden cardiac death by chest wall impact (commotio cordis). *Clinical Journal of Sport Medicine, 27*(1), 26–30. doi:10.1097/JSM.0000000000000297

Kwon, O., Jung, K., You, H., & Kim, H.-E. (2009). Determination of key dimensions for a glove sizing system by analyzing the relationships between hand dimensions. *Applied Ergonomics, 40*(4), 762–766. doi:10.1016/j.apergo.2008.07.003

LaBan, M. M., & Rapp, N. S. (1996). Low back pain of pregnancy: Etiology, diagnosis, and treatment. *Physical Medicine and Rehabilitation Clinics, 7*(3), 473–486. doi:10.1016/S1047-9651(18)30376-0

LaBat, K. L. (2007). Sizing standardization. In S. P. Ashdown, (Ed.), *Sizing in clothing: Developing effective sizing systems for ready-to-wear clothing* (pp. 88–107). Cambridge, UK: Woodhead Publishing.

LaBat, K. L., & DeLong, M. (1990). Body cathexis and satisfaction with fit of apparel. *Clothing and Textiles Research Journal, 8,* 97–102. doi:10.1177/0887302X9000800206

LaBat, K. L., & Sokolowski, S. (1999). A three-stage design process applied to an industry-university textile product design project. *Clothing and Textiles Research Journal, 17*(1), 11–20. doi:10.1177/0887302X9901700102

LaBat, K. L., & Sokolowski, S. L. (2012). Olympic dress, uniforms, and fashion. In J. B. Eicher & P. G. Tortora (Eds.), *Berg encyclopedia of world dress and fashion: Global perspectives.* doi:10.2752/BEWDF/EDch10412

LaBat, K. L., Ryan, K. S., & Sanden-Will, S. (2016). Breast cancer survivors' wearable product needs and wants: A challenge to designers. *International Journal of Fashion Design, Technology and Education, 10*(3), 308–319. doi:10.1080/17543266.2016.1250289

Laing, R. M., & Sleivert, G. G. (2002). Clothing, textiles, and human performance. *Textile Progress, 32*(2), 1–122. doi:10.1080/00405160208688955

Lamb, S. E., Marsh, J. L., Hutton, J. L., Nakash, R., Cooke, M. W., & Collaborative Ankle Support Trial (CAST Group). (2009). Mechanical supports for acute, severe ankle sprain: A pragmatic, multicentre, randomised controlled trial. *The Lancet, 373*(9663), 575–581. doi:10.1016/S0140-6736(09)60206-3

Langan, L. M., & Watkins, S. M. (1987). Pressure of menswear on the neck in relation to visual performance. *Human Factors, 29*(1), 67–71. doi:10.1177/001872088702900107

Langley, L. L., Telford, I. R., & Christensen, J. B. (1980). *Dynamic anatomy and physiology* (5th ed.). New York, NY: McGraw-Hill Book.

Larson, P., & Katovsky, B. (2012). *Tread Lightly: Form, footwear, and the quest for injury-free running.* New York, NY: Skyhorse Publishing.

Lattimer, C. R., Azzam, M., Kalodiki, E., Makris, G. C., & Geroulakos, G. (2013). Compression stockings significantly improve hemodynamic performance in post-thrombotic syndrome irrespective of class or length. *Journal of Vascular Surgery, 58*(1), 158–165. doi:10.1016/j.jvs.2013.01.003

Laven, J. S., Haverkorn, M. J., & Bots, R. S. (1988). Influence of occupation and living habits on semen quality in men (scrotal insulation and semen quality). *European Journal of Obstetrics & Gynecology and Reproductive Biology, 29*(2), 137–141. doi:10.1016/0028-2243(88)90140-2

Lawson, L., & Lorentzen, D. (1990) Selected sports bras: Comparisons of comfort and support. *Clothing and Textiles Research Journal, 8*(4), pp. 55–60. doi:10.1177/0887302X9000800409

Layton, A. M., Garber, C. E., Thomashow, B. M., Gerardo, R. E., Emmert-Aronson, B. O., Armstrong, H. F., ...& Bartels, M. N. (2011). Exercise ventilatory kinematics in endurance trained and untrained men and women. *Respiratory Physiology & Neurobiology, 178*(2), 223–229. doi:10.1016/j.resp.2011.06.009

Lebaric, J. E., Adler, R. W., & Limbert, M. E. (2001). Ultra-wideband, zero visual signature RF vest antenna for man-portable radios. Paper presented at the *Military Communications Conference, 2001. MILCOM 2001. Communications for Network-Centric Operations: Creating the Information Force. IEEE, 2* 1291-1294. Retrieved from http://hdl.handle.net/10945/41130 doi:10.1109/MILCOM.2001.986063

Lee, D. (2006). *U.S. Patent No. 7,037,284*. Washington, DC: U.S. Patent and Trademark Office.

Lee, D. (2011). *The pelvic girdle: An integration of clinical expertise and research* (4th ed.). New York, NY: Elsevier.

Lee, D., & Vleeming, A. (2007). An integrated therapeutic approach to the treatment of pelvic girdle pain. In A. Vleeming, V. Mooney, & R. Stoeckart (Eds.), *Movement, stability & lumbopelvic pain* (2nd ed., pp. 621–638). London, England: Churchill Livingstone Elsevier.

Lee, H. Y., Hong, K., & Kim, E. A. (2004). Measurement protocol of women's nude breasts using a 3D scanning technique. *Applied Ergonomics, 35*(4), 353–359.

Lee, W. C., Wong, W. Y., Kung, E., & Leung, A. K. (2012). Effectiveness of adjustable dorsiflexion night splint in combination with accommodative foot orthosis on plantar fasciitis. *Journal of Rehabilitation Research and Development, 49*(10), 1557–1564. http://dx.doi.org/10.1682/JRRD.2011.09.0181

Lee, W., Jung, H., Bok, I., Kim, C., Kwon, O., Choi, T., & You, H. (2016). Measurement and application of 3D ear images for earphone design. *Proceedings of the Human Factors and Ergonomics Society Annual Meeting, 60*(1), 1052–1056. doi:10.1177/1541931213601244

Lee, Y. W., Lee, J., & Warwick, W. J. (2008). The comparison of three high-frequency chest compression devices. *Biomedical Instrumentation & Technology, 42*(1), 68–75. doi:10.2345/0899-8205(2008)42[68:TCOTHC]2.0.CO;2

Lefaucheur, J.-P., Antal, A., Ayache, S. S., Benninger, D. H., Brunelin, J., Cogiamanian, F., . . . Paulus, W. (2017). Evidence-based guidelines on the therapeutic use of transcranial direct current stimulation (tDCS). *Clinical Neurophysiology, 128*(1), 56–92. doi:10.1016/j.clinph.2016.10.087

Legg, S. J., & Mahanty, A. (1985). Comparison of five modes of carrying a load close to the trunk. *Ergonomics, 28*(12), 1653–1660. doi:10.1080/00140138508963301

Leon, G. R., Koscheyev, V. S., Fink., B., Ciofani, P., Warpeha, J., Gernhardt, M. L., & Skytland, N. G. (2009). Subjective perception of thermal and physical comfort in three liquid cooling garments. *SAE Technical Paper Series 2009-01-2516*. Warrendale, PA: SAE International. doi:10.4271/2009-01-2516

Leppänen, M., Aaltonen, S., Parkkari, J., Heinonen, A., & Kujala, U. M. (2014). Interventions to prevent sports related injuries: A systematic review and meta-analysis of randomised controlled trials. *Sports Medicine, 44*(4), 473–486. doi:10.1007/s40279-013-0136-8

Lester, T. (2012). *Davinci's ghost: Genius, obsession, and how Leonardo created the world in his own image*. New York, NY: Simon and Schuster.

Levangie, P. K., & Norkin, C. C. (2005). *Joint structure and function: A comprehensive analysis* (4th ed.). Philadelphia, PA: F. A. Davis Company.

Levin, R. J. (2003). The ins and outs of vaginal lubrication, *Sexual and Relationship Therapy, 18*(4), 509–513. doi:10.1080/14681990310001609859

Leyland, N., Casper, R., Laberge, P., Singh, S. S., Allen, L., Arendas, K., . . . & Senikas, V. (2018). Endometriosis: Diagnosis and management. *Journal of Endometriosis, 2*(3), 107–134. doi:10.1177/228402651000200303

Li, X., & Tokura, H. (1996). The effects of two types of clothing on seasonal heat tolerance. *European Journal of Applied Physiology and Occupational Physiology, 72*(4), 287–291. doi:10.1007/BF00599686

Lieberman, D. E. (2012). What we can learn about running from barefoot running: An evolutionary medical perspective. *Exercise and Sport Sciences Reviews, 40*(2), 63–72. doi:10.1097/JES.0b013e31824ab210

Link, M. S. (2012). Commotio cordis: Ventricular fibrillation triggered by chest impact-induced abnormalities in repolarization. *Circulation: Arrhythmia and Electrophysiology, 5*(2), 425–432. doi:10.1161/CIRCEP.111.962712

Liu, K., Wang, J., Tao, X., Zeng, S., Bruniaux, P., & Kamalha, E. (2016). Fuzzy classification of young women's lower body based on anthropometric measurement. *International Journal of Industrial Ergonomics, 55*, 60–68.

Liu, R., Guo, X., Lao, T. T., & Little, T. (2017). A critical review on compression textiles for compression therapy: Textile-based compression interventions for chronic venous insufficiency. *Textile Research Journal, 87*(9), 1121–1141. doi:10.1177/0040517516646041

Lloyd, J., Crouch, N. S., Minto, C. L., Liao, L. M., & Creighton, S. M. (2005). Female genital appearance: "Normality" unfolds. *BJOG: An International Journal of Obstetrics and Gynaecology, 112*(5), 643–646. doi:10.1111/j.1471-0528.2004.00517.x

Lombardi, D. A., Pannala, R., Sorock, G. S., Wellman, H., Courtney, T. K., Verma, S., & Smith, G. S. (2005). Welding related occupational eye injuries: A narrative analysis. *Injury Prevention, 11*(3), 174–179. doi:10.1136/ip.2004.007088

Lübke, K. T., & Pause, B. M. (2015). Always follow your nose: The functional significance of social chemosignals in human reproduction and survival. *Hormones and Behavior, 68*, 134–144. doi:10.1016/j.yhbeh.2014.10.001

Lucas, P. A., Page, P. R. J., Phillip, R. D., & Bennett, A. N. (2014). The impact of genital trauma on wounded servicemen: Qualitative study. *Injury, 45*(5), 825–829. doi:10.1016/j.injury.2013.12.009

Ludewig, P. M., Cook, T. M., & Nawoczenski, D. A. (1996). Three-dimensional scapular orientation and muscle activity at selected positions of humeral elevation. *Journal of Orthopaedic & Sports Physical Therapy*, 24(2), 57–65. doi:10.2519/jospt.1996.24.2.57

Luk, N., & Yu, W. (2016). Bra fitting assessment and alteration. In W. Yu (Ed.), *Advances in women's intimate apparel technology* (pp. 109–133). Cambridge, UK: Woodhead Publishing. doi:10.1016/B978-1-78242-369-0.00007-4

Lukacz, E. S., Sampselle, C., Gray, M., MacDiarmid, S., Rosenberg, M., Ellsworth, P., & Palmer, M. H. (2011). A healthy bladder: A consensus statement. *International Journal of Clinical Practice*, 65(10), 1026–1036. doi:10.1111/j.1742-1241.2011.02763.x

Lun, V. M., Wiley, J. P., Meeuwisse, W. H., & Yanagawa, T. L. (2005). Effectiveness of patellar bracing for treatment of patellofemoral pain syndrome. *Clinical Journal of Sport Medicine, 15*(4), 235–240. doi:10.1097/01.jsm.0000171258.16941.13

Lund, P. J., Krupinski, E. A., & Brooks, W. J. (1996). Ultrasound evaluation of sacroiliac motion in normal volunteers. *Academic Radiology, 3*(3), 192–196. doi:10.1016/S1076-6332(96)80438-7

Luximon, A., & Luximon, Y. (2013). Shoe-last design and development. In R. S. Goonetilleke (Ed.), *The science of footwear* (pp. 193–212). Boca Raton, FL: CRC Press. doi:10.1201/b13021-13

Lyle, C., & Hubbard, M. (1983, May). Optimal ski boot stiffness for the prevention of boot top fracture. *Skiing Trauma and Safety: Fifth International Symposium.* Symposium conducted at Keystone, CO. In R. J. Johnson, & C. D. Mote, (Eds.), *Skiing trauma and safety: Fifth international symposium,* ASTM special technical publication 860 (1985), pp. 173–181. Philadelphia, PA: American Society for Testing and Materials. doi.10.1520/STP46636S

Lyu, S. (2016). Posture modification effects using soft materials structures (Unpublished doctoral dissertation). University of Minnesota, Minneapolis, MN.

Lyu, S., & LaBat, K. L. (2016). Effects of natural posture imbalance on posture deviation caused by load carriage. *International Journal of Industrial Ergonomics, 56,* 115–123. doi:10.1016/j.ergon.2016.09.006

MacDonald, N. M. (2010). *Principles of flat pattern design* (4th ed.). New York, NY: Fairchild Books.

MacRae, B. A., Laing, R. M., & Partsch, H. (2016). General considerations for compression garments in sports: Applied pressures and body coverage. In F. Engel, & B. Sperlich (Eds.), *Compression Garments in Sports: Athletic Performance and Recovery* (pp. 1–32). Cham, Switzerland: Springer. doi:10.1007/978-3-319-39480-0_1

Madziyire, M. G., Magure, T. M., & Madziwa, C. F. (2018). Menstrual cups as a menstrual management method for low socioeconomic status women and girls in Zimbabwe: A pilot study. *Women's Reproductive Health, 5*(1), 59–65. doi:10.1080/23293691.2018.1429371

Mafart, S. (2007). Hallux valgus in a historical French population: Paleopathological study of 605 first metatarsal bones. *Joint Bone Spine, 74*(2), 166–170. doi:10.1016/j.jbspin.2006.03.011

Magnusson, M., Enbom, H., Johansson, R., & Pyykkö, I. (1990). Significance of pressor input from the human feet in anterior-posterior postural control: The effect of hypothermia on vibration-induced body-sway. *Acta Oto-Laryngologica, 110*(3–4), 182–188. doi:10.3109/00016489009122535

Mahan, L. K., & Escott-Stump, S. (2008). Chapter 1: Digestion, absorption, transport, and excretion of nutrients, In L. Kathleen Mahan, & Sylvia Escott-Stump (Eds.), *Krause's food & nutrition therapy* (12th ed.). St. Louis, MO: Elsevier Saunders.

Mailler, E. A., & Adams, B. B. (2004). The wear and tear of 26.2: Dermatological injuries reported on marathon day. *British Journal of Sports Medicine, 38*(4), 498–501. doi:10.1136/bjsm.2004.011874

Mallouris, A., Yiacoumettis, A., Thomaidis, V., Karayiannakis, A., Simopoulos, C., Kakagia, D., & Tsaroucha, A. (2012). A record of skin creases and folds. *European Journal of Plastic Surgery, 35,* 847–854. doi:10.1007/s00238-012-0774-3

Manley, J. W. (1997). Protective clothing fitting considerations for pregnant women. In J. O. Stull, & A. D. Schwope (Eds.), *Performance of protective clothing: Sixth volume,* ASTM Publication Code Number (PCN) 04-012730-55. (pp. 293–302), West Conshohocken, PA: ASTM International. doi:10.1520/STP19911S

Marchi, N., Bazarian, J. J., Puvenna, V., Janigro, M., Ghosh, C., Zhong, J., . . . Janigro, D. (2013). Consequences of repeated blood-brain barrier disruption in football players. *PLoS ONE, 8*(3), e56805. doi:10.1371/journal.pone.0056805

Mason, B. P., Page, K. A., & Fallon, K. (1999). An analysis of movement and discomfort of the female breast during exercise and the effects of breast support in three cases. *Journal of Science & Medicine in Sport,* 2(2), 134–144. doi:10.1016/S1440-2440(99)80193-5

Masters, W. H., & Johnson, V. E. (1966). *Human sexual response.* Boston, MA: Little, Brown.

Matic, P., Hubler, G. K., Sprague, J. A., Simmonds, K. E., Rupert, N. L., Bruno, R. S., ... & Peksoz, S. (2006). QuadGard arm and leg protection against IED's. Naval Research Lab, Washington, DC, Material Science and Technology Div. Retrieved from https://www.dtic.mil/tr/fulltext/u2/a522792.pdf

MayoClinic.org. (2017a) Diseases and conditions: Burns. http://www.mayoclinic.org/diseases-conditions/burns/basics/symptoms/con-20035028?p=1

MayoClinic.org. (2017b) Diseases and conditions: Contact dermatitis. http://www.mayoclinic.org/diseases-conditions/contact-dermatitis/basics/definition/con-20032048

McCool, F. D. (2006). Global physiology and pathophysiology of cough: ACCP evidence-based clinical practice guidelines. *CHEST Journal, 129*(1), Suppl, 48S–53S. doi:10.1378/chest.129.1_suppl.48S

McCrory, P., Bell, S., & Bradshaw, C. (2002). Nerve entrapments of the lower leg, ankle and foot in sport. *Sports Medicine, 32*(6), 371–391. doi:10.2165/00007256-200232060-00003

McCrory, P., Meeuwisse, W. H., Aubry, M., Cantu, B., Dvořák, J., Echemendia, R. J., ... Turner, M. (2013). Consensus statement on concussion in sport: The 4th international conference on concussion in sport held in Zurich, November 2012. *British Journal of Sports Medicine, 47*(5), 250–258. doi:10.1136/bjsports-2013-092313

McCullough, E. A. (2009). Evaluation of cold weather clothing using manikins. In J. T. Williams (Ed.), *Textiles for cold weather apparel* (pp. 244–255). Cambridge, UK: Woodhead Publishing. doi:10.1533/9781845697174.2.244

McDowell, C. (1989). *Shoes: Fashion and fantasy.* New York, NY: Rizzoli International Publications.

McFarland, S. M. (2016, July 10–14). *Z-2 space suit: A case study in human spaceflight public outreach.* Paper presented at the 46th International Conference on Environmental Systems, Vienna, Austria. Retrieved from https://ttu-ir.tdl.org/ttu-ir/

McGhee, D. E., & Steele, J. R. (2006). How do respiratory state and measurement method affect bra size calculations? *British Journal of Sports Medicine, 40*(12), 970–974. doi:10.1136/bjsm.2005.025171

McGinn, C. (2015). *Prehension: The hand and the emergence of humanity.* Cambridge, MA: MIT Press.

McHenry, R. D., Arnold, G. P., Wang, W., & Abboud, R. J. (2015). Footwear in rock climbing: Current practice. *The Foot, 25*(3), 152–158. doi:10.1016/j.foot.2015.07.007

McKenzie, J. (1955). The foot as a half-dome. *British Medical Journal, 1*(4291), 1068–1070.

McKeon, P. O., & Hertel, J. (2007). Diminished plantar cutaneous sensation and postural control. *Perceptual Motor Skill, 104*(1), 56–66. doi:10.2466/pms.104.1.56-66

McKeon, P. O., Hertel, J., Bramble, D., & Davis, I. (2015). The foot core system: A new paradigm for understanding intrinsic foot muscle function. *British Journal of Sports Medicine, 49*(5), 290. doi:10.1136/bjsports-2013-092690

McKinley, M. P., & O'Loughlin, V.D. (2006). *Human anatomy.* New York, NY: McGraw-Hill.

McKinney, E. (2007). *Towards a three-dimensional theory of pattern drafting: Relationship of body measurements and shapes to pattern measurements and shapes* (Unpublished doctoral dissertation). University of Minnesota, Minneapolis, MN.

McKinney, E. C., Bye, E., & LaBat, K. (2012). Building patternmaking theory: A case study of published patternmaking practices for pants. *International Journal of Fashion Design, Technology and Education, 5*(3), 153–167. doi:10/1080/17543266.2012.666269

McKinney, E., Gill, S., Dorie, A., & Roth, S. (2017). Body-to-pattern relationships in women's trouser drafting methods: Implications for apparel mass customization. *Clothing and Textiles Research Journal, 35*(1), 16–32. doi:10.1177/0887302X16664406

McPoil, T. G., Jr. (1988). Footwear. *Physical Therapy, 68*(12), 1857–1865. PMID:3057521

McQuerry, M., Barker, R., & DenHartog, E. (2018). Functional design and evaluation of structural firefighter turnout suits for improved thermal comfort: Thermal manikin and physiological modeling. *Clothing and Textiles Research Journal, 36*(3), 165–179. doi:10.1177/0887302X18757349

McSweeney, S. C., & Cichero, M. (2015). Tarsal tunnel syndrome—a narrative literature review. *The Foot, 25*(4), 244–250. doi:10.1016/j.foot.2015.08.008

Mease, P. J. (2008). The management of psoriatic arthritis. In P. J. Mease, & P. S. Helliwell, (Eds.), *Atlas of psoriatic arthritis* (pp. 81–97). London, UK: Springer-Verlag.

Mehta, S., & Hill, N. S. (2001). Noninvasive ventilation. *American Journal of Respiratory and Critical Care Medicine, 163*(2), 540–577. doi:10.1164/ajrccm.163.2.9906116

Mellström, G., & Boman, A. (2005). Gloves: Types, materials, and manufacturing. In A. Boman, T. Estlander, J. E. Wahlberg., & H. I. Maibach (Eds.), *Protective gloves for occupational use* (2nd ed., pp. 15–28). Boca Raton, FL: CRC Press.

Meng, J., Zhang, S., Bekyo, A., Olsoe, J., Baxter, B., & He, B. (2016). Noninvasive electroencephalogram based control of a robotic arm for reach and grasp tasks. *Scientific Reports, 6*, 38565. doi:10.1038/srep38565

Meunier, P., Tack, D., Ricci, A., Bossi, L., & Angel, H. (2000). Helmet accommodation analysis using 3D laser scanning. *Applied Ergonomics, 31*(4), 361–369. doi:10.1016/s0003-6870(00)00006-5

Michael, R. P., Bonsall, R. W., & Warner, P. (1974). Human vaginal secretions: Volatile fatty acid content. *Science, 186*(4170), 1217–1219. doi.10.1126/science.186.4170.1217

Miller, M. R., Hankinson, J., Brusasco, V., Burgos, F., Casaburi, R., Coates, A., . . . & ATS/ERS Task Force. (2005). Standardisation of spirometry. *The European Respiratory Journal, 26*(2), 319–338. doi:10.1183/09031936.05.00034805

Millodot, M. (2004). *Dictionary of optometry and visual science* (6th ed.). Philadelphia, PA: Butterworth-Heinemann.

Milsom, I., Coyne, K. S., Nicholson, S., Kvasz, M., Chen, C. I., & Wein, A. J. (2014). Global prevalence and economic burden of urgency urinary incontinence: A systematic review. *European Urology, 65*(1), 79–95. doi:10.1016/j.eururo.2013.08.031

Minott, J. (1974). *Pants and skirts: Fit for your shape*. Minneapolis, MN: Burgess Publishing Company.

Minott, J. (1978). *Fitting commercial patterns: The Minott method*. Minneapolis, MN: Burgess

Miranda, H., Viikari-Juntura, E., Martikainen, R., & Riihimäki, H. (2002). A prospective study on knee pain and its risk factors. *Osteoarthritis and Cartilage, 10*(8), 623–630. doi:10.1053/joca.2002.0796

Mobley, J. P., Zhang, C., Soli, S. D., Johnson, C., & O'Connell, D. (1995). *U.S. Patent No. 5,467,784*. Washington, DC: U.S. Patent and Trademark Office.

Mobolaji-Lawal, M., & Nedorost, S. (2015). The role of textiles in dermatitis: An update. *Current Allergy and Asthma Reports, 15*(4), 17. doi:10.1007/s11882-015-0518-0

Mochimaru, M., Kouchi, M., & Dohi, M. (2000). Analysis of 3D human foot forms using the Free Form Deformation method and its application in grading shoe lasts. *Ergonomics, 43*(9), 1301–1313. doi:10.1080/001401300421752

Moody, R. O., Chamberlain, W. E., & Van Nuys, R. G. (1926). Visceral anatomy of healthy adults. *Developmental Dynamics, 37*(2), 273–288. doi:10.1002/aja.1000370205

Moore, K. L., Agur, A. M. R., & Dalley, A. F. (2011). *Essential clinical anatomy* (4th ed.). Baltimore, MD: Lippincott Williams & Wilkins.

Moore, K. L., Dalley, A. F. II., & Agur, A. M. R. (2014). *Clinically oriented anatomy* (7th ed.). Baltimore, MD: Lippincott Williams & Wilkins.

Moreno, J. C., Figueiredo, J., & Pons, J. L. (2018). Exoskeletons for lower-limb rehabilitation. In R. Colombo & V. Sanguineti (Eds.), *Rehabilitation robotics* (pp. 89–99). London, UK: Elsevier. doi:10.1016/B978-0-12-811995-2.00008-4

Morris, D., Coyle, S., Wu, Y., Lau, K. T., Wallace, G., & Diamond, D. (2009). Bio-sensing textile based patch with integrated optical detection system for sweat monitoring. *Sensors and Actuators B: Chemical, 139*(1), 231–236. doi:10.1016/j.snb.2009.02.032

Morris, K., Park, J., & Sarkar, A. (2017). Development of a nursing sports bra for physically active breastfeeding women through user-centered design. *Clothing and Textiles Research Journal, 35*(4), 290–306. doi:10.1177/0887302X17722858

Moseley, A. L., Carati, C. J., & Piller, N. B. (2006). A systematic review of common conservative therapies for arm lymphoedema secondary to breast cancer treatment. *Annals of Oncology, 18*(4), 639–646. doi:10.1093/annonc/mdl182

Mosti, G., & Partsch, H. (2014). Improvement of venous pumping function by double progressive compression stockings: Higher pressure over the calf is more important than a graduated pressure profile. *European Journal of Vascular and Endovascular Surgery, 47*(5), 545–549. doi:10.1016/j.ejvs.2014.01.006

Motawi, W. (2015). *How shoes are made*. Lexington, KY: https://www.SneakerFactory.net

Motosko, C. C., Bieber, A. K., Pomeranz, M. K., Stein, J. A., & Martires, K. J. (2017). Physiologic changes of pregnancy: A review of the literature. *International Journal of Women's Dermatology, 3*(4), 219–224. doi:10.1016/j.ijwd.2017.09.003

Moyer, R. F., Birmingham, T. B., Bryant, D. M., Giffin, J. R., Marriott, K. A., & Leitch, K. M. (2015). Biomechanical effects of valgus knee bracing: A systematic review and meta-analysis. *Osteoarthritis and Cartilage, 23*(2), 178–188. doi:10.1016/j.joca.2014.11.018

Muallem, M., & Rubeiz, N. G. (2006). Physiological and biological skin changes in pregnancy. *Clinics in Dermatology, 24*(2), 80–83. doi:10.1016/j.clindermatol.2005.10.002

Mueller, M. J., & Maluf, K. S. (2002). Tissue adaptation to physical stress: A proposed "Physical Stress Theory" to guide physical therapist practice, education, and research. *Physical Therapy, 82*(4), 383–403. doi:10.1093/ptj/82.4.383

Mueller, W. H., & Wohlleb, J. C. (1981). Anatomical distribution of subcutaneous fat and its description by multivariate methods: How valid are principal components? *American Journal of Physical Anthropology, 54*(1), 25–35. doi:10.1002/ajpa.1330540104

Munro, M. G. (2012). Classification of menstrual bleeding disorders. *Reviews in Endocrine and Metabolic Disorders, 13*(4), 225–234. doi:10.1007/s11154-012-9220-x

Munteanu, S. E., Scott, L. A., Bonanno, D. R., Landorf, K. B., Pizzari, T., Cook, J. L., & Menz, H. B. (2015). Effectiveness of customised foot orthoses for Achilles tendinopathy: A randomised controlled trial. *British Journal of Sports Medicine, 49*(15), 989–994. doi:10.1136/bjsports-2014-093845

Nácher, B., Alemany, S., González, J. C., Alcantara, E., García-Hernández, J., Heras, S., & Juan, A. (2006). A footwear fit classification model based on anthropometric data. *SAE Technical Paper, 2006-01-2356*. doi:10.4271/2006-01-2356

Napier, J. (1970). *The roots of mankind*. Washington, DC: Smithsonian Institution Press.

NASA (2012, May 11). *Cool spacesuits*. NASA summer of innovation, Life Science—Survival Teacher Lesson, Aerospace Education Services Project. Retrieved from https://www.nasa.gov/sites/pdf/544886main_LS5_Cool-Space-Suits_C1.pdf

NASA (2014). *Human integration design processes (HIDP) TP-2014-218556*. NASA: Human Health and Performance Directorate. Retrieved from http://ston.jsc.nasa.gov/collections/TRS

National Center on Birth Defects and Developmental Disabilities, Centers for Disease Control and Prevention. (2011). Joint range of motion video—hemophilia—videos. https://www.cdc.gov/ncbddd/hemophilia/video/intro.html

National Collaborating Centre for Women's and Children's Health (UK). (2018). Heavy menstrual bleeding. NICE Guideline, No. 88. London, UK: RCOG Press. Available from: https://www.ncbi.nlm.nih.gov/books/NBK493300/ or https://www.nice.org.uk/guidance/ng88

National Heart, Lung, and Blood Institute; National Institutes of Health; U.S. Department of Health and Human Services. *Health Topics/ Raynaud's*. Retrieved from https://www.nhlbi.nih.gov/health-topics/raynauds

National Operating Committee on Standards for Athletic Equipment. (2017). *Standard test method and performance specification used in evaluating the performance characteristics of chest protectors for commotio cordis* (NOCSAE Doc (ND) 200-17a m17a). Overland Park, KS: NOCSAE. Retrieved from http://nocsae.org/wp-content/files_mf/1494961061ND20017am17aCommotioCordisTestMethod.pdf

Navy Environmental Health Center: Bureau of Medicine and Surgery. (2007). Prevention and Treatment of Heat and Cold Stress Injuries, *Technical Manual NEHC-TM-OEM 6260.6A June 2007*, Portsmouth, VA: Navy Environmental Health Center. Retrieved from https://www.med.navy.mil/sites/nmcphc/Documents/policy-and-instruction/oem-prevention-and-treatment-ofheat-and-cold-stress-injuries.pdf

Netter, F. H. (1959). In E. Oppenheimer (Ed.). *The CIBA collection of medical illustrations: Volume 3: Digestive system: Part I upper digestive tract*. West Caldwell, NJ: CIBA Pharmaceutical Company. ISBN 0-914168-03-7

Netter, F. H. (1979). In R. K. Shapter & F. F. Yonkman (Eds.), *The CIBA collection of medical illustrations: Volume 6: Kidneys, ureters, and urinary bladder*. West Caldwell, NJ: CIBA Pharmaceutical Company. ISBN 0-914168-08-8

Netter, F. H. (1980). In M. B. Divertie & A. Brass (Eds.), *The CIBA collection of medical illustrations: Volume 7: Respiratory system* (2nd ed.). Summit, NJ: CIBA Pharmaceutical Company. ISBN 0-914168-09-6

Netter, F. H., 1986. In A. Brass (Ed.), *The CIBA collection of medical illustrations: Volume 1 nervous system: Part I anatomy and physiology*. West Caldwell, NJ: CIBA Pharmaceutical Company. ISBN 0-914168-10-X

Neumann, D. A. (2017). Fundamental and Clinical Considerations of the Muscles of the Hip. In J. C. McCarthy, P. C. Noble, & R. N. Villar (Eds.), *Hip joint restoration* (pp. 35–51). New York, NY: Springer. doi:10.1007/978-1-4614-0694-5_5

Niamtu, J. (2002). Eleven pearls for cosmetic earlobe repair. *Dermatologic Surgery, 28*(2), 180–185. doi:10.1046/j.1524-4725.2002.01052.x

Nikolopoulos, I., Osman, W., Haoula, Z., Jayaprakasan, K., & Atiomo, W. (2013). Scrotal cooling and its benefits to male fertility: A systematic review. *Journal of Obstetrics and Gynaecology, 33*(4), 338–342. doi:10.3109/01443615.2012.758088

Nonfoux, L., Chiaruzzi, M., Badiou, C., Baude, J., Tristan, A., Thioulouse, J., . . . Lina, G. (2018). Impact of currently marketed tampons and menstrual cups on Staphylococcus aureus growth and TSST-1 production in vitro. *Applied and Environmental Microbiology, 84*(12). AEM.00351. doi:10.1128/AEM.00351-18

Norkin, C., & White, D. (2009). *Measurement of joint motion: A guide to goniometry* (4th ed.). Philadelphia: F.A. Davis.

Notarnicola, A., Maccagnano, G., Pesce, V., Tafuri, S., Mercadante, M., Fiore., A., & Moretti, B. (2015). Effect of different types of shoes on balance among soccer players. *Muscles, Ligaments and Tendons Journal, 5*(3), 208–213. doi:10.11138/mltj/2015.5.3.208

Noyes, N., Cho, K.-C., Ravel, J., Forney, L. J., & Abdo, Z. (2018). Associations between sexual habits, menstrual hygiene practices, demographics and the vaginal microbiome as revealed by Bayesian network analysis. *PloS One., 13*(1), e0191625. doi:10.1371/journal.pone.0191625

O'Brien, A. (2005). Maternity dress. In V. Steele, (Ed.), *Scribner library of daily life, encyclopedia of clothing and fashion* (pp. 394–396). Detroit, MI: Thomson Gale.

O'Connor, K., Bragdon, G., & Baumhauer, J. (2006). Sexual dimorphism of the foot and ankle. *Orthopedic Clinics of North America, 37*(4), 569–574. doi:10.1016/j.ocl.2006.09.008

O'Hearn, B. E., Bensel, C. K., & Polcyn, A. F. (2005). *Biomechnical analyses of body movement and locomotion as affected by clothing and footwear for cold weather climates.* NATICK Technical Report, NATICK/TR-05/013, Natick, MA: U. S. Army Research, Development and Engineering Command, Natick Soldier Center.

O'Keefe, L. (1996). *Shoes: A celebration of pumps, sandals, slippers and more.* New York, NY: Workman.

O'Keefe, L. (2012). *Shoes: Instant expert.* New York, NY: Princeton Architectural Press.

Occupational Safety and Health Administration (OSHA). (2017). *Eye and face protection eTool,* Washington DC: Occupational Safety and Health Administration. Retrieved from United States Department of Labor, Occupational Safety and Health Administration: https://www.osha.gov/SLTC/etools/eyeandface/

Occupational Safety and Health Administration (OSHA). (2017). Respiratory Protection. In OSHA, *OSHA Technical Manual, Section VIII, Chapter 2* (OSHA Instruction TED 01-00-015). Washington DC: Occupational Safety and Health Administration. Retrieved from United States Department of Labor, Occupational Safety and Health Administration: https://www.osha.gov/dts/osta/otm/otm_viii/otm_viii_2.html

O'Followell, L., & Lion, D. G. (1908). In A. Maloine (Ed.), *Le corset, histoire, médecine, hygiène.. étude médicale, avec une préface du Dr. Lion.* Paris, France: A. Maloine. Retrieved from http://gallica.bnf.fr/ark:/12148/bpt6k6468254j/f21.image

Oggiano, L., Brownlie, L., Troynikov, O., Bardal, L. M., Sæter, C., & Sætran, L. (2013). A review on skin suits and sport garment aerodynamics: Guidelines and state of the art. *Proceedia Engineering, 60,* 91–98. doi:10.1016/j.proeng.2013.07.018

Okabe, K., & Sugimoto, H. (2007). A study on how to design comfortable maternity trousers which adjust to body changes during pregnancy. *Journal of Home Economics of Japan, 58*(12), 763–770.

Oliver, R., Thakar, R., & Sultan, A. H. (2011). The history and usage of the vaginal pessary: A review. *European Journal of Obstetrics & Gynecology and Reproductive Biology, 156*(2), 125–130. doi:10.1016/j.ejogrb.2010.12.039

Opperman, R. A., Waldie, J. M. A., Natapoff, A., Newman, D. J., & Jones, J. A. (2010). Probability of spacesuit-induced fingernail trauma is associated with hand circumference. *Aviation, Space, and Environmental Medicine, 81*(10), 907–913. doi:10.3357/ASEM.2810.2010

Oster, E., & Thornton, R. (2011). Menstruation, sanitary products, and school attendance: Evidence from a randomized evaluation. *The American Economic Journal, 3*(1), 91–100. doi:10.1257/app.3.1.91

Östgaard, H., Zetherström, G., Roos-Hansson, E., & Svanberg, B. (1994). Reduction of back and posterior pelvic pain in pregnancy. *Spine, 19*(8), 894–900.

Ott, K., Serlin, D., & Mihm, S. (Eds.). (2002). *Artificial parts, practical lives: Modern histories of prosthetics.* New York, NY: New York University Press.

Palastanga, N., & Soames, R. (2012). *Anatomy and human movement: Structure and function* (6th ed.). Edinburgh, Scotland: Elsevier/Churchill Livingstone.

Palastanga, N., Field, D., & Soames, R. (2002). *Anatomy and human movement: Structure and function* (4th ed.). Oxford, England: Butterworth-Heinemann.

Palmitier, R. A., An, K. N., Scott, S. G., & Chao, E. Y. (1991). Kinetic chain exercise in knee rehabilitation. *Sports Medicine, 11*(6), 402–413. doi:10.2165/00007256-199111060-00005

Palosuo, T., Antoniadou, I., Gottrup, F., & Phillips, P. (2011). Latex medical gloves: Time for a reappraisal. *International Archives of Allergy and Immunology, 156*(3), 234–246. doi:10.1159/000323892

Pan, C. S., Chiou, S., Kau, T.-Y., Bhattacharya, A., & Ammons, D. (2009). Effects of foot placement on postural stability of construction workers on stilts. *Applied Ergonomics, 40*(4), 781–789. doi.10.1016/j.apergo.2008.08.003

Pandya, S., & Moore, R. G. (2011). Breast development and anatomy. *Clinical Obstetrics and Gynecology, 54*(1), 91–95. doi:10.1097/GRF.0b013e318207ffe9

Panicker, V. V., Riyaz, N., & Balachandran, P. K. (2017). A clinical study of cutaneous changes in pregnancy. *Journal of Epidemiology and Global Health, 7*(1), 63–70. doi:10.1016/j.jegh.2016.10.002

Pannier, F., Hoffmann, B., Stang, A., Jöckel, K. H., & Rabe, E. (2007). Prevalence and acceptance of therapy with medical compression stockings. *Phlebologie, 36*(05), 245–259. doi:10.1055/s-0037-1622192

Paquet, E., & Viktor, H. L. (2014). Segmentation and classification of anthropometric data for the apparel industry. In D. Gupta, & N. Zakaria (Eds). *Anthropometry, apparel sizing and design.* Cambridge, UK: Woodhead Publishing, Ltd.

Park, J. (2013). Gauging the emerging plus-size footwear market an anthropometric approach. *Clothing and Textiles Research Journal, 31*(1), 3–16. doi:10.1177/0887302X12469291

Park, J., & Curwen, L. G. (2013). No pain, no gain?: Dissatisfied female consumers' anecdotes with footwear products. *International Journal of Fashion Design, Technology and Education, 6*(1), 18–26. doi:10.1080/17543266.2012.757369

Park, H.-S., & Lee, J.-J. (2007). Study on features that pregnant women find important and desirable when choosing maternity wear. *Journal of Korean Society of Design Science, 20*(2), 41–52.

Park, H., Park, J., Lin, S.-H., & Boorady, L. (2014). Assessment of firefighters' needs for personal protective equipment. *Fashion and Textiles, 1*(1)8. doi:10/1186/s40691-014-0008-3

Park, H., Trejo, H., Miles, M., Bauer, A., Kim, S., & Stull, J. (2015). Impact of firefighter gear on lower body range of motion. *International Journal of Clothing Science and Technology, 27*(2), 315–334. doi:10.1108/IJCST-01-2014-0011

Parks, M., & Block, J. C. (2010). *U.S. Patent No. 7682347 B2*. Washington, DC: U.S. Patent and Trademark Office.

Parry, I., Hanley, C., Niszczak, J., Sen, S., Palmieri, T., & Greenhalgh, D. (2013). Harnessing the transparent face orthosis for facial scar management: A comparison of methods. *Burns, 39*(5), 950–956. doi:10.1016/j.burns.2012.11.009

Parsons, M., Tissot, W., Cardozo, L., Diokno, A., Amundsen, C. L., & Coats, A. C. (2007). Normative bladder diary measurements: Night versus day. *Neurourology and Urodynamics, 26*(4), 465–473. doi:10.1002/nau.20355

Pause, B. M. (2017). Human chemosensory communication. In: A. Buettner (Ed.), *Springer handbooks: Springer handbook of odor* (pp. 129–130). Cham, Switzerland: Springer doi:10.1007/978-3-319-26932-0_52

Peat, G., McCarney, R., & Croft, P. (2001). Knee pain and osteoarthritis in older adults: A review of community burden and current use of primary health care. *Annals of the Rheumatic Diseases, 60*(2), 91–97.

Peate, W. (2007). Work-related eye injuries and illnesses. *American Family Physician, 75*(7), 1017–1022.

Pechter, E. A. (1998). A new method for determining bra size and predicting post-augmentation breast size. *Plastic and Reconstructive Surgery, 102*(4), 1259–1265. doi:10.1097/00006534-199809040-00056

Pehlivan, A. U., Sergi, F., Erwin, A., Yozbatiran, N., Francisco, G. E., & O'Malley, M. K. (2014). Design and validation of the RiceWrist-S exoskeleton for robotic rehabilitation after incomplete spinal cord injury. *Robotica, 32*(08), 1415–1431. doi:0.1017/S0263574714001490

Pei, J., Park, H., Ashdown, S. P., & Vuruskan, A. (2017). A sizing improvement methodology based on adjustment of interior accommodation rates across measurement categories within a size chart. *International Journal of Clothing Science and Technology, 29*(5), 716–731. doi:10.1108/IJCST-03-2017.0024

Peña, I., Viktor, H. L., & Paquet, E. (2012). Who are our clients: Consumer segmentation through explorative data mining. *International Journal of Data Mining, Modelling, and Management* 4(3), 286–308.

Pendergrass, P. B., Reeves, C. A., & Belovicz, M. W. (1991). A technique for vaginal casting utilizing vinyl polysiloxane dental impression material. *Gynecologic and Obstetric Investigation, 32*(2), 121–122. doi:10.1159/000293010

Persson, P. R., Hirschfeld, H., & Nilsson-Wikmar, L. (2007). Associated sagittal spinal movements in performance of head pro- and retraction in healthy women: A kinematic analysis. *Manual Therapy, 12*(2), 119–125. doi:10.1016/j.math.2006.02.013

Phang, K., Bowman, M., Phillips, A., & Windsor, J. (2014). Review of thoracic duct anatomical variations and clinical implications. *Clinical Anatomy, 27*(4), 637–644. doi:10.1002/ca.22337

Pheasant, S., & Haslegrave, C. M. (2006). *Bodyspace: Anthropometry, ergonomics and the design of work* (3rd ed.). Boca Raton, FL: CRC Press.

Piérard-Franchimont C., Piérard G.E., & Hermanns-Lê T. (2017). Sweat gland histophysiology. In: P. Humbert, F. Fanian, H. Maibach, & P. Agache (Eds.), *Agache's measuring the skin* (pp. 617–621). Cham, Switzerland: Springer. doi:10.1007/978-3-319-32383-1_84

Pilsl, U., & Anderhuber, F. (2010). The chin and adjacent fat compartments. *Dermatologic Surgery, 36*(2), 214–218. doi:10.1111/j.1524-4725.2009.01424.x

Pinter, M., Eckelt, M., & Schretter, H. (2010). Evaluation of ski boot fitting characteristics by means of different pressure distribution measurements. *Procedia Engineering, 2*(2), 2875–2880. doi:10.1016/j.proeng.2010.04.081

Pipes, A. (2009). *Introduction to design* (2nd ed.). Upper Saddle River, NJ: Pearson Education, Inc.

Poblet, E., Jimenez, F., Escario-Travesedo, E., Hardman, J. A., Hernández-Hernández, I., Agudo-Mena, J. L., . . . & Paus, R. (2018). Eccrine sweat glands associate with the human hair follicle within a defined compartment of dermal white adipose tissue. *British Journal of Dermatology, 178*(5), 1163–1172. doi:10.1111/bjd.16436

Postacchini, F., Gumina, S., De Santis, P., & Albo, F. (2002). Epidemiology of clavicle fractures. *Journal of Shoulder and Elbow Surgery, 11*(5), 452–456. doi:10.1067/mse.2002.26613

Postolache, G., Carvalho, H., Catarino, A., & Postolache, O. A. (2017). Smart clothes for rehabilitation context: technical and technological issues. In O. A. Postolache, S. C. Mukhopadhyay, K. P. Jayasundera, & A.K. Swain (Eds.), *Sensors for everyday life. Smart sensors, measurement and instrumentation* (Vol. 22). *Sensors for everyday life: Healthcare settings* (pp. 185–219). Cham, Switzerland: Springer International. doi:10.1007/978-3-319-47319-2_10

Poston, A. (2000). *Human engineering design data digest: Human factors standardization systems.* Washington, DC: Department of Defense Human Factors Standardization.

Pouliot, C. J. T. (1980). *Pants alteration by graphic somatometry techniques* (Unpublished master's thesis). Iowa State University, Ames, Iowa.

Price, C. I., & Pandyan, A. (2001). Electrical stimulation for preventing and treating post-stroke shoulder pain: A systematic Cochrane review. *Clinical Rehabilitation, 15*(1), 5–19. doi:10.1191/026921501670667822

Prime, D. K.; Geddes, D. T., & Hartmann, P.E. (2007). Oxytocin: Milk ejection and maternal-infant well-being. In T. Hale, & P. E. Hartmann, (Eds.), *Textbook of human lactation* (pp. 141–158). Amarillo: Hale Publishing.

Raanan, A. (2017). *U.S. Patent No. 9,629,399.* Washington, DC: U.S. Patent and Trademark Office.

Rahman, T., Sample, W., & Seliktar, R. (2004). 16 design and testing of WREX. In Z. Z. Bien, & D. Stefanov (Eds.), *Advances in rehabilitation robotics, lecture notes in control and information science, 306* (pp. 243–250). Berlin, Germany: Springer-Verlag. doi:10.1007/10946978_16

Raj, M., Patel, S., Lee, C. H., Ma, Y., Banks, A., McGinnis, R., . . . & Ghaffari, R. (2016). Multifunctional epidermal sensor systems with ultrathin encapsulation packaging for health monitoring. In J. A. Rogers, R. Ghaffari, & D. H. Kim (Eds.), *Stretchable bioelectronics for medical devices and systems* (pp. 193–205). Springer. doi:10.1007/978-3-319-28694-5

Rajulu, S., & Corner, B. (2013). 3D surface scanning. In R. S. Goonetilleke (Ed.), *The science of footwear* (pp. 127–145). Boca Raton, FL: CRC Press.

Ramsay, D. T., Kent, J. C., Hartmann, R. A. and Hartmann, P. E. (2005). Anatomy of the lactating human breast redefined with ultrasound imaging. *Journal of Anatomy, 206*(6), 525–534. doi:10.1111/j.1469-7580.2005.00417

Ramsey, S., Sweeney, C., Fraser, M., & Oades, G. (2009). Pubic hair and sexuality: A review. *The Journal of Sexual Medicine, 6*(8), 2102–2110. doi:10.1111/j.1743-6109.2009.01307.x

Randell, A. G., Nguyen, T. V., Bhalerao, N., Silverman, S. L., Sambrook, P. N., & Eisman, J. A. (2000). Deterioration in quality of life following hip fracture: A prospective study. *Osteoporosis International, 11*(5), 460–466. doi:10.1007/s001980070115

Randers-Pehrson, J. D. (1960). *The surgeon's glove*. Springfield, IL: Charles C. Thomas, Publisher.

Raudrant, D., Frappart, L., De Haas, P., Thoulon, J. M., & Charvet, F. (1989). Study of the vaginal mucous membrane following tampon utilisation; aspect on colposcopy, scanning electron microscopy and transmission electron microscopy. *European Journal of Obstetrics and Gynecology and Reproductive Biology, 31*(1), 53–65. doi:10.1016/0028-2243(89)90026-9

Raudrant, D., Landrivon, G., Frappart, L., De Haas, P., Champion, F., & Ecochard, R. (1995). Comparison of the effects of different menstrual tampons on the vaginal epithelium: A randomised clinical trial. *European Journal of Obstetrics and Gynecology and Reproductive Biology, 58*(1), 41–46. doi:10.1016/0028-2243(94)01977-F

Redwood, M. (2016). *Gloves and glove-making*. London, Great Britain: Shire Publications.

Reece, M., Herbenick, D., Monahan, P., Sanders, S. A., Temkit, M., & Yarber, W. (2008). Breakage, slippage and acceptability outcomes of a condom fitted to penile dimensions. *Sexually Transmitted Infections, 84*(2), 143–149. doi:10.1136/sti.2007.028316

Ren, L., Simon, D., & Wu, J. (2018). Meaning in absence: The case of tampon use among Chinese women. *Asian Journal of Women's Studies, 24*(1), 28–46. doi:10.1080/12259276.2017.1421291

Richards, C. E., Magin, P. J., & Callister, R. (2009). Is your prescription of distance running shoes evidence-based? *British Journal of Sports Medicine, 43*(3), 159–162. doi:10.1136/bjsm.2008.046680

Ridgway, J. L., Parson, J., & Sohn, M.-H. (2017). Creating a more ideal self through the use of clothing: An exploratory study of women's perceptions of optical illusion garments. *Clothing and Textile Research Journal, 35*(2), 111–127. doi:10.1177/0887302X16678335

Riello, G., & McNeil, P. (Eds.). (2006). *Shoes: A history from sandals to sneakers*. New York, NY: Berg.

Rigoard, P. (2017). The common fibular nerve. In P. Rigoard (Ed.), *Atlas of Anatomy of the Peripheral Nerves* (pp. 264–279). Cham, Switzerland: Springer. doi:10.1007/978-3-319-43089-8_17

Rinehart, J. (1975). *How to make men's clothes*. Garden City, New York, NY: Doubleday & Company, Inc.

Roaas, A., & Andersson, G. B. (1982). Normal range of motion of the hip, knee and ankle joints in male subjects, 30–40 years of age. *Acta Orthopaedica Scandinavica, 53*(2), 205–208. doi:10.3109/17453678208992202

Roach, M. (2010). *Packing for Mars: The curious science of life in the void*. New York, NY: W. W. Norton.

Roach-Higgins, M. E., Eicher, J. B., Johnson, K. P. (1995). *Dress and identity*. New York, NY: Fairchild Publications.

Robinette, K. M., Annis, J. F., Anthropology Research Project Inc. Yellow Springs OH (1986). *A nine-size system for chemical defense gloves.* Technical Report (AAMRL-TR-86-029) (AD A173 193). Harry G. Armstrong Aerospace Medical Research Library. Wright-Patterson Air Force Base, OH. Retrieved from https://www.dtic.mil/dtic/tr/fulltext/u2/a173193.pdf

Robinovitch, S. N., Evans, S. L., Minns, J., Laing, A.C., Kannus, P., Cripton, P. A., ... & Lauritzen, B. (2009). Hip protectors: Recommendations for biomechanical testing—an international consensus statement (part I). *Osteoporosis International, 20*(12), 1977–1988. doi:10.1007/s00198-009-1045-4

Rodgers, G. B., & Topping, J. C. (2012). Safety effects of drawstring requirements for children's upper outerwear garments. *Archives of Pediatrics and Adolescent Medicine, 166*(7), 651–655. doi:10.1001/archpediatrics.2011.1269

Rodrigues, S., Domingues, G., Ferreira, I., Faria, L., & Seixas, A. (2017, April). Influence of different backpack loading conditions on neck and lumbar muscles activity of elementary school children. In P. M. Arezes, J. S. Baptista, M. P. Barroso, P. Carneiro, P. Cordeiro, N. Costa, . . . & G. Perestrelo, (Eds.), *Occupational Safety and Hygiene V: Selected Proceedings of the International Symposium on Occupational Safety and Hygiene (SHO 2017), Guimarães Portugal, April 10–11, 2017* (pp. 485–489). Boca Raton, FL: CRC Press. ISBN 1351675249, NCT 02725645

Roebuck, J. A. (1995). *Anthropometric methods: Designing to fit the human body.* Santa Monica, CA: Human Factors and Ergonomics Society.

Roemer, F. W., Jomaah, N., Niu, J., Almusa, E., Roger, B., D'Hooghe, P., . . . & Guermazi, A. (2014). Ligamentous injuries and the risk of associated tissue damage in acute ankle sprains in athletes: A cross-sectional MRI study. *American Journal of Sports Medicine, 42*(7), 1549–1557. doi:10.1177/0363546514529643

Rohrich, R. J., & Pessa, J. E. (2007). The fat compartments of the face: Anatomy and clinical implications for cosmetic surgery. *Plastic and Reconstructive Surgery,119*(7), 2219–2227. doi:10.1097/01.prs.0000265403.66886.54

Rønning, R., Rønning, I., Gerner, T., & Engebretsen, L. (2001). The efficacy of wrist protectors in preventing snowboarding injuries. *The American Journal of Sports Medicine, 29*(5), 581–585. doi:10.1177/03635465010290051001

Roos, E. M., & Arden, N. K. (2016). Strategies for the prevention of knee osteoarthritis. *Nature Reviews Rheumatology, 12*(2), 92–101. doi:10.1038/nrrheum.2015.135

Rosciam, C. (2010). The evolution of catcher's equipment. *The Baseball Research Journal, 39*(1), 104–111. Retrieved from https://www.sabr.org/research/evolution-catchers-equipment

Rose, W., Bowser, B., McGrath, R., Salerno, J., Wallace, J., & Davis, I. (2011, August). Effect of footwear on balance. Poster session presented at the *American Society of Biomechanics 35th Annual Meeting.* Long Beach, CA. http://www.asbweb.org/conferences/2011/pdf/344.pdf

Rosenblad-Wallin, E. (1987). An anthropometric study as the basis for sizing anatomically designed mittens. *Applied Ergonomics, 18*(4), 329–333. doi:10.1016/0003-6870(87)90141-4

Roses, D. F. (2005). *Breast Cancer* (2nd ed.). Philadelphia, PA: Elsevier.

Ross, A., Rhodes, R., Graziosi, D., Jones, B., Lee, R., Haque, B. Z. G., & Gillespie, J. W. (2014, July). *Z-2 prototype space suit development.* Paper presented at the 44th International Conference on Environmental Systems, Tucson, AZ, 13–17. Retrieved from https://ttu-ir.tdl.org/

Rossi, W. (1977). *The sex life of the foot and shoe.* London: Routledge & Kegan Paul.

Rossi, A. B. R., & Vergnanini, A. L. (2000). Cellulite: A review. *Journal of the European Academy of Dermatology and Venereology, 14*(4), 251–262. doi:10.1046/j.1468-3083.2000.00016.x

Rowen, T. S., Gaither, T. W., Awad, M. A., Osterberg, E. C., Shindel, A. W., & Breyer, B. N. (2016). Pubic hair grooming prevalence and motivation among women in the United States. *JAMA Dermatology, 152*(10), 1106–1113. doi:10.1001/jamadermatol.2016.2154

Ruggiu, A., & Cancedda, R. (2015). Bone mechanobiology, gravity and tissue engineering: Effects and insights. *Journal of Tissue Engineering and Regenerative Medicine, 9*(12), 1339–1351. doi:10.1002/term.1942.

Runeman, B. (2008). Skin interaction with absorbent hygiene products. *Clinics in Dermatology, 26*(1), 45–51. doi:10.1016/j.clindermatol.2007.10.002

Ryan, E. L. (2000). Pectoral girdle myalgia in women: A 5-yr. study in a clinical setting. *Clinical Journal of Pain, 16*(4), 298–303.

Ryan, K. (2006). *Aesthetically unique, specially sized clothing for women with osteoporotic posture changes* (Unpublished master's thesis). University of Minnesota, St. Paul, MN.

Ryan, T. J. (2017). Care of oedematous skin in a resource-poor environment: A commentary of practice strategies to address a global community need. *Wound Practice & Research: Journal of the Australian Wound Management Association, 25*(3), 134.

Ryan, T. J., & Narahari, S. R. (2012). Reporting an alliance using an integrative approach to the management of lymphedema in India. *The International Journal of Lower Extremity Wounds, 11*(1), 5–9. doi:10.1177/1534734612438548

Saint, S., Kaufman, S. R., Rogers, M. A. M., Baker, P. D., Ossenkop, K., & Lipsky, B. A. (2006). Condom versus indwelling urinary catheters: A randomized trial. *Journal of the American Geriatrics Society, 54*(7), 1055–1061. doi:10.1111/j.1532-5415.2006.00785.x

Sakamoto, H., Yajima, T., Nagata, M., Okumura, T., Suzuki, K., & Ogawa, Y. (2008). Relationship between testicular size by ultrasonography and testicular function: Measurement of testicular length, width, and depth in patients with infertility. *International Journal of Urology, 15*(6), 529–533. doi:10.1111/j.1442-2042.2008.02071.x

Saladin, K. (2007). *Anatomy and physiology: The unity of form and function* (4th ed.). New York, NY: McGraw-Hill.

Saladin, K. S. (2014). *Human anatomy* (4th ed.). New York, NY: McGraw-Hill.

Saunders, C. G., & Chang, J. (2013). 3D shape capture of human feet and shoe lasts. In R. S. Goonetilleke (Ed.), *The science of footwear* (pp. 113–125). Boca Raton, FL: CRC Press.

Saxton, T. K., Lyndon, A., Little, A. C., & Roberts, S. C. (2008). Evidence that androstadienone, a putative human chemosignal, modulates women's attributions of men's attractiveness. *Hormones and Behavior, 54*(5), 597–601. doi:10.1016/j.yhbeh.2008.06.001

Schieber, R. A., Branche-Dorsey, C. M., Ryan, G. W., Rutherford Jr., G. W., Stevens, J. A., & O'Neil, J. (1996). Risk factors for injuries from in-line skating and the effectiveness of safety gear. *New England Journal of Medicine, 335*(22), 1630–1635. doi:10.1056/NEJM199611283352202

Schlievert, P. M., Tripp, T. J., & Peterson, M. L. (2004). Reemergence of staphylococcal toxic shock syndrome in Minneapolis-St. Paul, Minnesota, during the 2000-2003 surveillance period. *Journal of Clinical Microbiology, 42*(6), 2875–2876. doi:10.1128/JCM.42.6.2875-2876.2004

Schneider, T., Sperling, H., Lümmen, G., Syllwasschy, J., & Rübben, H. (2001). Does penile size in younger men cause problems in condom use? A prospective measurement of penile dimensions in 111 young and 32 older men. *Urology, 57*(2), 314–318. doi:10.1016/S0090-4295(00)00925-0

Schofield, N. A., & LaBat, K. L. (2005). Defining and testing the assumptions used in current apparel grading practice. *Clothing and Textiles Research Journal, 23*(3), 135–150. doi:10.1177/0887302X0502300301

Schofield, N. A., Ashdown, S. P., Hethorn, J., LaBat, K., & Salusso, C. J. (2006). Improving pant fit for women 55 and older through an exploration of two pant shapes. *Clothing and Textiles Research Journal, 24,* 147–160. doi:10.1177/08873 02X0602400208

Schrödter, E., Brüggemann, G.-P., Hamill, J., Rohr, E., & Willwacher, S. (2016). Footwear-related variability in running. *Footwear Science, 8*(1), 23–31. doi:10.108 0/19424280.2016.1142002

Scrafton, S., Stainer, M. J., & Tatler, B. W. (2017). Coordinating vision and action in natural behavior: Differences in spatiotemporal coupling in everyday tasks. *Canadian Journal of Experimental Psychology, 71*(2), 133–145. doi:10.1037/ cep0000120

Selfe, J., Janssen, J., Callaghan, M., Witvrouw, E., Sutton, C., Richards, J., . . . & Dey, P. (2016). Are there three main subgroups within the patellofemoral pain population? A detailed characterisation study of 127 patients to help develop targeted intervention (TIPPs). *British Journal of Sports Medicine, 50*(14), 873–880. doi:10.1136/bjsports-2015-094792

Selius, B. A., & Subedi, R. (2008). Urinary Retention in adults: Diagnosis and initial management. *American Family Physician, 77*(5), 643–650.

Seppänen, T. M., Alho, O., Vakkala, M., Alahuhta, S., & Seppänen, T. (2017). Continuous postoperative respiratory monitoring with calibrated respiratory effort belts: Pilot study. Paper presented at the *9th International Joint Conference on Biomedical Engineering Systems and Technologies, Rome, Italy, February 21–23, 2016.* Published in A. Fred, & H. Gamboa (Eds.), *Biomedical Engineering Systems and Technologies, BIOSTEC 2016. Communications in Computer and Information Service, 690* (pp. 340–359). Cham, Switzerland: Springer. doi:10.1007/978-3-319-54717-6_19

Serlin, D. (2002). *Replaceable you: Engineering the body in postwar America.* Chicago, IL: The University of Chicago Press.

Severinghaus, J. W., & Honda, Y. (1987). History of blood gas analysis. VII. Pulse oximetry. *Journal of Clinical Monitoring, 3*(2), 135–138. doi:10.1007/BF00858362

Shands, K. N., Schmid, G. P., Dan, B. B., Blum, D., Guidotti, R. J., Hargrett, N. T., … & Fraser, D. W. (1980). Toxic-Shock Syndrome in menstruating women: Association with tampon use and staphylococcus aureus and clinical features in 52 cases. *New England Journal of Medicine, 303*(25), 1436–1442. doi:10.1056/ NEJM198012183032502

Shea, J. D., & McClain, E. J. (1969). Ulnar-nerve compression syndromes at and below the wrist. *Journal of Bone and Joint Surgery, 51*(6), 1095–1103.

Shealy, J. E., & Ettlinger, C. F. (1985, April). The in-boot fracture. *Skiing and Trauma Safety: Sixth International Symposium.* Symposium conducted at Naeba, Japan. In C. D. Mote, & R. J. Johnson (Eds.), *Skiing and Trauma Safety: Sixth International Symposium,* ASTM special technical publication 938 (1987), pp. 113–126. Philadelphia, PA: American Society for Testing and Materials. doi:10.1520/ STP23180S

Shelbourne, K. D., & Nitz, P. A. (1991). The O'Donoghue triad revisited: Combined knee injuries involving anterior cruciate and medial collateral ligament tears. *American Journal of Sports Medicine, 19*(5), 474–477. doi:10.1177/036354659101 900509

Sheldon, W. H. (1940). *The varieties of human physique*. New York, NY: Harper and Brothers Publishers.

Shin, K. (2014). Intimate apparel: Designing intimate apparel to fit different body shapes. In M.-E. Faust, & S. Carrier (Eds.), *Designing apparel for consumers: The impacts of body shape and sizes* (pp. 273–29). Oxford, UK: Woodhead. doi:10.1533/9781782422150.2.273

Shoemake, J. J. (2014). Men's underwear with fitted frontal pouch and removable ergonomic gel pack for testicular cooling. US Patent Application 20140137316 A1. Retrieved from https://patents.google.com/patent/US20140137316?oq =US+Patent+Application+20140137316+A1

Shorter, K. A., Kogler, G. F., Loth, E., Durfee, W. K., & Hsiao-Wecksler, E. T. (2011). A portable powered ankle-foot orthosis for rehabilitation. *Journal of Rehabilitation Research & Development, 48*(4), 459–472.

Shwom, I. M. (2011). This time it's personal: Provide exceptional patient care by offering new ophthalmic lens technologies uniquely crafted for each patient. *Review of Optometry, 148*(4), 33–38.

Sibonga, J. D., Spector, E. R., Johnston, S. L., & Tarver, W. J. (2015). Evaluating bone loss in ISS astronauts. *Aerospace Medicine and Human Performance, 86*(12), A38–A44. doi:10.3357/AMHP.EC06.2015

Siegel, R. L., Miller, K. D., & Jemal, A. (2016). Cancer statistics, 2016. *CA: A cancer journal for clinicians, 66*(1), 7–30. doi 10.3322/caac.21332

Simmons, K. P. (2002). *Body shape analysis using three-dimensional body scanning technology* (Unpublished doctoral dissertation). North Carolina State University, Raleigh, NC.

Simmons, K. P., & Istook, C. L. (2003). Body measurement techniques: Comparing 3D body-scanning and anthropometric methods for apparel applications. *Journal of Fashion Marketing and Management, 7*(3), 306–2026. doi:10.1108/13612020310 484852

Simoes, I. (2013). Viewing the mobile body as the source of the design process. *International Journal of Fashion Design, Technology and Education, 6*(2), 72–81. doi:1 0.1080/17543266.2013.793742

Simon, C., Dunne, L., Zeagler, C., Martin, T., & Pailes-Friedman, R. (2014). *NASA Wearable technology CLUSTER2013-2014 Report*. Retrieved from https://ntrs. nasa.gov/

Singer Sewing Library. (1988). *Sewing for children*. Minnetonka, MN: Cy DeCosse Incorporated.

Sizer, P. S., Jr., & James, C. R. (2008). Considerations of sex differences in musculoskeletal anatomy. In J. J. Robert-McComb, R. Norman, & M. Zumwalt (Eds.), *The active female: Health issues throughout the lifespan* (pp. 25–54). Totowa, NJ: Humana Press. doi:10.1007/978-1-59745-534-3

Škerlj, B., Brožek, J., & Hunt, E. E., Jr., (with Chen, K.-P., Carlson, W., Bronczyk, F., & Baker, P.) (1953). Subcutaneous fat and age changes in body build and body form in women. *American Journal of Physical Anthropology, 11*(4), 577–600. doi:10.1002/ ajpa.1330110406

Skoog, A. I. (2013). Space suits. In C. Norberg (Ed.), *Human spaceflight and exploration* (pp. 209–254). Heidelberg, Germany: Springer-Praxis. doi:10.1007/978-3-642-23725-6_6

Smidt, G. L., McQuade, K., Wei, S.-H., & Barakatt, E. (1995). Sacroiliac kinematics for reciprocal straddle positions. *Spine, 20*(9), 1047–1054.

Smith, A. M. A., Jolley, D. D., Hocking, J., Benton, K., & Gerofi, J. (1998). Does penis size influence condom slippage and breakage? *International Journal of STD and AIDS, 9*(8), 444–447. doi:10.1258/0956462981922593

Smith, C. J., & Havenith, G. (2012). Body mapping of sweating patterns in athletes: A sex comparison. *Medicine & Science in Sports & Exercise, 44*(12), 2350–2361. doi:10.1249/MSS.0b013e318267b0c4

Smith, C. J., Machado-Moreira, C. A., Plant, G., Hodder, S., Havenith, G., & Taylor, N. (2013). Design data for footwear: Sweating distribution on the human foot. *International Journal of Clothing Science and Technology, 25*(1), 43–58. doi:10.1108/09556221311292200

Smith, R. O., & Okamoto, G. A. (1981). Checklist for the prescription of slings for the hemiplegic patient. *American Journal of Occupational Therapy, 35*(2), 91–95. doi:10.5014/ajot.35.2.91

Smith, R. P., Turek, P. J., & Netter, F. H. (2011). *The Netter collection of medical illustrations. Volume 1, reproductive system* (2nd ed.). Philadelphia, PA: Elsevier Saunders.

Smits, R. (2011). *The puzzle of left handedness*. London: Reaktion Books.

Snijders, C. J., Vleeming, A., & Stoeckart, R. (1993). Transfer of lumbosacral load to iliac bones and legs: Part 1: Biomechanics of self-bracing of the sacroiliac joints and its significance for treatment and exercise. *Clinical Biomechanics, 8*(6), 285–294. doi:10.1016/0268-0030(93)90002-Y

Sohn, M.-H., & Bye, E. (2012). Visual analysis of body shape changes during pregnancy. *International Journal of Fashion Design, Technology and Education, 5*(2), 117–128. doi:10.1080/17543266.2011.649792

Sohn, M.-H., & Bye, E. (2014). Exploratory study on developing a body measurement method using motion capture. *Clothing and Textiles Research Journal 32*(3), 170–185. doi:10.1177/0887302X14526302

Sohn, M.-H., & Bye, E. (2015). Pregnancy and body image: Analysis of clothing functions of maternity wear. *Clothing and Textiles Research Journal, 33*(1), 64–78. doi:10.1177/0887302X14557809

Sokolowski, S. L. (1999). *Development of a methodology to describe the morphology of the foot for footwear: Application for women's footwear.* (Unpublished doctoral dissertation). University of Minnesota, Minneapolis, MN.

Sokolowski, S. L., Hansen, N., & Roether, J. (2011). *U.S. Patent No. 7,934,325 B2.* Washington, DC: U.S. Patent and Trademark Office.

Sokolowski, S., Griffin, L., Carufel, R., & Kim, N. (2018). Drawing hands for glove design: Does the data match-up? In *Proceedings of the 9th International Conference on Applied Human Factors and Ergonomics*, Orlando, FL.

Song, H. K., & Ashdown, S. P. (2012). Development of automated custom-made pants driven by body shape. *Clothing and Textile Research Journal, 30*(4), 315–329. doi:10.1177/0887302X12462058

Song, S. J., Beard, C. A., & Ustinova, K. I. (2016). The effects of wearing a compression top on trunk and golf club motions during golf swing. *Clothing and Textiles Research Journal, 34*(1), 48–60. doi:10.1177/0887302X15602096

Sonnenburg, J., & Sonnenburg, E. (2015). *The good gut: Taking control of your weight, your mood, and your long-term health.* New York, NY: Penguin Press.

Soper, D. E., Brockwell, N. J., & Dalton, H. P. (1991). Evaluation of the effects of a female condom on the female lower genital tract. *Contraception, 44*(1), 21–29. doi:10.1016/0010-7829(91)90103-M

Sosin, M., Weissler, J. M., Pulcrano, M., & Rodriguez, E. D. (2015). Transcartilaginous ear piercing and infectious complications: A systematic review and critical analysis of outcomes. *The Laryngoscope, 125*(8), 1827–1834. doi:10.1002/lary.25238

Soyer, F. (2001, May 1). The evolution of baseball gloves. *Popular Mechanics*. Hearst Publications.

Spalding, K. L., Arner, E., Westermark, P. O., Bernard, S., Buchholz, B. A., Bergmann, O., . . . & Concha, H. (2008). Dynamics of fat cell turnover in humans. *Nature, 453*(7196), 783–787. doi:10.1038/nature06902

Spina Bifida Association. (2015). *LATEX in the home & community*. Retrieved from http://spinabifidaassociation.org/wp-content/uploads/2015/07/latex-in-the-home-and-community-eng.pdf

Squadroni, R., Rodano, R., Hamill, J., & Preatoni, E. (2015). Acute effect of different minimalist shoes on foot strike pattern and kinematics in rearfoot strikers during running. *Journal of Sports Sciences, 33*(11), 1196–1204. doi:10.1080/02640414.2014.989534

Staker, M., Ryan, K., & LaBat, K. (2009). Medicine and design investigate residual limb volume fluctuations. *Australasian Medical Journal, 1*(12), 156–161. doi:10.4066/AMJ.2009.92

Stall-Meadows, C. (2004). *Know your fashion accessories*. New York, NY: Fairchild Publications.

Stanton, C. (2017). Guideline for prevention of venous thromboembolism. *AORN Journal, 106*(3), P7–P9. doi:10.1016/S0001-2092(17)30730-5

Stanton-Hicks, M., & Salamon, J. (1997). Stimulation of the central and peripheral nervous system for the control of pain. *Journal of Clinical Neurophysiology, 14*(1), 46–62.

Starr, C. L., Cao, H., Peksoz, S., & Branson, D. H. (2015). Thermal effects of design and materials on QuadGard™ body armor systems. *Clothing and Textiles Research Journal, 33*(1), 51–63. doi:10.1177/0887302X14556151

Stergiou, G. S., Parati, G., Vlachopoulos, C., Achimastos, A., Andreadis, E., Asmar, R., . . . & O'Brien, E. (2016). Methodology and technology for peripheral and central blood pressure and blood pressure variability measurement: Current status and future directions—position statement of the European Society of Hypertension Working Group on blood pressure monitoring and cardiovascular variability. *Journal of Hypertension, 34*(9), 1665–1677. doi:10.1097/HJH.0000000000000969

Steven, J., Katz, E. G., Huxley, R. R. (2010). Associations between gender, age, and waist circumference. *Journal of Clinical Nutrition, 64*, 6–15.

Stevens, R. E., & Cooper, J. (1999). Radiotherapy for in situ, stage I and stage II breast cancer. In D. F. Roses (Ed.), *Breast cancer* (pp. 385–415). Philadelphia, PA: Harcourt Brace.

Stewart, A. D., Nevill, A. M., Stephan, R., & Young, J. (2010). Waist size and shape assessed by 3D photonic scanning. *International Journal of Body Composition Research, 8*(4), 123–129.

Stifani, B. M., Plagianos, M., Vieira, C. S., & Merkatz, R. B. (2018). Factors associated with nonadherence to instructions for using the Nestorone®/ethinyl estradiol contraceptive vaginal ring. *Contraception, 97*(5), 415–421. doi:10.1016/j.contraception.2017.12.011

Stolzenberg, D., Siu, G., & Cruz, E. (2012). Current and future interventions for glenohumeral subluxation in hemiplegia secondary to stroke. *Topics in Stroke Rehabilitation, 19*(5), 444–456. doi:10.1310/tsr1905-444

Stone, L. W. (1976). *Men's tailoring.* Tampa, FL: Louise, W. Stone.

Sturm, R. A., & Frudakis, T. N. (2004). Eye colour: Portals into pigmentation genes and ancestry. *Trends in Genetics, 20*(8), 327–332. doi:10.1016/j.tig.2004.06.010

Sueki, D., & Brechter, J. (2010). *Orthopedic rehabilitation: Clinical advisor.* Maryland Heights, MO: Mosby.

Sun, G. (Ed.). (2016). *Antimicrobial textiles. Woodhead Publishing Series in Textiles: No. 180.* Cambridge, MA: Woodhead Publishing.

Sweeney, D. H., & Taber, M. J. (2014). Cold-water immersion suits. In F. Wang & C. Gao (Eds.). *Protective clothing: Managing thermal stress* (pp. 39–69). Amsterdam, The Netherlands: Woodhead Publishing. doi:10.1533/9781782420408.1.39

Tadisina, K. K., Frojo, G., Plikaitis, C. M., & Bernstein, M. L. (2016, October). Basics of bra sizing: Essentials for the plastic surgery resident. *Plastic and Reconstructive Surgery, 138*(4), 780e–781e. doi:10.1097/PRS.0000000000002593

Taherali, F., Varum, F., & Basit, A. W. (2018). A slippery slope: On the origin, role and physiology of mucus. *Advanced Drug Delivery Reviews, 124,* 16–33. doi:10.1016/j.addr.2017.10.014

Talbot, B. S., Gange, C. P. Jr., Chaturvedi, Apeksha, Klionsky, N., Hobbs, S. K., & Chaturvedi, Abhishek (2017). Traumatic rib injury: Patterns, imaging pitfalls, complications, and treatment. *Radiographics, 37*(2), 628–651. doi:10.1148/rg.2017160100

Tang, C. Y., Tang, N., & Stewart, M. C. (1998a). Facial measurements for frame design. *Optometry and Vision Science, 75*(4), 288–292.

Tang, C. Y., Tang, N., & Stewart, M. C. (1998b). Ophthalmic anthropometry for Hong Kong Chinese adults. *Optometry and Vision Science, 75*(4), 293–301.

Tanner, J. M. (1962). *Growth at adolescence.* Oxford, UK: Blackwell Scientific Publications.

Tasron, D. N., Thurston, T. J., & Carré, M. J. (2015). Frictional behavior of running sock textiles against plantar skin. *Procedia Engineering, 112,* 110–115. doi:10.1016/j.proeng.2015.07.184

Taylor, D., Miaskowski, C., & Kohn, J. (2002). A randomized clinical trial of the effectiveness of an acupressure device (relief brief) for managing symptoms of dysmenorrhea. *Journal of Alternative & Complementary Medicine, 8*(3), 357–370. doi:10.1089/10755530260128050

Taylor, N. A., & Machado-Moreira, C. A. (2013). Regional variations in transepidermal water loss, eccrine sweat gland density, sweat secretion rates and electrolyte composition in resting and exercising humans. *Extreme Physiology & Medicine, 2*(1), 4. doi:10.1186/2046-7648-2-4

Taylor, R. W., Grant, A. M., Williams, S. M., & Goulding, A. (2010). Sex differences in regional body fat distribution from pre- to post-puberty. *Obesity, 18*(7), 1410–1416. doi:10.1038/oby.2009.399

Tedeschi Filho, W., Dezzotti, N. R. A., Joviliano, E. E., Moriya, T., & Piccinato, C. E. (2012). Influence of high-heeled shoes on venous function in young women. *Journal of Vascular Surgery, 56*(4), 1039–1044. doi:10.1016/j.jvs.2012.01.039

Teitel, A. S. (2016, June 10). A brief history of menstruating in space. *Popular Science, POPSCI.com/blogs.* Retrieved from https://www.popsci.com/brief-history-menstruating-in-space

Tennstedt, S. L., Link, C. L., Steers, W. D., & McKinlay, J. B. (2008). Prevalence of and risk factors for urine leakage in a racially and ethnically diverse population of adults: The Boston Area Community Health (BACH) Survey. *American Journal of Epidemiology, 167*(4), 390–399. doi:10.1093/aje/kwm356

Teton Data Systems, & Primal Pictures Ltd. (2001). *Anatomy.tv: The World's Most Detailed 3D Model of Human Anatomy Online*, STAT!Ref (Internet version). Jackson, WY: Teton Data Systems.

Thanassoulis, G., Massaro, J. M., Hoffmann, U., Mahabadi, A., Vasan, R. S., O'Donnell, C. J., & Fox, C. S. (2010). Prevalence, distribution and risk factor correlates of high pericardial and intra-thoracic fat depots in the Framingham heart study. *Circulation: Cardiovascular Imaging*. doi:10.1161/CIRCIMAGING.110.956706

Thelen, M. D., Dauber, J. A., & Stoneman, P. D. (2008). The clinical efficacy of kinesio tape for shoulder pain: A randomized, double-blinded, clinical trial. *Journal of Orthopaedic & Sports Physical Therapy, 38*(7), 389–395. doi:10.2519/jospt.2008.2791

Thomas, K. S., & McMann, H. J. (2012). *U. S. space suits* (2nd ed.). New York, NY: Springer.

Thompson, I. M., Flaherty, S. F., & Morey, A. F. (1998). Battlefield urologic injuries: The Gulf War experience. *Journal of the American College of Surgeons, 187*(2), 139–141. doi:10.1016/S1072-7515(98)00120-3

Thompson, S., Mesloh, M., England, S., Benson, E., & Rajulu, S. (2010, January). The effects of extravehicular activity (EVA) glove pressure on tactility (JSC-CN-20034). *54th Annual Meeting of the Human Factors and Ergonomics Society*; San Francisco, CA. Human Factors and Ergonomics Society.

Thornton, M. J. (1943). The use of vaginal tampons for the absorption of menstrual discharges. *American Journal of Obstetrics and Gynecology, 46*(2), 259–265. doi:10.1016/S0002-9378(15)32918-5

Thryft, A. R. (2018). 3D printing in space: Telescope mirrors, Mars space suits, and space structures. *Design News*. Massachusetts: UBM Americas. Retrieved from https://www.designnews.com/

Tiberio, D. (1988). Pathomechanics of structural foot deformities. *Physical Therapy, 68*(12), 1840–1849. PMID: 3194451

Tiggemann, M., & Lacey, C. (2009). Shopping for clothes: Body satisfaction, appearance investment, and functions of clothing among female shoppers. *Body Image, 6*(4), 285–291. doi:10.1016/j.bodyim.2009.07.002

Tipton, M. J. (1989). The initial responses to cold-water immersion in man. *Clinical Science, 77*(6), 581–588.

Torrens, G., Campbell, I., & Tutton, W. (2012). Chapter 7: Design issues in military footwear and handwear. In E. Sparks (Ed.), *Woodhead Publishing Series in Textiles: Number 122. Advances in military textiles and personal equipment* (pp. 139–164). Cambridge, MA: Woodhead Publishing Limited.

Toxqui, L., Pérez-Granados, A. M., Blanco-Rojo, R., Wright, I., & Vaquero, M. P. (2014). A simple and feasible questionnaire to estimate menstrual blood loss: Relationship with hematological and gynecological parameters in young women. *BMC Women's Health, 14*(1), 71. doi:10.1186/1472-6874-14-71

Trujillo, M. E., & Scherer, P. E. (2006). Adipose tissue-derived factors: Impact on health and disease. *Endocrine Reviews, 27*(7), 762–778. doi:10.1210/er.2006-0033

Trulock, E. P., III. (1990). Arterial blood gases. In H. K. Walker, W. D. Hall & J. W. Hurst (Eds.), *Clinical methods: The history, physical, and laboratory examinations* (3rd ed., pp. 254–257). Boston, MA: Butterworths.

Trussell, J. (2011). Contraceptive failure in the United States. *Contraception, 83*(5), 397–404. doi:10.1016/j.contraception.2011.01.021

Tsiaras, A. (2004). *The architecture and design of man and woman: The marvel of the human body, revealed.* New York, NY: Doubleday.

Turner, C. H., & Pavalko, F. M. (1998). Mechanotransduction and functional response of the skeleton to physical stress: The mechanisms and mechanics of bone adaptation. *Journal of Orthopaedic Science, 3*(6), 346–355. doi:10.1007/s007760050064

Tyrrell, W., & Carter, G. (2009). *Therapeutic footwear: A comprehensive guide.* Edinburgh; New York, NY: Churchill Livingstone.

U.S. Consumer Product Safety Commission (CPSC). (1979, 1991). *16 CFR Part 1501, Method for identifying toys and other articles intended for use by children under 3 years of age which present choking, aspiration, or ingestion hazards because of small parts.* Retrieved from https://www.cpsc.gov/PageFiles/111656/regsumsmallparts.pdf

U.S. Consumer Product Safety Commission (CPSC). (2011). *16 CFR Part 1120.3, Substantial product hazard list: Children's upper outerwear in sizes 2T to 12 with neck or hood drawstrings and children's upper outerwear in sizes 2T to 16 with certain waist or bottom drawstrings.* Retrieved from https://www.cpsc.gov/Business--Manufacturing/Business-Education/Business-Guidance/Drawstrings-in-Childrens-Upper-Outerwear/Frequently-Asked-Questions-FAQs

U.S. Consumer Product Safety Commission (CPSC). (2013). *The Regulated products handbook.* Bethesda, MD: U. S. Consumer Product Safety Commission, Office of Compliance and Field Operations. Retrieved from https://www.cpsc.gov/s3fs-public/RegulatedProductsHandbook.pdf

U.S. Consumer Product Safety Commission (CPSC). (2014). *Which helmet for which activity* (Publication # 349: 072014). Retrieved from http://www.cpsc.gov/en/safety-education/safety-guides/sports-fitness-and-recreation/bicycles/which-helmet-for-which-activity/

U.S. Consumer Product Safety Commission (CPSC). (2017). *CPSC Approves New Federal Safety Standard for Infant Sling Carriers* (Release number 17-069). Retrieved from https://cpsc.gov/newsroom/news-releases/2017/cpsc-approves-new-federal-safety-standard-for-infant-sling-carriers U.S. Food and Drug Administration (FDA) (2017). Alivecor letter of approval. Retrieved from https://www.accessdata.fda.gov/cdrh_docs/pdf17/K171816.pdf

Urban, J. P., & Roberts, S. (2003). Degeneration of the intervertebral disc. *Arthritis Research and Therapy, 5*(3), 120–130. doi:10.1186/ar629

U.S.A. Boxing (2013). *U.S.A. Boxing official rule book: Competition rules.* U.S.A. Boxing Publishers.

Vail, J. (2007). The rectal rocket: A two-day treatment for hemorrhoids. *International Journal of Pharmaceutical Compounding, 11*(3), 194–199.

Van Amber, R. R., Wilson, C. A., Laing, R. M., Lowe, B. J., & Niven, B. E. (2015). Thermal and moisture transfer properties of sock fabrics differing in fiber type, yarns, and fabric structure. *Textile Research Journal, 85*(12), 1269–1280. doi:10.1177/0040517514561926

van Andel, C. J., Wolterbeek, N., Doorenbosch, C. A. M., Veeger, D. H. E. J., & Harlaar, J. (2008). Complete 3D kinematics of upper extremity functional tasks. *Gait & posture, 27*(1), 120–127. doi:10.1016/j.gaitpost.2007.03.002

Van Haarst, E. P., Heldeweg, E. A., Newling, D. W., & Schlatmann, T. J. (2004). The 24-h frequency volume chart in adults reporting no voiding complaints: Defining reference values and analysing variables. *BJU International, 93*(9), 1257–1261. doi:10.1111/j.1464-4096.2004.04821.x

Van Langenhove, L. (2013). Smart textiles for protection: An overview. In R. A. Chapman (Ed.), *Smart textiles for protection* (pp. 3–31). Cambridge, UK: Woodhead Publishing. doi:10.1533/9780857097620.1.333

Van Tiggelen, D., Wickes, S., Coorevits, P., Dumalin, M., & Witvrouw, E (2009). Sock systems to prevent foot blisters and the impact on overuse injuries of the knee joint. *Military Medicine, 174*(2), 183–189.

Van Tiggelen, D., Witvrouw, E., Roget, P., Cambier, D., Danneels, L., & Verdonk, R. (2004). Effect of bracing on the prevention of anterior knee pain—a prospective randomized study. *Knee Surgery, Sports Traumatology, Arthroscopy, 12*(5), 434–439. doi:10.1007/s00167-003-0479-z

van den Hurk, C. J., Peerbooms, M., van de Poll-Franse, L. V., Nortier, J. W., Coebergh, J. W. W., & Breed, W. P. (2012). Scalp cooling for hair preservation and associated characteristics in 1411 chemotherapy patients—results of the Dutch scalp cooling registry. *Acta Oncologica, 51*(4), 497–504. doi:10.3109/0284186X.2012.658966

Van den Oord, M. H., Steinman, Y., Sluiter, J. K., & Frings-Dresen, M. (2012). The effect of an optimised helmet fit on neck load and neck pain during military helicopter flights. *Applied Ergonomics, 43*(5), 958–964. doi:10.1016/j.apergo.2012.01.004

Van der Putte, R. C. J. (1991). Anogenital "sweat" glands: Histology and pathology of a gland that may mimic mammary glands. *American Journal of Dermatopathology, 13*(6), 557–567.

Van der Velden, S. K., Pichot, O., van den Bos, R. R., Nijsten, T. E. C., & De Maeseneer, M. G. R. (2015). Management strategies for patients with varicose veins (C2–C6): Results of a worldwide survey. *European Journal of Vascular and Endovascular Surgery, 49*(2), 213–220. doi:10.1016/j.ejvs.2014.11.006

Vaneechoutte, M. (2017). The human vaginal microbial community. *Research in Microbiology, 168*(9–10), 811–825. doi10.1016/j.resmic.2017.08.001

VanLeeuwen, C., & Torondel, B. (2018). Improving menstrual hygiene management in emergency contexts: Literature review of current perspectives. *International Journal of Women's Health,10*,169–186. doi:10.2147/IJWH.S135587

Varshney, S., Malhotra, M., Gupta, P., Gairola, P., & Kaur, N. (2015). Cavernous sinus thrombosis of nasal origin in children. *Indian Journal of Otolaryngology and Head & Neck Surgery, 67*(1), 100–105. doi:10.1007/s12070-014-0805-4

Vinger, P. F. (2000). A practical guide for sports eye protection. *The Physician and Sportsmedicine, 28*(6), 49–69. doi:10.1080/00913847.2000.11439513

Wade, R., Paton, F., & Woolacott, N. (2016). Systematic review of patient preference and adherence to the correct use of graduated compression stocking to prevent deep vein thrombosis in surgical patients. *Journal of Advanced Nursing, 73*(2), 336–348. doi:10.1111/jan.13148

Waldie, J. M., & Newman, D. J. (2011). A gravity loading countermeasure skinsuit. *Acta Astronautica, 68*(7–8), 722–730. doi:10.1016/j.actaastro.2010.07.022

Walford, J. (2005). Shoes. In V. Steele (Ed.), *Encyclopedia of Clothing and Fashion* (pp. 168–169). Farmington Hills, MI: Thomson Gale.

Walker, N., Bohannon, R. W., & Cameron, D. (2000). Discriminant validity of temporomandibular joint range of motion measurements obtained with a ruler. *Journal of Orthopaedic & Sports Physical Therapy, 30*(8), 484–492.

Wang, J., Thornton, J. C., Bari, S., Williamson, B., Gallagher, D., Heymsfield, S. B., … & Pierson, R. N. (2003). Comparisons of waist circumferences measured at 4 sites. *American Society for Clinical Nutrition, 77*(2), 379–384.

Wang, Y., Wang, Z., Zhang, X., Wang, M., & Li, J. (2015). CFD simulation of naked flame manikin tests of fire proof garments. *Fire Safety Journal, 71,* 187–193. doi:10.1016/j.firesaf.2014.11.020

Warren, J. (1997). Catheter-associated urinary tract infections. *Infectious Disease Clinics of North America, 11*(3), 609–622. doi:10.1016/S0891-5520(05)70376-7

Warren, J. (2001). Catheter-associated urinary tract infections. *International Journal of Antimicrobial Agents, 17*(4), 299–303. doi:10.1016/S0924-8579(00)00359-9

Warwick, W. J., & Hanson, L. G., Regents of the University of Minnesota, "Chest compression apparatus," United States Patent 5,056,505A, October 15, 1991.

Wasserman, M. A., & McGee, M. F. (2017). Preoperative considerations for the ostomate. *Clinics in Colon and Rectal Surgery, 30*(3), 157–161. doi:10.1055/s-0037-1598155

Watkins, S. M. (2005). Space suit. In V. Steele (Ed.), *Encyclopedia of clothing and fashion* (Vol. 3, pp. 200–202). Detroit, MI: Thompson Gale.

Watkins, S. M., & Dunne, L. E. (2015). *Functional clothing design: From sportswear to spacesuits.* New York, NY: Bloomsbury.

Watnick, D., Keller, M. J., Stein, K., & Bauman, L. (2018). Acceptability of a tenofovir disoproxil fumarate vaginal ring for HIV prevention among women in New York city. *AIDS and Behavior, 22*(2), 421–436. doi:10.1007/s10461-017-1962-8

Webb, P., & Annis, J. F. (1967). *The principle of the space activity suit.* NASA-CR-973. Prepared for Langley Research Center by Webb Associates, Inc., Yellow Springs, OH. Retrieved from https://ntrs.nasa.gov/

Wedekind, C. (2002). "Good" and "bad" body odours. In O. P. Kreyden, R. Böni, G. Burg (Eds.), *Hyperhidrosis and botulinum toxin in dermatology, current problems in dermatology* (Vol. 30, pp. 23–29). Basel: Karger. doi:10.1159/000060692

Wei, P., Morey, B., Dyson, T., McMahon, N., Hsu, Y.-Y., Gazman, S., … & Rafferty, C. (2013). A conformal sensor for wireless sweat level monitoring. *SENSORS, 2013, Conference 3-6 Nov 2013, Baltimore, MD*: IEEE, 1–4. doi:10.1109/ICSENS.2013.6688376

Weiss, L., DeForest, B., Hammond, K., Schilling, B., & Ferreira, L. (2013). Reliability of goniometry-based Q-angle. *PM&R, 5*(9), 763–768. doi:10.1016/j.pmrj.2013.03.023

West, C. R., Campbell, I. G., Shave, R. E., & Romer, L. M. (2012). Effects of abdominal binding on cardiorespiratory function in cervical spinal cord injury. *Respiratory Physiology & Neurobiology, 180*(2), 275–282. doi:10.1016/j.resp.2011.12.003

Whitaker, J. (2016). Non-freezing cold injury, lessons from history for future prevention. *Trauma, 18*(3), 178–185. doi:10.1177/1460408615617789

Whitbourne, S. K. (2002). *The aging individual: Physical and psychological perspectives* (2nd Ed.). New York, NY: Springer.

White, A. A., III, & Panjabi, M. M. (1978). The basic kinematics of the human spine: A review of past and current knowledge. *Spine, 3*(1), 12–20.

White, T. D., & Folkens, P. A. (2005a). Thorax: Sternum & ribs. In *The human bone manual* (pp. 181–192). Boston, MA: Elsevier.

White, T. D., & Folkens, P. A. (2005b). The skeletal biology of individuals & populations. In *The human bone manual* (pp. 359–418). London, UK: Elsevier.

Wiberg, G. (1941). Roentgenographs and anatomic studies on the femoropatellar joint: With special reference to chondromalacia patellae. *Acta Orthopaedica Scandinavica, 12*(1–4), 319–410. doi:10.3109/17453674108988818

Wiesemann, F., Adam, R., & Proctor and Gamble Service GmbH, Germany (2011). Absorbent products for personal health care and hygiene. In V. T. Bartels (Ed.), *Handbook of medical textiles* (pp. 316–335). Cambridge, UK: Woodhead Publishing. doi:10.1533/9780857093691.3.316

Wigg, J., & Lee, N. (2014). Redefining essential care in lymphoedema. *British Journal of Community Nursing, 19*(Sup4), S20–S27. doi:10.12968/bjcn.2014.19.Sup4.S20

Wilcox-Levine, M. (2014). Kid gloves. *The Ladies' Paradise: Artifact study*, paper 7. Retrieved from http://digitalcommons.uri.edu/ladies_paradise/7

Wilke, K., Martin, A., Terstegen, L., & Biel, S. S. (2007). A short history of sweat gland biology. *International Journal of Cosmetic Science, 29*(3), 169–179. doi:10.111/j.1467-2494.2007.00387.x

Williams, G. N., Jones, M. H., & Amendola, A. (2007). Syndesmotic ankle sprains in athletes. *American Journal of Sports Medicine, 35*(7), 1197-1207. doi:10.1177/0363546507302545

Windebank, A. J., & Grisold, W. (2008). Chemotherapy-induced neuropathy. *Journal of the Peripheral Nervous System, 13*(1), 27–46. doi:10.1111/j.1529-8027.2008.00156.x

Wisdom, K. M., Delp, S. L., & Kuhl, E. (2015). Use it or lose it: Multiscale skeletal muscle adaptation to mechanical stimuli. *Biomechanics and Modeling in Mechanobiology., 14*(2), 195–215. doi:10.1007/s10237-014-0607-3

Witana, C. P., Goonetilleke, R. S., Xiong, S., & Au, E. Y. L. (2009). Effects of surface characteristics on the plantar shape of feet and subjects' perceived sensations. *Applied Ergonomics, 40*(2), 267–279. doi:10.1016/j.apergo.2008.04.014

Wittens, C., Davies, A. H., Bækgaard, N., Broholm, R., Cavezzi, A., Chastanet, S., . . . & Kakkos, S. (2015). Editor's choice-management of chronic venous disease: Clinical practice guidelines of the European Society for Vascular Surgery (ESVS). *European Journal of Vascular and Endovascular Surgery, 49*(6), 678–737. doi:10.1016/j.ejvs.2015.02.007

Wittlinger, G., & Wittlinger, H. (2004). *Textbook of Dr. Vodder's manual lymph drainage: Volume 1: Basic course* (7th ed.). Stuttgart, Germany: Georg Thieme Verlag.

Woodruff, A. W. (1950). Unilateral spinal accessory nerve palsy caused by arm sling. *British Medical Journal, 1*(4657), 821–822.

Woolford, B. & Mount, F. (2006). Human space flight. In G. Salvendy (Ed.), *Handbook of human factors and ergonomics* (3rd ed., pp. 929–944). Hoboken, NJ: John Wiley.

World Health Organization (WHO). (2011). *Waist circumference and waist-hip ratio*. Geneva, Switzerland: Report of a WHO Expert Consultation, 8–11 December, 2008. Available from the WHO web site: https://www.who.int

World Health Organization Department of Reproductive Health and Research (WHO/RHR) & Johns Hopkins Bloomberg School of Public Health/Center for Communication Programs (CCP), INFO Project (2007). *Family planning: A global handbook for providers: Evidence-based guidance developed through worldwide collaboration*. Baltimore, MD and Geneva, Switzerland: CCP and WHO.

Wright, R. (2002). Interview with Sally K. Ride. *NASA Johnson Space Center Oral History Project (Edited oral history transcript)*. Houston, TX: NASA Johnson Space Center. Retrieved from https://www.jsc.nasa.gov/history/oral

Wright, R. W., & Fetzer, G. B. (2007). Bracing after ACL reconstruction: A systematic review. *Clinical Orthopaedics and Related Research, 455*, 162–168. doi:10.1097/BLO.0b013e31802c9360

Wright, T. (1922). *The romance of the shoe*. London: C. J. Farncombe & Son.

Wu, G., Siegler, S., Allard, P., Kirtley, C., Leardini, A., Rosenbaum, D., ... & Stokes, I. (2002). ISB recommendation on definitions of joint coordinate system of various joints for the reporting of human joint motion—part I: Ankle, hip, and spine. *Journal of Biomechanics, 35*(4), 543–548. doi.org/10.1016/S0021-9290(01)00222-6

Wu, G., Van der Helm, F. C. T., Veeger, H. E. J. D., Makhsous, M., Van Roy, P., Anglin, C., . . . & Buchholz, B. (2005). ISB recommendation on definitions of joint coordinate systems of various joints for the reporting of human joint motion—Part II: Shoulder, elbow, wrist and hand. *Journal of Biomechanics, 38*(5), 981–992. doi:10.1016/j.jbiomech.2004.05.042

Wyndaele, J. J. (1998). The normal pattern of perception of bladder filling during cystometry studied in 38 young healthy volunteers. *Journal of Urology, 160*(2), 479–481. doi:10.1016/S0022-5347(01)62929-X

Xiao, M., Luximon, Y., & Luximon, A. (2013). Foot structure and anatomy. In R. S. Goonetilleke (Ed.), *The science of footwear* (pp. 3–18). Boca Raton, FL: CRC Press.

Xiong, S., Rodrigo, A., & Goonetilleke, R. S. (2013). Foot characteristics and related empirical models. In R. S. Goonetilleke (Ed.), *The science of footwear* (pp. 47–77). Boca Raton, FL: CRC Press.

Yamashita, M. H. (2005). Evaluation and selection of shoe wear and orthoses for the runner. *Physical Medicine and Rehabilitation Clinics of North America, 16*(3), 801–829. doi:10.1016/j.pmr.2005.02.006

Yang, C., Vora, H. D., & Chang, Y. (2018). Behavior of auxetic structures under compression and impact forces. *Smart Materials and Structures, 27*(2), 025012. doi:10.1088/1361-665X/aaa3cf

Yang, P. J., LaMarca, M., Kaminski, C., Chu, D. I., & Hu, D. L. (2017). Hydrodynamics of defecation. *Soft Matter, 13*(29), 4960–4970. doi:10.1039/c6sm02795d

Yang, P. J., Pham, J., Choo, J., & Hu, D. L. (2014). Duration of urination does not change with body size. *Proceedings of the National Academy of Sciences, 111*(33), 11932–11937. doi:10.1073/pnas.1402289111

Yang, Z., Gu, W., Zhang, J., & Gui, L. (2017). *Force control theory and method of load carrying exoskeleton suit*. Berlin, Germany: Springer. doi:10.1007/978-3-662-54144-9

Yao, K. T., Lin, C. C., & Hung, C. H. (2009). Maximum mouth opening of ethnic Chinese in Taiwan. *Journal of Dental Sciences, 4*(1), 40–44. doi:10.1016/S1991-7902(09)60007-6

Yip, J., & Yu, W. (2006). Intimate apparel with special functions. In W. Yu, J. Fan, S. C. Harlock, & S. P. Ng (Eds.), *Innovation and technology of women's intimate apparel* (pp. 171–195). Cambridge, England: Woodhead Publishing.

Young, C. C., Niedfeldt, M. W., Morris, G. A., & Eerkes, K. J. (2005). Clinical examination of the foot and ankle. *Primary Care: Clinics in Office Practice, 32*(1), 105–132. doi:10.1016/j.pop.2004.11.002

Yu, H.-L., Chase, R. A., & Strauch, B. (2004). *Atlas of Hand Anatomy and Clinical Implications*. St. Louis, MO: Mosby.

Yu, W., & Zhou, J. (2016). Sports bras and breast kinetics. In W. Yu (Ed.), *Advances in women's intimate apparel technology* (pp. 135–146). Cambridge, MA: Woodhead Publishing. doi:10.1016/B978-1-78242-369-0.00008-6

Yu, W., Wang, J-P., & Shin, K. (2006). Bra pattern technology. In W. Yu, J. Fan, S. C. Harlock, & S. P. Ng (Eds.), *Innovation and technology of women's intimate apparel* (pp. 76–113). Cambridge, England: Woodhead Publishing.

Zeagler, C. (2017). Where to wear it: Functional, technical, and social considerations for wearable technology in on-body location for wearable technology 20 years of designing for wearability. *Proceedings of the 2017 ACM/ISWC International Symposium on Wearable Computers*, September 11–15, 2017, Maui, Hawaii (pp. 150–157). doi:10.1145/3123021.3123042

Zhang, J., & Stringer, M. D. (2010). Ophthalmic and facial veins are not valveless. *Clinical & Experimental Ophthalmology, 38*(5), 502–510. doi:10.1111/j.1442-9071.2010.02325.x

Zheng, R., Yu, W., & Fan, J. (2006). Breast measurement and sizing. In W. Yu, J. Fan, S. C. Harlock, & S. P. Ng. (Eds.), *Innovation and technology of women's intimate apparel* (pp. 28–58). Cambridge, England: Woodhead Publishing.

Zhou, J., Yu, W., & Ng, S.-P. (2012). Identifying effective design features of commercial sports bras. *Textile Research Journal 83*(14), 1500–1513. doi:10.1177/0040517512464289

Zhuang, Z., & Bradtmiller, B. (2005). Head-and-face anthropometric survey of U.S. respirator users. *Journal of Occupational and Environmental Hygiene, 2*(11), 567–576. doi:10.1080/15459620500324727

Zhuang, Z., Shu, C., Xi, P., Bergman, M., & Joseph, M. (2013). Head-and-face shape variations of U.S. civilian workers. *Applied Ergonomics, 44*(5), 775–784. doi:10.1016/j.apergo.2013.01.008

Zong, Y., & Lee, Y-A. (2011). An exploratory study of integrative approach between 3D body scanning technology and motion capture systems in the apparel industry. *International Journal of Fashion Design, Technology and Education, 4*(2), 91–101. doi:10.1080/17543266.2010.537281

Zuckerman, J. D. (1996). Hip fracture. *New England Journal of Medicine, 334*(23), 1519–1525. doi:10.1056/NEJM199606063342307

Zuther, J. E., & Norton, S. (2013). (With collaborators Norton, S., Armer, J. M., et al.). *Lymphedema management: The comprehensive guide for practitioners* (3rd ed.). New York, NY: Thieme.

# Index